ERGOT

ERGOT

The Genus *Claviceps*

Edited by

Vladimír Křen
Institute of Microbiology
Academy of Sciences of the Czech Republic
Prague, Czech Republic
and
Ladislav Cvak
Galena Pharmaceutical Company
Opava, Czech Republic

CRC PRESS

Boca Raton   London   New York   Washington, D.C.

FIRST INDIAN REPRINT, 2012

This book contains information obtained from authentic and highly regarded sources. Reprinted material is quoted with permission, and sources are indicated. A wide variety of references are listed. Reasonable efforts have been made to publish reliable data and information, but the author and the publisher cannot assume responsibility for the validity of all materials or for the consequences of their use.

Visit the CRC Press Web site at www.crcpress.com

Printed and bound in India by
Replika Press Pvt. Ltd.

ISBN 10 : 90-5702-375-X
ISBN 13 : 978-90-5702-375-0

FOR SALE IN SOUTH ASIA ONLY.

CONTENTS

PREFACE TO THE SERIES

There is increasing interest in industry, academia and the health sciences in medicinal and aromatic plants. In passing from plant production to the eventual product used by the public, many sciences are involved. This series brings together information which is currently scattered through an ever increasing number of journals. Each volume gives an in-depth look at one plant genus, about which an area specialist has assembled information ranging from the production of the plant to market trends and quality control.

Many industries are involved such as forestry, agriculture, chemical, food, flavour, beverage, pharmaceutical, cosmetic and fragrance. The plant raw materials are roots, rhizomes, bulbs, leaves, stems, barks, wood, flowers, fruits and seeds. These yield gums, resins, essential (volatile) oils, fixed oils, waxes, juices, extracts and spices for medicinal and aromatic purposes. All these commodities are traded world-wide. A dealer's market report for an item may say "Drought in the country of origin has forced up prices".

Natural products do not mean safe products and account of this has to be taken by the above industries, which are subject to regulation. For example, a number of plants which are approved for use in medicine must not be used in cosmetic products.

The assessment of safe to use starts with the harvested plant material which has to comply with an official monograph. This may require absence of, or prescribed limits of, radioactive material, heavy metals, aflatoxin, pesticide residue, as well as the required level of active principle. This analytical control is costly and tends to exclude small batches of plant material. Large scale contracted mechanised cultivation with designated seed or plantlets is now preferable.

Today, plant selection is not only for the yield of active principle, but for the plant's ability to overcome disease, climatic stress and the hazards caused by mankind. Such methods as *in vitro* fertilisation, meristem cultures and somatic embryogenesis are used. The transfer of sections of DNA is giving rise to controversy in the case of some end-uses of the plant material.

Some suppliers of plant raw material are now able to certify that they are supplying organically-farmed medicinal plants, herbs and spices. The Economic Union directive (CVO/EU No. 2092/91) details the specifications for the

obligatory quality controls to be carried out at all stages of production and processing of organic products.

Fascinating plant folklore and ethnopharmacology leads to medicinal potential. Examples are the muscle relaxants based on the arrow poison, curare, from species of *Chondrodendron,* and the antimalarials derived from species of *Cinchona* and *Artemisia.* The methods of detection of pharmacological activity have become increasingly reliable and specific, frequently involving enzymes in bioassays and avoiding the use of laboratory animals. By using bioassay linked fractionation of crude plant juices or extracts, compounds can be specifically targeted which, for example, inhibit blood platelet aggregation, or have antitumour, or antiviral, or any other required activity. With the assistance of robotic devices, all the members of a genus may be readily screened. However, the plant material must be **fully** authenticated by a specialist.

The medicinal traditions of ancient civilisations such as those of China and India have a large armamentaria of plants in their pharmacopoeias which are used throughout South East Asia. A similar situation exists in Africa and South America. Thus, a very high percentage of the World's population relies on medicinal and aromatic plants for their medicine. Western medicine is also responding. Already in Germany all medical practitioners have to pass an examination in phytotherapy before being allowed to practise. It is noticeable that throughout Europe and the USA, medical, pharmacy and health related schools are increasingly offering training in phytotherapy.

Multinational pharmaceutical companies have become less enamoured of the single compound magic bullet cure. The high costs of such ventures and the endless competition from me too compounds from rival companies often discourage the attempt. Independent phytomedicine companies have been very strong in Germany. However, by the end of 1995, eleven (almost all) had been acquired by the multina-tional pharmaceutical firms, acknowledging the lay public's growing demand for phytomedicines in the Western World.

The business of dietary supplements in the Western World has expanded from the Health Store to the pharmacy. Alternative medicine includes plant based products. Appropriate measures to ensure the quality, safety and efficacy of these either already exist or are being answered by greater legislative control by such bodies as the Food and Drug Administration of the USA and the recently created European Agency for the Evaluation of Medicinal Products, based in London.

In the USA, the Dietary Supplement and Health Education Act of 1994 recognised the class of phytotherapeutic agents derived from medicinal and aromatic plants. Furthermore, under public pressure, the US Congress set up an Office of Alternative Medicine and this office in 1994 assisted the filing of several Investigational New Drug (IND) applications, required for clinical trials of some Chinese herbal preparations. The significance of these applications was that each Chinese preparation involved several plants and yet was handled as a **single** IND. A demonstration of the contribution to efficacy, of **each** ingredient of

each plant, was not required. This was a major step forward towards more sensible regulations in regard to phytomedicines.

My thanks are due to the staff of Harwood Academic Publishers who have made this series possible and especially to the volume editors and their chapter contributors for the authoritative information.

Roland Hardman

PREFACE

Ergot *(Claviceps purpurea)* is best known as a disease of rye and some other grasses. However, it is probably the most widely cultivated fungus and it has become an important field crop.

The main reason for its importance is ergot alkaloids, which are extensively used in medicine. No other class of compounds exhibits such a wide spectrum of structural diversity, biological activity and therapeutic uses as ergot derivatives. Currently, ergot alkaloids cover a wide spectrum of therapeutic uses as the drugs of high potency in the treatment of uterine atonia, postpartum bleeding, migraine, orthostatic circulatory disturbances, senile cerebral insufficiency, hypertension, hyperprolactinemia, acromegaly and parkinsonism.

Ergot—once dreaded pest and cause of epidemic intoxications has now become a profitable crop for farmers. However, the danger of intoxication and crop damage still persists. The fungus was already well known in the middle ages, causing outbreaks of ergotism or "epidemic gangrene" called for example, St Anthony's fire.

Ergot alkaloids are traditionally obtained by extraction of ergot sclerotia artificially cultivated on cereals. The parasitic cultures are not able to produce some, e.g., clavine alkaloids necessary for most semisynthetic drugs. Crop fluctuations and market demands lead to the development of submerged cultivation in production plants. Present trends in ergot cultivation are the development of saprophytic cultivation processes and improvement of field production by, for example, introduction of new hosts and ergot strains. Even though there is a constant effort to prepare ergot alkaloids synthetically their bio-production is still much more competitive. In the contemporary economical crisis of agriculture, especially in Europe, the ergot is a good and profitable alternative crop for farmers. Thanks to the new advanced technologies it experiences a real renaissance.

Various strains of *Claviceps* served as models for study of the fungal metabolism, biogenesis, physiological and genetic aspects of ergot alkaloids production. This interest continues because of good perspectives of submerged and field production of ergot alkaloids.

The volume on the *Claviceps* genus should provide readers with both biotechnological aspects of ergot alkaloid production, genetic and physiological data but also with newly emerging dangers of toxicology and environmental risks of ergot infection and contamination of food and forage. Chemistry and pharmacology of ergot alkaloids will demonstrate both their use as classical drugs and their newly discovered pharmacological applications.

Vladimír Křen

NAMES OF ERGOT IN VARIOUSCOUNTRIES OF THE WORLD

Anyarozs	Hungary
Bakkaku, Ergot	Japan
Çacdar Mahmuzu	Turkey
Centeio erspigado	Brazil
Cornezuelo de centeno	Argentina
Cornezuelo de centeno	Chile
Cornezuelo de centeno	Paraguay
Cornezuelo de centeno	Spain
Cravagem de anteio	Brazil
Cravagem de anteio	Portugal
Cuernicillo de centeno	Mexico
Ergot	England
Ergot	United States
Ergot de seigle	Belgium
Ergot de seigle	France
Erperao de anteio	Brazil
Erüsi bôdês briza	Greece
Grano speronato	Italy
Meldröje	Denmark
Meldröye	Norway
Mjöldryga	Finland
Moederkoorn	The Netherlands
Mutterkorn	Austria
Mutterkorn	Germany
Námel	Czech Republic
Razema glavnica	Yugoslavia
Secara cornuta	Romania
Segale cornuto	Italy
Sporyn'ja	Russia
Sporyzs	Poland

CONTRIBUTORS

Martin Buchta
Galena Pharmaceutical Company
74770 Opava 9
Czech Republic
e-mail:
MARTIN_BUCHTA@IVAX.COM

Petr Bulej
Galena Pharmaceutical Company
74770 Opava 9
Czech Republic e-mail:
PETR_BULEJ@IVAX.COM

Nicoletta Crespi-Perellíno
Department of Pharmaceutical
Sciences
University of Bologna
Via Belmeloro 6
40126 Bologna
Italy

Ladislav Cvak
Head of R&D
Galena Pharmaceutical Company
747 70 Opava 9
Czech Republic
e-mail:
LADISLAV_CVAK@IVAX.COM

Eckart Eich
Institut für Pharmazie II
Freie Universität Berlin
Koenigin-Luise-Strasse 2
14195 Berlin (Dahlem)
Germany

Anna Fišerová
Department of Immunology and
Gnotobiology
Institute of Microbiology
Academy of Sciences of the Czech
Republic
Vídeňská 1083
142 20 Prague 4

Czech Republic
e-mail: fiserova@biomed.cas.cz

Alexandr Jegorov
Galena Pharmaceutical Company
Research Unit
Branišovská 31
37005 České Budějovice
Czech Republic
e-mail: husakm@marvin.jcu.cz

Ullrich Keller
Max-Volmer-Institut für
Biophysikalische Chemie und
Biochemie
Fachgebiet Biochemie und Molekulare
Biologie
Technische Universität Berlin
Franklinstrasse 29
10587 Berlin-Charlottenburg
Germany
e-mail: kellghbe@mailszrz.zrz.TU-
Berlin.de

Anatoly G.Kozlovsky
Laboratory of Biosynthesis of
Biologically Active Compounds
Institute of Biochemistry and
Physiology of Microorganisms
Russian Academy of Sciences
142292 Pushchino
Moscow Region
Russia
e-mail: kozlovski@ibpm.serpukhov.su

Vladimír Křen
Laboratory of Biotransformation
Institute of Microbiology
Academy of Sciences of the Czech
Republic
Vídeňská 1083
142 20 Prague 4

Czech Republic
e-mail: kren@biomed.cas.cz
Zdeněk Malinka
Galena Pharmaceutical Company
74770 Opava 9
Czech Republic
e-mail:
ZDENEK_MALINKA@IVAX.COM

Anacleto Minghetti
Department of Pharmaceutical
Sciences
University of Bologna
Via Belmeloro 6
40126 Bologna
Italy
e-mail: ming@kaiser.alma.unibo.it

Éva Németh
University of Horticulture and Food
Industry
Villányi str. 29–43
1114 Budapest
Hungary
e-mail: h11531ber@ella.hu

Douglas P.Parbery
Faculty of Agriculture, Forestry and
Horticulture
University of Melbourne
Parkville 3052
Australia
Sylvie Pažoutová
Institute of Microbiology
Academy of Sciences of the Czech
Republic
Vídeňská 1083
14220 Prague 4
Czech Republic
e-mail: pazouto@biomed.cas.cz

Heinz Pertz
Institut für Pharmazie II
Freie Universität Berlin

Koenigin-Luise-Strasse 2
14195 Berlin (Dahlem)
Germany
Miloslav Pospíšil
Department of Immunology and
Gnotobiology
Institute of Microbiology
Academy of Sciences of the Czech
Republic
Vídeňská 1083
142 20 Prague 4
Czech Republic
e-mail: pospisil@biomed.cas.cz

Richard A.Shelby
Department of Plant Pathology
209 Life Sciences
Auburn University
AL 36849
USA
e-mail: rshelby@earthlink.net

Klaus B.Tenberge
Institut für Botanik
Westfälische Wilhelms-Universität
Schlossgarten 3
48149 Münster
Germany
e-mail: tenberg@uni-muenster

Paul Tudzynski
Institut für Botanik
Westfälische Wilhelms-Universität
Schlossgarten 3
48149 Münster
Germany
e-mail: tudzyns@uni-muenster.de

1.
THE HISTORY OF ERGOT

ANACLETO MINGHETTI and NICOLETTA CRESPI-
PERELLINO

*Department of Pharmaceutical Sciences, University of Bologna, Via
Belmeloro 6, 40126 Bologna, Italy*

1.1.
THE EARLIER ALKALOIDS: ISOLATION AND STRUCTURES

Years ago a review on ergot alkaloids (EA) appeared with the title: "The biosynthesis of ergot alkaloids; the story of the unexpected" (Floss, 1980). I don't believe that any other title could be more appropriate since the entire history of EA research, from the discovery of ergotamine almost a century ago until the present, has truly been the history of the unexpected.

The beginning of modern ergot alkaloid research dates back to 1918 when A.Stoll isolated in crystalline form ergotamine (Stoll, 1945), an alkaloid present in the sclerotia of the *Claviceps purpurea* fungus and patented it. In 1917 the Sandoz pharmaceutical company of Basel granted Stoll, then a young Swiss chemist and student of R.Willstaetter, already distinguished in the field of natural products, the responsibility of setting up a laboratory and developing new drug research (Stoll, 1965). Stoll proposed the goal of isolating the oxytocic active principle present in *Claviceps* sclerotia, universally used in post partum hemorrhages and now known as ergometrine. He hoped to do exactly as Sertumer had done a century earlier in isolating the active principle morphine from opium. Unfortunately, unlike morphine, ergometrine was not easily extractable with solvents due to its tendency to remain in the aqueous phase and, above all, because it was present in scarce quantities in the mixture of alkaloids produced by the *Claviceps* sclerotia: often one tenth in comparison with the production of ergotamine (Hofmann, 1964). This explains how Stoll ended up finding ergotamine, the major and the most lipophilic alkaloid in the extracted mixture, while looking for ergometrine. Nevertheless, ergotamine was used for some time as an oxytocic drug but with poor results. In fact, the crude drug (ground sclerotia) was used for many years in spite of serious dosage problems. Stoll was credited with being able to isolate in the pure state the first alkaloid of a series of almost one hundred products the majority of which were present in traces in the sclerotia collected in the Black Forest region. Stoll isolated from the mother liquor of ergotamine, also ergotaminine, a more liposoluble alkaloid of the same elementary composition as ergotamine, but which was dextrorotatory.

At the beginning of research on EA scientists had the following means for characterizing a product: elemental analysis, melting point, characteristic chromatic reactions and measurement of optical rotation. Elemental analyses were the most reliable data while the melting point could vary by a few units due to impurities or traces of the solvent present in the crystals used. The characteristic chromatic reactions included entire classes of compounds, and two reagents, Keller's (1896) and Van Urk's (1929), specific for the indoles substituted in position 4, were used for EA. Optical rotation, largely used in the characterization of isolated compounds, in the case of EA initially provided more confusion than help. It is now known that all EA are levorotatory in nature and that during extraction and isolation procedures these compounds can turn to be dextrorotatory as a result of temperature, light or pH. It was for this reason that Stoll, after having isolated ergotamine, found in the mother liquor a notable quantity of ergotaminine which was for a long time believed to be a natural product. In addition, dextrorotatory EA, when treated with acids as in the case of the transformation of bases to salts, can retroisomerize to the levorotatory state. Hence, the discrepancy which exists in literature between the previously reported values of optical rotation for various isolated products can be explained.

Other researchers had been trying to isolate the alkaloids present in *Claviceps* for 50 years, and upon Stoll's success/failure with ergotamine, this goal was reached step by step. In 1875 C.Tanret, in the hopes of isolating ergometrine, had obtained the so-called "ergotinine cristallisée" which was, contrary to its name, a mixture of almost all the solvent extractable alkaloids and composed mainly of peptide alkaloids (Barger, 1931). Tanret also dedicated himself to the systematic research of the compounds present in *C. purpurea* sclerotia which contain a significant quantity of lipids. He isolated several fatty acids and identified two "sterines" in the non-saponifiable fraction, one with a higher melting point which he called ergosterol and the other with the lower melting point, phytosterol. These products were later found to be ubiquitous in the plant kingdom. Regarding nitrogenous compounds, Tanret, isolated ergothioneine (the betaine of thiolhistidine) histidine, choline, and betaine. Trisaccharides such as clavicepsine, made of glucose and mannitol, as well as disaccharides such as trehalose and free mannitol were also identified from the sclerotia (Barger, 1931). Another researcher who was very active in the study of the components of *C. purpurea* was F.Kraft, a Swiss pharmacist, who was attracted by the purple color of the mature sclerotia and identified several pigments including ergochrisine and ergoflavine as derivatives of secalonic acid (Frank *et al.,* 1973). In addition, Kraft extracted a fraction composed mainly of ergotoxine group alkaloids which he called hydroergotinine (Hofmann, 1964). Subsequent progress in isolating alkaloids produced by *C. purpurea* was made by G.Barger and F.Carr (1907). These researchers isolated ergotoxine, which was at first considered a pure product, but later was recognized as a mixture of three alkaloids (ergocornine, ergocristine and ergocryptine). After the isolation of ergotoxine, a mixture which was rarely constant, the interesting pharmacological activities of other EA different from ergometrine, began to be discovered.

However, the problem concerning the oxytocic activity of *Claviceps* remained unsolved because *in vitro* pharmacological tests to follow the activity in the various steps of the extraction were not available. Stoll, after having realized that ergotamine was not the active principle of *Claviceps,* ceased to work on this topic for 16 years. But he suddenly woke up when he knew that an English physician, C.Moir, had found the oxytocic activity to be present in the exhausted aqueous phase after extraction of the alkaloids with organic solvents. Since that moment Stoll devoted himself to the search of new Claviceps alkaloids. He carried out this task so thoroughly that, after having isolated in less than one decade four alkaloid couples (ergometrine/ inine, ergocristine/inine, ergokryptine/inine and ergocornine/inine), he is now considered by everybody as "the father" of the EA.

As reported above, C.Moir (1932) made the unexpected discovery that the *Claviceps* oxytocic activity was present in the aqueous phase which remained after the extraction with solvents of the main alkaloids. It should not be surprising that in all that time many researchers had continued to look for ergometrine in the organic extracts after having thrown it down the drain with the aqueous phase. This is a rather frequent occurrence in research. As soon as Moir realized his discovery, he published it and, with the help of the chemist H.W.Dudley, devoted himself to the compound's extraction. He achieved this goal three years later (Dudley *et al.,* 1935), and along with defining the chemical-physical characteristics gave this new product the name ergometrine (Dudley *et al.,* 1935) with obvious reference to its activity on the endometrium. However, three other laboratories were also working independently on the isolation of this compound and as a result four papers describing a product isolated from *Claviceps* with oxytoxic activity appeared in the literature in 1935. The authors include M.Kharash and W.Legault from Chicago who named the product ergotocine (Kharasch *et al.,* 1935), M.Thompson from the Johns Hopkins University in Baltimore who called his product ergostetrine (Thompson, 1935) and, finally, A.Stoll from Sandoz in Basel who, being a chemist and not a physician, baptized his long-sought-after compound ergobasine (Stoll *et al.,* 1935) in reference to its basic characteristics. The chemical-physical data for the isolated compounds reported by the various authors were similar enough to infer that they referred to the same molecule. In any case, after a great deal of polemics documented in several articles (Thompson, 1935; Stoll, 1935) in order to settle the controversy, the four researchers exchanged their respective products for chemical and pharmacological comparison. The four groups came to the unanimous conclusion that the slight differences in melting point and optical rotation were due to the differing degrees of purity and that they had all found the same compound. In short, ergometrine, ergotocine, ergostetrine and ergobasine were synonyms (Kharasch *et al.,* 1936). However, which name to adopt for this molecule so as to avoid confusion in the literature remained to be decided. The American authors opted for ergonovine as a substitution for ergotocine and ergostetrine but the English and Swiss researchers didn't agree and so today, even after sixty years, the three synonyms ergonovine, ergobasine and ergometrine (now the prevailing name) can be found in the literature. This international race to identify such a compound demonstrates how great the need was at the time for a drug which could save many mothers from dying for post partum hemorrhage. After the isolation of ergometrine the classification of the EA into "liposoluble alkaloids", including those products with a peptide chain such as ergotamine, and "water soluble alkaloids" such as ergometrine was adopted. Given, however, that in reality ergometrine is not much more water soluble than ergotamine, it is now preferred to define the first group as ergopeptines and the second as "simple amides of lysergic acid". The discovery of ergometrine gave a significant boost to ergot alkaloid research, especially in the Anglo-Saxon scientific world which was more interested in natural products than its Germanic counterpart in which synthetic chemistry research was at that time at its zenith.

The second half of the thirties was one of the most productive periods regarding EA structural elucidation. From Tanret's ergotinine English researchers S.Smith and G.M.Timmis isolated a new peptide alkaloid, ergosine (Smith et al., 1937). It is noteworthy that Smith in his first report, assigned no name to his new alkaloid, (the third after ergotamine and ergometrine) "in order to prevent possibly later unnecessary complications" (Smith et al., 1936). In fact, the debate on ergometrine had just ceased. Almost contemporary Stoll isolated from ergotoxine ergocristine (Stoll et al., 1937; Stoll et al., 1951). In the same years also a great development in EA chemistry occurred. Smith and Timmis while trying to understand the structure of EA refluxed ergotinine, ergotoxine (Smith et al., 1932), ergotamine and ergotaminine (Smith et al., 1932) with metanol in IN KOH and obtained, from all four products, a crystallized basic compound which they named ergine. Although ergine had a much lower molecular weight than the starting materials and no biological activity, it showed all the reactions characteristic of the Claviceps alkaloids (Keller's and Van Urk's) and therefore it was guessed that it had to be the fundamental nucleus of all these compounds. The same researchers found ergine to be also the degradation product of ergometrine which had a slightly higher molecular weight (Smith et al., 1935). From this discovery it was clearly understood that ergometrine was composed of ergine and a short side chain which lacked nitrogen. Contemporaneously W.A.Jacobs and L.C.Craig, researchers at the Rockefeller Institute in New York, carried out the degradation of ergometrine and found 2-amino-propanol (Jacobs et al., 1935).

Since the structure of EA was still unknown, the same authors performed nitric oxidation, acidic and basic hydrolysis, and pyrolysis of ergotinine in order to elucidate its structure. It is not clear as to why these authors chose to use ergotinine, known to be a mixture, rather than ergotamine, the only pure alkaloid known at the time, perhaps because ergotamine was patented while ergotinine was readily available. Nitric oxidation brought about the possible existence of a methylindole ring as well as of a nucleus of methylhydroquinoline in the structure of ergotinine (Jacobs et al., 1932). Alkaline degradation was more fruitful and, instead of ergine, gave an acid which crystallized easily and which was named lysergic acid (Jacobs et al., 1934). Elemental analysis of this acid revealed that its structure corresponded to that of ergine, minus a nitrogen atom. When subjected to identical alkaline hydrolysis conditions ergine gave lysergic acid and ammonia, thus demonstrating that ergine was the amide of lysergic acid and that it contained three of the five nitrogens of ergotinine (Jacobs et al., 1934). Lysergic acid was shown to contain only one carboxylic group and a-N-CH$_3$ group and gave all the chromatic reactions characteristic of ergotinine (Jacobs et al., 1935). At the same time Smith and Timmis also isolated lysergic acid from ergine (Smith et al., 1934) and two years later isolysergic acid (Smith et al., 1936; Smith et al., 1936), thus indicating where in the EA structure the isomerization point was located. Stoll later admitted that the American researchers were more bold than him since he, knowing the sensitivity of the EA, would have never thought of treating them with an aqueous solution of boiling 7% potash (Stoll, 1965).

The structure of lysergic acid, which relation with tryptophan was already evident from the Ehrlich and Van Urk reactions, remained to be established. Jacobs and Craig from the fragments obtained by ergine degradation inferred the tetracyclic ergoline structure (Jacobs et al., 1936) which they confirmed with an elegant chemical synthesis (Jacobs et al., 1937; Jacobs et al., 1939). In order to have the definitive structure of lysergic acid, the positions of the carboxyl group (Jacobs et al., 1938) and of the double bond needed to be defined. Position 4 was assigned to the carboxyl group, as reported in Figure 1. Later position 7 was considered, but experimental data finally confirmed that this group was located in position 8 (Uhle et al., 1945). The double bond had to be assigned to the position conjugated with the indole aromatic ring, therefore between carbons 10–5 or 9–10 where rings C and D join. Although the former position was initially believed to correct (Craig et al., 1938), the latter was finally confirmed (Uhle et al., 1945). These structure-defining experiments led to the logical attempts to chemically synthesize lysergic acid, a goal which was achieved 20 years later, with extremely discouraging yields (Stoll et al., 1954; Kornfeld et al., 1954; Kornfeld et al., 1956). In fact, in spite of its high cost and demand as an intermediate for new drugs, the only source of lysergic acid is up to now the natural one.

The elucidation of the structure of the side chain of the alkaloids present in ergotinine or ergotoxine was performed mainly by Jacobs and Craig who in a few years produced a lot of work on this topic. After alkaline hydrolysis of ergotinine, by which lysergic acid was removed, the building blocks of the aminoacidic moiety were found to be: proline, phenylalanine, pyruvic and α-hydroxyisovaleric acids (Jacobs et al., 1935). The residue from alkaline hydrolysis

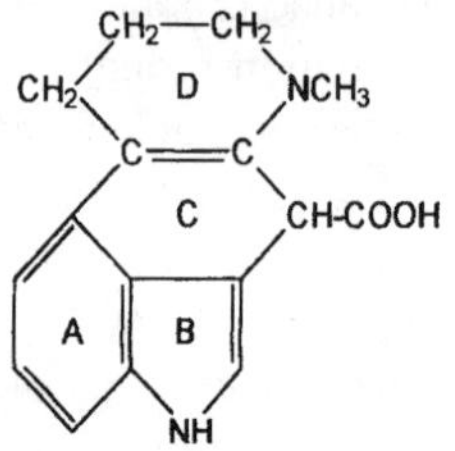

Figure 1 The first structure of lysergic acid as reported by Jacobs *et al.* (1936)

was treated with hydrochloric acid and was found to contain a dipeptide which after acidic hydrolysis gave an equimolecular mixture of proline and phenylalanine (Jacobs *et al.,* 1935). This dipeptide originated from ergocristine which was probably the major component of the starting material. After hydrogenation of ergotinine the same authors found isobutyric, formic, and α-hydroxyvaleric acid (Jacobs *et al.,* 1938). The isolation of all these compounds meant a solution to the ergotinine, or rather ergotamine, composition puzzle, but not in a correct sequence, as reported in Figure 2.

Research on new alkaloids present in *C. purpurea* was being carried out in parallel with EA structure research. Stoll, besides ergotamine, isolated two other peptide alkaloids from ergotinine: ergocornine and ergokryptine (Stoll *et al.,* 1943), the latter as its name indicates being the most difficult to isolate since it was a minor component in the mixture. In 1943 Stoll published the isolation and the structures of these two new alkaloids (Stoll *et al.,* 1943), perhaps with the thought that he had almost exhausted the series of alkaloids present in *Claviceps* and also being thankful for having worked in the peaceful island known as Switzerland. Surely he never imagined that at that time a new and unexpected chapter in EA history was being opened in war-torn England. Here indeed, the era of microbial fermentation had begun.

Nineteen forty-three was a historic year in EA research thanks to the unexpected discovery of the hallucinogenic properties of lysergic acid diethylamide (LSD) a synthetic derivative of lysergic acid. LSD had already been synthesized in 1938 in a screening of compounds with oxytocic activity, and tested in comparison with ergometrine, but it was impossible to note its effects on the central nervous system since trials were carried out on laboratory animals. It was found only to have less oxytocic activity than ergometrine. In the spring of 1943 A. Hofmann, in the Sandoz labs in Basel, resynthesized LSD in order to further investigate its analeptic activity owing its structural analogy to nikethamide (Stoll *et al.,* 1943). Without realizing it, Hofmann contaminated himself with the product. Returning home from work he noticed that he was having reactions and sensations completely new and unusual for an orderly and methodical person like himself. He understood immediately that what he was experiencing was the result of something he had come into contact with at work

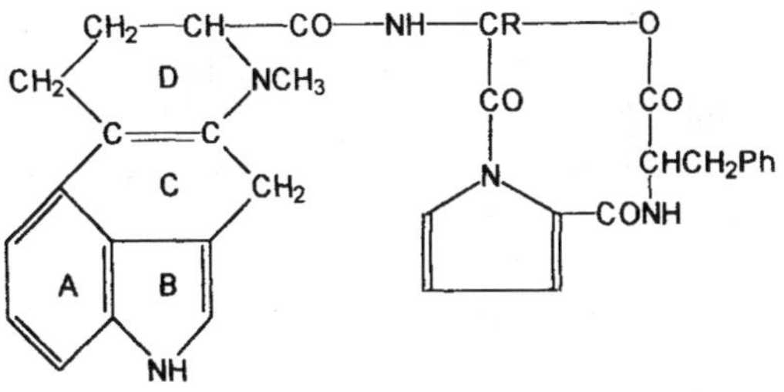

Figure 2 The first structure of ergotamine as reported by Jacobs *et al.* (1938)

and since he hadn't voluntarily inflicted anything on himself he concluded that traces of a product from the laboratory were causing the strange effects. It is not unusual for a chemist, even one as precise as Hofmann, to become contaminated by working at the lab bench. In fact, the cases of self-contamination are much more frequent than one would expect but are rarely evident because of the low biological activity of the product as well as the dose, usually less than a milligram, taken. The effect Hofmann experienced was so strong with respect to the dose ingested that he, like any curious researcher, couldn't keep himself from discovering which product had caused the sensations. He therefore separately ingested every product in small doses he had handled that day. He began by carefully taking 0.25 mg of each substance, and after ingesting the product labeled LSD 25 the hallucinations and feelings that he had experienced before burst inside him again. In his personal trial he took a quantity of LSD which corresponded to 5 times the active dose. It would be funny to think about how things would have been if Hofmann, had stopped at a bar with his friends, the habit of many researchers, instead of going directly home. He would have attributed the observed anomalies to something else and we would have never known about LSD, whose effects on the central nervous system would have been felt only by the few laboratory animals treated in 1938 and 1943. LSD remained a molecule used only in psychiatry and under strict medical control for 20 years. This was due mainly to the limited availability of lysergic acid which was obtainable only from *C. purpurea* sclerotia. However, its numerous and peculiar biological activities increased the reputation of lysergic acid year by year for its use by the pharmaceutical industries interested in new drug research.

1.2.

THE FERMENTATION ERA

The discovery of penicillin and its production in large scale caused the explosion of microorganism fermentation technology in the major pharmaceutical companies which, in the second half of the 1940's, began research on new antibiotics. The earliest attempts to obtain EA from *C. purpurea* in saprophytic conditions were performed by several authors, all in academic institutions. Their aim was to investigate the best growth conditions in connection with the production of alkaloids (Bonns, 1922; Kirchhoff, 1929; McCrea, 1931; Baldacci, 1946). Despite the luxuriant growth of *Claviceps* in some cases, only traces of alkaloids were detected, visualized by color reactions in crude extracts but never confirmed by isolation of the pure products. The decreasing of the ergotamine added to the medium led to the conclusion that, even if alkaloids were produced in artificial conditions, they were destroyed by the microorganism (Michener *et al.*, 1950). After obtaining these results none of the authors above mentioned went more deeply into the question. However, over the years the biological activities of natural and semisynthetic derivatives of lysergic acid attracted more and more interest.

The first ergot alkaloid produced in saprophytic conditions was achieved in 1949 by Matazô Abe, a Japanese scientist working at the Takeda Research Laboratories in Tokyo. He cultured a strain of *Claviceps* in surface cultures grown on *Agropyrum semicostatum,* a plant common in his area. At first he performed large flask surface cultures of the fungus, whose filtrate gave an evident and constant positive reaction with Van Urk's reagent (Abe, 1949). After extraction of the culture medium he found an alkaloid of a new class having an ergolinic structure but being unexpectedly not a lysergic derivative. He named it agroclavine (Abe *et al.,* 1951; Abe *et al.,* 1953). In spite of Abe's joy in discovering a new alkaloid produced under saprophytic conditions, he was quite disappointed when pharmacologically tested pure agroclavine showed none of the biological activities of the known lysergic derivatives produced by *C. purpurea*. He continued to look for new alkaloids in several *Claviceps* strains grown on different plants and later he isolated a new alkaloid, belonging to the same class as agroclavine, which he named elymoclavine (Abe *et al.,* 1952). It was produced by a *Claviceps* strain collected on a grass, *Elymus mollis,* growing on Kurili island. Elymoclavine differs from agroclavine in that it bears an alcoholic group instead of a methyl group in C-8 position. Having found the conditions for producing alkaloids on surface and submerged cultures, Abe continued to isolate new alkaloids such as festuclavine and penniclavine (Abe *et al.,* 1954) in *Claviceps* on *Festuca rubra;* molliclavine from *Claviceps* on *Elymus mollis* (Abe *et al.,* 1955); pyroclavine and costaclavine from *Claviceps* grown on *Agropyrum semicostatum* (Abe *et al.,* 1956). Since these alkaloids belonged to the same class as agroclavine they were of no therapeutic value. In his research on EA from *Claviceps* of different origins he finally succeeded in isolating a true lysergic derivative with an unusual peptide side chain from a Spanish strain. This new compound named ergosecaline can be considered as an ergocornine analogue lacking proline (Abe *et al.,* 1959). However, it was not mentioned again and was never reported among the alkaloids isolated from *Claviceps*. Abe worked for more than two decades with *Claviceps* and may be considered the most unlucky researcher in EA history. Nevertheless, we are indebted to him for having found the media suitable for culturing alkaloid-producing *Claviceps* strains in good yield.

At the end of the nineteen fifties Abe was not the only researcher interested in producing lysergic acid derivatives from *Claviceps purpurea* grown in saprophytic cultures. Several other laboratories in both universities and industries in Europe and the United States were working on this problem. Such a widespread interest could be attributed to both the great development in the techniques for microorganism fermentation, mainly for antibiotic production, and the high cost of lysergic acid, which was at that time about $500 per gram. The production of lysergic acid and especially of its natural and semi-synthetic derivatives was practically monopolized at that time by the Sandoz Pharmaceutical Company of Basel which, since Stoll's discovery of ergotamine, had produced these alkaloids by large scale cultivations of *Claviceps*-infected rye. At that point a person of fundamental importance in EA history entered the scene: Ernest Boris Chain.

E.Chain was awarded the Nobel prize in 1945 along with A.Fleming and H.Florey for their discovery of penicillin. Chain was a Jewish chemist who had studied and worked in Berlin as a young man during the nineteen thirties. Because of the nazi racial laws he moved to Oxford and worked as a researcher at the Sir William Dunn School of Pathology. It was here that Chain resumed the studies on penicillin that Fleming had interrupted in 1929. Chain devoted himself first to studying the production of penicillin by means of fermentation and then to the elucidation of its structure. During these years and with scarce means Chain acquired solid experience in fermentation which complemented the chemical knowledge he had obtained in Berlin. He was a brilliant researcher, a great mind, and had a vast experience in chemistry and biology as well as a strong drive to achieve his goals, but he was a difficult man. Until a few years ago, in Oxford, the door of the study of professor R.Robinson (one of the most brilliant English chemists of his time) bearing a large ink spot had been preserved. The spot was formed when Sir Robinson threw an ink bottle at Chain at the end of a discussion on the structure of that "diabolical linkage of reactive groups" otherwise known as penicillin (Sheehan, 1982).

In the middle of the nineteen fifties E.Chain moved to Italy to the Istituto Superiore di Sanità (ISS) in Rome, attracted by the possibility to set up and direct the International Center of Microbiological Chemistry (CICM). The new and well-equipped center was furnished with good chemistry and microbiology laboratories and a modern pilot plant with fermenters of up to 100L. Chain continued his studies on the fermentation of penicillin and its analogues and under his direction in 1958 at the CICM, 6-amino penicillanic acid was isolated for the first time. Chain directed the research on other applicative topics in addition to that of penicillin and its analogues. Among these topics was EA research aimed, however, not at speculative objectives but at the production of lysergic acid. With the help of microbiologist A.Tonolo and the use of small fermenters "hand made" by CICM technicians, Chain was able to transfer various *Claviceps* strains in submerged cultures. As in all studies of this type, he started with colonies obtained from sclerotia of different origins grown in dishes on solid media which were then transferred to liquid media in 300 mL Erlertmeyer flasks. The cultures which gave a sign of alkaloid production were transferred to small 2 L fermenters composed of a large glass tube and numerous electrodes in which an air flow constantly agitated the medium. Alkaloid production was directly assayed on the culture filtrates in the flask with Van Urk's reagent. The mycelium was never taken into consideration because at that time it was a common notion that all the known secondary metabolites produced by microorganisms were released into the medium and not retained in the cells. This was verified for Abe's clavines, penicillin, and many other antibiotics discovered in those years. It is very likely that this procedure had caused the discovery of one of the few water soluble alkaloids produced by *Claviceps,* the methylcarbinol amide of lysergic acid (MCA). The hypothesis that some ergopeptine-producing strains could have been cultured and then discarded cannot be excluded, given the numerous strains of *C. purpurea* assayed.

After more than one year of attempts Tonolo was able to obtain with Van Urk's reagent an evident coloration of a culture filtrate, thus indicating an alkaloid production greater than 10 mg/L as ergotamine. This culture originated from a sclerotium collected by Tonolo himself from a graminaceous plant *(Paspalum distichum)* growing in the Roman countryside. The microorganism, named *Claviceps paspali,* grew abundantly in submerged culture in pale-colored filamentous aggregates, and when the strain exhibited a stable alkaloid production, the research was focused exclusively on it and the other strains were given up. This strain was labelled F550 and all the strains used afterward derived from it. Chain's goal at the time was not to produce a specific alkaloid, known or unknown, but rather to get by fermentation whatever lysergic acid derivative in high yield from which pure lysergic acid could be obtained. The first cultures of *C. paspali* which gave a positive reaction with Van Urk reagent, were analyzed by paper chromatography because at that time thin layer chromatography (TLC) was still in its infancy. The chromatograms showed five fluorescent spots four of which, reacting with the Van Urk reagent, had the characteristic UV spectra of true lysergic derivatives (quite different from those of clavines). In order to confirm the presence of the lysergic acid in their structures the crude alkaloids were hydrolyzed, extracted and analyzed. IR analysis, the most reliable technique at the time, revealed an exact overlapping of the bands of the obtained product with those of a pure sample of lysergic acid. In December 1959, the possibility to produce lysergic acid by saprophytic cultures of *C. paspali* was at last achieved (Arcamone *et al.,* 1960).

Chain did not intend to repeat his experience with penicillin: a lot of honor but no economic benefits. In fact, the production of penicillin by English laboratories was submitted to payment of royalties to the American laboratories who owned the patents, dated before the isolation of penicillin itself, on the producer microorganisms. Chain had a position in a public institution in Rome, the ISS, so he could not be either the owner of a patent on the production of lysergic acid derivatives (LAD) or sell it to anyone. In order to bypass the problem he agreed with the head of the ISS to create a foundation which was allowed to give financial support to the research performed in the ISS. A patent was issued on the production of LAD in submerged cultures by *C. paspali* and was transferred to a pharmaceutical firm, leader in the field of industrial fermentations, the Farmitalia Company in Milan. Farmitalia sent a team of five young researchers to work in Chain's laboratories in Rome, each specialized in one of the following fields: chemical synthesis, extraction, fermentation process, microbiology and biochemical analysis, covering all the fields concerning the cultivation and fermentation of microorganisms as well as the isolation and identification of their metabolites. The researcher considered to be the head of the group, F.Arcamone, had already isolated a few new antibiotics including the antitumor agent adriamycin.

Research was carried out in two directions: identification of the alkaloids produced and increase of their production. As far as the identification of the alkaloids is concerned, it was surprisingly found that they did not correspond to any known LAD, either water soluble like ergometrine or liposoluble like the ergopeptines. Hydrolysis of the crude extracts gave no amino acids except traces of 2-aminopropanol, a sign of the presence of ergometrine, but less than 5% of the total alkaloids. It soon became clear that the four compounds corresponded to only two alkaloids, both present in two isomeric form. One of them was the lysergic acid amide, ergine, already known as a degradation product of ergotinine, which was classified as a natural product from ergot. Later, its presence in nature was excluded for biosynthetic reasons. The other one, during purification, constantly released small amounts of ergine, and thus was revealed to be a labile derivative of ergine. This alkaloid gave acetaldehyde upon treatment with diluted sulfuric acid and was identified as the methyl carbinol amide of lysergic acid (Arcamone *et al.,* 1961). The fifth spot observed on the chromatograms, Van Urk reagent negative, was identified as 2,3-dihydroxy benzoic acid, a compound related to the alkaloid production. In fact, biosynthetic studies confirmed that it derived from the catabolism of tryptophan, the earliest EA precursor (Arcamone *et al.,* 1961). In conclusion, the four spots observed on the chromatograms of the crude extracts all derived from a mere MCA which formed ergine by degradation. Both compounds then isomerized into the respective dextrorotatory form. The five Farmitalia researchers remained in Rome for more than a year during which Chain was very interested in their progress. Every afternoon he came to CICM to discuss the results, to express hypotheses and to plan the next experiments. The following day he appeared without fail to inquire about the results of the experiments even if they were not yet finished. Very seldom did a fermenter or flask fermentation arrive to its natural or scheduled end: the data from each experiment being immediately used as starting point for new ones under different conditions. Chain had surprising and almost always correct perceptions which resulted from his vast and long experience in industrial fermentation. He became very interested in measuring the oxygen consumption and concentration inside the medium, at that time very difficult to perform. The cost of the fermentation both in terms of time and money was another constant problem for him. After one year production of LAD reached 1 g/L with a very simple and cheap medium consisting of mannitol, succinic acid and ammonia. A medium so "clean" which avoided complex components such as corn steep or starch was very useful in both studying the fermentation process and in the recovering the alkaloids. Chain had been working in Rome until 1964 when, together with the head of the ISS, he was involved in a lawsuit. A violent press campaign against the ISS management, desired and fomented by political factions, compelled him to move to England. There he was appointed to the head of the Biochemical Department of the Imperial College of London where he never ceased his research on EA.

As the work on LAD progressed in Rome the Farmitalia company in Milan started its production on large scale. C.Spalla, the head of the Department of Industrial Microbiology, studied the strain F550 as soon as it was available, and in a short time a production of over 2 g/L was achieved. Spalla was also supervisor of the Farmitalia industrial fermentation plant in Settimo Torinese, near Turin, where fermenters with capacities up to 50,000 L and 100,000 L were used for antibiotics and vitamins production. He was credited with having largely increased alkaloid production but mainly for his efforts and success in transferring the good results obtained in flasks in large fermenters.

With strain F550 and large-scale fermentation Farmitalia soon became the major producer of lysergic acid, and was able to satisfy every request until something unexpected happened: the discovery of paspalic acid (Kobel *et al.,* 1964). The fact that Abe had found clavine-producing *Claviceps* sp. in Japan and Chain had found MCA-producing *Claviceps* in Italy led to the realization that other strains able to produce new alkaloids may exist. This idea prompted some companies to search all around the world for sclerotia and to analyze their alkaloid production. The company most interested in this pursuit was Sandoz. Sandoz's luck brought the company to Africa, to the Portuguese colony of Angola, where a strain of *Claviceps* which produced a new ergot alkaloid, paspalic acid, was found. Paspalic acid has a structure which is an intermediate between the clavines and lysergic acid differing from the latter only in the presence of the double bond in position 8,9 instead of 9,10. But the fact of utmost importance is that, upon mild alkaline treatment, the double bond shifts to position 9,10, thus making the conversion to lysergic acid. Sandoz was a pharmaceutical company with a synthetic and extractive history; in fact its main drugs were ergot alkaloids and digitalic glycosides. Nevertheless, the firm had no experience in fermentation. After the discovery of paspalic acid, Sandoz acquired a small company with fermentation plants in Austria and thus started producing paspalic acid which was easily transformed into lysergic acid. In this field Sandoz became a competitor of Farmitalia but the latter maintained the advantage due to its high producing strains and the potential of its fermenters.

As far as the investigation of the production of sclerotia collected in different regions is concerned, a surprising pattern became evident. In Spain and Portugal *Claviceps* sclerotia were found to produce mainly ergotoxine, while in central Europe they produced ergotamine, in east Europe ergometrine, in Japan clavines and in North America chanoclavines were found in maize-parasitizing strains. In addition to this linear positioning of strain structure complexity along a geographic pathway, some anomalies also existed. MCA-producing sclerotia were found only in Rome and in Australia. Their location in two such distant places puzzled Tonolo who, after consulting botanical texts on Italian flora, found that *Claviceps* parasitizing *Paspalumdistichum* had never been reported. One day, upon returning to where he had picked the sclerotia, he learned from the local people that New Zealander troops had been stationed in that area for a long time during the second world war.

1.3.
THE ERGOPEPTINES

As soon as the problem of the production of lysergic acid by fermentation was resolved, EA research efforts turned to the production of ergopeptines. At the beginning of the 60's the known ergopeptines (ergotamine, ergosine, ergocornine, ergocristine and ergocryptine) were extracted from ripe sclerotia (Hofmann, 1964) or produced semisynthetically from lysergic acid (Hofmann *et al.*, 1961). Nevertheless, it was evident that the production of these compounds by fermentation would have captured the market. At the same time a dozen research groups in both Europe (Gröger, 1959; Plieninger, 1961; Agurell, 1962; Floss, 1967) and the USA (Abou-Chaar, 1961; Pacifici, 1962; Taber, 1967), almost all in Universities, were working on this topic. Each of these groups had a strain of *C. purpurea* which produced small quantities of a complex mixture of ergopeptines often as minor alkaloids among a lot of clavines. The major problems were to first find producer strains of the desired alkaloid, and then to enhance the production by 10–100 times.

A project involving the same researchers who had been working in Rome with Chain was started in 1961 in the Farmitalia R & D Laboratories in Milan. In those years ergotamine was the most popular alkaloid on the market. By exploiting the previously mentioned knowledge on the production of different alkaloids according to the place of origin, sclerotia were collected in central Europe where ergotamine-producing *Claviceps* were present. Colonies obtained from sclerotia, often quite different from one another in spite of their common origin, were cultured on solid medium. Each colony was homogenized and utilized as starting material for a submerged culture. After 10 days the cultures were checked for alkaloid production with Van Urk's reagent and by TLC. In one year's time a few hundred strains were examined but no constant producer of EA was found. Some cultures produced a few mg/L, but the alkaloids always disappeared in subculturing. At this point a microbiologist, Alba M.Amici, joined the group. She had never worked on *Claviceps,* but to the great surprise of her colleagues and herself, she observed considerable alkaloid production in her first isolate. Her satisfaction was greatly enhanced when TLC analysis indicated that the strain produced mainly ergotamine. Furthermore, the mycelium of the culture, discarded but quickly retrieved from the garbage can, still contained a good amount of alkaloids. This strain, labelled FI 275 with a lasting production of 150 mg per liter, was the starting point for the industrial production of ergotamine (Amici *et al.,* 1966). Attempts to optimize the medium, the same one adopted for *C. paspali* but containing glucose instead of mannitol, gave inconsistent results. Phosphate was the only component whose concentration showed to be crucial, within limits, for a good production. A dramatic improvement took place when Amici remembered that the sclerotia in nature grow on the nectar in rye flowers. She performed an experiment by increasing the amount of sucrose in the medium up to 20%, a quantity that she considered quite high. Surprisingly, analyses revealed that the amount of the alkaloids produced increased proportionally to the amount of sucrose added. A second experiment with sucrose amounts ranging from 5 to 50% allowed for the determination that the optimum quantity fell between 30 and 35%. With this new medium EA production rose several fold. This result represented the most characteristic and unexpected among the data reported on EA production in *Claviceps* (Amici *et al.,* 1967). No medium with such a high concentration of sucrose had been previously reported in the literature for the production of a secondary metabolite. A medium containing 300 kg of sucrose per m^3 couldn't be initially considered for use as an industrial medium because it was too expensive, but the value of the alkaloids produced made up for the expense. When this medium was published, another researcher pointed out an "error"; he thought that the reported figure of 300 g/L was an error in the transcription from the manuscript in place of the more likely value of 30.0 g/L. Analyses at the end of fermentation indicated that only a part of the sucrose was consumed, the great majority being converted into oligosaccharides formed by a molecule of glucose linked to a linear sequence of fructose (Arcamone *et al.,* 1970). It was later found that 10% of the sucrose could be substituted with KCI, a significantly cheaper compound.

From that moment on, strategies for increasing production were no longer based on the assessment of the nutrients, but on the selection of cell lines endowed with the highest potency. The procedure was very simple and was based on cultures plated on solid agar medium. The undersides of colony-containing plates were inspected by UV light in order to identify the most fluorescent colonies presumed to contain the largest amount of alkaloids. Those colonies were picked up, grown onto slants and divided in two halves. The first half was fermented in submerged culture and analyzed, the second half preserved at low temperature. The second half of the most productive colonies was homogenized, plated again, then submitted to an analogous treatment. Little by little, with each selection step, the production of ergotamine increased from a few hundred to a few thousand mg/L. In spite of this ten-fold improvement, the internal ratio of the produced alkaloids never changed significantly. The researchers involved in this project were frequently asked at international meetings how these results were achieved. Although they always explained this strategy in detail, nobody ever believed them. Everyone suspected the existence of some trick which could not be revealed; on the contrary it was the pure and simple truth. Perhaps people did not realize that such an endeavor, which involved the fermentation of thousands cultures and their analyses, could have been performed only at an industrial level and not in a University laboratory, even one well equipped.

During these years Chain and Tonolo, while working at the Imperial College of London, also found a strain which produced ergotamine in addition to other alkaloids, but they couldn't obtain more than a few hundred mg/L (Basset *et al.*, 1973). After ergotamine the Farmitalia's researchers succeeded in producing, with the same technique and in a short time, the other known ergopeptines: ergocryptine (Amici *et al.*, 1969), ergocristine (Minghetti *et al.*, 1971), ergocornine and ergosine (Amici *et al.*, 1971). Usually the strains produced two major alkaloids, one belonging to the ergotamine group and one to that of ergotoxine. This fact was explained when the biosynthetic pathways of the overproduced amino acids were considered. The identification and the quantification of each alkaloid in a crude extract were determined as follows: alkaloids were divided in the three groups of ergotamine, ergoxine and ergotoxine by TLC. The groups were eluted and submitted to acid hydrolysis and the resulting amino acids were determined by an amino acid analyzer. With this procedure, only the amino acids present in position 2 and 3 (proline) were detected (see Figure 3). The amino acid in position 2 characterizes the alkaloid in the group. When several ergopeptines were present in the same group, the total molecular amount of amino acids (valine, leucine, isoleucine and phenylalanine) had to be equal to the amount of proline. This procedure was very useful in identifying some minor alkaloids. For example, the presence of isoleucine revealed β-ergocryptine production. Stepwise selection, as previously described, led to the attainment of a strain with a high production of this alkaloid (Bianchi *et al.*, 1976).

In some cases proline was found in smaller amounts than expected. This fact together with the discovery of large amounts of lysergylvaline methylester in the crude extracts revealed the production of "open" ergopeptines (ergopeptams) in which the side chain was not completely cyclized. The first compound of this series was discovered by Sandoz researchers who isolated lysergylvaline methylester and considered it a natural product (Schlientz *et al.*, 1963). Later the same authors realized that it was an artifact resulting from the methanolysis which had occurred during the extraction of an unexpected ergopeptine with a structure similar to ergocristine, in which valine was not linked to proline by an oxygen bridge (Stütz *et al.*, 1973). In such a structure valine is hydrolyzed in acidic conditions to a free amino acid instead of being degraded, as in the case of

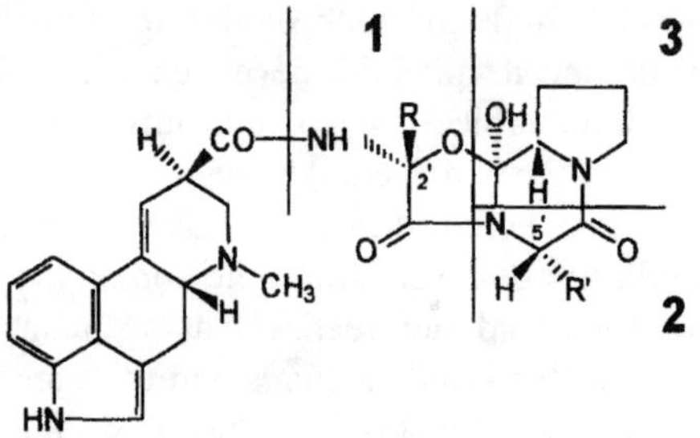

	R	Ergotamine group	Ergoxine group	Ergotoxine group
R'		R = CH$_3$	R = CH$_2$-CH$_3$	R = CH(CH$_3$)$_2$
-CH$_2$-C$_6$H$_5$		ERGOTAMINE	ERGOSTINE	ERGOCRISTINE
-CH$_2$-CH(CH$_3$)$_2$		α-ERGOSINE		α-ERGOCRYPTINE
-CH(CH$_3$)-C$_2$H$_5$				β-ERGOCRYPTINE
-CH$_2$-CH(CH$_3$)$_2$				ERGOCORNINE

Figure 3 The first known natural ergopeptines

ergocristine. This alkaloid open a new series of ergopeptines: the ergopeptams which exactly overlap the structures reported in Figure 3. Another finding included the presence of small amounts of other amino acids on the chromatographic panel. These amino acids were at first neglected because they were considered impurities, but their constant presence caused some suspicion since they were the amino acids typically present in the different ergopeptines.

The insertion of all the known alkaloids in a grid where each compound has its own place according to the side chain of either the first and the second amino acid, as reported in Figure 3, was very useful in predicting the existence of other new missing alkaloids. In fact, three alkaloids were missing in the ergoxine group: α and β-ergoptine and ergonine. An investigation was undertaken to find these alkaloids in the crude extract of the strain FI 231, mainly an ergotoxine producer. During the isolation of these compounds a paper appeared, in which Sandoz researchers described the isolation and characterization of ergonine and ergoptine from sclerotia, two alkaloid already obtained by synthesis (Stütz *et al.*, 1970). However, the Farmitalia researchers found small amounts of alanine and α-aminobutyric acid (ABA) in the chromatographic fractions containing the two previously mentioned alkaloids. Alanine was considered an impurity and ABA was believed to have arisen from the ergopeptams corresponding to ergonine and ergopeptine. Nevertheless, analysis of the spectroscopic data of one isolated compound revealed that it was an authentic ergopeptine in which ABA was the second amino acid. This compound was named ergobutine and originated a new series of ergopetines (Bianchi *et al.,* 1982). Following the discovery of ergobutine it was not difficult to find another member of this series, ergobutyrine, which corresponded to the ergotoxine group. The last missing alkaloid, ergobine, which belonged to the ergotamine group, owing to its very low production went on be discovered later by the same authors (Crespi-Perellino *et al.,* 1993). Only two alkaloids, β-ergoptine and β-ergosine, remained to be included in the previous mentioned grid. During the research on these products two other unexpected alkaloids were found which corresponded to ergocryptine and ergocornine and bore a methoxy instead of a hydroxy group in 12' position (Crespi-Perellino *et al.,* 1987). They could have been considered the first members of a new series, but no similar alkaloid has since been described. This was the second time that, looking for a well defined alkaloid, the researchers found some different ones. Recently, researchers of Galena (Czech Republic) have isolated an ergopeptine, ergogaline, bearing homoisoleucine in position 2 (Cvak *et al.,* 1994), which could also be considered the first member of a new series of ergopeptines, as for ABA. The presence of alanine among the amino acids in the panel of the hydrolyzed alkaloids led to the idea that it could also be present in the second position of the ergopeptines' side chain. In spite of very careful investigations no alkaloid was found with such a structure. It was analogously investigated whether the amino acid in 1 position of the ergopeptines could be one other than the usual alanine, ABA or valine. The most likely candidate would be leucine, but experiments with high quantities of added leucine as well as with labeled leucine excluded this hypothesis.

After having worked for more than a decade on EA, researchers wondered why only those specific alkaloids were produced and if it was possible to obtain others. If one considers all ergopeptines reported in Figure 4, the question arises as to why only alanine, aminobutyric acid, valine, leucine, isoleucine, phenylalanine and proline are involved in ergopeptine biosynthesis. Comparing these amino acids with all the other natural ones it became evident that the first group had a lipophilic side chain structure while the others had a more polar one. Since the lipophilicity of the amino acid side chains cannot be sacrificed because of its incorporation into the ergopeptines, unnatural amino acids were fed to the cultures of all four high producing *Claviceps* strains (Bianchi *et al.,* 1982). The following amino acids were tested: p-fluorophenylalanine, p-nitrophenylalanine, p-methoxyphenylalanine, norleucine and norvaline. All the strains gave a consistent amount (10–40%) of ergopeptines with the unnatural amino acid present as second amino acid of the side chain. A few analogues of such ergopeptines were produced and pharmacologically tested, but they showed no promise for the future.

In addition to the production by fermentation of the natural alkaloids, the biosynthesis of these compounds was also investigated. As soon as the strain F 550, a producer of simple lysergic amides, was available studies aimed at explaining the biosynthesis of the lysergic acid moiety were initiated. A

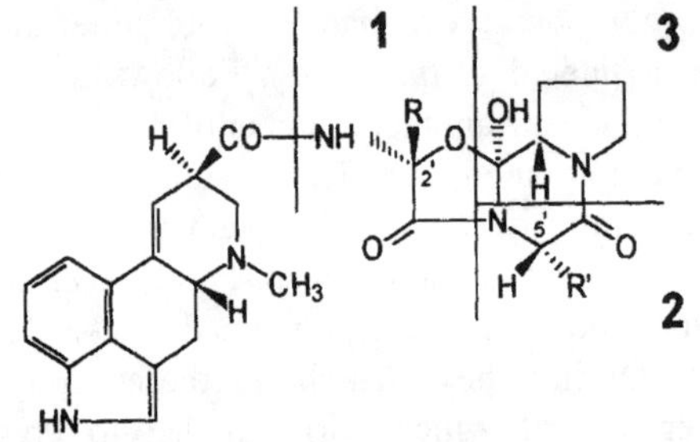

	R	Ergotamine group	Ergoxine group	Ergotoxine group	
R'		R = CH₃	R = CH₂-CH₃	R = CH(CH₃)₂	R = CH₂-CH(CH₃)₂
-CH₂-C₆H₅		ERGOTAMINE	ERGOSTINE	ERGOCRISTINE	_____
-CH₂-CH(CH₃)₂		α-ERGOSINE	α-ERGOPTINE	α-ERGOCRYPTINE	_____
-CH(CH₃)-C₂H₅		β-ERGOSINE*	β-ERGOPTINE*	β-ERGOCRYPTINE	_____
-CH₂-CH(CH₃)₂		ERGOVALINE	ERGONINE	ERGOCORNINE	_____
-CH₂-CH₃		ERGOBINE	ERGOBUTINE	ERGOBUTYRINE	_____
-CH₃		_____	_____	_____	_____

not yet found in nature.

Figure 4 Ergopeptines produced by *Claviceps purpurea*

few years earlier, in 1956, mevalonic acid had been discovered and it became clear that the building blocks of lysergic acid were tryptophan, mevalonic acid and a methyl group. All the steps in the biosynthetic pathway from tryptophan to lysergic acid were soon demonstrated also in details (Floss, 1976). However, the origin of the ergometrine side chain, the aminopropanol moiety, still remains obscure. The biosynthesis of ergopeptines, as a serial addition to the lysergic moiety of amino acids assembled in the cyclol structure, initially seemed easy to demonstrate. The early discovery of lysergylvaline methylester strengthened this hypothesis (Schlientz *et al.,* 1963), but what was expected to be clear in a short time needed thirty years to be explained. In fact only one year ago it has been demonstrated, by using the FI 275 strain, that the tripeptide was assembled starting from the last amino acid proline, bound to the enzyme and, only when linked to the lysergic acid, was the entire ergopeptine released in free form: an unusual and unexpected mechanism of biosynthesis (Riederer *et al.,* 1996).

Now, after almost forty years since the first fermentation of EA, some unexpected findings have taken place: among these ergobalansine, a new ergopeptine, produced by the endophytic fungus *Balansia* living on sandbur grass *(Cenchrus echinata)* and the sedge *Cyperus virens* has been found. This alkaloid may be considered an analogue of ergotamine in which proline is substituted by alanine and its isomer ergobalansinine has been isolated also from the seeds of *Ipomoea piurensis* (Jenett-Siems *et al.,* 1994). The isolation of ergobalansine and other unusual alkaloids reveals that the matter may not be so clear cut as one might have thought. Many other unknown alkaloids still lay hidden in sclerotia of *Claviceps* sp. or in some endophytes fungi like *Balansia* sp., Sphacelia sorghi (Mantel *et al.,* 1968; Mantel *et al.,* 1981), and *Aspergillus fumigatus. (Spilsbury et al.,* 1961) Also higher plants (Convoluvulaceae) have been recognized to be good producers of EA. Ergobalansinine in *Ipomoea piurensis* (Jenett-Siems *et al.,* 1994), ergosine, ergosinine and agroclavine in *Ipomoea argyrophylla* (Stauffacher *et al.,* 1966), lysergol in *Ipomoea bederacea,* ergine, isoergine and clavines in *Rivea corymbosa* (Hofmann *et al.,* 1960), ergine, isoergine and clavines in tropical wood roses *Argyreia nervosa,* and *Ipomoea tuberosa* (Hylin *et al.,* 1965) have been found. The finding of EA in higher plants was completely unexpected because until then they had been searched for and considered to be present exclusively in *Claviceps* sp.

At this point a question arises: why is the mechanism of EA biosynthesis, so complex and exclusive, spread both in microorganisms and in higher plants? Considering that all the alkaloids are produced by higher plants it might be that the genes encoding the enzymes responsible for their biosynthesis had been present first in plants. Later a parasitic microorganism like Claviceps could have picked up them and evolved increasing the production. This is the case for taxol, a macrocyclic diterpene produced by the tree *Taxus brevifolia* and now found also in some endophytic microorganisms (Stierle *et al.,* 1993). This hypothesis has never been considered but this story, not concerning the few past decades but the millions of years elapsed during the evolution, if investigated, might lead to truly unexpected conclusions.

REFERENCES

Abe, M. (1949) Ergot fungus. IX . Separation of an active substance and its properties. *J. Agr. Chem. Soc. Japan,***22,**2–3.

Abe, M., Yamano, T., Kozu, Y. and Kusumoto, M. (1952) A new water-soluble ergot alkaloid, elimoclavine. *J. Agric. Chem. Soc.,***25,**458–459.

Abe, M., Yamano, T., Kozu, Y. and Kusumoto, M. (1951) Ergot fungus. XVIII. Production of ergot alkaloids in submerged cultures. *J. Agr. Chem. Soc. Japan,***24,** 416–422.

Abe, M., Yamano, T., Kozu, Y. and Kusumoto, M. (1953) Ergot fungus. XX. Isolation of a mutant which produces good yields of agrociavine even in submerged cultures. *J. Agr. Chem. Soc. Japan,***27,**18–23.

Abe, M., Yamano, T., Yamatodani, S., Kozu, Y., Kusumoto, M., Kamatsu, H. and Yamada, S. (1959) On the new peptide-type ergot alkaloid, ergosecaline and ergosecalinine. *Bull. Agric. Chem. Soc. (Japan),***23,**246–248.

Abe, M. and Yamatodani, S. (1954) Isolation of two further water-soluble ergot alkaloid. *J. Agr. Chem. Soc. (Japan),***28,**501–502.

Abe, M. and Yamatodani, S. (1955) A new water-soluble ergot alkaloid, molliclavine. *Bull. Agric. Chem. Soc. (Japan),***19,**161–162.

Abe, M., Yamatodani, S., Yamano, T. and Kusumoto, M. (1956) Isolation of two new water-soluble alkaloids pyroclavine and costaclavine. *Bull. Agric. Chem. Soc.(Japan),* **20,**59–60.

Abou-Chaar, C.I., Brady, L.R. and Tyler, V.E. (1961) Occurrence of lysergic acid in saprofytic cultures of *Claviceps. Lloydia,***24,**89–93.

Agurell, S. and Ramstad, E. (1962) Biogenetic interrelationships of ergot alkloids. *Arch.Biochem. Biophys.,***98,**457–470.

Amici, A.M., Minghetti, A., Scotti, T., Spalla, C. and Tognoli, L. (1966) Production of ergotamine by a strain of *Claviceps purpurea* (Fr.) Tul.*Experientia,***22,**415–418.

Amici, A.M., Minghetti, A., Scotti, T., Spalla, C. and Tognoli, L. (1967) Ergotamine production in submerged cultures and physiology of *Claviceps purpurea.Appl.Microbiol.,***15,**597–602.

Amici, A.M., Minghetti, A., Scotti, T., Spalla, C. and Tognoli, L. (1969) Fermentative process for producing ergocriptine. U.S. Pat. Off. 3,485,722.

Amici, A.M., Minghetti, A. and Spalla, C. (1971) Fermentative process for the preparation of ergocornine and ergosine. U.S. Pat. Off. 3,567,584.

Arcamone, F., Barbieri, W., Cassinelli, G., Pol, C., (1970) Structure of the oligosaccharides synthetized by a strain of *Claviceps purpurea.Carbohydr. Res.,***14,** 65–71.

Arcamone, F., Bonino, C., Chain, E.B., Ferretti, A., Pennella, P., Tonolo, A. and Vero, L. (1960) Production of lysergic acid derivatives by a strain of *Claviceps paspali* in submerged culture. *Nature (London),***187,**238–239.

Arcamone, F., Chain, E.B., Ferretti, A., Minghetti, A., Pennella, P., Tonolo, A. and Vero, L. (1961) Production of a new lysergic acid derivative in submerged cultures by a strain of *Claviceps paspali* Stivens & Hall*Proc. of the Royal Soc., B,***155,**26–54.

Arcamone, F., Chain, E.B., Ferretti, A. and Pennella, P. (1961) Formation of 2, 3dihydroxybenzoic acid in fermentation liquor during the submerged culture production of lysergic acid hydroxyethylamide by *Claviceps paspali.Nature,***192,**552–553.

Baldacci, E. (1946) Produzione di alcaloidi della segale cornuta nelle colture di agar. Coltivazione *in vitro. Farmaco Sci. E. Tec.,***1,**110–115.

Barger, G. (1931) *Ergot and ergotism,*Gurney and Jackson, London.

Barger, G. and Carr, F.H. (1907) *J. Chem. Soc. (London),***91,**337.

Basset, R.A.Chain, E.B. and Corbett, K. (1973) Biosynthesis of ergotamine by *Clavicepspurpurea* (Fr.) Tul.*Biochem. J.,***134,**1–10.

Bianchi, M., Minghetti, A. and Spalla, C. (1976) Production of b-ergocriptine by a strain of *Claviceps purpurea* (Fr.) Tul. In submerged culture. *Experientia,***32,**145–147.

Bianchi, M.L., Cattaneo, P.A., Crespi-Perellino, N., Guicciardi, A., Minghetti, A. and Spalla, C. (1982) Mechanisms of control in the qualitative biosynthesis of lysergic acid alkaloids. In Krumphanzl, V., Sikyta, B. and Vanek, Z. (eds.), *Overproduction* of *microbial products,*Academic Press, London, pp. 436–442.

Bianchi, M.L., Crespi-Perellino, N., Gioia, B. and Minghetti, A. (1982) Production by *Claviceps purpurea* of two new ergot peptide ergot alkaloids belonging to a new series containing α-aminobutyric acid. *J. Nat. Prod.,***45,**191–196.

Bonns, W.W. (1922) A preliminary study of *Claviceps purpurea* in culture. *Amer. Jour.Bot.,* **9,** 339–353.

Craig, L.C., Shedlovsky, T., Gould, R.G. and Jacobs, W.A. (1938) The ergot alkaloids. XIV. The position of the double bond and the carboxyl group in lysergic acid and its isomer. The structure of the alkaloids. *J. Biol. Chem.,* **125,** 289–298.

Crespi-Perellino, N., Ballabio, M., Gioia, B. and Minghetti, A. (1987) Two unusual ergopeptins produced by saprophitic cultures of *Claviceps purpurea.J. Nat. Prod.,* **50,** 1065–1074.

Crespi-Perellino, N., Malyszko J., Ballabio, M., Gioia, B. and Minghetti, A. (1993) Identification of ergobine, a new natural peptide ergot alkaloid. *J. Nat. Prod.,* **56,** 489–493.

Cvak, L., Jegorov, A., Sedmera, P., Havlicek, V., Ondracek, J., Husak, M., Pakhomova, S., Kratochvil, B. and Granzin, J. (1994) Ergogaline, a new ergot alkaloid, produced by *Claviceps purpurea*—isolation, identification, crystal-structure and molecularconformation. *J. Chem. Soc. Perkin Trans. 2,* **8,** 1861–1865.

Dudley, H.W. and Moir, C. (1935) New active principle of ergot. *Science,* **81,** 559–560.

Dudley, H.W. and Moir, C. (1935) The substance responsible for the traditional clinical effect of ergot. *Brit. Med. J.,* **1,** 520–523.

Floss, H.G. (1976) Biosynthesis of ergot alkaloids and related compounds. *Tetrahedron,* **32,** 873–912.

Floss, H.G. (1980) The biosynthesis of ergot alkaloids (or the story of the unexpected). *Annual Proc. Phytochem. Soc. Eur.,* 249–270.

Floss, H.G., Günter, H., Mothes, U and Becker, I. (1967) Isolierung von ElymoclavineO-β-D-fruktosid aus Kulturen des Mutterkornpilzes. *Z. Naturforschg.,* **22b,** 399–402.

Frank, B. and Flash, H. (1973) Die ergochrome (Pysiologie, Isolierung, Struktur und Biosynthese). In *Forsch. Chem. Org. Naturstoffe,* **30,** pp. 151–206.

Gröger, D. (1959).Über die Bildung von Clavin-Alkaloide in Submers-Cultur. *Archiv.Pharm.,* **292/64,** 389–397.

Hofmann, A. (1964) Analytik der Mutterkornalkaioide. In *Die Mutterkorn alkaloide,* Ferdinand Enke, Verlag, Stuttgard, pp. 113–130.

Hofmann, A. (1964) Die Chemie der Mutterkoralkaloide. In *Die Mutterkorn alkaloide,* Ferdinand Enke, Verlag, Stuttgard, pp. 14–175.

Hofmann, A., Frey, A.J. and Ott, H. (1961) Ergot alkaloids. L. Total synthesis of ergotamine. *Experientia,* **17,** 206–207.

Hofmann, A. and Tscherter, H. (1960) Isolierung von Lysergsaure-Alkaloiden aus der mexikanischen Zauberdroge Ololuiqui (*Rivea corymbosa* (L.) Hall. f.). *Experientia,* **16,** 414–416.

Hylin, J.W. and Watson, D.P. (1965) Ergoline Alkaloids in Tropical Wood Roses. *Science,* **148,** 499–500.

Jacobs, W.A. and Craig, L.C. (1932) The ergot alkaloids. 1. The oxidation of ergotinine. *J.Biol. Chem.,* **97,** 739–743.

Jacobs, W.A. and Craig, L.C. (1934) The ergot alkaloids. 2. The degradation of ergotinine with alkali. Lysergic acid. *J. Biol. Chem.,* **104,** 547–551.

Jacobs, W.A. and Craig, L.C. (1934) The ergot alkaloids. 3. Lysergic acid. *J. Biol. Chem.,* **106,** 393–399.

Jacobs, W.A. and Craig, L.C. (1935) Alkaloids from ergot. *Science,* **82,** 16–17.

Jacobs, W.A. and Craig, L.C. (1935) Ergot alkaloids. *Science,* **81,** 256–257.

Jacobs, W.A. and Craig, L.C. (1935) Ergot alkaloids. V. Hydrolysis of ergotinine. *J. Biol.Chem.,***110,**521–530.

Jacobs, W.A. and Craig, L.C. (1935) Ergot akaloids. VI. Lysergic acid. *J. Biol. Chem.,* **111,**455–465.

Jacobs, W.A. and Craig, L.C. (1936) Ergot alkaloids. IX. Structure of lysergic acid. *J. Biol. Chem.,***113,**767–778.

Jacobs, W.A. and Craig, L.C. (1936) Ergot alkaloids. XI. Isomeric dihydrolysergic acids and the structure of lysergic acid. *J. Biol. Chem.,***115,**227–238.

Jacobs, W.A. and Craig, L.C. (1937) The ergot alkaloids. XII. Synthesis of substances related to lysergic acid. *J. Biol. Chem.,***120,**141–150.

Jacobs, W.A. and Craig, L.C. (1938) Ergot alkaloids. XIII. Precursors of pyruvic and isobutyrylformic acids. *J. Biol. Chem.,***122,**419–423.

Jacobs, W.A. and Craig, L.C. (1938) The position of the carboxyl group in lysergic acid. *J. Am. Chem. Soc.,***60,**1701–1702.

Jacobs, W.A. and Gould, R.G. (1939) Ergot alkaloids. XVIII. Production of a base from lysergic acid and its comparison with synthetic 6,8-dimethylergoline. *J. Biol. Chem.,* **130,**399–405.

Jenett-Siems, K., Kaloga, M. and Eich, E. (1994) Ergobalansine/ergobalansinine, a proline-free peptide-type alkaloid of the fungal genus Balansia is a constituent of *Ipomoea piurensis.J. Nat. Prod.,***57,**1304–1306.

Keller, C.C. (1896) *Scbweiz. Wschr. Chem. Pharm.,***34,**65.

Kharasch, M.S., King, H., Stoll, A. and Thompson, M. (1936) New ergot alkaloid. *Nature,* **137,**403–405.

Kharasch, M.S. and Legaul, R.R. (1935) Ergotocin: the active principle of ergot responsible for the oral effectiveness of some ergot preparations on human uteri. *J. Am. Chem. Soc.,***57,**956–957.

Kharasch, M.S. and Legaul, R.R. (1935) New active principle(s) of ergot. *Science,***81,** 614–615.

Kirchhoff, H. (1929) Beitrage zur Biologie und Physiologie des Mutterkornes. *Zentralbl. Bakt. Parasitenk. und Infectionskr.,***77,**310–369.

Kobel, H., Schreier, E. and Rutschmann, J. (1964) 6-methyl-$\Delta^{8,9}$-ergolen-carbonsäure, ein neues Ergolinderivat aus Kulturen eines Stammes von *Claviceps paspali* Stevens et Hall. *Helv. Chim. Acta,***47,**1054–1064.

Kornfeld, E.C., Fornefeld, E.J., Bruce Kline, G., Mann, M.J., Jones, R.G. and Woodward, R.B., (1954) The total synthesis of lysergic acid and ergonovine. *J. Am. Chem. Soc.,***76,** 5256–5257.

Kornfeld, E.G., Fornefeld, E.J., Bruce Kline, G., Mann, M.J., Morrison, D.E., Jones, R.G. and Woodward, R.B. (1956) The total synthesis of lysergic acid. *J. Am. Chem. Soc.,***78,** 3087–3114.

Mantle, P.G. and Atwell, S.M. (1981) Hydroxydihydroergosine, a new ergot alkaloid analogue from directed biosynthesis by *Sphacelia sorghi. Experientia,***37,**1257–1258.

Mantle, P.G. and Waight, E.S. (1968) Dihydroergosine: a new naturally occurring alkaloid from the sclerotia of *Sphacelia sorghi* (McRea). *Nature,***218,**581–582.

McCrea, A. (1931) The reactions of *Claviceps purpurea* to variations of environment. *Amer. Jour. Bot.,***18,**50–78.

Michener, H.D. and Snell, N. (1950) Studies on cultural requirements of *Clavicepspurpurea* and inactivation of ergotamine. *Amer. Jour. Bot.,***37,**52–59.

Minghetti, A., Spalla, C. and Tognoli, L. (1971) Fermentative process for producing ergocristine. U.S. Pat. Off. 3,567,582.

Moir, C. (1932) A new constituent of ergot. *Brit. Med. J.,* 1119–1123.

Pacifici, L.R., Kelleher, W.J. and Schwarting, A.E. (1962) Production of lysergic acid derivatives in submerged culture. I. Fermentation studies. *Lloydia,* **25,** 37–45.

Plieninger, H., Fischer, R., Keilich, G. and Orth, H.D. (1961) Untersuchungen zur Biosyntese der Clavin-Alkaloide mit Deuterierten Verbindungen. *Liebigs Ann.Chem.,* **642,** 214–224.

Riederer, B., Han, M. and Keller, U. (1996) *d*-Lysergyl peptide synthetase from the ergot fungus *Claviceps purpurea.J. Biol. Chem.,* **271,** 27524–27530.

Schlientz. W., Brunner, R. and Hofmann, A. (1963) *d*-Lysergyl-L-valin-methylester, eine neues natürliches Mutterkornalkaloide. *Experientia,* **19,** 397–398.

Sheehan, J.C. (1982) The early history of penicillin. In *The enchanted ring. The untoldstory of penicillin.* MIT press, Cambridge Massachusetts, pp. 18–43.

Smith, S. and Timmis, G.M. (1932) Alkaloids of ergot III. Ergine, a new base obtained by the degradation of ergotoxine and ergotinine. *J. Chem. Soc.,* 763–766.

Smith, S. and Timmis, G.M. (1932) Alkaloids of ergot IV. Complex group common to ergotoxine and ergotamine. *J. Chem. Soc.,* 1543–1544.

Smith, S. and Timmis, G.M. (1934) Alkaloids of ergot V. Nature of ergine. *J. Chem. Soc.,* 674–675.

Smith, S. and Timmis, G.M. (1935) Alkaloid of ergot. *Nature,* **136,** 259–263.

Smith, S. and Timmis, G.M. (1936) Alkaloids of ergot. Isoergine and isolysergic acid. *J. Chem. Soc. (London),* 1440–1444.

Smith, S. and Timmis, G.M. (1936) New alkaloid of ergot. *Nature,* **137,** 111–113.

Smith, S. and Timmis, G.M. (1937) Alkaloids of ergot. VIII. New alkaloids of ergot: ergosine and ergosinine. *J. Chem. Soc.,* 396–401.

Spilsbury, J.F. and Wilkinson, S. (1961) The isolation of festuclavine and two new clavine alkaloids from *Aspergillus fumigatus* Fres. *J. Chem. Soc.,* **398,** 2085–2091.

Stauffacher, D., Tscherter, H. and Hofmann, A. (1965) Isolierung von Ergosin und Ergosinin neben Agroclavin aus den Samen von *Ipomoea argyrophylla* Vatake (Convolvulaceae) . *Helv. Chim. Acta,* **48,** 1379–1380.

Stierle, A., Strobel, G. and Stierle, D. (1993) Taxol and taxane production by Taxomyces andreanae, an endophytic fungus of pacific yew. *Science,* **260,** 214–216.

Stoll, A. (1935) The new ergot alkaloid. *Science,* **82,** 415–417.

Stoll, A. (1945) Über Ergotamin. *Helv. Chim. Acta,* **28,** 1283–1308.

Stoll, A. (1965) Ergot—a treasure house for drugs. *The Pharm. J.,* **194,** 605–613.

Stoll, A. and Burckhard, E. (1935) Ergobasine, a new water-soluble alkaloid from Seigel ergot. *Compt. Rend. Acad. Sci.,* **200,** 1680–1682.

Stoll, A. and Burckhard, E. (1937) Ergocristin und Ergocristinin, ein neues Alkaloidpaar aus Mutterkorn. *Hoppe Seiler's Z. Phisiol. Chem.,* **250,** 1–6.

Stoll, A. and Hofmann, A. (1943) Die alkaloide der Ergotoxingruppe: Ergocristin, Ergokryptin und Ergocornin. *Helv. Chim. Acta,* **26,** 1570–1601.

Stoll, A. and Hofmann, A. (1943) Partialsynthese von Alkaloiden vom Typus des Ergobasins. *Helv. Chim. Acta,* **26,** 944–965.

Stoll, A. and Hofmann, A. (1951) Die Konstitution der Mutterkornalkaloide. Struktur des Peptidteils. III. *Helv. Chim. Acta,* **34,** 1544–1576.

Stoll, A., Hofmann, A. and Becker, B. (1943) Die Spaltstucke Ergocristin, Ergokryptin und Ergocornin. *Helv. Chim. Acta,* **26,** 1602–1613.

Stoll, A., Petrzilka, Th., Rutschmann, J., Hofmann, A. and Günthard, H. (1954) Über die Stereochemie der Lysergsäuren. Und der Dihydro-lysergsäuren. *Helv. Chim.Acta,***37,** 2039–2057.

Stütz, P., Brunner, R. and Stadler, P.A. (1973) Ein neues Alkaloid aus dem Mycel eines *Claviceps purpurea-Stammes*. Zur Biogenese von Ergocristin. *Experientia,***29,**936–937.

Stütz, P., Stadler, P.A. and Hofmann, A. (1970) Syntese von Ergonin und Ergoptin, zweier Mutterkornalkaloid-Analoga der Ergoxin-Gruppe. *Helv. Chim. Acta.,***53,** 1278–1285.

Taber, W.A. (1967) Fermentative production of hallucinogenic indole compounds. *Lloydia,***30,**39–66.

Thompson, M.R. (1935) The active constituents of ergot. A pharmacological and chemical study. *J. Am. Pharm. Assoc.,***24,**24–38, 185–196.

Thompson, M.R. (1935) The active principles of ergot. *Science,***81,**636–639.

Thompson, M.R. (1935) The active principles of ergot. *Science,***82,**62–63.

Uhle, F.C. and Jacobs, W.A. (1945) The ergot alkaloids. XX. The synthesis of dihydro-*dl*-lysergic acid. A new synthesis of 3-substituted quinolines. *J. Org. Chem.,***10,**76–86.

Van Urk, H.W. (1929) Een nieuwe gevoelige reactie op de moederkoornalkaloiden ergotamine, ergotoxine en ergotinine en de toepassig voor het onderzoek en de colorimetrische bepaling in moederkoornpreparaten. *Pharm. Weekbl.,***66,**473–481.

2.
BIOLOGY AND LIFE STRATEGY OF THE ERGOT FUNGI

KLAUS B.TENBERGE

Institut für Botanik, Westfälische Wilhelms-Universität,Schloßgarten 3, D-48149 Münster, Federal Republic of Germany

2.1.
INTRODUCTION

Claviceps species are the causal agents of the ubiquitous ergot disease. About thirty-six different filamentous fungi constitute this genus of phytopathogenic ascomycetes. They parasitize more than 600 monocotyledonous plants of the families *Poaceae, Juncaceae* and *Cyperaceae* (Bové, 1970), including forage grasses and the leading cereals worldwide: wheat, rice, corn, barley, sorghum, oats, rye, millets (Baum *et al.,* 1992). Being epidemic to a greater extent in semi-arid regions than in temperate zones, ergot is of increasing importance in India and Africa, where pearl millet and sorghum are essential crops (Frederickson *et al.,* 1993). Although the fungi cause harvest losses due to replacement of host ovaries with the parasite's resting structures, the ergotcalled sclerotia, the main problem is not a severe loss in seed quantity but arises from complete ruin of grain quality due to the alkaloid content of the sclerotia. Admittedly, ergot alkaloids are secondary metabolites of high pharmacological value and are, therefore, produced worldwide on a large scale, nevertheless, these toxins cause highly dangerous or even deadly ergotism when contaminated grains are fed to animals or are consumed by man. These are the reasons for a continuous interest for ages in ergot fungi and their persistent importance (see chapter 1 in this volume), which will remain valid as long as the main ubiquitous nutritional basis to man and herbivorous livestock is concerned. Worldwide reduction in grain yield and quality causes the permanent necessity for an expensive cleaning of attacked cereals to maintain a minimum of purity standard. A contamination of crops with ergots higher than 0.3% by weight spoils the grain even for feeding (Agrios, 1988). Specific measures for reliable control as well as utilization of positive capacities of ergot fungi closely depend on an overall understanding of host- and pathogen biology.

This article deals with general biology and histopathology of ergot fungi pointing to similarities and differences in the scale of biological forms of *Claviceps* species associated with variable host types and numbers. In contrast to the limited knowledge of fundamental biology in many plantparasite systems

(Mims, 1991), numerous investigations, most of them from Mantle's group and mainly based on the few most important ergot species, *C. purpurea* (Fries ex Fries) Tulasne, *C. fusiformis* Loveless, *C. sorghi* Kulkarni, Seshadri and Hegde, *C. africana* Frederickson, Mantle & De Milliano, and *C. paspali* Stevens and Hall, add to a considerable body of research on *Claviceps* species reviewed by Taber (1985) and, focused on *C. purpurea,* by Tudzynski *et al.* (1995). This article emphasizes on recent advances in general biology, epidemiology and control as well as histopathology and molecular cytology in the genus *Claviceps*. Although much further work is needed in this field of research on both, the pathogen and the targeted host organ, substantial knowledge and modern methods in fine structural analysis of interaction-specific reactions *in situ* open the opportunity to address unsolved hypotheses in this specific ergot-grass relationship and therewith contribute to general understanding of molecular mechanism in the interaction of hosts and pathogens.

2.2.
LIFE-CYCLE

In nature, the parasitic lifes of ergot fungi start with windborne ascospores landing on susceptible hosts in spring. All arising stages of their life-cycle can develop from one single spore, therefore, the ergot fungi are homothallic as shown by Esser and Tudzynski (1978) (see chapter 4 in this volume) for *C.purpurea* (Figure 1). Typically, spores attach and germinate on the pistil surfaces of blooming host florets and initiate a specific pathogenesis pattern with little variation between ergot species (Parbery, 1996). Hyphae invade and colonize the ovary, grow down to the tip of the ovary axis, the rachilla, and establish a specific and persisting host-parasite frontier. The fungi never invade any part further down in the host but proliferate above this site. A sphacelial stroma grows profusely in the ovary, producing masses of anamorphic spores which are exuded into a syrupy fluid (Figure 1). With this honeydew, the conidiospores are transferred to other blooming florets by rainsplash, head-to-head contact or insect vectors. Thereby ergot fungi spread spatially in the field having used the plant gynoecia for their own proliferation.

A few ergot fungi, e.g., *C. africana, C. fusiformis, C. cynodontis* Langdon, *C. paspali* and *C. sorghi,* produce two types of anamorphic spores. They cover a wide range in size and mostly divide into microconidia, measuring about 6 × 2.5 μm, and macroconidia, measuring about 16 × 4 μm. Firstly, the honeydew contains macroconidia, often microconidia as well, which are able to germinate in the honeydew just below the syrup surface. Secondly, conidiophores emerge and differentiate "secondary conidia" outside the liquid in a secondary conidiation cycle. Masses of conidia, mostly microconidia, whiten the surfaces of sticky colourless honey dew droplets one day after their exudation. Since both conidia types can initiate infection, these ergot fungi spread in the field by a second airborne inoculum in addition to the transmittance of macroconidia with

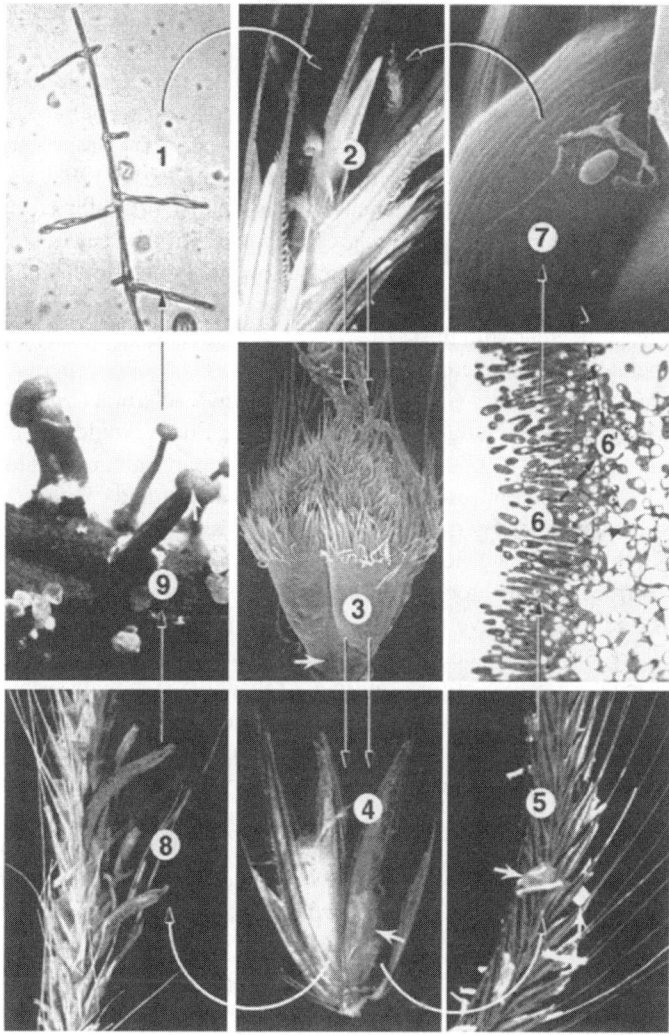

Figure 1 Life-cycle of ergot fungi, shown for *C. purpurea*. The different stages depicted are 1, germinating ascospore; 2, a rye floret at anthesis exposing the stigma between the opened glumes; 3, an infected rye ovary during the endophytic colonization phase with withered stigma and style, long ovary cap hairs and the rachilla (arrow); 4, a biflorescented spikelet of rye after selective inoculation with *C. purpurea* which has formed a sphacelium (arrow) in the infected right floret next to a rye seed developing in the neighbouring uninfected floret; 5, a rye ear with honeydew (arrows) flowing out of infected florets; 6, a sphacelial stroma with phialidic conidiophores producing many anamorphous spores; 6', pointing to the additional microcycle producing airborne microconidia in some other ergot species; 7, germinating conidiospore on the host ovary cap with subcuticular hyphal growth towards the cellular junction; 8, a mature rye ear with several sclerotia; 9, germinating sclerotium with stromata that differentiate perithecia (arrow) in the head periphery containing asci with ascospores (Figures 1 and 9 courtesy of P.Tudzynski)

the honeydew (Luttrell, 1977; Frederickson *et al.,* 1989, 1993; Parbery, D.G. pers. communication).

Next, honeydew production and conidiation usually cease when the formation of sclerotia starts. Sclerotia mature in about five weeks (Figure 1). Finally, during autumn, instead of a caryopsis, a ripe sclerotium leaves the spike, therewith making ergot a replacement tissue disease (Luttrell, 1980). The hard compact ergot consists of a plectenchymatous whitish medulla consisting of special storage cells and a typically pigmented outer cortex. It serves for sexual reproduction and as a resting structure to survive unfavourable conditions, e.g., in temperate zones for overwintering after having fallen to the ground or having been harvested together with the seed.

In temperate zones, sclerotia germinate in spring after a period of low temperature, which favours germination of *C. paspali* sclerotia (Luttrell, 1977). For *C. purpurea,* a temperature of 0°C for at least 25 days would be optimal for germination (Kirchhoff, 1929). Possibly, low temperatures are needed to activate enzymes for lipid mobilization in sclerotia (Cooke and Mitchell, 1967). Optimally, at about 20°C, germination can occur above or just beneath of the soil surface (Kirchhoff, 1929). Germination results in one to sixty clavicipitaceous stromata, formed of mushroom-like stalks with spherical capitula (Figure 1). Both, stalk and capitula have species-specific pigmentation (Frederickson *et al.,* 1991) and are 0.5 to 3.8 cm long, growing positively phototrophic (Hadley, 1968) to reach the air. Except for *C. paspali* (Taber, 1985), female ascogonia and male antheridia develop in the periphery of the capitula and fuse to form dikaryotic ascogenous hyphae. The hyphae surrounding the fertilized ascogonia build flask-shaped perithecia within which karyogamy and meiosis occur, producing asci with thin, needle-like, hyaline, nonseptate ascospores. *Claviceps* ascospores appear to be comparable in different species (Frederickson *et al.,* 1991) and measure 40 to 176×0.4 to 1.2 µm. Under suitable moist conditions, eight ascospores are forcibly ejected through the apical pores of asci, which emerge through the ostiole of perithecia in *C. paspali* (Luttrell, 1977). About four weeks after sclerotia germination, these ascospores represent the airborne primary inoculum and give rise to new infection foci.

2.3.
HOST RANGE AND SPECIFICITY

2.3.1.
Host Range

As mentioned above, ergot species are common on cereals, cultivated forage grasses and many wild grasses, in addition, ergot infects sedges and a few rushes (Taber, 1985). Although *Claviceps* infects over 600 host species worldwide, most of the thirty-six different *Claviceps* species have a monogeneric host range, a few are tribe-specific (Loveless, 1971), and some are species-specific (Parbery, 1996). Some ergot fungi are widely endemic, e.g., *C. africana,* living solely in eastern and south-eastern Africa, or *C. sorghi,* confined to India (Frederickson *et al.,* 1994). In conclusion, endemism is not due to spatial host restrictions, since both species share the same host type.

   C. purpurea, however, parasitizes mainly rye, wheat and barley as well as numerous forage- and roadside grasses (Campbell, 1957; Loveless, 1971). All in all, *C. purpurea* attacks about 400 species of grasses throughout the world (Taber, 1985). Such wide host range is unique in the genus, and rises the question whether physiological races have been evolved. Although substantially investigated, the existence of *formae specialis* has not been verified thus far (Stäger, 1922; Campbell, 1957; Loveless, 1971; Darlington *et al.,* 1977; Frauenstein, 1977; Taber, 1985; Tudzynski *et at.,* 1995). Nevertheless, a high variation in random amplified polymorphic DNA (RAPD) pattern with considerable strain-specificity among 29 field isolated of *C. purpurea* indicated an unusual high degree of genetic diversity in this ergot species; host specificity is indicated to some extend, since most strains from specific host plants group together in preliminary tree analysis (Tudzynski and Tudzynski, 1996; Jungehülsing and Tudzynski, 1997; see chapter 4 in this volume). With few exceptions and regardless of locations, most strains of *C. purpurea* isolated from one host can pass over onto another and vice versa (Campbell, 1957), in particular individuals from ryegrass can infect rye (Mantle, 1967). More aggressive strains can replace other ergot strains having already settled in the ovary (Swan and Mantle, 1991). This multidirectional infection is not without consequences for epidemiology and control (see below).

   Notably, all natural ergot-plant associations share two conspicuous host features, which obviously reflects special adaptations of the antagonist (see below): (a) all hosts are anemophilous monocotyledons and (b) the fungal objectives are host gynoecia solely.

2.3.2.
Organ Specificity

In all *Claviceps* species studied thus far, infection is confined to host ovaries (Parbery, 1996). In general, organ specificity is poorly understood (Schäfer, 1994). In ergot, in particular, this phenomenon is a matter of speculations, because the biological function of the targeted host organ intended for sexual reproduction and its distinctive adaptations thereupon offer some additional concepts, however, are not conceived at their molecular function itself.

   Although the inoculum reaches most likely every host surface area, in nature, successful ergot infection is strictly specific to florets. Florets can prevent infection completely by denying access of inoculum to their pistils due to tightly closed bracts (see Chapter Ergot Virulence and Host Susceptibility). However, one can artificially induce sclerotia development by wounding and inoculating young tissue, e.g., on stalks (Stoll and Brack, 1944), on the shoot apex of rye seedlings (Lewis, 1956), or on nodes and internodes of rye (Garay, 1956). Jointly, these observations point to unique features of pistil surfaces which appear to be indispensable for the establishment of infection. Additionally, imitation of specific pollen-stigma interaction has been suggested. Therefore, molecular cytological investigations of the pistil surface and the interaction-specific reactions are necessary (see Chapter Histo- and cytopathology of infection). Mechanisms might be analogous to pollen adhesion and penetration processes that have been examined in grasses (Heslop-Harrison and Heslop-Harrison, 1980, 1981; Heslop-Harrison et al., 1984, 1985). There are some striking similarities between the process of fungal colonization and natural fertilization, for the plant's ovary mainly serves as a host for the male gametophyte in order to favour and to guide pollen tube tip-growth. Furthermore, both invaders interfere with each other when landing on their objective at the same time, in particular, simultaneous pollination favours fungal penetration, e.g., in C. purpurea (Williams and Colotelo, 1975), however, pollen tubes grow much quicker reaching the ovule in wheat in about 30 min (You and Jensen, 1985). Admittedly, the ovary appears to be dispensable for fungal development subsequent to primary infection, because sclerotia are formed even after artificial inoculation of florets, the ovaries of which have been removed previously (Cherewick, 1953) or after advanced kernel growth (Mantle, 1972). Nevertheless, mimicry of pollen tube growth might also occur in the ovary colonization process.

A second reason for organ specificity might be the exceptional molecular architecture of the monocotyledonous cell wall (Carpita and Gibeaut, 1993) certainly with some additional unique cell wall modifications in different pistil tissues (Tenberge et at., 1996a, b; see below), e.g., pistil epidermis with stigma hairs, ovary mesophyll, transmitting tissue and integuments.

Host floret biology is fundamentally important in every respect. Since ergot hosts are anemophilous, floret morphology and the time course of anthesis determines pollination. Likewise, access of fungal spores, especially of airborne primary inoculum, is subject to floret biology. In most grasses, single florets gape for several hours, then close tightly, sometimes leaving the stigmas exposed between the glumes. After pollination, stigmas wither directly in pearl millet, quickly in rye or remain turgid over a period of two month, even after infection, in sorghum. Regarding complete ears, florets open, starting near the top of the ear and progressing in a series basipetally over several days, as in sorghum (Frederickson and Mantle, 1988) or rye. Hence, cereal fields are in bloom for about two weeks.

2.4.
HISTO- AND CYTOPATHOLOGY OF INFECTION

Different ergot species exhibit little variation in overall pathogenesis and their cytopathology of infection, which has recently been reviewed with emphasis on

C. purpurea (Tudzynski *et al.,* 1995), is basically identical (Luttrell, 1980; Parbery, 1996). In the few species investigated at the microscopical level, so far, starting with the first detailed description of infection by Tulasne (1853), there exists both, some peculiar findings and some conflicting data still open to question.

2.4.1.
Host Infection

Infection Site and Route

Infection of a single host plant is naturally induced by spores landing on the pistil of open florets for which less than ten conidia are sufficient (Puranik and Mathre, 1971). The precise site of spore germination and the resulting infection route, i.e., either via stigma or style with help of the pollen tube path in contrast to ovary wall or ovary base, has been controversially discussed (Engelke, 1902; Kirchhoff, 1929; Campbell, 1958; Luttrell, 1980; Shaw and Mantle, 1980a; Tudzynski *et al.,* 1995). Conflicts arose because the penetration site was deduced from hyphal locations, i.e., in the transmitting tissue after inoculation, but was not shown in micrographs itself. Rye stigmas were shown to be penetrated by *C. purpurea* (Luttrell, 1980), however, the use of squash mounts is a questionable approach, since ergot fungi will grow into almost every young tissue after its epidermis has been wounded. Thin sectioning appears to be not feasible, but, employing scanning microscopy, spore adhesion, spore germination and host cuticle penetration could be documented to occur on either part of the pistil surface (Figures 1 and 2) (Tenberge, 1994; Tenberge and Tudzynski, 1995). Therefore, germination of *C. purpurea* is not restricted to the stigma; it does obviously not depend on stigmatic fluid. Likewise, regarding other *Claviceps* species, each one of the penetration site has been reported, in particular the ovary wall, and in some cases visualized (Luttrell, 1977; Thakur and Williams, 1980; Willingale and Mantle, 1987a; Frederickson and Mantle, 1988), but never has proof been presented for a distinct epidermal region of the pistil to be resistant to ergot penetration. In conclusion, ergot fungi most likely are able to penetrate the pistil epidermis anywhere.

Following penetration in the outer epidermal wall of the pistil, the hyphae keep on growing towards the rachilla (Figure 2). They grow either down the style in the transmitting tissue following the pollen tube path outside the ovule and leaving this way at the micropylar region in direction of the rachilla or, after lateral entrance into the ovary, in the carpel mesophyll to the ovary basis (Luttrell, 1980; Shaw and Mantle, 1980a; Tudzynski *et al.,* 1995). The ovary wall gets completely colonized after about 2 days post inoculation (dpi) in *C. paspali,* 4 dpi in *C. fusiformis* or 8 dpi in *C. sorghi* depending on temperature, geographical location (Willingale and Mantle, 1987a) and species. However, the

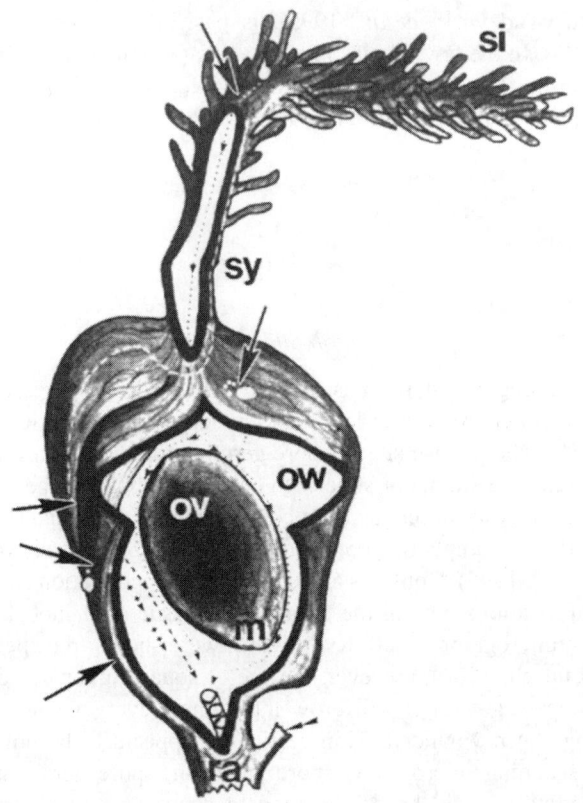

Figure 2 Illustration of different conidia germination places and penetration sites (long arrows) with resultant infection routes of *Claviceps purpurea* in a cereal pistil......, spore germination and infection via stigma or style or ovary cap partly corresponding to the pollen tube path; +++, spore germination and infection at the base of the ovary wall; ———, spore germination on the stigma and infection at the base of the ovary wall suggested by Kirchhoff (1929). Double-arrow, indicating filament base; m, micropylar region; ov, ovule; ow, ovary wall; ra, rachilla; si, stigma; sy, style

ovule first remains uninvaded (Kirchhoff, 1929) due to the integuments which appear to form a temporary barrier to the fungus (Campbell, 1958; Willingale and Mantle, 1987a). Ergot of dallisgrass is much quicker and colonizes the ovule already after 2 dpi (Luttrell, 1977). Sorghum ovules are additionally invaded through the chalazal region (Frederickson and Mantle, 1988). The same occurs in rye. Integuments typically collapse during development of caryopses. However, possibly due to chitinases or chitinbinding lectins found in rye seeds (Raihkel *et al.*, 1993; Yamagami and Funatsu, 1996), growing kernels remain noncolonized after late infection.

Fungal cells colonize the entire ovary wall but in the ovarian axis the hyphae of all ergot species stop to spread in the plant tissue. No hyphae emerge beyond the rachilla tip. Thus, a narrow frontier between the fungal stroma and the noncolonized host tissue develops, which is finished approximately six days after infection with *C. purpurea* and persists throughout the remaining life-span (Luttrell, 1980; Shaw and Mantle, 1980a; Tudzynski *et al.,* 1995). The completion of this frontier coincides with the exudation of honeydew, 4 to 10 dpi depending on the species, and indicates the begin of the sphacelial phase (see below). Honeydew presents the first macroscopic evidence for infection in most host grasses but not in pearl millet, where browning, withering and constriction of stigmas and stylodia is obvious already after 36 h post inoculation (hpi) (Willingale and Mantle, 1987a). Stoppage of fungal growth in the rachilla of *C. purpurea* has been shown to be most likely caused by host phenolics that accumulate during infection at this site (Mower and Hancock, 1975; Shaw and Mantle, 1980a; Hambrock, 1996) and might inhibit fungal pectin-degrading enzymes (Mendgen *et al.,* 1996).

The route which is usually used, however, depends on floral biology and spore vehicles. Short floret gaping followed by stigma exposure between tightly closed glumes causes the pollen tube path to be of importance in nature, such is valid for *C. paspali* in dallisgrass (Luttrell, 1977) and *C. fusiformis* in pearl millet with very large feathery stigmas (Thakur and Williams, 1980; Willingale and Mantle, 1985). Additional routes in rye, however, are obvious from successful infection with *C. purpurea* after previous ovary colonization with the bunt fungus *Tilletia caries* (Willingale and Mantle, 1987b) or after fertilization with advanced kernel development (Mantle, 1972), since in both cases the pollen tube path is blocked. This shows that, regardless of the infection route, the growth of ergot fungi is strongly directed at the vascular tissue supply of the ovary, which itself represents the entrance but is dispensable (Luttrell, 1980; Willingale and Mantle, 1987b) although incidentally consumed for overall nutrition.

Spore Adhesion, Infection Structures and the Infection Process

The attachment process of ergot spores is not investigated precisely. Since the attaching force of ungerminated spores of *C. fusiformis* increases during the first 8h following inoculation (Willingale and Mantle, 1987a), one may infer adhesive strategies of spores to the plant surface. But in *C. sorghi* conidia appear to become detached from the surface during germination and penetration (Frederickson and Mantle, 1988). On stigma or style, the stigmatic fluid might offer hydrophilic conditions or, in case of honeydew-mediated transmittance, the syrupy fluid may support adhesion to the host surface. After attachment, conidia germination starts with the formation of one to several germ-tubes, supported by dew periods, and is accomplished very quickly in some species, e.g., in *C. paspali* within 4 hpi (Luttrell, 1977), in *C. fusiformis* within 12–16 hpi (Willingale and Mantle, 1987a), in *C. sorghi* within 16–48 hpi (Frederickson and Mantle, 1988).

In the most important ergot fungus, *C. purpurea,* early infection events are not described and growth modes of ergot fungi are not analyzed functionally. To study the penetration and colonization mechanisms of ergot fungi, early events of the infection process of *C. purpurea* were first documented and then cytochemically analyzed in detail in our laboratory. Spores attached everywhere on the pistil epidermis. Its outer epidermal wall comprises a faint cuticular membrane, measuring about 15 nm in thickness, and forms a continuous outer barrier of the host ovary. Germination results in one or two germ-tubes, e.g., on the ovary cap (Figure 1). Sometimes a limited external mycelium is formed. Next, the faint plant cuticle of the outer epidermal wall is directly penetrated. An indirect entry via natural openings is unimportant because the pistil is free of stomata and natural wounding was never observed. Infection hyphae originate either directly from the germ-tube or from the external mycelium (Figure 1). Since no changes in hyphal shape were apparent, it appears that *C. purpurea* penetrates without specialized infection structures (K.B.Tenberge, unpubl.). At suitable sites, infection hyphae pass through the outer epidermal cell layer growing intercellularly into the anticlinal epidermal walls. However, hyphae may as well pass through the outer epidermal cell wall away from cellular junctions (Shaw and Mantle, 1980a; Tudzynski *et al.,* 1995).

So far investigated, all ergot fungi penetrate directly into the anticlinal walls between epidermal host cells, sometimes after a period of subcuticular growth similar to that of *C. purpurea* (Figure 1), e.g., *C. sorghi* (Frederickson and Mantle, 1988). However, the hyphae of some species develop special morphological structures prior to penetration. *C. gigantea* produces an appressorium (Osada Kawasoe, 1986). In *C. fusiformis,* an external mycelium is formed and bulbous infection structures arise at the tips of several germtubes of a single macrospore (Willingale and Mantle, 1987a). Whether these structures fulfil appressorial function is not described so far.

Mechanisms of Adhesion and Penetration

The mechanism of cuticle penetration still needs to be elucidated, although, in most ergot fungi, the direct push of a infection hypha into the epidermal wall obviously matches the more simple type of penetration as classified by Mims (1991). Infection structures of phytopathogenic fungi are specialized hyphae adapted for the invasion of the host and show considerable variations in penetration strategy (Mendgen and Deising, 1993; Mendgen *et al.,* 1996). The bulbous structures of *C. fusiformis* suggest the utilization of turgor pressure, the one used for penetration is mediated by functional appressoria and has been shown to be obligatory in the rice blast pathogen (Howard and Valent, 1996). Turgor pressure itself is essential for fungal tip growth (Wessels, 1994), however, it is an open question whether it is sufficient for penetration. According to Mendgen *et al.* (1996), directly penetrating fungi that do not form appressoria clearly need cell wall degrading enzymes for penetration. During penetration of grass stigma cuticles, tip growth of pollen tubes is mediated by cutin-degrading enzymes (Baum *et al.,* 1992; Heslop-Harrison and Heslop-Harrison, 1981), however, in case of ergot fungi, the secretions of such enzymes have not been demonstrated so far.

Recently, genes coding for hydrophobin-type proteins have been isolated from a *Claviceps* sp. (Arntz and Tudzynski, 1997) and from *C. purpurea* (V.Garre and P.Tudzynski, pers. communication, see chapter 4 in this volume), in addition, hydrophobin-type proteins have been identified and purified from corresponding axenic cultures (O.de Vries, S.Moore, C.Arntz, J.G.H.Wessels and P.Tudzynski, pers. communication). Hydrophobins can act as adhesive to the host cuticle (Wessels, 1994) and have been shown to play a crucial role in formation, adhesion and infection court preparation of phytopathogenic fungi (Beckerman and Ebbole, 1996; Talbot *et at.,* 1996). Therefore, the presence of such proteins suggests similar functions in ergot. In addition, one can speculate that they might mediate the intimate contact and abundant wrapping of superficial ergot hyphae with the host cuticle demonstrated by TEM (Tudzynski *et al.,* 1995; K.B.Tenberge, unpubl.) and therewith might cause a mechanical disruption of the thin cuticle itself. Cytological expression analysis of the hydrophobin gene with *in situ* hybridization technique showed that transcripts were located in external and penetrating hyphae as well as in *C. purpurea* conidiophores (Tenberge *et al.,* 1998). This localization indicates another function of hydrophobins that are thought to be essential for the formation of aerial hyphae and fruit bodies (Wessels, 1994) to be met in *C. purpurea.*

2.4.2.
Fungal Mechanisms for Host Colonization

Ectotrophic Growth in the Host Ovary with Limited
Endotrophism

After penetration, ergot fungi live inside the ovary, i.e., endophytically, during the colonization phase. Subcuticular hyphae, in fact, are located within the outer epidermal cell wall and then the fungi usually grow between epidermal cells into the host apoplast. Before tapping the vascular traces, fungal growth during the colonization phase has been reported to be exclusively intercellular, i.e., ectotrophic, in all ergot fungi investigated (Luttrell, 1977, 1980; Shaw and Mantle, 1980a; Willingale and Mantle, 1987a; Frederickson and Mantle, 1988). However, a limited intracellular growth has been documented electron microscopically in *C. purpurea* (Tenberge and Tudzynski, 1994; Tudzynski *et al.,* 1995). Therefore, at least the mycelium of this ergot species is ectotrophic but with limited endotrophism (Figure 3). The vegetative hyphae exhibit ultrastructural features typical for ascomycetous fungi and well preserved during processing for TEM. The thin fungal cell wall and the host cell wall build up an

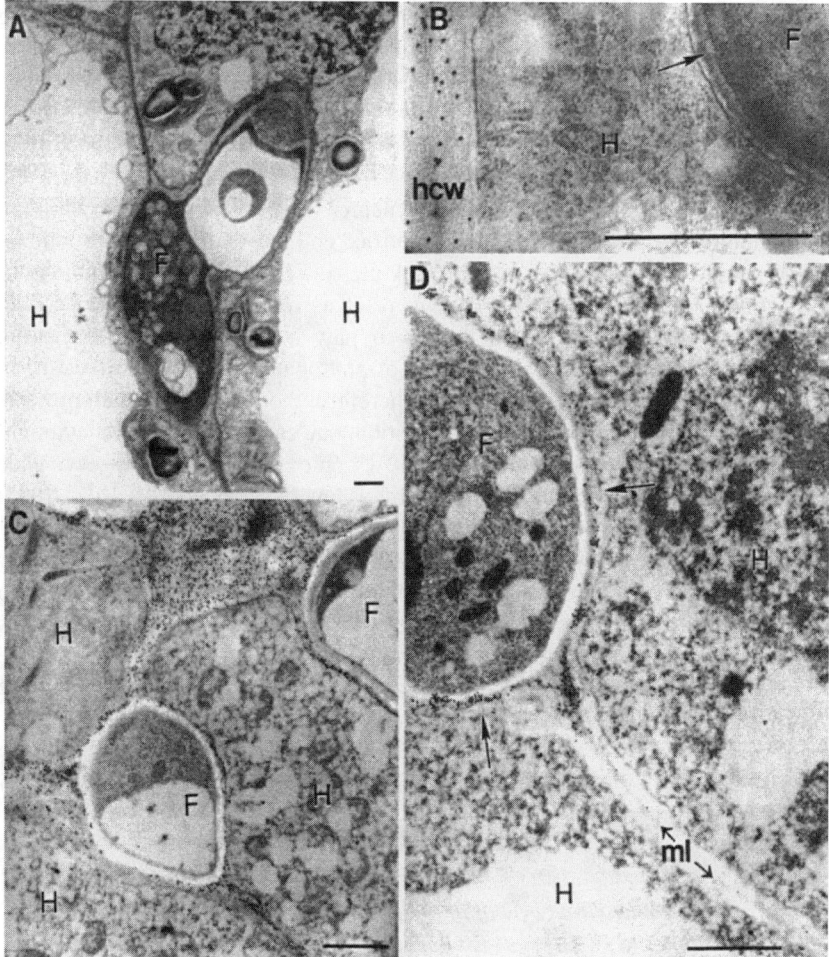

Figure 3 TEM micrographs of rye ovaries during the colonization phase with *C. purpurea* showing inter- and intracellular growth in host mesophyll cells. **A,** An intercellularly growing hyphal cell (F) actively penetrating a living host mesophyll cell (H). **B,** Enzyme-gold localization of β-1,4-glucan with a cellulase-gold sol showing gold label in the host cell wall (hcw) distant from the fungus (F) while no label is visible immediately at the interface of the intracellular hypha (arrow). C and D, Immunogold localization of homogalacturonan epitopes in pectin with a monoclonal antibody JIM5 specific for non-methyl-esterified polygalacturonic acid visualized with gold-linked secondary antibody. Distant from intercellular hyphae (F) in the ovary mesophyll (H), low JIM 5 label was restricted to the middle lamella (ml) **(D);** however, JIM 5 label was high throughout host walls after demethylation with sodium carbonate **(C).** At the interface of hyphae, high JIM 5 label was present above the entire host wall (arrows) and above the altered middle lamella zone **(D). A, B,** Glutaraldehyde fixation, osmication, epoxy resin embedding. **C, D,** Formaldehyde-glutaraldehyde fixation, no osmication, LR White embedding. Scale bars=1 μm

intimate zone of contact while both, host and pathogen, appear to be healthy (Tudzynski *et al.*, 1995). This is particularly valid for intracellular hyphae and the penetrated host cells and points to haustorial function. For this purpose, the endotrophic mycelium is well positioned in the chalazal region that serves for the host ovule nutrition. The interface of the intracellular hyphae, which are completely encapsulated by the host plasma membrane (Figure 3B), has developed special adaptations (Tenberge *et al.*, 1996a; Müller *et al.*, 1997), however, nutrient uptake into hyphal cells is not yet investigated.

Mechanisms of Fungal Growth in the Host Tissue

The chemical composition of the infection court is a matter of speculation if assuming that the cell wall type of grass leaves (Carpita and Gibeaut, 1993) is also valid for the ovary neglecting the ovary's specific function. We currently are analysing the molecular architecture of the host-parasite association with emphasis on interaction specific reactions, e.g., polymer alterations and protein secretion, at the electron microscopical level. This molecular cytological study is intensely co-ordinated with a molecular genetical approach (see chapter 4 in this volume) in order to elucidate fungal mechanisms utilized for infection court preparation, penetration and further ecto- and endotrophic host colonization, because functional studies are very limited on ergot pathogenicity.

Although only very low pectin content is expected in grass cell walls according to Carpita and Gibeaut (1993), host cell wall loosening during subcuticular and intercellular growth indicated that actions of pectolytic enzymes, which Shaw and Mantle (1980a) have proved to be active in culture, in honeydew and in parasitic tissue extracts, play a role in parasitism. Then, the simultaneous presence of both pectin types, non-methyl-esterified and methyl-esterified galacturonan, in the cell walls along the usual infection path in healthy carpels has been documented, using the two monoclonal antibodies JIM 5 and JIM 7 (Tenberge *et al.*, 1996a, b). With the same experimental design applied to parasitic culture of ergot on rye, a local molecular pectin modification as well as degradation have been demonstrated for the host cell wall and the middle lamella zone at the interface of subcuticularly and intercellularly growing hyphae *in situ* (Figure 3D). Chemical demethylation and immunogold labelling indicated a high total content of galacturonan that, in late infection phases, was completely absent, emphasizing its use for nutrition together with other plant polysaccharides. The observed host wall alterations provide evidence for the secretion and activity of extracellular pectinolytic enzymes *in planta*. From on-section saponification studies, the local reactions specific for interaction were concluded to comprise an enzymatic demethylation mediated by pectin-methylesterases, converting the pectin into the appropriate substrate of endo-polygalacturonases for final degradation (Figure 3, C and D). While the analysis of the pectin-methylesterase is a matter of future research, two genes, putatively encoding two endo-polygalacturonases, have already been isolated from *C.*

purpurea and were shown to be expressed during infection of rye (Tenberge *et al.*, 1996a). This strongly indicates the fungal origin of the pectinolytic activities found in infected ovaries by Shaw and Mantle (1980a).

The cellular junctions were shown to consist of high amount of unesterified pectin. Therefore, polygalacturonase activity seems to be a well fitting means to enable an entry into the middle lamella from the intercellular spaces which is not continuous along the infection route towards the rachilla. The outer cellular junctions in epidermal cells were rich in homo-polygalacturonan, too, clearly representing the right conditions for an entry into the anticlinal epidermal walls, which were usually selected for the primary penetration of the host dermal tissue as concluded from carefully directed hyphal branching above those sites (Tudzynski *et al.*, 1995). During primary penetration, degradation of host pectin could be detected (Tenberge *et al.*, 1996a, b), however, elicitation of host defence reactions causing incompatible interactions has not been observed. Since pollen tubes secrete pectic enzymes during tip-growth (Derksen, 1996), too, one can speculate that modifications of host pectin by fungal pectinases does not betray the pathogen presence.

The ergot fungus actively penetrates plant cell walls to produce an endotrophic mycelium during the primary penetration of the epidermis, the colonization of the mesophyll (Figure 3A) and later for the tapping of xylem vessels. At the interface of intracellular hypha, the host cell wall is obviously lacking, as seen in TEM (Tudzynski *et al.*, 1995). Since grasses have developed a special cell wall type containing low amounts of pectins and considerably high amounts of xylans in addition to the major polysaccharide portion of cellulose (Carpita and Gibeaut, 1993), xylanases and cellulases as well as pectinases are expected to be necessary for breaking down the major cell wall components during infection. At the host-pathogen interfaces of intracellular hyphae (Figure 3B), and additionally of intercellular hyphae, a lack of β-1,4-glucan in host cell walls has been found with the use of a specific enzyme-gold probe, pointing to the enzymatic action of cellulases in ergot infection (Tenberge and Tudzynski, 1994). Correspondingly, a putative cellobiohydrolase gene (*cel*1) has recently been isolated from *C. purpurea* and was found to be induced during the first days of infection of rye (Müller *et al.*, 1997; see chapter 4 in this volume). Therefore, this cellobiohydrolase may be involved in the penetration and degradation of host cell walls by depolymerising plant β-1,4-glucan as is characteristic of true cellulolytic fungi.

β-1,4-Xylan, i.e., the substrate of the fungal β-1,4-xylanase used in the enzyme-gold technique, has been localized in rye ovary cell walls throughout the infection route (Giesbert *et al.*, 1998), confirming that this major cell wall component in grass leaves is in fact a structural compound in ovary cell walls. The β-1,4-xylan is expected to represent only the backbone of the typical grass heteropolysaccharid, glucuronoarabinoxylan (GAX). Arabinofuranosyl epitopes, one of the possible side chains in GAX, were localized in ovary cell walls (Giesbert *et al.*, 1998). Giesbert and Tudzynski (1996) isolated a xylanase gene (*xyl*1), cloned another putative xylanase gene (*xyl*2) and proved their expression during parasitic culture of *C. purpurea* of rye (Giesbert *et al.,* 1998; see chapter 4 in this volume). Using three different heterologous antibodies in tissue printing experiments, the secretion of ergot xylanases in axenic culture and during infection of rye has been localized *in situ* (Giesbert *et al.,* 1998). Currently, the assumed xylan alteration during infection is being investigated.

While in necrotrophs the release of cell wall-degrading enzymes results in tissue maceration and host cell death immediately ahead of invading hyphae (Parbery, 1996), in the biotrophic *C. purpurea,* secretion of these enzymes causes no such drastic effects but only limited damage to the host during colonization. To restrict the enzymatic action to an adequate but limited area, the fungus might control the physico-chemical properties of the interface (Tenberge *et al.,* 1996a).

In conclusion, these different cell wall-degrading enzymes appear to be essential for the preparation of the infection court and the establishment of infection. The cell wall material is thought to be important for nutrition during colonization of the ovary supported by their complete use and also because cell wall extracts of ears stimulated growth in culture (Garay, 1956). In order to evaluate the importance of enzymes for ergot pathogenicity, deficient mutants have been created by targeted gene replacement (Giesbert *et al.,* 1998; see chapter 4 in this volume) which are currently being investigated microscopically. A novel finding in fungal phytopathology and particularly in ergot was the detection of a fungal catalase secreted in axenic culture of *C. purpurea* and most likely during infection of rye (Tenberge and Tudzynski, 1995; Garre *et al.,* 1998a, b). Catalase activity has been measured in axenic culture of *C. purpurea* (Tenberge and Tudzynski, 1995). Isoelectric focusing together with diaminobenzidine (DAB)-mediated activity staining showed the presence of a specific catalase in parasitic culture of *C. purpurea* as well as in infected ovaries and in honeydew and that it is likely induced during infection (Garre *et al.,* 1998b). Electron dense deposits have been found *in situ* in multivesicular bodies as well as in cell walls of hyphae from axenic culture (Figure 4) and during infection of rye if using ultrastructural enzyme-activity staining with DAB but not after inhibition with aminotriazole, indicating catalase activity (Garre *et al.,* 1998b). Putative catalase proteins have been immunogold localized in axenic (Figure 4) and parasitic culture using different heterologous antibodies (Tenberge and Tudzynski, 1995; Garre *et al.,* 1998b). It has been shown therewith that the detected antigen is secreted via fungal multivesicular bodies into the fungal cell wall diffusing further into the adjacent host apoplast at the host-pathogen interface exclusively. Moreover, one catalase gene comprising a putative signal sequence has been isolated from *C. purpurea* and shown to be expressed during infection of rye (Garre *et al.,* 1998a; see chapter 4 in this volume) providing further evidence for the fungal origin of the detected catalase and the possibility of a functional analysis by gene replacement experiments. Since infection-induced H_2O_2 production has been shown to occur outside host cells

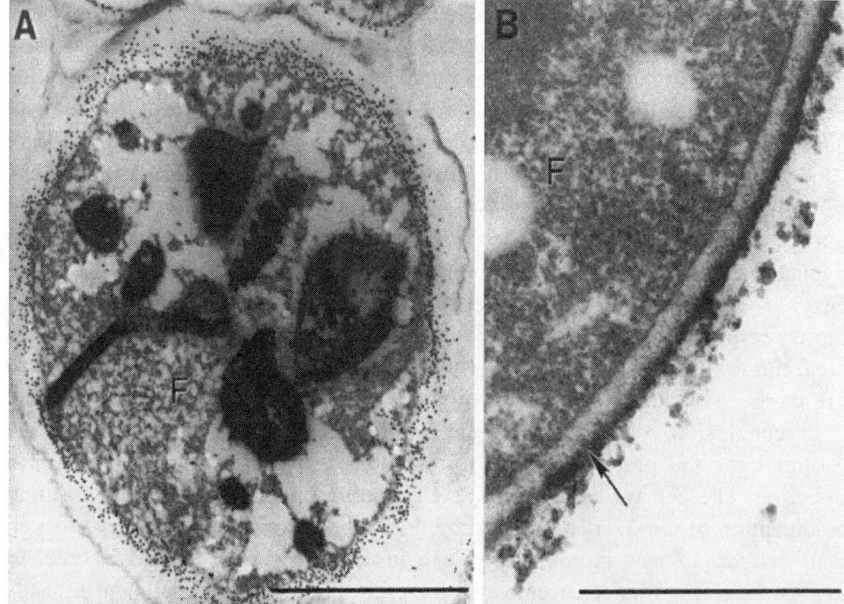

Figure 4 TEM micrographs of *C. purpurea* showing hyphal cells from axenic culture. **A,** Immunogold localization with a polyclonal anti-catalase antibody, raised against native sunflower catalase (Tenberge and Eising, 1995), and protein A-gold showing gold label for catalase-like proteins in the fungal cell wall. Preimmune controls lack gold particles (not shown). **B,** Activity staining for catalase with diaminobenzidine showing moderate electron dense deposits throughout the fungal wall and strong depositions at the wall periphery (arrow). Controls with aminotriazole for catalase inhibition lack electron dense deposition (not shown). Scale bars=1 μm

(Mehdy, 1994), we suggest a multiple function of fungal catalase in pathogenesis due to its hydrogen peroxide decomposing activity: (a) Cytotoxic effects to fungal cells are prevented, (b) Mechanical barrier formation during host defense reaction, i.e., H_2O_2-mediated cross-linkage of cell wall components during oxidative burst or lignification, are suppressed, (c) In particular, in grass cell walls, a phenolic cross-linkage of polysaccharides is supposed to occur while cell walls expand during their ontogeny causing high cell wall rigidity (Carpita and Gibeaut, 1993). By suppression of those reactions, ergot fungi might maintain a convenient habitat for colonization. In the rachilla, infection induced cross-linkage of phenolics (see above) forestalls fungal colonization and, therefore, might efficiently control fungal growth.

Direct Tapping of the Vascular Nutrition Supply

Plant synthates, primarily intended for the developing seed, are the main nutrition source to the fungus (Mower and Hancock, 1975) which obviously is exploited at about 5 dpi depending on the ergot species. To use this natural sink, several enzymes are secreted such as the cell-wall bound inducible fructosyltransferase (invertase) (Bassett *et al.,* 1972; Taber, 1985; Tudzynski *et al.,* 1995) and the fungal foot is developed structurally for attaching and absorbing (Luttrell, 1980). While intense exudation of honeydew is reported to occur without penetration of vascular cells, e.g., in *Pennisetum americanum* infected by *C. fusiformis* (Willingale and Mantle, 1987a), tapping of the vascular traces in rye by intracellular hyphae of *C. purpurea* has been documented (Luttrell, 1980; Tudzynski *et al.,* 1995).

The cytological basis of assimilate flow is an ectotrophic mycelium with limited endotrophism in the host phloem, which appears to be unchanged in vitality. In sharp contrast to uninfected ovaries, however, common phloem callose was not found in infected ovaries at all or was distinctly reduced as outlined by Tudzynski *et al.* (1995). This unblocking of sieve elements may be the reason for increased flow of assimilates to the infected floret causing limited growth of neighbouring seeds. The current opinion of the mechanisms is that ergot fungi enzymatically degrade the phloem callose by secreting *β*-1,3-glucanases, which have been purified from axenic cultures of *C. purpurea* (Dickerson and Pollard, 1982; Brockmann *et al.,* 1992). Using immunofluorescence, Dickerson and Pollard (1982) localized *β*-l,3-glucanase in the sphacelium but not before the tenth day after inoculation. Recently, the callase has been immunogold localized throughout the colonization phase. Detection of antigens in the fungal secretion pathway proved the fungal origin of the *β*-1,3-glucanase activity found in infected ovaries and honeydew (Tenberge *et al.,* 1995). Furthermore, immunogold electron microscopy documented that the secreted enzyme is diffusing into the host apoplast up to the host periplasmic area, which is the site of host callose deposition. Cross-reactivity of the antiserum produced by Dickerson and Pollard (1982) with callase from *C. fusiformis* (Willingale and Mantle, 1987b) indicates that this enzymatic action might be used in other ergot fungi, too. In addition, it is likely used to suppress callose deposition during potential defence reaction of the host, because only limited aniline blue fluorescence for callose has been detected in the ovary wall during its colonization with *C. purpurea* (Hambrock *et al.,* 1992).

2.4.3.

Sphacelial Stromata for Secondary Propagation

The sphacelial stroma is evident between 4 dpi, e.g., in *C. fusiformis,* and 6 dpi depending on ergot species and sporulation ceases approximately 11 dpi (Luttrell, 1980). The sphacelial plectenchyma is formed intercalarly (Tulsane,

1853) and accumulates lipids later in this phase. These filamentous hyphae are of the type normally found in axenic culture. At the base of the ovary, proliferation of the fungal cell starts (Kirchhoff, 1929); hyphae accumulate beneath the host cortical layers and break through the epidermis towards the ovarian outer surface (Luttrell, 1980). Finally, fungal cells cause ovary replacement, which is not necessary but a consequence of acropetal development (Willingale and Mantle, 1987a). Phialidic conidiophores emerge from the sphacelial stroma possibly favoured by ergot hydrophobins (see above). Numerous oblong conidia are produced (Figure 1). These conidia do not germinate in the honeydew, which is excreted simultaneously, due to high osmotic pressure (Kirchhoff, 1929; Taber, 1985). Some isolates of *C. purpurea* do not produce normal exudates of honeydew, only a few spores were detected later. However, these strains produced normal sclerotia (Mantle, 1967) indicating that abundant formation of honeydew is important for secondary infection but appears not to be a necessary prerequisite for sclerotial formation. In some species, macroconidia are produced that are able to germinate in honeydew and subsequently conidiophores emerge for the production of microspores (see above).

2.4.4.
Ergot Sclerotia

After the secondary conidiation has ceased, the sclerotia start growing. Within about three, e.g., *C. paspali* (Luttrell, 1977), up to five weeks post inoculation, e.g., *C. purpurea* (Kirchhoff, 1929), the maturity of sclerotia is achieved. In general, morphology and anatomy of sclerotia is species-specific. They measure 2–50 mm in length and a few millimetres in diameter. In *C. purpurea,* they are oblong, in *C. fusiformis* spherical with large variation at different location (Thakur *et al.,* 1984; Chahal *et al.,* 1985). Sclerotia clearly grow epiphytically on top of the ovary stalk and they mostly emerge out of the florets. Since the sclerotia are no longer enclosed between the glumes, energy must be provided to protect them from desiccation, UV radiation and mycoparasitism (Parbery, 1996). The outer rinds get naturally pigmented resulting into different sclerotia colours, such as dark-purple in *C. purpurea* (Luttrell, 1980), red-brownish in *C. sorghi* and *C. africana* (Frederickson *et al.,* 1991), dark-brown in *C. fusiformis* (Thakur *et al.,* 1984), except those of *C. paspali,* which are white-to-brown (Luttrell, 1977).

The differentiation mode of the sclerotium varies between *Claviceps* species. In *C. purpurea,* first at several places within the sphacelium, sphacelial hyphae differentiate into sclerotial hyphae, later at the sclerotial base, sclerotial hyphae are formed directly (Shaw and Mantle, 1980b). The sclerotial plectenchyma develops intercalarly above the stromatic fungal foot by a generative zone as previously did the sphacelium (Campbell, 1958). *C. sorghi* forms an proximal plectenchyma below an extended sphacelial stroma, within which elongation is evident from a thin red core, representing cortical tissue enclosing some of the first differentiating sclerotial medulla (Frederickson *et al.,* 1991). Remnants of sphacelial stroma with conidia as well as the ovary cap may persist on top of the growing sclerotia, but no internal conidiogenous locules were found. Hence, in these species, the sclerotial hyphae are newly formed. *C. africana,* on the other hand, largely differentiates the sclerotium be transformation of the spherical sphacelium with little change in total size (Frederickson *et al.,* 1991).

Purple pigmentation is the first sign of sclerotial development in *C. purpurea* about 12 dpi (Shaw and Mantle, 1980b), but the trigger for the change from sphacelial into sclerotial growth is unknown (Parbery, 1996). It has been speculated that nutrition is a major factor, supported by the transition effect of certain amino acids found in axenic culture (Mantle and Nisbet, 1976). Changes in cytology of the cells coincide with increasing levels of lipids which are predominantly triglycerides with the fatty acid ricinoleate (Corbett *et al.,* 1974). This increase in total lipid content from 10% to 30% of the dry weight is the first metabolic indicator of the morphogenesis of sclerotial cells (Bassett *et al.,* 1972). The youngest fungal cells of the sclerotial stroma are longitudinally organized, distinct, frequently septate hyphae forming the prosenchymatous region at the proximal end of the sclerotium. Lipid content is evident in these storage cells, thus they are packed with osmiophilic globules in contrast to the differentiating hyphae of the generative zone in the lower ergot region. In the medulla, distal from the prosenchymatous region, fungal cells form a region of a compact plectenchyma which is build out of bulbous storage cells interspersed with narrower hyphae (Shaw and Mantle, 1980b). The absorbing hyphae, which connect the sclerotium to the ovary stalk and form the stable host-parasite frontier, however, are neither of the sclerotial type nor typically sphacelial cells. They exhibit no parallel orientation, lack lipids and contain large vacuoles (Shaw and Mantle, 1980b). A special function of the fungal foot is also suggested by xylanase activity which has been localized in the sclerotial phase at this host-parasite frontier exclusively, using tissue printing experiments (Giesbert *et al.,* 1998).

Sclerotia are the only ergot structure containing alkaloids (Ramstad and Gjerstad, 1955) and the pigmentation of the sclerotial cortex might protect these light sensitive alkaloids (Taber, 1985). In axenic culture, differentiation of the sphacelial-like hyphae into sclerotial-like cells occurs (Kirchhoff, 1929). They accumulate up to 40% of triglyceride/ricinoleate, but cells show no pigmentation and, as in the parasitic state, do not always produce alkaloids (Bassett *et al.,* 1972).

2.5.

ERGOT VIRULENCE AND HOST SUSCEPTIBILITY

Different *Claviceps* species cause ergot disease in a scale of crops, e.g., wheat, rye, sorghum and millet with various levels of virulence. To date, knowledge about resistance in the hosts is fragmentary. Resistance is reported only in a few cases, such as spring and durum wheat (Platford and Bernier, 1970). In the experiments with four different *C. purpurea* strains, however, obscure ovary necrosis occurred only in some of the florets, others produced sclerotia in these hosts. Necrotic ovary response also has been found after double inoculation with two different *C. purpurea* strains (Swan and Mantle, 1991). In sorghum, some ergot resistant lines had been identified (Musabyimana *et al.,* 1995). In pearl millet, *Pennisetum glaucum,* a female nuclear genetic factor contributes to high resistance to *C. fusiformis* (Rai and Thakur, 1995). So far, only limited evidence for a gene-determined somatic resistance has been discovered in diverse hosts. In pearl millet, some lines are resistant to ergot, other genotypes expressed considerably reduced susceptibility to *C. fusiformis* (Willingale *et al.,* 1986). Resistance in these lines, however, is not based on a specific gene-for-gene interaction but on host floret biology. Distinctive ecological and morphological adaptations to pollination determine gaping and therewith affect susceptibility to ergot drastically. Firstly, pollination induces closing of florets and withering of stigma and style. Hence, competition arises between inoculation and pollination, although, admittedly, pollen landing aside of spores on the pistil surface can stimulate conidia germination in *C. purpurea* (Williams and Colotelo, 1975). Since settlement of inoculum on the pistil surface can initiate infection exclusively, the period of exposure determines the infection success. Susceptibility is highest at anthesis and declines rapidly afterwards in sorghum (Frederickson and Mantle, 1988); but in rye, wheat or barley infection with *C. purpurea* remains possible at a lower rate even after fertilization (Kirchhoff, 1929; Campbell and Tyner, 1959; Puranik and Mathre, 1971). Therefore, floret biology is important for epidemiology and control (see below). The smaller the time window of gaping is, the lower the susceptibility of the host is, e.g., in sorghum (Musabyimana *et al.,* 1995). While host male-sterility or protogyny enhances gaping duration and consequently susceptibility (Willingale and Mantle, 1985), cleistogamy is most effective in avoiding ergot, as has been shown for sorghum (Frederickson *et al.,* 1994).

In addition to gaping control, pearl millet (Willingale and Mantle, 1985) and maize (Heslop-Harrison *et al.,* 1985) evolved a pollination-induced constriction of host stylodia. Constriction occurs about 6 h after pollination, rapidly after passage of compatible pollen tubes and after 4 to 5 days of ageing of unpollinated pistils. This controlled mechanical barrier, based on unique pistil anatomy, effectively directs fertilization and produces passive resistance to *C. fusiformis* in some lines of *Pennisetum americanum*. Passive resistance is efficiently accomplished by early pollination and immediate stigmatic constriction response, which is quicker than ergot establishment and, therefore, always blocks infection (Willingale *et al.,* 1986; Willingale and Mantle, 1987a). In contrast, the susceptibility of wheat and rye to *C. purpurea* persists at least a few days after fertilization and withering of the stigmas (Willingale *et al.,* 1986).

The limited reports on resistance could point to special adaptations of the ergot fungi to the monocotyledonous host ovaries. Assuming that the ergot fungi indeed mimic pollen tube growth with the genius of using components of the specific signal exchange of the pollen-stigma interaction, they possibly grow unrecognized in the ovary and completely avoid host defense reaction of any type. This would suggest that the ergot fungi have coevolved, as assumed for anthracnose fungi's use of their host ripening hormone (Flaishman and Kolattukudy, 1994). One can speculate that successful defense reactions producing host resistance might interfere with the basic function of the ovary and, therefore, do not evolve, since no progenies arise.

2.6.
TROPHISM AND ECOLOGY

The more precisely fungal-plant associations are studied, the less clear do the original distinctions between biotrophs, necrotrophs and saprophytes become, as profoundly outlined in an outstanding review by Parbery (1996). Since *C. purpurea,* in contrast to some biotrophic rusts, is able to survive for a short time outside living host tissue while growing saprophytically in axenic culture, *C. purpurea* and other *Claviceps* spp. were distinctively classified as belonging to the hemibiotrophs. Since this is conflicting with widely accepted terminology, according to which hemibiotrophs live part of their lifes in living tissue and part in tissue that they have subsequently killed (Luttrell, 1974), the trophism type of *Claviceps* needs to be reconsidered. In addition, other fungal pathogens, such as *Cladosporium fulvum,* grow easily in axenic culture, although belonging to the biotrophs (van den Ackerveken and de Wit, 1995).

Ergot fungi never kill host cells in advance of colonizing with the intention to draw nutrition from the killed cells. Hence, they are clearly no necrotrophs but are biotrophs according to Münch (1929) which colonize and draw their nutrients from living host tissue (see Chapter Histo- and cytopathology of infection). Admittedly, throughout the ergot-infected pistil, host cells die not in advance but only subsequent to fungal exploitation of living tissue and possibly some due to induced senescence. In addition, due to the unique pathogenesis pattern, the ergot fungi proliferate intercalarly in the ovary basis that inevitably results into an separation of the host ovary cap. Nevertheless, the separated and colonized tissue stays alive for a while, possibly with nutritional support from the sugary honeydew. Furthermore, already fertilized ovules can develop on top of sphacelia into mature seeds of normal function (Mantle, 1972). Thus, ergot-grass interactions are classified as belonging to a pathosystem free of necrosis in fully susceptible hosts; cell death is not intended but inevitably induced after a while, similar to early host senescence by other true biotrophs (Parbery, 1996). The onset of fruit ripening and host senescence does not trigger necrotrophic vegetative growth of *Claviceps* species, which would be typical for necrotrophs and hemibiotrophs, but coincides with sclerotia formation for sexual reproduction and survival, hence matching another feature of holobiotrophic fungi (Parbery, 1996). In conclusion, ergots are true holobiotrophs which are, following Luttrell's (1974) terminology, ecologically obligate parasites and in nature obtain nutrients only from living host tissue (Mims, 1991) while managing to maintain host cell viability for extended periods, and serve as sink for plant metabolites (Mendgen and Deising, 1993).

Claviceps species appear to be most efficient biotrophs. They don't establish themselves in living tissue in order to wait for host death to use the remainings. They don't establish pathogenic sinks in organs actually intended for synthate export. Instead, *Claviceps* species likely take advantage of the most common source-sink system for synthate in a host by directly tapping the host's nutrition supply network and exploiting the plant resources in a working sink (see Chapter Direct tapping of the vascular nutrition supply). However, the fungus likely utilizes principles to maintain the phloem synthate flow. In addition, the parasite might enhance the sink, which is supported by the suppression of seed development in noninfected florets of the same ear. Likewise, the inverse correlation of sclerotia size and number indicates a competition of sinks in one ear.

While being true holobiotrophic parasites, *Claviceps* species are symbiotrophic at the same time. In exchange for a habitat and nutrition, they protect their hosts and thereby themselves against grazing animals. The introduction of *C. paspali* in Italy caused its host, *Paspalum distichum,* to become a widespread weed. In contrast to earlier days the grass was avoided by animal, which might have recognized ergot infected grasses by its typical smell, reminding the animal of negative experience. This is due to the toxic ergot alkaloids which, although representing dispensable secondary metabolites in the pathogenesis, benefit the fungal's survival by supporting its host's survival (Parbery, 1996).

In this respect and regarding the infection path, there exists great resemblance between *Claviceps* species and some relatives in the *Clavicipitaceae,* e.g., *Acremonium* and some other balansoid genera, which are grass endophytes using toxic alkaloids for defense against herbivory, too (Taber, 1985; Cheeke, 1995). Both fungal groups infect florets and get access to the host ovary probably via the stigma. While ergot fungi sometime enter and later destroy the ovule incidentally, some balansoids, e.g., *Acremonium lolii,* immediately get into the ovule but do not interfere with seed development to ensure systemic inoculation (Parbery, 1996). Since they never leave the plant, i.e., they are living totally endophytic, they are systemic symptomless endophytes. However, some of these symbiotrophs are able to grow and conidiate in axenic culture. Between the two ecological forms, pathogenic ergot fungi on the one hand and symptomless protective commensals on the other, there exist intermediates among the balansoids with great similarities in each direction. Like *Claviceps* species, they first grow endophytic and symptomless in the host, but in order to multiply, they produce damaging epiphytic stromata (White *et al.,* 1992). This is accompanied by suppression of anthesis which is easily tolerated by the host but this phytopathogenic potential characterizes these symbiotrophs to be holobiotrophic.

The trophism type varies with the host in some species (White *et al.,* 1992). In summary, there is a great body of evidence for the fact that biotrophic pathogenesis offers several ecological advantages, conserved in obviously similar pathogenesis pattern across different taxa. However, it is controversial whether biotrophism is the origin (Parbery, 1996) or a secondary, younger development derived from saprophytism (Luttrell, 1974; Mendgen *et al.,* 1996) during the evolution of trophism types.

The strategies for the parasite's survival covers several highly specialized ecological adaptations interconnected with the biology of the host, (a) Sclerotia dormancy, germination and ejection of airborne primary inoculum are perfectly located and timed, in particular in case of monogeneric host specificity, corresponding to the brief receptive phase of blooming florets of the annual anemophilous host plant, (b) The fungi are clearly able to penetrate the outer epidermal wall of the host pistil solely, (c) The fungi might mimic the pollen tube growth, including the mechanisms and mutual recognition, (d) Fungal growth mechanisms are likely adapted to the specialized molecular architecture of monocotyledonous cell walls. Colonization is endophytic, mainly ectotrophic with limited endotrophic hyphae and very cautious, hence nearly symptomless. (e) Ergot fungi directly utilize an already working synthate sink. This exclusive niche might compensate the limitations caused by the narrow time window open for infection, (f) Quickly after infection, sphacelial stromata produce anamorphic conidia for secondary infection to ensure the expansion of the fungal population in the field within the small flowering period of the host. In nature, the infectious inoculum consists of several spores from various strains, resulting in sclerotia of heterogeneous composition. Admittedly, *C. purpurea* is homothallic, however, the mixed inoculum might facilitate the hybridization of strains (Swan and Mantle, 1991) and have some evolutive advantages, e.g., adaptation to host range, (g) Alkaloid content creates a mutualistic interaction assuring the survival of the host together with the parasite.

Although the strategies are not fully perceived on the molecular functional level, it is clear that they bring the ergot fungi in full use of three fundamental advantages of parasitism given by Parbery (1996). (a) There is little competition in the host ovary. In addition, *C. purpurea,* for example, can settle on wheat and barley already infected with smuts like *Ustilago tritici* or *U. nuda* (Cherewick, 1953). (b) The habitat is more than large enough with sufficient nutrition to multiply, to spread and to produce a succeeding generation within six weeks, (c) The preservation of exclusiveness and supply of renewable substrate is guaranteed. Ergot fungi are seldom that highly virulent strains causing a complete suppression of host reproduction; some florets remain noninfected and fertile in every ear. Even if epidemic in the field, the fungi only mix sclerotia in the seed. The susceptibility of hosts evolved to an degree that maintains a certain level of infection with ergot. Symbiotrophism is likely the reason for tolerance and is antagonistic to evolution of resistant species. In conclusion, ergot disease appears not to be a severe problem to the plant but to animal and mostly to man in the need for clean crops.

2.7.
EPIDEMIOLOGY AND CONTROL

2.7.1.
Occurrence and Spread of Ergot Disease

As already mentioned, ergot disease is common throughout the world. Even single species, e.g., *C. purpurea,* are ubiquitous and several forms may occur in one locality (Swan and Mantle, 1991), producing a mixed inoculum. However, there are recent reports on immigration of individual ergot species into areas not occupied, so far, and, occasionally, ergot epidemics flame up for some reasons. Alternate hosts for *C. purpurea* proliferated on field borders or due to shut-down of arable land; small sclerotia which can not be shifted out during harvest remained in the seed. Increase in ergot severity or natural epidemics were correlated with reduced male-fertility or malesterility in sorghum (Frederickson *et al.,* 1993), in wheat after limited pollen supply (Mantle and Swan, 1995) and in pearl millet (Rai and Thakur, 1995). However, the highly susceptible male-sterile plants were used for hybrid breeding on a large scale. Highly protogynous hosts, e.g., pearl millet, are particularly susceptible to ergot by exposing the stigmas for two to five days before pollination.

In 1991, *C. paspali* was newly recognized in France after ergotism on cattle (Raynal, 1996). In 1995, *C. africana* was discovered on sorghum in Brazil for the first time, causing a widespread and economically important epidemic in commercial forage and hybrid seed production fields (Vasconcellos, 1996), and in 1996, *C. africana* was detected on sorghum in Australia for the first time (Ryley *et al.,* 1996). Ergot is severely epidemic on sorghum in Eastern Africa, i.e., Ethiopia-Swaziland, and in southern India (Frederickson and Mantle, 1988). In the USA, ergot is one of the most serious diseases in Kentucky bluegrass (*Poa pratensis* L.) (Johnston *et al.,* 1996) with up to 504 sclerotia per gram seed, which is 47% ergot by weight in one strain. Between 1991 and 1994, seed replacement amounts up to 0.44% at year's average, making 9% seed loss during cleaning to meet purity standards (Alderman *et al.,* 1996). In 1994, *C. purpurea* was epidemic in eastern Germany with average 10.4 sclerotia per rye ear and up to 10% seed loss during cleaning (Amelung, 1995).

In spite of its large host range, epidemics are in most cases spatially restricted with *Claviceps* species, they don't even threaten nearby cereal crops (Frederickson *et al.,* 1989, 1993). Host floral biology (see above) largely determines epidemics, nevertheless, this indicates rather unfavourable climatic conditions or limited inoculum transmission only at rather low distances. Though maximum ascospore release is perfectly timed with respect to host anthesis

(Mantle and Shaw, 1976) and this primary inoculum is wind distributed, it induces only the primary infection focus in the field but is not supposed to contribute much to the disease spread (Mantle, 1988). Spatial distribution of ergot disease is achieved by a secondary inoculum and is thought to be transmitted together with honeydew by rain-splash, insect vectors, physical head-to-head contact between ears and dripping down onto florets inserted below in the same ear (Tulasne, 1853; Engelke, 1902; Kirchhoff, 1929; Swan and Mantle, 1991). However, conidia transmittance, in particular, by insect vehicles is not investigated in detail (Luttrell, 1977; Mantle, 1988) but a matter of current research (von Tiedemann, pers. communication). On the contrary, recent aerobiological investigations into the spread of ergot in semi-arid zones showed that it are not the more infectious macroconidia but the microconidia produced outside of honeydew during microcycle are effective windborne propagules, which provide the principle epidemiological agent in the transmittance of *C. africana* in sorghum fields (Frederickson *et al.*, 1989, 1993). Higher humidities and lower temperature favour conidiation and drying-off periods support spore distribution by air (Marshall, 1960; Frederickson *et al.*, 1993).

2.7.2.
Control Measures

Although the ergot attack is a minor problem or even beneficial to the host, it is a great disadvantage in agriculture and, for certain reasons as discussed above, ergot disease gained strength in recent years. To enhance quantity and, more important, quality of grain and to minimize cleaning expenditure, control of ergot is needed, which has mainly been based on sanitary and cultural procedures (Kirchhoff, 1929) since many years. Protective fungicides have been ineffective or technically infeasible in controlling ergot (Puranik and Mathre, 1971; Thakur and Williams, 1980), because appropriate amounts must reach the ovary surfaces directly before spores landing. Unfortunately, the same is true for the fungicide benomyl, therewith the more feasible application of one systemic fungicide failed, so far. Recently, triazole fungicides have been identified as highly efficient for the control in experimental trials and in seed production fields (Vasconcellos, 1996). Hyperparasites growing on sclerotia of *Claviceps* species, e.g., *Fusarium heterosporum,F. roseum* and *F. acuminatum* are not yet developed for biological control (Chelkowski and Kwasna, 1986; Ali *et al.*, 1996).

Actually executed control measures are the removal of sclerotia from seeds before sowing combined with postharvest deep plowing and crop-rotation (Agrios, 1988). Usually, ergots do not live longer than one year (Taber, 1985). Normally, they do not survive if buried deep in the ground and, 1–1.5 inch below the soil surface, they do not produce functional stromata (Mantle and Shaw, 1976). Sometimes, however, even one and the same ergot can produce functional perithecia in a second and a third season, although their germination rate is much reduced (Kirchhoff, 1929). Cutting of intermediate hosts at field borders, particularly ryegrass or blackgrass, before flowering is recommended (Mantle, 1967; Mantle and Shaw, 1976). In USA, field burning is traditionally used but the control effect is controversial, because sclerotia viability is only markedly reduced above soil by temperatures higher than 100°C (Johnston *et al.*, 1996). Since host floral biology widely determines passive resistance and susceptibility, floral features antagonistic to ergot, e.g., pollination-induced host stigmatic constriction, has been exploited in the breeding of ergot-resistant strains for commercial use (Willingale *et al.*, 1986). Additionally, provision of pollen donor lines in the protogynous pearl millet efficiently reduces ergot attack (Thakur *et al.*, 1983). Inoculation with an ergot strain, which itself turned out to be relatively incompetent in sclerotia formation, could reduce the rate of successful infection of other strains which were already in place (Swan and Mantle, 1991).

2.8.
ACKNOWLEDGEMENTS

I wish to acknowledge Dr. B.Brockmann, Dr. V.Garre and Prof. Dr. P.Tudzynski for sharing of results prior to publication, Dr. R.Eising for providing anti-catalase sera, Prof. Dr. K.Roberts (John Innes Centre, Norwich) for the gift of JIM 5 and JIM 7 antibodies. I would like to thank Prof. Dr. P.Tudzynski, Dr. S.Giesbert and Dr. U.Müller for substantial discussions, inspirations and ideas through the years, and appreciate their comments on this review after critical reading. I am indebted to Mrs. B.Berns for technical and secretarial assistance, Mrs. W.Höfer for photomechanical support over several years, and Mrs. P.Stellamanns for reading the English. Our experimental research was partly financed by the Deutsche Forschungsgemeinschaft (DFG), Germany.

I.
REFERENCES

Agrios, G.N. (1988) *Plant pathology.*3rd. ed., Academic Press, San Diego.

Alderman, S.C., Coats, D.D. and Crowe, F.J. (1996) Impact of ergot on Kentucky bluegrass grown for seed in northeastern Oregon. *Plant Dis.,***80,**853–855.

Ali, H., Backhouse, D. and Burgess, L.W. (1996) *Fusarium heterosporum* associated with paspalum ergot in Eastern Australia. *Australiasian Plant Pathol.,***25,**120–125.

Amelung, D. (1995) Zum Auftreten von Mutterkorn im Jahr 1994. *Phytomedizin,***25**(3), 15.

Arntz, C. and Tudzynski, P. (1997) Identification of genes induced in alkaloid-producing cultures of *Claviceps* sp.*Curr. Genet.,*31, 357–360.

Bassett, R.A., Chain, E.B., Corbett, K., Dickerson, A.G.F. and Mantle, P.G. (1972) Comparative metabolism of *Claviceps purpurea in vivo* and *in vitro. Biochem. J.,***127,** 3P-4P.

Baum, M., Lagudah, E.S. and Appels, R. (1992) Wide crosses in cereals. *Annu. Rev.Plant Physiol.Plant Mol. Biol.,***43,**117–143.

Beckerman, J.L. and Ebbole, D.J. (1996) MPG1, a gene encoding a fungal hydrophobin of *Magnaporthe grisea,* is involved in surface recognition. *Mol. Plant-Microbe Interact.,***9,**450–456.

Bové, F.J. (1970) *The story of Ergot.*S.Karger, Basel.

Brockmann, B., Smit, R. and Tudzynski, P. (1992) Characterization of an extracellular β-1,3-glucanase *of Claviceps purpurea.Physiol. Mol. Plant Pathol.,***40,**191–201.

Campbell, W.P. (1957) Studies on ergot infection in gramineous hosts. *Can. J. Bot.,***35,** 315–320.

Campbell, W.P. (1958) Infection of barley by *Claviceps purpurea.Can. J. Bot.,***36,** 615–619.

Campbell, W.P. and Tyner, L.E. (1959) Comparison of degree and duration of susceptibility of barley to ergot and loose smut. *Phytopathology,***49,**348–349.

Carpita, N.C. and Gibeaut, D.M. (1993) Structural models of primary cell walls in flowering plants: consistency of molecular structure with the physical properties of the walls during growth. *Plant J.,***3,**1–30.

Chahal, S.S., Rao, V.P. and Thakur, R.P. (1985) Variation in morphology and pathogenicity in *Claviceps fusiformis,* the causal agent of pearl millet ergot. *Trans.Br. mycol. Soc.,***84,**325–332.

Cheeke, P.R. (1995) Endogenous toxins and mycotoxins in forage grasses and their effects on livestock. *J. Animal Science,***73,**909–918.

Chelkowski, J. and Kwasna, H. (1986) Colonization of *Claviceps purpurea* sclerotia on Triticale by fusaria in 1985. *J. Phytopath.,***117,**77–78.

Cherevick, W.J. (1953) Association of ergot with loose smut of wheat and of barley. *Phytopathology,***43,**461–463.

Cooke, R.C. and Mitchell, D.T. (1967) Germination pattern and capacity for repeated stroma formation in *Claviceps purpurea. Trans. Br. Mycol. Soc.,***50,**275–283.

Corbett, K., Dickerson, A.G. and Mantle, P.G. (1974) Metabolic studies on *Clavicepspurpurea* during parasitic development on rye. *J. Gen. Microbiol.,***84,**39–58.

Darlington, L.C., Mathre, D.E. and Johnston, R.H. (1977) Variation in pathogenicity between isolates of *Claviceps purpurea. Can. J. Plant Sci.,***57,**729–733.

Derksen, J. (1996) Pollen tubes: a model system for plant cell growth. *Bot. Acta,***109,** 341–345.

Dickerson, A.G., Mantle, P.G., Nisbet, L.J. and Shaw, B.I. (1978) A role for β-glucanases in the parasitism of cereals by *Claviceps purpurea. Physiol. Plant Pathol.,***12,**55–62.

Dickerson, A.G. and Pollard, C.M.D. (1982) Observations on the location of β-glucanase and an associated β-glucosidase in *Claviceps purpurea* during its development on rye. *Physiol. Plant Pathol.,***21,**179–191.

Engelke, C. (1902) Neue Beobachtungen über die Vegetations-Formen des Mutterkornpilzes *Claviceps purpurea* Tulasne. *Hedwigia,***41,**221–222.

Esser, K. and Tudzynski, P. (1978) Genetics of the ergot fungus *Claviceps purpurea.* I. Proof of a monoecious life-cycle and segregation patterns for mycelial morphology and alkaloid production. *Theor. Appl. Genet.,***53,**145–149.

Flaishman, M.A. and Kolattukudy, P.E. (1994) Timing of fungal invasion using host's ripening hormone as a signal. *Proc. Natl. Acad. Sci. USA,***91,**6579–6583.

Frauenstein, K. (1977) Studies on the host plant range of *Claviceps purpurea* (Fr.) Tul.— Influence of host plant on shape and size of conidia. *Proceedings of the XIIIInternational Farmland Congress,*Leipzig, 18–27 May. Akademie Verlag, Berlin, pp. 1277–1279.

Frederickson, D.E. and Mantle, P.G. (1988) The path of infection of sorghum by *Claviceps sorghi*. *Physiol. Mol. Plant Pathol.,***33,**221–234.

Frederickson, D.E., Mantle, P.G. and De Milliano, W.A.J. (1989) Secondary conidiation of *Sphacelia sorghi* on sorghum, a novel factor in the epidemiology of ergot disease. *Mycol. Res.,***93,**497–502.

Frederickson, D.E., Mantle, P.G. and De Milliano, W.A.J. (1991) *Claviceps africana* sp. nov.; the distinctive ergot pathogen of sorghum in Africa. *Mycol. Res.,***95,**1101–1107.

Frederickson, D.E., Mantle, P.G. and De Milliano, W.A.J. (1993) Windborne spread of ergot disease *(Claviceps africana)* in sorghum A-lines in Zimbabwe. *Plant Pathol.,***42,** 368–377.

Frederickson, D.E., Mantle, P.G. and De Milliano, W.A.J. (1994) Susceptibility to ergot in Zimbabwe of sorghums that remained uninfected in their native climates in Ethiopia and Rwanda. *Plant Pathol.,***43,**27–32.

Garay, A.St. (1956) The germination of ergot conidia as affected by host plant, and the culture of ergot on excised roots and embryos of rye. *Physiol. Plant.,***9,**350–355.

Garre, V., Müller, U. and Tudzynski, P. (1998a) Cloning, characterization and targeted disruption of *cpcat1,* coding for an in planta secreted catalase of *Clavicepspurpurea*. Mol. Plant-Microbe Interact., *(in press)*.

Garre, V., Tenberge, K.B. and Eising, R. (1998b) Secretion of a fungal extracellular catalase by *Claviceps purpurea* during infection of rye: Putative role in pathogenicity and suppression of host defense. Phytopathology, **88***(in press)*.

Giesbert, S., Lepping, H.-B., Tenberge, K.B. and Tudzynski, P. (1998) The xylanolytic system of *Claviceps purpurea:* cytological evidence for secretion of xylanases in infected rye tissue and molecular characterization of two xylanase genes. Phytopathology, (accepted for publication).

Giesbert, S. and Tudzynski, P. (1996) Molecular characterization of the xylanolytic system from the fungal pathogen *Claviceps purpurea*.ECFG 3, Münster, *Fun.Genet. Newslett.,***43:B,**79.

Hadley, G. (1968) Development of stromata in *Claviceps purpurea. Trans. Br. Mycol.Soc.,***51,**763–769.

Hambrock, A. (1996) *Untersuchungen über pflanzliche Abwebrreaktionen bei derInteraktion von Claviceps purpurea (Fr.) Tul. mit Secale cereale L.—Licht-undelektronenmikroskopische sowie biochemische Analysen.* Westfälische WilhelmsUniversität Münster, Ph.D. thesis.

Hambrock, A., Peveling, E. and Tudzynski, P. (1992) Lokalisation von Callose in verschiedenen Stadien der Entwicklung von *Claviceps purpurea* auf *Secale cereale.* In Haschke, H.-P. and Schnarrenberger, C. (eds.), *Botanikertagung 1992: Berlin,* 13.-19. September 1992, Deutsche Botanische Gesellschaft, Vereinigung für angewandte Botanik, Akademie Verlag, Berlin, p. 403.

Heslop-Harrison, J. and Heslop-Harrison, Y. (1980) The pollen-stigma interaction in the grasses. 1. Fine-structure and cytochemistry of the stigmas of *Hordeum* and *Secale. Acta Bot. Neerl.,***29,**261–276.

Heslop-Harrison, J. and Heslop-Harrison, Y. (1981) The pollen-stigma interaction in the grasses. 2. Pollen-tube penetration and the stigma response in *Secale. Acta Bot.Neerl.,* **30,**289–307.

Heslop-Harrison, Y., Heslop-Harrison, J. and Reger, B.J. (1985) The pollen-stigma interaction in the grasses. 7. Pollen-tube guidance and the regulation of tube number in *Zea mays* L. *Acta Bot. Neerl.,***34,**193–211.

Heslop-Harrison, Y., Reger, B.J. and Heslop-Harrison, J. (1984) The pollen-stigma interaction in the grasses. 6. The stigma ('silk') of *Zea mays* L. as host to the pollens of *Sorghum bicolor* (L.) Moench and *Pennisetum americanum* (L.) Leeke. *Acta Bot.Neerl.,***33**,205–227.

Howard, R.J. and Valent, B. (1996) Breaking and entering: host penetration by fungal rice blast pathogen*Magnaportbe grisea. Annu. Rev. Microbiol.,***50**,491–512.

Johnston, W.J., Golob, C.T., Sitton, J.W. and Schultze, T.R. (1996) Effect of temperature and postharvest field burining of Kentucky bluegrass on germination od sclerotia of *Claviceps purpurea. Plant Dis.,***80**,766–768.

Jungehülsing, U. and Tudzynski, P. (1997) Analysis of genetic diversity in *Clavicepspurpurea* by RAPD markers. *Mycol. Res.,***101**,1–6.

Kirchhoff, H. (1929) Beitrage zur Biologie und Physiologie des Mutterkornpilzes. *Centralbl. Bakteriol. Parasitenk. Abt. II,***77**,310–369.

Lewis, R.W. (1956) Development of conidia and sclerotia of the ergot fungus on inoculated rye seedlings. *Phytopathology,***46**,295–296.

Loveless, A.R. (1971) Conidial evidence for host restriction in *Claviceps purpurea. Tram. Br. my col Soc.,***56**,419–434.

Luttrell, E.S. (1974) Parasitism of fungi on vascular plants. *Mycologia,***66**,1–15.

Luttrell, E.S. (1977) The disease cycle and fungus-host relationships in dallisgrass ergot. *Phytopathology,***67**,1461–1468.

Luttrell, E.S. (1980) Host-parasite relationships and development of the ergot sclerotium in *Claviceps purpurea. Can. J. Bot.,***58**,942–958.

Mantle, P.G. (1967) Emergence and phytopathological properties of a new strain of *Claviceps purpurea* (Fr.) Tul. on rye. *Ann. Appl. Biol.,***60**,353–356.

Mantle, P.G. (1972) An unusual parasitic association between *Claviceps purpurea* and rye. *Trans. Br. My col. Soc.,***59**,327–330.

Mantle, P.G. (1988) *Claviceps purpurea*. In *European Handbook of Plant Diseases.* Blackwell Scientific Publications, Oxford, pp. 274–276.

Mantle, P.G. and Nisbet, L.J. (1976) Differentiation of *Claviceps purpurea* in axenic culture. *J. Gen. Microbiol.,***93**,321–334.

Mantle, P.G. and Shaw, S. (1976) Role of ascospore production by *Claviceps purpurea* in aetiology of ergot disease in male sterile wheat. *Trans. Br. Mycol. Soc.,***67**,17–22.

Mantle, P.G. and Swan, D.J. (1995) Effect of male sterility on ergot disease spread in wheat. *Plant Pathol.,***44**,392–395.

Marshall, G.M. (1960) The incidence of certain seed-borne diseases in commercial seed-samples. II. Ergot, *Claviceps purpurea* (Fr.) Tul. in cereals. *Ann. Appl. Biol.,***48**,19–26.

Mehdy, M.C. (1994) Active oxygen species in plant defense against pathogens. *PlantPhysiol.,***105**,467–472.

Mendgen, K. and Deising, H. (1993) Tansley Review No. 48. Infection structures of fungal plant pathogens—a cytological and physiological evaluation. *New. Phytol.,***124**, 193–213.

Mendgen, K., Hahn, M. and Deising, H. (1996) Morphogenesis and mechanisms of penetration by plant pathogenic fungi. *Annu. Rev. Phytopathol.,***34**,367–386.

Mims, C.W. (1991) Using electron microscopy to study plant pathogenic fungi. *Mycologia,***83**,1–19.

Mower, R.L. and Hancock, J.G. (1975) Mechanism of honeydew formation by *Claviceps* species. *Can. J. Bot.,***53**,2826–2834.

Müller, U., Tenberge, K.B., Oeser, B. and Tudzynski, P. (1997) *Cel*1, probably encoding a cellobiohydrolase lacking the substrate binding domain, is expressed in the initial infection phase of *Claviceps purpurea* on *Secale cereale.Mol. Plant-MicrobeInteract.*, **10**,268–279.

Münch, E. (1929) Über einige Grundbegriffe der Phytopathologie. *Z. Pflanzen-krankheit.*, **39**,276–286.

Musabyimana, T., Sehene, C. and Bandyopadhyay, R. (1995) Ergot resistance in sorghum in relation to flowering, inoculation technique and disease development. *Plant Pathol.*, **44**,109–115.

Osada Kawasoe, S., Fucikovsky Zak, L.Ortega Delgado, M.L. and Engleman, E. (1986) Study on the germination of the ergot on maize and host parasit interaction *Zeamays* and *Claviceps gigantea. Agrociencia O*,**66**,57–70.

Parbery, D.G. (1996) Trophism and the ecology of fungi associated with plants. *Biol.Rev.*, **71**,473–527.

Platford, R.G. and Bernier, C.C. (1970) Resistance to *Claviceps purpurea* in spring and durum wheat. *Nature,***226**,770.

Puranik, S.B. and Mathre, D.E. (1971) Biology and control of ergot on male sterile wheat and barley. *Phytopathology,***61**,1075–1080.

Rai, K.N. and Thakur, R.P. (1995) Ergot reaction of pearl millet hybrids affected by fertility restoration and genetic resistance of parental lines. *Euphytica,***83**,225–231.

Raikhel, N.V., Lee, H.-I. and Broekaert, W.F. (1993) Structure and function of chitinbinding proteins. *Annu. Rev. Plant Physiol. Plant Mol. Biol.,***44**,591–615.

Ramstad, E. and Gjerstad, G. (1955) The parasitic growth of *Claviceps purpurea* (Fries) Tulasne on rye and its relation to alkaloid formation. *J. Am. Pharm. Assoc. Sci. Ed.,***44**, 741–743.

Raynal, G. (1996) Presence in France of *Claviceps paspali* Stev. et Hall on *Paspalumdistichium* L. and of the corresponding ergotism on cattle. *Cryptogamie, Mycol.,***17**,21–31.

Ryley, M.J., Alcorn, J.L., Kochman, J.K., Kong, G.A. and Thompson, S.M. (1996) Ergot of *Sorghum* spp. in Australia. *Australasian Plant Pathol.,***25**,214.

Schäfer, W. (1994) Molecular mechanisms of fungal pathogenicity to plants. *Annu. Rev.Phytopathol.,***32**,461–477.

Shaw, B.I. and Mantle, P.G. (1980a) Host infection by *Claviceps purpurea. Trans. Br.Mycol. Soc.,***75**,77–90.

Shaw, B.I. and Mantle, P.G. (1980b) Parasitic differentiation of *Claviceps purpurea. Trans. Br. Mycol. Soc.,***75**,117–121.

Stäger, R. (1922) Beitrage zur Verbreitungsbiologie der *Claviceps* Sklerotien. *Zentralbl. Bakt. Abl. II,***56**,329–339.

Stoll, A. and Brack, A. (1944) Über die Entstehung von Sklerotien des Mutterkornpilzes *(Claviceps purpurea)* an den obersten Halmknoten des Roggens. *Ber.Schweiz. Bot. Ges.,***54**,252–254.

Swan, D.J. and Mantle, P.G. (1991) Parasitic interactions between *Claviceps purpurea* strains in wheat and an acute necrotic host response. *Mycol. Res.,***95**,807–810.

Taber, W.A. (1985) Biology of *Claviceps.* In Demain, A.L. and Solomon, N.A. (eds.), *Biology of industrial microorganisms.*The Benjamin Cummings Publishing Company, London, Amsterdam, Don Mills, pp. 449–486.

Talbot, N.J., Kershaw, M.J., Wakley, G.E., de Vries, O.M.H.,Wessels, J.G.H. and Hamer, J.E. (1996) *MPG1* encodes a fungal hydrophobin involved in surface interactions during infection-related development of *Magnaporthe grisea*. *PlantCell,***8,**985–999.

Tenberge, K.B. (1994) Infection by the ergot fungus *Claviceps purpurea:* ultrastructure and cytochemistry of a host-pathogen relationship. *Fifth International MycologicalCongress (IMC 5),*August 14–21, 1994, Vancouver, British Columbia, Canada, p. 219.

Tenberge, K.B., Brockmann, B. and Tudzynski, P. (1995) Die extrazelluläre *β*-1, 3Glucanase von *Claviceps purpurea:* cytologische und molekularbiologische Analysen zur Bedeutung in der Wirt-Parasit-Beziehung. *Molekularbiologie der Pilze II,Tagung in Gosen bei Berlin 5.10.–8.10. 1995,*p. 76.

Tenberge, K.B. and Eising, R. (1995) Immunogold labelling indicates high catalase concentrations in amorphous and crystalline inclusions of sunflower *(Helianthusannuus* L.) peroxisomes. *Histochem. J.,***27,**184–195.

Tenberge, K.B., Homann, V., Oeser, B. and Tudzynski, P. (1996a) Structure and expression of two polygalacturonase genes of *Claviceps purpurea* oriented in tandem and cytological evidence for pectinolytic enzyme activity during infection of rye. *Phytopathology,***86,**1084–1097.

Tenberge, K.B., Homann, V. and Tudzynski, P. (1996b) Der Mutterkornpilz *Clavicepspurpurea* auf Roggen: Struktur und Funktion zweier Polygalakturonase-Gene in Relation zu Veränderungen von Pektinen in der Wirtszellwand. *Phytomedizin,***26**(3), 55–56.

Tenberge, K.B., Stellamanns, P., Plenz, G. and Robenek, H. (1998) Nonradioactive in situ hybridization for detection of hydrophobin mRNA in the phytopathogenic fungus *Claviceps purpurea* during infection of rye. *Eur. J. Cell Biol.,***75,**265–272.

Tenberge, K.B. and Tudzynski, P. (1994) Early infection of rye ovaries by *Clavicepspurpurea* is inter- and intracellular. *BioEng. Sondernummer,***10**(3), 22–22.

Tenberge, K.B. and Tudzynski, P. (1995) Der Mutterkornpilz *Claviceps purpurea* auf Roggen: Affinitätscytochemische Untersuchungen einer Wirt-Parasit-Beziehung. *Phytomedizin,***25**(3), 51–52.

Thakur, R.P., Rao, V.P. and Williams, R.J. (1984) The morphology and disease cycle of ergot caused by *Claviceps fusiformis* in pearl millet. *Phytopathology,***74,**201–205.

Thakur, R.P. and Williams, R.J. (1980) Pollination effects on pearl millet. *Phytopathology,***70,**80–84.

Thakur, R.P., Williams, R.J. and Rao, V.P. (1983) Control of ergot in pearl millet through pollen management. *Ann. App. Biol.,***103,**31–36.

Tudzynski, P., Tenberge, K.B. and Oeser, B. (1995) *Claviceps purpurea.*In Kohmoto, K., Singh, U.S. and Singh, R.P. (eds.), *Pathogenesis and host specificity in plantdiseases: histopathological, biochemical, genetic and molecular bases,*vol. IIEukaryotes, Pergamon, Elsevier Science Ltd., pp. 161–187.

Tudzynski, P. and Tudzynski, B. (1996) Genetics of phytopathogenic fungi. *Progressin Botany,***57,**235–252.

Tulasne, L.-R. (1853) Mémoire sur l'ergot des glumacées. *Ann. Set. Nat. (PartieBotanique),***20,**5–56.

van den Ackerveken, G.F.J.M. and de Wit, P.J.G.M. (1995) The Cladosporium fulvumtomato interaction, a model system for fungus-plant specificity. In Kohmoto, K., Singh, U.S. and Singh, R.P. (eds.), *Pathogenesis and host specificity in* *plantdiseases: histopathological, biochemical, genetic and molecular bases,*vol. IIEukaryotes, Pergamon, Elsevier Science Ltd, pp. 145–160.

Vasconcellos, J.H. (1996) Ergot on *Sorghum. Intern. Newslett. Plant Pathol.,*26, 1.

Wessels, J.G.H. (1994) Developmental regulation of fungal cell wall formation. *Annu.Rev. Phytopathol.,***32,**413–437.

White, J.F., Hakisky, P.M., Sun, S.C., Morgan-Jones, G. and Funk, C.R. (1992) Endophyte-host associations in grasses. XVI. Patterns of distribution in species of the tribe Agrostidae. *Amer. J. Bot.,***79,**472–477.

Williams, R.J. and Colotelo, N. (1975) Influence of pollen on germination of conidia of *Claviceps purpurea. Can. J. Bot.,***53,**83–86.

Willingale, J. and Mantle, P.G. (1985) Stigma constriction in pearl milltet, a factor influencing reproduction and disease. *Ann. Bot.,***56,**109–115.

Willingale, J. and Mantle, P.G. (1987a) Stigmatic constriction in pearl millet following infection by *Claviceps fusiformis. Physiol. Mol. Plant Pathol.,***30,**247–257.

Willingale, J. and Mantle, P.G. (1987b) Interactions between *Claviceps purpurea* and *Tilletia caries* in wheat. *Tram. Br. Mycol. Soc.,***89,**145–153.

Willingale, J., Mantle, P.G. and Thakur, R.P. (1986) Postpollination stigmatic constriction, the basis of ergot resistance in selected lines of pearl millet. *Phytopathology,***76,**536–539.

Yamagami, T. and Funatsu, G. (1996) Limited proteolysis and reduction-carboxymethylation of rye seed chitinase-a: role of the chitin-binding domain in its chitinase action. *Biosci. Biotechnol. Biochem.,***60,**1081–1086.

You, R. and Jensen, W.A. (1985) Ultrastructural observations of the mature megagametophyte and the fertilization in wheat *(Triticum aestivum). Can. J. Bot.,***63,** 163–178.

3.
THE TAXONOMY AND PHYLOGENY OF *CLAVICEPS*

SYLVIE PAŽOUTOVÁ[1] and DOUGLAS P.PARBERY[2]

[1]*Institute of Microbiology, Academy of Sciences of the CzechRepublic, Vídeňská 1083, 142 20 Prague, Czech Republic*
[2]*Faculty of Agriculture, Forestry and Horticulture,University of Melbourne, Parkville 3052, Australia*

3.1.
THE TAXONOMIC POSITION OF *CLAVICEPS*

The evolution and to some extent the taxonomy of the parasitic fungi belonging to the genus *Claviceps* are linked to the evolution of their host plants (grasses, rushes, and sedges). Although the recent Internet version of the Index of Fungi (http://nt.ars_grin.gov/indxfun/frmlndF.htm) lists 38 recorded species of *Claviceps,* at least 7 species are missing and *C. oryzae-sativae* and *C. virens* were removed from *Claviceps,* so that there are 43 known species.

The genus *Claviceps* is in the family Clavicipitaceae which was initially placed in the order Hypocreales. During the 50's, however, doubts about this based on comparative study of conidiogenous stroma development resulted in its transfer into the Xylariales (Luttrell, 1951) and then to the Clavicipitales, an order close to Hypocreales and erected specifically to accommodate clavicipitaceous fungi (Gäumann, 1952). More recently, application of DNA analysis has been used to test the relationships between different members of Hypocreales and has confirmed the initial placement of the monophyletic Clavicipitaceae as family belonging to the order Hypocreales. Molecular phylogenies suggest that the genus *Claviceps* was the first group derived from a common ancestor line, then *Epichloe/Neotyphodium* followed by *Atkinsonella* and with the last clade containing species of *Balansia* and *Myriogenospora* (Spatafora and Blackwell, 1993; Rehner and Samuels, 1994; Glenn *et al.,* 1996). This contradicts the formerly held hypothesis that *Balansia* contains the most primitive clavicipitoids.

3.2.
TAXONOMIC MARKERS

The taxonomic criteria used to deliminate *Claviceps* species are: the color, size and shape of sclerotia, the color of stipes and capitula, the presence or absence of loose hyphae on the stroma, the size and shape of perithecia, asci, ascospores and conidia (Langdon, 1942).

3.2.1.
Sclerotium

Sclerotium size is largely dependent on host. For example, *C. purpurea* sclerotia produced in florets of *Poa annua* or *Phalaris tuberosa* are about 1–2 mm long, those formed in florets of *Secale cereale* are up to 50 mm. Similarly, there is a threefold difference also between the *C. gigantea* sclerotia formed in *Zea mays* and *Z. mexicana*. Sclerotia contain lipid reserves that are consumed during germination (Mitchell and Cooke, 1968). In the wet tropics, where hosts may flower for most of the year, the selective pressure for sclerotium production is so low for some ergot species that they produce only few sclerotia erratically or even not at all (*Sphacelia* spp.). For example, although some African mature sclerotia of *C. africana* were successfully germinated (Frederickson *et al.*, 1991), this ergot is worldwide found in the sphacelial state. Other ergots, such as *Sphacelia tricholaenae* are known only for its anamorph.

Langdon (1954) described three basal types of sclerotia:

1. primitive, irregularly globose, where the mycelium emerges from the infected ovary and envelops parts of the spikelet(s) into pseudosclerotia resembling those of balansioid genera. Species producing these primitive sclerotial forms, *C. diadema* and *C. flavella,* occur only on panicoid hosts with C3 type of photosynthesis in tropical forest regions
2. subglobose to elongated, usually light-colored, (*C. paspali, C. queenslandica, C. hirtella* and *C. orthocladae*), occurring on Panicoids
3. elongated sclerotia, ovoid to cylindric in shape, dark coloured. On the distal tip of this sclerotium type, there is usually a cap formed by the remnants of sphacelial tissue. Species forming this type are found on members of all gramineous subfamilies. Their most advanced representative is *C. purpurea* which exhibits intercalary growth in the proliferative zone distal to the sclerotial foot (site of contact with the plant vascular system). This category, however, should be subdivided, because elongated sclerotia differ greatly in the resistance to drought, frost and long storage as well as in dormancy and germination requirements.

3.2.2.
Stroma

Stroma (also clava) consists of stipe and capitulum. The coloration of clavae is either straw to yellow or in shades of brown-purple to black-purple, with the exception of *C. viridis* which is green. The "yellow" clavae are mostly encountered in species from panicoid hosts. The young capitulum is smooth, the ostioles of the perithecia appear as darker pores. During the maturation the ostioles enlarge and become papillate, very prominently in some species *(C. ranunculoides, C. fusiformis)*. This, however, appears to be due more to the shrinkage of the tissue of the outer capitulum cortex than to the emergence of perithecia beyond it although the end result would be the same.

The size and shape of perithecia is dependent on their degree of maturation. Young perithecia are small and oblong to oval, whereas fully matured ones are often ellipsoid to pyriform. Once the filamentous asci which form inside perithecia are mature, they protrude through the ostiole prior to discharging their ascospores. Loveless (1964) considered the length of asci an unreliable criterion, because asci at different stage of maturity and therefore different length may occur in a single perithecium and the stage of maturity is difficult to determine. The length of filiform ascospores should be measured after their discharge from asci.

3.2.3.
Conidia

The size and shape of conidia is less dependent on environmental factors than are ascospores and are valuable traits for species determination. Loveless (1964) mapped the range in size and shape of conidia in fresh honeydew and from dried sclerotial material collected from Rhodesian grasses. Thirteen conidial types were defined, six of them belonged to known species *C. paspali,C. digitariae, C. sulcata, C. maximensis, C. pusilla* and *C. cynodontis,* seventh group with large falcate or fusiform conidia occurring on *Cenchrus ciliaris* and *Pennisetum typhoideum* was probably *C. fusiformis,* described three years later by the same author (Loveless, 1967). Conidial characters together with considerations of host range are generally sufficient to determine if a sphacelial ergot is the anamorph of an already named species or is representative of a species new to science. For e.g., the tenth conidial group described from *Sorghum caffrorum* resembled anamorph of *C. africana* described much later by Frederickson *et al.* (1991).

A study of conidia of English *C. purpurea* collections revealed the similarity in conidial shape among the samples from certain host groups suggesting the existence of *C. purpurea* host races (Loveless, 1971).

3.2.4.
Chemotaxonomic Markers

The qualitative analysis of alkaloid content as a possible marker was made in only seven species from the 43 described, these being *C. purpurea* (peptide alkaloids: ergotamine, ergocornine, ergocristine), *C. fusiformis* (agro-, elymoand chanoclavine), *C. paspali* (lysergic acid amides), *C. gigantea* (festuclavine) (for detailed review see Flieger *et al.,* 1997). Mantle (1968) detected the peptide alkaloid dihydroergosine as the main component in *C. africana* and traces of agroclavine in *C. sorghi.* Tanaka and Sugawa (1952) and Yamaguchi *et al.* (1959) found the peptide alkaloids, ergometrine and agroclavine in sclerotia of *C. imperatae.* Porter *et al.* (1974) described the occurrence of ergometrine related alkaloids in ergotized *Cynodon dactylon,* but did not identify the *Claviceps* species which was probably *Claviceps cynodontis*.

In other species, evidence of alkaloid production was obtained colorimetrically using vanUrk's reagent which gives blue coloration with ergoline compounds. Prior to describing new species, Tanda assayed sclerotial extracts for alkaloids as well as testing their toxicity to mice. Only in *C. bothriochloae,* some positive colorimetric reaction was found (Tanda, 1991), whereas in *C. microspora* and *C. yanagawaensis,* the weight loss and other signs of mice toxicity were not apparently connected with alkaloid content (Tanda, 1981) and may therefore have been caused by other toxic metabolites. Taber and Vining (1960) detected alkaloid production colorimetrically in shake cultures of *C. maximensis* and the ergot isolated from wild rice, that could possibly be *C. zizaniae*. As yet unidentified alkaloids have been found in the shaken cultures *of C. grohii* and *C. sulcata* (Pažoutová *et al.,* unpublished).

An indirect proof of alkaloid production by *C. cinerea* was the occurrence of abortions in cattle in North Mexico grazing *Hilaria mutica,* colonized by this ergot (Zenteno-Zevada, 1958).

Walker (personal communication) found no alkaloids in *C. phalaridis* sclerotia. Similarly, HPLC analysis at the Inst. Microbiology CAS, Prague failed to detect alkaloids at the ppm level. Preliminary results confirmed the presence of a group of unknown metabolites. No alkaloids were also detected in the sclerotia of new species of ergot *C. citrina* from *Distichlis spicata* (Pažoutová *et al.,* in press).

Another attempt to find chemotaxonomic markers for distinguishing *Claviceps* species was presented by Mower and Hancock (1975). The sugar composition of the thick liquid honeydew secreted during the sphacelial stage to protect and disseminate conidia, was determined for nine known species and possibly five undescribed species collected on specified grasses and rushes.

The Claviceps and Sphacelia species differed in the amount and representation of 23 mono-, di- and oligosaccharides with two basic types being identified; those with prevalent glucose and fructose (*C. purpurea, C. fusiformis, C. nigricans, C. grohii* and *Sphacelia* isolates from *Pennisetum,Setaria* and *Juncus*) and those with arabinitol and mannitol *(C. gigantea,C. cinerea, C. tripsaci)*. An intermediary group containing both sugars and sugar alcohols included *C. paspali, C. uleana, C. zizaniae* and a *Sphacelia* from *Andropogon tener*.

3.3.
THE CHARACTERIZATION OF SOME CLAVICEPS
SPECIES

3.3.1.
C. paspali

According to Hitchcock (1950) and Langdon (1952), this species originates from South America (Uruguay or Argentina) and was introduced to USA about 1850. In the years 1927 to 1937 it was reported from Australia and New Zealand, 1947–1948 from the Mediterranean region (where it followed the introduction of *Paspalum distichum* in 1929) and now is found worldwide where *Paspalum* species grow. Infection is followed 4–5 days later with the production of abundant honeydew containing primary and secondary (microcycle) conidia. Subglobose light brown sclerotia encompass the remnants of disintegrated ovary (Luttrell, 1977). Cold and moist storage is required for sclerotium germination, the germination ability is lost after dry storage and exposure to temperatures under 0°C (Cunfer and Marshall, 1977). Similar species, however, endemic to Australia and detected there before the introduction of *C. paspali* is *C. queenslandica* (Langdon, 1952).

3.3.2.
C. phalaridis

Walker (1957) discovered this Australian endemic ergot, that persists as systemic endophyte similarly to *Epichloe/Neotyphodium* and forms sclerotia in florets of the diseased plants, rendering them sterile. Intercellular hyphae are present in tillers, stems, leaf sheaths and blades. It was found on pooid grasses including *Lolium rigidum* and species of *Phalaris, Vulpia, Dactylis,* as well as on arundinoid grasses in the genus *Danthonia* (Walker, 1970). A white fungal mass envelopes young anthers and the ovary. Infected florets are later incorporated in the mature subglobose sclerotium. The sclerotia were able to germinate after 5-month storage either dry at room temperature or humid at 6°C (Walker, 1970). Ascospore formation in this species differs from that of either *C. purpurea* and *E. typhina* (Decker, 1980). Clay (1988) considered this species as possible intermediate between Balansiae and *Claviceps,* but our current investigations of DNA sequences do not support this view (Pažoutová, unpublished).

3.3.3.
C. viridis

C. virdis was found on *Oplismenus compositus* in India (Padwick and Azmatullah, 1943; Thomas *et al.,* 1945) and on O. *undulatifolius* recently in Japan (Tanda, 1992), in forest regions with monsoon rains. Its sclerotia are cylindrical, blackish with a tinge of green and require a dormancy period prior germination (Tanda, 1992).

3.3.4.
C. gigantea

C. gigantea is found on maize *(Zea mays)* in the high valleys of Central Mexico (Fuentes *et al.,* 1964). Its occurrence seems to be limited by the annual mean temperature of 13–15°C (Fucikovsky and Moreno, 1971). The formation of sclerotia (Fuentes *et al.,* 1964) is quite different from that in *C. purpurea* and *C. paspali* as described by Luttrell (1977, 1980 respectively). It starts as a soft and hollow structure that produces many conidia suspended in honeydew. The resting structure develops later in the internal cavity as firm pseudoparenchymatous tissue surrounded by thin dark rind. Maturation proceeds from the interface with the host progressing distally. The surface of the mature sclerotium is covered with dried remnants of the sphacelia. With respect to climatic conditions, the sclerotia survive short frost periods and the dry winter.

3.3.5.
C. africana

C. africana colonizes sorghum *(Sorghum bicolor)* and is endemic in eastern and southern Africa, especially in Kenya, Zimbabwe and South Africa. However, *C. africana* is spreading to Southeast Asia, Japan (Bandyopadhyay, 1992) South America (Reis *et al.,* 1996) and Australia (Ryley *et al.,* 1996), presenting danger for the male sterile A-lines of sorghum and seed production. In 1997, it reached Texas, USA. Infection ability is enhanced by the extensive secondary conidiation on the surface of excreted honeydew that presents source of windborne propagules (Frederickson *et al.,* 1993). *C. africana* sclerotia were able to germinate after 1 year of dry storage at ambient temperature (15–30°C) (Frederickson *et al.,* 1991). The most often encountered alternative host is *S. halepense* (Johnson grass). A common feature of *C. africana* and *C. gigantea* is the production of dihydrogenated ergoline alkaloids.

3.3.6.
Claviceps pusilla

C. pusilla is a widespread ergot of the Mediterranean region, Africa, India, Australia and probably China, colonizing seventeen andropogonoid Old World species (among them *Bothriochloa, Dichanthium, Capillipedium,Cymbopogon, Heteropogon, Vetiveria* and *Themeda*) (Langdon, 1954; Loveless, 1964). Its sclerotia are probably of the advanced type, and a characteristic marker is the triangular shape of its conidia.

3.3.7.
Claviceps cynodontis

C. cynodontis is one of the few ergots colonizing chloridoid grasses (Langdon, 1954; Loveless, 1965). It grows in the florets of *Cynodon dactylon* (Bermuda grass) and it is distributed from southern Europe to Africa, India, Burma and Philippines. It often occurs in sphacelial state without developing sclerotia.

3.3.8.
Claviceps sulcata

This species was originally found in Rhodesia (Zimbabwe) on a *Brachiaria* species (Langdon, 1952), an important forage grass in warm regions. Since then, it has been found in other areas of southern Africa, but recently, the sphacelial stage was reported as widespread from Brazil (Fernandes *et al.,* 1995). It is characterized by large conidia and prominent broad furrows running along each side of the sclerotium.

3.3.9.
Claviceps fusiformis

C. fusiformis is widespread in semi-arid regions of Africa and India. It occurs on pearl millet *(Pennisetum americanum)* which was domesticated in Sudanian region in Africa and introduced to India (for review see de Wet, 1992). Distribution of *C. fusiformis* extends from the Transvaal to Equatorial Africa (Loveless, 1967) and widely within India. Despite its typical fusiform elongated conidia, Thirumalachar (1945) concluded that it was *C. microcephala* (syn. *C. purpurea*) and this misidentification persisted until the 70's (Sundaram *et al.,* 1972). Rapid spread and high infection rates are enhanced by the formation of secondary conidia and its sclerotia are drought-resistant (Thakur *et al.,* 1984). Loveless (1964) found the conidia of *C. fusiformis* as the falcate type on the species of *Cenchrus ciliaris, P. typhoideum* and *P. maximum*. Ramakrishnan (1952) successfully cross-inoculated *C. fusiformis* from six species of *Pennisetum* to species of *Cenchrus* and *Urochloa*.

3.3.10.
Claviceps purpurea

C. purpurea occurs in all temperate regions and has undoubtedly the widest host range of any *Claviceps* species. As a result, its name has been suspected of being a blanket term for several distinct taxa. Morphology descriptions of the species reflect this variability. Sprague (1950) described *C. purpurea* as having purple-black sclerotia and flesh-colored capitula, Dickson (1956) characterized its stromatal heads as pale-fawn, while Tanda (1979a) reported pale purple to blackish brown sclerotia and light orange to pale red capitula.

Sclerotium morphology is influenced by the size of host florets and climatic differences (for e.g., the sclerotial size ranges from 1–50 mm). However, for precisely these reasons it happens that almost any ergot with elongated dark brown to black sclerotia and purplish capitulum is classified as *C. purpurea* whereas it could represent yet another undescribed species. A good example of such misdetermination is the above mentioned designation of *C. fusiformis* from Indian *Pennisetum typhoideum* as *C. microcephala* which persisted for many years. *C. grohii* (Groves, 1943) was described as a new species mostly because of its colonization of *Carex,* which was considered too great an extension of its host range for *C. purpurea.* One other discriminative marker distinguishing the two species was the insect damage of sclerotia in *C. grohii* collections when *C. purpurea* sclerotia remained untouched. Later it was found (Langdon, 1952) that conidia of *C. grohii* differ in the shape from those of *C. purpurea.* However, *C. purpurea* was recently isolated from *Carex* and its identity confirmed by RAPD (Jungehülsing, 1995). Thus, both these species occur on *Carex hosts.*

It has been noted, that isolates of *C. purpurea* from certain hosts infect other grass species under laboratory conditions with varying results. The successful infection is highly dependent on the length of the period of anthesis and the inoculation technique used. Stäger (1903, 1905, 1908, 1922, cf. Campbell, 1957) used three methods: spraying the heads with a conidial suspension, prying the glumes apart and spraying the florets and dipping grass heads in conidial suspension. These methods succeeded unless the florets were closed or waxy or haired glumes protected them.

Stäger found three races of *C. purpurea* in Europe that should infect only certain groups of hosts and later, together with other authors seven races was identified:

P1, *Bromus sterilis, Festuca elatior, Hordeum,* four species of *Poa, Secale,Triticum*

P2, *Brachypodium sylvaticum*

P3, *Bromus erectus, Lolium* sp.

P4, *Bromus erectus, Festuca arundinacea, Lolium perenne* (Mastenbroek and Oort, 1941)

P5, *Lolium* spp., *Secale* (Baldacci and Forlani, 1948)

P6, *Aira, Molinia, Nardus, Phragmites* (Arundinoideae)

P7, *Poa annua* only (also considered *C. microcephala*)

Other authors, however, came to different conclusions when they transferred ergot from *Lolium* sp. to rye (Békésy, 1956) or to wheat (Bretag and Merriman, 1981). Kybal and Brejcha (1956) succeeded in infecting rye with ergot from *Phragmites* and *Molinia,* whereas Campbell (1957) found no host preferences at all in Canadian isolates.

Campbell removed the glume tips at anthesis and sprayed the heads with conidial suspension but when infecting *Hordeum,* it was necessary to inoculate the heads just emerging from the leaf sheath. Inoculation at anthesis was not successful. Inoculations were repeated until the first signs of sphacelial development appeared.

In the first part of his experiments, Campbell took the conidial suspensions from ergots collected on 38 host species in 19 genera from three subfamilies (as defined by Watson and Dalwitz, 1996):

Pooideae

Agropyron, Agrostis, Arrhenatherum, Avena, Bromus, Calamagrostis,Dactylis, Elymus, Festuca, Glyceria, Hordeum, Lolium, Phleum, Poa,Secale, Triticum

Arundinoideae

Stipa

Chloridoideae

Calamovilfa, Spartina

and inoculated rye, wheat and barley. Every isolate infected each cereal with the exception of one isolate from *Glyceria borealis*. Other isolates from *G. borealis*, however, were infectious.

In a reverse procedure, conidial inoculum of rye ergot to was applied to plants in the following subfamilies and genera:

Pooideae:

Agropyron, Agrostis, Alopecurus, Bromus, Calamagrostis, Dactylis,Elymus, Festuca, Koeleria, Phalaris, Phleum, Poa, Polypogon, SitanionSphenopholis

Arundinoideae

Stipa, Danthonia

Chloridoidae:

Eragrostis, Sporobolus

Panicoideae

Setaria

Again, all plants became infected both in the field and in the greenhouse. These experiments confirmed that with the use of a suitably aggressive inoculation technique, *C. purpurea* is capable of colonizing a wide range of not only pooid, but also chloridoid and some panicoid hosts. Natural occurrences of *C. purpurea* on the species of Setaria and Pennisetum were reported from South Africa (Doidge, 1950). Brewer and Loveless (1977) took honeydew from *Pennisetum macrourum* and inoculated wheat getting a mean of 1.6 sclerotia per head of wheat compared with 3.9 sclerotia when they transferred honeydew from wheat to wheat.

However, the question of host races and specificity of *C. purpurea* continues. Loveless (1971) found that the host-specific differences in conidial size and shape in English collections were retained even in laboratory cultures and corresponded to some extent to the Stäger's groups. Following numerous cross-inoculation experiments, Tanda allocated Japanese isolates of *C. purpurea* from different pooid and arundinoid hosts to four varieties: *C. purpurea* var. *purpurea*, var. *alopecuri*, var. *phalaridis* and var. *sasae* (summarized in Tanda, 1979a, b). The latter variety occurs on a bambusoid host and could be probably species different from *C. purpurea*.

It seems that the races of *C. purpurea* are associated with regions and most probably they cannot be world-wide generalized. The host ranges of the Stägerian groups are not the same as the ones listed for Tanda's varieties, and even other European authors came to different conclusions. Also in North America, *C. purpurea* seems to have other preferences.

C. purpurea sclerotia contain peptide alkaloids in varying ratios. Kobel and Sanglier (1978) after screening of alkaloid content in sclerotia collected for years in Europe and North America on rye and wild grasses found 10 chemoraces:

Combination	% of samples
Ergotamine	13.1
Ergocristine	7.3
Ergosine	2.2
Ergotamine + ergosine	7.3
Ergocornine + ergocryptine	22.6
Ergocornine	1.5
Ergocristine + ergotamine	4.4
Ergocristine + ergosine	20.4
Ergocryptine α	1.5
Ergocryptine $\alpha + \beta$	1.5

The authors observed that the strains containing the combination ergocristine-ergocornine-ergokryptine were unstable and dissociated in the culture into ergocristine strains and ergocornine+ergocryptine strains.

Gröger (1979) found 6 chemoraces among strains capable of alkaloid production in laboratory culture. These strains were isolated from sclerotia found on rye, triticales or fescue:

Ergotamine
Ergocristine
Ergosine
Ergocornine+ergosine (1:1)
Ergotamine+ergocryptine (1:1)
Ergocornine+ergocryptine (3:2)

In our laboratory, the measurements of alkaloid content in natural sclerotia collected on wild grasses and cereals in Europe and in USA were made. We observed that the European strains tend to belong either to ergotamine or to ergotoxine group, whereas the American strains contained various combinations of ergotamine with ergotoxines, covering almost the whole spectrum.

3.4.
CLAVICEPS INTRASPECIFIC VARIABILITY ASSESSED
BY DNA ANALYSIS

3.4.1.
Variable *C. purpurea*

Jungehülsing (1995) addressed the problem of intraspecific variability of *C. purpurea* using the methods of molecular genetics (RAPD analysis and electrophoretic karyotyping). Her RAPD results were probably the most variable of any RAPD's found among filamentous fungi studied so far, although one or several species-specific bands was observed with each primer. The isolates originated mainly from Germany and England, the host being species of *Agropyron, Agrostis, Dactylis, Elymus, Festuca, Holcus, Lolium, Molinia,Phleum, Secale,* and non-grass host Carex. The analysis of data pooled from nine primers showed that the *Secale* isolates from various parts of Germany were related and that isolates from the same locality were the most similar. Bootstrap analysis confirmed a marked tendency towards relatedness of the isolates from the same region followed by relatedness according to host preference.

Jungehülsing also found possible reason for the extreme variability even among the isolates from the same host species and locality (see also Tudzynski, this volume). The chromosome number and size in *C. purpurea* isolates was found to be variable. Despite this, crosses were possible between parents with different sized chromosomes and translocations resulted in progeny of different karyotypes.

RAPD fingerprinting of different *Claviceps* species has also been done in the Institute of Microbiology CAS. When testing various *C. purpurea* isolates with the primer RP2 (AAGGATCAGA) (Lehmann *et al.,* 1992), we obtained similar results as Jungehülsing (1995) confirming that the strains collected in the same region are more related than the geographically distant isolates (Figure 1). Variability among natural isolates from the same locality differed. Single sclerotium isolates from the same rye field (strains M, S, J, A) showed fewer differences than the isolates from a seed producing field of *Poa annua* (strains C/A, C/B, C/D). This variability was due to the presence of a population of distinct strains on the given locality and not to variations in the progeny of a single strain, because the banding pattern of the strain BRD used to inoculate one cultivar of *Poa pratensis* and reisolated from the field sclerotia in two subsequent years (CS 94, CS 95) remained unchanged. Also the mutation-and-screening procedure did not change the RAPD patterns, shown here for the ergotoxine producing strain Pepty 695/S, its clavineproducing blocked mutant Pepty 695/ch isolated by Schumann *et al.* (1982) and another isolate PD59 isolated in our laboratory and differing slightly in morphology from the parent Pepty 695/S strain.

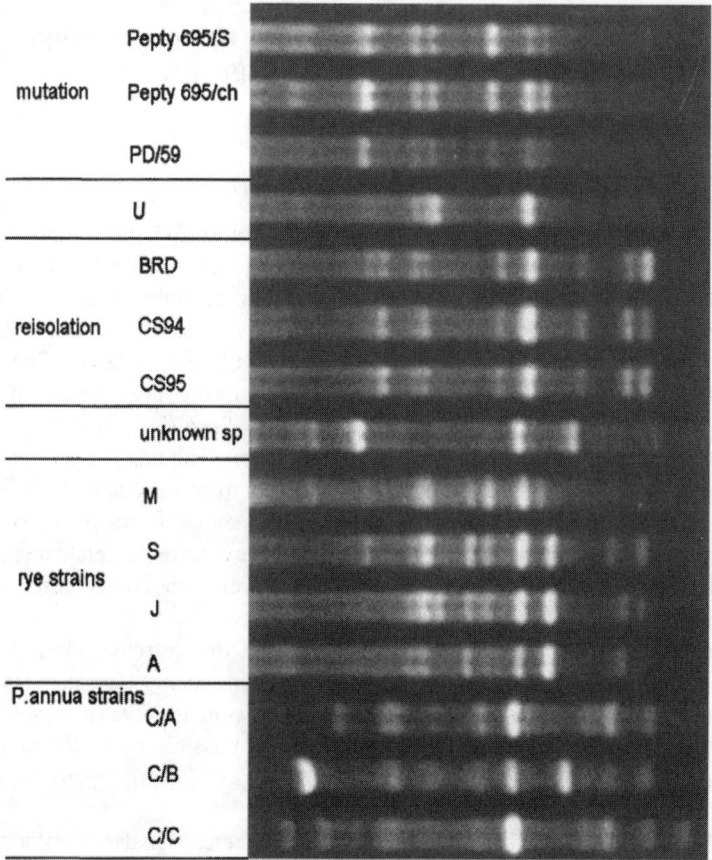

Figure 1 RAPD patterns of *C. purpurea* strains obtained with the primer RP2. The strain Pepty 695/S and its two mutants, strain U isolated from *Lolium multiflorum* in South Africa, strain BRD used for field infection and two strains CS94 and CS95 reisolated from the resulting sclerotia in two subsequent years, independent isolates M, S, J, A from the spontaneously infected rye field and independent isolates from spontaneously infected *Poa annua* field are shown. The species specific band is marked by black arrow

Another experiment assessed the degree of variability among isolates from sclerotia collected from the same flower head, on different flower heads of the same plant and from the different plants of the same locality. Although the colony morphology of the isolates was almost identical, we found that even the isolates from the same flower head of *Lolium* sp. exhibited different RAPD patterns (data not shown).

In contrast to the RAPD's, the sequence of the ribosomal region containing 5. 8S rDNA and spacers ITS1 and ITS2 is less variable. We sequenced the 556 bp fragments obtained from Middle-European cereal isolate *C. purpurea* P695/S

(Schumann *et al.,* 1982) and from *C. purpurea* isolated in our laboratory from sclerotia from *Phalaris* sp. collected in Australia. The sequences were compared to two American *C. purpurea* sequences. U57669 (GenBank) from isolate GAM 12885 collected on *Dactylis glomerata* near Athens, Georgia, USA (Glenn and Bacon, 1996; unpublished) and the sequence of the isolate 109 from *Festuca arundinacea,* Lexington, Kentucky (Schardl *et al.,* 1991).

The European and Australian isolates differed only in a single base out of 556 localized in ITS2. The American isolates GAM 12885 and 109 differed from the European Pepty 695/S each in 5 and 6bp respectively, all in ITS1 spacer, but not the identical ones (Figure 2), whereas the difference between them was in 10 positions of ITS1.

This result seems to support the hypothesis that *C. purpurea* is a relatively recent introduction to Australia together with cereals from Europe. The

```
Pepty 695/S ATCATTACCG AGTTTACAAC TCCCAAACCC ACTGTGAACT TATACCC-AA AACGTTGCCT
Australia   .......... .......... .......... .......... .......... ..........
GAM 12885   .......... .......... .......... .T........ .......... ..........
109         .......... .......... .......... .......... .......G.. ..........


Pepty 695/S CGGCGGGCAC AGCGGTACCC GAGCCCCC-G CAAGGGAG-C AGAGGCGCC- CGCCCGCCAG
Australia   .......... .......... .......... .......... .......... ..........
GAM 12885   ......A... .......... .......... .......... .A....... .C ..T......
109         .........- .......... ........C. ........G. .........G ..........


Pepty 695/S GGGACCAAAA CTCTTCTGTA TACCCATAGC GGCATGTCTG AGTGGATTTA AAAACAAAT
Australia   .......... .......... .......... .......... .......... ..........
GAM 12885   .......... .......... .......... .......... .......... .........
109         .......... .......... .......... .......... .......... C........
```

Figure 2 The differences in ITS1 spacer sequence among *C. purpurea* isolates from different regions. P695/S—European isolate; Australia—Australian isolate from *Phalaris* sp.; GAM 12885—*D. glomerata,* Georgia, USA; 109—*F. arundinacea,* Kentucky, USA

differences between European and American isolates could be attributed to the presence of indigenous strains of *C. purpurea* that evolved independently of the European populations.

3.4.2.
Other Claviceps Species

We made RAPD's of other Claviceps species with the primer 206 (TCAACAATGTCGGCCTCCGT) (Figure 3). Isolates of *C. paspali* and *C. fusiformis,* although originating from distant regions, had almost the same banding patterns. The patterns of three Mexican *C. gigantea* isolates were identical (only one representative shown) as well as the patterns of two *C. africana* strains from Bolivia and Australia. Probably, *C. africana* is spreading too quickly to develop genome differences among distant isolates. The patterns with the primer 206 were markedly species-specific and that could enable the identification of sphacelial stages once the teleomorph RAPD was determined. We used the RAPD pattern as the first criterion for excluding the possibility of *D. spicata* ergot *(C. citrina)* being *C. purpurea.*

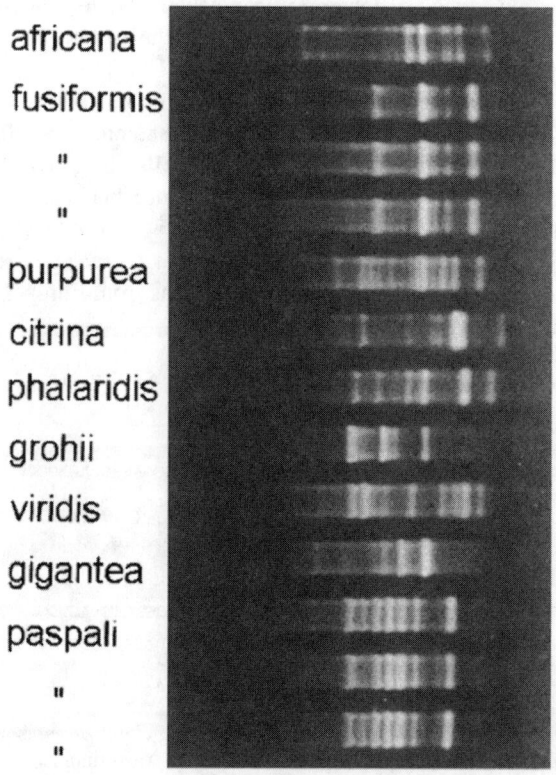

africana

fusiformis

"

"

purpurea

citrina

phalaridis

grohii

viridis

gigantea

paspali

"

"

Figure 3 RAPD patterns of different *Claviceps* species with the primer 206. Note the almost identical patterns of *C. paspali* and *C. fusiformis* isolates

3.5.
THE CLAVICEPS PHYLOGENY

Its main course is connected with the evolution of grasses and the global climatic changes, in the side branches the incidence of endemic species is correlated with ecological niches.

3.5.1.
Grass Evolution

The evolution of grasses is summarized in Brown and Smith (1972) and Jones (1991). The grasses probably appeared during the Jurassic period (Mesozoic), in the wet tropics. The Centothecoideae, Bambusoideae, and Arundinoideae with

C3 type of photosynthesis are supposed to be the most primitive. Their descendants, the subfamilies Chloridoideae and Panicoideae developed C4 photosynthetic pathways and thus acquired competitive advantage in warmer regions.

Chloridoideae are adapted to stressful arid or saline habitats (sand dunes, coastal marshes, semi-deserts). Their widespread distribution with the radiation center in South Africa suggests that they acquired the C4 photosynthesis and expanded before the breakup of the Pangea (Middle Cretaceous).

One section of the Panicoideae (the tribe Paniceae) acquired C4 photosynthesis shortly before South America was separated from the rest of Gondwana in the Late Cretaceous. In the then isolated South America, Paniceae became dominant. C3 photosynthesis was retained in apparently more primitive shade loving species (*Oplismenus, Icnanthus,* and some species of *Panicum*) whereas C4 species colonized open habitats in the tropical and subtropical zones (e.g. *Paspalum, Cenchrus, Brachiaria, Pennisetum*).

In southern Asia, the Andropogoneae *(Sorghum, Zea, Saccharum, Bothriochloa, Imperata)* developed from some C4 ancestor and rapidly occupied savanna habitats and dry open woodlands. It is supposed that this occurred some 25–30% Mya (million years ago). Andropogonoid species form 30–40% of all grass species in India and Africa. The transfer of Andropogoneae to America probably occurred via southern Europe, before their separation in the Tertiary period while they probably reached Australia via island chains.

The subfamily Pooidae retained C3 photosynthesis, as the C4 type does not provide any advantage in cooler climates. It expanded mainly in the northern hemisphere. The southern hemisphere does not support its climatic needs except New Zealand, southern Australia, and South America. As distinct from other subfamilies, Pooidae possess large chromosomes.

The grass phylogenies derived from plastid as well as nuclear DNA sequences show two main groups. In the PACC group (including Panicoideae, Arundinoideae, Chloridoideae, and Centothecoideae) chloridoids and arundinoids are sister groups to the ancestor of the panicoids, which correlates well with the theory based on the geographical distribution of grass subfamilies. The group BOP contains Bambusa-Oryza and Pooids (Mathews and Sharrock, 1996).

3.5.2.
Ergot Distribution

Knowledge of the geographic occurrence of *Claviceps* species is influenced by the presence of the grass hosts and also by the human factor. With the exception of Brazil and West India at the end of the 19th century (Möller, 1901; Hennings, 1899), the information concerning the occurrence of tropical forest ergots is scarce. The best information about geographical and host distribution is available for the ergots that colonize cereals and pasture grasses.

The distribution of ergot species throughout the world has several interesting features. First, there is a striking difference in the number of ergot species colonizing the chloridoid, pooid, and panicoid subfamilies. The only chloridoid-specializing species known so far are *C. cynodontis (Cynodon)C. yanagawaensis (Zoysia)* and *C. cinerea (Hilaria)* and we add new species *C. citrina* found in Central Mexico on *Distichlis spicata*. Also the occurrence of *C. purpurea* on recent *Spartina* hybrid in England was documented (Gray *et al.*, 1990).

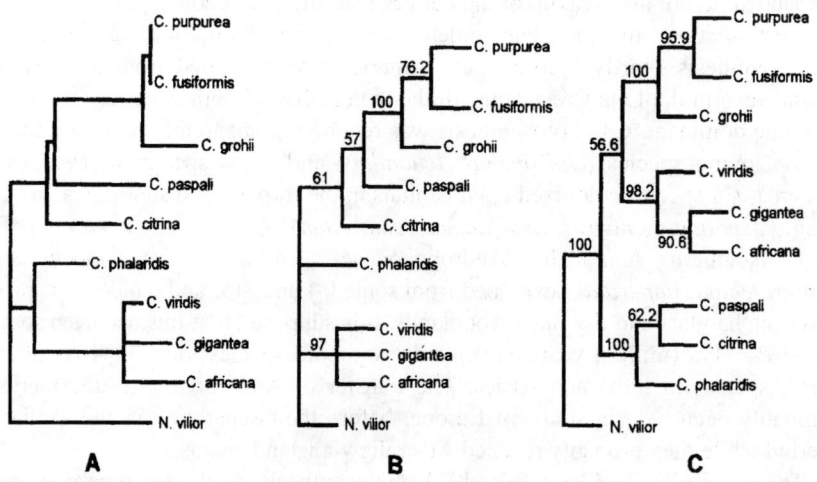

A   **B**   **C**

Figure 4 Phylogenetic relationships among *Claviceps* species based on rDNA ITS1 region.
A—maximum likelihood tree;
B—neighbor–joining tree (Jukes–Cantor distance matrix and 500x bootstrap);
C—maximum parsimony tree (500x bootstrap); Strains used for sequencing:
C. purpurea Pepty 695/S; *C. fusiformis* SD-58; *C. grohii* strain 124.47 (CBS, Barn); *C. paspali*—our conidial isolate (*P. dilatatum*, Alabama); *C. citrina* sp. nova, our isolate (*D. spicata*, Central Mexico); *C. phalaridis*—our isolate (*Phalaris,* sample DAR 69619 Australia); *C. viridis* strain 125.63 (CBS, Barn); *C. gigantea*—our isolate (*Zea mays,* Toluca Valley, Mexico); *C. africana*—our isolate (*Sorghum,* Bolivia)

The pooid ergot is represented only by *C. purpurea* with Laurasian distribution and the unusual endophytic *C. phalaridis* endemic to Australia. Both species colonize also arundinoid species like *Danthonia* and have wide host ranges, although *C. purpurea* is considered advanced species and *C. phalaridis* a rather primitive one. There is also *C. litoralis* occurring on the northern Japanese islands and Sakhalin on *Elymus* and *Hordeum* (Kawatani, 1944), but some mycologists doubt its species status and consider it a variety of *C. purpurea*.

Some morphologic similarity to *C. purpurea* is seen in *C. grohii, C. cyperi* and *C. nigricans* that colonize sedges in the North temperate regions.

In the genus *Claviceps,* the panicoid species with monogeneric *(C. paspali,C. viridis, C. gigantea)* to polygeneric host ranges *(C. fusiformis, C. pusilla)* predominate. The most primitive species *C. orthocladae, C. flavella* and *C. diadema* (undifferentiated sclerotium encompassing the flower parts, germinating directly on the host) are found on *Orthoclada* (Centothecae) and panicoids in South America tropics, whereas none are found in wet tropical and subtropical forests of Africa and South Asia. Also none of the South American ergots was found in the other regions of the world except *C. paspali* which has been spread by human influence. On the other hand, several species are common in Africa and India and some also in Australia. No colonization of arundinoid grasses has been documented for panicoid ergots.

It seems, that species of *Claviceps* should be considered predominantly parasites of panicoids. The colonization of chloridoid and pooid hosts is somewhat obscured by the vast occurrence of *C. purpurea,* but in fact it is as marginal as the occurrence on cyperaceous hosts. The distribution of *Claviceps* species suggests, that the genus origin probably dates from the period of the expansion of panicoids from South America in the Late Cretaceous. The speciation then continued independently on all continents of the former Gondwana. The exchange among African and Indian species was enabled probably at the same time as the spreading of andropogonoid grasses.

3.5.3.
Claviceps Phylogenetic Tree

In our laboratory, we made phylogenetic analysis of the nine *Claviceps* species based on the comparison of internal transcribed spacers ITS1. *Nectriavilior* (Glenn *et al.,* 1996) was used as an outgroup species. The trees (Figure 2) acquired with different algorithms contained in PHYLIP (Felsenstein, 1995) differed, but the segregation of two distinct groups was always supported over 95% of bootstraps:

1. *C. purpurea, C. fusiformis, C. grohii*
2. *C. viridis, C. gigantea, C. africana*

The remaining species *C. phalaridis, C. citrina and C. paspali* were separated by the maximum parsimony as the third group (100%) containing more, "primitive" species. Also, these species share a marked deletion of 38 bp.

On the other side, maximum likelihood tree (out of 211 tested) placed Australian *C. phalaridis* basal to the group 2, whereas American *C. paspali* and *C. citrina* were closer to the ancestors of group 1. The branches with zero length were collapsed to polytomies. The neighbor-joining tree resembles the ML tree (with the exception of *C. phalaridis*) and their log likelihoods were almost identical (ML=−990.245; NJ =−990.256). The NJ tree branches supported by less than 50% bootstraps were reduced to polytomies.

The advanced species of the group 1, with drought and cold resistant sclerotia may be related to the group of ergots with prevailing glucose and fructose as described by Mower and Hancock (1975). The high similarity of ITS1 DNA of these species otherwise so different in their habitats and host specificity is rather puzzling, but the 95.6% identity in 960 bp fragment of 28S rDNA of *C. purpurea* and *C. fusiformis* was also observed by Rehner and Samuels (1995). The evolution course in this branch is unclear, but we very tentatively speculate about the Laurasian origin from more primitive ancestors related to *C. paspali* and *C. citrina* and relatively recent transfer to Africa.

The second group, encompassing two andropogonoid ergots together with *C. viridis* from rather primitive C3 panicoid from Asian monsoon regions, clearly correlates with the radiation origin of andropogonoids in Southern Asia. It also shows that the andropogonoids took their ergots with them on their way to Africa and North America and were not colonized by the local ergot species. The ergots with the prevalent mannitol and arabinitol in their honeydew (Mower and Hancock, 1975) should be expected in this branch. Together with related *C. phalaridis* it represents probably the genuine ergots that developed in the Old World warm regions.

REFERENCES

Baldacci, E. and Forlani, R. (1950) Ricerche sulla spezializzazione della *Clavicepspurpurea. Phytopathol. Zschr.,***17,**81–84.

Bandyopadhyay, R. (1992) Sorghum ergot. In W.A.J. De Milliano, R.A.Frederiksen, and G.D.Bengston (eds.), *Sorghum and Millets Diseases; A Second World Review,*ICRISAT, Pantacheru, Andhra Pradesh, India, pp. 235–244.

Békésy, N. (1956) Ein Beitrag zur Biologie des Mutterkorns. *Phytopathol, Zschr.,***26,** 49–56.

Bretag, T.W. and Merriman, P.R. (1981) Epidemiology and cross-infection of *C. purpurea. Trans. Br. Mycol. Soc.,***77,**211–213.

Brewer, J. and Loveless, A.R. (1977) Ergot of *Pennisetwn macrourum* in South Africa. *Kirkia,***10,**589–600.

Brown, W.V. and Smith, B.N. (1972) Grass evolution and the Kranz syndrome, C^{13}/C^{12} ratios and continental drift. *Nature,***239,**345–346.

Campbell, W.P. (1957) Studies of ergot infection in gramineous hosts. *Can. J. Bot.,***35,** 315–320.

Clay, K. (1988) Clavicipitaceous endophytes of grasses. Coevolution and the change from parasitism to mutualism. In D.Hawksworth and K.Pirozynski (eds.), *Coevolution of Fungi with Plants and Animals,*Academic Press, London, UK, pp. 79–105.

Cunfer, M.B. and Marshall, D. (1977) Temperature and moisture requirements for germination of *C. paspali* sclerotia. *Mycologia,***69,**1137–1141.

de Wet, J.M.J. (1992) The three phases of cereal domestication. In G.P.Chapman (ed.), *Grass evolution and domestication,*Cambridge University Press, Great Britain, pp. 176–198

Dickson, G.J. (1956) *Diseases of Field Crops,*McGraw-Hill, New York.

Doidge, E.M. (1950) The South African fungi and lichens. *Bothalia,*5, 1–1094.

Felsenstein, J. (1995) PHYLIP (Phylogeny Inference Package) Version 3.5c. University of Washington, Scattle, USA.

Fernandes, C.D., Fernandes, A.T.F. and Bezerra, J.L. (1995) "Mela": Uma nova doença em sememes de *Brachiaria* spp. no Brasil. *Fitopatol. Bras.,* **20,**501–503.

Flieger, M., Wurst, M. and Shelby, R. (1997) Ergot alkaloids—sources, structures and analytical methods. *Folia Microbiol.,* **42,**3–30.

Frederickson, D.E. and Mantle, P.G. (1988) The path of infection of sorghum by *C. sorghi. Physiol. Mol. Pathol.,* **33,**221–234.

Frederickson, D.E., Mantle, P.G. and De Milliano, W.A.J. (1989) Secondary conidiation of *Sphacelia sorghi* on sorghum, a novel factor in the epidemiology of ergot disease. *Mycol. Res.,* **93,**497–502.

Frederickson, D.E., Mantle, P.G. and De Milliano, W.A.J. (1991) *Claviceps africana* sp. nov.; the distinctive ergot pathogen of sorghum in Africa. *Mycol. Res.,* **65,**1101–1107.

Frederickson, D.E., Mantle, P.G. and De Milliano, W.A.J. (1993) Windborne spread of ergot disease *(C. africana)* in sorghum A-lines in Zimbabwe. *Plant Pathol.,* **42,** 368–377.

Fučikovský, L. and Moreno, M. (1971) Distribution of *C .gigantea* and its percent attack on two lines of corn in the state of Mexico, Mexico. *Plant Dis. Rep.,* **55,**231–233.

Fuentes, S.F., de la Isla, M.-L., Ullstrup, A.J. and Rodriguez, A.E. (1964) *Clavicepsgigantea,* a new pathogen of maize in Mexico. *Phytopathology,* **54,**379–381.

Gäumann, E.A. (1952) *The Fungi,*Hafner Publishing Co., N.Y.

Glenn, A.E., Bacon, C.W., Price, R. and Hanlin, R.T. (1996) Molecular phylogeny of *Acremonium* and its taxonomic implications . *Mycologia,* **88,**369–383.

Gray, A.J., Drury, M. and Raybould, A.F. (1990) *Spartina* and the ergot fungus *Clavicepspurpurea*—a singular contest? In J.J.Burdon and S.R.Leather (eds.) *Pests, Pathogens and Plant Communities,*Blackwell Scientific Publications, Oxford, pp. 63–79.

Groves, J.W. (1943) A new species of *Claviceps* on *Carex. Mycologia,* **35,**604–609.

Gröger, D. (1979) Saprophytische Gewinnung von Mutterkornalkaloiden. *Pharmazie,* **34,** 278.

Hennings, P. (1899) *Xylariodiscus* nov. gen. und einige neue brasilianische Ascomycetes E. Ule'schen Herbars. *Hedwigia,*Beibl. **38,**63–65.

Hitchcock, A.S. (1950) *Manual of grasses of the United States,*Ed. 2.

Jones, C.A. (1991) *C4 Grasses and Cereals, Growth, Development and Stress Response.*John Wiley, New York, Chichester, Brisbane, Toronto, Singapore, pp. 39–54.

Jungehülsing, U. (1995) Genomanalyse bei *C. purpurea. Bibliotheca Mycologica,* **161,**J. Cramer, Berlin, Stuttgart.

Kawatani, T. (1944) Clavicipiti species nova parasitica ad Eiymum mollem Trin. *Bull.Nat. Inst. Hyg. Sci.,* **65,**81–83.

Kobel, H. and Sanglier, J.J. (1978) Formation of ergotoxine alkaloids by fermentation and attempts to control their biosynthesis. In R.Hütter *et al.* (eds.), *Antibiotics andOther Secondary Metabolites. Biosynthesis and Production,*FEMS Symp. **5,**Academic Press.

Kybal, J. and Brejcha, V. (1955) Problematik der Rassen und Stämmen des Mutterkorns *Claviceps purpurea* Tul. *Die Pharmazie,* **10,**752–755.

Langdon, R.F.N. (1942) Ergots on native grasses in Queensland. *Proc. Roy. Soc. Qld.,* **54,** 23–32.

Langdon, R.F.N. (1952) *Studies on ergot.,*PhD. Thesis, University of Queensland.

Langdon, R.F.N. (1954) New species of *Claviceps. Univ. Queensland Papers (Botany)*3, 37–40.

Langdon, R.F.N. (1954) The origin and differentiation of *Claviceps* species. *Univ.Queensland Papers (Botany)*,**3**,61–68.

Lehmann, P.F, Lin, D. and Lasker, B.A. (1992) Genotypic identification and characterization of species and strains within the genus *Candida* using random amplified polymorphic DNA. *J. Clin. Microbiol.*,**30**,3249–3254.

Loveless, A.R. (1964) Studies on Rhodesian ergots 1. *C. maximensis* Theis and *C. pusilla* Cesati. *Kirkia*,**4**,37–41.

Loveless, A.R. (1964) Use of the honeydew state in the identification of ergot species. *Trans. Brit. Mycol. Soc.*,**47**,205–213.

Loveless, A.R. (1965) Studies on Rhodesian ergots 4. *Claviceps cynodontis* Langdon. *Kirkia*,**5**,25–29.

Loveless, A.R. (1967) *C. fusiformis* sp. nov., the causal agent of agalactia of sows. *Trans. Brit. Mycol. Soc.*,**50**,13–18.

Loveless, A.R. (1971) Conidial evidence for host restriction in *C. purpurea*. *Trans. Brit.Mycol. Soc.*,**56**,419–434.

Luttrell, E.S. (1951) Taxonomy of Pyrenomycetes. *Univ. Missouri Studies,***24**,1–120.

Luttrell, E.S. (1977) The disease cycle and fungus-host relationships in dallisgrass ergot. *Phytopathology,***67**,1461–1468.

Luttrell, E.S. (1980) Host parasite relationship and development of the ergot sclerotium in *C. purpurea*. *Can. J. Bot.*,**58**,942–958.

Mantle, P. (1968) Studies on *Sphacelia sorghi* McRae, an ergot of *Sorghum vulgare* Pers. *Annals Appl. Biol.*,**62**,443–449.

Mastenbroek, C. and Oort, A.J.P. (1941) Het voorkomen van moederkoren *(Claviceps)* oop granen en grassen en de specialisatie van de moederkorenschimmel. *Tijdschr.Plziekt.*,**47**,165–185.

Mathews, S. and Sharrock, R.A. (1996) The phytochrome gene family in grasses (Poaceae), a phylogeny and evidence that grasses have a subset of the loci found in dicot angiosperms. *Mol Biol. Evol.*,**13**,1141–1150.

Mitchell, D.T. and Cooke, R.C. (1967) Water uptake, respiration pattern and lipid utilization in sclerotia of *C. purpurea* during germination. *Trans. Brit. Mycol. Soc.*,**51**, 731–736.

Mower, R.L. and Hancock, J.G. (1975) Sugar composition of ergot honeydews. *Can.J. Bot.*, **53**,2813–2825.

Möller, A. (1901) Phycomyceten und Askomyceten von Brasil. In A.F.W.Schimper (ed.) *Bot. Mittheilungen aus Tropen,*Jena, Germany, pp. 1–319.

Padwick, G.W. and Azmatullah, M. (1943) *Claviceps purpurea* and the new species from Simla. *Curr. Sci.,***12**,257.

Pažoutová, S., Fučíkovský, L., Leyva-Mir, S.G. and Flieger, M. (1998) *Claviceps citrina* sp. nov.; the parasite of halophytic grass *Distichlis spicata* from central Mexico. *Mycol. Res.,***102**,850–854.

Porter, J.K., Bacon, C.W. and Robbins, J.D. (1974) Major alkaloids of a *Claviceps* isolated from toxic Bermuda grass. *Agric. Food. Chem.*,**22**,838–841.

Rehner, S.A. and Samuels, G.J. (1995) Molecular systematics of the Hypocreales, a teleomorph gene phylogeny and the status of their anamorphs. *Can. J. Bot.*,**73** (Suppl.), S816-S823.

Reis, E.M. (1966) First report on the Americas of sorghum ergot disease, caused by a pathogen diagnosed as *C. africana*. *Plant Dis.*,**80**,463.

Ryley, M.J., Acorn, J.L., Kochman, J.K., Kong, G.A. and Thompson, S.M. (1996) Ergot on sorghum in Australia. *Australasian Plant Patbol.,***25**,214.

Schardl, C.L., Liu, J.S., White, J.F., Finkel, R.A., An, Z. and Siegel, M.R. (1991) Molecular phylogenetic relationships of nonpathogenic grass mycosymbionts and clavicipitaceous plant pathogens. *Plant Syst. Evol.,***178**,27–41.

Schumann, B., Erge, D., Maier, W. and Gröger, D. (1982) A new strain of *C. purpurea* accumulating tetracyclic clavine alkaloids. *Planta Med.,***45**,11–14.

Spatafora, J.W. and Blackwell, M. (1993) Molecular systematics of unitunicate perithecial ascomycetes. The Clavicipitales-Hypocreales connection. *Mycologia,***85**, 912–922.

Sprague, R. (1950) *Diseases of Cereals and Grasses in North America,*Ronald Press, New York, pp. 59–67.

Sundaram, N.V., Palmer, L.T., Nagarajan, K. and Prescott, D. (1972) Disease survey of sorghum and millets in India . *Plant Dis. Rep.,***56**,740–743.

Taber, W.A. and Vining, L.C. (1960) A comparison of isolates of *Claviceps* spp. for the ability to grow and to produce alkaloids on certain nutrients. *Can. J. Microbiol.,***6**, 355–365.

Tanaka, K. and Sugawa, T. (1952) *J. Pharm. Soc. Japan,***72**,616–620.

Tanda, S. (1979a) Mycological studies on the ergot in Japan (6) A physiologic race of *C. purpurea* Tul. var. *alopecuri* Tanda collected from *Trisetum bifidum* Ohwi. *Journal of Agric. Sci., Tokyo Nogyo Daigaku,***23**,207–214.

Tanda, S. (1979b) Mycological studies on the ergot in Japan (9) Distinct variety of *C. purpurea* Tul. on *Phalaris arundinacea* L. and *P. arundinacea* var. *picta* L. *Journal of Agric. Sci., Tokyo Nogyo Daigaku,***24**,79–95.

Tanda, S. (1981) Mycological studies on the ergot in Japan (20) *Clavicepsyanagawaensis* Togashi parasitic on *Zoysia japonica* Steud. *Journal of Agric. Sci.,Tokyo Nogyo Daigaku,***26**,193–199.

Tanda, S. (1991) Mycological studies on ergot in Japan (26) A new ergot, *Clavicepsbothriochloae* parasitic on *Bothriochloa parviflora. Journal of Agric. Sci., TokyoNogyo Daigaku,***36**,36–42.

Tanda, S. (1992) Mycological studies on the ergot in Japan (28). An ergot new to Japan, *Claviceps viridis* parasitic on *Oplismenus. Trans. Mycol. Soc. Japan,***33**,343–348.

Thakur, R.P., Rao, V.P. and Williams, R.J. (1984) The morphology and disease cycle of ergot caused by *Claviceps fusiformis* in pearl millet. *Phytopathology,***74**,201–205.

Thirumalachar, M.J. (1945) Ergot on *Pennisetum hohenackeri,* Hochst. *Nature,***156**,754.

Thomas, K.M., Ramakrishnan, T.S. and Srinivasan, K.V. (1945) The natural occurrence of ergot in south India. *Proc. Ind. Acad. Sci., Sec. B,***21**,93–100.

Uecker, F.A. (1980) Cytology of the ascus of *C. phalaridis. Mycologia,***72**,270–278.

Walker, J. (1957) A new species of *Claviceps* on *Phalaris tuberosa* L. *Proc. Linn. Soc.New South Wales,***82**,322–327.

Walker, J. (1970) Systemic fungal parasite of *Phalaris tuberosa* in Australia. *Search,***1**, 81–83.

Watson, L. and Dallwitz, M.J. (1992 onwards) *Grass Genera of the World,* Descriptions, Illustrations, Identification, and Information Retrieval; including Synonyms, Morphology, Anatomy, Physiology, Phytochemistry, Cytology, Classification, Pathogens, World and Local Distributions and References. Version: 30th April 1998. URL http://biodiversity.uno.edu/delta.

Yamaguchi, K., Ito, H., Ito, M., Kawatani, T. and Kashima, T. (1959) *Bull. Nat. Inst.Hyg. Set.,***77**,235–239.

Zevada-Zenteno, M. (1958) Estudios sobre hongos parasites de Gramineas de la republica Mexicana. II. *Memorios del Primer Congreso Nacional de Entomología yFitopatología,*Chapingo, Mexico, pp. 501–519.

4.
GENETICS OF *CLAVICEPS PURPUREA*

PAUL TUDZYNSKI

Institut für Botanik, Westf. Wilhelms-Universität,Schlossgarten 3,

D-48149 Münster

For about 70 years genetic analyses of *Claviceps purpurea* have been performed (e.g. Kirchhoff, 1929); nevertheless, genetic data available are still rather scarce, mainly due to intrinsic experimental problems, which place this fungus far apart from being a genetic model organism: the generation time is extremely long and —since it involves parasitic culture of the fungus—the conditions of obtaining sexual progeny are rather complex; biotechnologically relevant strains normally are imperfect and even do not produce conidia; and, last but not least, *C. purpurea* strains tend to be instable during prolonged vegetative cultivation. Nevertheless, due to the development of molecular genetic approaches, and due to growing interest in *C. purpurea* as a pathogen, in recent years genetic research in *C. purpurea* has been intensified. In some recent review articles on *C. purpurea* also genetic data have been included (e.g. Socic and Gaberc-Porekar, 1992; Tudzynski *et al.,* 1995; Lohmeyer and Tudzynski, 1997).

4.1.
CLASSICAL GENETICS

Compared to other biotrophic parasites, *C. purpurea* can be rather easily handled: most laboratory strains and field isolates grow well and sporulate abundantly in axenic culture. Since conidia normally are mononuclear, mutants can be easily obtained by standard procedures; heterokaryons are readily formed and can be used for parasexual approaches; and sexual crosses are possible, though time consuming.

4.1.1.
Mutants

Numerous stable mutants of *C. purpurea* have been obtained, mainly by UV-mutagenesis. They include various auxotrophic mutants (e.g. Strnadova, 1964; Tudzynski *et al.,* 1982), for only a few of which the metabolic block was determined exactly, e.g. for 5 *arg⁻* mutants (Kus *et al.,* 1987) and several *pyr⁻* mutants (Smit and Tudzynski, 1992). The latter mutants were obtained in an "intelligent" screening approach, using resistance against the toxic analogue 5-

fluoro-orotate (FOA) as selective principle; FOA-resistant mutants were obtained after UV-mutagenesis at a rate of 3.2×10^{-6}; from these, 12.5% were pyrimidine auxotrophs, about half of these were shown to be defective in the orotidine-monophosphate-carboxylase (OMPD) gene (the rest probably was defective in the OMP pyrophosphorylase gene). Also several fungicide resistant mutants have been described, e.g. against benomyl (Kus et al., 1986, Jungehülsing, 1995) or cercobin (Tudzynki et al., 1982). Recently also recombinant DNA technology has been applied to mark strains with antibiotic resistance genes and use them in crosses (Jungehülsing, 1995).

The main effort over the years, however, has been laid on mutagenesis programs concerning alkaloid biosynthesis, often using successive steps of various mutagens; various mutant strains showing increased or decreased amount or a different spectrum of alkaloids have been described (e.g. DidekBrumec et at., 1987; Kobel and Sanglier, 1978).

Recently a novel and promising approach to generate mutants by random integration of transforming vector DNA has been successfully used to obtain auxotrophic and pathogenicity mutants in C. purpurea (Voß, Stehling and Tudzynski, unpubl., see below).

Despite these numerous successfull reports on mutagenesis in C. purpurea, we found that several field isolates of this fungus were unexpectedly highly recalcitrant against mutagenesis. Determination of relative DNA content per nucleus by DAPI staining showed a high level of variation among different isolates, indicating various degrees of aneuploidy in these field isolates; haploidization experiments using low doses of benomyl applied on two different isolates yielded derivatives with only 50% of relative DNA content, confirming the idea that the original strains were not haploid (Jungehülsing, 1995; Müller, 1996), and that benomyl treatment represents a good tool for the generation of bona-fide haploid strains for mutagenesis programs and certain molecular studies (see below).

4.1.2.

Heterokaryosis

Most field isolates and industrial strains of C. purpurea have been found to be heterokaryotic, inspite of the mononuclear status of their conidia; reasons for this heterogenicity could be mixed infections in the field strains and generally a high mutation rate. When we analysed selfing F1 progeny of field isolates, we found a high degree of morphological variation, in one case about 15 different phenotypes (Tudzynski, unpubl.). Already Kirchhoff (1929) found segegration within an F1 progeny of albino sclerotia, confirming a heterokaryotic status of these sclerotia (see also Bekesy, 1956). Despite this high level of heterokaryosis in field isolates, heterokaryosis is not required for completion of the life-cycle: Strnadova and Kybal (1974) and Esser and Tudzynski (1978) could prove that homokaryotic strains can perform the parasitic cycle and produce ascospores; C.

purpurea (at least the isolates tested sofar) obviously is homothallic. Esser and Tudzynski (1978) could demonstrate that homokaryotic mycelia can produce alkaloids, in contrast to earlier observations (e.g. Amici *et al.,* 1967) that only heterokaryotic strains were good producers. Obviously these different data reflect differences in the genetic background of the strains used in these studies. It could be shown by several groups that formation of heterokaryons between strains of different genetic background could indeed significantly increase or generally alter alkaloid biosynthesis; here probably effects like heterosis, complementation, gene dosis, etc. are important. Spalla *et al.* (1976) produced a forced heterokaryon (which had to be kept on minimal medium) between *C. purpurea* strains producing ergocristine and ergocornine/ergokryptine, respectively; the heterokaryon produced all 3 alkaloids. Even interspecies fusions have been successfully used to obtain different spectra of alkaloids, e.g. between an ergotamin-producing *C. purpurea* strain and a clavine producing *C. fusiformis* strain (Robbers, 1984), the fusion product synthesizing both types of alkaloids.

Brauer and Robbers (1987) could demonstrate that in heterokaryotic strains parasexual processes take place, leading to recombination. Therefore, fusion of strains and generation of defined heterokaryons represent a powerful tool for improvement of the (normally imperfect) alkaloid production strains.

4.1.3.
Sexual Cycle and Recombination

For completion of the sexual cycle in *C. purpurea* the formation and germination of sclerotia (see chapter 3 "Biology of *Claviceps*") is obligatory. Since—despite intensive efforts of several groups—so far induction of sclerotia formation in axenic culture never has been achieved, genetic analysis of *C. purpurea* requires facilities to grow *Secale cereale* (or another host plant) under defined ± sterile conditions. Matings are obtained by inoculating young rye florets with mixed conidial suspensions of both strains; inspite of the homothallism mentioned above, such mixed infections lead to a high rate of mixed sclerotia yielding recombinant ascospores (e.g. Tudzynski *et al.,* 1982; Taber, 1984). A major technical problem hampering such genetic analyses is the low germination rate of sclerotia; conditions for induction of this germination process have been studied in detail (e.g. Kirchhoff, 1929; Bekesy, 1956; Taber, 1984), nevertheless the process cannot be considered as being easily reproducible in any lab. In short, sclerotia are harvested, surface-sterilized by $HgCl_2$ or formaline treatment, placed on wet sand or vermiculite, and kept for at least 3 months at about 0°C. After transfer to 14–17°C (white light) sclerotia (hopefully) germinate to form perithecial stromata. Ejected ascospores are trapped on agar-covered glass slides and isolated; they germinate within a few hours. Isolation of tetrads is not possible, since the needle-like ascospores are tightly intertwined and cannot be isolated without breakage from an intact ascus.

Another problem is that virulence/pathogenicity is an essential prerequisite for successful matings; since e.g. most auxotrophic mutants turned out to be non-pathogenic (Tudzynski *et al.*, 1982), the use of auxotrophic markers in crosses is limited.

In our lab recently the construction of a (preliminary) genetic map was achieved by analysing the progeny of a cross between two fungicide/antibiotic resistant strains by molecular methods (see below). Therefore the sexual cycle and meiotic recombination could form a valuable basis for molecular genetic approaches, inspite of the pitfalls and technical problems of the system.

4.2.
MOLECULAR GENETICS

Molecular genetic techniques now readily available also for fungal systems have been successfully applied to *C. purpurea* in recent years and have considerably widened the options for genetic research: they allow detailed molecular characterization of strains—this is important for "typing" of strains and systematic analyses—; they facilitate and speed up the evaluation of crosses and the establishment of a genetic map considerably; they provide the tools for detailed regulatory studies, especially of the alkaloid pathway; and, last not least, they open the possibility to create recombinant strains, by enhancing expression of genes of interest or by creating defined deletion mutants.

4.2.1.
Molecular Characterization of Strains

Molecular methods like restriction-fragment-length-polymorphism (RFLP) analysis, polymerase-chain-reaction (PCR) based techniques using random primers (random-amplified-polymorphic DNA, RAPD) or specific primers (e.g. for ribosomal DNA), and pulsed field gel electrophoresis (PFGE) for the separation of fungal chromosomes have been widely used to characterize fungal isolates with respect to aspects of evolution, population biology, definition of compatibility groups/subspecies, disease diagnosis, etc. (see recent compilation by Tudzynski and Tudzynski, 1996, which lists corresponding data for more than 40 species of phytopathogenic fungi). All these techniques have been successfully applied also for the genus *Claviceps* (see Pazoutova, this volume).

(a)*RFLP:* The first investigations on RFLPs in *C. purpurea* used mitochondrial DNA (mtDNA): it could be shown that the restriction pattern of mtDNA is strain specific, i.e. allows the identification of specific isolates (Tudzynski *et al.,* 1983; Tudzynski and Düvell, 1985). Mt plasmids present in most field isolates, however, are not strictly strain-specific: identical plasmids can occur in strains from different geographic origin (Tudzynski and Esser, 1986).

Recently in our lab we started a genomic RFLP program with the aim to establish a genetic map of *C. purpurea* (Luerweg and Tudzynski, unpubl.).

Genomic DNA from two non-related *secale-derived* field isolates, digested with four different restriction endonucleases, was blotted and hybridized to randomly chosen lambda clones from a genomic library of *C. purpurea* strain *T5* (Smit and Tudzynski, 1992). 50% of the clones checked obviously contained repetitive DNA and could not be used for genetic analyses (nevertheless, could be valuable for fingerprint studies!). Only 6% of the single copy clones yielded no RFLPs, the rest showed one or more RFLPs. The two strains, which carried resistance markers (benomyl resistance, and a transformation introduced phleomycin resistance, respectively), were crossed and the progeny (56 randomly isolated ascospore lines) were evaluated by RFLP, using the randomly selected lambda clones and several cloned genes from *C. purpurea*. Most of them segregated in a Mendelian way (1:1).

Figure 1 shows the segregation pattern of an RFLP detected by the *cpxyl* gene of *C. purpurea,* coding for a xylanase (Giesbert and Tudzynski, unpubl., see Table 1).

Altogether 72 RFLP markers (and the two resistance markers) were used for a linkage analysis. A preliminary genetic map was obtained (see Figure 2), comprising 13 linkage groups (and two unlinked markers). Inclusion of additional markers and combination with PFGE (see below) will allow to fill in the gaps and to assign these linkage groups to specific chromosomes.

(b)*RAPD:* This widely applied technique is based on the polymerase-chainreaction (PCR) using single random nona- or dekanukleotides as primers (Williams *et al.,* 1991). RAPD allows a fingerprint-like characterization of strains, depending on the primers used. Jungehülsing and Tudzynski (1996) tested 58 different random decamer primers using as template genomic DNA from 29 field isolates of *C. purpurea* from various host-plants; 16 of these primers yielded a reproducible RAPD pattern, 9 were used for comparative analysis, since they showed significant differences. With some of these primers each of the 29 isolates showed a unique RAPD pattern, confirming that this technique can be used for the identification of *C. purpurea* strains. A preliminary tree analysis based on these 9 RAPD primers gave indications for some degree of host specificity or beginning sub-species formation (though this data must be interpreted cautiously). RAPD without any doubt is a very useful tool for the typing of strains and for systematic analyses in *C. purpurea* (see also chapter 3, this volume). We tested the use of RAPD bands as molecular markers in genetic crosses; it turned out that more than 50% of the RAPD fragment were derived from repetitive DNA, and that others yielded unusual segregation patterns; nevertheless, about a quarter of the RAPD markers tested segregated 1:1 and could be used as genetic markers (Jungehülsing, 1995). Therefore, RAPD can be useful for the construction of a genetic map; each RAPD marker, however, has to be checked carefully for its Mendelian behaviour.

(c)*PFGE:* Pulsed-field-gel-electrophoresis has turned out to be an effective method to separate fungal chromosomes, due to their comparatively small sizes (for review see Skinner *et al.,* 1991; Tudzynski and Tudzynski, 1996).

Jungehülsing (1995) successfully optimized electrophoresis conditions for the separation of *C. purpurea* chromosomes (see also Tudzynski *et al.,* 1995). Unexpectedly the karyotypes of the field isolates differed considerably, showing numerous chromosome length polymorphisms (the size varying

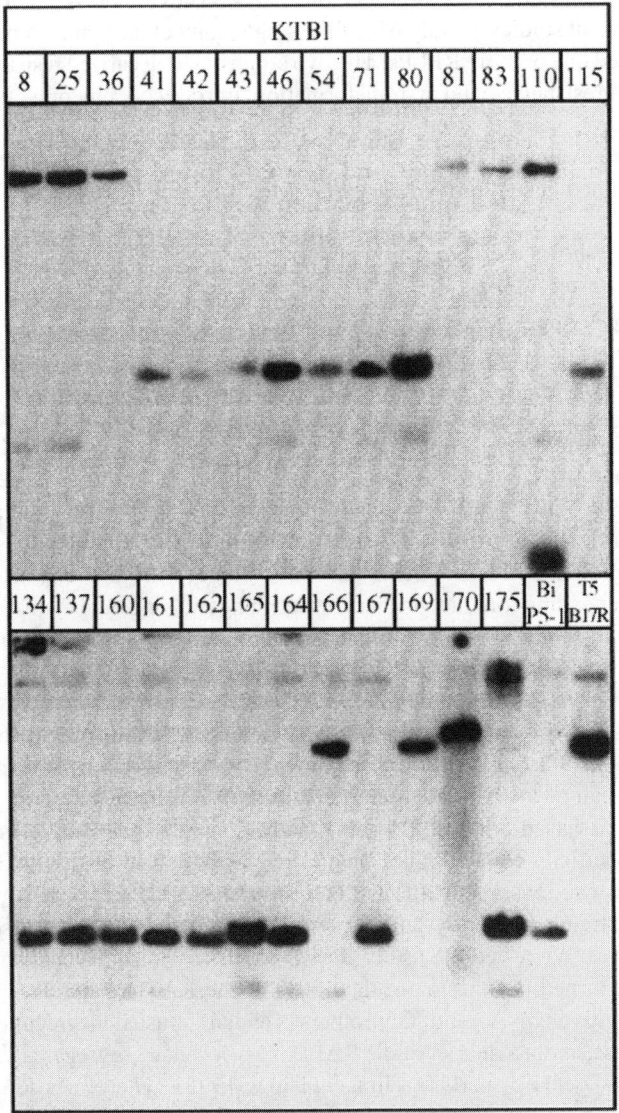

Figure 1 Example for an RFLP analysis in *C. purpurea.* Genomic DNA of the parent strains (T5 B17R and BiP5-1) and F1 progeny strains (KTB) was digested with BamHI, separated in an agarose gel, blotted and hybridized with a labelled fragment of the *cpxyl* gene of *C. purpurea* (see Table 1)

Table 1 Genes (or fragment of genes) available from Claviceps sp.

Gene	Strain	(*putative*) *Gene product*	*Method of cloning*	*Reference*
rrn ITS 1/2	from Festucca	rDNA	PCR	Schardl *et al.*, 1991
pyr4	T5	orotidine-5'-mono-phosphate-decarboxylase	heterol. probe (*N. crassa* pyr4)	Smit and Tudzynski, 1992
gpd	T5	glyceraldehyde-3-phosphate-dehydrogenase	heterol. probe (*A. nidulans* gpdA)	Jungehülsing *et al.*, 1994
dma	ATCC 26245	dimethylallyl-tryptophan synthase	reverse genetics	Tsai *et al.*, 1995
cpa1	ATCC 26245	dimethylallyl-tryptophan synthase	diff. cDNA[1]	Arntz and Tudzynski, 1996
cpa2	ATCC 26245	general stress protein? (ccg1-homologue)	diff. cDNA[1]	Arntz and Tudzynski, 1996
cpa3	ATCC 26245	hydrophobin-like protein	diff. cDNA[1]	Arntz and Tudzynski, 1996
cp605/608	ATCC 34501	peptide synthetase	PCR	Panaccione, 1996
pg1/pg2	T5	(endo-) polygalacturonase	heterol. probe (*A. niger* pgaII)	Tenberge *et al.*, 1996
cel1	T5	cellobiohydrolase	heterol. probe (*Trichoderma ressei* cbhI)	Müller *et al.*, 1996
cpxyl1/2	T5	xylanases	heterol. probe (*C. carbonum* xyl1/*M. grisea* xyn33)	Giesbert in press
cpcat1	T5	catalase	heterol. probe (*A. niger* catR)	Garre and Tudzynski, in press
cpg1	T5	heat-shock protein? (hsp70 homologue)	random cDNA[2]	Garre and Tudzynski, unpubl.
cpg2	T5	hydrophobin-like protein(cpa3-homologue)	random cDNA[2]	Garre and Tudzynski, unpubl.
cpg3	T5	transl. elongation factor? (*P. anserina* EF-1α homologue)	random cDNA[2]	Garre and Tudzynski, unpubl.
cpg4	T5	stress protein? (*N. crassa* ccg6 homologue)	random cDNA[2]	Garre and Tudzynski, unpubl.

1: differential screening of a cDNA library from an alkaloid-producing culture with cDNA from a non-producing (cDNA⁻) and a producing culture (cDNA⁺).
2: cDNA-clones obtained (from a cDNA library from a glucose-derepressed culture) during the screening for other genes.

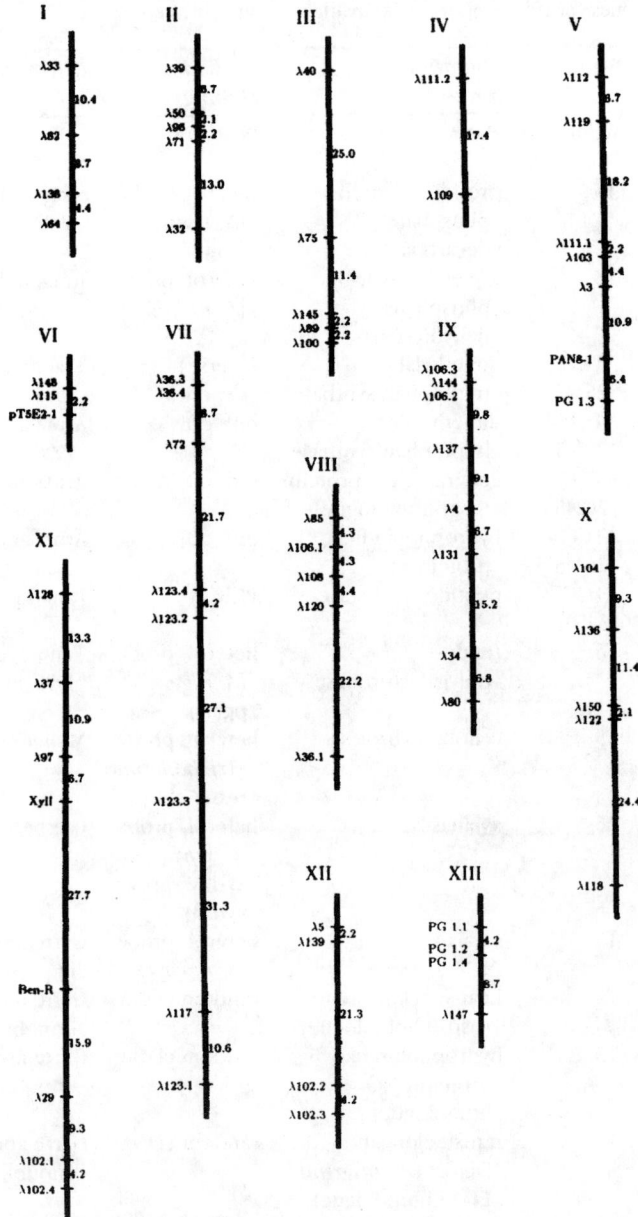

Figure 2 Preliminary genomic map of *C. purpurea*. Linkage groups are designed with roman numbers, the distance of the various markers (random lambda clones from a EMBL library from *C. purpurea* T5 and various genes of *C. purpurea* as listed in Table 1) is indicated in map units (Frank Luerweg, unpubl. data)

between 7 and 1.8 Mb) and differences in numbers (3–7). Actually each *C. purpurea* strain tested sofar showed a unique karyotype, indicating a high degree of interchromosomal recombinations (translocations, etc.). Interestingly, mating between strains with different karyotypes yielded viable progeny, showing recombinant karyotypes including a high degree of "new" chromosome sizes. This indicates that meiotic recombination considerably contributes to the heterogenicity in the karyotypes of field isolates.

A major advantage of PFGE from the genetic point of view is the option to perform Southern hybridization experiments with PFGE gels; this opens the possibility to assign cloned genes or specific RFLP markers to single chromosomes, allowing to establish a link between the genetic and the physical genomic map.

4.2.2.
Isolation of Genes

The first gene that has been cloned from a *C. purpurea* strain was *pyr4,* coding for a orotidine-monophosphate-decarboxylase (OMPD); it was cloned from a genomic *EMBL3* library of the rye isolate *T5,* using the *pyr4* gene of *N. crassa* as heterologous probe (Smit and Tudzynski, 1992). Its function was confirmed by complementation analyses using defined OMPD-deficientmutants of *A. niger* and *C. purpurea* as recipients. It can be used as an efficient selection marker in transformation experiments (see below).

Since then a large number of genes from *C. purpurea* have been cloned and characterized, mostly by using heterologous probes, but also by cDNA screening and PCR techniques (see Table 1). Some of these genes have been cloned because of their potential usefulness in transformation systems (*pyr4,gpd,* etc.), most of the others stem from molecular analyses of the pathogen *C. purpurea,* e.g. genes coding for plant cell wall degrading enzymes, etc. (see also Chapter 2 by Tenberge, this volume). Recently, however, also studies on the molecular genetics of alkaloid biosynthesis were initiated, and the first genes have been characterized. In the group of Schardl (Tsai *et al.,* 1995) the gene for the first specific enzyme of the alkaloid pathway, the dimethylallyltryptophansynthase (DMATS) was cloned, based on oligonucleotides derived from a partial aminoacid sequence of the purified enzyme; the identity of the gene was verified by expression in yeast.

In our group a differential cDNA screening approach was initiated (Arntz and Tudzynski, 1996), using a cDNA library (lambda *Zap*) from an alkaloid producing culture of strain ATCC 26245 (the same strain as used by Tsai *et al.,* 1995) and cDNA preparations from an alkaloid producing and non-producing culture as probes. Among the clones identified so far by this screening as being selectively expressed during alkaloid biosynthesis were clones coding for a DMATS (confirming that this appproach is useful for the isolation of genes of the alkaloid pathway), a hydrophobin-like putative cell-wall protein showing high homology to other fungal hydrophobins, e.g. *Qid3* from *Trichoderma viride* (Lora *et al.,* 1994), and a protein showing homology to the *ccg1* gene product of *N. crassa* (a circadian clock regulated general stress protein, Loros *et al.,* 1989). The latter two clones indicate that this approach also yields genes that might be correlated with the physiological stress and different hyphal morphology (sclerotial growth) typical for alkaloid producing cultures. These preliminary data open up interesting new perspectives for the molecular analyses of ergot alkaloid biosynthesis: (1) the identification of the DMATS gene in this differential cDNA screening proves that at least this gene is regulated at the transcriptional level. This allows a detailed analysis of the different regulatory circuits influencing alkaloid biosynthesis (phosphate and nitrate regulation, tryptophan induction, etc.). (2) probably this approach will allow the identification of further genes of this pathway. Recently also parts of a cyclic peptidase from *C. purpurea* have been characterized which might be involved in the last step of the pathway (Panaccione, 1996). (3) The identification of genes which might be correlated with the different hyphal morphology of producing cultures could lead to an understanding of the relation between this switch in morphology and induction of alkaloid synthesis.

4.2.3.
Transformation

Several transformation systems have been developed for *C. purpurea.* Van Engelenburg *et al.* (1989) used the bacterial phleomycin resistance gene *(phleo^R)* driven by the *A. nidulans trpC* promoter for the first successfull transformation experiments; the transformation rate was poor (~ 1 transformant/μg DNA), and the integration modus was complex (multicopy/multisite integration). Comparable results were obtained by Comino *et al.* (1989), who used two other selection systems: the acetamidase *(amds)* gene of *A. nidulans,* and the prokaryotic hygromycin B-resistance gene unter the control of the *A. nidulansgpd* promoter. The phleomycin selection system was further improved in our group by modifications of the selection process and by the use of stronger promoters. Good results were obtained with vector *pAN 8–1* (Mattern *et al.,* 1988), in which the *pbleo^R* gene is controlled by the *A. nidulans gpd* promoter. Even better transformation rates were achieved using the homologous promoter of the *C. purpurea gpd* gene: when an DNA fragment spanning 1.4 kb of the non-coding upstream region of this gene was fused to the *phleo^R* gene, up to 300 transformants/μg DNA were obtained (Jungehülsing *et al.,* 1994).

In addition, a highly efficient homologous transformation system was developed based on the complementation of auxotrophic recipient strains. The *C. purpurea pyr4* gene (see above) was used to transform OMPD-deficient mutants of strain *T5* to prototrophy, at a rate of more than 400 transformants/μg DNA (Smit and Tudzynski, 1992). The *pyr4* transformation system also proved to be very efficient in co transformation experiments using the bacterial *β-*

glucuronidase gene *(uidA)* as a reporter gene, which was efficiently expressed during the parasitic cycle of *C. purpurea* on rye: honeydew produced by plants infected with *pyr4/uidA* co-transformants was shown to contain significant levels of β-glucuronidase activity (Smit and Tudzynski, 1992). These data show that co-transformation is effective in *C. purpurea,* and that transformation technology can be efficiently used to monitor the fungal development in planta.

Homologous integration of transforming DNA in the genome obviously is a rare event in *C. purpurea;* Smit and Tudzynski (1992) found a rate of ~10% at the *pyr4* locus. Nevertheless, targeted gene inactivation by transformation is possible: Müller (1996) inactivated the *cell* gene of strain *T5* by a gene replacement approach, using the *phleo^R* expression cassette (see above) flanked by the 5' part of *cell* and the downstream non-coding region. Three of 178 transformants obtained showed gene replacement at the *cell* locus, leading to gene inactivation. Comparable results were obtained by Giesbert (in press) using the *cpxyll* gene of *C. purpurea;* also this gene was successfully inactivated by a gene replacement approach.

Gene deletions of this type allow a functional analysis of genes which are suspected to be pathogenicity factors (see also Chapter 2 by Tenberge, this volume); they could also be efficient tools for biotechnological purposes (see below).

Recently another promising application for transformation technology has been developed for filamentous fungi; the so-called REMI (restriction enzyme mediated integration) technique (Kuspa and Loomis, 1992). The method mimics the transposon tagging approach in bacteria and plants. Though transposons occur in fungi (Oliver, 1992), so far no transposon tagging method could be developed for hyphal fungi. REMI is a transformation-based approach which leads to mutants tagged by an integrated plasmid. As outlined above, ectopic (non-homologous) integration is frequent in *C. purpurea* as in other fungi. REMI involves the use of a transforming vector without major homology to the host's genome. The transformation assay includes significant amounts of a restriction enzyme, which randomly cuts the genomic DNA, leading to a random integration of vector DNA at these sites. The genomic DNA of transformants having a desired mutant phenotype is digested with an restriction enzyme having no site within the vector DNA, then the integrated vector plus parts of the tagged gene is excised, can be re-ligated, recovered after transformation in *E. coli,* and sequenced. The link between this "tag" and the observed mutant phenotype can be confirmed by using the rescued vector in knock out experiments with the recipient strain or by genetic linkage analysis. In *C. purpurea* recently a REMI library of more than 1000 transformants has been established and so far 10 pathogenicity mutants have been identified; in addition, 4 of about 500 tested mutants were auxotrophic, confirming the usefulness of this method (Voß, Stehling and Tudzynski, unpubl.).

4.3.
PERSPECTIVES

The combination of classical and molecular genetic techniques opens interesting new perspectives of the analysis of ergot alkaloid biosynthesis in *Claviceps* and for the development of strain improvement programs. The available molecular techniques should allow in the near future the cloning of important genes of the pathway and the analysis of the targets and the mode of action of the regulatory circuits involved. A combination of classical genetic analyses and RFLP/PFGE could answer the question if genes of the pathway are clustered (comparable to the penicillin biosynthesis genes in several fungi). Several options are brought up by the availability of transformation technology in *Claviceps:*

(a) the introduction of multiple copies of a "bottle neck" gene of the alkaloid pathway could lead to a yield improvement; since the initial gene of the pathway has been cloned (see above), such experiments are to be expected for the next future;

(b) targeted gene inactivation represents an important tool to identify so far unkown (and by sequence analysis not indentifyable) genes of the pathway, obtained by "black box" approaches like differential cDNA screening; if the deleted gene belongs to the pathway, the alkaloid synthesis should be blocked, and the new endproduct (intermediate) should be identifyable.

(c) Random insertion mutagenesis could be used to isolate blocked mutants and to identify the (hopefully tagged) mutated gene.

(d) *In vitro* mutagenesis of promoter regions or fusion of strong inducible promoters to important genes of the pathway could be used to construct better industrial strains.

In addition, the availability of several interesting genes of *C. purpurea* (e.g. for the surface-active hydrophobins or for a secreted catalase, see Table 1) opens new perspectives for alternative biotechnological processes.

In conclusion, the field is now open for a concerted breeding program in *Claviceps purpurea*.

ACKNOWLEDGEMENT

I wish to thank my co-workers S.Giesbert, U.Müller, F.Luerweg, T.Voß, M.Stehling for providing me with data prior to publication and A.Kammerahl for typing the manuscript. The experimental work performed in my laboratory was funded by the Deutsche Forschungsgemeinschaft (DFG, Bonn).

REFERENCES

Amici, A.M., Scotti, T., Spalla, C. and Tognoli, L. (1967) Heterokaryosis and alkaloid production in *Claviceps purpurea* . *Appl. Microbiol.*,**15**,611–615.

Arntz, C. and Tudzynski, P. (1997) Identification of genes induced in alkaloidproducing cultures of *Claviceps* sp.*Curr. Genet.*,**31**,357–360.

Bekesy, N. (1956) Über die vegetative und generative Übertragung von Mutterkorneigenschaften. *Z. Pflanzenzüchtung,***35**,461–496.

Brauer, K.L. and Robbers, J.E. (1987) Induced parasexual processes in *Claviceps* sp. strain SD58. *Appl. Environ. Microbiol.*,**53**,70–73.

Comino, A., Kolar, M., Schwab, H. and Socic, H. (1989) Heterologous transformation of *Claviceps purpurea. Biotechnol. Lett.*,**11**,389–392.

Didek-Brumec, M., Puc, A., Socic, H. and Alacevic, M. (1987) Isolation and characterization of a high yielding ergotoxines producing *Claviceps purpurea* strains. *FoodTechnol. Biotechnol. Rev.*,**25**,103–109.

Engelenburg, F.van, Smit, R., Goosen, T., Broek, H. van den and Tudzynski, P. (1989) Transformation of *Claviceps purpurea* using a bleomycin resistance gene. *Appl.Microbiol. Biotechnol.*,**30**,364–370.

Esser, K. and Tudzynski, P. (1978) Genetics of the ergot fungus *Claviceps purpurea*. I. Proof of a monoecious life-cycle and segregation patterns for mycelial morphology and alkaloid production. *Theor. Appl. Genet.*,**53**,145–149.

Garre, V., Müller, U. and Tudzyuski, P. (1998) Cloning, characterization and targeted disruption of cpcatl, coding for an in planta secreted catalase of Claviceps purpurea. *Molec. Plant Microbe Interact.*, in press.

Gesbert, S. (1998) Die Xylanasen des phytopathogenen Axomyceten Claviceps purpurea: Molekülarbiologisde und biochemische charakterisierung. Bibliotheca Mycologica, J. Cramer, Berlin-Stuttgart, in press.

Jungehülsing, U. (1995) Genomanalyse bei *Claviceps purpurea. BibliothecaMycologica,*Vol. 161. J. Cramer, Stuttgart.

Jungehülsing, U. and Tudzynski, P. (1997) Analysis of genetic diversity in *Clavicepspurpurea* by RAPD. *Mycol. Res.*,**101**,1–6.

Jungehülsing, U., Arntz, C., Smit, R. and Tudzynski, P. (1994) The *Claviceps purpurea* glyceraldehyde-3-phosphate dehydeogenase gene: cloning, characterization and use for improvement of a dominant selection system. *Curr. Genet.*,**25**,101–106.

Kirchhoff, H. (1929) Beitrage zur Biologie und Physiologic des Mutterkornpilzes. *Centralbl. Bakteriol. Parasitenk. AH II,***77**,310–369.

Kobel, H. and Sanglier, J.J. (1978) Formation of ergotoxine alkaloids by fermentation and attempts to control their biosynthesis. In Hütter, R., Leisinger, T., Nüesch J. and Wehrli, W. (eds.) *Antibiotics and other secondary metabolites,*Academic Press, New York, pp. 233–242.

Kus, B., Didek-Brumec, M., Alacevic, M. and Socic H. (1986) Benomyl resistant mutant of *Claviceps purpurea* strains. *Biol. Vestnik,***34**,43–52.

Kus, B., Didek-Brumec, M., Socic, H.Gaberc-Porekar, V. and Alacevic, M. (1987) Determination of metabolic blocks in *Claviceps purpurea arg* mutants. *FoodTechnol. Biotechnol. Rev.*,**25**,127–131.

Kuspa, A. and Loomis, W.F. (1992) Tagging development genes in *Dictyostelium* by restriction enzyme-mediated integration of plasmid DNA. *Proc. Nat. Acad. Sci. USA,* **89**,8803–8807.

Lohmeyer, M. and Tudzynski, P. (1997) *Claviceps* alkaloids. In Anke, T. (ed.) *FungalBiotechnology.*Chapman and Hall, Weinheim, pp. 173–185.

Lora, J.M., de la Cruz, J., Benitez, T., Llobell, A. and Pintor-Toro, J.A. (1994) A putative catabolite-repressed cell wall protein from the mycoparasitic fungus *Trichodermaharzianum. Mol. Gen. Genet.,***242,**461–466.

Loros, J.J., Denome, S.A. and Dunlap, J.C. (1989) Molecular cloning of genes under control of the circadian clock. *Neurospora Science,***243,**383–388.

Mattern, I.E., Punt, P.J. and van den Hondel, C.A.M.J.J. (1988) A vector for *Aspergillus* transformation conferring phleomycin resistance. *Fungal Genet. Newsletter,***35,**25.

Müller, U. (1997) Struktur, Expression und gezielte Inaktivierung von *cel1,* einem vermutlich cellobiohydrolase-codierenden Gen von *Claviceps purpurea. BibliothecaMycologica.,*Vol. 166. J.Cramer, Stuttgart.

Müller, U., Tenberge, K.B., Oeser, B. and Tudzynski, P. (1996) *Cel1,* probably encoding a cellobiohydrolase lacking the substrate binding domain, is expressed in the initial infection phase of *Claviceps purpurea* on *Secale cereale. Molec. Plant Microb.Interact.,***10,**268–279.

Oliver, R.P. (1992) Transposons in filamentous fungi. In Stahl, U. and Tudzynski, P. (eds.) *Molecular Biology of Filamentous Fungi.*Verlag Chemi, Weinheim, New York, pp. 3–12.

Panaccione, D.G. (1996) Multiple families of peptide synthetase genes from ergopeptine producing fungi. *Mycol. Res.,***100,**429–436.

Robbers, J.E. (1984) The fermentative production of ergot alkaloids. In Mizrahi, A. and Wezel, A.L.(eds.) *Advances in Biotechnological Processes,*Vol. 3, Alan R.Liss, New York, pp. 197–239.

Schardl, C.L.Liu, J.S., White, J.F., Finkel, R.A., An, Z. and Siegel, M.R. (1991) Molecular phylogentic relationships of nonpathogenic grass mycosymbionts and clavicipitaceous plant pathogens. *Pl. Syst. Evol.,***178,**27–41.

Skinner D.Z., Budde, A.D. and Leong, S.A. (1991) Molecular karyotype analysis of fungi. In Bennet, J.W. and Lasure, L.L.(eds.) *More Gene Manipulation in Fungi.*Academic Press, New. York, p. 86–98.

Smit, R. and Tudzynski, P. (1992) Efficient transformation of *Claviceps purpurea* using pyrimidine auxotrophic mutants: cloning of the OMP decarboxylase gene. *Molec.Gen. Genet.,***234,**297–305.

Socic, H. and Gaberc-Porekar, V. (1992) Biosynthesis and physiology of ergot alkaloids. In Arora, D.K., Elander R.P. and Mukerji, K.G. (eds.) *Handbook of AppliedMycology,*Marcel Dekker, New York, pp. 475–516.

Spalla, C., Guicciardi, A., Marnati, M.P. and Oddo, N. (1976) Protoplasts and production of alkaloids with a heterokaryotic strain of *Claviceps purpurea.* Abstracts. *Fifth Int. Ferment. Symposium,*Berlin, p. 192.

Strnadova, K. (1964) Methoden zur Isolierung und Ermittlung auxotropher Mutanten by *Claviceps purpurea* (fr.) Tul.*Z. Pflanzenzüchtung,***51,**167–169.

Strnadová, K. and Kybal, J. (1974) Ergot Alkaloids. V.Homokaryosis of the sclerotia of *Claviceps purpurea. Folia Microbiol.,***19,**272–280.

Taber, W.A. (1985) Biology of *Claviceps.* In Demain, A.L. and Nadine, A.S. (eds.) *Biotechnology Series,*Vol. 6. *Biology of Industrial Microorganisms.*The BenjaminCummings Publishing Company Inc., New York, pp. 449–486.

Tenberge, K.B., Homann, V., Oeser, B. and Tudzynski, P. (1996) Structure and expression of two polygalacturonase genes of *Claviceps purpurea* oriented in tandem and cytological evidence for pectinolytic enzyme activity during infection of rye. *Phytopathology,***86,**1084–1097.

Tsai, H.F., Wang, J.H., Gebler, C.J. and Schardl, C.L. (1995) The *Claviceps purpurea* gene encoding dimethylallytryptophan synthase, the committed step for ergot alkaloid biosynthesis. *Biochem. Biophys. Research Comm.,***216,**119–125.

Tudzynski, P. and Düvell, A. (1985) Molecular aspects of mitochondrial DNA and mitochondrial plasmids in *Claviceps purpurea.* In Quagliariello, E., Slater, E.C., Palmieri, F., Saccone C. and Kroon, A.M. (eds.) *Achievements and perspectivesof mitochondrial research. Biogenesis,*Vol. II, Elsevier, Amsterdam, New York, Oxford, pp. 249–256.

Tudzynski, P. and Esser, K. (1986) Extrachromosomal genetics of *Claviceps purpurea.* II. Mt plasmids in various wild strains and integrated plasmid sequences in mitochondrial genomic DNA. *Curr. Genet.,***10,**463–467.

Tudzynski, P. and Tudzynski, B. (1996) Genetics of phytopathogenic fungi. *Prog. Bot.,***57,** 235–252.

Tudzynski, P., Düvell, A. and Esser, K. (1983) Extrachromosomal genetics of *Clavicepspurpurea.* I. Mitochondrial DNA and mitochondrial plasmids. *Curr. Genet.,***7,** 145–150.

Tudzynski, P., Esser, K. and Gröschel, H. (1982) Genetics of the ergot fungus *Clavicepspurpurea.* II. Exchange of genetic material via meiotic recombination. *Theor. Appl.Genet.,***61,**97–100.

Tudzynski, P., Tenberge, K. and Oeser, B. (1995) *Claviceps purpurea.* In Singh, U. (ed.) *Pathogenesis and host parasite specificity in plant diseases: histopathological, biochemical, gentic and molecular bases.*Vol. IV, *Eukaryotes.* Pergamon Press, New York, pp. 161–187.

Williams, J.G.K., Kubelik, A.R., Rafalski, J.A. and Tingey, S.V. (1991) Genetic analysis with RAPD markers. In Bennet, J.W. and Lasure, L.L. (eds.) *More Gene Manipulations in Fungi.*Academic Press, New York, pp. 431–439.

5.
BIOSYNTHESIS OF ERGOT ALKALOIDS

ULLRICH KELLER

Max-Volmer-Institut für Biophysikalische Chemie und Biochemie,Fachgebiet Biochemie und Molekulare Biologie, TechnischeUniversität Berlin, Franklinstrasse 29, D-10587Berlin- Charlottenburg, Germany

5.1.
INTRODUCTION

5.1.1.
Historical Background and Importance of Ergot Alkaloids

The ergot alkaloids, metabolites formed during sclerotial growth of the ergot fungus *Claviceps* sp. *(secale cornutum),* occupy an important chapter in the history of natural drugs. As toxins they were responsible for the mass poisonings in the middle ages caused by consumption of bread made from rye contaminated with sclerotia (leading to symptoms known as ergotism) and for the accompanying socio-cultural implications of ergotism in the medieval society (Barger, 1931). On the other hand, the use of ergot as a medical drug is old and appears to have developed during the same times. It is documented as early as 1582 in a German herbarium (Lonitzer, 1582) when sclerotia wrongly considered as malformed grains were recommended as a remedy for use in gynaecology. However, it took centuries to recognise that ergot *per se* was responsible for the mass poisonings and that it was a fungus. Thus, the complete life cycle was not described until the middle of the 19th century (Tulasne, 1853). Production of ergot alkaloids in plant-parasitizing fungi is still a risk today; less important for man but of significant extent for livestock breeding in agriculture. These facts indicate the need of constant control of the spread of ergot alkaloid producing fungi in the environment (Tudzynski *et al.,* 1995; Latch, 1995).

The era of rational medicinal use of ergot drugs in the form of alkaloid mixtures extracted from sclerotia of *Claviceps purpurea* began only about one hundred years ago (Tanret, 1875). However, doubts on the identity of the alkaloids as the active principle of the ergot drug remained until 1918, when Stoll succeeded to crystallise ergotamine and was able to assign the specific action of the ergot drug to that pure substance (Stoll, 1945). This was the starting point for the chemical investigation of the ergot alkaloids which led to knowledge of the structures of a large number of d-lysergic acid derived peptides and amides (Hofmann, 1964). The value of d-lysergic acid derivatives in therapy and the need to investigate their structure-action relationships in their various different effects on the adrenergic, noradrenergic and serotoninergic receptor families strongly stimulated research in the chemistry of the ergot alkaloids. These efforts culminated in the structural elucidation and the chemical total synthesis of d-lysergic acid and ergotamine (Kornfeld *et al.,* 1954; Hofmann *et al.,* 1963). Modifications of compounds and many semisynthetic approaches led to new compounds exerting useful effects or unexpected activities such as in the case of LSD (reviewed by Stadler and Stütz, 1975; Arcamone, 1977). As result of the progress in alkaloid chemistry various ergot alkaloids with different structures have been introduced into therapy such as in the treatment of migraine, high blood pressure, cerebral circulatory disorders, parkinsonism, acromegalia or prolactin-disorders (reviewed by Berde and Stürmer, 1978; Stadler and Giger, 1984). The fact that no other group of substances exhibits such a wide spectrum of biological actions as the d-lysergic acid peptides makes it likely that new derivatives may provide novel unexpected "hidden structures" of the d-lysergic acid and clavine pharmacophores (Hofmann, 1972) (see also Chapter 14 in this book). This potential recommends to explore further the ergot alkaloid substance pool as a class of versatile compounds for future developments of new drugs and also for use as leads in rational drug design for synthesis of new natural compound mimetics (Eich and Pertz, 1994; Clark, 1996).

5.1.2.
Development of Alkaloid Biochemistry

Although the technology of field cultivation and classical strain breeding afforded a solid basis for the large scale production of ergot peptide alkaloids, there was increasing need to explore the molecular basis of their formation both in respect of developing new compounds and laboratory cultures. In the 1950s, the enormous progress of microbial chemistry and microbiology of the β-lactams (Abraham, 1991) had set novel standards of microbial metabolite production and stimulated general interest in the biosynthesis of secondary metabolism products. The beginning studies on the biosynthesis of ergot alkaloids thus aimed at identifying the biogenetic precursors and elucidating the mechanisms of their incorporation. Concomitantly with the onset of research in the biosynthesis of d-lysergic acid-derived alkaloids, the discovery of a hitherto unknown second class of ergot alkaloids, the clavines—simpler in their structures than the classical ergot peptide alkaloids—enormously enlarged the working basis for biosynthetic studies because they proved to be intermediates of the ergot peptide alkaloid synthetic pathway (Abe, 1948a-c, 1952). Moreover, the progress in ergot microbiology such as the development of saprophytic and submerged cultures of ergot, the isolation and selection of *Claviceps* strains valuable for production of compounds and suitable for biosynthetic studies of clavines and of d-lysergic acid-derived alkaloids beneficially supported these efforts (Stoll *et al.,* 1954; Abe and Yamatodani, 1961; Arcamone *et al.,* 1960, 1961). Both academic and industrial researchers working in the fields of microbiology and biochemistry of the ergot fungus have elaborated a solid biogenetic basis of ergoline ring synthesis and of d-lysergic acid-derived alkaloids (reviewed in Weygand and Floss, 1963; Floss, 1976; Vining and Taber, 1979; Floss and Anderson, 1980; Boyes-Korkis and Floss, 1993). In parallel, detailed research in ergot microbiology contributed much to understand the physiology and regulation of alkaloid formation (reviewed in Udvardy, 1980; Esser and Düwell, 1984; Kobel and Sanglier, 1986; Socic and Gaberc-Porekar, 1992) (see also Chapter 6 in this book).

However, enzymology of ergot alkaloid formation was a slowly progressing field as it appears to be extremely difficult to establish stable cell-free systems in *Claviceps*. With the exception of dimethylallyltryptophan synthase (DMAT synthase), enzymatic reactions of the ergoline ring pathway had poorly been investigated in cell-free conditions and—if at all—they served to confirm results that were obtained in *in vivo* precursor studies. Instability of most of the enzymes involved in ergoline ring synthesis prevented their purification and the study of their actions and structures. Advanced enzymology of the biosynthetic enzyme by means of gene cloning, gene expression and testing gene products in suitable biosynthetic backgrounds is as yet only possible with DMAT synthase. The field of ergot alkaloid biosynthesis, therefore, still contains important challenges for both enzymologists and geneticists for future research. This chapter summarises the essential steps of biogenesis of both clavine and peptide alkaloid biosynthesis from a present-day view and will describe previous and most recent accomplishments in the elucidation of their enzymatic steps. It is not the intention of this chapter, of course, to compile and present structures of all previously known and recently isolated novel structures which represent more or less modifications of the intermediates or main products of the ergoline and d-lysergyl peptide pathway. For a compilation of ergoline ring-derived structures and their sources see the Chapters 7 and 18 in this book.

5.2.

STRUCTURES AND CLASSIFICATION OF ERGOT ALKALOIDS

The ergot alkaloids are indole derivatives substituted at their 3,4 positions. The tetracyclic ergoline ring system (1) is the basic structure of most ergot alkaloids (Figure 1). In these alkaloids the N-6 is methylated, ring D contains a $\Delta^{8,9}$ or $\Delta^{9,10}$ double bond and most importantly, C-8 is substituted with a one-carbon unit (methyl, hydroxymethyl or carboxyl) which according to its oxidation state may serve as attachment site of characteristic side chains. Additional substitutions with hydroxy groups can occur in few other positions of the ergoline ring system. Variations in structures can also arise by the formation of epimers at C-8

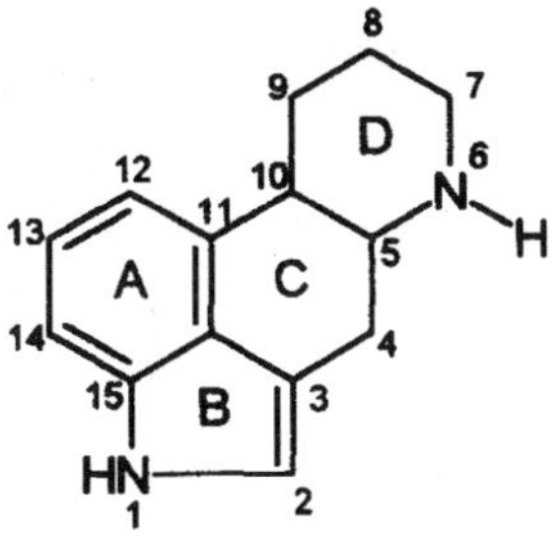

Figure 1 Structure of ergoline

or C-10 or double bond reduction in ring D. Open form of ring D in the ergoline ring system creates an additional subclass of the ergot alkaloids.

The formation of the tetracyclic ergoline ring and derivatives thereof is not restricted to the ergot fungus *Claviceps purpurea* and other *Claviceps* species. Ergoline ring synthesis and both pathway-specific and non-specific modification may occur in numerous other fungi and even in higher plants (see Chapter 18 in this book). *Convolvulaceae* have been shown to be the main plant source of clavines and d-lysergic acid amides and peptides (Hofmann and Tscherter, 1960; Stauffacher *et al.,* 1965). However, in fungi the occurrence of d-lysergic acid derivatives is restricted to the genus *Claviceps* and—of more recent importance— to some *Acremonium* and *Balansia* species living endophytically in grasses (Bacon and White, 1994).

5.2.1.
d-Lysergic Acid and its Derivatives

According to the oxidation state of the C-17 substituent located at C-8 of the ergoline ring, ergot alkaloids are classified into two series of compounds. The first and more important one is the class of amide derivatives of d-lysergic acid ($\Delta^{9,10}$–6-methyl-ergolene-8-carboxylic acid) (Figure 2). The amide portions may consist of a short peptide or may be represented by an amino alcohol. d-Lysergic acid **(2)** and its derivatives can readily epimerize at C-8 which leads to co-occurrence of d-lysergic acid-containing alkaloids with their corresponding d-isolysergic acid **(4)**-containing antipodes (in the case of the latter indicated by the ending -inine instead of -ine). Hydrogenation of d-lysergic acid and its derivatives leads to the corresponding dihydroergot alkaloids containing dihydrolysergic acid **(3)** instead of d-lysergic acid. In contrast to d-lysergic or -isolysergic acid, dihydrolysergic acid does not epimerize at C-8 (Hofmann, 1964). Paspalic acid **(5),** also called $\Delta^{8,9}$ d-lysergic acid, differs from d-lysergic acid in the position of the double bond in ring D. In aqueous solution, paspalic

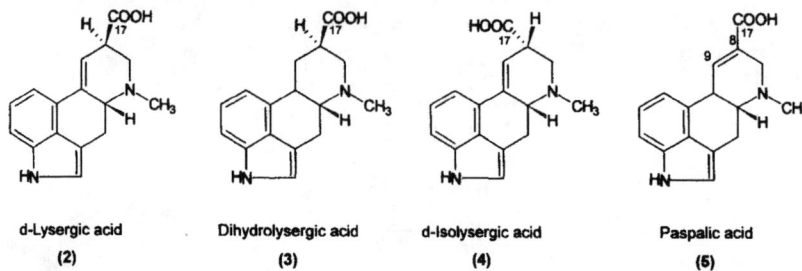

d-Lysergic acid | Dihydrolysergic acid | d-Isolysergic acid | Paspalic acid
(2) | (3) | (4) | (5)

Figure 2 Structures of 6-methyl-ergoline- and -ergolene-8-carboxylic acids

acid can spontaneously isomerize to d-lysergic acid (for ergot alkaloid chemistry and modifications see Chapters 7, 8, and 9).

The naturally occurring derivatives of d-lysergic acid can be arranged—according to the complexity of the structures of their amide portions—in a hierarchical order (Table 1). The ergopeptines—the classical ergot peptide alkaloids produced by *C. purpurea*—are d-lysergyltripeptides with the tripeptide portion arranged in a bicyclic cyclol-lactam structure (Hofmann, 1964). Their sequence is characterised by two non-polar amino acids followed always by proline (position III). The first amino acid (next to d-lysergic acid, position I) has a hydroxy group located at the α-C which is esterified via a cyclol bridge to the carboxy group of proline which formally is in its ortho form, a rare example for an ortho amino acid in nature. That same carboxy group has undergone simultaneous condensation with the peptide nitrogen of its amino-sided amino acid (position II) to give the lactam ring. The numerous ergopeptines differ from each other on a combinatorial basis by amino acid substitutions in the positions I and II. Based on this they can be arranged into three main groups—the ergotamine, ergoxine and ergotoxine groups—each containing various members (Table 2).

Ergopeptams are the non-cyclol versions of the ergopeptines (Stadler, 1982). They still posses the lactam ring and the same amino acid sequence arrangement as their corresponding ergopeptines but differ from the latter in that the amino acid in position I is not hydroxylated and hence no cyclol bridge is present (Table 1). Furthermore, the proline residue is always in the D-configuration while in the ergopeptines it is always L-configured. Ergopeptams co-occur with their corresponding ergopeptines. However, the ergopeptam family of ergot peptide alkaloids has, as yet, much less members than the ergopeptines (for the structures see the Chapter 7).

d-Lysergyl dipeptides are arranged in a structure more similar to the ergopeptines than to the ergopeptams. The structure of ergosecaline **(6)** isolated from Spanish ergot (Abe *et al.,* 1959)—which as yet is the only member of the family—shows that it is a d-lysergyl-dipeptide lactone containing the first two amino acids of an ergopeptine with proline (position III) missing. Hence, no lactam can form. However, the compound still contains the a-hydroxyamino acid

Table 1 Structures of the naturally occurring d-lysergic acid amides arranged on a combinatorial basis considering the multiplicity of amide-bond forming building blocks in their side chains

Ergopeptines
(Cyclol-tripeptide)

Ergopeptams
(Tripeptide lactams)

Ergosecaline (6)
(Dipeptide lactone)

Ergometrine (7)

Lysergyl α-hydroxyethylamide (8)

Table 2 Combinatorial system of the naturally occurring ergopeptines

		Ergotamines	Ergoxines	Egotoxines
	Position I	Alanine	α-Amino butyric acid	Valine
Position II				
Phenylalanine		Ergotamine	Ergostine	Ergocristine
Leucine		α-Ergosine	α-Ergoptine	α-Ergokyptone
Isoleucine		β-Ergosine	β-Ergoptine*	β-Ergokryptine
Valine		Ergovaline	Ergonine	Ergocornine
α-Amino butyric acid		Ergobine	Ergobutine	Ergobutyrine

* not yet found in nature

in position I which is esterified to the carboxyterminal end of the peptide (Table 1).

In the d-lysergyl alkylamides such as ergometrine (ergonovine **(7)**) or d-lysergyl α-hydroxyethylamide **(8),** d-lysergic acid is connected with only one building block and these compounds may be regarded as d-lysergyl monopeptide-related compounds which instead of a single amino acid contain amino alcohols such as α-hydroxyethylamine or α-hydroxy-β-aminopropane (alaninol) linked to d-lysergic acid in amide linkage. d-Lysergyl α-hydroxyethylamide **(8)** is produced by strains *C. paspali*, ergometrine **(7)** has been found in both *C. paspali* and *C. purpurea* (Hofmann, 1964). *C. paspali* is also the source of d-lysergyl amide, ergine. Ergine is supposed to be a non-natural compound arising by spontaneous decomposition of d-lysergyl α-hydroxyethylamide (Kleinerová and Kybal, 1973).

d-Lysergic acid **(2)** itself does not occur in a free form (Floss, 1976). By the contrast, the closely related paspalic acid usually does occur in the free form and sometimes is an abundant metabolite in the alkaloid mixture of *C. paspali* strains (Kobel *et al.,* 1964). 9,10-Dihydrolysergic acid **(3)** has an importance as the constituent of several semisynthetic dihydroergopeptines (produced by hydrogenation of the natural ones) used in therapy (Berde and Stürmer, 1978). The natural occurrence of dihydrolysergic acid is restricted to one single dihydroergopeptine (dihydroergosine) (Mantle and Waight, 1968; Barrows *et al.,* 1974).

5.2.2.
Clavines and Secoergolines

The clavines are the second series of ergot alkaloids which have been discovered by Abe and coworkers when examining ergot fungi parasitizing on the wild grasses from Japan and other parts of Asia (Abe, 1948a-c). The structures of some representative clavines are shown in Figure 3. They are simpler than d-lysergic acid-derived compounds and they have the same ergoline ring system with a methyl or hydroxymethyl group (and occasionally a carboxy group) at C-8. Agroclavine (9), the prototype of this class of compounds, was found in the saprophytic cultures of the grass ergot (Abe, 1948a-c). Elymoclavine (10) differs from agroclavine by the presence of a hydroxymethyl substituent at C-8 instead of a methyl group (Abe *et al.,* 1952). Clavines can have $\Delta^{8,9}$ double bonds but a $\Delta^{9,10}$ double bond (ergolenes) is found also in a number of compounds and there are also examples of hydrogenated compounds (ergolines). Additional diversity is created in some compounds by hydroxylations in the apparently readily available positions C-8, C-9 and occasionally C-10 of the ergoline ring system. Clavines have been isolated from many *Claviceps* sp. including *C. purpurea* and non-ergot fungi (Floss, 1976; Floss and Anderson, 1980; Flieger *et al.,* 1997).

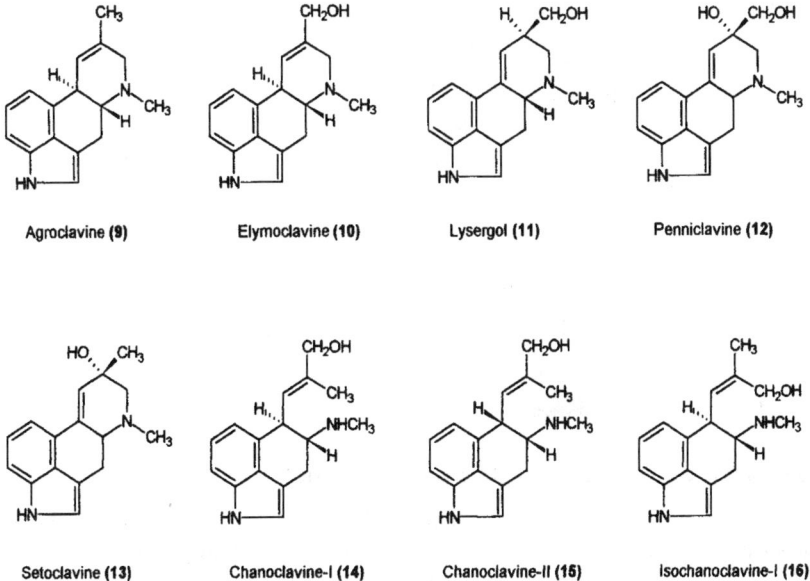

Figure 3 Structures of representative clavines and 6,7 secoergolenes

The 6,7-seco ergolenes, also called chanoclavines, are a subclass of the clavines. They are formed by *Claviceps* sp. and also by fungi outside this genus (see Chapters 7 and 18) and have ring D of the tetracyclic ergoline ring system in an open form. Chanoclavines-I (14) and -II (15) are stereoisomers having hydrogens *trans* or *cis,* respectively, at positions 5 and 10. Isochanoclavine-I (16) is the (E)-, (Z)-isomer of chanoclavine-I while racemic chanoclavine-II is a mixture of (+) and (−)-chanoclavine-II (Stauffacher and Tscherter, 1964). Variations in structures of the chanoclavines can arise by, e.g., hydrogenation (secoergolines), by isomerization of the double bond or by oxidation of the hydroxymethyl group to the aldehyde or carboxy group. Structures of some secoergolines (chanoclavines) are shown in Figure 3.

5.3.
BIOSYNTHESIS OF THE ERGOLINE RING

5.3.1.
Precursors of the Ergoline Ring

The fact that the indol ring system is a part of the ergoline ring nucleus of the ergot alkaloids led Mothes and coworkers to propose that tryptophan and a C5-isopren unit would serve as building blocks in the assembly of the ergoline backbone (Figure 4) (Mothes *et al.*, 1958; Gröger *et al.*, 1959). Feeding sclerotia

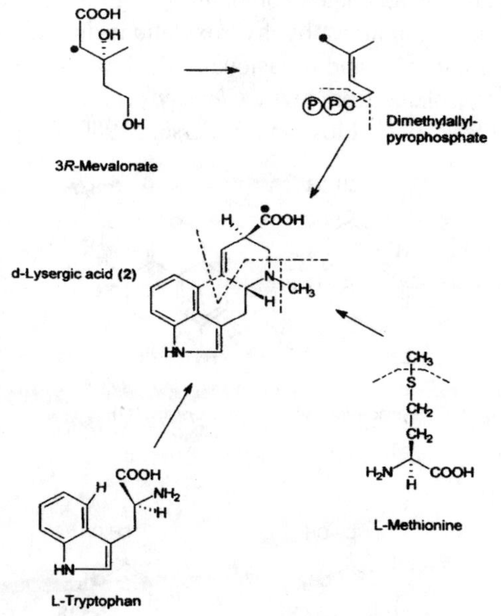

Figure 4 Biogenetic precursors of the ergoline ring system

of *Claviceps* growing on the intact plant with [^{14}C]tryptophan resulted in the low but specific incorporation of radioactivity into the ergoline part of ergometrine and ergopeptine (Mothes *et al.*, 1958). Mycelium of saprophytically grown *Claviceps* gave better results with 10% to 39% specific incorporation into elymoclavine after administration of [2–^{14}C]-tryptophan (Gröger *et al.*, 1959). Furthermore, feeding ergot cultures with tryptophan specifically labelled in the benzene ring or side chain revealed that during conversion to alkaloid all of the hydrogen atoms of the amino acid are retained except the one residing at C-4 of the indol ring which was lost (Plieninger *et al.*, 1967; Bellati *et al.*, 1977). Not only the carbon skeleton but also the α-amino nitrogen was found to become incorporated into alkaloid indicating that tryptophan was incorporated without rearrangement (Floss *et al.*, 1964).

Incorporation of [^{14}C]-acetate into ergoline was demonstrated by Gröger *et al.*, (1959). In extending this observation, several groups reported specific incorporation of [2–^{14}C]-mevalonate into the ergoline moiety of alkaloids (Gröger *et al.*, 1960; Birch *et al.*, 1960; Taylor and Ramstad, 1961; Baxter *et al.*, 1961; Bhattacharji *et al.*, 1962). Further evidence indicated that the mode of mevalonate incorporation was similar to that into terpenes as was documented by the finding that $[1-^{14}C]$-mevalonate was not incorporated (Baxter *et al.*, 1961).

The question of the origin of the methyl group at N-6 of the ergoline ring was addressed by feeding labelled methionine or formate. Methionine labelled in the methyl group simultaneously with both ^{14}C and tritium gave a constant ratio of both labels with high incorporation rate which was by far exceeding that of $[^{14}C]$-formate (Baxter *et al.*, 1964). This suggested that L-methionine in the form of S-adenosyl-L-methionine can serve as direct precursor of the methyl group at N-6 of the ergoline ring of elymoclavine or d-lysergic acid.

5.3.2.
The Pathway of Ergoline Ring Synthesis

The structural similarity between the clavines and d-lysergic acid-derived alkaloids in their tetracyclic ring systems implies that their biosyntheses have similar or same reaction steps in common. The presence of methyl, hydroxymethyl and carboxyl groups located exclusively at the same C-8, and apparently incomplete structure represented by the 6,7-secoergolines led to suggest that clavine alkaloids such as chanoclavine-I, agroclavine or elymoclavine were themselves not only metabolic end products *per se* but in other biosynthesis backgrounds represented intermediates in the pathway leading to the complex d-lysergic acid amides produced in *C. purpurea*. Thus, the biosynthetic sequence chanoclavine (**14**)→agroclavine (**9**)→elymoclavine (**10**) →d-lysergic acid (**2**) was postulated already by Rochelmeyer in the 1950s (Rochelmeyer, 1958). Evidence for the validity of that proposal came further by the result that clavine alkaloids—after their initial discovery in grassparasitizing *Claviceps*—were subsequently found accompanying the d-lysergic acid alkaloids in sclerotia of rye ergot (Voigt, 1962).

The findings that mevalonate entered into the ergoline structure in some form of an isoprene unit and that labelled 4-(γ,γ-dimethylallyl)pyrophosphate (DMAPP) was incorporated intact into alkaloids (Plieninger *et al.*, 1967) established 4-(*γ,γ*-dimethylallyl)tryptophan (DMAT, 17) as the structural basis of the ergoline ring system. Consequently, the conversion to agroclavine as an end product of the ergoline ring pathway is a series of successive modifications of this molecule required for two ring closures with the right stereochemistry.

Analysis of the essential steps of the pathway was done by feeding isotopically labelled precursors to ergot cultures and by analysing label distribution in the various alkaloids formed. In addition, chemical and enzymatic synthesis of various isotopically labelled putative intermediates and following their incorporations into tri- and tetracyclic ergolenes allowed to describe the steps of ring assembly and also the route leading from the clavines to the d-lysergic acid amides. In addition, enzymatic conversions with crude cell-extracts and partially purified enzyme fractions from *Claviceps* helped to confirm the identity of intermediates formed and to confirm reaction mechanisms of several steps of the pathway. These precursor studies led to the pathway shown in Figure 5.

After isoprenylation of tryptophan the resulting DMAT (17) is methylated with S-adenosylmethionine (AdoMet) to give N-methyl-DMAT (MeDMAT, 18). Next steps involve conversion to chanoclavine-I (14) under closure of ring C. 14 is then oxidised to chanoclavine-I aldehyde (19) which in turn is converted to agroclavine (9) by a closure of ring D. The formation of ring C and ring D each is accompanied by a *cis-trans* isomerization of the allylic double bond of the isoprene moiety. Two successive oxygenations at the C-17-methyl group of agroclavine lead to elymoclavine (10) and then to d-lysergic acid (2).

5.3.3.
The First Step: Isoprenylation of Tryptophan

Methylated or decarboxylated derivatives of tryptophan were not incorporated into elymoclavine (10) whereas 4-(γ,γ-dimethylallyl)tryptophan (DMAT, 17) was an efficient precursor (Baxter et al., 1961; Floss and Gröger, 1963, 1964; Plieninger et al., 1962, 1964; Agurell, 1966a). These experiments suggested that decarboxylation and methylation of the tryptophan residue must occur later in the biosynthetic sequence and that an alkylation at C-4 of tryptophan with the C5-isopren unit in the biosynthetically active form of 4(γ,γ-dimethylallyl)-1-pyprophosphate (DMAPP) is the first pathway-specific reaction in ergot alkaloid biosynthesis. Studies of the stereochemistry of incorporation revealed that mevalonate is used exclusively in the R-configuration as in other systems (Floss et al., 1968; Seiler et al., 1970a, b; Abou-Chaar et al., 1972). Other reports consistently demonstrated that after administration of [2–^{14}C]-mevalonate the majority of label resided into the C-17 of the tetracyclic alkaloids with only slight scrambling of 7–10% of label in the C-7 position indicating somehow incomplete stereospecificity in the DMAT formation (Gröger et al., 1960; Birch et al., 1960; Taylor and Ramstad, 1961; Baxter et al., 1961; Bhattacharji et al., 1962). Furthermore, feeding experiments with D- and DL-tryptophan indicated that L-tryptophan is incorporated with the complete retention of the α-hydrogen (Floss et al., 1964). This is particularly interesting in view of the fact that the C-5 of the ergoline ring system (identical with C-2 of tryptophan) has R-configuration which corresponds to a D-configured tryptophan side chain. This inversion occurs in a later step and it is of mechanistic importance as it is under full retention of hydrogen. Later, in agreement with these findings, DMAT (17) was detected as free intermediate in *Claviceps* mycelium when total synthesis of alkaloids was inhibited by depriving *Claviceps* cultures from supply with oxygen or adding ethionine that finally established a direct correlation between DMAT formation and alkaloid synthesis (Robber and Floss, 1968; Agurell and Lindgren, 1968).

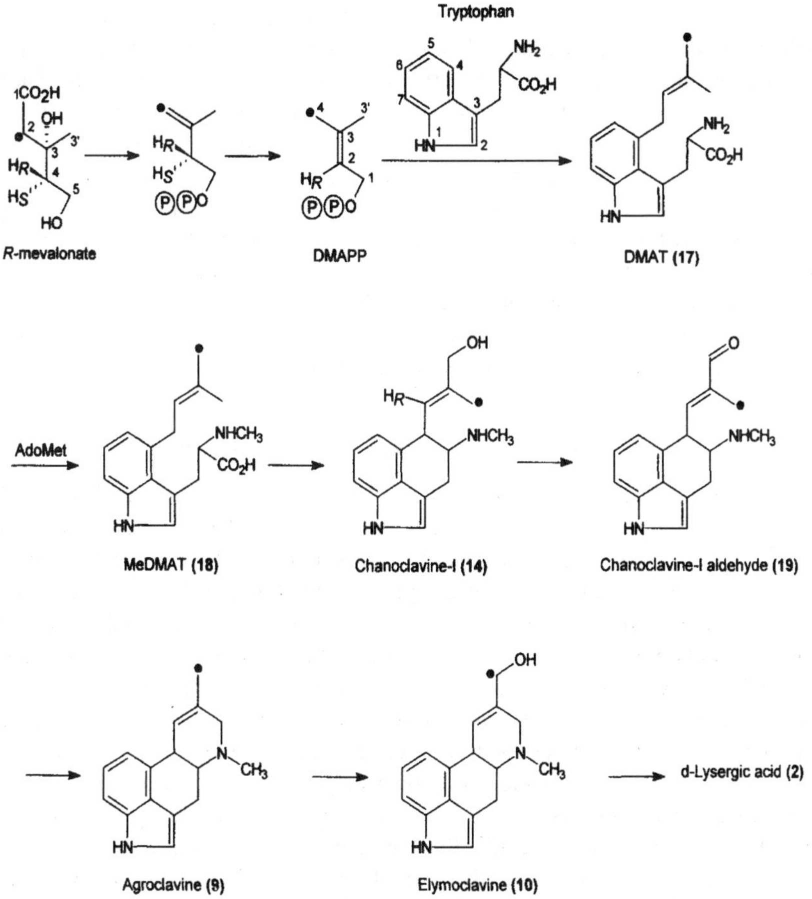

Figure 5 Biosynthetic pathway of ergoline ring formation in *Claviceps* sp. • denotes label (e.g. ^{14}C) originally present in C-2 of *R*-mevalonate

Stereochemistry of DMAT Formation

Isolation of the enzyme responsible for synthesis of DMAT **(17)**, 4-(γ,γdimethyllally)tryptophan synthase (DMAT synthase), has greatly assisted in studying the isoprenylation of tryptophan (Lee *et al.*, 1976). Indirect evidence from feeding experiments with whole cells of *Claviceps* had suggested that the configuration of the allylic double bond of **(17)** should be the same as that of DMAPP from which it is derived (Pachlatko *et al.*, 1975). Enzymatic synthesis of **(17)** from tryptophan and DMAPP deuterated specifically in the Z-methyl group (i.e. (Z)-3-methyl-2-[4–^{2}H$_3$] butenyl pyrophosphate) gave a product in which—as proved by NMR—the intensity ratio between the (Z)and (E)-methyls was the same as in the labelled DMAPP (Shibuya *et al.*, 1990). This indicates that the (Z)-methyl group of DMAPP gives rise to the (Z)-methyl group of DMAT. Thus the geometry of the allylic double bond is retained in the DMAT synthesis reaction as shown initially for the (E)-methyl group in Figure 5. Furthermore, these findings assigned the scrambling of the label between the C-17 and C-7 in the tetracyclic alkaloids after feeding [2–^{14}C]mevalonate to the ergot cultures to the one of the reaction steps prior to the DMAT synthase reaction converting the mevalonate into DMAPP. This is unusual because the reaction sequence from mevalonic acid to DMAPP in other organisms appears to be stereospecific (Popják and Cornforth, 1966). From the four enzymes involved in that conversion, isopentenyl pyrophosphate (IPP) isomerase is considered to be most likely responsible for this effect and thus this enzyme in *Claviceps* is obviously not absolutely stereospecific (Shibuya *et al.,* 1990).

Mechanism of the Isoprenylation Reaction

The chemical mechanism of the reaction leading to DMAT—which leads to the loss of hydrogen at C-4 of tryptophan (Plieninger *et al.,* 1967)—has been subject of a number of investigations to solve the problem that C-4 of an indole is not the most favoured position for an electrophilic substitution (Floss and Anderson, 1980). These studies—performed by feeding precursors labelled in various positions and determining their distributions in the ergoline ring backbone—have not yielded any evidence for activation of the C-4 or for the migration of the isoprene moiety from another position to C-4 (Floss *et al.,* 1965). From studying the enzyme-catalysed reaction, it was concluded that the enzyme promoted a direct attack at the C-4 position by proper alignment of the two substrates in a sequential ordered process. However, it was observed in the previous feeding experiments with stereospecifically tritiated (5R-^{3}H)- and (5S-^{3}H)-[2–^{14}C] mevalonate that that there was also scrambling of label between the two allylic hydrogens at C-l of the isoprenoid moiety (Seiler *et al.,* 1970a; Abou-Chaar *et al.,* 1972) (besides the scrambling of the label from [2–^{14}C]-mevalonate into the (Z)- and (E)methyl group-derived positions of the tetracyclic alkaloids). This indicated incomplete stereospecificity of the alkylation of tryptophan DMAT synthase. Shibuya *et al.* (1990) determined the steric course of isoprenylation reaction at C-1 of DMAPP during the reaction catalysed by DMAT synthase. For this purpose enzyme was reacted with (1R)- or (1S)-[1–^{3}H]DMAPP and unlabeled L-tryptophan. Analysis by chemical and enzymatic degradation of the reaction products for their configuration at the allylic-benzylic carbon atom indicated that DMAT generated from (1S)-[1–^{3}H]DMAPP had retained about 80% of the radioactivity in the *pro-R* position whereas that one produced from (1R)-[1–^{3}H]DMAPP had retained some 20% of total radioactivity in the *pro-R* hydrogen (Figure 6). These results clearly proved inversion at the C-1 of DMAPP in the course of the isoprenylation reaction but indicated that the observed scrambling of label in the allylic hydrogens of DMAT was most

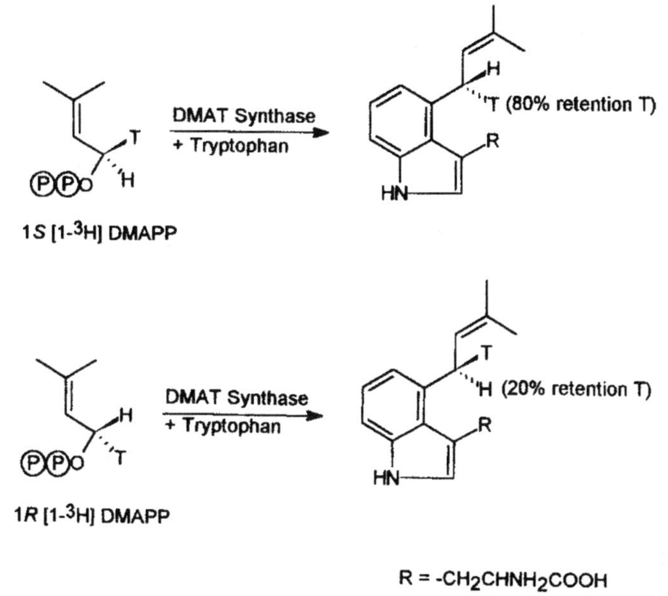

Figure 6 Retention of radioactivity in the pro-*R*-positions of DMAT after reaction of tryptophan with (1*R*)- or (1*S*)-[1–³H]DMAPP catalysed by DMAT synthase according to Shibuya *et al.* (1990)

probably due—at least in part—to the incomplete stereospecificity of the DMAT synthase reaction itself.

The conclusion of Shibuya *et al.* (1990) from these experiments was that isoprenylation of L-tryptophan involved displacement of the pyrophosphate moiety of DMAPP by the indole residue with inversion at C-1 of the isoprene moiety. In the ternary enzyme substrate complex, dissociation of the C-1/ pyrophosphate bond of DMAPP would yield an ion pair or an allylic carbocation stabilised by the countercharge of the enzyme-bound pyrophosphate as shown in Figure 7. From C-4 of enzyme-bound L-tryptophan on the face opposite to the pyrophosphate, attack would create DMAT with inversion at C-1 and retention of double bond geometry. If the scrambling between allylic hydrogens occurred during these event, this can be readily accounted for by rotation around the C-1/ C-2 bond of the allylic carbocation in a fraction of the molecules prior to bond formation with the indole. These findings characterise the prenyl transfer of DMAT synthase as an electrophilic aromatic

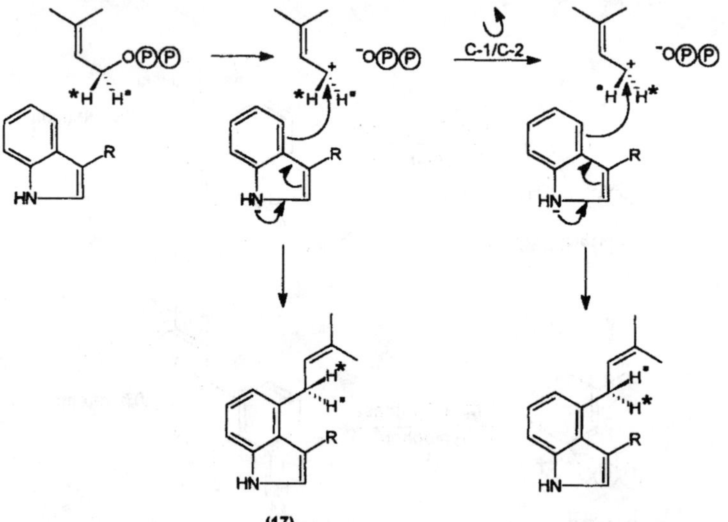

(17)
main product

(17)
minor product

Figure 7 Reaction mechanism of DMAT formation catalysed by DMAT synthase

substitution, mechanistically similar to the electrophilic alkylation catalysed by farnesyl diphosphate synthase (Song and Poulter, 1994).

5.3.4.
The Second Step: Methylation of Dimethylallyltryptophan

For a time, methylation at N-6 was believed to take place in the course of formation of ring C because both N-demethylchanoclavine-I and -II in feeding experiment were not incorporated into tetracyclic ergolines (Cassady *et al.,* 1973). However, Barrow and Quigley (1975) showed the presence of a new indole amino acid in *C. fusiformis* which they identified as N-methyl-dimethylallyltryptophan (MeDMAT, **18**). These authors showed also the incorporation of [N-[14]CH$_3$]MeDMAT into clavine alkaloids. Otsuka *et al.* (1979) fed *Claviceps* sp. SD 58 with chemically synthesised MeDMAT and its decarboxylation product N-methyldimethylallyltryptamine **(20)** both trideuterated in the N-methyl group and additionally carrying ^{15}N in the N-methyl nitrogen (Figure 8). Massspectrometric analysis of elymoclavine obtained after feeding MeDMAT to ergot cultures revealed intact incorporation of the double-labelled MeDMAT in the tetracyclic ergoline. By contrast, **20** was not incorporated that was consistent with the previously observed failure of incorporation of dimethylallyltryptamin into chanoclavine-I. This suggested that decarboxylation must

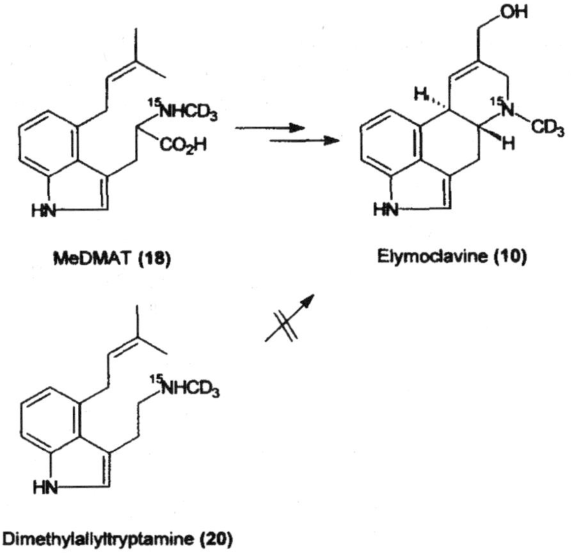

MeDMAT (18)

Elymoclavine (10)

Dimethylallyltryptamine (20)

Figure 8 Intact incorporation of [N-^{14}CH$_3$]MeDMAT into elymoclavine establishes methylation of DMAT as the second step of the ergoline synthesis pathway

be a later step than methylation (Plieninger *et al.*, 1967). Later, Otsuka *et al.* (1980) have isolated an enzyme which catalyses transfer of the methyl group of S-adenosylmethionine to DMAT as substrate which indicates that methylation of 17 is the second step of the ergoline pathway.

5.3.5.
The Third Step: Closure of Ring C and Formation ofChanoclavine-I

Of the four stereoisomers of chanoclavine, chanoclavine-I **(14)** is the only one to be converted into the tetracyclic ergolines (Gröger *et al.*, 1966). On the other hand, Plieninger *et al.* (1978) showed that chanoclavine-I **(14)** is formed from DMAT **(17)** (via MeDMAT **(18)**) which establishes chanoclavine-I as an intermediate between DMAT and the tetracyclic ergoline alkaloids. As can be seen from the structure of chanoclavine-I **(14),** its formation from MeDMAT **(18)** involves changes of the oxidation states at one of the methyl groups and the two C-5 and C-10 carbons involved in ring C closure. This requires two oxidative steps (Figure 9). However, these events and their mechanisms are not yet completely understood.

Cis-Trans Isomerization at the Double Bond of the Isoprenoid Moiety

One of the most interesting features of the conversion of MeDMAT **(18)** to chanoclavine-I **(14)** is the fact that the (Z)- and (E)-methyls of the isoprenoid moiety of DMAT have opposite configuration in the isoprenoid moiety of chanoclavine-I. Feeding ergot cultures with [2–^{14}C)-mevalonate and isolation

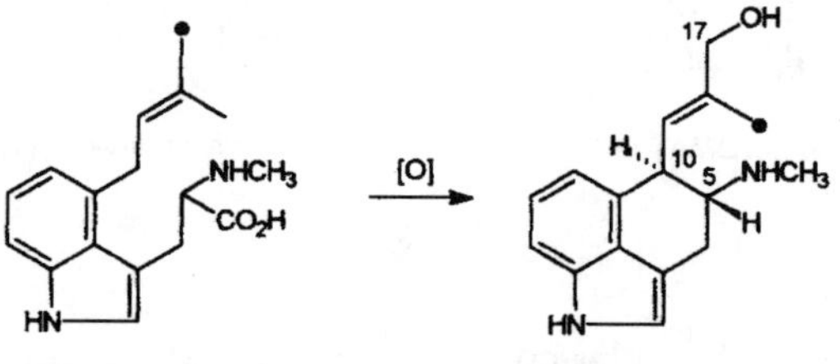

MeDMAT (18)   Chanoclavine-I (14)

Figure 9 Stereochemical course of the conversion of MeDMAT to chanoclavine-I

of the chanoclavine-I formed revealed the majority of the radiolabel to be present in C-methyl group ((Z)-methyl) of the isoprenoid moiety (Fehr et al., 1966). This indicates that isomerization at the double bond of the isoprenoid moiety must have taken place in the course of chanoclavine synthesis from DMAT because C-2 of mevalonate gives rise to the (E)-methyl group of DMAPP via the isopentenyl pyrophosphate isomerase reaction (Figure 9). This finding was confirmed when DMAT labelled in the (E)-methyl group was fed to chanoclavine-I-producing ergot cultures which resulted in the exclusive labelling of the (Z)-methyl group of chanoclavine-I (Plieninger et al., 1978).

Hydroxylation Reactions in the Course of Chanoclavine-I Synthesis

Early experiments suggested that introduction of the hydroxyl group into the (Z)-methyl group of MeDMAT **(18)** (i.e. (E)-methyl group of chanoclavine-I) must take place prior to ring closure of ring C because deoxychanoclavine **(21)** is not incorporated into agroclavine (Fehr, 1967). However, hydroxylation of the (Z)-methyl group of **18** as the first step could be ruled out because of the finding that chemically synthesised (E)-OH-DMAT **(22)** and (Z)-OH-DMAT **(23)** labelled with ^{14}C in their 3-methyl groups were both incorporated by cultures of ergot into elymoclavine **(10)** but not into agroclavine **(9)** (Plieninger et al., 1971; Pachlatko et al., 1975). Surprisingly, the analysis of radiolabel distribution in the elymoclavine formed revealed that label resided in C-7 of elymoclavine **(10)** irrespective of which of the two potential precursors was administered (Figure 10). The interpretation of these results was that the (E)-OH-isomer of **22** is an unnatural intermediate which is accepted by the cyclizing enzymes as if the

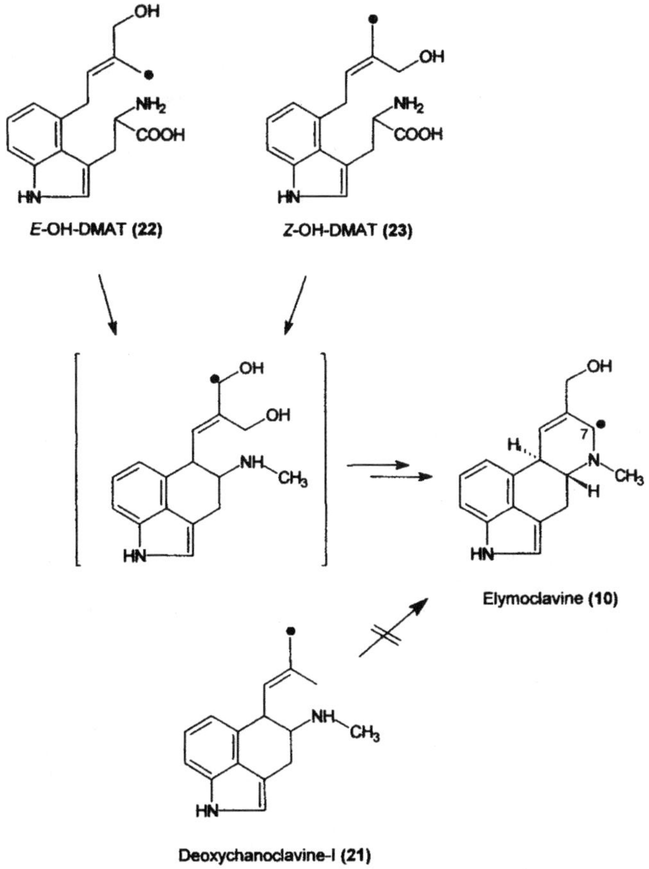

Figure 10 Conversion of *(E)*-OH-DMAT and *(Z)*-OH-DMAT into elymoclavine

hydroxyl group were not present. This leads, of course, directly to **10** because the (E)-hydroxymethyl group of the educt contains a preformed hydroxyl group. The elymoclavine formed here would represent a "pseudoagroclavine". The fact that the (Z)-OH-isomer also yielded **10** with the radioisotope present in C-7 of ring D as did the (E)-isomer could only be explained by assuming conversion of the (Z)-isomer to the (E)-isomer in the cell prior to incorporation (Pachlatko *et al.*, 1975). E-OH-DMAT **(22)** was later detected as a metabolite in cultures of *Claviceps* species. Its formation was assigned to action of non-specific hydroxylases (Anderson and Saini, 1974; Floss and Anderson, 1980).

These data and work of Seiler *et al.* (1970a,b) and Abou-Chaar *et al.* (1972) concerning the loss of hydrogen at C-10 (i.e. position C-1 of the isoprene moiety) during conversion, led to envisage initial attack by hydroxylation at C-10 of **18** instead at the (Z)-methyl group of MeDMAT **(18)** (Figure 11). Following route

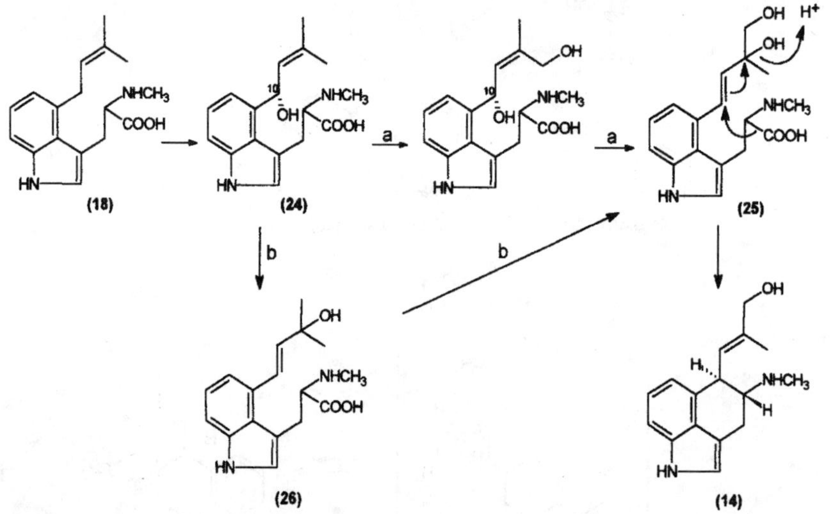

Figure 11 Early scheme of chanoclavine-I synthesis from MeDMAT initiated by hydroxylation at C-10 of MeDMAT involving intermediacy of a diol. Note that the structures have been modified for the presence of the methyl group in the amino nitrogen of MeDMAT. DMAT was considered previously as precursor of chanoclavine-I

designated "a" the resultant 10-OH-MeDMAT **(24)** could be hydroxylated further at the (Z)-methyl group and finally converted with allylic rearrangement into the 8,17-diol **(25)**. Alternatively, following the route designated "b", the 8-OH-derivative of **26** could be formed from **24** by allylic rearrangement. An additional hydroxylation of **26** would then result in formation of the same 8,17-diol **(25)** as in the route "a" (Figure 11). Pachlatko *et al.* (1975) proposed an alternative mechanism (Figure 12) suggesting formation of a hydroperoxide of **18** with subsequent conversion to the corresponding epoxide **(27)**. This could be cyclized with simultaneous decarboxylation via an S_N2 mechanism (Figure 12).

More recent data from Kozikowski's and Floss' groups clarified some steps of chanoclavine-I synthesis summarised in Figure 13. They showed that the diol **(25)** was not incorporated into **10** (Kozikowski *et al.*, 1988). As a more plausible precursor of chanoclavine-I Kozikowski *et al.* (1993) envisaged then 8-hydroxy-MeDMAT **(26)** as intermediate. In fact, 8-OH-MeDMAT (deuterated in its methyl group) became incorporated into **10** with up to 33% specific incorporation. However, experiments to trap **26** in *Claviceps* cultures failed, indicating that **26** was not a normal intermediate. In these trapping experiments they, however, detected the diene **(29)** as a new metabolite. The diene **(29)**, chemically synthesised and labelled with deuterium in the methyl group, was incorporated by ergot cultures with much higher specific incorporation (>80%) than **26**. These findings suggest that 8-OH-MeDMAT **(26)** had been incorporated

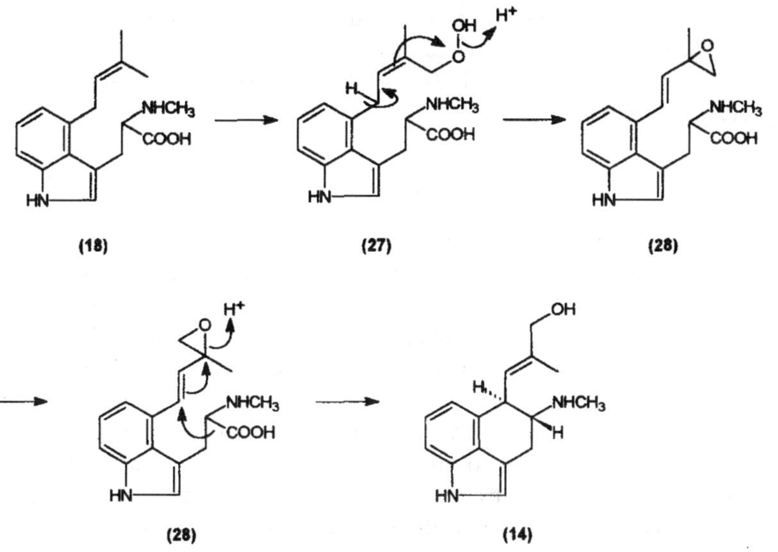

Figure 12 Hydroperoxide formation as first step in chanoclavine-I synthesis (Pachalatko *et al.*, 1975). Structures have been modified for the presence of the methyl group in the amino nitrogen of DMAT

into 10 via 14 by virtue of nonenzymatic dehydration to the diene **(29)**. This implicates that the epoxide **(28)** formed from the diene—which had been suggested earlier by Pachlatko *et al.* (1975) (Figure 12) as plausible species to undergo cyclization—is most likely to be the next intermediate. However, instead via the hydroperoxide, the results of Kozikowski *et al.* (1993) suggest that **28** is generated by epoxidation of the diene **(29)** that is consistent with the observation that the hydroxyl oxygen of chanoclavine-I **(14)** originates from molecular oxygen and not from water (Kobayashi and Floss, 1988). The data also suggest that the diene **(29)** is not formed directly from the 8-OHMeDMAT derivative **(26)**. Instead, it may be formed from **18** by hydroxylation at the benzylic carbon C-10 of the isoprene moiety to yield **24** followed by a direct 1,4-dehydration. The scheme in the Figure 13 involving the diene **(29)** as an intermediate may also provide an explanation for the observed *cis-trans* isomerization observed in chanoclavine-I synthesis which—facilitated by free rotation between C-8/C-9 of the isoprene moiety—may be dictated by the steric course of the dehydration step and/or steric constraints in the active site of the epoxidizing enzyme. The conversions of both E-OH-DMAT and Z-OH-DMAT into elymoclavine (Pachlatko *et al.*, 1975) shown in the Figure 10 point to this mechanism because formation of the diene **(29)** from (E)-OH-DMAT or (Z)-OH-DMAT would in both cases involve the C-methyl groups but not the hydroxymethyl groups which contain the preformed hydroxyl groups of the elymoclavine. Thus, one can envisage that isomerization would probably take place at the stage of the diene and this can explain the efficient incorporation of both E-OH-DMAT **(22)** and Z-OH-DMAT **(23)** into the above-mentioned "pseudoagroclavine".

Clavicipitic Acids

The structures of the clavicipitic acids **(30, 30a)** (Figure 14), a pair of diastereomeric compounds, initially detected in ethionine-inhibited *Claviceps* cultures (Robbers and Floss, 1969) and later isolated in normal conditions from *C. fusiformis* (King *et al.,* 1973) have raised the attention of investigators as possible products of a side reaction in chanoclavine-I formation. Apparently, clavicipitic acid—which is not converted into tetracyclic ergolines— arises from DMAT **(17)** by a hydroxylation at C-10 which subsequently leads to a substitution with the amino nitrogen of the tryptophan moiety. It was found that microsomal fractions from *Claviceps* convert DMAT into clavicipitic acid under consumption of oxygen with the formation of hydrogenperoxide (Bajwa *et al.,* 1975). This indicated no direct incorporation of oxygen into DMAT but facilitating in some way substitution at C-10. The formation of clavicipitic acids has been attributed to nonenzymatic cyclization of 10-OH-DMAT. 10-OH-DMAT may be formed under methionine limitation by non-specific hydroxylation of accumulating DMAT by the same enzyme that normally hydroxylates MeDMAT (Floss and Anderson, 1980). 10-OH-DMAT has not been detected in *Claviceps,* because it may be too reactive. By contrast, 10-OH MeDMAT **(24)** has a fate different from that of DMAT by reacting to the diene **(29)** with subsequent ring closure between C-5 and C-10. This may illustrate the importance of the methyl group at the amino nitrogen of DMAT for substrate recognition by the enzymes involved in chanoclavine-I synthesis. These considerations characterize the clavicipitic acids as a dead end of the ergoline pathway.

The Decarboxylation Step

Before N-methylation of DMAT has been established as the second step of the ergoline ring synthetic pathway, the reaction models presented in Figures 11 and 12 were based on results obtained with DMAT and not MeDMAT as substrate and the dominating hypothesis was that the decarboxylation of DMAT should proceed in a pyridoxalphosphate-dependent mechanism (Floss and Anderson, 1980). However, the presence of the methyl group at the amino nitrogen of MeDMAT **(18),** makes the participation of a pyridoxal phosphate cofactor unlikely (Otsuka *et al.,* 1980). Steric constraints will interfere with the coplanar rearrangement of Schiff base imine double bond with the pyridine ring which is a prerequisite for decarboxylations and also for the subsequent stabilisation of a carbanion structure at the α-C of the tryptophan moiety. This was initially postulated to be necessary for attack at C-10 and ring closure. However, the data

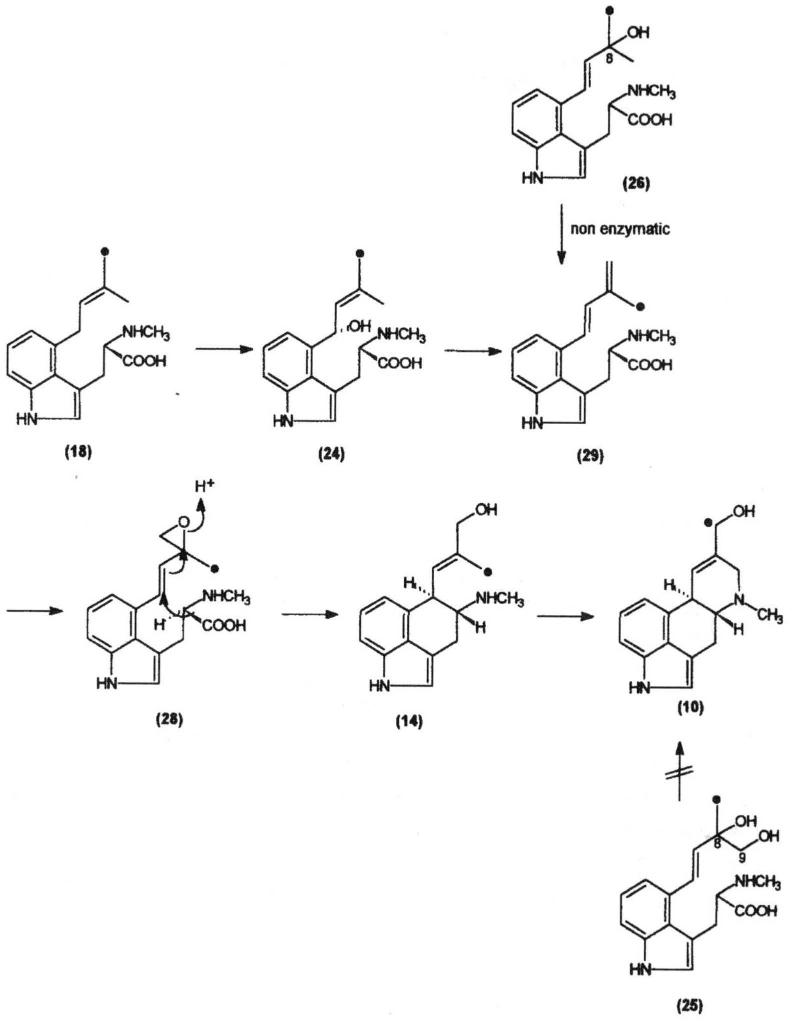

Figure 13 Scheme of reactions leading from MeDMAT to chanoclavine-I involving hydroxylation at C-10 leading to the diene isomer of MeDMAT (Kozikowski *et al.*, 1993)

suggest formation of a carbanion structure at C-5 by a cofactor independent decarboxylation with attack a benzylic carbocation structure at C-10 (Otsuka *et al.*, 1980; Kozikowski *et al.*, 1993). The fact, that the formation of chanoclavine-I from MeDMAT proceeds with inversion and retention of hydrogen at C-5 suggests that decarboxylation is concerted with ring closure between C-5 and C-10 on the face opposite to the departing carboxyl group (Figure 13).

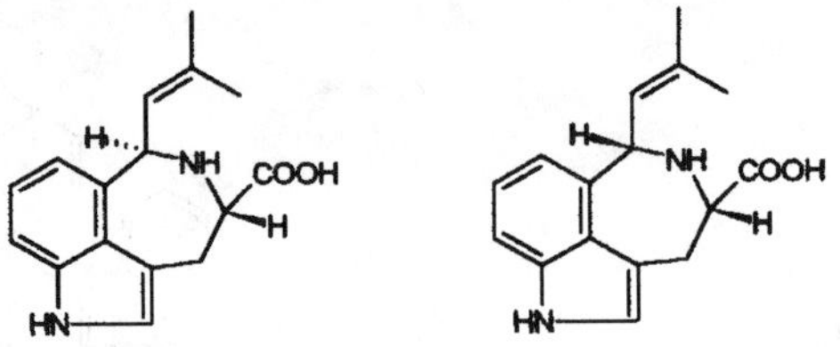

(30)   **(30a)**

Clavicipitic acids

Figure 14 Structures of the clavicipitic acids

5.3.6.
The Fourth Step: Formation of Ring D of Tetracyclic Ergot Alkaloids

A Second Cis-Trans Isomerization in Ergoline Ring Synthesis

Feeding [17–^{14}C]chanoclavine-I or [7–^{14}C]chanoclavine-I to the ergot cultures and degradation of the resulting agroclavine and/or elymoclavine had shown that the hydroxymethyl group of chanoclavine-I gave rise to C-7 and the C-methyl group to C-17 of agroclavine and elymoclavine (Fehr *et al.*, 1966; Floss *et al.*, 1968). These results indicate that cyclization of ring D is accompanied by another *cis-trans* isomerization as indicated in the scheme of the pathway shown in Figure 4. Feeding (E)-[methyl-^{14}C]- or -[methyl-^{13}C]-DMAT **(17)** and determination of label in agroclavine **(9)** and elymoclavine **(10)** formed showed that it is converted into agroclavine and elymoclavine with no or an even number of *cis-trans* isomerizations. Together with the established isomerization between MeDMAT and chanoclavine-I it demonstrates two double bond isomerizations in the ergoline pathway (Pachlatko *et al.*, 1975; Plieninger *et al.*, 1978).

Chanoclavine-I Aldehyde

Loss of tritium in the C-7 of elymoclavine obtained after feeding *Claviceps* strain SD 58 with [17–^{3}H, 4–^{14}C] chanoclavine-I indicated an oxidation step at the hydroxymethyl group of chanoclavine-I. Chanoclavine-I aldehyde **(19)** was postulated to be the intermediate produced by this oxidation (Naidoo *et al.*, 1970). Later, it was shown, that chemically synthesised chanoclavine-I aldehyde **(19)** was much better incorporated into elymoclavine **(10)** than chanoclavine-I **(14)** (Floss *et al.*, 1974). This led to a scheme of agroclavine **(9)** formation via chanoclavine-I aldehyde **(19)** and its isomer isochanoclavine-I aldehyde **(31)**. Intramolecular cyclization of **31** and reduction of the intermediate would yield agroclavine **(9)**. The intermediacy of **31** would explain the isomerization during closure of ring D and from a mechanistic point of view, isomerization would

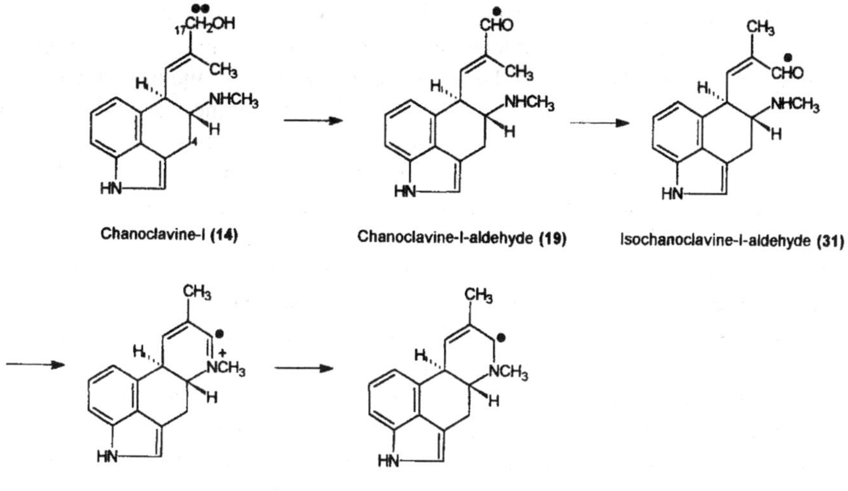

Chanoclavine-I (14)   Chanoclavine-I-aldehyde (19)   Isochanoclavine-I-aldehyde (31)

(9)

Figure 15 Closure of ring D: chanoclavine-I cyclization. •• and • indicate loss of tritium diring the conversion of chanoclavine-I to chanoclavine-I-aldehyde

substantially facilitate reaction between the N-methylamino group and the aldehyde to yield the bi-unsaturated system in ring D (Figure 15).

Further evidence for the involvement of chanoclavine-I aldehyde **(19)** was obtained by the preparation of a mutant of *C. purpurea* blocked in the formation of ergometrine and ergotoxine (Maier *et al.,* 1980). This mutant designated Pepty 695/ch—accumulated chanoclavine-I **(14)** and chanoclavine-I aldehyde **(19)** instead of producing d-lysergic acid-derived alkaloids. This supports the idea that **19** is indeed a natural intermediate in tetracyclic ergoline ring formation. Moreover, when fed with d-lysergic acid or elymoclavine, the strain Pepty 695/ch produced ergometrine and ergotoxine like the parent strain indicating that it was a true blocked mutant apparently unable to perform the cyclization step which converts chanoclavine-I aldehyde to agroclavine **(9)**. The presence of this mutation also provides indirect proof for a two-step-reaction of conversion of chanoclavine-I **(14)** into agroclavine **(9)**. Formation of the aldehyde is apparently not affected in the mutant while activity catalysing the cyclization step is absent suggesting that either isomerization of chanoclavine-I aldehyde to isochanoclavine-I aldehyde or reduction of the 1,4 iminium double bond in ring D are affected (Figure 15).

Mechanism of Cis-Trans Isomerization and Closure of Ring D

The mechanism of isochanoclavine-I aldehyde formation and cyclization are as yet unclear. Isochanoclavine-I aldehyde **(31)** has not been detected in *Claviceps* cultures (Floss *et al.*, 1974b). Despite the isolation of an enzyme activity capable to catalyse conversion of chanoclavine to agroclavine/elymoclavine (Gröger and Sajdl, 1972; Erge *et al.*, 1973) the available few data about the mechanism stem from *in vivo* feeding experiments. C-9-tritiated chanoclavine-I or [2–^{14}C, 4–^{3}H] mevalonate were converted to elymoclavine with 70% tritium retention which indicated hydrogen recycling at the active site of the enzyme responsible for isomerization of chanociavine-I aldehyde to isochanoclavine-I aldehyde (Floss *et al.*, 1974b). Possibly this enzyme could also be involved in the subsequent step of cyclization and reduction. Direct proof for the recycling came from feeding a 1:1 mixture of [2–^{13}C] mevalonate and [4–^{2}H] mevalonate to *Claviceps* strain SD 58. Analysis of the chanoclavine-I **(14)** and elymoclavine **(10)** formed revealed that chanoclavine-I consisted of a mixture of either deuterated or ^{13}C-labelled molecules. By contrast, the elymoclavine sample contained besides the single-labelled species about 7% double-labelled molecules which indicated that the population had been produced by hydrogen exchange with enzyme molecules which were loaded with a fraction of the abstracted tritium from the previous rounds of turnover (Floss *et al.*, 1974). Retentions in chanoclavines and clavines clearly indicated reversibility of reactions from chanoclavine-I up to the hydrogen elimination step and low rates of alkaloid production correlated with low rates of tritium retention. The abstraction and addition of hydrogen at C-9 indicate an acid-base catalytic mechanism involving addition and abstraction of hydrogen and of a catalytic basic group of the enzyme. As a result of addition of hydrogen, the allylic double bond of chanoclavine-I aldehyde is lifted to allow rotation. After isomerization and subsequent elimination of hydrogen by the enzyme from C-9 of the intermediate, the formed isochanochlavine-I aldehyde **(31)** has the same steric configuration as agroclavine which facilitates its cyclisation. Cyclisation may be spontaneous as in the biosynthesis of proline from glutamate semialdehyde. Alternatively, an enzyme-assisted carbinolamine formation in the course of isomerization has been discussed (Floss *et al.*, 1974b). In both cases, reduction of the resultant bi-unsaturated system would yield agroclavine (Figure 15). Purification of the enzyme catalysing chanoclavine-I cyclization would allow to study this mechanisms in more detail.

5.4.
PATHWAY-SPECIFIC AND -NONSPECIFIC
MODIFICATIONS OF THE ERGOLINE RING SYSTEM

5.4.1.
The Ergoline Ring as a Substrate for Hydroxylation
Reactions

Agroclavine is the precursor of the various tetracyclic clavines and the d-lysergyl moiety of the d-lysergic acid derived alkaloids (Agurell, 1966e; Agurell and

Ramstad, 1962). The conversion of this compound thus requires oxydations/ hydroxylations at the C-17-methyl or—if necessary—at positions C-8, C-9 or C-10. These conversions may be accompanied by double bond isomerizations and dehydrogenations. Conversion of chanoclavines to the various related ergosecalines may proceed by similar mechanisms. However, one has to distinguish between pathway-specific and pathway non-specific modifications. Pathway-specific conversions should be correlated with alkaloid production during cultivation while non-specific modifications need not and they can occur in ergoline non-producing fungal species. Hydroxylation of agroclavine and elymoclavine at the C-8 position leads to setoclavine/isosetoclavine (13) and penniclavine/isopenniclavine (12), respectively (Agurell, 1966b; Agurell and Ramstad, 1962). This reaction is catalysed by peroxidase the activity of which is widely distributed in fungi (Ramstad, 1968; Taylor *et al.,* 1966; Shough and Taylor, 1969). A system of thioglycolate, Fe^{2+} and oxygen catalyzes the same conversion which indicates that in strains where these alkaloids are not major products could be nonenzymatic (Bajwa and Andersons, 1975). The microsomal cytochrome P-450 mono-oxygenase system from mammalian liver converts agroclavine to noragroclavine (i.e. demethylated agroclavine (9)) and to a lesser extent to elymoclavine and other hydroxylation products formed in alkaloid producers (Wilson *et al.,* 1971). Addition of inducers of the liver microsomal oxygenase such as phenobarbital and polycyclic hydrocarbons to the culture medium of *Claviceps* increased alkaloid production and the level of cytochrome P-450 in the microsomal fraction (Ogunlana *et al.,* 1969; Ambike and Baxter, 1970; Ambike *et al.,* 1970). These results indicate that a broad specificity inducible cytochrome P-450 similar to the liver enzyme is present in *Claviceps* and that it may be able to catalyse one or more steps in the pathway of ergot alkaloid biosynthesis as well transformation of compounds not related to the tetracyclic ergolines. These broad specificity-hydroxylases could lead to conflicting results such as in the case of the isolation of 4[E-4-hydroxy-3-methyl-Δ^2-butenyl]-tryptophan ((E)-OH-DMAT, 22) from *Claviceps* (Anderson and Saini, 1974). Cell-free formation of (E)-OH-DMAT has been observed in the cytosol fraction of *Claviceps* sp. (Petroski and Kelleher, 1978; Saini and Anderson, 1978). Since the previous data exclude OH-DMAT as a regular intermediate in the ergoline biosynthesis pathway, the formation of OH-DMAT may be due to broad-specificity hydroxylases and thus classifies this compound either as a dead end product of the pathway such as the clavicipitic acids or as an intermediate of a separate pathway leading from DMAT directly to elymoclavine (Floss and Anderson, 1980) (for more details on ergot alkaloid biotransformations see Chapter 9).

5.4.2.
Hydroxylation of Agroclavine

The available evidence strongly suggests that independent from the various broad-specificity conversions there is one main pathway leading from agroclavine to elymoclavine and to d-lysergic acid in *Claviceps* (Agurell and Ramstad, 1962). In this pathway elymoclavine is the direct oxidation product of agroclavine in a mixed-type oxygenase reaction. The hydroxyl oxygen comes from molecular oxygen and not from water as is the case for the hydroxyl group of chanoclavine (Floss *et al.*, 1967). Alkaloid-producing strains of *Claviceps* species carried out the conversion of agroclavine to elymoclavine wheras non-producing strains did not. Other types of fungi lacked this activity (Tyler *et al.*, 1965).

5.5.
ENZYMES INVOLVED IN ERGOLINE RING ASSEMBLY AND IN CONVERSION OF ERGOLENES INTO ERGOLENE-8-CARBOXYLIC ACIDS

5.5.1.
Dimethylallyltryptophan Synthase

The enzyme responsible for condensation of L-tryptophan and DMAPP is dimethylallylpyrophosphate: L-tryptophan dimethylallyl transferase (DMAT synthase). It has been isolated by the group of Floss and these investigators also purified it to homogeneity and performed the first characterisations (Heinstein *et al.*, 1971; Lee *et al.*, 1976). Initially, the enzyme was reported to be a monomeric species of 73 kDa size. Cress *et al.* (1981) later obtained the enzyme as a dimer of 34 kDa subunit size which they were able to crystallise. More recent work of Gebler and Poulter (1992) identified the enzyme as a α_2 dimer of 105 kDa. The large discrepancies in the molecular weight reported appear to be due to different levels of proteolysis in the different preparations and strains. Proteolysis which is generally strong in *C. purpurea* is a main problem in the enzymology of ergot alkaloid biosynthesis.

Gebler and Poulter (1992) reported that proteinase inhibitors excellently stabilised the enzyme. Characterisation of the enzyme revealed that the enzyme unlike other prenyltransferases was active in the absence of Ca^{2+}-ions (Cress *et al.*, 1981; Gebler and Poulter, 1992). These authors also showed that the enzyme is stimulated more than twofold in the presence of Ca^{2+} and Mg^{2+}. At low substrate concentrations and in metal-free conditions enzyme displayed negative cooperativity and at 4 mM Ca^{2+} or at high substrate concentrations displayed typical Michaelis-Menten kinetics. These findings are indicative for a role of Ca^{2+} as a positive allosteric effector which deregulates the enzyme (Gebler and Poulter, 1992; Cress *et al.*, 1981). Lee *et al.* (1976) reported mild stimulation upon addition of other metals such as Na^+, Li^+, Cu^{2+} and Fe^{2+}. They also reported that the DMAT synthase reaction proceeds by a random or ordered sequential rather than a ping-pong mechanism. The mechanism of action of DMAT synthase has been established only indirectly by the analysis of labelled reaction products which indicate that the enzyme catalyses the electrophilic alkylation of tryptophan (Shibuya *et al.*, 1990).

Molecular Cloning of DMAT Synthase Gene

Molecular cloning of the DMAT synthase gene from *Claviceps purpurea* marks a milestone in the biochemistry and genetics of ergot alkaloid biosynthesis because it makes possible for the first time the detailed investigation of a step of ergot alkaloid biosynthesis at the genetic level (Tsai *et al.*, 1995). The gene was cloned by a screening of a gene bank of *C. purpurea* DNA with oligonucleotides derived from peptide sequences of the protein and sequenced. From the deduced amino acid sequence, the gene codes for a protein of 51.8 kDa which is in agreement with the previous data on DMAT synthase identifying the protein as a homodimer of 105 kDa. The entire gene was expressed in *S. cerevisiae* and shown to produce a catalytically active enzyme. Surprisingly, the DMAT synthase gene displayed nearly no similarity with other prenyltransferase sequences with the exception of a possible prenyl diphosphate motif (DDSYN) at a position 113–117 of the amino acid sequence. This motif is also conserved in other farnesyl diphosphate and geranylgeranyl diphosphate synthases (Song and Poulter, 1994). The cloned DMAT synthase gene will facilitate the future investigation of regulation of the gene at both the transcriptional and translational level.

Dimethylallyltryptophan N-Methyltransferase

In the view of the precursor role of MeDMAT (see above) in the ergoline biosynthesis, Otsuka *et al.* (1980) performed studies on the enzymatic methylation of DMAT as the second pathway-specific step in ergoline biosynthesis and succeeded in purifying to some extent a DMAT N-methyltransferase from *Claviceps* strain SD 58. The enzyme, which is not located in the particulate fraction, has relatively high K_m-values for DMAT and S-adenosyl-L-methionine of 0.11–0.27 mM and 0.5–1.6 mM, respectively. The methylation reaction is rather specific in respect of the natural substrate DMAT. Structural analogues such as tryptophan or 4-methyltryptophan were much less active (c. 10%). Interestingly β,γ-dihydro-DMAT is a better substrate for the enzyme than DMAT itself, presumably producing the N-methyl derivative. Recent work on the stereospecificity of the methylation reaction indicated that the methylation proceeded under inversion suggesting a S_N2 mechanism of N-alkylation. (Gröger *et al.*, 1991). The DMAT N-methyltransferase was related to the age and alkaloid production of *Claviceps* cultures establishing this enzyme as catalysing the second step in ergoline ring biosynthesis pathway.

5.5.2.
Enzymatic Conversion of Tryptophan into Chanoclavine-I and -II

Cavender and Anderson (1970) found that the synthesis of chanoclavine-I and -II from tryptophan and isopentenyl pyrophosphate was catalysed by the 60–80% ammonium sulfate precipitate obtained from the 105000 g-supernatant of a homogenate of *C. purpurea*. The reaction was dependent on the presence of isopentenyl pyrophosphate, methionine and ATP. Addition of a liver concentrate to the reaction was necessary, presumably to provide additional cofactors.

5.5.3.
Chanoclavine-I Cyclase

The enzyme system catalysing the conversion of chanoclavine-I into agroclavine or elymoclavine has been characterised to some extent (Gröger and Sajdl, 1972; Erge *et al.*, 1973; Floss, 1976). It requires NADPH/NADP and Mg^{2+}, but not oxygen or FAD. Dependence on ATP was reported (Ogunlana *et al.*, 1970). The conversion rates were 10–30%. Chanoclavine-I cyclase also converts chanoclavine-I aldehyde, but not isochanoclavine or dihydrochanoclavine into agroclavine (Erge *et al.*, 1973). The enzyme is stabilised by glycerol but heavy losses of activity during purification prevented its identification and further characterisation.

5.5.4.
Agroclavine 17-hydroxylase

Kim *et al.* (1980) were seeking for enzymes specifically involved in the conversion of agroclavine into elymoclavine. In their study they characterised a specific cytochrome P-450 mono-oxygenase (non-inducible with liver enzyme inducers) catalysing exclusively the conversion of agroclavine to elymoclavine. The enzyme is located in the 15000–105000 g-fraction. NADPH and molecular oxygen were required for its activity. Under optimised condition 16% conversion was reached within one hour of incubation. The enzyme did not catalyse the conversion of DMAT to OH-DMAT that is apparently a non-specific reaction leading to elymoclavine catalysed by broad specificity oxygenases of *Claviceps* (Petroski and Kelleher, 1977; Saini and Anderson, 1978). Agroclavine hydroxylase activity was inhibited by carbon monoxide, but not by cyanide or EDTA that is typical for a cyctochrome P-450 monooxygenase. There is also a good correlation between alkaloid production and the enzyme activity. Practically comparable results concerning agroclavine hydroxylase (which they named agroclavine 17-mono-oxygenase) were obtained by Maier *et al.* (1988).

They showed that agroclavine hydroxylase is inhibited by elymoclavine in a feed-back mechanism consistent with similar observation concerning feedback regulation of DMAT synthase and of other enzymatic conversions in the tetracyclic ergoline synthesis pathway (Cheng *et al.*, 1980).

5.5.5.
Elymoclavine 17-mono-oxygenase

Likewise agroclavine 17-hydroxylase an enzyme catalysing the conversion of elymoclavine **(10)** to paspalic acid **(5)** was found to be located in the microsomal fraction of ergotamine-producing *Claviceps purpurea* PCCE 1 (Kim *et al.*, 1983). The enzyme, named elymoclavine 17-mono-oxygenase, is dependent on NADPH and molecular oxygen. Consistently, the carbonyl oxygen of d-lysergic acid in the ergopeptines was shown to stem exclusively from molecular oxygen in contrast to the carbonyl oxygens of the amino acids of the peptide because of the different biosynthetic origin of their carboxy groups (Quigley and Floss, 1981). The enzyme was inhibited by sulfhydryl inactivating agents such as *p*-chloromercuribenzoate. Potassium rhodanide like carbon monoxide, which is known to convert cytochrome P-450 into cytochrome P-420, inhibited elymoclavine-17-mono-oxygenase as well as agroclavine hydroxylase which leaves no doubt about the nature of the two enzymes (Maier *et al.*, 1988). Elymoclavine 17-mono-oxygenase seems to be highly specific since lysergol **(11)**, the $\Delta^{9,10}$-isomer of elymoclavine, is not converted to d-lysergic acid indicating that the shift of the 8,9 double bond takes place exclusively during conversion of paspalic acid **(5)** to d-lysergic acid **(2)** (Kim *et al.*, 1983; Maier *et al.*, 1988) (Figure 16). Attempts to solubilize the clavine hydroxylase for the purpose of purification led to inactive material (Maier *et al.*, 1988). A further clue to the significance of elymoclavine 17-mono-oxygenase was that Kim *et al.* (1983) could not detect the enzyme in extracts of *Claviceps* strain SD 58 which does not produce ergopeptines but only clavines. This indicates that *Claviceps* strain sp. SD 58 lacks elymoclavine mono-oxygenase as a part of the d-lysergic acid synthesis pathway. This led to conclude that elymoclavine 17-mono-oxygenase is a key enzyme in the

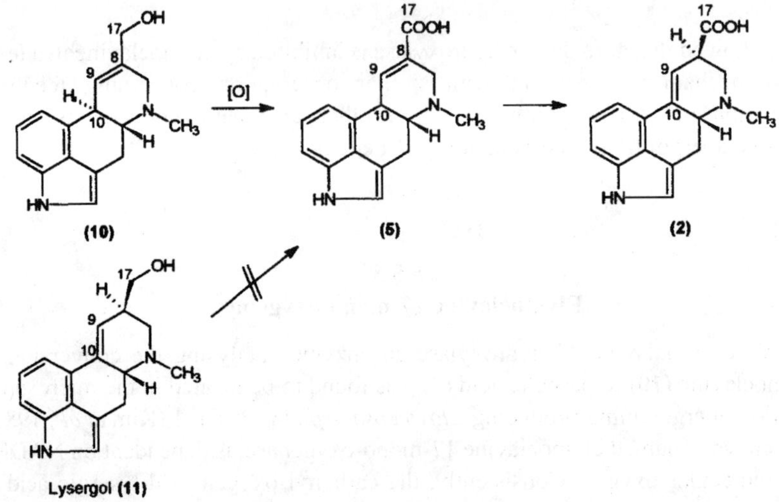

Figure 16 Conversion of elymoclavine to paspalic acid catalyzed by elymoclavine 17-mono-oxygenase

biosynthesis of paspalic acid necessary for supplying of the ergpeptinesynthesizing enzyme system with d-lysergic acid (Keller, 1986).

5.6.

BIOSYNTHESIS OF CYCLOL-TYPE d-LYSERGIC ACID PEPTIDES

5.6.1.

Origin of d-Lysergic Acid

Elymoclavine (**10**) is converted to d-lysergic acid amides *in vivo* with a high conversion rate and was considered to be the immediate precursor of d-lysergic acid (Mothes *et al.,* 1962). Similarly, paspalic acid (**5**), the $\Delta^{9,10}$lysergic acid, was converted into d-lysergic acid amides with comparable efficiency (Agurell, 1966d; Ohashi *et al.,* 1970). This indicates that the pathway is in the order elymoclavine (**10**)→paspalic acid (**5**)→d-lysergic acid (**2**) and—as mentioned previously—the shift of the double bond from $\Delta^{8,9}$- into $\Delta^{9,10}$-position occurs after the introduction of the carboxyl function into the ergoline ring system (Figure 16). However, the fact that paspalic acid isomerizes to d-lysergic acid spontaneously and also that paspalic acid—in contrast to d-lysergic acid—is often present in a free form in *Claviceps* cultures left the possibility that conversion may have occurred nonenzymatically. This prompted to propose an alternative pathway leading from elymoclavine to d-lysergic involving an aldehyde stage of elymoclavine leading to d-lysergylcoenzyme A ester via the corresponding coenzyme A thiohemiacetal (Lin *et al.*, 1973). In this hypothesis, d-lysergyl coenzyme A ester would represent the activated form of d-lysergic acid entering into the processes of d-lysergic acid amide/peptide assembly. However, the failure to demonstrate the d-lysergyl coenzyme A thioester in ergot cultures (Lin *et al.*, 1973) and the previously established role of both free d-lysergic acid and dihydrolysergic acid in the assembly of their corresponding peptide derivatives makes d-lysergylcoenzyme A thioester as a precursor of the d-lysergyl moiety of d-lysergyl amides and peptides unlikely (Keller *et al.*, 1988; Riederer *et al.*, 1996; Walzel *et al.*, 1997). All of the available evidence suggests that paspalic acid is a regular intermediate in the pathway and not a product of a side reaction.

5.6.2.
Origin of the Building Blocks of Ergopeptines

The formation of ergopeptines in *C. purpurea* has been investigated by short-and long-term feeding of labelled amino acids and d-lysergic acid and following their incorporation into the respective compounds. Besides measuring the extent of incorporation into individual ergopeptines, degradation of the products formed by acid or alkaline hydrolysis helped to determine the specific incorporations of the radiolabelled amino acids administered. Cleavage products which are obtained upon acid and alkaline hydrolysis of ergopeptines such as e.g. ergotamine **(32)** are L-phenylalanine and D-proline (which is always obtained when L-proline containing acyl-diketopiperazine structures are hydrolysed) or d-lysergic acid, phenylalanine, D-proline and pyruvate (Stoll *et al.*, 1943). In this way, it was possible to demonstrate the specific incorporation of L-phenylalanine (Bassett *et al.*, 1973), L-proline (Gröger and Erge, 1970; Bassett *et al.*, 1973), L-valine (Maier *et al.*, 1971; Floss *et al.*, 1971b) and L-leucine (Maier *et al.*, 1971) into the corresponding positions of the respective alkaloids formed by various *Claviceps* species.

Moreover, degradation analysis of ergopeptines in respect of the hydroxyamino acid moiety such as the hydroxyvaline in ergotoxines showed that externally added radiolabelled valine labelled the hydroxyl-valine moiety in position I with high efficiency comparable to the simultaneous incorporation of the radiolabelled valine into the position II in the ergocornine component (Table 2) of the alkaloid mixture (Maier *et al.*, 1971; Floss *et al.*, 1971b). Similar experiments with the aim to show the incorporation of [U-^{14}C]-L-alanine into ergotamine demonstrated high incorporation of the administered radiolabel into the α-hydroxy-alanine portion of the alkaloid (Floss *et al.*, 1971a). These data clearly showed that the precursor of the hydroxyamino acids in the ergopeptines are the corresponding free nonhydroxylated ones and that these amino acids stem from the free amino acids of the cellular pool.

Similarly to the results of feeding experiments with labelled amino acids, evidence was obtained that free d-lysergic acid **(2)** is precursor of the d-lysergic

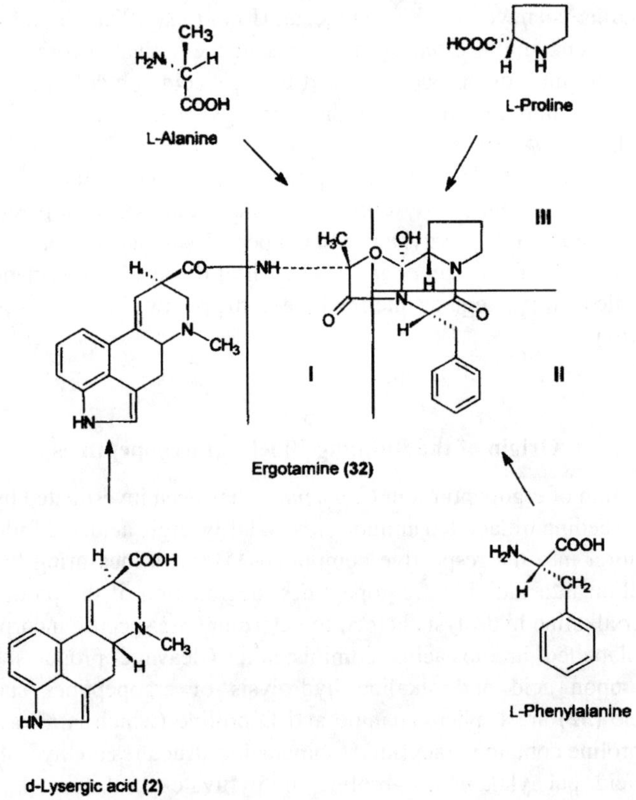

Figure 17 Biogenetic precursors in the assembly of ergopeptines (ergotamine (32))

acid alkaloids (Agurell, 1966b, c; Minghetti and Arcamone, 1969; Bassett *et al.,* 1973). Furthermore, adding non-labelled d-lysergic acid to both mycelium and protoplasts of *Claviceps purpurea* substantially stimulated incorporation of radiolabelled amino acids into ergotamine **(32)** suggesting the d-lysergic acid as a free intermediate in the process of ergopeptine assembly (Keller *et al.,* 1980). Anderson *et al.* (1979) approached the role of d-lysergic acid as a precursor in the ergopeptine assembly by feeding dihydrolysergic acid as a structural analogue of d-lysergic acid to the cultures of ergotamme-producing *Claviceps purpurea* PCCE1. This resulted in the formation of substantial amounts of dihydroergotamine and suggested direct incorporation of the administered dihydroiysergic acid. Figure 17 shows the precursor scheme derived from the results of the feeding studies with precursors of ergopeptines (ergotamine **(32)** taken as example).

5.6.3.
Early Models of Ergopeptine Biosynthesis

A central issue in the pre-enzymological era of research in the enzymatic formation of ergopeptine was to demonstrate their assembly in a step-wise fashion from preformed peptide intermediates. Early hypotheses postulated

Figure 18 d-Lysergyl peptide lactam as intermediate in ergopeptine synthesis and mode of conversion to ergopeptine (Agurell, 1966e; Ramstad, 1968)

a d-lysergyl peptide lactam as product of the assembly process (Agurell, 1966e). Ramstad (1968) extended this by proposing that after enzymatic hydroxylation of the hypothetical d-lysergyl peptide lactam, spontaneous (nonenzymatic) cyclization would yield the corresponding d-lysergylcyclollactam (Figure 18). In view of the structure of the d-lysergyl peptide lactam Agurell (1966a, e) and later Abe (Abe *et al.,* 1971) proposed that d-lysergyl-L-alanine **(33)** should be a precursor of ergotamine or in an extension of this d-lysergyl-L-valine or its methyl ester a precursor of ergocristine, respectively. However, d-lysergyl-L-alanine labelled in the alanine moiety when fed to the ergot cultures, appeared to be cleaved into d-lysergic acid and alanine and then incorporated which did not support the d-lysergyl amino acid hypothesis (Floss *et al.,* 1971c). Similar experiments with d-lysergyl L-valine labelled in the valine moiety led to the same conclusions, in fact the level of its incorporation was that of valine itself indicating cleavage of the d-lysergyl amino acid into its components prior to incorporation (Floss *et al.,* 1971b). Since the possibility could not be ruled out that the d-lysergyl amino acid was cleaved during entry into the *Claviceps* cell into d-lysergic acid and that the labelled d-lysergyl monopeptide did not reach the cellular location of ergot peptide alkaloid synthesis, Gröger and coworkers tested the fate of d-lysergylalanyl ester and d-lysergylvalyl methylesters in both intact mycelia and cell-free extracts (Maier *et al.,* 1974). They found the compounds to be cleaved to d-lysergic acid and alanine and valine, respectively. In addition, trapping

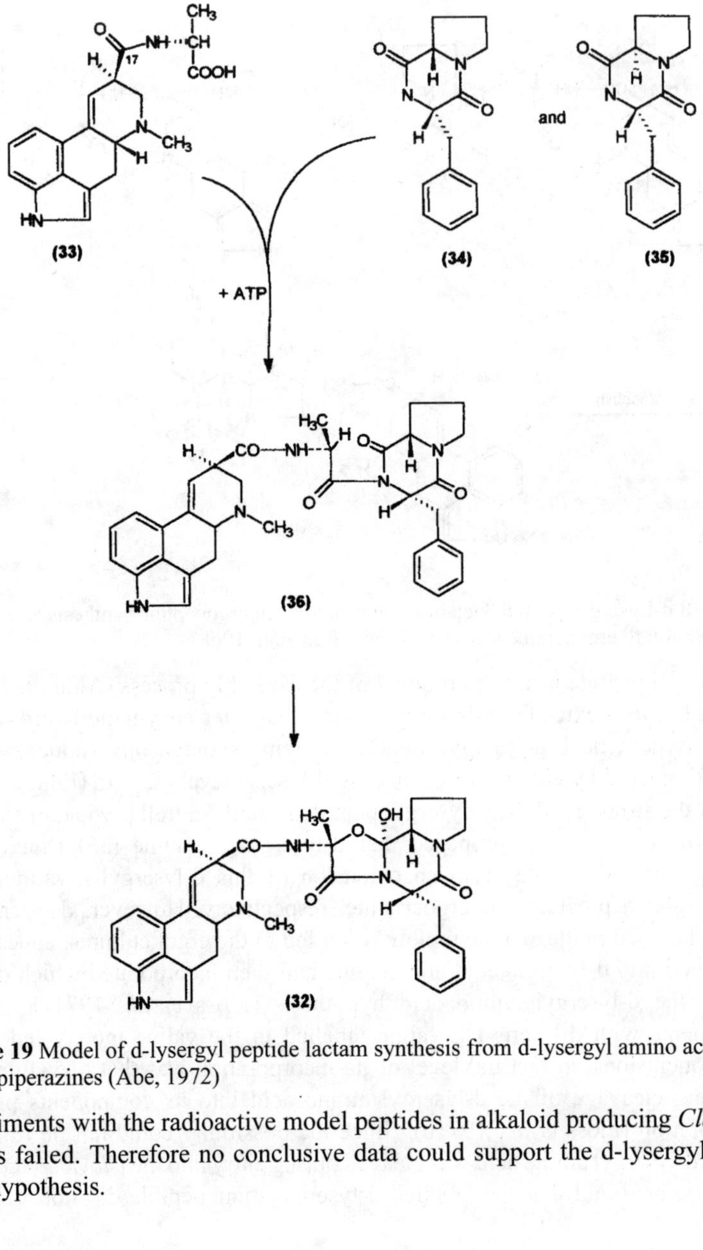

Figure 19 Model of d-lysergyl peptide lactam synthesis from d-lysergyl amino acids and diketopiperazines (Abe, 1972)

experiments with the radioactive model peptides in alkaloid producing *Claviceps* strains failed. Therefore no conclusive data could support the d-lysergyl amino acid hypothesis.

The finding that cultures of *C. purpurea* strains besides ergopeptines accumulate diketopiperazines consisting of proline and its amino-sided amino acid in the ergopeptine sequence (position II of the ergopeptine) led Abe and coworkers to a postulate that diketopiperazines may serve as building blocks in the assembly of the d-lysergyl peptide lactam (Abe *et al.,* 1971; Abe, 1972). In the diketopiperazines, the lactam ring structure of the d-lysergyl peptide lactam is clearly visible, so that their combination with d-lysergyl amino acids would directly yield the proposed structures (Figure 19). However, the model suffered from the fact, that diketopiperazines such as cyclo (leucyl-proline) or cyclo (phenyl-alanyl-proline) (34, 35) isolated from the cultures were always present as pairs of diastereomers containing L-proline and D-proline, respectively. Both diastereomeric compounds were incorporated in equal amounts into ergokryptine/inine or ergotamine/inine (Abe, 1972). However, the optical configuration of proline in ergopeptines is the L-configuration and in view of the preference of proline-containing diketopiperazines to stay in the D-proline configuration (Hofmann *et al.,* 1963; Ott *et al.,* 1963) a mechanism of cyclodipeptide condensation with d-lysergyl amino acids under simultaneous epimerization was hardly to imagine. These results stimulated extensive feeding experiments with various labelled diketopiperazines and open-chain tripeptides (Gröger and Johne, 1972; Floss *et al.,* 1974) as well as d-lysergyl di- and tripeptides (Baumert *et al.,* 1977). Degradation analysis of the alkaloids formed showed in all cases that these peptide had been cleaved into the individual amino acids and then incorporated.

In summary, all of these data virtually exclude the presence of any free intermediate in the formation of d-lysergyl peptides. It was, therefore, suggested that this must take place in a mechanism where all intermediates of d-lysergyl peptide formation remain covalently bound on the surface of a large multienzyme until completion of the final product (Floss *et al.,* 1974a). Such a mechanism would resemble the non-ribosomal synthesis on multifunctional protein templates operating in the synthesis of the peptide antibiotics and related compounds originally elaborated by Lipmann and colleagues (Lipmann, 1971, 1973, 1982). Peptide synthetases catalyse activation and condensations of amino acyl building blocks resulting in peptide formation. Such an enzyme could also operate in the case of the putative d-lysergyl tripeptide lactams. The discovery of the first ergopeptam, d-lysergyl-L-valyl-L-phenylalanyl-D-proline lactam (Figure 20, **(37)**) (Stütz *et al.,* 1973) and of analogous compounds in cultures of several *Claviceps* strains supported this model. The ergopeptams are thought to arise by the readily occurring epimerization from the corresponding L-isomers which were considered as the intermediates of the corresponding ergopeptines (Agurell, 1966e). However, L-isomers of the ergopeptams have never been detected in cultures of *Claviceps* which indicates that they—if they were normal intermediates—are short-lived being converted immediately to the corresponding ergopeptines by hydroxylation of the amino acid in position I. From these considerations, Stadler (1982) concluded that the ergopeptams are dead-end products of the pathway unavailable for the converting enzyme due to the presence of the D-proline in the lactam ring.

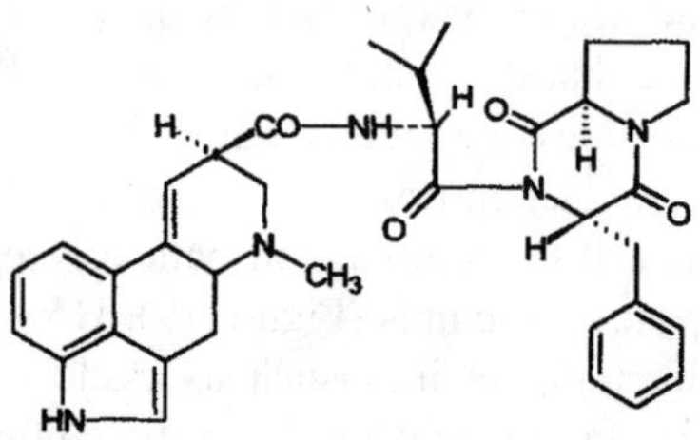

Ergocristam (37)

Figure 20 Structure of d-lysergyl-L-valine-L-phenylalanyl-D-proline lactam (ergocristam (37))

5.6.4.
Non-ribosomal Peptide Synthesis as Model forErgot Peptide Alkaloid Assembly

Non-ribosomal peptide synthesis is catalysed by multifunctional peptide synthetases which activate their amino acid substrates as amino acyl adenylates and bind them as thioesters (Figure 21). Peptide synthetases are huge enzymes which have been recognized early as consisting of amino acid-activating units— rather than of subunits—each responsible for activation and covalent binding in thioester linkage of one individual amino acid of the peptide synthesized (Lipmann, 1973). Originally, the activating units were postulated to be arranged peripheral to a central 4′-phosphopantetheine cofactor covalently attached to an ACP-region of the enzyme operating in analogy to the 4′-phosphopantetheine "swinging arm" involved in fatty acid synthesis catalysed by fatty acid synthase and polyketide synthases (Wakil *et al.,* 1982; Hopwood and Sherman, 1990; Katz and Donadio, 1993). The number of specific amino acid activating units and their sequential arrangement in the enzyme would dictate the number and sequence of amino acids

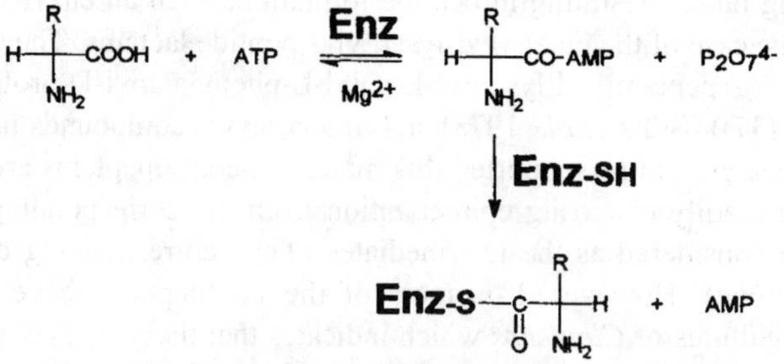

Figure 21 Two step activation of an amino acid substrate in non-ribosomal biosynthesis as aminoacyl adenylate and enzyme thioester. Both steps are catalysed by the same enzyme

which catalyses amino acyladenylate formation (top) and covalently binds its amino acid substrate as thioester (bottom)

in the peptide product and each unit would correspond a 70 kDa equivalent (Lipmann, 1971).

Sequencing of a variety of peptide synthetase genes have revealed that the repeating units of all peptide synthetases are longer than initially believed and have a length of 1100 amino acids equivalent to about 120 kDa (Smith *et al.,* 1990; Turgay *et al.,* 1992). These repeating units (also called modules (Stein and Vater, 1996)) share homology with each other in their amino acids sequences which is particularly high in their central regions comprising some 600 amino acids and is less but still significantly conserved in the interdomain sequences (spacers) which comprise some further 500 amino acids (Figure 22). Biochemical data and comparison of the sequences of various peptide synthetases with the sequences of various acyl-adenylating enzymes identified the 600 amino acid-long most conserved region as responsible for the activation of amino acyl residues as adenylates (Marahiel, 1992). In addition, biochemical and sequence data revealed that peptide synthetases contain more than one 4'-phosphopantetheine cofactor. In fact, every amino acid-adenylating domain has a neighbouring distal region with a highly conserved sequence similar to the attachment site of 4'-phosphopantetheine in fatty acid and polyketide synthases (Schlumbohm *et al.,* 1989; Stein and Vater, 1996) (Figure 22). The amino acyl-adenylating domain in combination with that region is therefore referred to as the peptide synthetase domain (Zocher and Keller, 1997). In fungal peptide synthetases all peptide synthetase domains are assembled into one polypeptide whereas in bacteria they are distributed over more than one protein (Zocher and Keller, 1997). The 4'-phosphopantetheine cofactors serve as primary acceptors of the activated amino acids and as carrier arms transferring amino acyl residues to the next domain where the next amino acid—activated in the same fashion—is ready to react to the dipeptide. The dipeptide will be transferred to the next domain where it is elongated to the tripeptide and so on until the last condensation step will take place on the terminal domain. This reaction sequence is the normal route for peptide chain growth with the direction from the N-terminal to the carboxy-terminal end. Transfer reactions and epimerization reactions catalysed by peptide synthetases have been assigned to the spacer regions regions lying downstream of the peptide synthetase domains (Stachelhaus and Marahiel, 1995). Termination reactions—which are less well understood—proceed either through intramolecular cyclizations of thioesterbound peptide end-products or by release of the end-product with the assistance of thioesterases most probably operating in a similar manner as in fatty acid biosynthesis (Kleinkauf and von Döhren, 1990; Stachelhaus and Marahiel, 1995; Stein and Vater, 1996; Zocher and Keller, 1997).

5.6.5.
d-Lysergic Acid Activating Enzymes

Characteristic feature of the peptide synthetase systems is that the activation reaction of each amino acid substrate, i.e. amino acyl adenylate and thioester

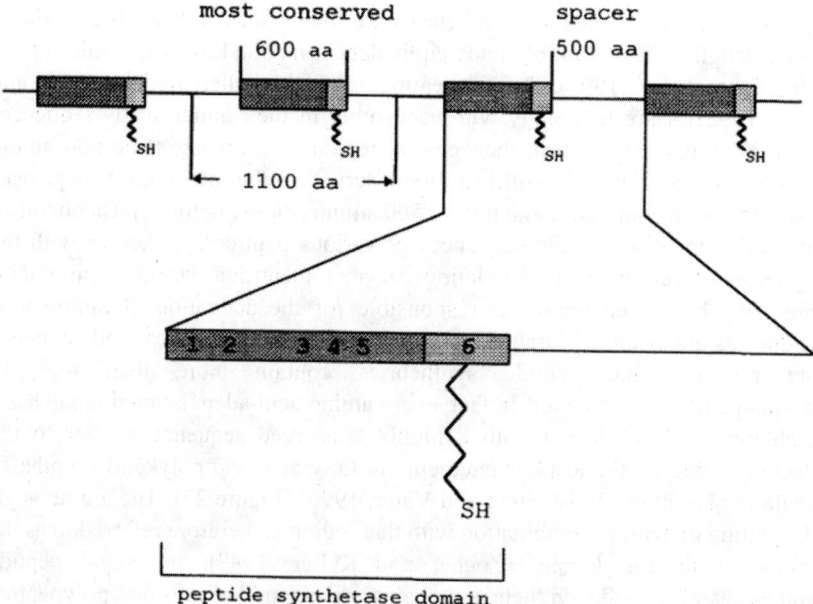

Figure 22 Schematic representation of the assembly of four peptide synthetase domains in a multienzyme peptide synthetase catalysing the synthesis of a tetrapeptide. The model protein consists of four repeating units (modules) which each contain a domain responsible for activation and covalent binding of an amino acid (peptide synthetase domain, shaded boxes). The various peptide synthetase domains share high similarity with each other and are characterized by 6 highly conserved amino acid sequence motifs from which motifs 1–5 are common to many acyl adenylate forming enzymes. The region containing motif 6 is found in all enzymes binding their amino acid substrates as thioesters (Marahiel, 1992). Motif 6 contains a conserved serine which is the attachment site for the covalently bound cofactor 4'-phosphopantetheine. The 4'-phosphopantetheine arms in each domain (indicated by wavy lines) are responsible both for covalent binding of substrate as well as carriers of peptidyl intermediates in the course of peptide synthesis from the N-terminal to C-terminal end. The spacer regions contain the sites responsible for acyl transfers (elongations) in the various condensation reactions. Release of products takes place on the terminal domains by action of cyclizing enzymes or thioesterase activities

formation, can be measured for each individual amino acid substrate separately which may facilitate the detection and isolation of peptide synthetases or peptide synthetase fragments in cell extracts even when a cell-free total synthesis of the corresponding peptide is not obtained (Figure 21). First attempts to characterise partial activities of the enzyme system responsible for assembly of d-lysergyl peptides were undertaken in respect of the activation of d-lysergic acid. As a prerequisite for peptide or amide bond formation, carboxy groups have to be activated prior to peptide bond formation. Therefore in all processes leading to d-lysergyl amides, d-lysergic acid has to be activated in a way presumably via the adenylate as enzyme-thioester as it is generally in the case of non-ribosomal peptide synthesis. An alternative way—in analogy to starter activation in chalcone and resveratrol biosynthesis (Schüz et al., 1983; Schröder et al., 1986) but less likely in a non-ribosomal peptide formation mechanism—would be activation as coenzyme A thioester (Figure 23). Maier et al. (1972) were the first to perform measurements of activation reactions in Claviceps and they reported that cell extracts of Claviceps can catalyse the ATP-dependent formation of d-lysergyl-coenzyme A. However, the level of the enzyme was little correlated with the alkaloid production and it was also present in the strains not producing d-lysergic acid-derived compounds. Purification of the enzyme preparation was not performed and it remained unclear whether the activity was due to one or two activating enzymes because the authors observed during their experiments—using an assay system based on hydroxamic acid formation—also a significant d-lysergic acid activation in the absence of coenzyme A. This is most likely due to activation of d-lysergic acid as adenylate because it is known that acyl- and amino acyl adenylates are reactive in the hydroxamate reaction as are acyl coenzyme A thioesters. The presence of hydroxylamine may strongly influence the specificity of activating enzymes (Eigner and Loftfield, 1974; Stadman, 1957). This observation has not been pursued and in retrospect it cannot be decided whether this activation in the absence of coenzyme A was elicited by the d-lysergic acid-coenzyme A ligase itself or by additional d-lysergic acid activating enzymes.

Keller et al. (1984) purified an enzyme from Claviceps purpurea that activates d-lysergic acid as adenylate as measured by the d-lysergic acid dependent ATP-pyrophosphate exchange (Figure 23). The enzyme did not activate dihydrolysergic acid and did not catalyse formation of d-lysergyl coenzyme A. No thioester formation between enzyme and d-lysergic acid was observed and none of the amino acid of the peptide portion of ergotamine was activated. The enzyme was partially purified and its native M_r was estimated to be in the range of 140000. In the light of the efficient in vivo dihydrolysergic acid incorporation into dihydroergotamine it is difficult to assign this enzyme a function in d-lysergyl peptide assembly. In an extension of this study, d-lysergic acid-activating enzyme was later purified to homogeneity (Keller et al., 1988). In contrast to the previous findings, enzyme displayed a native molecular mass 240–250 kDa instead of the earlier observed value of 140 kDa which was hard to understand. However, under denaturing conditions enzyme was identified as a 62 kDa band in SDS gels which suggested that in native conditions it is a tetramer. In the retrospect it cannot be decided whether the 62 kDa protein is a breakdown product of the previous 140 kDa protein because no size determinations of the latter protein have been performed under denaturing conditions (Keller et al.,

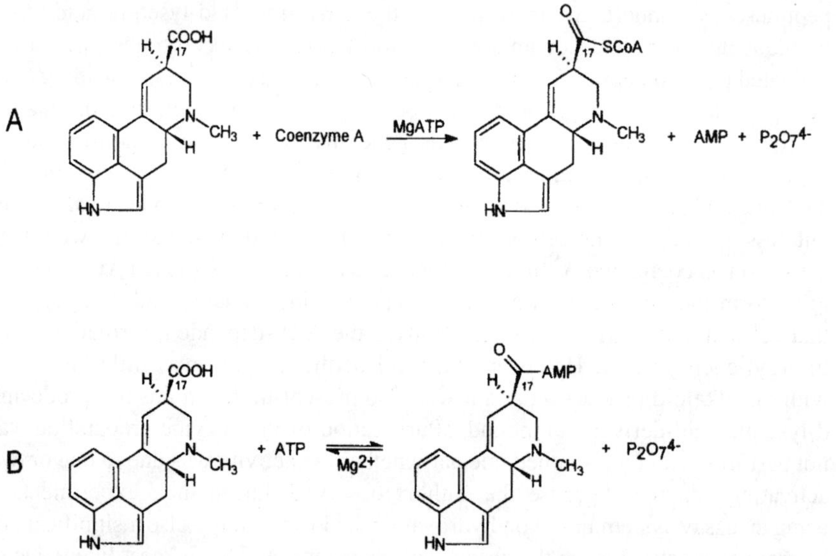

Figure 23 Activation of d-lysergic acid as d-lysergyl coenzyme A ester (A). Activation of d-lysergic acid as d-lysergyl adenylate (B)

1984). The d-lysergic acid-activating enzyme in its appearance during cultivation was little correlated with alkaloid production and it was found in comparable specific activities in strains selected for high production as well as in their wild type parent suggesting that the d-lysergic acid-activating enzyme may not be subject to the mechanisms governing regulation of ergot alkaloid production in these strains. The specific activity was significantly lower but still present when the strains were grown in the media in which alkaloid production does not take place (Keller *et al.,* 1988; Riederer and Keller, unpublished data). These data suggested that the enzyme—like the one catalysing d-lysergyl coenzyme A thioester (Maier *et al.,* 1972)—could have an unknown function in d-lysergic acid metabolism in *Claviceps purpurea.*

5.6.6.

Enzymatic Synthesis Systems of Ergopeptines andd-Lysergyl Peptide Lactams

First reports on the cell-free assembly of ergopeptines came from Abe's group who demonstrated the formation of ergokryptine and ergotamine in extracts from *C. purpurea* from preformed d-lysergyl amino acids and diketopiperazines. The enzyme system was not characterised and thus the protein components and the mechanism of formation remained obscure (Abe *et al.,* 1971; Abe, 1972; Ohashi *et al.,* 1972).

In a series of reports, Gröger's group demonstrated the ATP-dependent incorporation of individual amino acids such as L-[U-^{14}C]-leucine into ergosine/-inine in the 15000 g-supernatant of a cell homogenate obtained from ergosine-producing *Claviceps purpurea* Mut 168/2 (Maier *et al.*, 1981a). Remarkably, incorporation of the amino acid into product was strongly stimulated by the presence of elymoclavine and even more by agroclavine rather than by addition of d-lysergic acid suggesting a direct role of the clavines in the formation of the peptide alkaloid and hence, in the formation of the activated form of d-lysergic acid. Comparable results were later obtained in respect of incorporation of L-[U-^{14}C]-phenylalanine and, to a much lesser extent of L-[U-^{14}C]-alanine into ergotamine by the extracts from an ergotamine-producing *Claviceps* strain (Maier *et al.*, 1981b). Purification of the ergotamine-synthesising activity from the latter strain by ammonium sulphate precipitation, ion exchange chromatography, gel permeation chromatography and chromatography on hydroxylapatite revealed that the activity in each purification step always resided in one single peak. The enzyme system kept its dependence on agroclavine or elymoclavine up to the highest purification step (Maier *et al.*, 1985). These findings suggested co-purification of a clavine-converting activity and of the enzyme responsible for alanine hydroxylation residing in a multienzyme complex (cyclol synthetase). The final purification of the enzyme was 172 fold. The size of the multienzyme complex was estimated to be 195000±5000 by gel filtration (Maier *et al.*, 1983). Size and purity of the polypeptides in denatured form from this protein fraction were not examined. Concerns about the significance of this multienzyme complex are based on the fact that—from theoretical considerations—the enzyme is too small to account for so many functions involved in the synthetic process. Apart from the clavine-converting activities the mere presence of four activation domains each having 100–120 kDa size equivalent and additional enzyme activities would predict a size of 500–600 kDa instead of the observed 195 kDa. Furthermore, the reaction conditions as described indicate that neither cell-free extract not the purified enzyme need any cofactor such as NADH or NADPH normally necessary for mixed-type oxygenations of agroclavine/ elymoclavine and for formation of the α-hydroxy amino acid of cyclol-portion (Floss *et al.*, 1967; Belzecki *et al.*, 1980; Quigley and Floss, 1981; Kim *et al.*, 1983; Maier *et al.*, 1988). These unusual substrate requirements as well as the obvious failure of d-lysergic acid incorporation implicates a too much complicated reaction mechanism for a suitable working model of ergot peptide assembly.

Our working hypothesis was based on the assumption that d-lysergyl peptides are assembled from the free amino acids and free d-lysergic acid (Keller *et al.*, 1984; Keller, 1985; Keller *et al.*, 1988). We detected the ATP-dependent incorporation of d-lysergic acid, alanine, phenylalanine and proline into two d-lysergic acid containing peptides catalysed by an enzyme fraction from ergotamine-producing *C. purpurea* (Keller *et al.*, 1988). Neither of the two peptides was identical with ergotamine and they were not formed when d-lysergic acid was omitted. The structures of the compounds were determined by alkaline and acid hydrolysis after cell-free incorporation of ^{14}C-L-alanine, ^{14}C-L-phenylalanine or ^{14}C-L-proline. Both compounds contained radioactive L-alanine, L-phenylalanine and D-proline, respectively. From this it was concluded that the two compounds **36** and **38** were non-cyclol d-lysergyl tripeptides because if they were ergopeptines one should have obtained radioactive pyruvate instead of the alanine (Figure 24). Furthermore, the exclusive recovery of D-proline in the hydrolysate suggested that in both compounds the proline was likely to be part of a lactam structure.

Base-catalysed methanolysis of the compounds as has been done previously in the analysis of the first isolated ergopeptam, ergocristam (Stütz *et al.,* 1973) showed that both d-lysergyl peptides yielded exclusively L-phe-D-pro diketopiperazine which strongly suggested that both compounds were stereoisomeric d-lysergyl peptide lactams with L- or D-proline in the lactam rings, respectively (Figure 24). The possibility that the two compounds were different epimers in respect of the C-8 of the d-lysergic acid moiety (d-lysergic acid and isolysergic acid derivatives) could be ruled out because the analogous compounds were also formed when dihydrolysergic acid was used for their synthesis instead of d-lysergic acid. Dihydrolysergic acid, however, does not epimerize at C-8 which virtually provides strong evidence that the two compounds are epimeric in respect to the chiral centre in the peptide portion and not in the ergoline ring. Thus, these results firmly established the structures of these compounds as d-lysergyl-L-alanyl-L-phenylalanyl-L-proline lactam **(36)** and d-lysergyl-L-alanyl-L-phenylalanine-D-proline lactam **(38)** (Figure 24). The exclusive formation of the compound in the cell-free system also suggested that hydroxylation of the peptide is a later step in ergopeptine synthesis catalysed by a separate modifying enzyme.

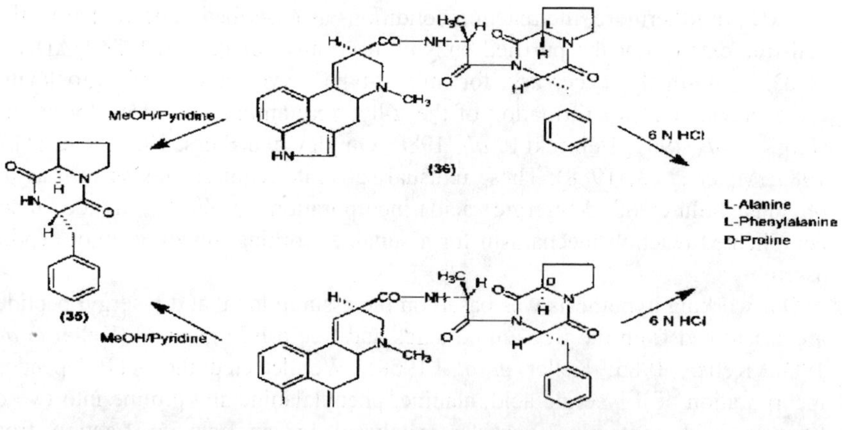

Figure 24 Analysis of *in vitro* products of d-lysergyl peptide synthetase

The formation of the D-proline epimer by cell-free preparation poses the question whether it is formed in the course of enzymatic synthesis or whether it is derived from the L-proline containing isomer. Experiments revealed that the ratio of the amounts of the two compounds formed was dependent on the quality of the preparations. Predominant formation (90%) of the L-proline isomer has been seen only in a few rare cases when the crude enzyme preparation had an exceptional high level of synthetic activity (Keller, unpublished data). This suggests that the D-proline containing isomer may be formed as a result of suboptimal reaction conditions and this may have implications on formation of the compound *in vivo*.

5.6.7.
d-Lysergyl Peptide Synthetase, the Enzyme
CatalysingAssembly of d-Lysergyl Peptide Lactams

The *in vitro* system of d-lysergyl peptide lactam synthesis was the starting point to address the enzymatic basis and the reaction mechanism of that process. Purification of the d-lysergyl peptide lactam synthesising enzyme proved to be difficult when compared to other peptide synthetase systems from fungi such as enniatin synthetase or cyclosporin synthetase (Zocher *et al.*, 1982; Zocher *et al.*, 1986; Lawen and Zocher, 1990). The half-life time of enzyme was only some 10–20 hrs in the presence of high concentrations of glycerol which stabilised the enzyme. Riederer *et al.* (1996) reported an 18 fold partial purification of the enzyme activity which was achieved by fractionated polymin P, ammonium sulphate precipitation, gel filtrations and ion exchange chromatography. d-Lysergyl peptide lactam synthesising activity always eluted from columns as one single peak which suggested that the enzyme was one single polypeptide chain or a multienzyme complex. From the elution behaviour from gel filtration columns the native M_r was estimated to be much larger than 300 kDa. In addition to its ability to catalyse synthesis of the d-lysergyl peptide lactams, enzyme catalysed individual activation of d-lysergic acid (or dihydrolysergic acid) and of each of the amino acids of the peptide portion as adenylates (measured by the amino acid or d-lysergic acid dependent ATP-pyrophosphate exchange). Moreover, binding assays for each of the substrates showed that d-lysergic acid (or dihydrolysergic acid) and the amino acids were covalently bound to enzyme in thioester linkage. This suggested that the enzyme consists of four peptide synthetase domains (modules) each harbouring a covalently attached 4'-phosphopantetheine cofactor for binding of the relevant amino acids or d-lysergic acid, (see scheme in Figure 22).

The ability of the enzyme to bind covalently d-lysergic acid or dihydrolysergic and the amino acids of the peptide portion of the alkaloid cyclopeptides was used to label the polypeptide with radioactive amino acids and dihydrolysergic acid and to identify the corresponding band by autofluorography of polyacrylamide gels in SDS-polyacrylamide gels. The radioactive amino acids alanine, phenylalanine or proline significantly labelled a most prominent 370 kDa protein. By the contrast, radioactive dihydrolysergic acid labelled a 140 kDa protein distinct from the 370 kDa band, which clearly showed that the enzyme—which was designated d-lysergyl peptide synthetase (LPS)—consisted of two protein components. The larger 370 kDa polypeptide (LPS 1), activates and assembles alanine, phenylalanine and proline. The 140 kDa polypeptide is responsible for d-lysergic acid assembly (LPS 2). The total size of LPS of between 500 kDa and 520 kDa was in full accordance with the previous theoretical predictions derived from other non-ribosomal systems. Accordingly, analysis of LPS 1 and 2 revealed that they both contain 4'-phosphopantetheine as covalently bound cofactor identifying them as peptide synthetases assembling d-lysergyl peptide lactam in a non-ribosomal mechanism (Riederer et al., 1996).

5.6.8.
Substrate Specificities of d-Lysergyl Peptide Synthetase
(LPS)

The d-lysergic acid and amino acid activating domains of LPS differ considerably in their K_m-values under synthetis conditions (Table 4). The K_m-values for d-lysergic acid and dihydrolysergic were both 1.5 µM suggesting a very high affinity of the enzyme for the two ergoline-ring carboxylic acids. The K_m-values for the amino acids were higher and their differences reflect the different affinities of the three amino acid activating domains of the enzyme for their substrates. The K_m-value for phenylalanine was 15 µM and that for alanine and proline 200 µM and 120 µM, respectively. This appears to be a reason for the uneven labelling observed in the *in vivo* and *in vitro* incorporation experiments with radioactive L-alanine, L-phenylalanine and L-proline when radioactive substrates were not present at saturation (Keller et al., 1980; Keller et al., 1988).

To address the substrate specificities of each individual amino acid activation domain, LPS was tested to produce other d-lysergyl peptide lactams than d-lysergyl-L-alanyl-L-phenylalanine-L-proline lactam. Incorporation experiments with ^{3}H-dihydrolysergic acid, proline and various combinations of amino acids occurring in position I and II of ergopeptines revealed synthesis of various dihydrolysergyl peptide lactams. Thus, with phenylalanine being present as the substrate for position II, addition of alanine, aminobutyric acid or valine led to the formation of corresponding d-lysergyl peptide lactams of the ergotamine, ergoxine or ergotoxine group, respectively (Riederer et al., 1996). Similarly, enzyme synthesised an analogous series when phenylalanine was replaced by leucine in these incubations (Figure 25). Complex product mixtures were obtained in the case of combinations of valine/aminobutyric acid and valine/ alanine (Keller, unpublished data). Valine and aminobutyric acid both competed for incorporation into position I and position II which theoretically should yield 4 diastereomeric pairs of d-lysergyl peptide lactams. In the case of adding alanine and valine, valine competes with alanine for incorporation in position I which leads to two pairs of diastereomers. In summary, the data indicate a broad substrate specificity of the domain responsible for position I of the alkaloid

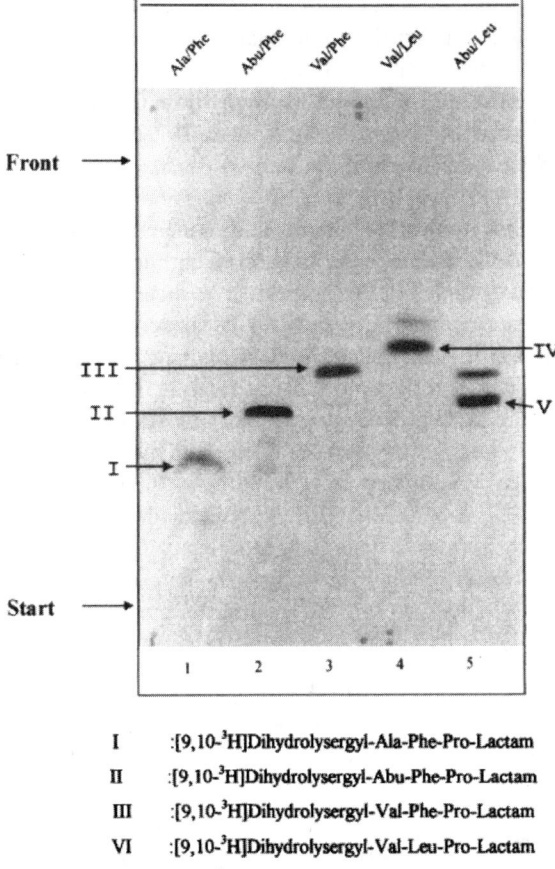

I	:[9,10-³H]Dihydrolysergyl-Ala-Phe-Pro-Lactam
II	:[9,10-³H]Dihydrolysergyl-Abu-Phe-Pro-Lactam
III	:[9,10-³H]Dihydrolysergyl-Val-Phe-Pro-Lactam
VI	:[9,10-³H]Dihydrolysergyl-Val-Leu-Pro-Lactam
V	:[9,10-³H]Dihydrolysergyl-Abu-Leu-Pro-Lactam

Figure 25 Synthesis of various d-lysergyl peptide lactams catalysed by d-lysergyl peptide synthetase (LPS). Compounds formed from [9–10-³H]dihydrolysergic acid, proline and different combinations of two amino acids occurring in the position of the natural ergopeptines were extracted from reaction mixtures, separated by thin-layer chromatography and analysed. The structures of the indicated bands are given. The indicated lanes contain products from reactions mixtures of LPS with [9–10-³H] dihydrolysergic acid, proline, ATP and alanine+phenylalanine (1), α-aminobutyric acid (Abu)+phenylalanine (2), valine+phenylalanine (3), valine+leucine (4), α-aminobutyric acid+leucine. All amino acids were present in saturating conditions

cyclopeptide. This also indicates that the spectrum of products formed *in vivo* is dependent on the actual concentrations of relevant amino acids in the cellular pool.

The domain for position II was found to have strong preference for phenylalanine which is incorporated efficiently already in the 10 μM range and displaces leucine, valine or α-aminobutyric acid from the enzyme. Leucine and valine have a higher K_m-value than phenylalanine which explains the strongly reduced incorporation of these two amino acids in the presence of phenylalanine. Nevertheless, good incorporation of the two amino acids at saturating concentrations into the position II of the corresponding dihydrolysergyl peptide lactam in the absence of phenylalanine was seen (Riederer et al., 1996).

Proline which is an invariable amino acid in the ergopeptines elaborated by C. purpurea has a K_m-value one order of magnitude higher than that of L-phenylanine. Nevertheless, the peptide synthetase domain catalysing its incorporation into product appears to be specific for proline. Structural homologues of proline such as sarcosine, azetidine-2-carboxylic acid or pipecolic acid could not replace proline in d-lysergyl peptide lactam synthesis (Walzelet et al., 1997). Examples for replacement of proline in ergopeptines by structural analogues are the in vivo incorporations of thiazolidine-4-carboxylic acid into position III of ergosine produced by C. purpurea (Baumert et al., 1982) and of 4-hydroxyproline into position III of dihydroergosine produced by Sphacelia sorghi (Atwell and Mantle, 1981). This suggests that a five-membered ring may be necessary for substrate recognition by the proline domain of LPS. It is worth to mention that ergobalansine, an ergopeptine from Balansia species contains alanine in the position III of the

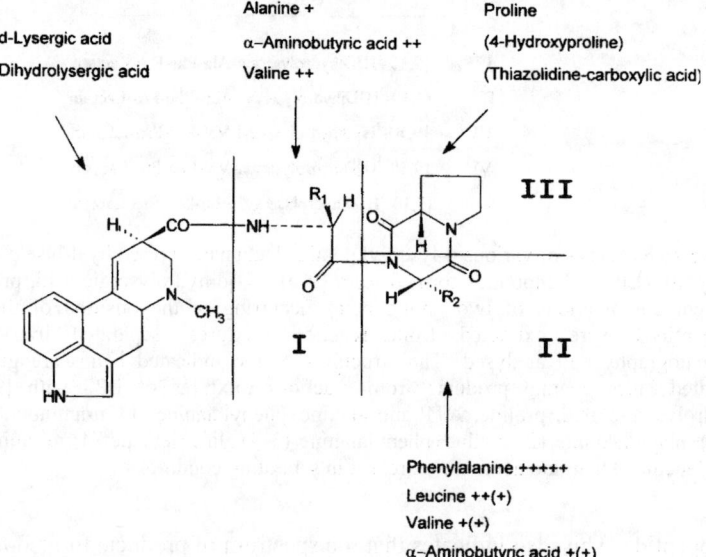

Figure 26 Substrates incorporated into d-lysergyl peptide lactams by d-lysergyl peptide synthetase and the relative affinities of the various domains for substrates on a K_{mapp}-basis in synthesis conditions

peptide portion instead of the proline (Powell *et al.,* 1990). In addition, the compound has alanine and leucine in positions I and II, respectively. It will be interesting to compare the two alanine activating domains of the corresponding peptide synthetase with the corresponding ones of LPS when the sequences of the enzymes will be available in the future.

In summary, the *in vitro* studies of Riederer *et al.* (1996) and further unpublished work of our group (Walzel and Keller, unpublished data) indicate that d-lysergyl peptide synthesis *in vivo* is governed by the actual concentrations of free d-lysergic acid and the free amino acid in the cellular pool which may explain the production of different ergopeptines in the various strains of *C. purpurea* strains. d-Lysergic acid, due to its low K_m, plays a pivotal role in the ergopeptine synthesis triggering (Figure 26).

5.6.9.
Constitutive Regulation of d-Lysergyl PeptideSynthetase
(LPS) in *Claviceps purpurea*

Riederer *et al.* (1996) measured the level of LPS in various derivatives of *C. purpurea* ATCC 20102 by Western blot analysis of cell extracts using anti-LPSl antibodies. Despite of the different ergotamine productivities of these strains ranging from 10 mg I^{-1} up to 1000 mg I^{-1} (2 weeks of cultivation), there were no remarkable differences in the amount of the immunoreactive material. A possible reason for the different productivities of the various strains was therefore supposed differences of the d-lysergic acid supply in the different strains. In fact, it was shown that ergotamine production of *C. purpurea* wild type strain ATCC 20102 (low producer) could be considerably stimulated by the external addition of d-lysergic acid to the protoplasts or to the intact mycelium whereas strain *C. purpurea* D1 (high producer) did not respond to d-lysergic acid in these experiments. This suggested that in this strain the intracellular level of d-lysergic acid was at saturation and therefore considerably higher than in its parental strain (Keller *et al.,* 1980; Riederer *et al.,* 1996). A constitutive expression of LPS1 (and most probably of LPS2) was inferred which suggests that regulation of alkaloid synthesis resides at some step of d-lysergic acid synthesis, presumably at the stage of DMAT synthase. DMAT formation is subject of regulation by tryptophan and phosphate analogously as is formation of ergot alkaloids themselves (Robbers *et al.,* 1972; Krupinski *et al.,* 1976). Confirming results were obtained when levels of LPS in *C. purpurea* grown in different media promoting or suppressing alkaloid formation were determined. The presence of LPS 1 in alkaloid-producing and non-producing conditions as well as in sclerotia-like cells and vegetative hyphae points to constitutive synthesis of LPS in *C. purpurea.*

5.6.10.
Reaction Mechanism of d-Lysergyl Peptide Assembly

To address the mechanism of d-lysergyl peptide formation, Walzel *et al.* (1997) isolated enzyme-bound radioactive reaction products formed after incubation of LPS with radiolabelled dihydrolysergic acid and with the successive addition of non-labeled alanine, phenylalanine and proline in the presence of ATP. After liberation from the enzyme by cleavage with alkali, HPLC analysis of the hydrolytic products and a comparison with chemically synthesized dihydrolysergyl peptides revealed the following steps of d-lysergyl peptide assembly.

Initiation of d-Lysergyl Peptide Synthesis

In the absence of the amino acids, LPS binds radioactive dihydrolysergic acid in substantial amounts as expected. In the additional presence of non-labeled alanine both radioactive dihydrolysergic acid and dihydrolysergyl-alanine were found to be attached to the LPS indicating that initiation of d-lysergyl peptide lactam synthesis starts with the activation of d-lysergic acid and the formation of d-lysergyl-alanine. (Figure 27). Furthermore, it could be demonstrated that the LPS is able to catalyze formation of enzyme-bound thioesters

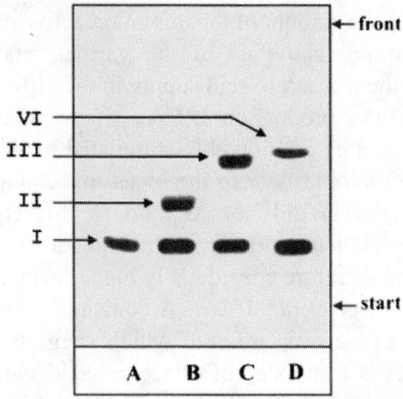

I   :[9,10-³H]Dihydrolysergic acid

II :[9,10-³H]Dihydrolysergyl-Alanine

III :[9,10-³H]Dihydrolysergyl-α-Aminobutyrate

IV :[9,10-³H]Dihydrolysergyl-Valine

Figure 27 Initiation of d-lysergyl peptide lactam synthesis catalysed by LPS. Formation of dihydrolysergyl amino acids of the ergotamine, ergoxine and ergotoxine groups of alkaloid cyclopeptides. Each track represents thin-layer chromatographic separation of accumulating compounds on the surface of LPS when incubated with [9–10–3H] dihydrolysergic acid, MgATP; B: in the additional presence of alanine; C: in the additional presence of α-aminobutyric acid; D: in the additional presence of valine. The structures of the dihydrolysergyl amino acids are as indicated, (with permission of Current Biology Ltd.)

of dihydrolysergyl-α-aminobutyrate and dihydrolysergyl-valine when incubated with dihydrolysergic acid and α-aminobutyric acid or valine, respectively (Figure 27). Their formation is consistent with the incorporation of α-aminobutyric acid and valine into the ergoxine and ergotoxine related peptide lactams catalyzed by LPS. Furthermore this indicates that the ergoxines and ergotoxines are synthesized by the same mechanism as the ergotamines.

Elongation of d-Lysergyl Peptide Synthesis

In the additional presence of the second amino acid of the ergot peptide portion (position II) such as phenylalanine, LPS was shown to contain covalently bound dihydrolysergic acid, dihydrolysergyl-alanine and dihydrolysergyl-phenylalanine. In the additional presence of proline in the reaction mixture much smaller amounts of the enzyme-bound peptide intermediates and nearly no dihydrolysergic acid was seen on the enzyme's surface whereas free dihydrolysergyl peptide lactams were formed at substantial levels. This suggested a low steady state concentration of accumulating intermediates when all substrates are present allowing synthesis of end product to progress.

Proline Incorporation and Termination Through Cyclization

To show that the linear d-lysergyl-L-alanyl-L-phenylalanyl-L-proline is the last intermediate prior to release of the tripeptide lactam, Walzel et al. (1997) used a partially deactivated LPS which is unable to catalyze dihydrolysergyl peptide lactam formation but still can activate d-lysergic acid or dihydrolysergic acid and the amino acids of the peptide portion. Interestingly, this enzyme accumulated besides dihydrolysergyl-alanine and dihydrolysergyl-alanylphenylalanine an additional dihydrolysergyl peptide when it was incubated with dihydrolysergic acid in the presence of alanine, phenylalanine and proline (Figure 28). This compound was considered to be dihydrolysergylalanyl-phenylalanyl-proline, which in turn was confirmed by the demonstration in parallel experiments of incorporation of ^{14}C-proline into the same compound. These data clearly indicated that the inability of the deactivated enzyme to catalyze peptide lactam formation was due to loss of its lactamcyclization activity. Therefore, the synthesis of alkaloid cyclopeptides proceeds via the lysergyl mono-, di-, and tripeptides and not in the opposite direction from the carboxy- to the N-terminal end as it was suggested earlier (Floss et al., 1974a).

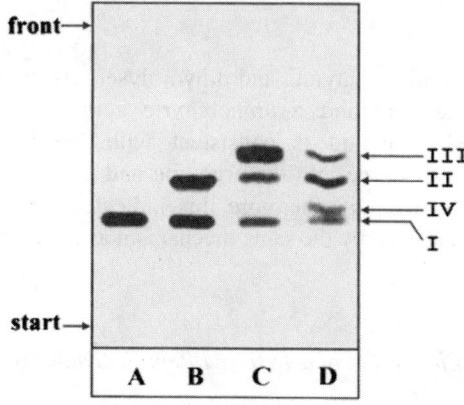

I   :[9,10-³H]Dihydrolysergic acid

II   :[9,10-³H]Dihydrolysergyl-Ala

III   :[9,10-³H]Dihydrolysergyl-Ala-Phe

IV   :[9,10-³H]Dihydrolysergyl-Ala-Phe-Pro

Figure 28 Intermediates accumulating on a modified d-lysergyl peptide synthetase unable to release d-lysergyl peptide lactam as end product. The indicated lanes are TLC separations of enzyme-bound peptide intermediates isolated from enzyme incubated with A. [9–10–³H]dihydrolysergic acid, MgATP; B. [9–10–³H]dihydrolysergic acid, MgATP and alanine; C. [9–10–³H]dihydrolysergic acid, MgATP, alanine and phenylalanine; D: [9–10–³H]dihydrolysergic acid, MgATP, alanine, phenyalanine and proline. The structures of the various substances are as indicated (with permission of Current Biology Ltd.)

Successive Acyl Transfers Between the d-Lysergic Acid and Amino Acid Activating Domains of LPS

Walzel *et al.* (1997) determined the specific functions of LPS 1 and LPS 2 by incubating the two enzymes with ³H-dihydrolysergic acid and different combinations of alanine, phenylalanine and proline. The enzyme subunits were then separated by sodium dodecyl suifate-polyacrylamide gel electrophoresis (SDS-PAGE) and the location of dihydrolysergic acid and intermediates was analyzed by autofluorography. Addition to LPS of ³H-dihydrolysergic acid alone led to the exclusive labeling of LPS 2, consistent with its role as the d-lysergic acid activating subunit of LPS. By contrast, the addition of nonlabelled alanine or alanine plus phenylalanine resulted in the additional labeling of LPS 1 indicating that the radioactive label from dihydrolysergic acid moved from LPS 2 to LPS 1 in an alanine-dependent acyltransfer reaction. Extent of labeling of LPS increased when phenylalanine was added together with alanine. Protein-bound label disappeared both from LPS 1 and LPS 2 when proline was added allowing completion of the reaction and release of the final product. These data led to the scheme of events catalyzed by LPS shown in Figure 29.

A. Initiation

B. elongation step 1

C. elongation step 2

D. lactamisation and product release

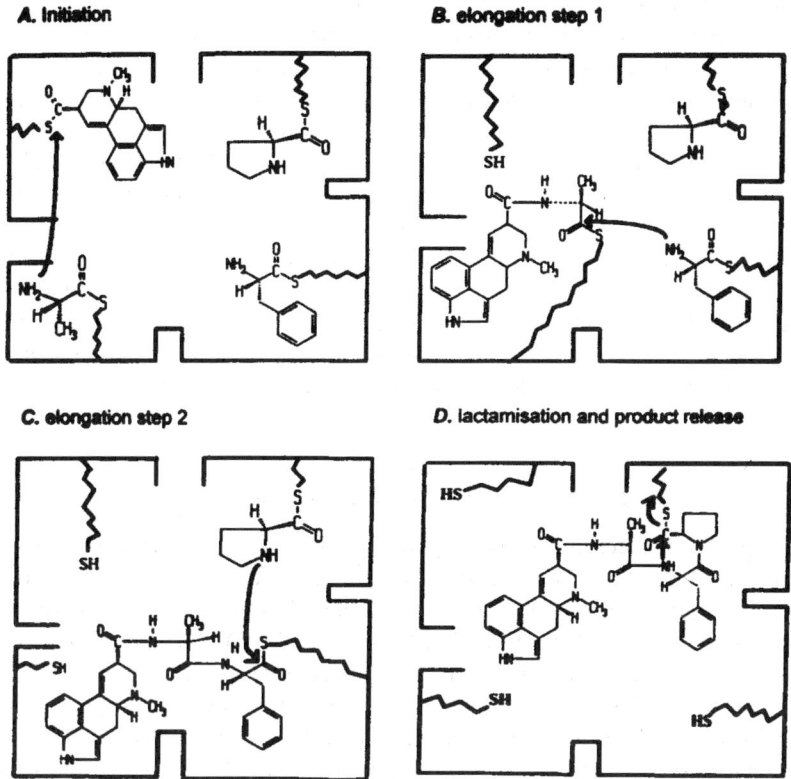

Figure 29 Scheme of the events catalysed by LPS in d-lysergyl peptide formation. The wavy lines indicate the 4'-phosphopantetheine arms of the various peptide synthetase domains of LPS. Note that the upper left corner in each diagram represents the active site of LPS 2 while the other three corners represent the three activation centers of LPS 1

Mechanistic Regulation of LPS

Summarizing these and other data of the work of Walzel *et al.* (1997) not explicitly mentioned here led to conclude that the synthesis of d-lysergyl peptide lactam is a sequential ordered process carried out by four domains which catalyze four separate activation reactions of d-lysergic acid and the three amino acids of the peptide portion. The domains cooperate in three acyltransfer reactions leading to thioester-bound peptide intermediates. In addition the enzyme (presumably in the last C-terminal domain) has an intrinsic lactamizing enzyme activity resembling the cyclizations in the biosynthesis of other microbial cyclopeptides and peptide lactones (Zocher and Keller, 1997). The irreversibility of peptide-thioester formation appears to dictate the direction of chain growth to form the end product. d-Lysergic acid is the primer of the peptide synthesis, in its absence no peptide bond formation between amino acids takes place. An uneven distribution of accumulating intermediates indicates constraints in the accessibility of the sites involved in the priming step due to steric hindrance of the bulky peptidyl intermediates formed. This may cause the drastically reduced binding of d-lysergic acid and of early peptide intermediates as the reaction is allowed to progress. The absence of covalently bound dihydrolysergic acid and intermediates in the presence of all substrates reveals a strict programming of the condensation reactions. This directs the process to end product formation and indicates that priming (entry) of d-lysergic acid into the reaction cycle occurs only after the previous round of d-lysergyl peptide lactam synthesis is completed. These data illustrate the role of d-lysergic acid in the programming of the events on LPS which fits with the previous observations of its role in the regulation of *in vitro* formation of alkaloid cyclopeptides.

5.6.11.
LPS—One Polypeptide or Two Subunits?

The gene sequences of fungal peptide synthetases which have been so far obtained indicate that these enzyme consist of one polypeptide chain while bacterial peptide synthetase systems are composed of more than one polypeptide chain (Zocher and Keller, 1997). Riederer *et al.* (1996) pointed out that LPS 1 and LPS 2 could be an exception from this rule. However, they mention the possibility that LPS 1 and LPS 2 have arisen by a proteolytic cleavage of a 500–520 kDa peptide synthetase. The 550 kDa HC-toxin synthetase from the phytopathogenic fungus *Cochliobolus carboneum* was found in protein extracts from the fungus in the form of two 200–300 kDa fragments (Scott-Craig *et al.,* 1992). Molecular cloning and sequencing of the HC-toxin synthetase revealed, however, that it encoded a 550 kDa enzyme consistent with proteolytic cleavage of the polypeptide into two fragments. In view of the possible generation of LPS2 through proteolysis from a 500 kDa polypepide it cannot be excluded that even the previously isolated d-lysergic acid activating enzyme (Keller *et al.,* 1984) is a product of further degradation of LPS 2. Both enzymes are most likely to be constitutively expressed. These questions, however, can be ultimately solved only when the gene sequences of these enzymes were available.

The molecular cloning of the gene of LPS has not been reported, yet. Panaccione (1996) has cloned DNA-fragments representing different peptide synthetase domains from *Claviceps purpurea* and from the grass-paraziting endophytic fungus *Acremonium coenophialum*. These sequences have been generated by PCR using oligonucleotides designed on the basis of highly conserved motif common to all known peptide synthetase domains. Surprisingly, hybridization screenings of a *C. purpurea* λ-gene bank with these fragments revealed that *C. purpurea* harbors different peptide synthetase genes which makes it difficult to assign which of them will code for LPS. One of the fragments hybridizes significantly with one of the peptide synthetase domains from *Acremonium* which makes this a candidate for the ergopeptine synthetase because the fungus produces ergopeptines as does *C. purpurea*. The results indicate that ergopeptine-producing fungi contain multiple families of peptide synthetase genes (Panaccione, 1996). Similar results have been obtained in our group (Riederer, Grammel, Schauwecker and Keller, unpublished data).

5.6.12.
Conversion of d-Lysergyl Peptide Lactams to Ergopeptines

Floss and coworkers have shown that introduction of the hydroxyl group into the amino acid at position I of the ergopeptines is by action of mixed-function oxygenases (Belzecki *et al.,* 1980; Quigley and Floss, 1981). It was shown that the oxygen of the cyclol bridge is derived from molecular oxygen and not from water. Cell-free cyclol synthesis has been reported by Gröger's group (Maier *et al.,* 1985), however the mechanism of introduction of the cyclol oxygen was not addressed. The fact, that LPS catalyses exclusively the formation of the d-lysergyl peptide lactam indicates that the oxygenase is a separate enzyme distinct from LPS (Walzel and Keller, unpublished data).

5.7.
BIOSYNTHESIS OF NON-CYCLOL d-LYSERGIC ACID PEPTIDES AND AMIDES

The features of non-ribosomal d-lysergyl peptide synthesis outlined in the previous section strongly suggest that prior to amide bond formation, d-lysergic acid has to be activated. Analysis of the reaction mechanism of LPS shows that in ergopeptine synthesis d-lysergic acid is activated as adenylate and thioester. The modular structure of LPS and the apparent combinatorial diversity of d-lysergic acid containing peptides and amides implicate a similar mechanism for the synthesis of other d-lysergic acid peptides and amides presented in Table 1. The d-lysergic acid activating domain of these systems should be structurally and functionally related to that one located on LPS 2. The fact that d-lysergic acid does not occur in free form suggests this kind of activation to be present in any *Claviceps* strain producing d-lysergic acid amides regardless whether the amide component is an amine or a peptide chain. Furthermore, high specificity of the LPS 2 activation domain for d-lysergic acid and dihydrolysergic acid of peptide alkaloid producing *Claviceps* suggests that this domain has exclusively evolved in the genera able to convert clavines to ergolene-8-carboxylic acids (i.e. paspalic acid as the precursor of d-lysergic acid). It also suggests that the mechanism of peptide bond formation would be similar and would require additional activating domains serving as templates for the amide components.

5.7.1.
d-Lysergyl Dipeptides—Ergosecaline

Ergosecaline **(6),** which was isolated by Abe *et al.* (1959)—as a shorter version of ergopeptines—would be an interesting model system to further elucidate the mechanisms of d-lysergyl peptide formation. Unfortunately, biosynthetic studies on that compounds have not been performed, yet, but in the view of its similarity with ergovaline it appears likely that it should be synthesized in a similar mechanism as that one catalyzed by LPS. LPS is able to synthesize the ergovaline-related d- lysergyl peptide lactam (Riederer *et al.,* 1996). The domain assembly would resemble a shortened d-lysergyl peptide synthetase with three domains for activation of d-lysergic acid, alanine (position I) and valine (position II) and catalyze formation of enzyme-bound d-lysergyl-alanylvaline as end product (Figure 30). Ergosecaline synthetase would have no functional proline domain or completely lacking it. This would imply that hydroxylation at the alanine moiety would occur

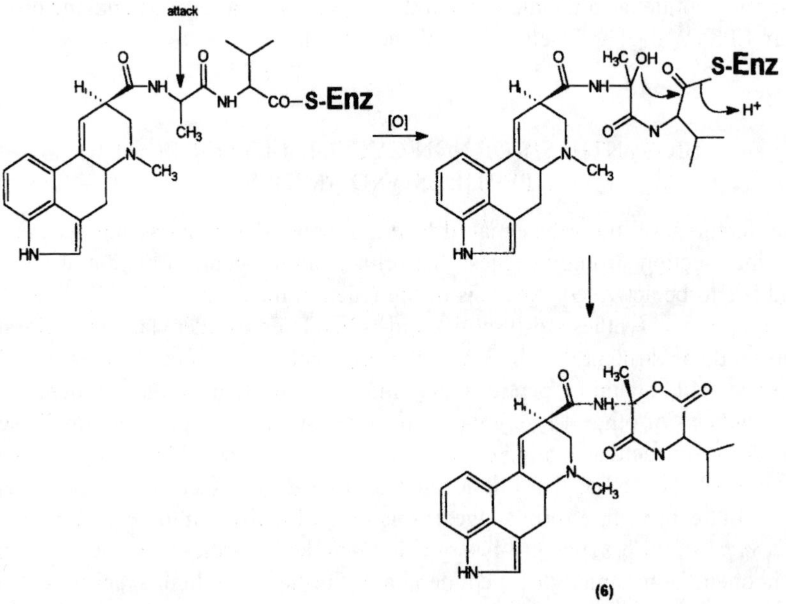

Figure 30 Model scheme of hydroxylation and cyclization events on a putative ergosecaline synthetase

in situ when the end product d-lysergyl-alanyl-valine is still bound to ergosecaline synthetase. The α-hydroxyl group of the alanine integrated in the d-lysergyl peptide chain could react with the thioester-activated carboxyl group in an intramolecular cyclization to yield the 6-membered lactone as shown in Figure 30 similar to the mechanism of other peptide lactonizing activities (Keller,

1995). This mechanism would also predict that the d-lysergyl peptide hydroxylase would be tightly associated with ergosecaline synthetase at least in the intact cell. At present the model of ergosecaline biosynthesis proposed here cannot be verified on an experimental level.

5.7.2.
d-Lysergyl Amide-Derived Compounds

Biogenetic Origin of Ergometrine and d-Lysergyl α-Hydroxyethylamide

Both d-lysergic acid amides ergometrine (7) and d-lysergyl α-hydroxyethylamide (8) have clearly recognizable d-lysergyl monopeptide structures suggesting d-lysergyl-alanine as their common precursor. Several groups have confirmed that both d-lysergic acid α-hydroxyethyl amide and ergometrine are derived from d-lysergic acid and alanine. Agurell (1966b, e) showed the non-incorporation of radiolabelled acetamide and ethylamine whereas feeding with L-[U-^{14}C]alanine to *C. paspali* cultures yielded exclusive labeling of the carbinolamide portion of d-lysergic acid α-hydroxyethylamide (Agurell, 1966e). Other groups confirmed this finding by showing the same labeling after administering L-[2–^{14}C]alanine and they showed that the amino nitrogen of ^{15}N-labelled alanine labeled the amide nitrogen of d-lysergic acid (Gröger *et al.*, 1968; Castagnoli *et al.*, 1970). By the contrast, L-[1–^{14}C]alanine was not incorporated suggesting that the carboxyl group of alanine is lost. Interestingly, incorporation studies using [2–^{14}C]pyruvate and [1–^{14}C]pyruvate gave essentially the same results like the two correspondingly labeled alanines confirming the role of alanine as a biosynthetic precursor of the carbinolamine moiety of d-lysergic acid α-hydroxyethylamide (Castagnoli *et al.*, 1970).

In the case of ergometrine, incorporation experiments have shown that like in the case of d-lysergyl α-hydroxyethylamide the alaninol side chain is derived from alanine (Nelson and Agurell, 1968; Minghetti and Arcamone, 1969). Alaninol was claimed to be a direct precursor, while other groups experiments to demonstrate alaninol incorporation failed (Minghetti and Arcamone, 1969; Majer *et al.*, 1967; Nelson and Agurell, 1968). Agurell proposed that d-lysergyl-alanine is an intermediate in the biosynthesis of the d-lysergyl amides (Agurell, 1966e). However, feeding d-lysergyl-L-[2–^{14}C]alanine to *C. paspali* gave only weak incorporation of this compound into the side chain of ergonovine (presumably due to the cleavage and metabolizing activity prior to incorporation). Trapping experiments with the aim to detect d-lysergyl-L-alanine as free intermediate remained unsuccessful (Basmadjan *et al.*, 1969; Floss *et al.*, 1971).

Models of d-Lysergic Acid Alkylamide Formation Based on a Non-ribosomal Mechanism

Although the cell-free formation of ergometrine or d-lysergyl a-hydroxyethylamide has as yet not been accomplished, data obtained with LPS and more recent novel insights in the mechanisms of post-assembly modifications of thioester-bound carboxyl groups provide a basis for new interpretations of the previous data establishing alanine as a precursor of both ergometrine and d-lysergic acid α-hydroxyethylamide. The formation of d-lysergyl-alanine as covalently enzyme-bound intermediate on LPS in the course of formation of d-lysergyl peptide lactam points to a mechanism of ergometrine formation which involves the formation of d-lysergyl-alanine covalently bound to a modified LPS or a specific peptide synthetase template with two domains arranged in a similar fashion as the d-lysergic acid- alanine-activating and peptide synthetase domains in LPS. Because of the above-mentioned specificity of the d-lysergic acid-activating domain of LPS 2 in *Claviceps* the formation of the d-lysergyl-alanine precursor via an alternative activation mechanism of d-lysergic acid activation appears unlikely. In the case of ergometrine one could envisage several possibilities of conversion of the d-lysergyl-alanine to ergometrine:

(a) Thioester-bound d-lysergyl-alanine would be released from the enzyme by action of a hydrolysing enzyme (e.g., thioesterase) as in the case of other linear peptides of non-ribosomal origin. Next step would involve immediate conversion of d-lysergyl-alanine to d-lysergyl-alanine-aldehyde in a similar mechanism to that operating in the synthesis of protease inhibitor peptides from *Streptomyces* such as leupeptins, antipain, chymostatins and elastatinal which are peptides aldehydes (Umezawa, 1975). They are formed from the corresponding peptide acids, which—like e.g. leupeptin—are released from the peptide synthetases responsible for their assembly and subsequently reduced to the corresponding aldehydes (Suzukake *et al.,* 1979). Formation of ergometrine would require further reduction of the peptide aldehyde to the alcohol.

(b) Direct conversion of d-lysergyl-alanine to d-lysergyl-alanine-aldehyde while still being attached to the 4′-phosphopantetheine cofactor of the peptide synthetase. This would be in a mode similar to that of reduction of α-aminoadipate to aminoadipate semialdehyde catalysed by aminoadipate reductase from *S. cerevisae* (Sagisaka and Shimura, 1960). Aminoadipate reductase (Lys2) is a 155 kDa protein with homology to the amino acid-activating peptide synthetases which like peptide synthetases activates α-aminoadipate at its δ-carboxy group as adenylate (Sinha and Bhattacharjee, 1971; Morris and Jinks-Robertson, 1991). Recent evidence strongly suggests that the enzyme binds α-aminoadipate as pantetheinyl thioester in a similar manner to the amino acid activation of peptide synthetases such as LPS (Lambalot *et al.,* 1996). Reduction of the pantetheinyl thioester with NAD(P)H yields aminoadipic acid-semialdehyde via the unstable thiohemiacetal. Enzyme bound d-lysergyl-

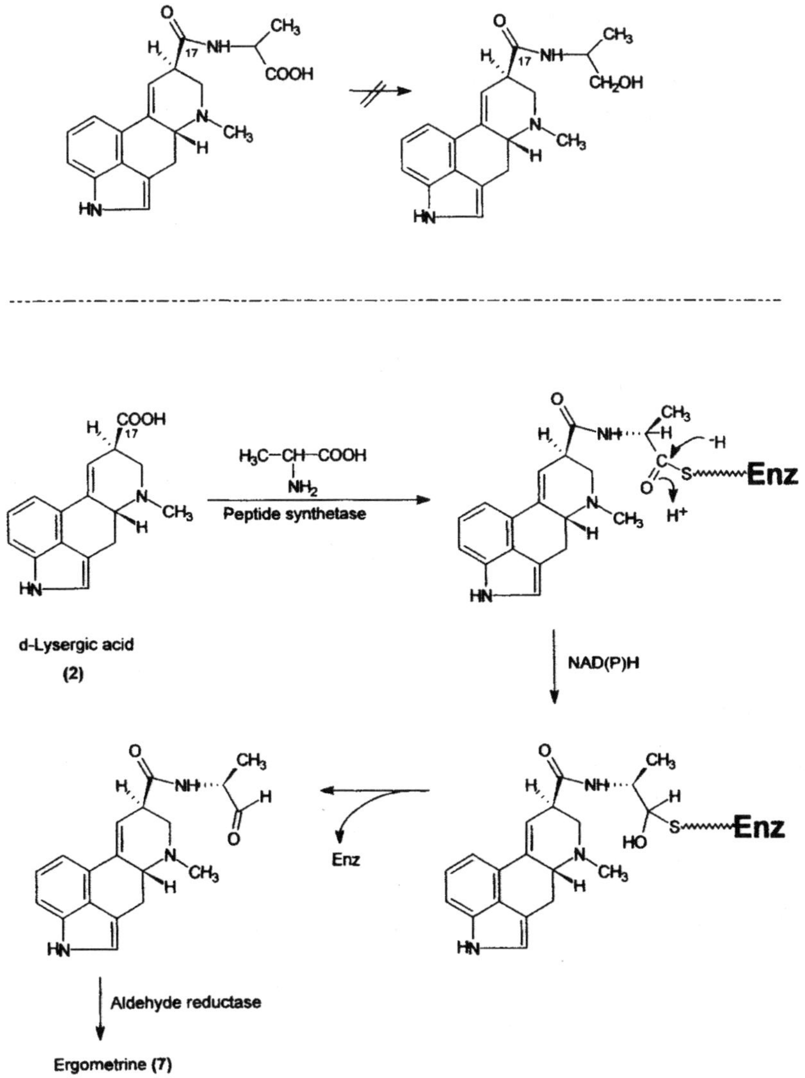

Figure 31 Possible pathway of ergometrine synthesis on a d-lysergyl peptide synthetase involving post-assembly modifications by *in situ* reduction of covalently bound d-lysergyl-alanine to d-lysergyl-alanine-aldehyde with subsequent reduction to ergometrine

alanine thioester could be converted to d-lysergyl-alaninealdehyde in a similar fashion. The aldehyde would then be converted to ergometrine (Figure 31). This model can explain that d-lysergyl-alanine is not a free intermediate in ergometrine synthesis.

The same mechanism such as activation of d-lysergic acid and alanine and subsequent formation of thioester-bound d-lysergyl-alanine catalysed by a peptide synthetase could also operate in the formation of d-lysergic acid α-hydroxyethylamide since d-lysergyl-alanine is not a free intermediate. Post-assembly modifications must convert the d-lysergyl-mono-peptide into the d-lysergic acid alkylamide. In any case the future investigation of the biosynthesis of the simple d-lysergic acid amides will give interesting new insights in the activation of d-lysergic acid and in the nature of the corresponding enzymes as peptide synthetases.

5.8.

FUTURE PROSPECTS OF ERGOT ALKALOID
BIOSYNTHESIS RESEARCH

5.8.1.

**Cloning and Expression of Ergot Alkaloid
BiosynthesisGenes for Applications in Enzymology and
Biotechnology**

The recent progress in cloning the gene of DMAT synthase and the progress in elucidating the enzymatic basis of ergot peptide assembly will stimulate further work on the enzymology and molecular genetics of ergot alkaloid synthesis. On the one hand, advanced techniques in protein isolation and purification have to be applied to identify and purify more enzymes involved in ergoline ring synthesis and of the assembly of d-lysergic acid amides. Alternatively, cloning genes of enzymes of biosynthetic steps by complementation of mutants or gene disruptions steps could lead to the identification of enzymes which cannot be purified with conventional techniques. The latter may apply in particular to the enzymes involved in ergoline ring cyclizations. The successful cloning of ergot alkaloid biosynthesis genes involved in ergoline ring and d-lysergic acid amide synthesis will enable one to express them in both homologous and heterologous hosts and studying the functions of the gene products in various biosynthetic backgrounds as well as at the cell-free level. The over-expression of alkaloid biosynthesis genes will faclitate the large-scale purification of enzymes for the purpose of structural studies and preparative *in vitro* systems.

The cloning and expression of alkaloid genes will also be of importance for the biotechnological production of alkaloids in genetically engineered alkaloid-high producing strains of *Claviceps* or in heterologous hosts with better rowth and yield of products. Cloning of ergot peptide alkaloid genes could also provide a means to develop technologies for the control of the production of ergot toxins in the environment such as of ergopeptines in endophytic fungi either on a genetic or biochemical basis.

5.8.2.
Combinatorial Biosynthesis of New d-LysergicAcid Containing Compounds

The apparent combinatorial basis of the structural diversity of d-lysergic acid peptides and amides provides a clue for the *in vitro* construction of novel enzymes. Construction of recombinant peptide synthetase containing the d-lysergic acid-activating peptide synthetase domains in combination with amino acid-activating domains from other fungi and bacteria should provide a means for the production of recombinant ergot drugs. Similar approaches in the field of peptide antibiotics have been realized recently (Stachelhaus *et al.,* 1995) when domain-coding regions of bacterial and fungal origin were combined with each other in hybrid genes that encoded peptide synthetases producing peptides with modified amino acid sequences. The introduction of new peptide building blocks in d-lysergic acid amides by domain exchange could lead to new or improved properties of the ergoline nucleus as a pharmacophore.

ACKNOWLEDGEMENTS

I appreciate the contributions of Brigitte Riederer, Bernd Walzel and Mehmet Han concerning the characterization and mechanism of LPS. I also thank Nicolas Grammel for preparing the illustrations of this article. The work performed in the author's laboratory has been supported by the Deutsche Forschungsgemeinschaft.

REFERENCES

Abe, M. (1948a) IX. Separation of an Active Substance and its properties. *J. Agric.Chem. Soc. Japan,***22,**2.

Abe, M. (1948b) XI. Separation of Agroclavine from Natural Ergot. *J. Agric. Chem. Soc.Japan,***22,**61–62.

Abe, M. (1948c) XIII. Position of Agroclavine in the group of Ergot Alkaloids. *J. Agric.Chem. Soc. Japan,***22,**85–86.

Abe, M. and Yamatodani, S. (1964) Preparation of Alkaloids by Saprophytic Culture of Ergot Fungi. *Progress in Industrial Microbiology,***V,**205.

Abe, M., Yamano, Y., Kozu, T. and Kusumoto, M.(1952) A new water-soluble ergot alkaloid, elymoclavine. Preliminary report. *J. Agric. Chem. Soc. Japan,***25,**458.

Abe, M., Ohashi, T., Ohmomo, S. and Tabuchi, T. (1971) Production of Alkaloids and Related Substances by Fungi. Part VIII. Mechanism of Formation of peptide-type Ergot Alkaloids. *J. Agr. Chem. Soc. Japan,***45,**6–10.

Abe, M. (1972) On the biosynthesis of Ergot Alkaloids. *4. Intern. Symp. Biochemie undPhysiologie der Alkaloide (Halle), Abh. dtsch. Akad. Berl.,***Bd. b,**Akademie-Verlag, Berlin, pp. 411–422.

Abe, M., Yamano, T., Yamatodani, S., Kozu, Y., Kusumoto, M., Komatsu, H. and Yamada, S. (1959) On the new peptide-type ergot alkaloids, ergosecaline and ergosecalinine. *Bull. Agr. Chem. Soc. (Japan),***23,**246–248.

Abou-Chaar C.I., Günther, F.H., Manuel, M.F., Robbers, J.E. and Floss, H.G. (1972) Biosynthesis of ergot alkaloids. Incorporation of (5R)- and (5S)-mevalonate-5-T into chanoclavines and tetracyclic ergolines. *Lloydia,***35,**272–279.

Abraham, E.P. (1991) From Penicillins to Cephalosporins. Kleinkauf, H. and von Döhren, H. (eds.), *50 Years of Penicillin Application,*Publica Press: Prague, pp. 7–23.

Acklin, W. and Arigoni, D. (1966) The Role of Chanoclavines in the Biosynthesis of Ergot Alkaloids. *J. Chem. Soc. Chem. Comm.,*801–802.

Agurell, S. (1966a) Biosynthesis of ergot alkaloids in *Claviceps paspali.* I. Incorporation of DL-4-dimethylallyltryptophan-[14]C. *Acta Pharm. Suec.,***3,**11–22.

Agurell, S. (1966b) Biosynthesis of ergot alkaloids in *Claviceps paspali.* II. Incorporation of labelled agroclavine, elymoclavine, lysergic acid and lysergic acid methyl ester. *Acta Pharm. Suec.,***3,**33–36.

Agurell, S. (1966c) Biosynthesis of ergot alkaloids in *Claviceps paspali.* II. Incorporation experiments with lysergic acid amide-^{3}H, isolysergic acid amide-^{3}H and ethylamine-[14]C. *Acta Pharm. Suec.,***3,**65–70.

Agurell, S. (1966d) Biosynthesis of ergot alkaloids in *Claviceps paspali.* 3. Incorporation of 6-methyl-Δ-8,9-ergolene-8-carboxylic acid-^{14}C and lysergic acid-8-[3]H. *ActaPharm. Suec.,***3,**65–70.

Agurell, S. (1966e) Biosynthetic studies on ergot alkaloids and related indoles. *ActaPharm. Suec.,***3,**71–100.

Agurell, S. and Ramstad, E. (1962) Biogenetic interrelationships of ergot alkaloids. *Arch. Biochem. Biophys.,***98,**457–470.

Agurell, S. and Lindgren, J.-E. (1968) Natural occurrence of 4-dimethylallyltryptophan—an ergot alkloid precursor. *Tetrahedron Lett.,***49,**5127–5128.

Ambike, S.H. and Baxter, R.M. (1970) Cytochrome P-450 and b$_5$ in *Claviceps purpurea:* Interconversion of P-450 and P-420. *Phytochemistry,***9,**1959–1962.

Ambike, S.H., Baxter, R.M. and Zahid, N.D. (1970) The relationship of cytochrome P-450 levels and alkaloid synthesis in *Claviceps purpurea. Phytochemistry,***9,** 1953–1958.

Anderson, J.A. and Saini, M.S. (1974) Natural occurrence of 4-[4-hydroxy-3-Methyl-Δ2-butenyl]-tryptophan in *Claviceps* purpurea PRL 1980. *Tetrahedron Lett.,*2107–2108.

Anderson, J.A., Kim, I.-S., Lehtonen, P. and Floss, H.G. (1979) Conversion of dihydrolysergic acid to dihydroergotamine in an ergotamine-producing strain of *Claviceps purpurea. J. Nat. Prod.,***42,**271–273.

Arcamone, F. (1977) The *Claviceps* fermentation and the development of new ergoline drugs. In Hems, D.A. (ed.), *Biologically Active Substances—Exploration andExploitation.*Wiley, Chichester, pp. 49–771.

Arcamone, F., Bonino, C., Chain, E.B., Ferretti, A., Pennella, P., Tonolo, A. and Vero, L. (1960) Production of lysergic acid derivatives by a strain of *Claviceps paspali* Stevens and Hall in submerged culture. *Nature,***187,**238–239.

Arcamone, F., Chain, E.B., Ferretti, A., Minghetti, A., Pennella, P., Tonolo, A. and Vero, L. (1961) Production of a new lysergic acid derivative in submerged cultures by a strain of *Claviceps paspali* Stivens & Hall. *Proc. Roy. Soc. London Ser B,***155,**26–54.

Atwell, S.M. and Mantle, P.G. (1981) Hydroxydihydroergosine, a new ergot alkloid analogue from directed biosynthesis by *Sphacelia sorghi. Experientia,***37,**1257–1258.

Bajwa, R.S. and Anderson, J.A. (1975) Conversion of agroclavine to setoclavine and isosetoclavine in cell-free extracts from *Claviceps* species SD 58 and in a thiolglycolate-iron(II) system. *J. Pharm. Sci.,***64,**343–344.

Bajwa, R.S., Kohler, R.-D., Saini, M.S., Cheng, M. and Anderson, J.A. (1975) Formation of clavicipitic acid in cell-free systems of *Claviceps* spec. *Phytochemistry,* **14,**735.

Barrow, K.D., Mantle, P.G. and Quigley, F.R. (1974) Biosynthesis of dihydroergot alkaloids. *Tetrahedron Lett.,* 1557–1560.

Barrow, K.D. and Quigley, F.R. (1975) Ergot alkaloids III: The isolation of N-Methyl-4-dimethylallyltryptophan from *Claviceps fusiformis. Tetrahedron Lett.,* **1975,**4269–4270.

Basmadjian, G.P., Floss, H.G., Gröger, D. and Erge, D. (1969) Biosynthesis of ergot alkaloids, lysergylalanine as precursor of amide-type Alkaloids. *J. Chem. Soc. Chem.Comm.,* 418–419.

Bassett, R.A., Chain, E.B. and Corbett, K. (1973) Biosynthesis of ergotamine by *Claviceps purpurea* (Fr.) Tul.*Biochem. J.,* **134,**1–10.

Baumert, A., Gröger, D. and Maier, W. (1977) Lysergylpeptides in the course of peptide ergot alkaloid formation. *Experientia,* **33,**881–882.

Baumert, A., Erge, D. and Gröger, D. (1982) Incorporation of thiazolidine-4-carboxylic acid into ergosine by *Claviceps purpurea. Planta med.,* **44,**122–123.

Bhattacharji, S., Birch, A.J., Brack, A., Hofmann, A., Kobel, H., Smith, D.C.C., Smith, H. and Winter, J. (1962) Studies in relation to biosynthesis. Part XXVII. The biosynthesis of ergot alkaloids. *J. Chem. Soc.,* 421–425.

Baxter, R.M., Kandel, S.I. and Okany, S.Y. (1961) Biosynthesis of ergot alkaloids. *Chemistry and Industry,* 1453–1455.

Baxter, R.M., Kandel, S.I. and Okany, S.Y. (1961) Studies on the mode of incorporation of mevalonic acid into ergot alkaloids. *Tetrahedron Lett.,* 596–600.

Baxter, R.M., Kandel, S.I. and Okany, S.Y. (1964) Biosynthesis of ergot alkaloids. The origin of the N-methyl group. *Can. J. Chem.,* **42,**2936–2938.

Bellatti, M., Casnati, G., Palla, G. and Minghetti, A. (1977) Fate of the benzene hydrogens in ergot alkloids synthesis. *Tetrahedron,* **33,**1821–1822.

Belzecki, C., Quigley, F.R., Floss, H.G., Crespi-Perellino, N. and Guicciardi, A. (1980) Mechanism of α-hydroxy-α-amino acid formation in the biosynthesis of peptide ergot alkaloids. *J. Org. Chem.,* **45,**2215–2217.

Berde, B. and Stürmer, E. (1978) Introduction to the pharmacology of ergot alkaloids and related compounds. In Berde, B. and Schild, H.O. (eds.), *Ergot alkaloids andrelated compounds.*Springer, Berlin-Heidelberg-New York, pp. 1–28.

Birch, A.J., McLoughlin, B.J. and Smith, H. (1960) Biosynthesis of ergot alkaloids. *Tetrahedron Lett.,* 1–3.

Cassady, J.M., Abou-Chaar, C.I. and Floss, H.G. (1973) Ergot alkaloids. Isolation of N-demethylchanoclavine-II from *Claviceps* strain SD 58 and the role of demethylchanoclavines in ergoline biosynthesis. *Lloydia,* **36,**390–396.

Castagnoli, N., Corbett, K.Chain, E.B. and Thomas, R. (1970) Biosynthesis of *N*-(α-hydroxy-ethyl)lysergamide, a metabolite of *Claviceps paspali. Biochem. J.,* **117,** 451–455.

Cavender, F.L. and Anderson, J.A. (1970) The cell-free synthesis of clavine alkaloids. *Biochem. Biophys. Acta,* **208,**345–348.

Clark, A.M. (1996) Natural products as a resource for new drugs. *Pharm. Res.,* **13,** 1133–1141.

Cheng, L.-J., Robbers, J.E. and Floss, H.G. (1980) End-product regulation of ergot alkaloid formation in intact cells and protoplasts of *Claviceps* Species, Strain SD 58. *J. Nat. Prod.,* **43,**329–339.

Cress, W.A., Chayet, L.T. and Rilling, H.C. (1981) Crystallization and partial characterization of dimethylallyl pyrophosphate: L-Tryptophan Dimethylallyltransferase from *Claviceps* sp. SD 58. *J. Biol. Chem.,***256,**10917–10923.

Eich, E. and Pertz, H. (1994) Ergot alkaloids as lead structures for different receptor systems. *Pharmazie,***49,**867–877.

Eigner, E.A. and Loftfield R.B. (1974) Kinetic techniques for the investigation of amino acid: tRNA ligases (aminoacyl-tRNA synthetases, amino acid activating enzymes). *Methods Enzymol.,***29,**601–619.

Erge, D., Maier, W. and Gröger, D. (1973) Untersuchungen über die enzymatische Umwandlung von Chanoclavine-I. *Biochem. Physiol. Pflanzen,***164,**234–247.

Esser, K. and Düvell, A. (1984) Biotechnological explotation of the ergot fungus *Claviceps purpurea*. *Process Biochem.,***8,**142–149.

Fehr, T. (1967) Untersuchungen über die Biosynthese der Ergotalkaloide. *DissertationETH Zürich No. 3967.*

Flieger, M., Wurst, M. and Shelby, R. (1997) Ergot alkaloids—sources, structures and analytical methods. *Folia Microbiol.,***42,**3–30.

Floss, H.G. (1976) Biosynthesis of ergot alkaloids and related compounds. *Tetrahedron,* **32,**873–912.

Floss, H.G. and Anderson, J.A. (1980) Biosynthesis of ergot toxins. In Steyn, P.S. (ed.), *The biosynthesis of mycotoxins. A study in secondary metabolism.* Academic Press, New York, pp. 17–67.

Floss, H.G. and Gröger, D. (1963) Über den Einbau von Nα-Methyltryptophan und N$_\omega$-Methyltryptamin in Mutterkornalkaloide vom Clavin-Typ. *Z. Naturforschg.,***18b,** 519–522.

Floss, H.G. and Gröger, D. (1964) Entmethylierung von Nα-Methyltryptophan durch *Claviceps* sp. *Z. Naturforschg.,***19b,**393–395.

Floss, H.G., Mothes, U. and Günther, H. (1964) Zur Biosynthese der Mutterkornalkaloide. Über den Einbau der Tryptophan-Seitenkette und den Mechanismus der Reaktion am α-C-Atom. *Z. Naturforschg.,***19B,**784–788.

Floss, H.G., Mothes, U., Onderka, D. and Hornemann, U. (1965) Nichtverwertung von 4-Hydroxytryptophan bei der Biosynthese der Mutterkornalkaloide. *Z. Naturforschg.,***20B,** 133–136.

Floss, H.G., Günther, H., Gröger, D. and Erge, D (1966) Biosynthesis of ergot alkaloids. Biogenetic relationships between clavine and lysergic acid derivatives. *Z.Naturforschg.,***21B,**128–131 (1966).

Floss, H.G., Günther, H., Gröger, D. and Erge, D. (1967) Biosynthesis of ergot alkaloids. Origin of the oxygens of chanoclavine-I and elymoclavine. *J. Pharm. Sci.,***56,** 1675–1677.

Floss, H.G., Hornemann, U., Schilling, N., Kelley, K., Gröger, D. and Erge, D. (1968) Biosynthesis of ergot alkaloids. Evidence for two isomerizations in the isoprenoid moiety during the formation of tetracyclic ergolines. *J. Am. Chem. Soc.,***90,**6500–6507.

Floss, H.G., Basmadjian, G.P., Tcheng, M., Spalla, C. and Minghetti, A. (1971a) Biosynthesis of peptide-type ergot alkaloids. Ergotamine. *Lloydia,***34,**442–445.

Floss, H.G., Basmadjian, G.P., Tcheng, M., Gröger, D. and Erge, D. (1971b) Biosynthesis of peptide-type ergot alkaloids. Ergocornine and ergocryptine. *Lloydia,***34,**446–448.

Floss, H.G., Basmadjian, G.P., Gröger, D. and Erge, D. (1971c) Biosynthesis of ergot alkaloids. Ergometrine. *Lloydia,***34,**449–450.

Floss, H.G., Tscheng-Lin, M., Kobel, H. and Stadler, P. (1974a) On the biosynthesis of peptide ergot alkaloids. *Experientia,* **30,** 1369–1370.

Floss, H.G., Tcheng-Lin, M., Chang, C.J., Naidoo, B., Blair, E., Abou-Chaar, C. and Cassidy, J.M. (1974b) Biosynthesis of ergot alkaloids. Studies on the mechanism of the conversion of chanoclavin-I into tetracyclic ergolines. *J. Am. Chem. Soc.,* **96,** 1898–1909.

Gebler, J.F. and Poulter, C.D. (1992) Purification and characterization of dimethylallyl tryptophan synthase from *Claviceps purpurea* *Arch. Biochem. Biophys.,* **296,** 308–313.

Gröger, D. and Sajdl, P. (1972) Enzymatic conversion of chanoclavine-I. *Pharmazie,* **27,** 188.

Gröger, D., Wendt, H.J., Mothes, K. and Weygand, F. (1959) Untersuchungen zur Biosynthese der Mutterkornalkaloide. *Z. Naturforschg.,* **14B,** 355–358.

Gröger, D., Mothes, K., Simon, H., Floss, H.G. and Weygand, F. (1960) Über den Einbau von Mevalonsäure in das Ergolinsystem der Clavin-Alkaloide. *Z. Naturforschg.,* **15B,** 141–143.

Gröger, D., Erge, D. and Floss, H.G. (1966) Biosynthese der Mutterkornalkaloide. Einbau von Chanoclavin I in Ergolin-Derivate. *Z. Naturforschg.,* **21B,** 827–832.

Gröger, D., Erge, D. and Floss, H.G. (1968) Biosynthesis of ergot alkaloids. On the origin of side chains in d-lysergic acid methylcarbinolamide. *Z. Naturforschg.,* **23B,** 177–180.

Gröger, D. and Johne, S. (1972) Über den Einbau von Dipeptiden in Ergotoxin Alkaloide. *Experientia,* **28,** 241–242.

Gröger, D. and Erge, D. (1970) Biosynthesis of the peptide alkaloids of *Claviceps purpurea.* *Z. Naturforschg.,* **25B,** 196–199.

Gröger, D., Johne, S. and Härtling, S. (1974) Über den Stoffwechsel von Oligopeptiden und Diketopiperazinen in *Claviceps-Arten.* *Biochem. Physiol. Pflanz.,* **166,** 33–43.

Gröger, D., Gröger, L., D'Amico, D., He, M.-X. and Floss, H.G. (1991) Steric course of the N-methylation in the biosynthesis of ergot alkaloids by *Claviceps purpurea.* *J. Basic Microbiol.,* **31,** 121–125.

Heinstein, P.P., Lee, S-L. and Floss, H.G. (1971) Isolation of dimethylallylpyrophosphate: tryptophan dimethylallyltransferase from the ergot fungus (*Claviceps* sp.). *Biochem. Biophys. Res. Commun.,* **44,** 1244–1251.

Hofmann, A., Ott, H., Griot, R., Stadler, P.A. and Frey, J. (1963) Die Synthese und Stereochemie des Ergotamins. *Helv. Chim. Acta,* **46,** 2306–2328.

Hofmann, A. (1964) *Die Mutterkornalkaloide,* Verlag Enke, Stuttgart, pp. 11–121.

Hofmann, A. (1972) Ergot-a rich source of pharmacologically active substances. In Swain, T. (ed.), *Plants in the development of modern Medicine.,* Harvard University Press, Cambridge, MA, pp. 235–260.

Hofmann, A. and Tscherter, H. (1960) Isolierung von Lysergsaure-Alkaloiden aus der mexikanischen Zauberdroge Ololiuqui (*Rivea corymbosa* (L.) Hall, f.) *Experientia,* **16,** 414.

Hopwood, D.A. and Sherman, D.H. (1990) Molecular Genetics of polyketides and its comparison to fatty acid biosynthesis. *Ann. Rev. Genet.,* **24,** 37–66.

Katz, L. and Donadio, S. (1993) Polyketide synthesis: prospects for hybrid antibiotics. *Annu. Rev. Microbiol.,* **47,** 875–912.

Keller, U., Zocher, R. and Kleinkauf, H. (1980) Biosynthesis of ergotamine in protoplasts of *Claviceps purpurea.* *J. Gen. Microbiol.,* **118,** 485–494.

Keller, U., Zocher, R., Krengel, U. and Kleinkauf, H. (1984) d-Lysergic acid-activating enzyme from the ergot fungus *Claviceps purpurea.* *Biochem. J.,* **218,** 857–862.

Keller, U. (1985) Ergot peptide alkaloid synthesis in *Claviceps purpurea*. In Kleinkauf, H., von Döhren, H., Dornauer, H. and Nesemann, G. (eds.), *Regulation of Secondary Metabolite Formation,*VCH Verlagsgesellschaft, Weinheim, Germany, pp. 157–172.

Keller, U., Han, M. and Stöffler-Meilicke, M. (1988) d-Lysergic acid activation and cell-free synthesis of d-lysergyl peptides in enzyme fractions from the ergot fungus *Claviceps purpurea. Biochemistry,*27,6164–6170.

Keller, U. (1995) Peptidolactones. In Vining, L.C. and Stuttard, C. (eds.), *Geneticsand Biochemistry of Antibiotic Production.*Butterworth-Heinemann, Boston, pp. 71–94.

Kim, I.-S., Kim, S.-U. and Anderson, J.A. (1981) Microsomal agroclavine hydroxylase of *Claviceps* species. *Photochemistry,*20,2311–2314.

Kim, S.-U., Cho, Y.-J., Floss, H.G. and Anderson, J.A. (1983) Conversion of elymoclavine to paspalic acid by a particulate fraction from an ergotamine-producing strain of *Claviceps* sp.*Planta Med.,*48,145–148.

King, G.S., Mantle, P.G., Sczyzbak, C.A. and Waight, E.S.(1973) A revised struture for Clavicipitic acid. *Tetrahedron Lett.,*215–218.

Kleinerová, E. and Kybal, J. (1973) Ergot alkaloids. IV. Contribution to the biosynthesis of lysergic acid amides. *Folia Microbiol.,*18,390–392.

Kleinkauf, H. and von Döhren, H. (1990) Non-ribosomal biosynthesis of peptide antibiotics. *Eur. J. Biochem.,*192,1–5.

Kobayashi, M. and Floss, H.G. (1987) Biosynthesis of ergot alkaloids: Origin of the oxygen atoms in chanoclavine-I and elymoclavine. *J. Org. Chem.,*52,4350–4352.

Kobel, H. and Sanglier, J.-J. (1986) Ergot Alkaloids. In Rehm, H.-J. and Reed, G. (eds.), *Biotechnology*4,VCH Verlagsgesellschaft, Weinheim, pp. 569–609.

Kobel, H., Schreier, E. and Rutschmann, J. (1964) 6-Methyl $\Delta^{8,9}$-ergolen-8-carbonsäure, ein neues Ergolinderivat aus Kulturen eines Stammes von *Claviceps paspali* Stevens and Hall. *Helv. Chim. Acta,*47,1052–1064.

Kornfeld, E., Fornefeld, E., Kline, G.B., Mann, M., Jones, R.G. and Woodward, R.B. (1954) The total synthesis of lysergic acid and ergonovine. *J. Am. Chem. Soc.,*76, 5256–5257.

Kozikowski, A.P., Okita, M., Kobayashi, M. and Floss, H.G. (1988) Probing ergot alkaloid biosynthesis: Synthesis and feeding of a proposed intermediate along the biosynthetic pathway. A new amidomalonate for tryptophan elaboration. *J. Org.Chem.,* 53,863–869.

Kozikowski, A.P., Chen, C., Wu, J.-P., Shibuya, M., Kim, C.-G. and Floss, H.G. (1993) Probing ergot alkaloid biosynthesis: Intermediates in the formation of ring C. *J. Am.Chem. Soc.,*115,2482–2488.

Krupinski, V.M., Robbers, J.E. and Floss, H.G. (1976) Physiological study of ergot: induction of alkaloid synthesis by tryptophan at the enzymic level. *J. Bacteriol.,*125, 158–165.

Lambalot, R.H., Gehring, A.M., Flugel, R.S., Zuber, P., LaCelle, M., Marahiel, M.A., Reid, R., Khosla, C. and Walsh, C.T. (1996) A new enzyme superfamily—the phosphopantetheinyl transferases. *Chem. Biol.,*3,923–936.

Latch, G.C.M. (1995) Endophytic fungi of grasses. In Khomoto, K., Singh, U.S. and Singh, R.P. (eds.), *Pathogenesis and host specificity in plant diseases*II. Pergamon Press, Oxford, pp. 265–275.

Lawen, A. and Zocher, R. (1990) Cyclosporin synthetase—the most complex peptide synthesizing multienzyme polypeptide so far described. *J. Biol. Chem.*265, 11355–11360.

Lonitzer, A. (1582) *Kreuterbuch,*Frankfurt.

Lin, C.C.L., Blair, G., Cassady, J.M., Gröger, D., Maier, W. and Floss, H.G. (1973) Biosynthesis of ergot alkloids. Synthesis of 6-methyl-8-acetoxy-9-ergolene and its incorporation into ergotoxine by *Claviceps. J. Org. Chem.,***38,**2249–2251.

Lipmann, F. (1971) Attempts to map a process evolution of peptide biosynthesis. *Science,* **173,**1435–1441.

Lipmann, F. (1973) Nonribosomal polypeptide synthesis on polyenzyme templates. *Acc. Chem. Res.,***6,**361–367.

Lipmann, F. (1980) Bacterial production of antibiotic polypeptides bithiol-linked synthesis on protein templates. *Adv. Microbiol. Physiol.,***21,**227–266.

Lee, S.L., Floss, H.G. and Heinstein, P. (1976) Purification and properties of dimethylallylpyrophosphate: tryptopharm dimethylallyl transferase, the first enzyme of ergot alkaloid biosynthesis in *Claviceps.* sp. SD 58. *Arch Biochem Biophys.,***77,**84–94.

Maier, W., Erge, D. and Gröger, D. (1971) Zur Biosynthese von Ergotoxinalkaloiden in *Claviceps purpurea. Biochem. Physiol. Pflanz.,***161,**559–569.

Maier, W., Erge, D., Schumann, B. and Gröger, D. (1972) Über Aktivierungsreaktionen bei *Claviceps. Biochem. Physiol. Pflanz.,***163,**432–442.

Maier, W., Erge, D. and Gröger, D. (1974) Über den Stoffwechsel von LysergylDerivaten in *Claviceps purpurea* (Fr.) Tulasne. *Biochem. Physiol. Pflanz.,***165,**479.

Maier, W., Baumert, A. and Gröger, D. (1978) Synthese und Stoffwechsel von Lysergylpeptiden in einem Ergotoxine bildenden Stamm. *Biochem. Physiol.Pflanz.,* **172,**15–26.

Maier, W., Erge, D. and Gröger, D. (1980) Mutational Biosynthesis in a strain of *Claviceps purpurea. Planta Med.,***40,**104–108.

Maier, W., Erge, D., Schumann, B. and Gröger, D. (1981a) Incorporation of L-[U-14C] leucine into ergosine by cell-free extracts of *Claviceps purpurea* (Fr.) Tul.*Biochem.Biophys. Res. Commun.***99,**155–162.

Maier, W., Erge, D. and Gröger, D. (1981b) Studies on the cell-free biosynthesis of ergopeptines in *Claviceps purpurea. FEMS Microbiol. Lett.,***12,**141–146.

Maier, W., Erge, D. and Gröger, D. (1983) Further studies on cell-free biosynthesis of ergotamine in *Claviceps purpurea. FEMS Microbiol. Lett.,***20,**223–236.

Maier, W., Krauspe, R. and Gröger, D. (1985) Enzymatic biosynthesis of ergotamine and investigation on some aminoacyl-tRNA synthetases in *Claviceps. J. Biotechnol.,***3,** 155–166.

Maier, W., Schumann, B. and Gröger, D. (1988) Microsomal oxygenases involved in ergoline alkaloid biosynthesis of various *Claviceps* strains. *J. Basic. Microbiol.,***28,** 83–93.

Majer, J., Kybal, J. and Komersová, I. (1967) Ergot alkaloids. I. Biosynthesis of peptide side chain*Folia Microbiol.,***12,**489–491.

Mantle, P.G. and Waight, E.S. (1968) Dihydroergosine: a new naturally occurring alkaloid from the sclerotia of *Sphacelia sorghi. Nature,***218,**581–582.

Marahiel, M.(1992) Multidomain enzymes involved in peptide synthesis. *FEBS Lett.,***307,** 40–43.

Minghetti, A. and Arcamone, F. (1969) Studies concerning the biogenesis of natural derivatives of d-Lysergic acid. *Experientia,***25,**926–927.

Morris, M.E. and Jinks-Robertson, S. (1991) Nucleotide sequence of the *LYS2* gene of *Saccharomyces cerevisiae:* homology to *Bacillus brevis* tyrocidine synthetase I. *Gene,* **98,**141–145.

Mothes, K., Weygand, F., Gröger, D. and Grisebach, H. (1958) Untersuchungen zur Biosynthese der Mutterkorn-Alkaloide. *Z. Naturforschg.,* **13B,**41–44.

Mothes, K., Winkler, K., Gröger, D., Floss, H.G., Mothes, U. and Weygand, F. (1962) Über die Umwandlung von Elymoclavin in Lysergsäurederivate durch Mutterkornpilze *(Claviceps)Tetrahedron Lett.,* **21,**933–937.

Naidoo, B., Cassady, J.M., Blair, G.E. and Floss, H.G. (1970) Biosynthesis of ergot alkaloids. Synthesis of chanoclavine-I-aldehyde and its incorporation into elymoclavine by *Claviceps. J. Chem. Soc. Chem. Comm.,*471–472.

Nelson, U. and Agurell, S. (1968) Biosynthesis of ergot alkaloids. Origin of the side chain of ergometrine. *Acta Chem. Scand.,* **23,**3393–3397.

Ogunlana, E.G., Ramstad, E. and Tyler, V.E. (1969) Effects of some substances on ergot alkaloid production. *J. Pharm. Sci.,* **58,**143–145.

Ogunlana, E.O., Wilson, B.J., Tyler, V.E. and Ramstad, E. (1970) Biosynthesis of ergot alkaloids. Enzymatic closure of ring D of the ergoline nucleus. *J. Chem. Soc. Chem.Commun.,*775–776.

Ohashi, T., Aoki, S. and Abe, M. (1970) Production of alkloids and related substances by fungi. V. Biogenetic relations among representative ergot alkloids. *J. Agr. Chem.Soc. Japan,* **44,**527–531.

Ohashi, T., Takahashi, H. and Abe, M. (1972) Production of alkaloids and related substances by fungi. X. Mechanism of biosynthetis of peptide-type alkloids ergocryptine-ergocryptinine*J. Agr. Chem. Soc. Japan,* **46,**535–540.

Otsuka, T., Anderson, J.A. and Floss, H.G. (1979) The stage of N-methylation as the second pathway-specific step in ergoline biosynthesis. *J. Chem. Soc. Chem. Comm.,* 660–662.

Otsuka, H., Quigley, F.R., Gröger, D., Anderson, J.A. and Floss, H.G. (1980) *In Vivo* and *in vitro* evidence for N-methylation as the second pathway-specific step in ergoline biosynthesis. *Planta Med.,* **40,**109–119.

Ott, H., Frey, A.J. and Hofmann, A. (1963) The stereospecific cyclolization of N-(α-hydroxyacyl)-phenylalanyl-proline lactams. *Tetrahedron,* **19,**1675–1684.

Pachlatko, P., Tabacik, Ch., Acklin, W. and Arigoni, D. (1975) Natürliche und unnatürliche Vorläufer in der Biosynthese der Ergotalkaloide. *Chimia,* **29,**526.

Panaccione, D.G. (1996) Multiple families of peptide synthetase genes from ergopeptine-producing fungi. *My col. Res.,* **100,**429–436.

Paradkar, A., Jensen, S.E. and Mosher, R.H. (1997) Comparative genetics and molecular biology of β-lactam biosynthesis. In Strohl, W.R. (ed.), *Biotechnology of Antibiotics.*Marcel Dekker, New York, pp. 241–277.

Petroski, R.J. and Kelleher, W.J. (1977) Biosynthesis of ergot alkaloids. Cell-free formation of 4-(E-4'-hydroxy-3'-methylbut-2'-enyl)-L-tryptophan. *FEBS Lett.,* **82,** 55–57.

Plieninger, H., Fischer, R., Keilich, G. and Orth, H.D. (1961) Untersuchungen zur Biosynthese der Clavin-Alkaloide mit deuterierten Verbindungen. *Liebigs Ann.Chem.,* **642,**214–224.

Plieninger, H., Fischer, R. and Liede, V. (1962) Dimethylallyl-tryptophan als Vorstufe der Clavin-Alkaloide. *Angew. Chem.,* **74,**430.

Plieninger, H., Fischer, R. and Liede, V. (1964) Untersuchungen zur Biosynthese der Mutterkornalkaloide. II. *Liebigs Ann. Chem.,* **672,**223.

Plieninger, H., Immel, H. and Völkl, A. (1967) Synthese und Einbau eines [14]C- und ^{3}H-markierten 4-Dimethylallyl-tryptophans und eines ^{14}C markierten 4-

Dimethylallyltryptamins sowie Einbau eines ^{14}C-markierten Dimethylallyl-pyrophosphates. *LiebigsAnn. Chem.,***706,**223–229.

Plieninger, H., Meyer, E., Maier, W. and Gröger, D. (1978) Über den Einbau von [4′-(E)-^{13}C]4-(3-Methyl-2-butenyl)tryptophan in Clavinalkloide und Lysergsaure. *Liebigs Ann. Chem.,***1978,**813–817.

Popják, G. and Cornforth, J.W. (1966) Substrate Stereochemistry in Squalene Biosynthesis. *Biochem. J.,***101,**553–568.

Powell, R.G., Planner, R.D., Yates, S.G., Clay, K. and Leuchtmann, J. (1990) Ergobalansin, a new ergot-type peptide alkaloid isolated from *Cenchrus echinatus* (Sandbur grass) infected with *Balansia obtecta* and produced in liquid culture of *B. obtecta* and *Balansia cyperi*. *J. Nat. Prod.,***53,**1272–1279.

Quigley, F.R. and Floss, H.G. (1981) Mechanism of amino acid α-hydroxylation and formation of the lysergyl moiety in ergotamine biosynthesis. *J. Org. Chem.,***46,** 464–466.

Ramstad, E. (1968) Chemistry of alkaloid formation in ergot. *Lloydia,*31, 327–341.

Riederer, B., Han, M. and Keller, U. (1996) d-Lysergyl peptide synthetase from the ergot fungus *Claviceps purpurea*. *J. Biol. Chem.,***271,**27524–27530.

Robbers, J.E. and Floss, H.G. (1968) Biosynthesis of ergot alkaloids: formation of 4-dimethylallyltryptophan by the ergot fungus. *Arch. Biochem. Biophys.,***126,**967–969.

Robbers, J.E. and Floss, H.G. (1969) Clavicipitic acid, a new 4-substituted indolic amino acid obtained from submerged cultures of the ergot fungus. *TetrahedronLett.,***23,** 1857–1858.

Robbers, J.E. and Floss, H.G. (1970) Physiological studies on ergot: Influence of 5-methyltryptophan on alkaloid biosynthesis and the incorporation of tryptophan analogs into protein. *J. Pharm. Sci.,***59,**702–703.

Robbers, J.E., Robertson, I.W., Hornemann, K.M., Jindra, J. and Floss, H.G. (1972) Physiological studies on ergot: further studies on the induction of alkaloid synthesis by tryptophan and its inhibition by Phosphate. *J. Bacterial.,***112,**791–796.

Rochelmeyer, H. (1958) Problem der biologischen Synthese der Mutterkornalkaloide. *Pharm. Ztg.,***103,**1269–1275.

Sagisaka, S. and Shimura, K. (1960) Mechanism of activation and reduction of α-aminoadipic acid by yeast enzyme. *Nature***188,**1189–1190.

Saini, M.S. and Anderson, J.A. (1978) Cell-free conversion of 4-γ, γ-dimethylallyltryptophan to 4-[4-hydroxy-methyl-Δ^2-butenyl]tryptophan in *Claviceps purpurea* PRL 1980. *Phytochemistry,***17,**799–800.

Schlumbohm, W., Stein, T., Ullrich, C., Vater, J., Krause, M., Marahiel, M.A., Kruft, V. and Wittmann-Liebold, B. (1991) An active serine is involved in covalent substrate amino acid binding at each reaction center of gramicidin S synthetase. *J. Biol.Chem.,* **266,**23135–23141.

Schröder, G., Brown, J.W.S. and Schröder, J. (1988) Molecular analysis of resveratrol synthase cDNA genomic clones and relationship with chalcone synthase. *Eur. J.Biochem.,***172,**161–169.

Schüz, R., Heller, W. and Hahlbrock, K. (1983) Substrate Specificity of chalcone synthase from *Petroselinum hortense*. *J. Biol. Chem.,***258,**6730–6734.

Schumann, B., Erge, D., Maier, W. and Gröger, D. (1982) A New strain of *Clavicepspurpurea* accumulating tetracyclic clavine alkaloids. *Planta Med.,***45,**11–14.

Scott-Craig, J.S., Panaccione, D.G., Pocard, J.-A. and Walton, J.D. (1992) The cyclic peptide synthetase catalysing HC-toxin production in the filamentous fungus

Cochliobolus carbonum is encoded by a 15.7-kilobase open reading frame. *J. Biol.Chem.,***267,**26044–26049.

Seiler, M., Acklin, W. and Arigoni (1970) Biosynthesis of ergot alkaloids from (5*R*)-and (5*S*)-[5–^{3}H$_1$]-Mevalonolactones. *J. Chem. Soc. Chem. Commun.,*1394–1395.

Seiler, M., Acklin, W. and Arigoni, D. (1970) Zur Biosynthese der Chanoclavine. *Chimia,***24,**449–450.

Shibuya, M., Chou, H.-M., Fountoulakis, M., Hassam, S., Kim, S.-U., Kobayashi, K., Otsuka, H., Rogalska, E., Cassady, J.M. and Floss, H.G. (1990) Stereochemistry of the isoprenylation of tryptophan catalyzed by 4-(γ, γ-dimethylallyl)tryptophan synthase from *Claviceps,* the first pathway-specific enzyme in ergot alkaloid biosynthesis. *J. Am Chem. Soc.,***112,**297–304.

Sinha, A.K. and Battacharjee, J.K. (1971) Lysine biosynthesis in *Saccharomycescerevisiae.* Conversion of α-aminoadipate into α-aminoadipic acid semialdehyde. *Biochem. J.,***125,**743–749.

Song, L. and Poulter, C.D. (1994) Yeast farnesyl-diphosphate synthase: site-directed mutagenesis of residues in highly conserved prenyltransferase domains I and II. *Proc. Natl. Acad. Sci. USA,***91,**3044–3048.

Shough, H.R and Taylor, E.H. (1969) Enzymology of ergot alkaloid biosynthesis. IV. Additional studies on the oxidation of agroclavine by horseradish peroxidase. *Lloydia,* **32,**315–326.

Smith, D.J., Alison, J.E. and Turner, G. (1990) The multifunctional peptide synthetase performing the first step of penicillin biosynthesis in *Penicillium chrysogenum* is a 421 073 dalton protein similar to *Bacillus brevis* peptide antibiotic synthetases. *EMBO J.,***9,** 2743–2750.

Stachelhaus, T. and Marahiel, M.A. (1995) Modular structure of genes encoding multifunctional peptide synthetases required for non-ribosomal peptide synthesis. *FEMS Microbiol. Lett.,***125,**3–14.

Stadler, P.A. (1982) New results of ergot alkaloid research. *Plant. Med.,*46, 131–144.

Stadler, P.A. and Stütz, P. (1975) Ergot Alkaloid. In R.H.F.Manske and H.L.Holmes (eds.), *The Alkaloids,***XV,**Academic Press, New York, pp. 1–40.

Stadler, P.A. and Giger, R. (1984) Ergot alkaloids and their derivatives in medical chemistry and therapy. In Krosgard-Larson, P., Christensen, C.H. and Kofod, H. (eds.), *Natural Products and Drug Development.*Munksgaard, Copenhagen., pp. 463–485.

Stadman, E.R. (1957) Preparation an Assay of acyl Coenzyme A and other Thiol Esters: Use of hydroxylamine. *Methods Enzymol.,***3,**931–941.

Stauffacher, D. and Tscherter, H. (1964) Isomere des Chanoclavins aus *Clavicepspurpurea* (Fr.) TUL.*(secale cornutum) Helv. Chim. Acta,***47,**2186–2194.

Stauffacher, D., Tscherter, H. and Hofmann, A. (1965) Isolierung von Ergosin und Ergosinin neben Agroclavin aus den Samen von *Ipomea argyrophylla* Vatke *(Convolvulaceae)Helv. Chim. Acta,***48,**1379–1380.

Stein, T. and Vater, J. (1996) Amino acid activation and polymerization at modular multienzymes in noribosomal peptide biosynthesis. *Amino Acids,***10,**201–227.

Stindl, A. and Keller, U. (1993) The initiation of peptide formation in the biosynthesis of actinomycin. *J. Biol. Chem.,***268,**10612–10620.

Stoll, A. (1945) Ergot alkaloids. *Helv. Chim. Acta,***28,**1283–1308.

Stoll, A. (1952) *Die spezifischen Inhaltstoffe des Mutterkorns.* Edition Cantor: Aulendorf.

Stoll, A., Hofmann, A. and Becker, B. (1943) Die Spaltstücke von Ergocristin, Ergokryptin und Ergocornin (8. Mitteilung über die Mutterkornalkaloide). *Helv. Chim.Acta,***26,**1602–1613.

Stoll, A., Brack, A., Kobel, H., Hofmann, A. and Brunner, R. (1954) Die Alkaloide eines Mutterkornpilzes von *Pennisetum typhoideum* Rich und deren Bildung in saprophytischer Kultur. *Helv. Chim. Acta,***37,**1815–1819.

Stütz, P., Brunner, R. and Stadler, P. (1973) Ein neues Alkaloid aus dem Mycel eines *Claviceps purpurea-Stammes*. Zur Biogenese von Ergocristin. *Experientia,***29,**936–937.

Suzukaze, K., Fuhyama, T., Hayashi, H., Hori, M. and Umezawa, H. (1979) Biosynthesis of leupeptin. II. Purification and properties of leupeptin acid synthetase. *J. Antibiotics,* **32,**523–530.

Taylor, E.H., Goldner, K.J., Pong, S.F. and Shough, H.R. (1966) Conversion of $\Delta^{8,9}$-ergolines to $\Delta^{9,10}$-8-hydroxyergolines in plant homogenates. *Lloydia,***29,**239–244.

Taylor, E.H. and Ramstad, E. (1960) Biogenesis of lysergic acid in ergot. *Nature,***188,** 494–495.

Tsai, H.-F., Wang, H., Gebler, J.C., Poulter, C.D. and Schardl, C.L. (1995) The *Clavicepspurpurea* gene encoding dimethylallyltryptophan synthase, the committed step for ergot alkaloid biosynthesis. *Biochem. Biophys. Res. Commun.,***216,**119–125.

Tulasne, L.R. (1853) Mémoire sur l'ergot des glumacées. *Ann. Sci. Nat. Botan. Biol.Végétate,***20,**5–56.

Tudzynski, P., Tenberge, K.B. and Oeser, B. (1995) *Claviceps purpurea*. In Khomoto, K., Singh, U.S. and Singh, R.P. (eds.), *Pathogenesis and host specificity in plantdiseases.***II.**Pergamon Press, Oxford, pp. 161–187.

Tyler, V.E., Erge, D. and Gröger, D. (1965) Biological conversions of agroclavine and elymoclavine. *Planta Med.,***13,**316–325.

Turgay, K., Krause, M. and Marahiel M.A. (1992) Four homologous domains in the primary structure of GrsB are related to domains in a superfamily of adenylateforming enzymes. *Molec. Microbiol.,***6,**529–546.

Udvardy, E.N. (1980) Consideration of the development of an ergot alkaloid fermentation process. *Process Biochemistry,***4,**5–8.

Umezawa, H. (1975) Structures and activities of protease inhibitors of microbial origin. *Methods Enzymol.,***XLI,**678–695.

Vining, L.C. and Taber, W.A. (1979) Ergot Alkaloids. In Rose, A.H. (ed.), *Secondaryproducts and metabolism.***3** (Economic Microbiology) Academic Press, London, pp. 389–415.

Voigt, R. (1962) Über Clavinalkloide im Roggenmutterkorn. *Pharmazie,***17,**101–106.

Wakil, S.J., Stoops, J.K. and Joshi, V.C. (1983) Fatty acid synthesis and its regulation. *Annu. Rev. Biochem.***52,**535–579.

Walzel, B., Riederer, B. and Keller, U. (1997) Mechanism of alkaloid cyclopeptide formation in the ergot fungus *Claviceps purpurea*. *Chem. Biol.,***4,**223–230.

Weygand, F. and Floss, H.G. (1963) Die Biogenese der Mutterkornalakloide. *Angew.Chem.,***75,**783–788.

Wilson B.J., Ramstad, E., Jansson, I. and Orrenius, S. (1971) Conversion of agroclavine by mammalian cytochrome P-450. *Biochim. Biophys. Acta,***252,**348–356.

Zocher, R., Keller, U. and Kleinkauf, H. (1982) Enniatin synthetase, a novel type of multifunctional enzyme catalyzing depsipeptide synthesis in *Fusarium oxysporum*. *Biochemistry,***21,**43–48.

Zocher, R., Nihira, T., Paul, E., Madry, N., Peeters, H., Kleinkauf, H. and Keller, U. (1986) Biosynthesis of cyclosporin A: Partial purification and properties of a multifunctional enzyme from *Tolypocladium inflatum*. *Biochemistry,***25,**550–553.

Zocher, R. and Keller, U. (1997) Thiol template peptide synthesis systems in bacteria and fungi. In Poole, R.K. (ed.), *Advances in Microbial Physiology.*Academic Press, San Diego,pp. 85–131.

6.
PHYSIOLOGICAL REGULATION OF ERGOT ALKALOIDPRODUCTION AND SPECIAL CULTIVATION TECHNIQUES

VLADIMÍR KŘEN

Laboratory of Biotransformation, Institute of

Microbiology,Academy of Sciences of the Czech Republic, Vídeňská

1083,CZ-142 20 Prague 4, Czech Republic

6.1.
REGULATION OF ERGOT ALKALOID BIOSYNTHESIS

The studies on the ergot physiology were mostly motivated by industrial process improvements. These studies nevertheless brought many basic findings from physiology of filamentous fungi.

The results of these studies that were performed with the strains not generally available can hardly be transferred to other strains and, therefore, general applicability or even reproducibility is sometimes questionable. Critical assessment of such results is always necessary. Therefore, all results obtained with, e.g. *Claviceps fusiformis* SD-58 that was made generally available by Prof. D.Gröger from Halle an der Saale (Germany) are especially valuable.

Deep knowledge of the regulation of the alkaloid synthesis and the metabolism of the producer is a prerequisite for successful design of biotechnological process of alkaloid production. Some products of primary metabolism serve in secondary metabolism as precursors. They, however, exhibit also regulatory effects (tryptophan) or act as structural elements (lipids) influencing thus the biosynthesis of secondary metabolites in a complex way.

6.1.1.
Carbon Sources and Saccharides

Saccharides are the main carbon and energy source for *Claviceps* both under parasitic and saprophytic conditions. Sucrose is an important component of phloem sap of the host plants (Basset *et al.*, 1972). The best substrates for alkaloid production are slowly metabolizable Saccharides, e.g., sucrose, maltose and polyols (mannitol, sorbitol) (Křen *et al.*, 1984).

Sucrose is metabolised faster at the very beginning of the submerse *C. purpurea* cultivation (Figure 1). Monosaccharides released are quickly utilized (Křen *et al.*, 1984). From the sucrose molecule, mostly glucose is utilized. Fructose, by the action of transglycosylating β-frucfuranosidase, is oligomerized starting with one sucrose molecule (Perényi *et al.*, 1968).

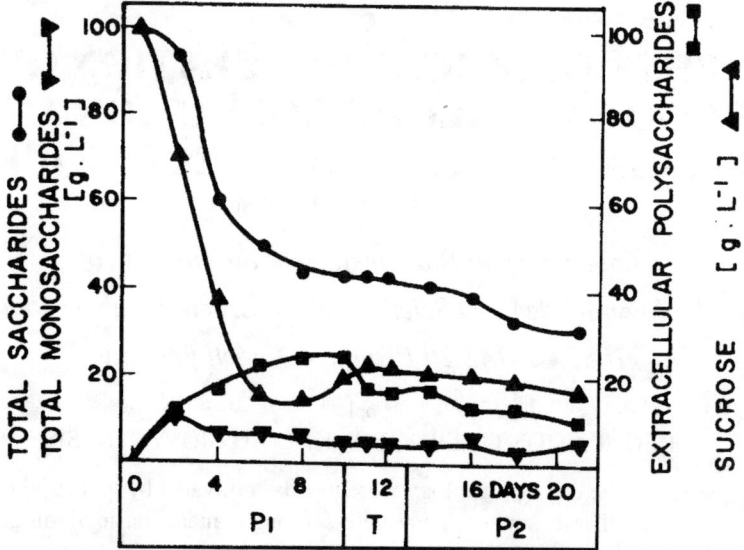

Figure 1 Sucrose metabolism in a submerged culture of *C. fusiformis* (former *purpurea*) 129/35. P_1, P_2 production phases of the culture; T transition (growth) phase (Křen *et at.*, 1984).

Glucose is not suitable as a sole carbon source because it supports mycelial growth and the alkaloid formation is virtually inhibited by glucose-catabolite repression (Křen *et al.*, 1987a). Glucose in some strains, e.g., *C. fusiformis*, also strongly supports extracellular polysaccharide (β-glucan) formation that increase viscosity and complicate the oxygen transport (Buck *et al.*, 1968). However, glucose, when added in an optimised proportion and at respective time, supports alkaloid production.

Secondary carbon-sources, i.e., intermediates of citrate or glyoxalate cycle are essential for good saccharide utilisation by saprophytically cultivated *Claviceps* strains.

C. purpurea growing on sucrose as sole C-source produces lactate and severe acidification of the medium diminishes activity of Krebs cycle enzymes, e.g., malatedehydrogenase and citratesynthase (Pažoutová and Řeháček, 1981). Addition of citrate or succinate to the medium maintains the citrate cycle active, ensures catabolism of pyruvate and by ATP production helps to control the rate of glycolysis.

Key regulatory enzyme of the glycolysis in *Claviceps* is phosphofructokinase that has rather atypical regulatory properties—feedforward regulation and hyperbolic saturation kinetics—(Křen and Řeháček, 1984).

Hexosomonophosphate pathway prevails in glucose breakdown during the vegetative phase of fermentation, the share of the glycolytic pathway becomes more pronounced during alkaloid synthesis (Gaberc-Porekar *et al.,* 1990). Supply of NADPH is important for ergot alkaloid biosynhesis, e.g., for cyt P-450-catalysed steps.

6.1.2.
Phosphate

Inorganic phosphate has an exquisite role among the inorganic nutrients in *C. purpurea* cultivation. In most of the microbial secondary metabolites (e.g., antibiotics) a suboptimal level of phosphate has a stimulative role for their production (Martín, 1977). In high alkaloid producing *Claviceps* strains the optimum level of phosphate in medium ranges from 1 to 4 mM. The phosphate is taken up by the fungus during first 2–4 days of the cultivation and it is accumulated in the mycelium. Higher phosphate concentrations stimulate vegetative growth of the mycelium, glucane production and inhibit alkaloid biosynthesis. Drop of the intracellular phosphate coincides with the stoppage of the mycelial growth and the onset of the alkaloid production (Pažoutová and Řeháček, 1984). High levels of phosphate also induce alkaloid degrading enzymes, lowering thus the yields (Robbers *et al.,* 1978).

Some strains as, e.g., *C. paspali,* demand for alkaloid production higher phosphate concentration.

6.1.3.
Lipids

Ergot alkaloid biosynthesis and lipid metabolism have some common regulatory points. Hydroxymethylglutaryl-coenzyme-A reductase is the key regulatory enzyme of isoprenyl unit production. The activity pattern of this enzyme in *Claviceps* (Křen *et al.,* 1986) reveals that the mevalonate distribution is shared at the beginning of the fermentation by sterol and alkaloid biosynthesis and later it is used solely for alkaloid building.

Substances known to modulate lipid biosynthesis (and cell lipid composition), e.g., chlorophenoxy acids are also able to influence ergot alkaloid biosynthesis and to amend their yields (Křen *et al.,* 1990a). Chlorophenoxypropionate in minute doses significantly stimulates alkaloid production in *C. purpurea* strain 59 (Figure 2).

6.1.4.
Medium Osmolarity

Claviceps in its parasitic stadium grows under high sucrose concentration of the phloem sap. This fact lead to assumption (Amici *et al.,* 1967) (see Chapter 1) that high sucrose concentration should be favourable for the high alkaloid production. Optimum sucrose concentration was identified to be 30–35%. At present, most media for saprophytic alkaloid production by *Claviceps* have at least 100 g sucrose per litre. Later it was found that part of sucrose (or manitol) can be replaced by NaCl (or KC1) to maintain high osmolarity (Puc and Sočič, 1977).

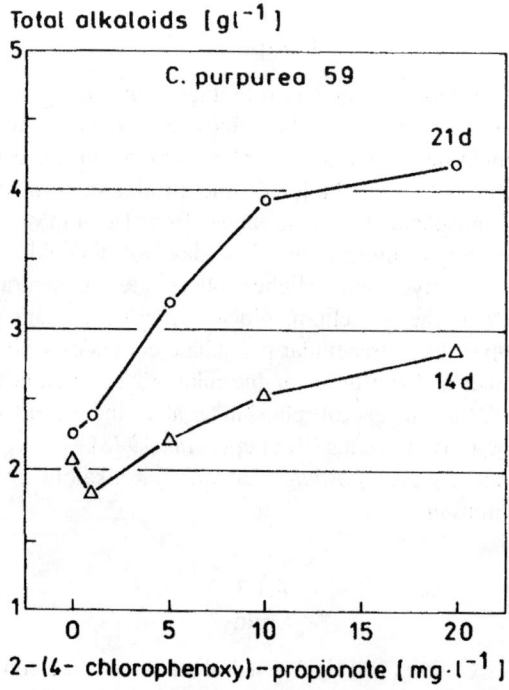

Figure 2 Effect of DL-2(4-chlorophenoxy)-propionate on ergot alkaloid production in *Claviceps purpurea* 59 on 14th (Δ) and 21st (o) day of cultivation (Křen *et al.*, 1990a).

6.2.
SEASONAL CYCLES IN ERGOT CULTIVATION

Most people working longer with *Claviceps* strains admit that there exist substantial difference in alkaloid production and strain growth during the year. These effects occur mostly in the Autumn season (September till beginning of December—in Europe) when the strains grow worse and the production is lower. Some people ascribe these effects to higher risk of aerial contamination in this period when, due to the fruit ripening, there exist high concentration of airborne germs. However, these effects occur also in noncontaminated cultures. This seasonal effect might have a correlation with time rhythms of the parasite growing in nature. Not all strains exert this effect and it is not always of the same magnitude.

To my knowledge, this observation has never been published as it should have been proved by a large statistical set. However, both people in research and in industry should bear this possibility in their minds when comparing the results from various seasons of the year.

6.3.
BIOTECHNOLOGY OF SUBMERGED CULTIVATION OF ERGOT

The prerequisite for successful, high yielding cultivation is a high-quality strain. There exist many *Claviceps* strains (Schmauder, 1982), some of them of unclear origin. Primarily, all the strains were isolated from the parasitic sclerotia and later were mutagenized and selected. Some of the strains are deposited in public collections, however, most of high-producing strains are in the possession of pharmaceutical companies (see also Chapter 12).

The *Claviceps* strains display high variability and they often tend to degeneration. Therefore continuous selection and standard storage techniques are integral part of the bioprocess. The best method for strain storage is deep freezing in liquid nitrogen or freeze drying in a preserving solution consisting of sucrose and skimmed milk.

Cultivation conditions for *Claviceps* strains do not basically differ from other cultivations of the filamentous fungi, e.g., *Penicillia*. A crucial step is the inoculum preparation. If the inoculum is of bad quality, the production stage cannot be usually recovered. Suitability of the inoculum can be judged also by morphological and bichemical parameters, e.g., RNA content (Sočič *et al.,* 1985). Production stage must be inoculated by at least 10% of inoculum. During the fermentation all the key parameters must be controlled, pH being 4.5–6.2 (optimum 5.4), aeration 0.5–1.5 l of air/l of medium, mixing rate 75–200rpm and temperature 24 °C. Some synthetic antifoam reagents diminish yields. Good results are obtained by use of plant oils (or ergot oil from sclerotia) that is partly utilised. The inoculum stage takes place for 6–10 days (or longer in multistage inoculum), the production stage takes place for 12–18 days. Long cultivation puts high demand on strict sterility of the process. Sometimes, low amount of broad spectrum antibiotics (chloramphenicol) could be added (Křen *et al.,* 1986c) to protect against prokaryotic contamination.

For most *Claviceps* strains synthetic media can be used. In industry sucrose or sorbitol is used (seldom other C-sources) in concentrations of 70–250 g/l. As a secondary C-source usually citrate in concentration of 5–10 g/l is used. salts, asparagine or salts serve as N-source and phosphate—*vide supra*—and other inorganic salts are supplemented. Inoculation medium differs from the production medium by higher phosphate concentration—up to 10 mM, different N-source, some additional vitamins (biotin, aneurine) and by complex nutrients added—yeast extract, corn steep, malt extract etc.

More than 120 various media with approx. 50 components were described for *Claviceps* strains cultivation. Choice of the suitable inoculation and production medium is rather important because each strain considerably differs in nutrient demands and the respective media optimization should be always performed.

6.4.
CULTIVATION OF ERGOT ON THE PLANT TISSUE CULTURES

Plant tissue culture technique provides completely controlled conditions to elucidate the growth and physiology of the cells and to study host-parasite interactions at the cellular level. There exists probably single paper on this interesting technique in ergot research on cultivation of *C. fusiformis* on a *Pennisetum typhoides* cell culture. The authors (Roy and Kumar, 1985) established the methodology and studied nutritional demands of the mixed culture with respect to the growth and alkaloid production.

6.5.
ALKALOID PRODUCTION BY IMMOBILISED CELLS OF *CLAVICEPS*

The use of immobilised cells of producing strain is a promising method for extending the production period of the ergot cultures.

Claviceps purpurea was immobilised in Ca-alginate by Kopp and Rehm (1983, 1984). The immobilised cells maintained the alkaloid production, for approximately 150 days under semicontinuous regime. Higher matrix (alginate) concentration promoted the overall production but mostly alkaloids in lower oxidation state were produced. The immobilised cells formed sklerotia-like structures suggesting the immobilisation (matrix) can simulate parasitic conditions.

Physiology of immobilised *C. fusiformis* during long-term semicontinuous cultivation was studied by Křen *et al.* (1987b). The immobilised cells maintained the alkaloid production for 770 days. The cells underwent profound morphological changes—vacuolisation, mitochondrial degeneration. The beads remained mechanically stable for the whole time.

Also *C. paspali* has been immobilized and semicontinuous production of simple lysergic acid derivatives has been achieved (Rozman *et al.*, 1989).

Immobilisation of *Claviceps* cells in other matrices (carrageenan, pectate) and their coimmobilisation with oxygen supplying systems was studied by Křen (1990). The immobilised *Claviceps* cells can be used not only for alkaloid production but also for biotransformations for e.g., stabilization of *C. purpurea* protoplasts (Komel *et al.*, 1985).

6.6.
PROTOCOL

Following procedure as a practical example describes submerged cultivation of *Claviceps fusiformis* SD-58 for clavine alkaloid production. The strain can be obtained, e.g., from American Type Culture Collection (ATCC).

The culture from a slant agar medium (Sabouraud) is inoculated into inoculation medium TI (60 ml/300 ml conical flasks) and aseptically cultivated on rotary shaker for 10 days. 5 ml of the mycelial suspension is transferred into same amount of production medium CS2 and shaken for 20 days (temperature 24°C). Ergot alkaloid production (agroclavine, elymoclavine, chanoclavine) is 0. 8–2.5 g/l. After biomass separation the liquid is alkalised by ammonia and extracted by ether. Alkaloid can be separated by TLC (SiO_2, $CHCl_3$: MeOH: NH_4OH=8:2:0.02) and the spots visualised by Ehrlich reagent. Media: TI, (g/1): sucrose, 100; L-asparagin, 10; $(NH_4)_2SO_4$, 10; KH_2PO_4, 0.25; $Ca(NO_3)_2$, 1; yeast extract, 0.1; L-cystein, 0.1; inorganic salts and trace elements, pH 5.5. CS2, (g/ 1): sucrose, 100; citric acid, 16.8; $(NH_4)_2SO_4$, 10; KH_2PO_4, 0.25; $CaCl_2$, 1; inorganic salts and trace elements, pH 5.5.

REFERENCES

Amici, A.M., Minghetti, A., Scotti, T., Spalla, C. and Tognoli, L. (1967) Ergotamine production in submerged cultures and physiology of *Claviceps purpurea*. *Appl.Microbiol.,***15,**597–602.

Basset, R.A., Chain, E.B., Dickerson, A.G. and Mantle, P.G. (1972) Comparative metabolism of *Claviceps purpurea in vivo* and *in vitro*. *Biochem. J.,***127,**3P-4P.

Buck, K.W., Chen, A.G., Dickerson, A.G. and Chain, E.B. (1968) Formation and structure of extracellular glucans produced by *Claviceps* species. *J. Gen. Microbiol.,***51,** 337–352.

Gaberc-Porekar, V., Didek-Brumec, M. and Sočič, H. (1990) Carbohydrate metabolism during submerged production of ergot alkaloids. *Appl. Microbiol. Biotechnol.,***34,** 83–86.

Kopp, B. and Rehm. H.-J. (1983) Alkaloid production by immobilized mycelia of *Claviceps purpurea. Eur. J. Appl. Microbiol. Biotechnol.,***18,**257–263.

Kopp, B. and Rehm, H.-J. (1984) Semicontinuous cultivation of immobilized *Claviceps purpurea. Appl. Microbiol. Biotechnol.,***19,**141–145.

Křen, V., Pažoutová, S., Rylko, V., Sajdl, P., Wurst, M. and Řeháček Z. (1984) Extracellular metabolism of sucrose in a submerged culture of *Claviceps purpurea* in relation to formation of monosaccharides and clavine alkaloids. *Appl. Environ.Microbiol.,***48,** 826–829.

Křen, V. and Řeháček, Z. (1984) Feedforward regulation of phosphofructokinase in a submerged culture of *Claviceps purpurea* producing clavine alkaloids. *Speculat.Sci. Technol.,***7,**223–226.

Křen, V., Pažoutová, S., Rylko, V. and Řeháček, Z. (1986) Saprophytic production of clavine alkaloids and activity of 3-hydroxy-3-methylglutaryl coenzyme-A reductase . *Folia Microbiol.,***31,**282–287.

Křen, V., Chomátová, S., Břemek, J., Pilát, P. and Řeháček, Z. (1986c) Effect of some broad-spectrum antibiotic on the high-production strain *Claviceps fusiformis* W1. *Biotechnol. Lett.*,**8**,327–332.

Křen, V., Mehta, P., Rylko, V., Flieger, M., Kozová, J. and Řeháček, Z. (1987a) Substrate regulation of elymoclavine formation by some saccharides. *Zentralblatt für Mikrobiol.*, **142**,71–85.

Křen, V., Ludvík, J., Kofroňová, O., Kozová, J. and Řeháček, Z. (1987b) Physiological activity of immobilized cells of *Claviceps fusiformis* during long-term semicontinuous cultivation. *Appl. Microbiol. Biotechnol.*,**26**,219–226.

Křen, V. (1990) Complex exploitation of *Claviceps* immobilized cells. In de Bont, J., Visser, J., Mattiasson, B. and Tramper, V. (eds.) *Physiology of immobilized cells,*Proc. Int. Symp. Wageningen, December, 10–13, 1989, Elsevier, Amsterdam, pp. 551–556.

Křen, V., Pažoutová, S., Rezanka, T., Viden, I., Amler, E. and Sajdl, P. (1990a) Regulation of lipid and ergot alkaloid biosynthesis in *Claviceps purpurea* by chlorophenoxy acids. *Biochem. Physiol. Pflanzen,***186**,99–108.

Komel, R., Rozman, D., Puc, A. and Socic, H. (1985) Effect of immobilization on the stability of *Claviceps purpurea* protoplasts. *Appl. Microbiol. Biotechnol.*,**23**,106–109.

Martin, J.F. (1977) Control of antibiotic synthesis by phosphate. *Adv. Biochem. Eng.*,**6**, 105–127.

Pažoutová, S. and Řeháček, Z. (1981) The role of citrate on the oxidative metabolism of submerged cultures of *Claviceps purpurea* 129. *Arch. Microbiol.*,**129**,251–253.

Pažoutová, S. and Řeháček, Z. (1984) Phosphate regulation of phosphatases in submerged cultures of *Claviceps purpurea* 129 producing clavine alkaloids. *Appl.Microbiol. Biotecbnol.*,**20**,389–392.

Perenyi, T., Udvardy-Nagy, E. and Novak, E.K. (1968) Studies on sucrose breakdown by invertase-producing *Claviceps* strains. *Acta Microbiol. Acad. Sci. Hung.*,**15**,17–28.

Puc, A. and Socic, H. (1977) Carbohydrate nutrition of *Claviceps purpurea* for alkaloid production related to the osmolarity of media. *Eur. J. Appl. Microbiol.*,**4**,283–287.

Robbers, J.E., Eggert, W.W. and Floss, H.G. (1978) Physiological studies on ergot. Time factor influence on the inhibitory effect of phosphate and the induction effect of tryptophan on alkaloid production. *Lloydia,***41**,120–129.

Roy, S. and Kumar, A. (1985) Production of alkaloids by ergot (*Claviceps fusiformis* Lov.) on *Pennisetum typhoides* (Burm.) Stapf and Hubb. *in vitro.* Neuman, K.H. and Bart, W. (eds.) *Primary and Secondary Metabolism of Plant Cell Cultures.*Springer-Verlag, Berlin, Heidelberg, pp. 117–123.

Rozman, D., Pertot, E., Komel., R. and Prošek, M. (1989) Production of lysergic acid derivatives with immobilized *Claviceps paspali* mycelium. *Appl. Microbiol.Biotecbnol.*,**32**,5–10.

Schmauder, H.-P. (1982) Ergot. Saprophytic production of ergot alkaloids. In Atal, C.K. and Kapur, B.M. (eds.) *Cultivation and Utilization of Medicinal Plants.*Council of Sci. and Ind. Res., Jammu-Tawi, India.

Socic, H., Gaberc-Porekar, V. and Didek-Brumec, M. (1985) Biochemical characterization of the inoculum of *Claviceps purpurea* for submerged production of ergot alkaloids. *Appl. Microbiol. Biotecbnol.*,**21**,91–95.

7.
ERGOT ALKALOIDS AND OTHER METABOLITESOF THE GENUS *CLAVICEPS*

MARTIN BUCHTA and LADISLAV CVAK

Galena a.s. Opava, 747 70 Czech Republic

7.1.
INTRODUCTION

Both parasitic and saprophytic ergot is a producer of vast number of compounds from which alkaloids represent the most important group. Over 80 ergot alkaloids (EA) have as yet been isolated from diverse natural material, mainly from *Claviceps* strains, but also from other fungi and higher plants. Besides alkaloids and some other secondary metabolites (pigments and mycotoxins), ergot produces also some unusual primary metabolites of the lipidic and sacharidic nature.

7.2.
ERGOT ALKALOIDS

Only alkaloids isolated from different *Claviceps* species are described in this chapter. EA produced by other fungi and higher plants are the subject of the Chapter 18. Beside this limitation, nearly 70 ergot alkaloids, produced by genus *Claviceps,* are discussed here.

7.2.1.
Ergot Alkaloid Structure

All the EA are biosynthetically derived from L-tryptophane, that means indole is the base of their structure. Although the most of EA contains a tetra-cyclic **ergoline** system (Figure 1), there are some exceptions whose structure is created only by two or three cycles. Natural alkaloids can be divided into three groups according to their structure: The first, structurally simplest group, contains alkaloids of clavine type, the second group are simple amides of lysergic and paspalic acids and the third, most complex group is created by the alkaloids of peptidic type. Biosynthetically, the peptidic EA can be understand as tetrapeptides containing lysergic acid as the first member of the peptidic chain. The other amino acids of the tetrapeptide are variable, which provides the basis for a great diversity of this group of EA. Structurally, two group of peptidic alkaloids can be distinguished: ergopeptines and ergopeptams.

7.2.2.
Ergot Alkaloid Nomenclature

EA nomenclature is very complex and systematic names are used only for semisynthetic derivatives or for exact chemical description. New natural compounds were usually given a trivial name by their discoverers which were later generally accepted.

Trivial names mark the milestones in the history of ergot research and their etymology unravels interesting facets of the ergot story. Most of the trivial names of EA stem from the botanical names of their producers or the host plant, e.g., agroclavine and pyroclavine *(Agropyrum),* elymoclavine and molliclavine *(Elymus mollis),* setoclavine and penniclavine *(Pennisetum),* paspalic acid *(Paspalum),* festuclavine *(Festuca),* ergosecaline *(Secale),* etc. Other trivial EA names are connected with the circumstances of their discovery, e.g., ergokryptine remained obscured (κρυπτός) for a long time (however, some authors and even official pharmacopaeias use incorrectly name "ergocriptine"), ergobasine was considered to be very basic or lysergic acid was a product of the **lys**is of **ergot** alkaloids. Still other names were connected with the pharmacological properties, e.g., ergotoxine or ergo**metri**ne (endo**metri**um uteri). Some scientists embodied parts of their private life in the EA names, e.g., Stoll's ergo**crist**ine according to a girl named **Crist**ine or Flieger's ergo**anna**me according to his daughter **Anna**. Sometimes, the discoverers wanted to make a compliment to a person or company closely associated with EA, e.g., ergo**lad**inine (**Ladi**slav Cvak) or ergo**gal**ine (**Gal**ena).

The rational approach to the nomenclature created hybrids bearing prefixes nor-, iso-, dihydro- and suffixes -inine, -ol, -am, etc. Other trivial names were coined also for some products of chemical modification, e.g., prefix lumifor the products of photochemically initiated water addition or aci- for ergopeptine isomers possessing acidic properties (see Chapter 8). Another such rational approach, but not generally accepted, is the use of the name **ergolene** for 8,9- or 9,10-didehydroergolines.

The systematic nomenclature of EA and their derivatives is based on general principles of chemical nomenclature. Three types of nomenclature are used. The first is the nomenclature according to Chemical Abstracts using the name **ergoline** (Figure 1) for the tetra-cyclic system which is present in most of EA and the name **ergotaman** (Figure 2) for the hepta-cyclic system of EA of peptidic type. The second type of systematic names, valid for peptidic alkaloids only, uses the name **ergopeptine** (Figure 3) for the whole hepta-cyclic system of these alkaloids including all the substituents present in natural EA of this type. The last one, the most complicated but the most rational, is the nomenclature based on IUPAC rules for nomenclature of heterocyclic compounds. According to this nomenclature the ergoline system can be described as 7-methyl4,6,6a,7,8, 9,10,10a-octahydro-indolo[4, 3-fg]quinoline. Examples of all those types of nomenclature can be found in the Chapter 13.

7.2.3. Ergot Alkaloid Stereochemistry

Although ergot alkaloids contain several chiral centres of various configuration, there is one fixed configuration in all the natural EA, i.e., 5*R* having relation to EA biosynthetic precursor L-tryptophane. Only EA prepared by

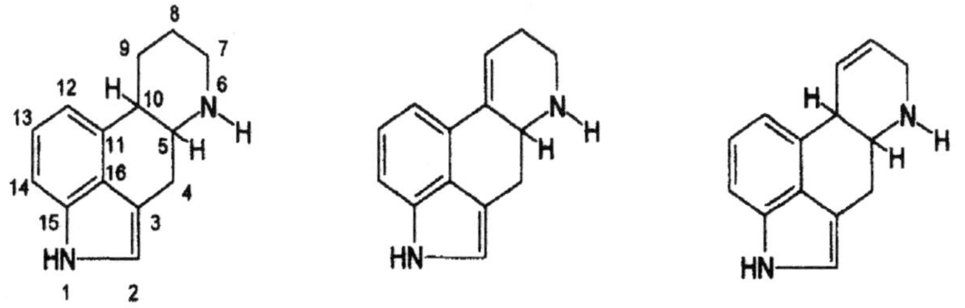

ergoline

$\Delta^{9,10}$-ergolene

9,10-didehydroergoline

$\Delta^{8,9}$-ergolene

8,9-didehydroergoline

Figure 1 Ergoline and ergolene structures and their numbering

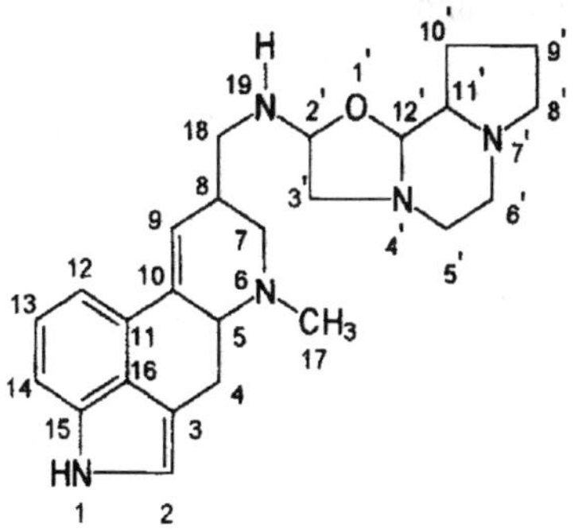

Figure 2 Ergotaman skeleton and its numbering[1]

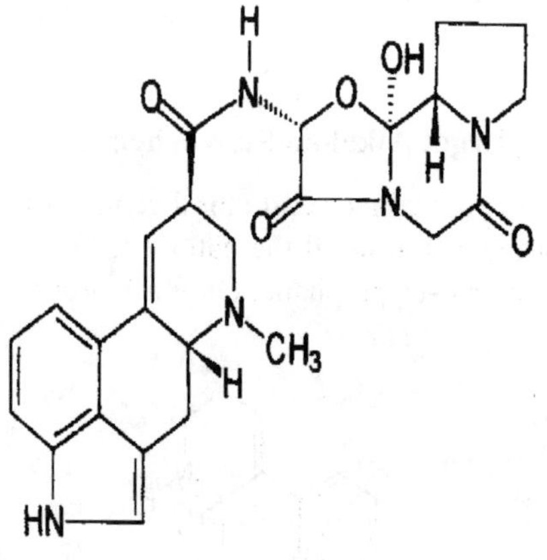

Figure 3 Ergopeptine skeleton

total synthesis or products prepared by isomerisation of natural alkaloids under harsh conditions (e.g. hydrolysis or hydrazinolysis) can have 5*S* configuration. Natural 5*R* EA are sometimes specified as D isomers, which is incorrect. The only *d*- and *l*-denomination for some derivatives e.g., *d*-lysergic acid can be accepted.

The prefix *iso*- in EA nomenclature is strictly confined to the position 8, i.e., lysergic acid for the 5*R*, 8*R* and isolysergic acid for the 5*R*, 8*S* derivatives. Stoll originally designated roman numbers for the stereochemistry at C5-C10: I for 5, 10-*trans*-isomers, e.g., dihydrolysergic acid-I and II for 5, 10-*cis*-isomers, e.g., dihydrolysergic acid-II. This system was used for many natural and semisynthetic alkaloids (e.g., chanoclavines), but in the case of agroclavine-I (5, 10-*cis*-isomer) it was used incorrectly and its name is misleading (see Chapter 18). The stereochemistry of lysergic and dihydrolysergic acids is demonstrated in the Figure 4.

Because of the ergoline system is only slightly puckered, the steroid chemistry nomenclature using designation α- for a substituent below and *β*- for that above the plane is convenient and is therefore used very often for semisynthetic derivatives e.g., 10α-methoxydihydrolysergol.

---

[1] Carbon on C-8 (e.g., carbonyl of lysergic acid) is usually numbered as C-17, however, according to Chemical Abstracts nomenclature C-17 is the Me on N-6, and the carbon on C-8 is numbered as C-18.

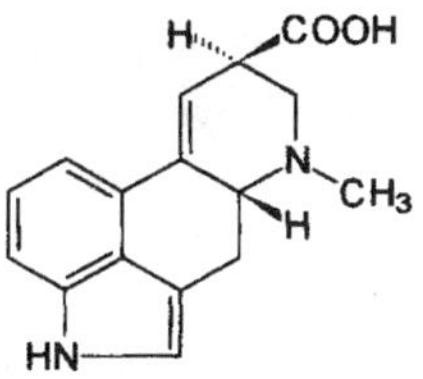

d-lysergic acid (5R, 8R)

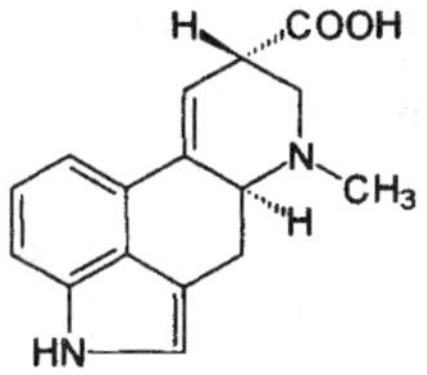

l-lysergic acid (5S, 8S)

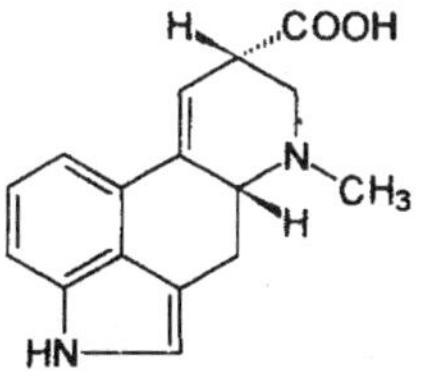

d-isolysergic acid (5R, 8S)

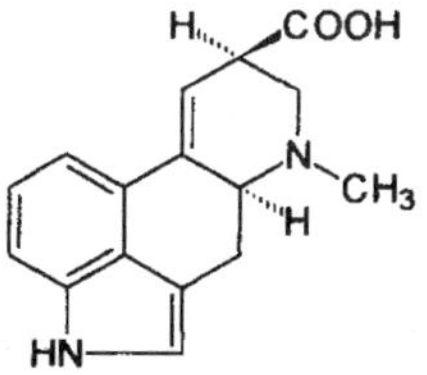

l-isolysergic acid (5S, 8R)

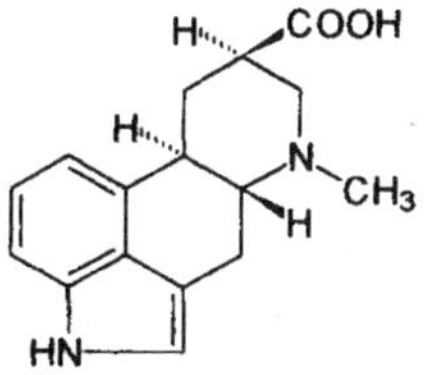

d-dihydrolysergic acid-I (5R, 8R, 10S)

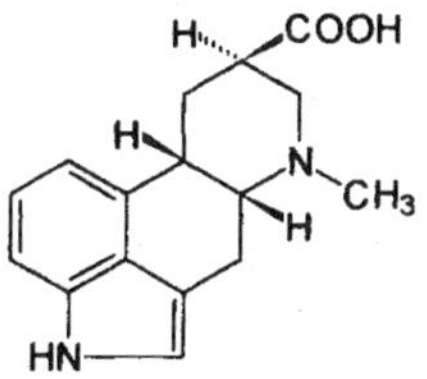

d-dihydrolysergic acid-II (5R, 8R, 10R)

Figure 4 Lysergic acid and dihydrolysergic acid stereoisomers (only two, most important dihydrolysergic acid stereoisomers are presented)

7.2.4.
Clavine Alkaloids

The clavine alkaloids are tricyclic (secoergolines) or tetracyclic (ergolines) compounds with relation to L-tryptophan. The ergoline (secoergoline) skeleton of clavine alkaloids is usually substituted in positions 8 and/or 9 with simple substituents e.g. methyl, hydroxyl, hydroxymethyl or a double bond is present in the positions 8,9 or 9,10. Some clavine alkaloids are the primary products of EA biosynthesis e.g. chanoclavine, agroclavine, elymoclavine and they serve as the biosynthetic precursors of other EA. Although some clavines are produced by various strains of *Claviceps* under parasitic conditions (e.g. chanoclavine-I, agroclavine, elymoclavine), most of the clavine alkaloids were isolated from the saprophytic cultures of some *Claviceps* species originating from the Far East. Particularly Abe and his group of Japanese scientists contributed substantially to the discovery of clavine alkaloids. A number of clavine alkaloids are produced outside of *Claviceps* genus, some of them are identical with that from *Claviceps,* some other are different—see Chapter 18. According to their structures the clavine alkaloids can be divided into two main groups, i.e., 6,7-secoergolines and ergolines and besides one smaller group containing alkaloids with modified ergoline structure.

6, 7-Secoergolines

The basic biosynthetic precursors of 6, 7-secoergolines are chanoclavine-I and chanoclavine-I-aldehyde. All the other secoergolines and $\Delta^{8,9}$–6,7-secoergolenes are the biosynthetic by-products or the products of their further biotransformations. All the natural secoergolines (including secoergolenes) and their occurrence in the nature are summarised in the Table 1. Structures of all the EA of this group are in the Figure 5.

Ergolines

Agroclavine is the biosynthetic precursor of the other ergolines and ergolenes, which is further transformed either to other $\Delta^{8,9}$-ergolenes (e.g. elymoclavine, molliclavine-I) or is isomerised to $\Delta^{9,10}$-ergolenes (e.g. lysergol, penniclavine, setoclavine), or alternatively is reduced to ergolines (e.g. festuclavine, dihydrolysergol). All the clavine alkaloids, with the structure of $\Delta^{8,9}$-ergolene, $\Delta^{9,10}$-ergolene and ergoline, isolated from different strains of *Claviceps* genus are summarised in Table 2. Structures of all clavine alkaloids of this group are on Figure 6.

Alkaloids with Modified Ergoline Structure

Many EA with modified ergoline skeleton have been found in the nature but most of them outside of *Claviceps* genus (e.g. aurantioclavine, cycloclavine, rugulovasine A and B). Only four such modified ergolines were isolated from ergot: 4-dimethylallyltryptophan and *N*-methyl-4-dimethylallyltryptophan are the biosynthetic precursors of the other EA, Clavicipitic acid is the biosynthetic by-product having still the carboxyl group of tryptophan and paspaclavine is a modified alkaloid of the clavine type. The occurrence of these alkaloids in ergot is summarised in the Table 3. Structures of all the modified ergolines isolated from *Claviceps* are in the Figure 7.

Table 1 6, 7-secoergolines

Alkaloid	Structure	Source	Growing	References
Chanoclavine-I	1	*Claviceps* sp. from *Pennisetum typhoideum*	saprophytic	Hofmann *et al.*, 1957
		Claviceps purpurea	parasitic, on rye	Stauffacher and Tscherter, 1964
		Claviceps paspali	saprophytic	Gröger, 1965
		Claviceps gigantea	parasitic, on maize	Agurell and Ramstad, 1965
Chanoclavine-II	2	*Claviceps purpurea*	parasitic, on rye	Stauffacher and Tscherter, 1964
Isochanoclavine-I	3	*Claviceps purpurea*	parasitic, on rye	Stauffacher and Tscherter, 1964
Dihydrochano-clavine-I	4	*Claviceps paspali*	saprophytic	Voigt and Zier, 1970
Dihydro-isochano-clavine-I	5	*Claviceps paspali*	saprophytic	Voigt and Zier, 1970
Norchano-clavine-II	6	*Claviceps purpurea*	saprophytic	Cassady *et al.*, 1973
Paliclavine	7	*Claviceps paspali* from *Paspalum dilatatum*	saprophytic	Tscherter and Haut, 1974
6,7-Secoagro-clavine	8	*Claviceps purpurea*	saprophytic	Horwell and Verge, 1979
Chanoclavine-I-aldehyde	9	*Claviceps purpurea*	saprophytic	Maier *et al.*, 1980a,b
Chanoclavine-I-monofructoside	10	*Claviceps fusiformis*	saprophytic	Flieger *et al.*, 1990
Chanoclavine-I-difructoside	11	*Claviceps fusiformis*	saprophytic	Flieger *et al.*, 1990

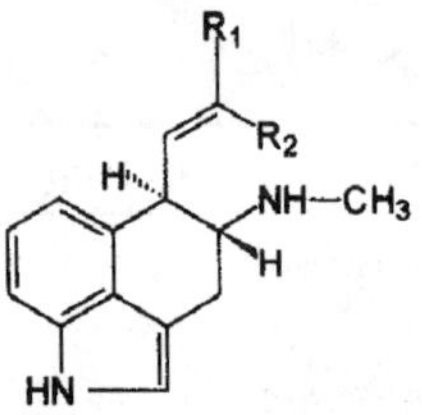

1 chanoclavine-I, R_1 = CH$_2$OH, R_2 = CH$_3$
3 isochanoclavine-I, R_1 = CH$_3$, R_2 = CH$_2$OH
8 secoagroclavine, R_1 = CH$_3$, R_2 = CH$_3$
9 chanoclavine-I aldehyde, R_1 = CHO, R_2 = CH$_3$
10 chanoclavine-I O-β-D-fructofuranoside
11 chanoclavine-I O-β-D-fructofuranosyl-(2→1)-O-β-D-fructofuranoside

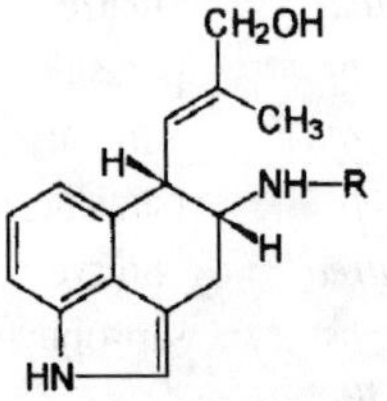

2 chanoclavine-II, R = CH$_3$
6 norchanoclavine-II, R = H

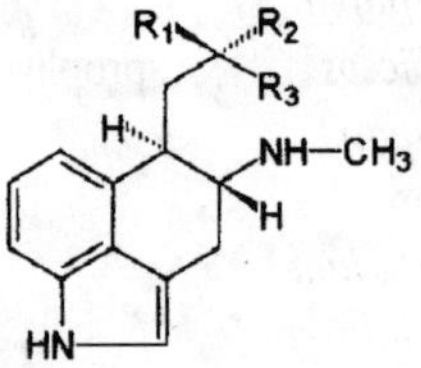

4 dihydrochanoclavine-I, R_1 = CH$_2$OH, R_2 = H, R_3 = CH$_3$
5 dihydroisochanoclavine-I, R_1 = H, R_2 = CH$_2$OH, R_3 = CH$_3$
7 paliclavine, R_1 = C=CH$_2$, R_2 = H, R_3 = OH
             |
            CH$_3$

Figure 5 Clavine alkaloids: 6,7-secoergolines

7.2.5.
Simple Derivatives of Lysergic and Paspalic Acids

Elymoclavine is the biosynthetic precursor of paspalic acid, which is further isomerised to lysergic acid. Lysergic acid can be transformed to its derivatives: simple amides and more complex derivatives—alkaloids of   **Table 2***(Continued)* peptidic type. Lysergic acid can be easily isomerised to isolysergic acid— Figure 8—and, therefore, most of its derivatives, both simple and peptidic ones,

Table 2 Ergolines

Alkaloid	Structure	Source	Growing	References
Agroclavine	12	*Clavicep* sp. from *Agropyrum semicostatum*	saprophytic	Abe, 1948; Abe *et al.*, 1951
		Claviceps sp. from *Pennisetum typhoideum*	saprophytic	Stoll *et al.*, 1954
Elymoclavine	13	*Claviceps* sp. from *Elymus mollis*	saprophytic	Abe *et al.*, 1952
Penniclavine	14	*Claviceps* sp. from *Festuca rubra*	saprophytic	Abe and Yamatodani, 1954
Festuclavine	15	*Claviceps* sp. from *Festuca rubra*	saprophytic	Abe and Yamatodani, 1954; Abe *et al.*, 1956
		Claviceps gigantea	parasitic, on maize	Agurell and Ramstad, 1965
Molliclavine-I	16	*Claviceps* sp. from *Elymus mollis*	saprophytic	Abe and Yamatodani, 1955a
Setoclavine	17	*Claviceps* sp. from *Elymus mollis*	saprophytic	Abe *et al.*, 1955b
		Claviceps sp. from *Pennisetum typhoideum*	saprophytic	Hofmann *et al.*, 1957
Isosetoclavine	18	*Claviceps* sp. from *Elymus mollis*	saprophytic	Abe *et al.*, 1955b
		Claviceps sp. from *Pennisetum typhoideum*	saprophytic	Hofmann *et al.*, 1957
Costaclavine	19	*Claviceps* sp. from *Agropyrum semicostatum*	saprophytic	Abe *et al.*, 1956
Pyroclavine	20	*Claviceps* sp. from *Agropyrum semicostatum*	saprophytic	Abe *et al.*, 1956
		Claviceps gigantea	parasitic, on maize	Agurell and Ramstad, 1965

Alkaloid	Structure	Source	Growing	References
Isopenniclavine	21	*Claviceps* sp. from *Pennisetum typhoideum*	saprophytic	Hofmann *et al.*, 1957
Lysergene	22	*Claviceps* sp. from *Agropyrum semicostatum*	saprophytic	Yamatodani, 1960; Abe *et al.*, 1961
Lysergol	23	*Claviceps* sp. from *Agropyrum semicostatum*	saprophytic	Yamatodani, 1960; Abe *et al.*, 1961
Lysergine	24	*Claviceps* sp. from *Agropyrum semicostatum*	saprophytic	Yamatodani, 1960; Abe *et al.*, 1961
Dihydrolysergol	25	*Claviceps gigantea*	parasitic, on maize	Agurell and Ramstad, 1965
Isolysergol	26	*Claviceps* sp. from *Pennisetum typhoideum*	saprophytic	Agurell, 1966a,b
Norsetoclavine	27	*Claviceps* sp. from *Pennisetum typhoideum*	saprophytic	Ramstad *et al.*, 1967
Elymoclavine-monofructoside	28	*Claviceps purpurea*	saprophytic	Floss *et al.*, 1967b
Dihydroseto-clavine	29	*Claviceps paspali* from *Paspalum dilatatum*	saprophytic	Tscherter and Haut, 1974
Elymoclavine-difructoside	30	*Claviceps fusiformis*	saprophytic	Flieger *et al.*, 1989a

occur in two isomeric forms: lysergic acid derivative bearing the basic trivial name, e.g., ergometrine or ergine and isolysergic acid derivative bearing the suffix -inine, e.g., ergometrinine or erginine.

The first simple derivative of lysergic acid, ergometrine, was isolated by three groups of researchers and thus it has two other alternative names—ergobasine and ergonovine. Later free lysergic acid, paspalic acid and many of their derivatives were found in the nature. All of them are summarised in **Table 4** *(Continued)* the Table 4 and their structures are in the Figure 9. Ergosecaline, mentioned as the last one in this group of EA is a transition between the simple

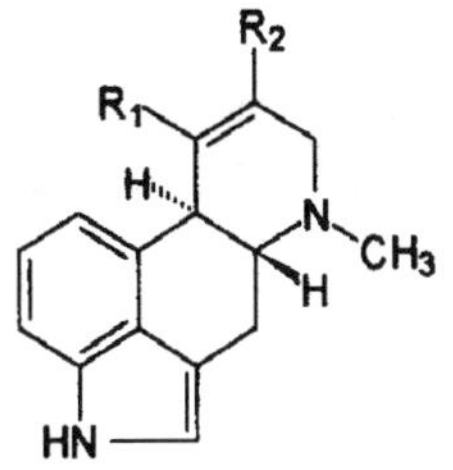

12 agroclavine, R_1 = H, R_2 = CH_3
13 elymoclavine, R_1 = H, R_2 = CH_2OH
16 molliclavine-I, R_1 = OH, R_2 = CH_2OH
28 elymoclavine O-β-D-fructofuranoside
30 elymoclavine O-β-D-fructofuranosyl-(2→1)-O-β-D-fructofuranoside

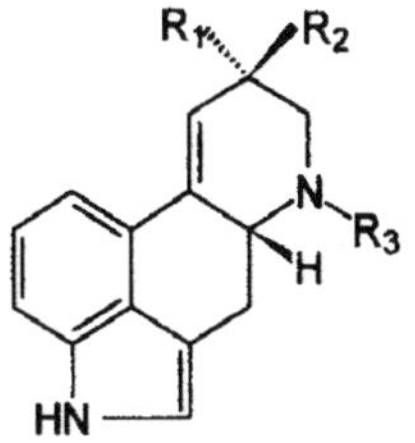

14 penniclavine, R_1 = OH, R_2 = CH_2OH, R_3 = CH_3
17 setoclavine, R_1 = OH, R_2 = CH_3, R_3 = CH_3
18 isosetoclavine, R_1 = CH_3, R_2 = OH, R_3 = CH_3
21 isopenniclavine, R_1 = CH_2OH, R_2 = OH, R_3 = CH_3
22 lysergene, R_1 + R_2 = CH_2, R_3 = CH_3
23 lysergol, R_1 = H, R_2 = CH_2OH, R_3 = CH_3
24 lysergine, R_1 = H, R_2 = CH_3, R_3 = CH_3
26 isolysergol, R_1 = CH_2OH, R_2 = H, R_3 = CH_3
27 norsetoclavine, R_1 = OH, R_2 = CH_3, R_3 = H

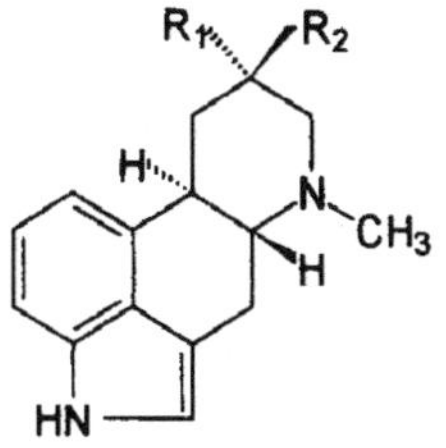

15 festuclavine, R_1 = H, R_2 = CH_3
20 pyroclavine, R_1 = CH_3, R_2 = H
25 dihydrolysergol, R_1 = H, R_2 = CH_2OH
29 dihydrosetoclavine, R_1 = OH, R_2 = CH_3

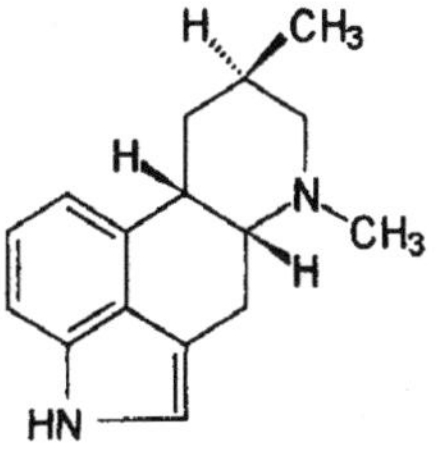

19 costaclavine

Figure 6 Clavine alkaloids: ergolines and ergolenes

Table 3 Alkaloids with modified ergoline skeleton

Alkaloid	Structure	Source	Growing	References
4-dimethylallyl-tryptophan	31	*Claviceps sp.* from *Pennisetum typhoideum*	saprophytic	Agurell and Lindgren, 1968
Clavicipitic acid	32	*Claviceps fusiformis*	saprophytic	King *et al.*, 1973
Paspaclavine	33	*Claviceps paspali* from *Paspalum dilatatum*	saprophytic	Tscherter and Haut, 1974
N-methyl-4-dimethyl-allyltryptophan	34	*Claviceps*	saprophytic	Agurell and Lindgren, 1968

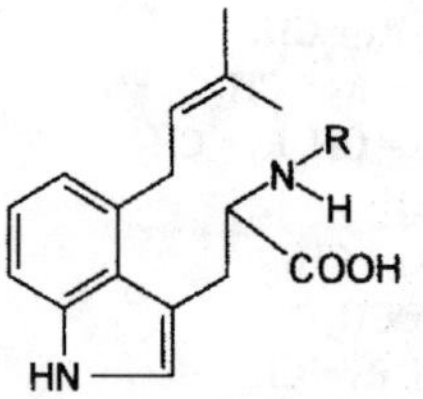

31 4-dimethylallyltryptophan, R = H
34 *N*-methyl-4-dimethylallyltryptophan, R = CH₃

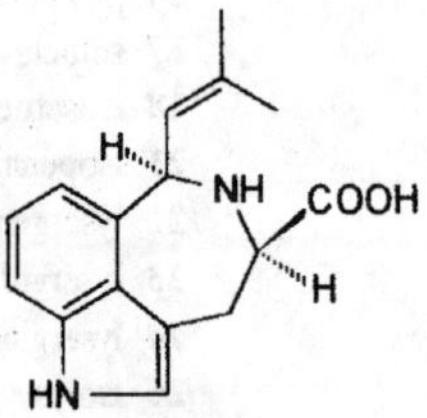

32 clavicipitic acid

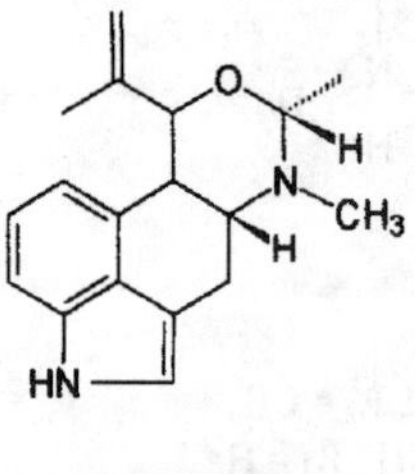

33 paspaclavine

Figure 7 Clavine alkaloids with modified ergoline skeleton

lysergic acid derivative   isolysergic acid derivative

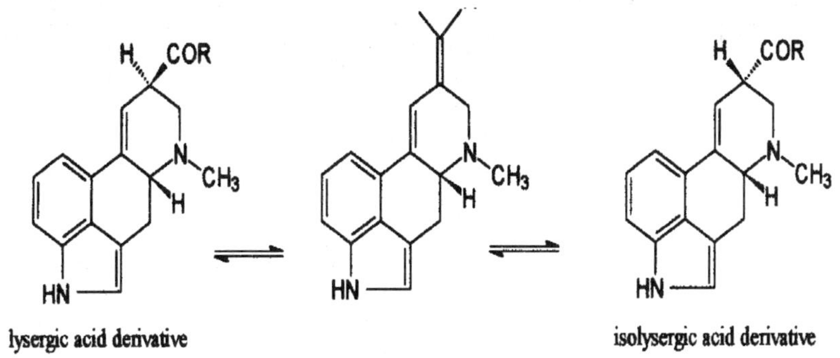

Figure 8 Epimerisation of lysergic acid derivatives

Table 4 Simple derivatives of lysergic and paspalic acids

Alkaloid	Structure	Source	Growing	References
Ergine	35	Claviceps paspali from Paspalum distichum	saprophytic	Arcamone et al., 1961
Ergometrine	36	Claviceps purpurea	parasitic, on rye	Dudley and Moir, 1935; Kharash and Legaul, 1935a, b Thompson, 1935a, b, c; Stoll and Burckhard, 1935
Ergosecaline	37	Claviceps purpurea	saprophytic	Abe et al., 1959
Lysergic acid α-hydroxy-ethylamide	38	Claviceps paspali from Paspalum distichum	saprophytic	Arcamone et al., 1961
		Claviceps paspali	saprophytic	Flieger et al., 1982
Lysergic acid	39	Claviceps paspali from Paspalum dilatatum,	saprophytic	Kobel et al., 1964
		Claviceps	saprophytic	Castagnoli and

Alkaloid	Structure	Source	Growing	References
Paspalic acid	40	*Claviceps paspali* from *Paspalum dilatatum*, *Paspalum distichum*	saprophytic	Kobel *et al.*, 1964
		Claviceps purpurea from *Spartina townsendii*	saprophytic	Castagnoli and Mantle, 1966
8α-Hydroxyergine	41	*Claviceps paspali*	saprophytic	Flieger *et al.*, 1989
10-Hydroxy-*cis*-paspalamide	42	*Claviceps paspali*	saprophytic	Flieger *et al.*, 1993
10-Hydroxy-*trans*-paspalamide	43	*Claviceps paspali*	saprophytic	Flieger *et al.*, 1993

derivatives of lysergic acid and alkaloids of peptide type, which are tetrapeptides, while ergosecaline is a tripeptide.

7.2.6.
Peptidic Ergot Alkaloids

Ergopeptines

Ergopeptines are tetrapeptides formed by lysergic acid and a tripeptide moiety, a unique tricyclic structure called cyclol—**cyclol alkaloids**. Some years ago, ergopeptines seemed to be a closed group of natural product, containing only a limited number of L-amino acids: in position 2 hydroxylated alanine (alkaloids of **ergotamine group),** 2-aminobutyric acid (alkaloids of **ergoxine group)** or valine (alkaloids of **ergotoxine group)** as the first amino acid, phenylalanine, valine, leucine and isoleucine as the second and proline as the third amino acid of the peptidic moiety. This variability formed a basis for existence of 12 natural ergopeptines (three groups, each having four alkaloids), of course each of them with its isolysergic acid derivative called **ergopeptinine**. Only 10 of these ergopeptines were found in *Claviceps:*

β-Ergosine was isolated from higher plant *Ipomoea argyrophylla* (Stauffacher *et al.,* 1965), β-ergoptine was prepared synthetically only. Some novel ergopeptines, containing unusual amino acids as the second amino acid of the peptidic part were isolated recently from both parasitic and saprophytic cultivated ergot: ergogaline with L-homoisoleucine and ergoladine (only isolysergic acid derivative—ergoladinine—was described, it was later transfer to

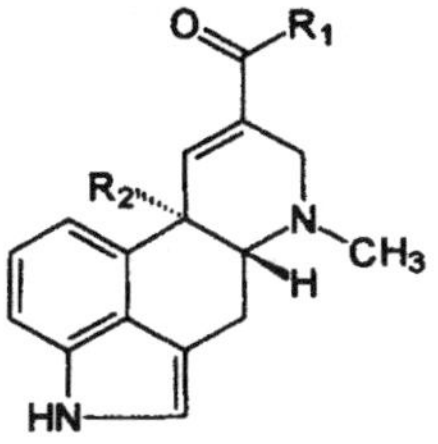

40 paspalic acid, R_1 = OH, R_2 = H
43 10α-hydroxy-*trans*-paspalamide, R_1 = NH₂, R_2 = OH

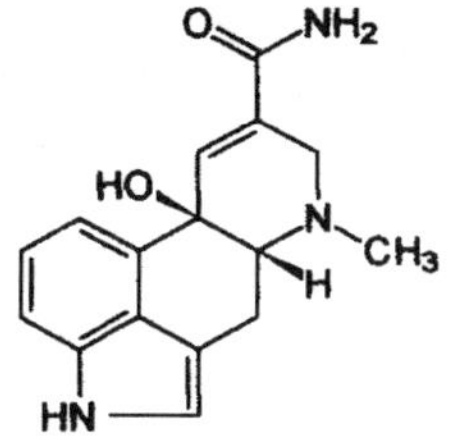

42 10β-hydroxy-*cis*-paspalamide

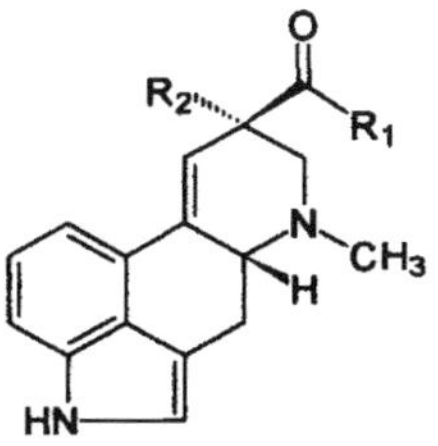

35 ergine, R_1 = NH₂, R_2 = H
36 ergometrine, R_1 = NH-CH(CH₃)CH₂OH, R_2 = H
38 lysergic acid α-hydroxyethylamide, R_1 = NH-CH(OH)CH₃, R_2 = H
39 lysergic acid, R_1 = OH, R_2 = H
41 8α-hydroxyergine, R_1 = NH₂, R_2 = OH

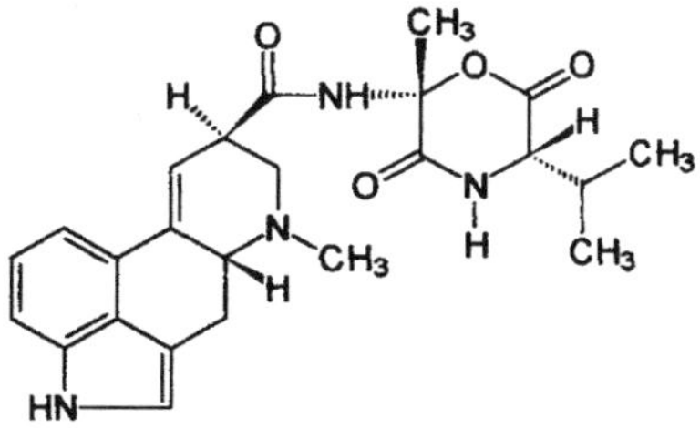

37 ergosecaline

Figure 9 Lysergic and paspalic acids and their simple derivatives

ergoladine—Cvak, personal communication) with L-methionine were isolated from parasitic ergot and ergobutine and ergobutyrine with L-2aminobutyric acid have been isolated from saprophytic culture of *Claviceps*. The last ergopeptine with L-2-aminobutyric acid in position 2 of peptidic part, ergobine, was prepared by directed biosynthesis only (Crespi-Perellino *et al.*, 1993), as well as many other novel ergopeptines. Both, natural and synthetic amino acids were used as biosynthetic precursors for the directed biosynthesis producing ergopeptines with the modified second amino acid of the peptidic moiety: 2-aminobutyric acid, nor-valine, nor-leucine, *p*-chloro-phenylalanine, *p*-fluoro-phenylalanine, 5, 5, 5-trifluoro-leucine, 3-hydroxy-leucine and D-isoleucine (Cvak *et al.*, 1996). The incorporation of D-isoleucine into *epi-β*-ergokryptine is very unusual (Flieger *et al.*, 1984). Many ergopeptine analogues were prepared by synthesis—these are not mentioned here. Finally a new alkaloid with hydroxylated isoleucine as the first amino acid was described by two groups of researchers. In accordance to the formerly accepted name β, α-ergoannam for the ergopeptam alkaloid with the same amino acids (Flieger *et al.*, 1984), the name of this new ergopeptine is β,α-ergoannine. This alkaloid is the first member of a new group of ergopeptines—*β*-**ergoainnine group,** having isoleucine as the first amino acid.

8α-Hydroxylated ergopeptines (8α-hydroxyergotamine and 8α-hydroxy-α-ergokryptine) are ergopeptine derivatives isolated from parasitic *Clavicepspurpurea*. The occurrence of other hydroxylated ergopeptines can be expected as well. Other ergopeptine derivatives isolated from saprophytic cultures are 12′-*O*-methyl ergopeptines (12′-methoxy ergopeptines).

Although many of the clavine alkaloids are ergolines without any double bond in the D cycle, dihydro-α-ergosine isolated from *Claviceps africana* (Mantle and Waight, 1968) is the only example of natural dihydrolysergic acid derivative.

Nearly all the new, recently described alkaloids were isolated by researchers from companies producing ergot alkaloids. The large quantities of processed material give them a chance to discover the minor alkaloids.

All the isolated natural ergopeptines and references about their occurrence are summarised in the Table 5 and their structures are in the Figures 10–12. Structures of some ergopeptines prepared by directed biosynthesis only are in the Figure 13.

Table 5 Ergopeptines

Alkaloid	Structure	Source	Growing	References
Ergotamine	44	*Claviceps purpurea*	parasitic	Stoll, 1945, 1952
α-Ergosine	45	*Claviceps purpurea*	parasitic	Smith and Timmis, 1936, 1937; Stoll, 1952
Ergocristine	46	*Claviceps purpurea*	parasitic	Stoll and Burckhard, 1937; Stoll, 1952
α-Ergokryptine	47	*Claviceps purpurea*	parasitic	Stoll and Hofmann, 1943; Stoll, 1952; Schlientz et al., 1968
Ergocornine	48	*Claviceps purpurea*	parasitic	Stoll and Hofmann, 1943; Stoll, 1952
Ergostine	49	*Claviceps purpurea*	parasitic	Schlientz et al., 1964
β-Ergokryptine	50	*Claviceps purpurea*	parasitic	Schlientz et al., 1968
Dihydro-α-ergosine	51	*Claviceps africana*	parasitic, on *Sorghum vulgare*	Mantle and Waight, 1968
Ergonine	52	*Claviceps purpurea*	parasitic, on rye	Brunner et al., 1979

Table 5*(Continued)*

Alkaloid	Structure	Source	Growing	References
α-Ergoptine	53	*Claviceps purpurea*	parasitic, on rye	Brunner *et al.*, 1979
Ergovaline	54	*Claviceps purpurea*	parasitic, on rye	Brunner *et al.*, 1979
8α-Hydroxy-ergotamine	55	*Claviceps purpurea*	parasitic, on rye	Krajíček *et al.*, 1979
Ergobutine	56	*Claviceps purpurea*	saprophytic	Bianchi *et al.*, 1982
Ergobutyrine	57	*Claviceps purpurea*	saprophytic	Bianchi *et al.*, 1982
12'-O-Methyl-ergocornine	58	*Claviceps purpurea*	saprophytic	Crespi-Perellino *et al.*, 1987
12'-O-Methyl-α-ergokryptine	59	*Claviceps purpurea*	saprophytic	Crespi-Perellino *et al.*, 1987
Ergogaline	60	*Claviceps purpurea*	parasitic, on rye	Cvak *et al.*, 1994a
β,α-Ergoannine	61	*Claviceps purpurea*	parasitic, on rye	Cvak *et al.*, 1994b; Szántay *et al.*, 1994
Ergoladinine	62	*Claviceps purpurea*	parasitic, on rye	Cvak *et al.*, 1996
8α-Hydroxy-α-ergokryptine	63	*Claviceps purpurea*	parasitic, on rye	Cvak *et al.*, 1997

Ergopeptams

Ergopeptams (non-cyclol alkaloids, lactam alkaloids) are a relatively new group of peptidic EA. The first member of the group—ergocristam was described by Stütz *et al* (1973). Two other natural ergopeptams, ergocornam and α-ergokryptam were isolated from field ergot by Flieger *et al.* (1981). The last three known ergopeptams (β-ergokryptam, α, β-ergoannam and β, β-ergoannam) were isolated as the products of directed biosynthesis aimed at the saprophytic production of β-ergokryptine by the feeding of isoleucine (Flieger *et al.*, 1984). The same authors suggested the new nomenclature of this new class of EA, using suffix -am to the name of ergopeptine containing the same amino acids.

Ergopeptams are minor products of the alkaloid biosynthesis. The probability of lactam formation decreases with the steric size of the side chain of the first amino acid. The lactam alkaloids originate as a result of easy racemisation of L-proline in *cis*-dioxopiperazines (biosynthetic precursors of ergopeptines—see Chapter 5) to the thermodynamically favoured *trans*-dioxopiperazines (Day *et al.,* 1985). The racemisation probably competes with hydroxylation of the amino acid attached to lysergic acid (Quigley and

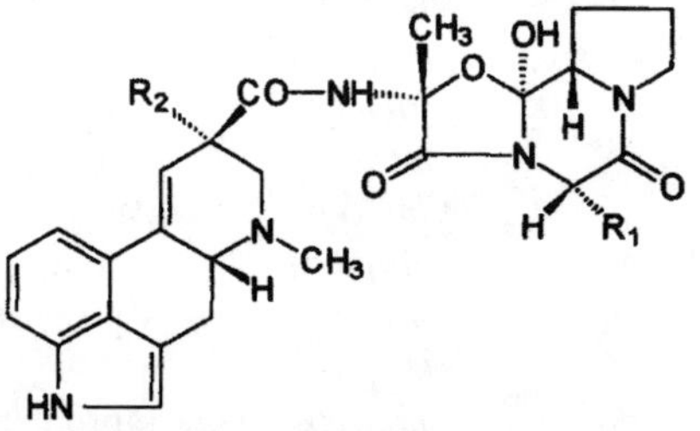

44 ergotamine, R$_1$ = benzyl, R$_2$ = H
45 α-ergosine, R$_1$ = isobutyl, R$_2$ = H
54 ergovaline, R$_1$ = isopropyl, R$_2$ = H
55 8α-hydroxyergotamine, R$_1$ = benzyl, R$_2$ = OH

Figure 10 Natural ergopeptines: ergotamine group

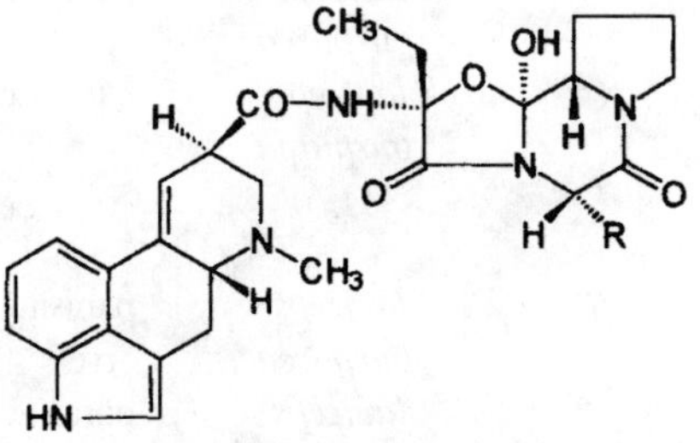

49 ergostine, R = benzyl
52 ergonine, R = isopropyl
53 α-ergoptine, R = isobutyl
56 ergobutine, R = ethyl

Figure 11 Natural ergopeptines: ergoxine group

Floss, 1981). The probability of this process also decreases with increasing size of the first amino acid side chain. Consequently, only ergopeptams with valine, leucine or isoleucine in position 1 of the peptidic moiety have been isolated yet.

The amidic bond between the first amino acid and the dioxopiperazine part of ergopeptams can be easily cleaved by nucleophiles. Lysergyl-valinamide is formed by aminolysis and lysergyl-valine methyl ester by methanolysis of ergocristam. This fact is necessary to take into consideration when ergopeptams are isolated or analysed. Previously described new natural alkaloid, lysergylvalin methyl ester (Schlientz *et al.,* 1963), was the artefact of ergocristam decomposition.

References about the isolation of all the natural ergopeptams are summarised in the Table 6 and their structures are in the Figure 14.

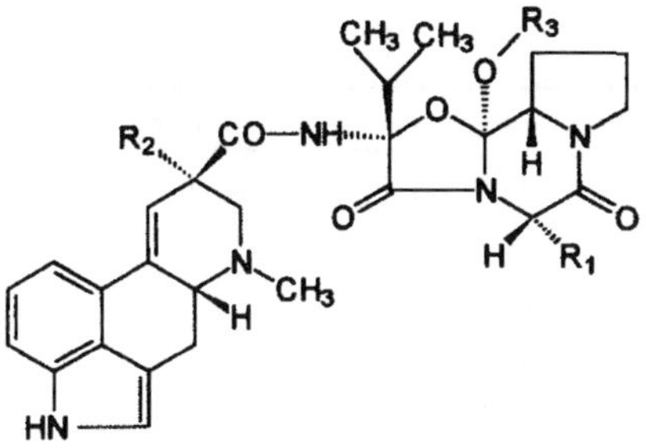

46 ergocristine, R_1 = benzyl, R_2 = R_3 = H
47 α-ergokryptine, R_1 = isobutyl, R_2 = R_3 = H
48 ergocornine, R_1 = isopropyl, R_2 = R_3 = H
50 β-ergokryptine, R_1 = sec-butyl, R_2 = R_3 = H
57 ergobutyrine, R_1 = ethyl, R_2 = R_3 = H
58 12'-O-methyl-ergocornine, R_1 = isopropyl, R_2 = H, R_3 = methyl
59 12'-O-methyl-α-ergpkryptine, R_1 = isobutyl, R_2 = H, R_3 = methyl
60 ergogaline, R_1 = 2-methyl-butyl, R_2 = R_3 = H
62 ergoladine, R_1 = methylthiomethyl, R_2 = R_3 = H
63 8α-hydroxy-α-ergokryptine, R_1 = isobutyl, R_2 = OH, R_3 = H

Figure 12 Natural ergopeptines: ergotoxine group

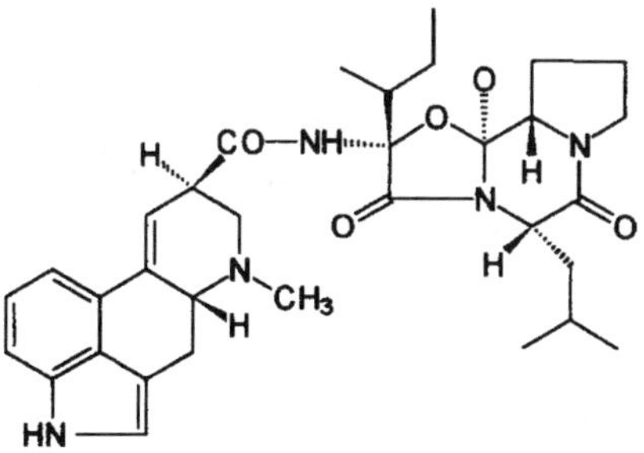

61 β,α-ergoannine

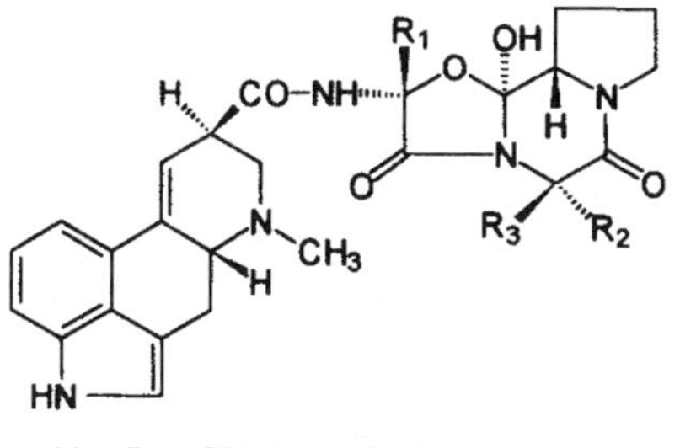

ergobine, R_1 = CH_3, R_2 = ethyl, R_3 = H
5'-epi-β-ergokryptine, R_1 = isopropyl, R_2 = H, R_3 = sec-butyl

Figure 13β, α-Ergoannine and some ergopeptines prepared by directed biosynthesis

Table 6 Ergopeptams

Alkaloid	Structure	Source	Growing	References
Ergocristam	64	*Claviceps purpurea*	saprophytic	Stütz *et al.*, 1973
			parasitic, on rye	Černý *et al.*, 1976
α-Ergokryptam	65	*Claviceps purpurea*	parasitic, on rye	Flieger *et al.*, 1981
				Stuchlík *et al.*, 1982
Ergocornam	66	*Claviceps purpurea*	parasitic, on rye	Flieger *et al.*, 1981
				Stuchlík *et al.*, 1982
β-Ergokryptam	67	*Claviceps purpurea*	saprophytic[*]	Flieger *et al.*, 1984
β,β-Ergoannam	68	*Claviceps purpurea*	saprophytic[*]	Flieger *et al.*, 1984

* directed biosynthesis.

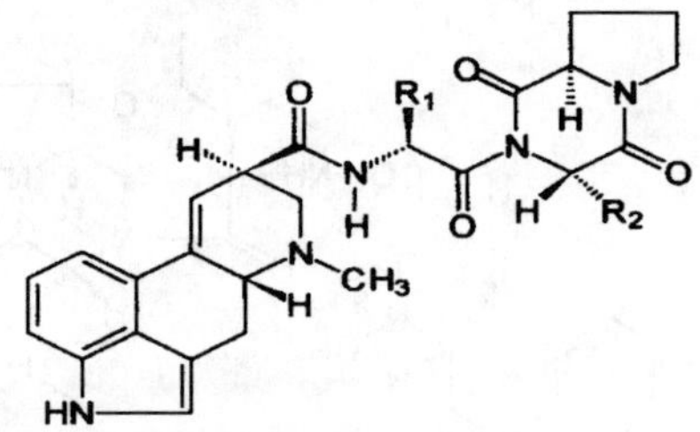

64 ergocristam, R_1 = isopropyl, R_2 = benzyl
65 α-ergokryptam, R_1 = isopropyl, R_2 = isobutyl
66 ergocornam, R_1 = isopropyl, R_2 = isopropyl
67 β-ergokryptam, R_1 = isopropyl, R_2 = *sec*-butyl
68 β,β-ergoannam, R_1 = *sec*-butyl, R_2 = *sec*-butyl

Figure 14 Ergopeptams (natural and prepared by directed biosynthesis)

7.3.
OTHER SECONDARY METABOLITES

7.3.1.
Pigments

Ergot contains 1–2% of pigments (Lorenz, 1979; Schoch *et al.*, 1985). Violet, yellow-red and yellow pigments were isolated from various *Claviceps* species (ApSimon *et al.*, 1965; Franck *et al.*, 1965a; Kornhauser *et al.*, 1965; Perenyi *et al.*, 1966; Franck, 1980). These pigments belong to several structural groups: anthraquinones, biphenyls and organic iron complexes (Šmíd and Beran, 1965).

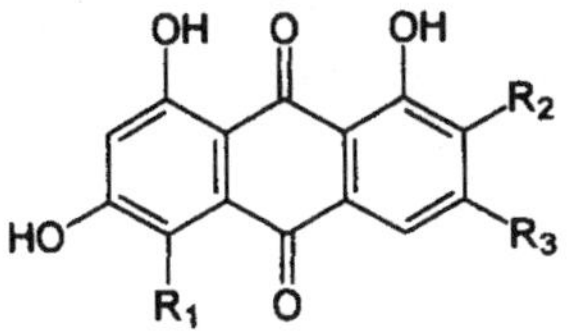

endocrocin, R_1 = H, R_2 = COOH, R_3 = CH3
clavorubin, R_1 = OH, R_2 = COOH, R_3 = CH3
emodin-2-carboxylic acid, R_1 = H, R_2 = H, R_3 = COOH

Figure 15 Anthraquinone pigments isolated from genus *Claviceps*

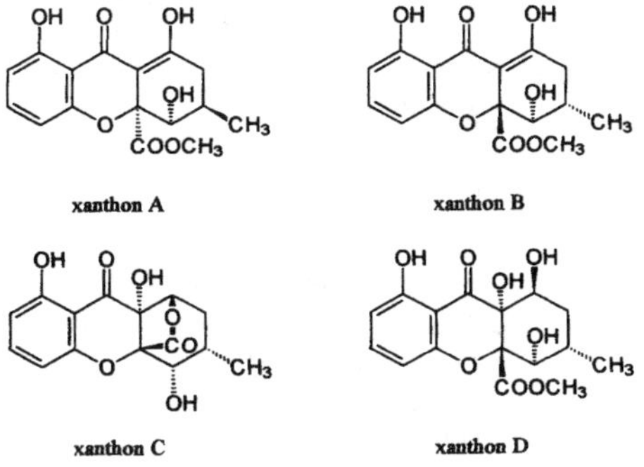

xanthon A

xanthon B

xanthon C

xanthon D

Figure 16 Tetrahydroxanthon units of biphenyl pigments

Three individual anthraquinone yellow-red pigments have been isolated from ergot: endokrocine, clavorubin and emodin-2-carboxylic acid—Figure 15— (Franck, 1960; Betina, 1988).

The yellow biphenyl pigments called ergochromes are dimers of tetrahydroxanthon units: four xanthon units, signed A, B, C and D were described by Franck *et al.,* (1965a, b)—Figure 16. Four particular dimers were described: ergoflavine (ergochrome CC 2, 2′), ergochrysine A (ergochrome AC 2, 2′), ergochrysin B (ergochrome BC 2, 2′) and secalonic acid A (ergochrome AA 4, 4′)—Figure 17. The secalonic acid A is the enantiomer of secalonic acid D the mutagenic and teratogenic metabolite produced by some species of *Penicillium* genus (Ciegler *et al.,* 1980; Betina, 1988).

Some species of *Claviceps* genus produce 2, 3-dihydroxybenzoic acid which forms violet complexes with Fe^{3+} (Kelleher *et al.,* 1971).

7.3.2.

Mycotoxins

Five tremorgenic mycotoxins, containing indole condensed with a diterpenoid unit, have been isolated from *Claviceps paspali* paraziting on the grass *Paspalum dilatatum* (Acklin *et al.,* 1977; Cole *et al,* 1977; Scott, 1984). Their structures are in the Figure 18.

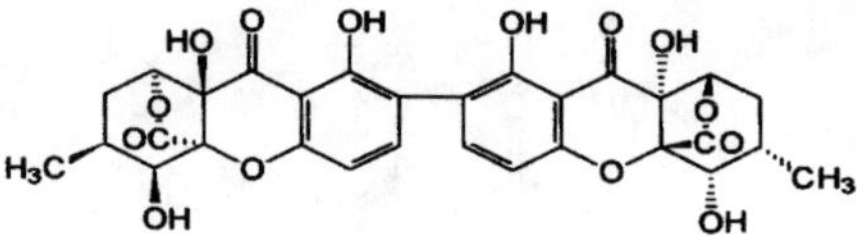

ergoflavin

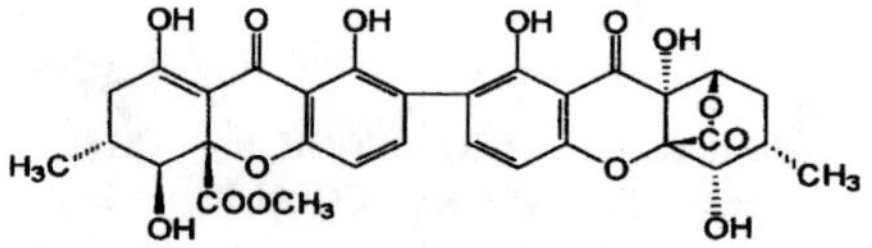

ergochrysin A

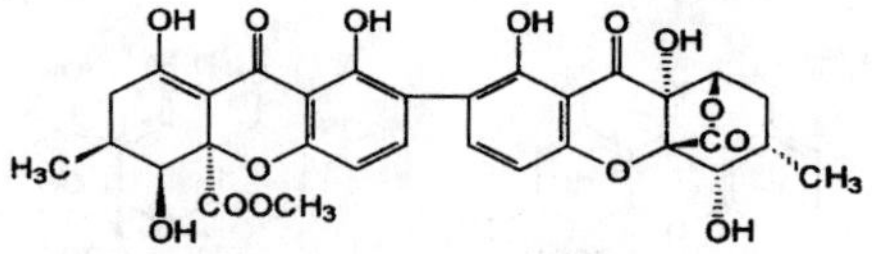

ergochrysin B

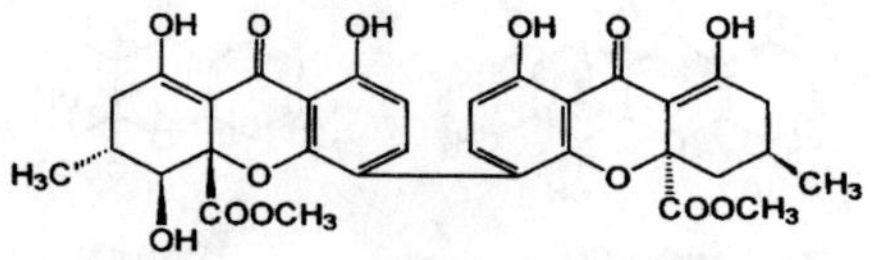

secalonic acid A

Figure 17 Biphenyl pigments (ergochromes) isolated from genus *Claviceps*

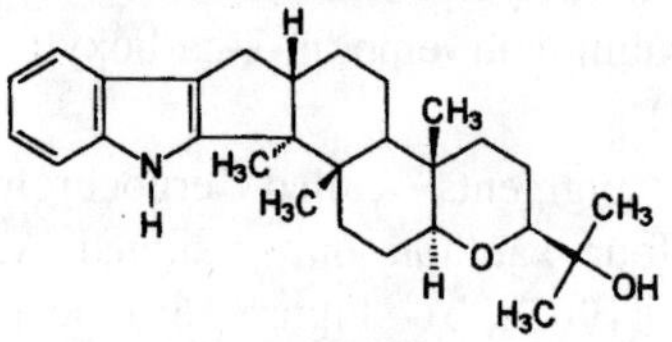

paspaline

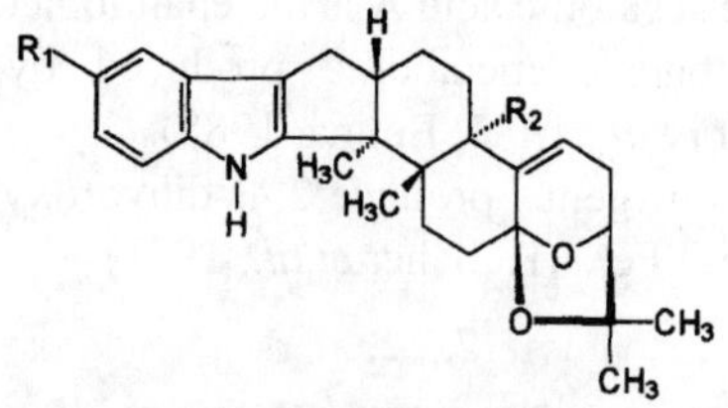

paspalicine, $R_1 = H$, $R_2 = H$
paspalinine, $R_1 = H$, $R_2 = OH$
paspalitreme A, $R_1 = CH_2CH{=}C(CH_3)_2$, $R_2 = OH$
paspalitreme B, $R_1 =$ *trans*$CH{=}CHC(OH)(CH_3)_2$, $R_2 = OH$

Figure 18 Mycotoxins isolated from genus *Claviceps*

7.4.
PRIMARY METABOLITES

7.4.1.
Lipids

Ergot contains usually 30–35% of oil, characterised by a high content of ricinoleic acid (12*(R)*-hydroxyoctadec-9*(Z)*-enoic acid)—up to 36%. Most of the oil is composed of triacylglycerols of estolide type in which, at least, one of the hydroxyl groups of glycerol is esterified by ricinoleic acid or its dimer or trimer (Morris and Hall, 1966; Mangold, 1967). The other hydroxyls of estolides are esterified by common higher fatty acids: palmitic, stearic, palmitoleic, oleic and linoleic (Batrakov and Tolkachev, 1997). Besides estolides the oil is created by common triacyl-, diacyl- and monoacyl-glycerols and free fatty acids. The cell membranes of ergot contain phospholipids (Anderson *et al.,* 1964).

Discovery of sterols began in 1889 when **ergosterol** was isolated from ergot (Tanret, 1889). Later, some other sterols and their precursors were isolated from ergot: fungisterol (Tanret, 1908), 7,22-ergostadienol (Heyl and Swoap, 1930; Heyl, 1932), squalene and cerevisterol (Wieland and Coutelle, 1941). Recently 15 sterols were identified in different *Claviceps* species, using GC-MS technique (Křen *et al,* 1986).

7.4.2.
Other Primary Metabolites

Formation and storage of saccharides depends on the nutrient source, e.g., the glucose rich medium supports the extracellular accumulation of polysacharides (Perlin and Taber, 1963). The saccharide typical for *Claviceps* genus is trehalose (Taber and Wining, 1963; Taber, 1964; Vining and Taber, 1964). Besides saccharides, 10–17% of polyols, mainly mannitol, is accumulated in the mycelium of *Claviceps* genus (Lewis and Smith, 1967). Also stable viscous glucan is produced by genus *Claviceps,* one consists of β-D-glucopyranosyl units, most of which constitute a (1→3)-linked main chain. Other units are attached as branches of the main chain by (1→6)-linkages and are distributed in a relatively uniform arrangement along the length of the polymer (Perlin and Taber, 1963; Buck *et al.,* 1968). This glucan could be autolysed by constitutive β (1→3)-glucanase and β-glucosidase to d-glucose (Dickerson *et al.,* 1970) or by exo type of laminarinase to D-glucose and gentobiose (Perlin and Taber, 1963). The content of polyphosphates in analysed saprophytic strains fluctuates between 0.1 and 2.0 in the dry basis of the mycelium (Taber and Wining, 1963; Kulaev, 1979).

REFERENCES

Abe, M. (1948) Researches on ergot fungus. Part IX. Separation of an active substance and its properties. *J. Agr. Chem. Soc. (Japan),***22,**2.

Abe, M., Yamano, T., Kozu, Y. and Kusumoto, M. (1951) Ergot fungus. XVIII. Production of ergot alkaloids in submerged cultures. *J. Agr. Chem. Soc. (Japan),***24,** 416–422.

Abe, M., Yamano, T., Kozu, Y. and Kosumoto, M. (1952) A new water-soluble ergot alkaloid, elimoclavine. *J. Agr. Chem. Soc. (Japan),***25,**458–459.

Abe, M. and Yamatodani, S. (1954) Isolation of two further water-soluble ergot alkaloid. *J. Agr. Chem. Soc. (Japan),***28,**501–502.

Abe, M. and Yamatodani, S. (1955a) On a new water-soluble ergot alkaloid, molliclavine. *Bull. Agr. Chem. Soc. (Japan),***19,**161–162.

Abe, M., Yamatodani, S., Yamano, T. and Kusumoto, M. (1955b) Communication. *Bull.Agr. Chem. Soc. (Japan),***19,**92.

Abe, M., Yamatodani, S., Yamano, T. and Kusumoto, M. (1956) Isolation of two new water-soluble alkaloids pyroclavine and costaclavine. *Bull. Agr. Chem. Soc. (Japan),* **20,**59–60.

Abe, M., Yamano, T., Yamatodani, S., Kozu, Y., Kusumoto, M., Komatsu, H. and Yamada, S. (1959) On the new peptide-type ergot alkaloids, ergosecaline and ergosecalinine. *Bull. Agr. Chem. Soc. (Japan),***23,**246–248.

Abe, M., Yamatodani, S., Yamano, T. and Kusumoto, M. (1961) Isolation of lysergol, lysergene and lysergine from the saprophytic cultures of ergot fungi. *Agr. Biol.Chem.,* **25,**594–595.

Acklin, W., Weibel, F. and Arigoni, D. (1977) Zur Biosynthese von Paspalin und verwandten Metaboliten aus *Claviceps paspali. Chimia,***31,**63.

Agurell, S. and Ramstad, E. (1965) A new ergot alkaloid from Mexican maize ergot. *Acta Pharm. Suecica,***2,**231–237.

Agurell, S. (1966a) Isolysergol from saprophytic cultures of ergot. *Acta Pharm.Suecica,***3,** 7–10.

Agurell, S. (1966b) Biosynthetic studies on ergot alkaloids and related indoles. *ActaPharm. Suecica,***3,**71–100.

Agurell, S. and Lindgren, J.-E. (1968) Natural occurence of 4-dimethylallyltryptophan— an ergot alkaloid precursor. *Terahedron Lett.,*5127–5128.

Anderson, J.A., Sun, F.K., McDonald, J.K. and Cheldelin, V.H. (1964) Oxidase activity and lipid composition of respiratory particles from *Claviceps purpurea* (Ergot fungus). *Arch. Biochem. Biophys.,***107,**37–50.

Ap Simon, J.W., Hannaford, A.J. and Whalley, W.B. (1965) The chemistry of fungi. Part XLIX. Aliphatic amides from ergot. *J. Chem. Soc.,*4164–4168.

Arcamone, F., Chain, E.B., Ferretti, A., Minghetti, A., Pennella, P., Tonolo, A. and Vero, A. (1961) Production of a new lysergic acid derivates in submerged culture by a strain of *Claviceps paspali* STEVENS & HALL. *Proc. Roy. Soc.,***155 B,**26–51.

Batrakov, S.G. and Tolkachev, O.N. (1997) The structures of triacylglycerols from sclerotia of the rye ergot *Claviceps purpurea* (FRIES) TUL. *Chem. Phys. Lipids,*86, 1–12.

Betina, V. (1988) "Mycotoxins-Chemical, Biological and Enviromental Aspects", 1st ed., Amsterdam.

Bianchi, M.L., Crespi-Perellino, N., Gioia, B. and Minghetti, A. (1982) Production by *Claviceps purpurea* of two new peptide ergot alkaloids belonging to a new series containing α-aminobutyric acid. *J. Nat. Prod.,* **45,**191–196.

Brunner, R., Stütz, P.L., Tscherter, H. and Stadler, P.A. (1979) Isolation of ergovaline, ergoptine, and ergonine, new alkaloids of the peptide type, from ergot sclerotia. *Can. J. Chem.,* **57,**1638–1641.

Buck, K.W., Chen, A.W., Dickerson, A.G. and Chain, E.B. (1968) Formation and structure of extracellular glucans produced by *Claviceps purpurea. J. Gen. Microbiol.,* **51,**337–352.

Cassady, J.M., Abon-Chaar, C.I. and Floss, H.G. (1973) Ergot alkaloids. Isolation of *N*-demethylchanoclavine-II from *Claviceps* strain SD 58 and the role of demethylchanoclavines in ergoline biosynthesis. *Lloydia,* **36,**390–396.

Castagnoli, N. and Mantle, P.G. (1966) Occurence of D-lysergic acid and 6-methylergol-8-ene-8-carboxylic acid in cultures of *Claviceps purpurea. Nature,* **211,** 859–860.

Ciegler, A., Hayes, A.W. and Vesonder, R.F. (1980) Production and biological activity of secalonic acid D, *Appl. Environ. Microbiol.,* **39,**285–287.

Cole, R.J., Dorner, J.W., Landsen, J.A., Cox, R.H., Pape, C., Cunfer, B., Nicholson, S.S. and Bedell, D.M. (1977) Paspalum staggers: Isolation and identification of tremorgenic metabolites from sclerotia of *Claviceps paspali. J. Agr. Food Chem.,* **25,**1197–1201.

Crespi-Perellino, N., Ballabio, M., Gioia, B. and Minghetti, A. (1987) Two unusual ergopeptins produced by saprophitic cultures of *Claviceps purpurea. J. Nat. Prod.,* **50,** 1065–1074.

Crespi-Perellino, N., Malyszko, J., Ballabio, M., Gioia, B. and Minghetti, A. (1993) Identification of ergobine, a new natural peptide ergot alkaloid. *J. Nat. Prod.,* **56,** 489–493.

Cvak, L., Jegorov, A., Sedmera, P., Havlíček, V., Ondráček, J., Hušák, M., Pakhomova, S., Kratochvíl, B. and Granzin, J. (1994a) Ergogaline, a new ergot alkaloid, produced by *Claviceps purpurea:* Isolation, identification, crystal structure and molecular conformation. *J. Chem. Soc. Perkin Trans.,* **2,**1861–1865.

Cvak, L., Jegorov, A., Sedmera, P., Havlíček, V., Ondráček, J., Hušák, M., Pakhomova, S. and Kratochvíl, B. (1994b) New ergopeptines isolated from ergot. In *Proc. 7thInternal. Congr. Mycology Div.",* Prague, 453.

Cvak, L., Minář, J., Pakhomova, S., Ondráček, J., Kratochvíl, B., Sedmera, P., Havlíček, V. and Jegorov, A. (1996) Ergoladinine, an ergot alkaloid. *Phytochemistry,* **42,**231–233.

Cvak, L., Jegorov, A., Pakhomova, S., Kratochvíl, B., Sedmera, P., Ondráček, J., Havlíček, V. and Minář, J. (1997) 8α-hydroxy-α-ergokryptine, an ergot alkaloid. *Phytochemistry,* **44,**365–369.

Černý, A., Krajíček, A., Spáčil, J., Beran, M., Kakáč, B. and Semonský, H. (1976) Isolation of *N*-[*N*-(D-lysergyl)-L-valyl]-cyclo(L-phenylalalnyl-D-prolyl) from the ergotoxine type of field ergot and some of its reactions. *Coll. Czech. Chem.Commun.,* **41,**3415–3419.

Day, R.O., Day, V.W., Wheeler, D.M.S., Stadler, P. and Loosli, H.-R. (1985) The structure of pyroergotamine. *Helv. Chim. Acta,* **68,**724–733.

Dickerson, A.G., Mantle, P.G. and Szczyrbak, C.A. (1970) Autolysis of extracellular glucans produced *in vitro* by a strain of *Claviceps fusiformis. J. Gen. Microbiol.,* **60,** 403–415.

Dudley, H.W. and Moir, C. (1935) The substance responsible for the traditional clinical effect of ergot. *Brit. Med. J.,***1,**520–523.

Flieger, M., Wurst, M., Stuchlík, J. and Řeháček, Z. (1981) Isolation and separation of new natural lactam alkaloids of ergot by high performance liquid chromatography. *J. Chromatogr.,***207,**139–144.

Flieger, M., Sedmera, P., Vokoun, J., Řičicová, A. and Řeháček, Z. (1982) Separation of four isomers of lysergic acid α-hydroxyethylamide by liquid chromatography and their spectroscopic identification. *J. Chromatogr.,***236,**453–459.

Flieger, M., Sedmera, P., Vokoun, J., Řeháček, Z., Stuchlík, J., Malinka, Z., Cvak, L. and Harazim, D. (1984) New alkaloid from saprophytic culture of *Claviceps purpurea.* *J.Nat. Prod.,***47,**970–976.

Flieger, M., Zelenkova, N.F., Sedmera, P., Křen, V., Novák, J., Rylko, V., Sajdl, P. and Řeháček, Z. (1989a) Ergot alkaloids glycosides from saprophytic cultures of *Claviceps.* I. Elymoclavine fructosides. *J. Nat. Prod.,***52,**506–510.

Flieger, M., Linhartová, R., Sedmera, P., Zima, J., Sajdl, P., Stuchlík, J. and Cvak, L. (1989b) New alkaloids of *Claviceps paspali. J. Nat. Prod.,***52,**1003–1007.

Flieger, M., Křen, V., Zelenkova, N.F., Sedmera, P., Novák, J. and Sajdl, P. (1990) Ergot alkaloid glycosides from saprophytic cultures of *Claviceps,* II. chanoclavine-I fructosides. *J. Nat. Prod.,***53,**171–175.

Flieger, M., Sedmera, P., Havlíček, V., Cvak, L. and Stuchlík, J. (1993) 10-hydroxy-*cis*- and 10-hydroxy-*trans*-paspalic acid amide: New alkaloids from *Claviceps paspali. J. Nat. Prod.,***56,**810–814.

Floss, H.G., Günther, H., Mothes, U. and Becker I. (1967b) Isolierung von ElymoclavinO-β-D-fruktosid aus Kulturen des Mutterkornpilzes. *Z. Naturforsch.,***22b,** 399–402.

Franck, B. (1960) Die Farbstoffe des Mutterkorns. *Planta Med.,***8,**420–429.

Franck, B., Baumann, G. and Ohnsorge, U. (1965a) Ergochrome, eine ungewöhnlich vollständige Gruppe dimerer Farbstoffe aus *Claviceps paspali* . *Tetrahedron Lett.,* 2031–2037.

Franck, B. and Zemer, I. (1965b) Mutterkorn-Farbstoffe, VIII. Konstitution und Synthese des Clavorubins. *Chem. Ber.,***98,**1514–1521.

Franck, B. (1980) in "The biosynthesis of Mycotoxins: A study in secondary metabolism", Steyn, P.S. ed., Academic Press, New York, 157–191.

Gröger D. (1965) Über die Bildung von Lysergsäurederivaten in Submerskultur von *Claviceps paspali.* 3. Mitteilung: Isolation von Chanoclavin-(I). *Pharmazie,***30,** 523–524.

Heyl, F.W. and Swoap, O.F. (1930) The sterols of ergot. II. The occurence of dihydro-ergosterol. *J. Am. Chem. Soc.,***52,**3688–3690.

Heyl, F.W. (1932) The sterols of ergot. III. The occurence of an isorner of alphadihydrolysergol. *J. Am.Chem. Soc.,***54,**1074–1076.

Hofmann, A., Brunner, R., Kobel, H. and Brack, A. (1957) Neue Alkaloide aus der saprophytischen Kultur des Mutterkornpilzes von *Pennisetum typhoideum* RICH. *Helv. Chim. Acta,***40,**1358–1373.

Horwell, D.C. and Verge, J.P. (1979) Isolation and identification of *6, 7-seco*agroclavine from *Claviceps purpurea. Phytochemistry,***18,**519.

Kelleher, W.J., Krueger, R.J. and Rosazza, J.P. (1971) The violet pigment of lysergic acid alkaloiD-producing cultures of *Claviceps paspali:* Fe(III) complex of 2, 3 dihydroxybenzoic acid *Lloydia,***34,**188–194.

Kharash, M.S. and Legaul, R.R. (1935a) Ergotocin: the active principle of ergot responsible for the oral effectiveness of some ergot preparations on human uteri. *J. Am. Chem. Soc.*,**57**,956–957.

Kharash, M.S. and Legaul, R.R. (1935b) New active principle(s) of ergot. *Science,***81**, 614–615.

King, G.S., Mantle, P.G., Szczyrbak, C.A., Waight, E.S. (1973) A revised structure for clavicipitic acid. *Tetrahedron Lett.,*215–218.

Kobel, H., Schreier, E. and Rutschmann, J. (1964) 6-methyl-$\Delta^{8,9}$-ergolen-carbonsäure, ein neues Ergolinderivat aus Kulturen eines Stammes von *Claviceps paspali* STEVENS *et* HALL. *Helv. Chim. Acta,***47**,1054–1064.

Kornhauser, A., Logar, S. and Perpar M. (1965) Über die Papier- und Dünnschichtchromatographie der Mutterkornfarbstoffe. *Pharmazie,***20**,447–449.

Krajíček, A., Trtík, B., Spáčil, J., Sedmera, P., Vokoun, J. and Řeháček, Z. (1979) 8-hydroxyergotamine, a new ergot alkaloid. *Collect. Czech. Chem. Commun.,***44**, 2255–2260.

Křen, V., Řezanka, T., Sajdl, P. and Řeháček, Z. (1986) Identification of sterols in submerged cultures of different *Claviceps* species. *Biochem. Physiol. Pflanzen,***181**, 505–510.

Kulaev, I.S. (1979) "The Biochemistry of Inorganic Polyphosphates". John Wiley and Sons, New York.

Lewis, D.H. and Smith, D.C. (1967) Sugar alkohols (polyols) in fungi and green plants. *New Phytol.,***66**,143–184.

Lorenz, K. (1979) Ergot on cereal grains. *Crit. Rev. Food Sci. Nutr.,***11**,311–354.

Maier, W., Erge, D. and Gröger, D. (1980a) Mutational biosynthesis in a strain of *Claviceps purpurea. Planta Med.,***40**,104–108.

Maier, W., Erge, D., Schmidt, J. and Gröger, D. (1980b) A blocked mutant of *Clavicepspurpurea* accumulating chanoclavine-I-aldehyde. *Experientia,***36**,1353–1354.

Mangold, H.K. (1967) Aliphatische Lipide. In "Stahl E.: Dünnschicht Chromatographie", 2. Aufl.—Berlin, Heidelberg, New York, Springer Verlag, 350–404.

Mantle, P.G. and Waight, E.S. (1968) Dihydroergosine: a new naturally occuring alkaloid from the sclerotia of *Sphacelia sorghi* (McRAE). *Nature,***218**,581–582.

Morris, L.J. and Hall, S.W. (1966) The Structure of the Glycerides of Ergot Oils. *Lipids,* **1**,188–196.

Perényi, T , Udvardy, E.N., Wack, G. and Boross, L. (1966) Über die Pigmentstoffe des saprophytisch kultivierten Roggen-Mutterkornes (*Claviceps purpurea* TUL.). *PlantaMed.,***14**,42–48.

Perlin, A.S. and Taber, W.A. (1963) A glucan produced by *Claviceps purpurea. Can. J.Chem.,***41**,2278–2282.

Quigley, F.R. and Floss, H.G. (1981) Mechanism of amino acid α-hydroxylation and formation of the lysergyl moiety in ergotamine biosynthesis. *J. Org. Chem.,***46**, 464–466.

Ramstad, E., Chan-Lin, W.-N., Shough, H.R., Goldner, K.J., Parikh, R.P. and Taylor, E.H. (1967) Norsetoclavine, a new clavine-type alkaloid from *Pennisetum* ergot. *Lloydia,***30**, 441–444.

Scott, P.M. (1984) in "Mycotoxins-Production, Isolation, Separation and Purification", Betina V. (ed.), Amsterdam, 463.

Schlientz, W., Brunner, R. and Hofmann, A. (1963) *d*-Lysergil-L-valin-methylester, eine neues natürliches Mutterkornalkaloide. *Experientia,***19**,397–398.

Schlientz, W., Brunner, R., Stadler, P.A., Frey, A.J., Ott, H. and Hofmann, A. (1964) Isolierung und Synthese des Ergostins, eines neuen Mutterkorn-Alkaloids. *Helv.Chim. Acta,* **47,** 1921–1933.

Schlientz, W., Brunner, R., Rüegger, A., Berde, B., Stürmer, E. and Hofmann, A. (1968) β-Ergokryptin, ein neues Alkaloid der Ergotoxin-Gruppe, *Pharm. Acta Helv.,* **43,** 497–509.

Schoch, U. and Schlatter, Ch. (1985) Gesundheitsrisiken durch Mutterkorn aus Getreide. *Mitt. Gebiete Lebensm. Hyg.,* **76,** 631–644.

Smith, S. and Timmis, G.M. (1936) New alkaloid of ergot. *Nature,* **137,** 111.

Smith, S. and Timmis, G.M. (1937) Alkaloids of ergot. VIII. New alkaloids of ergot: ergosine and ergosinine. *J. Chem. Soc.,* 396–401.

Stütz, P., Brunner, R. and Stadler, P.A. (1973) Ein neues Alkaloid aus dem Mycel eines *Claviceps purpurea-Stammes. Experientia,* **29,** 936–937.

Stauffacher, D. and Tscherter, H. (1964) Isomere des Chanoclavins aus *Clavicepspurpurea* (FR.) TUL. *(Secale cornutum). Helv. Chim. Acta,* **47,** 2186–2194.

Stauffacher, D., Tscherter, H. and Hofmann, A. (1965) Isolierung von Ergosin und Ergosinin neben Agroclavin aus den Samen von *Ipomoea argyrophylla* VATKE *(Convolvulaceae). Helv. Chim. Acta,* **48,** 1379–1380.

Stoll, A. and Burckhard, E. (1935) Ergobasine, a new water-soluble alkaloid from Seigel ergot. *Compt. Rend. Acad. Sci,* **200,** 1680–1682.

Stoll, A. and Burckhard, E. (1937) Ergocristin und Ergocristinin, ein neues Alkaloidpaar aus Mutterkorn. *Hoppe Seyler's Z. Physiol. Chem.,* **250,** 1–6.

Stoll, A. and Hofmann, A. (1943) Die Alkaloide der Ergotoxingruppe: Ergocristin, Ergokryptin und Ergocornin. *Helv. Chim. Acta,* **26,** 1570–1601.

Stoll, A. (1945) Über Ergotamin. *Helv. Chim. Acta,* **28,** 1283–1308.

Stoll, A. (1952) Recent investigations on ergot alkaloids. *Fortschr. Chem. Org. Naturst.,* **9,** 114–174.

Stoll, A., Brack, A., Kobel, H., Hofmann, A. and Brunner, R. (1954) Die Alkaloide eines Mutterkornpilzes von *Pennisetum typhoideum* RICH, und deren Bildung in saprophytischen Kultur. *Helv. Chim. Acta,* **37,** 1815–1827.

Stuchlík, J., Krajíček, A., Cvak, L., Spáčil, J., Sedmera, P., Flieger, M., Vokoun, J. and Řeháček, Z. (1982) Non-cyclol (lactame) ergot alkaloids. *Coll. Czech. Chem.Commun.,* **47,** 3312–3317.

Szántay, C. Jr., Bihari, M., Brlik, J., Csehi, A., Kassai, A. and Aranyi A. (1994) Structural elucidation of two novel ergot alkaloid impurities in α-ergokryptine and bromokryptine, *Acta Pharm. Hung.,* **64,** 105–108.

Šmíd, M. and Beran, M. (1965) Námelová barviva. *Českoslov. farm.,* **14,** 21–25.

Taber, W.A. and Wining, L.C. (1963) Physiology of alkaloid production by *Clavicepspurpurea* (FR.) TUL. Correlation with changes in mycelial polyol, carbohydrate, lipid, and phosphorus-containing compounds. *Can. J. Microbiol.,* **9,** 1–14.

Taber, W.A. (1964) Sequential formation and accumulation of primary and secondary shunt metabolic products in *Claviceps purpurea. Appl Microbiol.,* **12,** 321–326.

Tanret, M.C. (1889) Sur un nouveau principe immédiat de l'ergot de siegle, l'ergostérine. *C.R.Hebd. Seances Acad. Sci.,* **108,** 98–103.

Tanret, M.C. (1908) Sur l'ergostérine et la fongistérine. *C.R.Hebd. Seances Acad. Sci.,* **147,** 75–77.

Thompson, M.R. (1935a) The active constituents of ergot. A pharmacological and chemical study. *J. Am. Pharm. Assoc.,* **24,** 24–38.

Thompson, M.R. (1935b) The new active principle of ergot. *Science,***81,**636–639.

Thompson, M.R. (1935c) The new active principle of ergot. *Science,***82,**62–63.

Tscherter, H. and Haut, H. (1974) Drei neue Mutterkornalkaloide aus saprophytischen Kulturen von *Claviceps paspali* STEVENS *et* HALL. *Helv. Chim Acta,***57,**113–121.

Vining, L.C. and Taber, W.A. (1964) Analysis of the endogenous sugars and polyols of *Claviceps purpurea* (FR.) TUL. by chromatography on ion exchange resins. *Can. J.Microbiol.,***10,**647–657.

Voigt, R. and Zier, P. (1970) Zur Bildung der tetracyclischen Ergoline durch Hydrierung von Chanoclavin-(I) . *Pbarmazie,***25,**272.

Wieland, H. and Coutelle, G. (1941) Zur Kenntnis des Fungisterins und anderer Inhaltsstoffe von Pilzen. *Liebig Ann. Chem.,***548,**270–283.

Yamatodani, S. (1960) Researches on ergot fungus. Part XL. On the paper chromatography of water-soluble ergot alkaloids. *Ann. Rep. Takeda Res. Lab.,***19,**1–7.

8.
CHEMICAL MODIFICATIONS OF ERGOT ALKALOIDS

PETR BULEJ and LADISLAV CVAK

Galena a.s., Opava 74770, Czech Republic

8.1.
INTRODUCTION

Ergot alkaloids (EA) are called "dirty drugs", because they exhibit many different pharmacological activities (see Chapter 15). The objective of their chemical modifications is the preparation of new derivatives with higher selectivity for some types of receptors. The fact that many new drugs were developed by semisynthetic modification of natural precursors is a proof that this approach is fruitful.

Chemical modifications of EA were reviewed several times (Hofmann, 1964; Stoll and Hofmann, 1965; Bernardi, 1969; Semonský, 1970; Stadler and Stütz, 1975). The last and very extensive review was published by Rutschmann and Stadler (1978). Therefore, this chapter concentrates especially on derivatives described from 1978 to 1997. The papers and patents devoted to this topic, which appeared in this period, were too numerous to be included in our review. This is the reason why we selected only the most important contributions. Nevertheless, we believe that our survey covers the most important progress in this field.

Preparation of radiolabelled derivatives is also reviewed here. Total syntheses of the ergoline skeleton are not included, but they have been treated in a recent monograph (Ninomiya and Kiguchi, 1990).

8.2.
CHEMICAL MODIFICATIONS IN THE ERGOLINE SKELETON

Chemical modifications of individual positions of the ergoline skeleton (Figure 1) are described below.

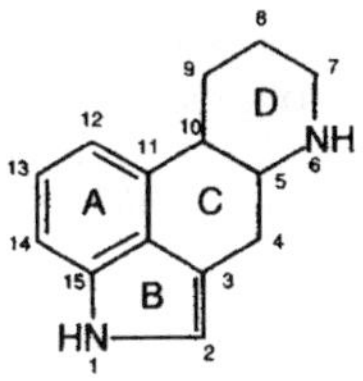

Figure 1 Ergoline numbering

8.2.1.
Modifications in Position 1

Alkyiation by alkyl halogenides in liquid ammonia (Troxler and Hofmann, 1957a), acylation by ketene and diketene, hydroxymethylation by formaldehyde, Mannich type reactions and Michaels addition of acrylonitrile affording *N*-1 cyanoethyl derivatives (Troxler and Hofmann, 1957b) are the well-known modifications in this position. Most of the modifications on the indole nitrogen were used in studies of structure—activity relationships. *N*-1Hydroxymethyl or aminomethyl derivatives were used for binding to proteins in immunological methods—see Chapter 11. Protection of the *N*-1 position in multi-steps syntheses can also be the reason of its modification. Many new reactions and *N*-1 substituted ergoline derivatives have therefore been described in the last two decades.

N-1-Alkyl Derivatives

Alkyiation in the *N*-1 position significantly changes EA activity. Some of *N*-l derivatives are used in therapy (nicergoline, metergoline, methysergide). That's why many new alkyl derivatives have been prepared and new alkylation procedures were investigated. Many processes using phase-transfer catalysis were developed, mainly for industrial methylation, and are used for production of nicergoline (Ručman, 1978; Cvak *et al.,* 1983; Gervais, 1986)—see Chapter 14. Šmidrkal and Semonský (1982a) developed a new procedure for alkylation by alkyl halogenides in dimethylsulfoxide in the presence of powdered NaOH. They prepared methyl, ethyl, n-propyl and phenyl derivatives of dihydrolysergic acid and some other ergolines. All the *N*-1 alkyl ergolines were less active in the test of prolactine secretion inhibition than the basic compounds. Eich and co-workers prepared many *N*-1 alkyl derivatives (up to C_8) of agroclavine and festuclavine by alkylation by primary alkyl halogenides in liquid ammonia in the presence of a strong base—sodium or potassium amide (Eich *et al.,* 1985; Eichberg and Eich, 1985). Dehalogenation of alkyl halogenides to alkenes was observed when higher alkyl halogenides were used. The longer the alkyl chain, the more preferred was the side reaction and less alkyl ergoline was formed. The yields of higher alkyl ergolines can be improved using alkyl tosylates instead of halides. Marzoni and Garbrecht (1987) prepared N-1 alkyl derivatives of dihydrolysergic acid by alkylation of dihydrolysergic acid in dimethylsulfoxide in the presence of powdered KOH with yields ranging from 80–95%, even with secondary alkyl tosylates (cyclopentyl, cyclohexyl).

A different approach must be adopted for introducing a tertiary alkyl group. *tert*-Butyl ergolines were prepared by reaction of ergoline with *tert-butyl* alcohol in the presence of trifluoroacetic anhydride (Temperilli *et al.*, 1987a; Beneš and Beran, 1989). Substitution in N-1 position is accompanied by substitution in positions 2, 13 and 14; a complicated mixture of products is therefore often obtained.

N-1-Aryl Derivatives

Aryl derivatives of ergolines were prepared by the treatment of ergoline derivatives with the appropriate aryl halogenides under phase-transfer catalysis (Sauer *et al.*, 1987).

N-1-Acyl and Sulfamoyl Derivatives

N-1-Acyl and N-1-sulfamoyl derivatives are formed by the reaction of appropriate chlorides with ergolines under phase-transfer catalysis (Lončarič and Ručman, 1984, Taimr and Křepelka, 1987). Dimethyl and diethylcarbamoyl derivatives were prepared in the same manner. N-1-Acetyl ergolines were also prepared by a reaction with acetic anhydride under catalysis with BF_3 etherate (Beneš, 1989). N-1 Formyl ergolines can be obtained by treatment with formic acid and a tertiary amine under palladium catalysis (Taimr *et al.*, 1987b).

N-1-Hydroxymethyl Derivatives

N-1-Hydroxymethyl ergolines can be prepared by refluxing the ergoline substrate with aqueous formaldehyde (Tupper *et al.*, 1993). Instead of expected 2-methoxyergolines, N-l-hydroxymethyl derivatives were obtained also by electrochemical reaction in methanol (Danieli *et al.*, 1983).

N-1-Carboxymethyl Derivatives

N-1-Carboxymethyl derivative resulted when dihydrolysergic acid was treated with ethyl bromoacetate and KOH in dimethylsulfoxide (Šmidrkal and Semonský, 1982b). It was further transformed to a hydroxyethyl derivative.

N-1-Trimethylsilyl Derivatives

Trimethylsilyl derivatives of EA are synthons suitable for the preparation of *N*-1 acylated or glycosylated ergolines. The trimethylsilyl group was introduced into the position 1 of some clavine alkaloids by the reaction with *N*-methyl-*N*-(trimethylsilyl)-trifluoroacetic amide in acetonitrile or by refluxing in hexamethyldisilazane in yields of 60 to 90% (Křen and Sedmera, 1996).

N-1-Glycosides of EA

Since *N*-1-glycosides of EA can be considered as nucleosides, some interesting activities can be expected. *β*-*N*-1-Ribofuranosides of clavine alkaloids were prepared by the reaction of *N*-1-trimethylsilyl derivatives with 1-O-acetyl-2, 3, 5tri-*O*-benzoyl-*β*-D-ribofuranose in dichloromethane under SnCl$_4$ catalysis in yields of 20–40% (Křen *et al.*, 1997a). Similarly a mixture of α and *β* anomers of *N*-1-deoxyribofuranosides of clavine alkaloids was prepared by reaction with 1-chloro-2-deoxy-3, 5-di-*O*-toluoyl-α-D-ribofuranose in acetonitrile. The two anomers were separated by preparative chromatography (Křen *et al.*, 1997b).

8.2.2.
Modifications in Position 2

Position 2 of the ergoline skeleton is highly suitable for synthetic modification of EA by both electrophilic and radical substitution. Many modifications have been reviewed by Rutschmann and Stadler (1978): chlorination, bromination and iodination, nitration and reduction of nitro derivatives to amino derivatives and reaction with 2-methoxy-1, 3-dithiolane affording an intermediate which can be desulfurised to a 2-methyl derivative. Troxler and Hofmann (1959) described the oxidation of lysergic acid diethylamide (LSD) to 2-oxo-3-hydroxy-2, 3-dihydrolysergic acid diethylamide.

One of the 2-bromo derivatives, 2-bromo-α-ergokryptine (bromokryptine), is used in therapy and many processes were therefore described for its production, mainly in patent literature—see Chapter 13.

2-Acyl Derivatives

The basic procedure for the preparation of 2-acylergolines consists in acylation of ergoline compounds with carboxylic acid anhydrides under catalysis by Lewis acid (Beneš and Křepelka; 1981a,b; Taimr *et al.*, 1987a). 2-Formyl derivatives were prepared from *N*-1 protected ergolines only by a reaction with dichloromethyl methyl ether in the presence of AlCl$_3$ (Sauer and Schröter, 1991). Using this procedure 8α-(3,3-diethylureido)-6-methylergoline2-carbaldehyde (2-formyl-terguride) was prepared from *N*-1-tosylterguride. The acylation of *N*-1 protected ergolines by acyl halogenides afforded 13-

acylergolines predominantly. 2-Acyl ergolines can be used for other transformations, e.g. to be subjected to aldolisation with aromatic aldehydes (Křepelka *et al.,* 1981).

2-Halogen Derivatives

The general method for introduction of a halogen group (Cl, Br, I) into the EA molecule is a reaction with *N*-halogen-succinimide (Troxler and Hofmann, 1957b). Many other processes, mainly for bromination of α-ergokryptine, have been developed: bromination by *N*-bromophthalimide, *N*-bromocaprolactam or dioxane-bromine complex in inert organic solvents (Flückiger *et al.,* 1971), usage of 3-bromo-6-chloro-2-methylimidazolo[1, 2-b]pyridazine-bromine complex, 2-pyrrollidinone-hydrotribromide or 2-piperidinone-hydrotribromide in dioxane containing peroxide or in the presence of 2, 2'-azo-bis-isobutyronitrile or dibenzoyl peroxide (Stanovnik *et al.,* 1981; Ručman *et al.,* 1983), the reaction with bromine in methylene chloride under hydrogen bromide (Börner *et al.,* 1984) or BF_3 etherate (Cvak *et al.,* 1992b) catalysis. A very interesting and efficient bromination and chlorination process using dimethylsulfoxide and trimethylbromo(chloro)silane was described by Megyeri and Keve (1989). An electrochemical halogenation method was also described (Palmisano *et al.,* 1987). The selectivity of halogenation in position 2 is an important question which must be taken into consideration. While in the case of 9, 10-didehydroergolines the regioselectivity into position 2 is high and dibrominated products are formed only when a large excess of the reagent or harsh reaction conditions are used, the parallel halogenation of ergoiines in positions 2 and 13 was observed yielding mixtures of products (Cvak, unpublished results).

Recently, 2-fluoroergolines have been prepared by an exchange reaction from 2-bromo derivatives via lithiated intermediates (Bohlmann *et al.,* 1993).

2-Hydroxymethyl Derivatives

Sauer *et al.* (1991) prepared 2-hydroxymethylergoline derivatives by treatment of terguride with paraformaldehyde under catalysis by $(CH_3)_2AlCl$.

2-Alkylthio Derivatives

The thioether group can be introduced into this position by reaction with sulfenyl chloride (Timms and Tupper, 1985; Tupper *et al.,* 1993). 2-(Methylthio)-agroclavine exhibits a strong antipsychotic effect. 2-Methylthio derivatives can be oxidised to corresponding sulfoxides and sulfones.

Electrochemical Functionalisation in Position 2

2-Alkoxyergolines were prepared by the electrolysis of EA in alcoholic solution of KOH (Seifert and Johne, 1980). The cyano group can be introduced by electrolysis of some EA in aqueous-methanolic solutions containing sodium cyanide (Seifert et al., 1983). Electrochemical oxidation of dihydrolysergol in aqueous methanol containing HBr gave, at a potential of 1.1 V, 2, 13-dibromo-dihydrolysergol and at 1.6V 12, 14-dibromo-2, 3-dihydro-8β(hydroxymethyl)-3β-methoxy-6-methylergolin-2-one (Seifert et al., 1992). Finally, the electrochemical oxidation of dihydrolysergol in acetonitrile gave a highly conjugated dimer (Figure 2) (Dankházi et al., 1993). Oxidative degradation and formation of a similar type of products can underlie the darkening of EA.

Electrophilic Substitution Using 2-Lithiated Ergoiines

Sauer and co-workers prepared many new 2-substituted ergoiines by a process including 2-lithiated ergoiines—Figure 3 (Sauer et al., 1985c, 1988a).

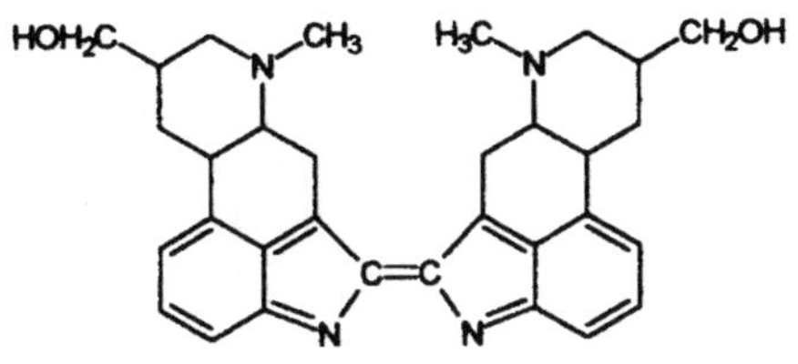

Figure 2 Ergoline dimer

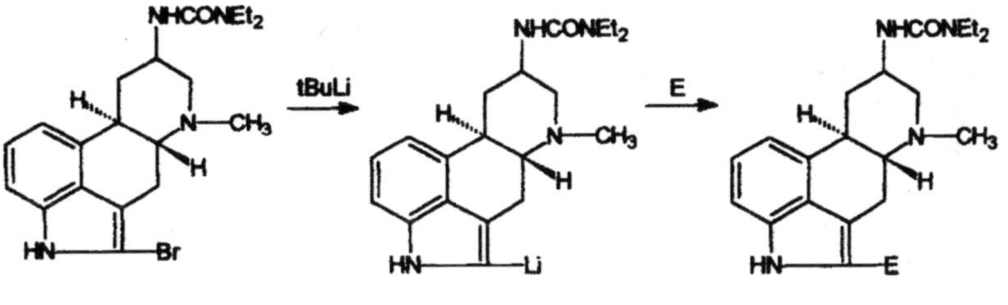

Figure 3 Electrophilic substitution of 2-lithiated ergolines

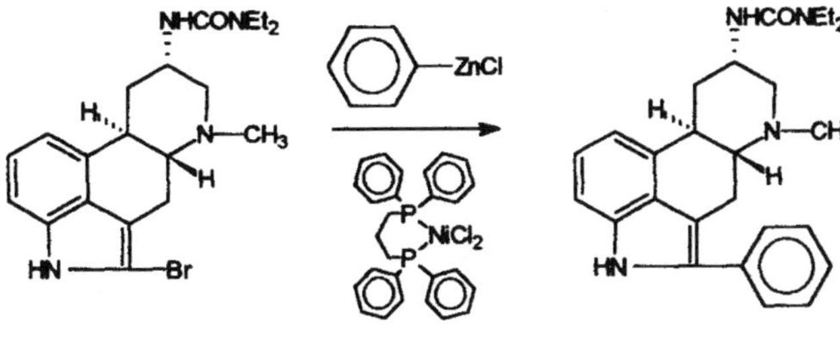

Figure 4 Nucleophilic substitution of 2-bromo derivatives

The starting 2-bromoergoline must be protected at *N*-1 position; using e.g. tert-butyldimethylsilyl group. The protected 2-bromoergoline is then transformed into a lithium derivative by tert-butyllithium and the lithiated ergoline is subjected to a reaction with an electrophilic reagent (ethylene oxide, methyl isocyanate, methyl isothiocyanate, carbon dioxide, alkyl halogenide, $(CH_3)_3Si—N=C=O$, etc.). Many derivatives of lisuride and terguride were prepared in this way.

Other Reactions Involving 2-Bromo Group Replacement— Nucleophilic Substitution

Phenyl group was introduced into position 2 by treatment of 2-bromoterguride with phenylzinc chloride under catalysis by palladium or nickel complexes— Figure 4 (Heindl *et al.*, 1989).

Oxidation in Position 2

2, 3-Dihydro-2-oxo-α-ergokryptine and 2, 3-dihydro-2-oxo-3-hydroxy-α-ergokryptine were found as the by-products in bromination of α-ergokryptine with bromine (Cvak *et al.*, 1992b). A procedure was developed for the preparation of 2-oxo derivatives of EA in 30–70% yield by treatment of EA by bromine in the presence of water (Cvak *et al.*, 1994). Also the oxidative cleavage of EA by sodium periodate giving aminoketone derivatives (Figure 5) can be considered as an oxidation in position 2. The aminoketones, prepared in this way were further transformed to depyrrolo analogues of EA (Bach *et al.*, 1980) or were used for the synthesis of 2-aza analogues of ergolines (Kornfeld and Bach, 1979; Stadler *et al.*,1981). Both depyrrolo and 2-aza analogues manifested a lower dopaminergic activity than their ergoline precursors. A similar type of oxidation was observed as a side reaction during bromination of some *N*-1-acylergolines by 2-pyrrolidone hydrotribromide when water was present (Lončarič and Ručman, 1984).

8.2.3.
Modifications in Position 3

All the reactions in this position affect also position 2, because they represent as a rule an addition to the 2, 3 double bond, followed in some cases by further transformation. Oxidation reactions leading to 2-oxo derivatives, in some cases substituted also in position 3, were described above.

First attempts at a direct reduction of EA in the 2, 3 position led to dimers (Bach and Kornfeld, 1973). The procedure for reduction of ergolines to 2, 3-dihydroergolines by Zn in HC1 and by $NaBH_4$ in trifluoroacetic acid was later described by Bach and Kornfeld (1976). Formation of 2, 3-dihydroergolines was also observed on hydrogenation of EA on a Raney-nickel catalyst (Cvak,

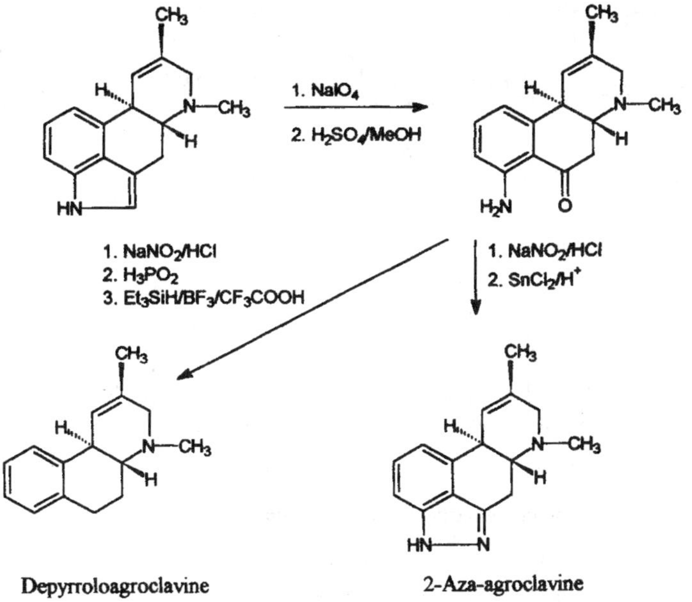

Figure 5 Oxidative cleavage of 1, 2-bond of ergoline skeleton and the use of the intermediates

unpublished results). Reduction of 2-bromoergolines by NaBH$_4$ in trifluoroacetic acid afforded 2, 3-dihydro derivatives in a high yield and without the formation of dimeric products (Sauer and Haffer, 1984).

Oxidation of 2, 3-dihydroergolines back to ergolines is an important process in some multi-step syntheses. It can be achieved by MnO$_2$, but this reaction is not satisfactorily reproducible and gives a poor yield. Better results were obtained with oxidation using some electrophilic reagents (*tert-butyl* hypochlorite, *N*-chlorosuccinimide, tosyl chloride, etc.) in nonpolar solvents (Sauer *et al.*, 1985a).

8.2.4.
Modifications in Positions 4 and 5

Despite the considerable efforts directed at chemical transformation of these positions (especially position 4 is a challenge for many chemists), no successful example has been described. The only transformation has been the racemisation of EA occurring during their hydrazinolysis (Stoll and Hofmann, 1937 and 1943; Stoll *et al.*, 1950).

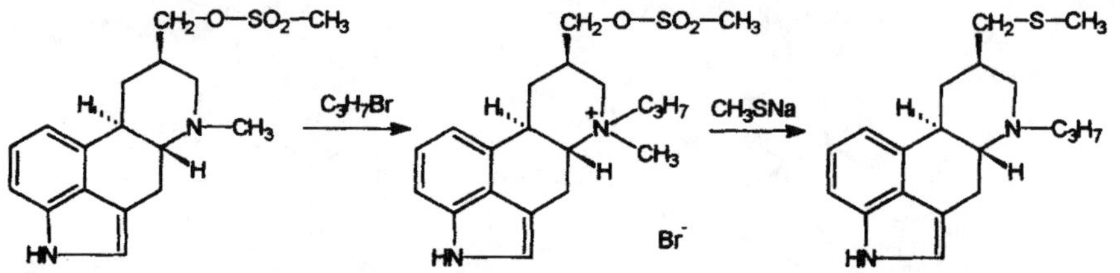

Figure 6 Pergolide manufacture

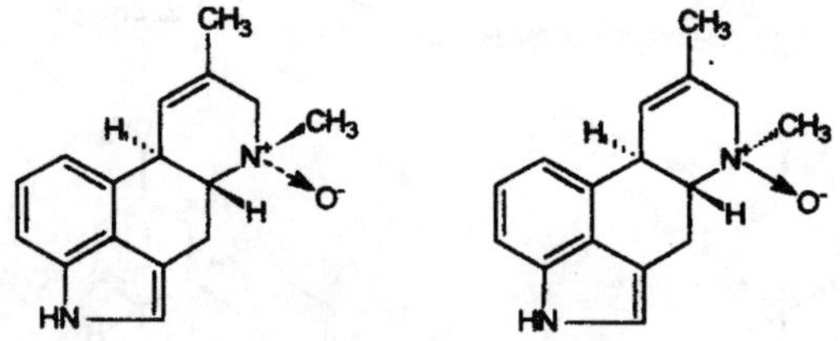

Figure 7 Diastereoisomers of agroclavine-N-6-oxides

8.2.5.
Modifications in Position 6

Demethylation

Classic von Braun *N*-demethylation used for the preparation of 6-norergolines consists in a reaction with BrCN followed by elimination of the cyanogen group by hydrolysis using Zn in acetic acid or KOH in diethylene glycol, or by hydrogenolysis on Raney-nickel (Rutschman and Stadler, 1978). Brumby and Sauer (1991) used organometallic reagents for CN removal, giving higher yield of 6-norergolines. An alternative process using 2, 2, 2-trichloroethyl-chloroformate as the demethylating agent was developed by Crider *et al.* (1981). Another possibility is the demethylation of *N*-6-oxides by strong bases (*n*- or *tert*-butyl lithium) in the presence of some Lewis acid (Sauer and Brumby, 1990). A very interesting, one-pot process for the synthesis of pergolide, including *N*-6 methyl group replacement as a key step, was developed by Misner (1985)—Figure 6.

N-6 Oxides

N-6-Oxides of EA were first described by Ponikvar and Ručman (1982) as by-products formed during the hydrogenation of EA in dioxane containing peroxides. They can be prepared by oxidation using hydrogen peroxide or 3-chloroperoxybenzoic acid (Mantegani *et al.*, 1988; Ballabio *et al.*, 1992). Křen *et al.* (1995) studied the chirality on *N*-6 of agroclavine and elymoclavine oxides—Figure 7. The ratio of diastereoisomers (6*R*/6*S*) was 2:3 in both cases.

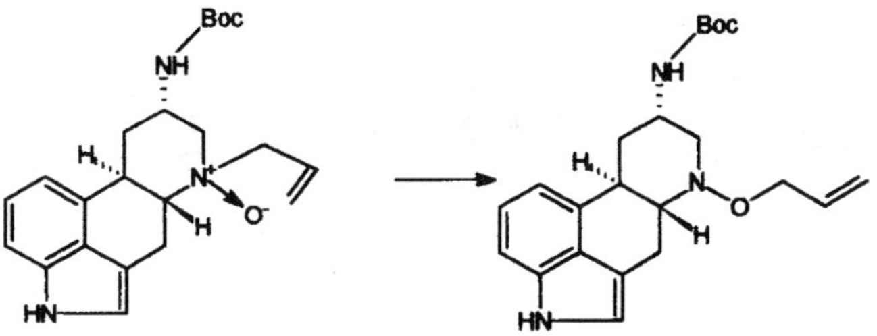

Figure 8 Meisenheimer's [2, 3]-sigmatropic rearrangement of 6-allylergoline-6-*N*-oxides

Other N-6 Derivatives

N-6-Demethylated ergolines were first allylated, then oxidised to 6-allyl oxides which were subjected to rearrangement to *N*-6-allyloxy derivatives—Figure 8 (Nordmann and Gull, 1986). Nordmann and Loosli (1985) prepared *N*-6-carboxamidines from 6-cyano-6-norergolines.

8.2.6.
Modifications in Position 7

The only modification in this position is formation of the $\Delta^{7,8}$ double bond (Stütz and Stadler, 1973) from *N*-6 oxides by Polonovski reaction.

8.2.7.
Modifications in Position 8

*Functional Derivatives of Lysergic, Isolysergic, Paspalic
and Dihydrolysergic Acids*

The review by Rutschmann and Stadler (1978) gives a survey of the derivatives and methods described until 1978. Only a few substantial news, particularly some new derivatives with new pharmacological activities, appeared since then. Esters of *N*-1-alkylated dihydrolysergic acid with ethylene glycol and 2, 3-butanediol were prepared by acid-catalysed esterification (Marzoni *et al.,* 1987). Another series of esters was prepared by the reaction of potassium salts of dihydrolysergic acid derivatives with cyclohexyl tosylates (Garbrecht *et al.,* 1988). In particular, *N*-1-isopropyl derivatives of these esters showed a high affinity for the 5HT$_2$ receptor. Misner *et al.* (1990) prepared a new series of cyclopentyl and cyclohexyl amides of *N*-1 alkylated dihydrolysergic acid. Condensation using 1.1′-carbonyldiimidazole as the coupling reagent and condensation via acyl chlorides prepared by the reaction with POCl$_3$/DMF gave the best yields. Also in this group the *N*-1-isopropyldihydrolysergic acid derivatives showed the highest affinity for the 5HT$_2$ receptor.

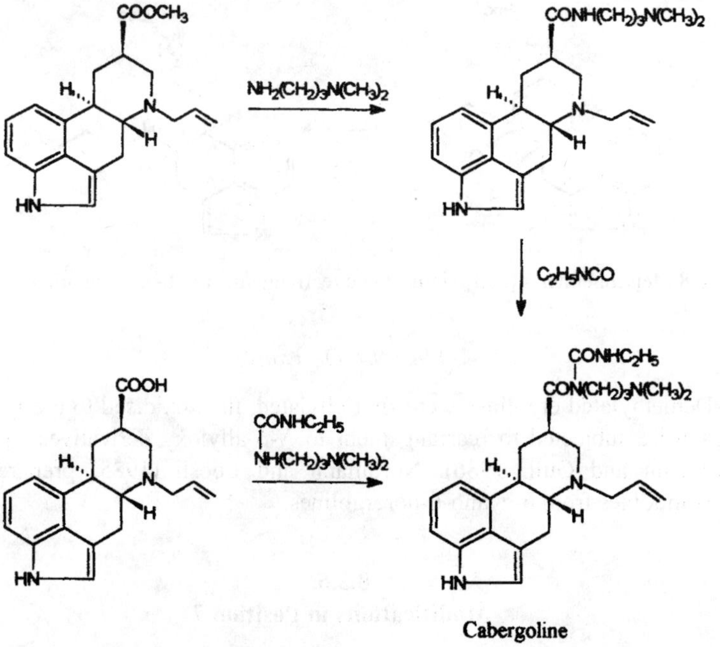

Figure 9 Two processes for cabergoline synthesis

Anilides of both lysergic and dihydrolysergic acids were prepared using trimethylaluminium as a coupling reagent (Neef *et al.*, 1982). Technologically interesting is the direct ester preparation by acid-catalysed esterification of some lysergic and isolysergic acid amides which was described by Sauer and Haffer (1983). New sugar esters of lysergic and dihydrolysergic acids were prepared by a reaction of these acids with protected (acetylated) 1-bromosugars under silver oxide catalysis in tetrahydrofuran (Seifert and Johne, 1979).

Cabergoline (see Chapter 13), a new long lasting dopamine agonist, was selected from a group of new derivatives of dihydrolysergic acid having the structure of acylurea. These derivatives were prepared by two ways—Figure 9. The first method (Brambilla *et al.*, 1989) represents the amidation of methyl 6-(2-propenyl)-9, 10-dihydrolysergate by 3-dimethylaminopropyl-1-amine followed by the reaction with ethyl isocyanate. In the alternative method, acylurea derivatives of dihydrolysergic acid can be synthesised by a reaction of dihydrolysergic acid or its derivatives with *N*-ethyl-*N'*-(3-dimethylamino)propyl-carbodiimide in the presence of triethylamine (Salvati *et al.*, 1981).

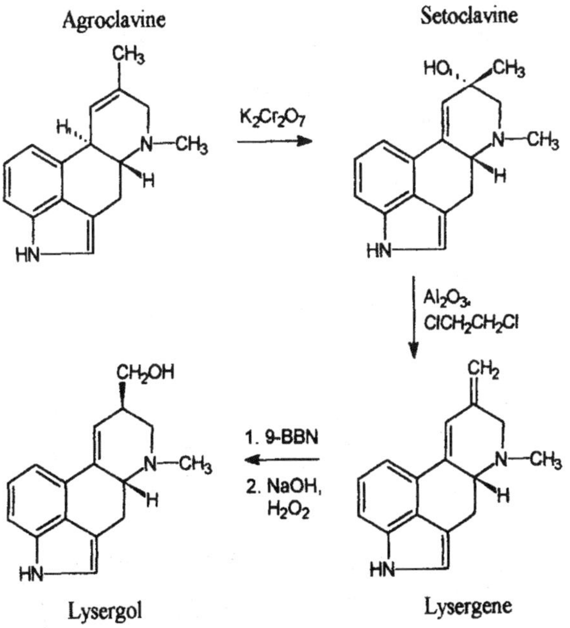

Figure 10 Agroclavine conversion to lysergol (9-BBN=9-borabicyclo[3, 3, 1]nonane)

Modification of the Side Chain in Position 8

The modifications of the side chain are the most frequently used in the studies of structure-activity relationships. The most important starting material for such syntheses is lysergol, elymoclavine and dihydrolysergol—for their industrial production see Chapter 13. They are transformed to mesylates or tosylates which are subjected to nucleophilic substitution using different nucleophilic agents. Another cheap precursor, produced by fermentation could be agroclavine, but its low reactivity prevents its wider use. Recently, Harris and Horvell (1992) described its transformation to lysergol in three steps, but the yield was not high —Figure 10. Only theoretically interesting is the preparation of agroclavine from 9, 10-didehydro-6-methylergolin-8-one (Wheeler, 1986). The starting 9, 10-didehydro-6-methylergolin-8-one was prepared from lysergic acid in five steps (Bernardi *et al.*, 1974). The last starting material for modification of the side chain can be lysergamine or dihydrolysergamine obtained by reduction of ergine or dihydroergine by LiAlH$_4$.

Many ergoline derivatives containing a heterocyclic moiety bonded to methylene group in position 8β were described—Figure 11. Bernardi *et al.* (1983) prepared ergolines with pyrazole, isoxazole, pyrimidine, 2-amino-pyrimidine and 2-phenylpyrimidine exhibiting antihypertensive and dopaminergic activity. Other heterocyclic derivatives were prepared by Bernardi *et al.* (1984), Rettegi *et al.* (1984), Mantegani *et al.* (1986) and Seifert *et al.* (1986). Japanese authors

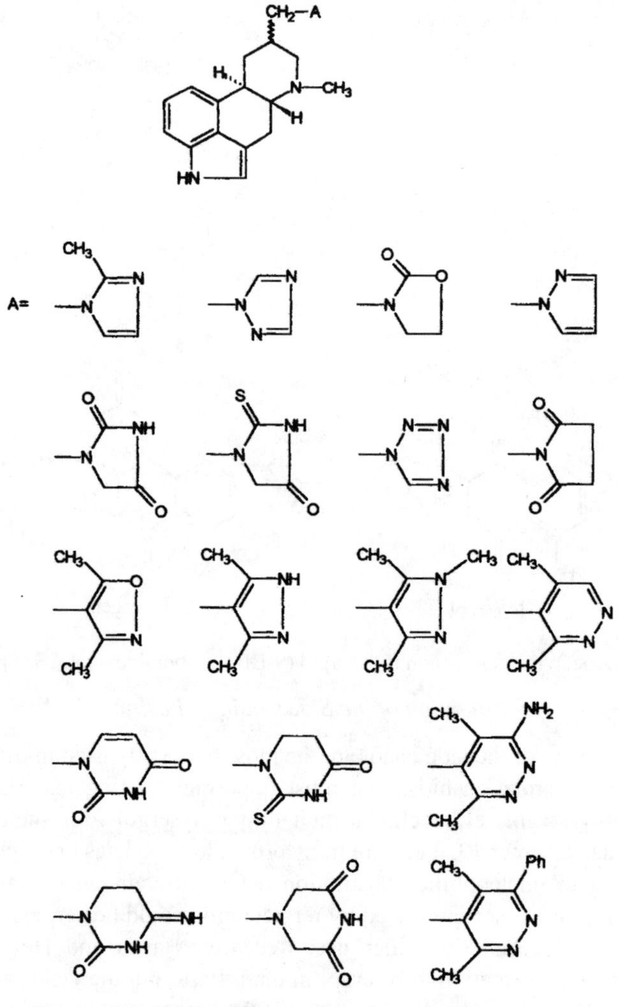

Figure 11 New ergoline derivatives containing a heterocyclic moiety bonded to methylene group in position 8

described similar derivatives (both 8α and 8β) with five-membered heterocycles (Ohno *et al.,* 1986, 1994a, b). Their syntheses use dihydrolysergol or dihydroisolysergol (8α-CH$_2$OH) mesylates or tosylates for further chemical substitution. Some of their derivatives exhibited dopaminergic activity higher than bromokryptine and pergolide and are in clinical trials as antiparkinsonic agents.

Using S nucleophiles, many thiodihydrolysergol derivatives were prepared from dihydrolysergol mesylate or tosylate (Gull, 1983, Kornfeld and Bach, 1978). One of them, pergolide (Misner, 1985; Misner *et al.,* 1997), is used in the

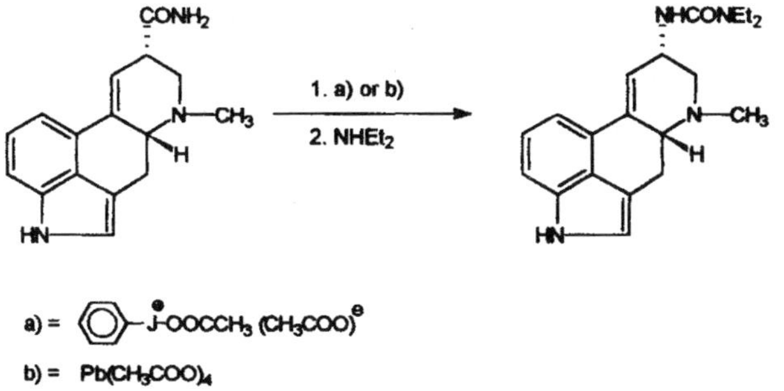

Figure 12 Conversion of erginine to lisuride

therapy (see Chapter 13). Derivatives of urea and thiourea were prepared from dihydrolysergamine or dihydroisolysergamine (8β- or 8α-aminomethyl-6-methylergoline) by a reaction with isothiocyanate or dithiocarbamate derivatives (Bernardi et al., 1982; Ruggieri et al., 1989). A number of recent papers were devoted to glycosides of EA possessing a hydroxy group—chanoclavine, elymoclavine, lysergol, dihydrolysergol and ergometrine (Křen et al., 1992, 1994, 1996, 1997c; Ščigelová et al., 1994). Both chemical and enzymatic syntheses were used for their preparation.

Prolongation of the Side Chain in the Position 8

Derivatives of 6-methylergolin-8β-yl-acetic and propionic acids were prepared by stepwise synthesis from dihydrolysergol mesylate or tosylate using NaCN or KCN (Beran and Benš, 1981; Brambilla et al., 1983). Bernardi et al. (1983a) condensed dihydrolysergol tosylate with malondiamide or malondinitrile and prepared a number of 6-methylergolin-8β-yl-propionic acid derivatives.

8α-Aminoergoline Derivatives

Lisuride and terguride are the therapeutically used derivatives of 8α-aminoergolene or ergoline, which were prepared by Curtius degradation of lysergic (dihydrolysergic) acid azides—see Chapter 13. Two new processes for lisuride preparation, both starting from isolysergic acid amide (erginine), were described. Sauer and Haffer (1981) used lead tetraacetate and Bulej et al. (1990) used iodosobenzene diacetate for erginine oxidation followed by Hofmann-like rearrangement giving 8α-isocyanate, which afforded lisuride in a reaction with diethylamine—Figure 12.

Isomerisation in the Position 8

It is well known that derivatives of lysergic acid epimerise easily to derivatives of isolysergic acid. On the other hand, the acquisition of dihydroisolysergic acid

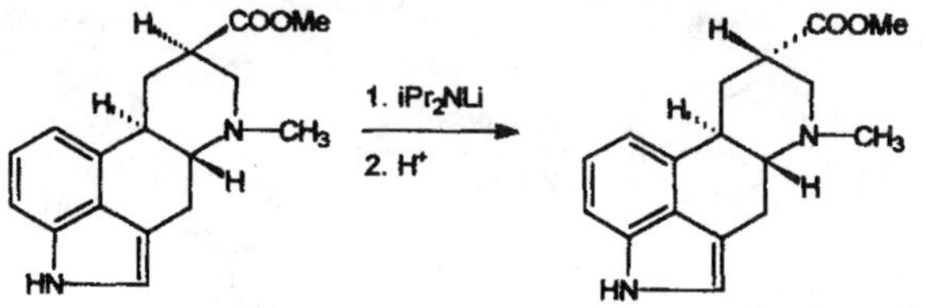

Figure 13 Isomerisation of methyl dihydrolysergate

derivatives is very complicated. That's why the new process for isomerisation of methyl dihydrolysergate described by Brich and Mühle (1981) brought a substantial progress into the synthetic modification of EA. Methyl dihydrolysergate is first deprotonated in position 8 by a strong base (lithium diisopropylamide) and the resulting anion is then hydrolysed under kinetic control giving 85% of methyl dihydroisolysergate—Figure 13.

8.2.8.
Modifications in Positions 9 and 10

The ergoline structure can be easily modified in these positions via addition reactions to $\Delta^{8,9}$ or $\Delta^{9,10}$ double bond. Hydrogenation, hydroboration, addition of mercury(II) salts in methanol and photochemically initiated addition of water or alcohols can therefore be found in older literature. The industrially used photochemical methoxylation of methyl lysergate and lysergol is discussed in Chapter 13.

Hydrogenation

Mayer and Eich (1984) described a new transfer-hydrogenation of both $\Delta^{8,9}$ and $\Delta^{9,10}$ ergolenes (agroclavine, elymoclavine, lysergol, ergotamine, ergocristine), giving high yields of stereospecifically hydrogenated products.

Classic hydrogenation of 8α-substituted ergolenes requires a high pressure and affords a mixture of 5,10 *cis* and *trans* products. One such industrially used process is the hydrogenation of lisuride to terguride (*trans*-dihydrolisuride)— Figure 14. When lisuride was reduced by lithium in liquid ammonia (Birch reduction), the reduction to terguride proceeded stereochemically in a yield over 90% (Sauer, 1981; Sauer *et al.,* 1986).

Electrophilic Substitution of 10-Lithiated Ergolines and $\Delta^{8,9}$-Ergolenes

The elimination of proton from C-10 of agroclavine can be achieved by butyl lithium, giving a carbanion. Because the C-10 carbanion can tautomerise to C-8 carbanion, a mixture of 10 and 8 substituted derivatives was obtained after

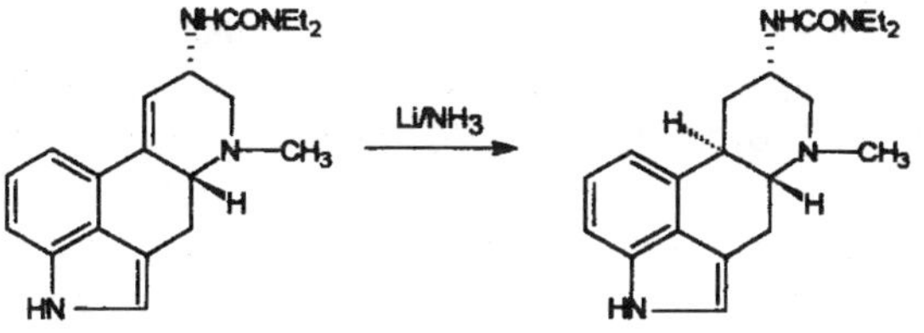

Figure 14 Birch reduction of lisuride to terguride

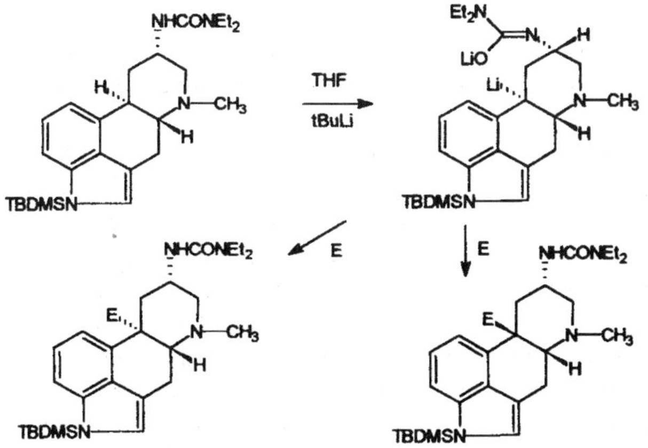

Figure 15 Electrophilic substitution of 10-lithiated terguride

treatment with an electrophilic agent (alkyl halogenides, chloroformates, methyl isocyanate, dimethylsulfide) (Timms *et al.,* 1989).

Sauer *et al.* (1988b) studied the lithiation of terguride. Reaction with *tert*-butyl lithium afforded bis-lithiated intermediate which was further treated with different electrophiles. Depending on the electrophilic reagent 10α- or 10β-substituted products or their mixture were obtained—Figure 15. The *N*-1 position must be protected, most conveniently by *tert*-butyldimethylsilyl group.

8.2.9.
Modifications in Positions 12, 13 and 14

These positions on the aromatic part of the ergoline skeleton are accessible to electrophilic substitution. The most reactive position for electrophilic substitution is position 2, less reactive is position 13. Positions 14 and 12 are substituted only under harsh conditions when usually complicated mixtures of products are obtained or some special techniques must be used. When 2, 3-dihydroergolines are subjected to substitution, position 12 reacts preferentially.

Sauer *et al.* (1985b) prepared 13-bromoterguride via 2, 13-dibromoterguride. Bromination of terguride by 2 equivalents of bromine gave 2, 13-dibromo derivative, which was then selectively debrominated by cobalt bromide (Bernardi *et al.*, 1975) or by hydrogenolysis using NaH_2PO_2 in the presence of a palladium catalyst (Boyer *et al.*, 1985). Similar approach can be used for preparation of other 13 substituted derivatives by electrophilic substitution of 2-bromoergolines and subsequent selective debromination of bromine in position 2. Many 13-acylergolines have been prepared in this way but the reaction is unfortunately accompanied by an exchange reaction giving 2-acyl-13-bromoergoline (Taimr *et al.*, 1987c).

12-Nitroergolines were prepared by Gull (1981a, b) by nitration of *N*-1 acetylated 2, 3-dihydroergolines. The nitro derivatives were further transformed to 12-amino, 12-hydroxy and 12-methoxy derivatives. The double bond was reintroduced into 2, 3 position by oxidation with air.

12-Bromoterguride was prepared by bromination of l-acetyl-2, 3-dihydroterguride. The oxidation of 12-bromo-2, 3-dihydroterguride was accomplished by *tert*-butylhypochlorite (Sauer *et al.*, 1985b).

The synthesis of clavine derivatives substituted in the C-14 position was described by Beneš and Beran (1989). A mixture of 1-*tert*-butyl-, 1-*O*-di-*tert*-butyl-, 2, 13-di-*tert*-butyl- and 2, 14-di-*tert*-butyl-elymoclavine resulted from the reaction of elymoclavine with tert-butylalcohol and trifluoroacetic anhydride. Sauer *et al.* (1990) prepared 13, 14-dibromo derivatives by bromination of 2-methyl-ergolines by different bromination reagents (bromine, pyridinehydrotribromide, pyrrolidone-hydrotribromide) in trifluoroacetic acid. When only one equivalent of bromine was used, 13-bromo derivative was preferentially formed.

The bromine in 12-bromo and 13-bromoergolines can be replaced by lithium and the lithiated ergolines subjected to further electrophilic substitution. The procedure was used to prepare 12- and 13- $CONH_2$, $COOCH_3$, CHO and OH ergolines (Sauer *et al.*, 1988a). Heindl *et al.* (1989) prepared 12- and 13- alkyl, alkenyl and alkinylergolines from the respective bromo derivatives by reaction with some organometallic compounds in the presence of a palladium catalyst.

8.2.10.
Changes of the Ergoline Skeleton

5, 6- and 6, 7-Secoergolines

When quarternary ammonium salts prepared from ergoline or ergolene precursors by the treatment with alkyl halogenides were reduced, the C-N bond in the D cycle of the ergoline skeleton was cleaved—Figure 16. While the reduction with alkali metals in liquid ammonia afforded 5, 6-secoergolines (Temperilli and Bernardi, 1980a) the catalytic hydrogenation gave 6, 7-secoergolines (Temperilli and Bernardi, 1980b).

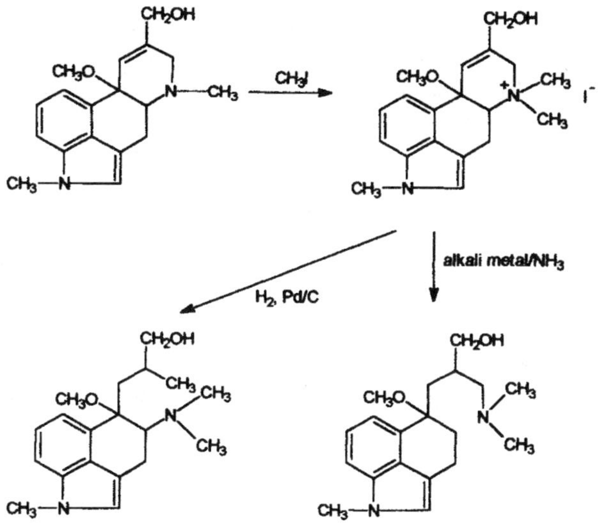

Figure 16 Synthesis of 5, 6 and 6, 7-secoergolines

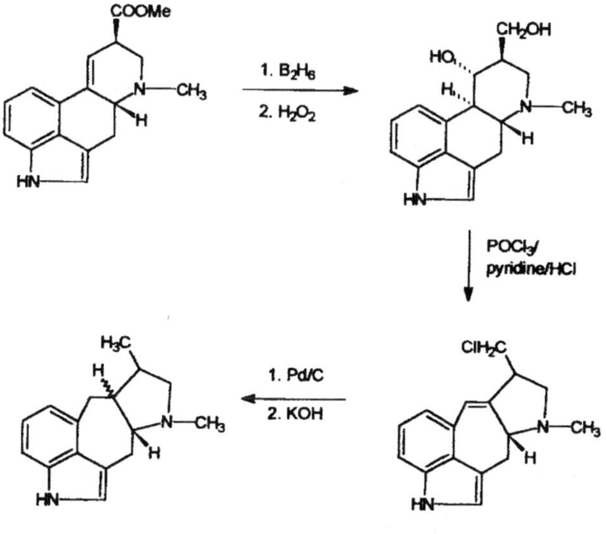

Figure 17 *(10→9)abeo-ergolines* synthesis

5(10→9)Abeo-ergolines

Hydroboration of $\Delta^{9,10}$ ergolenes affords 9α-hydroxy-ergolines (Bernardi *et al.,* 1976). When these derivatives are treated with $POCl_3$-pyridine hydrochloride, the Merweein-Wagner rearrangement takes place and 5(10→9)*abeo-ergolines* arise—Figure 17. Temperilli *et al.* (1980, 1987b) prepared a number of such derivatives.

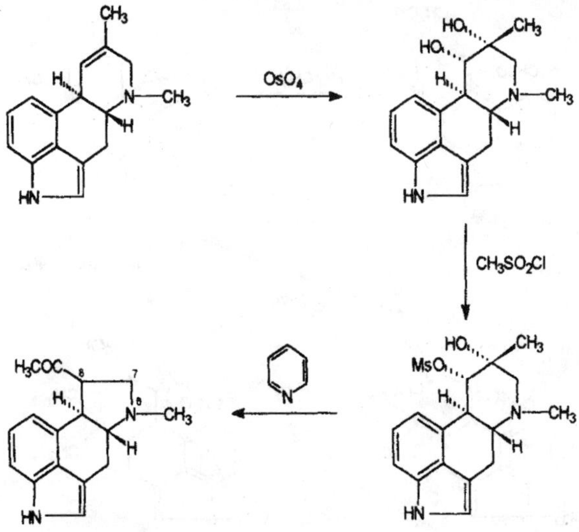

Figure 18 D-*nor*-7-ergolines synthesis

D-nor-7-Ergoline Derivatives

$\Delta^{8,9}$-Ergolenes can be oxidised by osmium tetroxide to 8β, 9β-dihydroxyergolines. When the 9-hydroxy group is activated (converted to a mesyloxy group) it can be eliminated under pinacol-pinacolone-like rearrangement giving D-nor-7-ergoline derivatives—Figure 18 (Hunter, 1985). A similar type of rearrangement starting from 8β-hydroxy-ergolines was described by Bernardi *et al.* (1987).

8.3.
CHEMICAL MODIFICATIONS IN THE PEPTIDIC
MOIETY OF ERGOPEPTINES

In contrast to the ergoline part of EA, not many chemical transformations of the peptidic moiety are known. Most of the described ergopeptine (EA of a peptidic type) derivatives and analogues modified in the peptidic part were prepared by the total synthesis or by directed biosynthesis, both using modified amino acids. Many other derivatives are known as products of metabolic biotransformations—see Chapter 9. Only positions 2', 6' and 12' are accessible for some types of regioselective chemical modification—Figure 19.

Aci-Derivatives

The hydroxyl group in position 12' of natural ergopeptines is engaged in a hydrogen bond to the ergoline part of the molecule and therefore it does not exhibit its acidic properties. When the configuration in position 2' changes, the

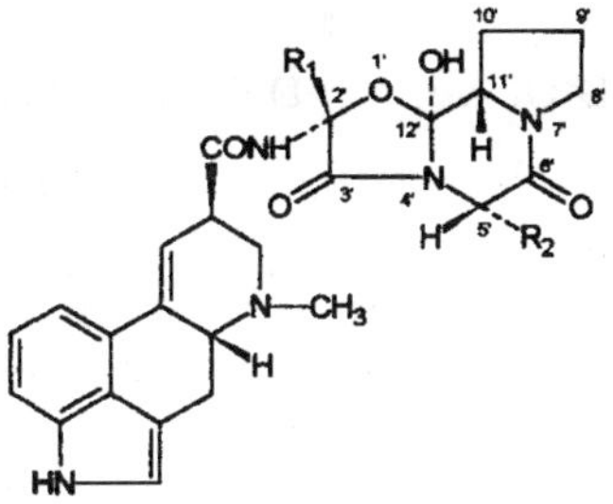

Figure 19 Ergopeptines—cyclol moiety numbering

ability to form the hydrogen bond disappears and the acidic character of the hydroxyl is demonstrated. Such derivatives with the alkyl group in position 2'α are soluble in alkaline aqueous solutions and are called *aci*-derivatives (*aci*-ergotamine, etc.). The *aci*-derivatives are formed in acidic aqueous solutions by a mechanism started by water elimination from the protonated 12'-hydroxy group (Ott *et al.,* 1966). Because the epimerisation in position 8 of the ergoline part of ergopeptine proceeds simultaneously, a mixture of four isomeric products is obtained: ergopeptine, *aci*-ergopeptine, ergopeptinine and *aci*-ergopeptinine. The dihydroergopeptines are epimerised only on C-2' and therefore only two products are obtained. The *aci*-derivatives of ergopeptines and dihydroergopeptines have no practical use but their occurrence must be monitored in the final products—both bulk substances and pharmaceutical products.

12'-Alkoxy-Derivatives

The mechanism of formation of 12'-alkoxy derivatives is the same as the one involved in the *aci*-rearrangement—acid-catalysed hydroxyl elimination in position 12', here followed by an alcohol addition (Schneider *et al.,* 1977; Ručman *et al.,* 1991). Similar to the *aci*-derivatives, also the 12'-alkoxy derivatives can be present as impurities in the salts of ergopeptines.

6'-Modified Derivatives

The stereochemically less hindered carbonyl group in position 6' can be selectively reduced. Bernardi and Bosisio (1977) prepared 6'-deoxo-$\Delta^{5'6'}$-didehydro9, 10-dihydroergopeptines by Birch reduction of dihydroergopeptines. Cvak *et al.* (1992a) prepared 6'-deoxo-ergopeptines and 6'-deoxo-9, 10-dihydroergopeptines by the selective reduction of ergopeptines or dihydroergopeptines by LiAlH$_4$.

8.4.

RADIOLABELLED DERIVATIVES

Whereas the structure-activity relationship was the main target of synthetic modification of the ergoline skeleton, syntheses with labelled compounds were aimed at obtaining identical molecules using suitable labelled intermediates. In most cases, the radiolabelled derivatives were used as tracers for the study of absorption, distribution, excretion and metabolic fate of EA. Nowadays, labelled compounds are often used also in competitive radioimmunoassays and for identification of binding sites of some neuroreceptors.

Among the numerous synthetic strategies, catalytic reduction with tritium gas of the 9, 10-double bond of EA is the simplest procedure ensuring specific activities of EA over 50 Ci/mmol. This procedure was used to prepare some dihydroergopeptines and the same method was used also for ^{3}H-terguride (Krause et al., 1991), ^{3}H-mesulergine (Voges, 1985) and its derivatives (Voges, 1988). A different synthetic strategy was used for ^{3}H-pergolide (Wheeler et al., 1990) and ^{3}H-cabergoline (Mantegani et al., 1991) where the 9, 10-double bond was not available. ^{3}H-Pergolide was synthesised by catalytic hydrogenation of an N-6-allyl intermediate and, similarly, ^{3}H-cabergoline from N-6-propargyl derivative. Since the majority of ^{3}H-EA was synthesised for the studies of their pharmacokinetics and metabolism directly in human volunteers (Maurer and Frick, 1984; Wyss et al., 1991; Krause et al., 1991), it is important to note that whereas 9 and 10 positions remain intact during the metabolism of EA, oxidative dealkylation at N-6 by cytochrome P-450 is a common metabolic pathway of EA (see Chapters 10 and 11).

In order to achieve a better sensitivity of detection and to avoid the possible exchange or metabolism of tritium-labelled derivatives, several ^{14}C-derivatives have been prepared. In contrast to the soft tritium label, ^{14}C-EA were used solely for animal studies, biotransformations in liver perfusates, and for in vitro experiments. Since the total synthesis of the ergoline moiety is a complex multistep procedure, ^{14}C-label was introduced by N-6-dealkylation to the nor-derivative and re-alkylation with a labelled alkylhalogenide. This procedure was used, e.g., for bromokryptine (Schreier, 1976) and pergolide (Wheeler et al., 1990). Since the N-6 position is metabolism-prone, several syntheses at the C8-atom have been described. A novel application of the ^{14}C-Wittig reagent with the 8-keto-EA intermediate has been developed for the preparation of ^{14}C-labelled 9, 10-didehydro-6, 8-dimethyl-2-methylthioergoline (Wheeler, 1988). Alternatively, the substitution of 8-mesyloxy- or 8-chlorogroups with [^{14}C]-sodium or potassium cyanides were used to introduce a series of C17 labelled derivatives (Wheeler et al., 1990; Mantegani et al., 1991; Angiuli et al., 1997). Lisuride and proterguride were prepared by introduction of ^{14}C-label into the carbonyl group of the urea moiety in the side chain (Toda and Oshino, 1981; Krause et al., 1993). 6-([^{11}C]-methyl)ergolines were prepared from corresponding 6-nor-derivatives by the reaction with labelled ^{n}CH$_3$I (Langström *et al.,* 1982). For special purposes, syntheses with other isotopes, leading to slight modification of original templates, have been described: ^{75}Se-pergolide (Basmadjian *et al.,* 1989) and 2-[^{125}I]-ergolines (Kadan and Hartig, 1988; Watts *et al.,* 1994; Bier *et al.,* 1996, Hartig *et al.,* 1985).

REFERENCES

Angiuli, P., Fontana, E. and Dostert, P. (1997) Synthesis of [17–[14]C] nicergoline. *J. Labelled Compd. Radiopharm.,***39,**331–337.

Börner, H., Haffer, G. and Sauer, G. (1984) Verfahren zur Herstellung von 2-Brom-8ergolinyl-Verbindungen. *EP Pat. 0141387.*

Bach, N.J. and Kornfeld, E.G. (1973) Dimerization of ergot derivatives. *TetrahedronLett.,* **1973,**3315–3316.

Bach, N.J. and Kornfeld, E.G. (1976) 2, 3-Dihydroergoline und Verfahren zu ihrer Herstellung. *DE Pat. 2601473.*

Bach, N.J., Kornfeld, E.G., Clemens, J.A. and Smalstig, E.B. (1980) Conversion of ergolines to hexahydro- and octahydrobenzo*[f]*quinolines (depyrroloergolines). *J. Med. Chem.,***23,**812–814.

Ballabio, M., Sbraletta, P., Mantegani, S. and Brambilla, E. (1992) Diastereospecific formation of 6-N-oxide ergoiines: A 1H NMR study of the configuration at nitrogen. *Tetrahedron,***48,**4555–4566.

Basmadjian, G.P., Sadek, S.A., Mikhail, E.A., Parikh, A., Weaver, A. and Mills, S.L. (1989) Structure biodistribution relationship of labeled ergolines: Search for brain imaging radiopharmaceuticals. *J. Labelled Compd. Radiopharm.,***27,**869–883.

Beneš, J. (1989) Ergolene acetylderivatives and the process for their preparation. *CSPat. Appl. 2967–89* (in Czech).

Beneš, J. and Beran, M. (1989) Elymoclavine *tert*-butylderivatives and the process for their preparation. *CS Pat. 275769* (in Czech).

Beneš, J. and Křepelka, J. (1981a) Ergoline 2-acylderivatives, their salts and the preparation thereof. *CS Pat. Appl. 2528–81* (in Czech).

Beneš, J. and Křepelka, J. (1981b) The process for the preparation of ergoline 2-acylderivatives. *CS Pat. 221421* (in Czech).

Bier, D., Dutschka, K. and Knust, E.J. (1996) Radiochemical synthesis of (^{123}I) 2-iodolisuride for dopamine D$_2$-receptor studies. *Nucl. Med. Biol.,***23,**373–376.

Beran, M. and Beneš, J. (1981) The process for the of synthesis of D-6-methyl-8β(2-cyanoethyl)ergoline-I. *CS Pat. 018536* (in Czech).

Bernardi, L. (1969) Recenti sviluppi della chimica degli alcaloidi dell' ergot. *ChimicaIndustria,***51,**563–569.

Bernardi, L. and Bosisio, G. (1977) Ergot alkaloids modified in the cyclitol moiety. *Experientia,***33,**704–705.

Bernardi, L., Gandini, E. and Temperilli, A. (1974) Ergoline derivatives-XIV. Synthesis of clavine alkaloids. *Tetrahedron,***30,**3447–3450.

Bernardi, L., Bosisio, G., Elli, C., Patelli, B., Temperilli, A., Arcari, G. and Glaesser, H.A. (1975) Ergoline derivatives. Note XIII. (—)-α-Adrenergic blocking drugs. *Il Farmaco,Ed. Sci.***30,**789–801.

Bernardi, L., Elli, C. and Temperilli, A. (1976) 5(10→9)*abeo*-ergolines from 9-hydroxyergolines. *J. Chem. Soc. Chem. Commun.,***1976,**570.

Bernardi, L., Temperilli, A., Ruggieri, D., Arcari, G. and Salvati, P. (1982) Verfahren zur Herstellung von neuen ergolinderivaten. *AT Pat. 381091*.

Bernardi, L., Bosisio, G., Mantegani, S., Sapini, O., Temperilli, A., Salvati, P., di Salle, E., Arcari, G. and Bianchi, G. (1983a) Antihypertensive ergolinepropionamides. *Arzneim.Forsch.,***33,**1094–1098.

Bernardi, L., Temperilli, A., Mantegani, S., Traquandi, G., Cornate, D. and Salvati, P. (1983b) Process for the synthesis of ergoline derivatives. *CS Pat. 236 874* (in Czech).

Bernardi, L., Chiodini, L., Mantegani, S.Ruggieri, D., Temperilli, A. and Salvati, P. (1984) Ergoline derivatives, processes for their preparation and pharmaceutical compositions containing same. *EP Pat. 0 126 968*.

Bernardi, L., Chiodini, L. and Temperilli, A. (1987) D-*Nor*-7-ergoline derivatives, process for preparing them, pharmaceutical composition and use. *EP Pat. 0 240 986*.

Bohlmann, R., Sauer, G. and Wachtel, H. (1993) Fluorierte ergoline. *DE Pat. 4 333 287*.

Boyer, S.K., Bach, J., McKenna, J. and Jagdmann, E. (1985) Mild hydrogen-transfer reductions using sodium hypophosphite. *J. Org. Chem.,***50,**3408–3411.

Brambilla, E., Chiodini, L., di Salle, E., Ruggieri, D., Sapini, O. and Temperilli, A. (1983) Ergoline derivatives, process for producing the ergoline derivatives and pharmaceutical compositions containing them. *EP Pat. 0 091 652*.

Brambilla, E., di Salle, E., Briatico, G., Mantegani, S. and Temperilli, A. (1989) Synthesis and nidation inhibitory activity of a new class of ergoline derivatives. *Eur.J. Med. Chem.,***24,**421–426.

Brich, Z. and Mühle, H. (1981) Verfahren zur Isomerisierung von Ergolinderivaten. *EPPat. 0 048 695*.

Brumby, T. and Sauer, G. (1991) Verfahren zur Herstellung von 6-H-Ergolinen. *DE Pat.4 114 230*.

Bulej, P., Cvak, L., Stuchlík, J., Markovič, L. and Beneš, J. (1990) Process for the synthesis of N-(D-6-methyl-8α-ergolenyl)-N',N'-diethylurea. *CZ Pat. 278 725* (in Czech).

Crider, M., Grubb, R., Bachmann, K.A. and Rawat, A.K. (1981) Convenient synthesis of 6-*nor*-9, 0-dihydrolysergic acid methyl ester. *J. Pharm. Sci.,***70**(12), 1319–1321.

Cvak, L., Stuchlík, J., Cerný, A., Křepelka, J. and Spáčil, J. (1983) Manufacture of 1-alkyl-derivatives of dihydrolysergol. *CS Pat. 234 498* (in Czech).

Cvak, L., Beneš, K., Pavelek, Z., Schreiberová, M., Stuchlík, J., Sedmera, P., Flieger, M. and Golda, V. (1992a) Ergopeptine 6'-deoxoderivatives and their synthesis. *CZ Pat.Appl. 2833–92* (in Czech).

Cvak, L., Stuchlík, J., Schreiberová, M., Sedmera, P. and Flieger, M. (1992b) Side reactions in bromination of α-ergocryptine. *Collect. Czech. Chem. Commun.,***57,** 565–572.

Cvak, L., Stuchlík, J., Schreiberová, M., Sedmera, P., Havlíček, V. and Flieger, M. (1994) 2, 3-dihydro-2-oxoergolene derivatives. *Collect. Czech. Chem. Commun.,***59 b,** 929–942.

Danieli, B., Fiori, G., Lesma, G. and Palmisano, G. (1983) On the alleged electrochemical methoxylation of ergolines. *Tetrahedron Lett.,***24**(8), 819–820.

Dankházi, T., Fekete, E., Paál, K. and Farsang, G. (1993) Electrochemical oxidation of lysergic acid-type ergot alkaloids in acetonitrile. Part 1. Stoichiometry of the anodic oxidation electrode reaction. *Anal. Chim. Acta,***282,**289–296.

Eich, E., Sieben, R. and Becker, Ch. (1985) N-1-, C-2- und N-6-monosubstituierte Agroclavine. *Arch. Pharm. (Weinheim, Ger.),***318,**214–218.

Eichberg, D. and Eich, E. (1985) N-1- und N-6-mono- und disubstituierte Festuclavine. *Arch. Pharm. (Weinheim, Ger.)*, **318**, 621–624.

Flückiger, D., Troxler, F. and Hofmann, A. (1971) 2-Bromo-α-ergocryptine. *CH Pat. 507 249.*

Garbrecht, W.L., Marzoni, G., Whitten, K.R. and Cohen, M.L. (1988) (8β)-Ergoline-8carboxylic acid cycloalkyl esters as serotonin antagonists: Structure-activity study. *J. Med. Chem.*, **31**, 444–448.

Gervais, Ch. (1986) Procédé de preparation des derives N-méthyles du lysergol et du méthoxy-10alpha lumilysergol. *Eur. Pat. Appl. 209 456.*

Gull, P. (1981a) Ergolinderivate, ihre Herstellung und Verwendung. *CH Pat. 645 894.*

Gull, P. (1981b) Ergolinderivate, ihre Herstellung und Verwendung. *CH Pat. 645 895.*

Gull, P. (1983) Mutterkornalkaloide, ihre Herstellung und Vervendung. *CH Pat. 657 366.*

Harris, J.R. and Horwell, D.C. (1992) Conversion of agroclavine to lysergol. *Synth. Commun.*, **22**, 995–999.

Hartig, P.R., Krohn, A.M. and Hirschman, S.A. (1985) Microchemical synthesis of the serotonin receptor ligand, ^{125}I-LSD. *Anal. Biochem.*, **144**, 441–446.

Heindl, J., Sauer, G. and Wachtel, H. (1989) 2'-, 12'- oder 13'-substituierte 3-(8'αErgolinyl)-1, 1-diethyl-harnstoffe, Ihre Herstellung und Verwendung in Arzneimitteln. *EP Pat. 0 351 351.*

Hofmann, A. (1964) *Die Mutterkornalkaloide.* F. Enke Verlag, Stuttgart.

Hunter, W.H. (1985) Pharmaceutical indoloindole compounds and their preparation. *GB Pat. 2 162 182.*

Kadan, M.J. and Hartig, P.R. (1988) Autoradiographic localization and characterization of (1–125) Lysergic acid diethylamide binding to serotonin receptors in Aplysia. *Neuroscience (Oxford)*, **24**, 1089–1102.

Kornfeld, E.C. and Bach, N.J. (1978) 6-N-propyl-8-methoxymethyl or methylmercaptomethylergolines and related compounds. *US Pat. 4 166 182.*

Kornfeld, E.C. and Bach, N.J. (1979) 2-Azaergolines and 2-Aza-8(or 9)-ergolenes. *USPat. 4 201 862.*

Krause, W., Kühne, G. and Seifert, W. (1991) Pharmacokinetics of ^{3}H-terguride in elderly volunteers. *Arzneim.-Forsch.*, **41**, 373–377.

Krause, W., Düsterberg, B., Jakobs, U. and Hoyer, G.-A. (1993) Biotransformation of proterguride in the perfused rat liver. *Drug Metab. Dtspos.*, **21**, 203–208.

Křen, V. and Sedmera, P. (1996) N-1-Trimethylsilyl derivatives of ergot alkaloids. *Collect. Czech. Chem. Commun.*, **61**, 1248–1253.

Křen, V., Sedmera, P., Havlíček, V. and Fišerová, A. (1992) Enzymatic galactosylation of ergot alkaloids. *Tetrahedron Lett.*, **33**, 7233–7236.

Křen, V., Sčigelová, M., Přkrylová, V., Havlíček, V. and Sedmera, P. (1994) Enzymatic synthesis of β-N-acetylhexosaminides of ergot alkaloids. *Biocatalysis,* **10**, 181–193.

Křen, V., Němeček, J. and Přikrylová, V. (1995) Assignment of nitrogen stereochemistry of agroclavine and elymoclavine 6-N-oxides. *Collect. Czech. Chem. Commun.*, **60**, 2165–2169.

Křen, V., Fišerová, A., Augé, C., Sedmera, P., Havlíček, V. and Šíma, P. (1996) Ergot alkaloid glycosides with immunomodulatory activities. *Bioorg. Med. Chem.*, **4**, 869–876.

Křen, V., Pískala, A., Sedmera, P., Havlíček, V., Přikrylová, V., Witvrouw, M. and De Clercq, E. (1997a) Synthesis and antiviral evaluation of N-β-D-ribosides of ergot alkaloids. *Nucleosides Nucleotides,* **16**, 97–106.

Křen, V., Olšovský, P., Havlíček V., Sedmera, P., Witvrouw, M. and De Clercq, M. (1997b) N-Deoxyribosides of ergot alkaloids: Synthesis and biological activity. *Tetrahedron,* **53,**4503–4510.

Křen, V. (1997c) Enzymatic and chemical glycosylations of ergot alkaloids and biological aspects of new compounds. *Top. Curr. Chem.,* **186,**45–65.

Křepelka, J., Vlčková, D. and Beneš, J. (1981) Process for the preparation of 2substituted ergoline-I derivatives. *CS Pat. 218 532* (in Czech).

Langström, B., Antoni, G., Halldin, C., Svärd, H. and Bergson, G. (1982) Synthesis of some ^{11}C-labeled alkaloids. *Chem. Scr.,* **20,**46–48.

Lončarič, S. and Ručman, R. (1984) Synthesis and chemical transformations of some new 1-acyl-ergolines. *Vestn. Slov. Kem. Drus.,* **31,**101–114.

Mantegani, S., Temperilli, A., Traquandi, G., Rossi, A. and Pegrassi, L. (1986) Piperazin1-yl-ergoline derivatives, process for preparing them and pharmaceutical compositions containing them. *EP Pat. 0 197 241.*

Mantegani, S., Brambilla, E., Temperilli, A., Ruggieri, D. and Salvati, P. (1988) Process for the synthesis of ergoline derivatives. *SU Pat. 1 634 137* (in Russian).

Mantegani, S., Brambilla, E., Ermoli, A., Fontana, E., Angiuli, P. and Vicario, G.P. (1991) Syntheses of tritium and carbon-14 labeled N-(3-dimethylaminopropyl)-N-(ethylamino-carbonyl)-6-(2-propenyl) ergoline-8β-carboxamide (cabergoline), a potent long lasting prolactin lowering agent. *J. Labelled Compd. Radiopharm.,* **29,**519–533.

Marzoni, G. and Garbrecht, W.L. (1987) N^1-Alkylation of dihydrolysergic acid. *Synthesis,* **1987,**651–653.

Marzoni, G., Garbrecht, W.L., Fludzinski, P. and Cohen, M.L. (1987) 6-Methylergoline8-carboxylic acid esters as serotonin antagonists: N^1-substituent effects on 5HT$_2$ receptor affinity. *J. Med. Chem.,* **30,**1823–1826.

Maurer, G. and Frick, W. (1984) Elucidation of the structure and receptor binding studies of the major primary metabolite of dihydroergotamine in man. *Eur. J. Clin.Pharmacol.,* **26,**463–470.

Mayer, K. and Eich, E. (1984) Raney-Nickel-katalysierte Transferhydrierung: Eine Methode zur Darstellung Ring D-gesättigter Ergot-Alkaloide. *Pharmazie,* **39,**537–538.

Megyeri, G. and Keve, T. (1989) Halogenation of indole alkaloids with dimethylsulfonium halogenides and halodimethylsulfoxonium halogenides. *Synth. Commun.,* **19,**3415–3430.

Misner, J.W. (1985) Decyanation of pergolide intermediate. *US Pat. 4 782 152.*

Misner, J.W., Garbrecht, W., Marzoni, G., Whitten, K.R. and Cohen, M.L. (1990) (8β)-6-Methylergoline amide derivatives as serotonin antagonists: N^1-Substituent effects on vascular 5HT$_2$ receptor activity. *J. Med. Chem.,* **33,**652–656.

Misner, J.W., Kennedy, J.H. and Biggs, W.S. (1997) Integration of a highly selective demethylation of a quaternized ergoline into a one-pot synthesis of pergolide. *Org.Process Res. Develop.,* **1,**77–80.

Neef, G., Eder, U., Saurer, G., Ast, G. and Schröder, G.(1982) Process for the preparation of ergoline derivatives. *CZ Pat. 229 948* (in Czech).

Ninomiya, I. and Kiguchi, T. (1990) Ergot Alkaloids. In A.Brossi (ed.), *The Alkaloids,* Vol. 38, Academic Press, New York, Chap. 1, pp. 1–156.

Nordmann, R. and Loosli, H.R. (1985) Synthesis and conformation of (5R, 8R, 10R)-8-(methylthiomethyl) ergoline-6-carboxamidine. *Helv. Chim. Acta,* 68, 1025–1032.

Nordmann, R. and Gull, P. (1986) Synthesis of (5*R*, 8*S*, 10*R*)-6-(allyloxy)- and (5*R*, 8*S*, 10*R*)-6-(propyloxy)- ergolines from the 6-methyl precursors. *Helv. Chim.Acta,***69**, 246–250.

Ohno, S., Adachi, Y., Koumori, M., Mizukoshi, K., Nagasaka, M., Ichihara, K. and Kato, E. (1994a) Synthesis and structure-activity relationships of new (5*R*, 8*R*, 10*R*)-ergoline derivatives with antihypertensive or dopaminergic activity. *Chem. Pharm. Bull.,***42**, 1463–1473.

Ohno, S., Koumori, M., Adaki, Y., Mizukoshi, K., Nagasaka, M. and Ichihara, K. (1994b) Synthesis and structure activity relationships of new (5*R*, 8*S*, 10*R*)-ergoline derivatives with antihypertensive or dopaminergic activity. *Chem. Pharm. Bull.,***42**,2042–2048.

Ohno, S., Ebihara, Y., Mizukoshi, K., Ichihara, K., Ban, T. and Nagasaka, M. (1986) Ergoline derivatives and salts thereof and pharmaceutical compositions thereof. *GBPat. 2 173 189.*

Ott, H., Hofmann, A. and Frey, A.J. (1966) Acid-catalyzed isomerization in the peptide part of ergot alkaloids. *J. Am. Chem. Soc.,***88**,1251–1256.

Palmisano, G., Danieli, B., Lesma, G. and Fiori, G. (1987) Electrochemical synthesis of 2-halo-ergolines. *Synthesis,***1987**,137–139.

Ponikvar, S. and Ručman, R. (1982) N-6-oxides of 9, 10-dihydroergot alkaloids. *Vestn.Slov. Kem. Drus.,***29**,119–128.

Rettegi, T., Magó, E., Toldy, L., Borsy, J., Berzétei, I., Rónai, A., Druga, A. and Cseh, G. (1984) Pyrazole derivatives with an ergoiine skeleton, a process for preparing them and pharmaceutical compositions containing these compounds. *EP Pat.0 128 479.*

Ručman, R. (1978) Process for the preparation of N-substituted 9,10-dihydrolysergic acid esters. *CS Pat. 216 231* (in Czech).

Ručman R., Kovšič, J. and Jurgec, M. (1983) A new synthesis of 2-bromo-α-ergocryptine and related ergot derivatives. *Il Farmaco,***38**,406–410.

Ručman, R., Kocjan, D., Grahek, R., Milivojevič, D. and Pflaum, Z. (1991) Isolation, synthesis and structure determination of 2-bromocryptine impurities. TRISOC, Trieste, 2.-5. April 1991.

Ruggieri, D., Arrigoni, C., Di Salle, E., Temperilli, A. and Giudici, D. (1989) Antiulcer and antisecretory ergoiine derivatives. *Il Farmaco,***44**,39–50.

Rutschmann, J. and Stadler, P.A. (1978) Ergot Alkaloids and Related Compounds. In B. Berde, H.O. Schield (eds), *Handbook of Experimental Pharmacology: NewSeries,*Vol. 49, Springer-Verlag, Berlin-Heidelberg-New York, Chap. II., pp. 29–85.

Salvati, P., Caravaggi, A.M., Temperilli, A., Bosisio, C., Sapini, O. and di Salle, E. (1981) Process for the synthesis of new ergoiine derivatives. *CS Pat. 221 828* (in Czech).

Sauer, G. (1981) Verfahren zur Herstellung von 8α-substituierten 6-Methylergolinen. *EP Pat. 0 032 684.*

Sauer, G. and Brumby, T. (1990) Verfahren zur Entalkylierung von Ergolinen. *DE Pat.4 034 031.*

Sauer, G. and Haffer, G. (1981) Process for the preparation of ergoiine derivatives. *DE Pat. 3 135 305.*

Sauer, G. and Haffer, G. (1983) Process for the preparation of lysergic acid esters. *CS Pat. 235 038* (in Czech).

Sauer, G. and Haffer, G. (1984) Verfahren zur Herstellung von 2, 3-Dihydroergolinen. *DE Pat. 3 411 981.*

Sauer, G. and Schröter, B. (1991) Verfahren zur Herstellung von 2- oder 13-Acylergolinen. *DE Pat. 4 113 609.*

Sauer, G., Biere, H., Haffer, G. and Huth, A. (1985a) Process for the preparation of ergoline derivatives. *DE Pat. 3 445 784.*

Sauer, G., Heindl, J., Schröder, G. and Wachtel, H. (1985b) Neue 12- und 13-BromErgolin-derivate. *DE Pat. 3 533 675.*

Sauer, G., Huth, A., Wachtel, H. and Schneider, H.H. (1985c) The way of preparation of new 2-substituted ergoiine derivatives. *DE Pat. 3 413 658.*

Sauer, G., Haffer, G. and Wachtel, H. (1986) Reduction of 8α-substituted 9,10-didehydroergolines. *Synthesis,* **12,**1007–1010.

Sauer, G., Biere, H. and Wachtel, H. (1987) Process for the preparation of new 1-arylergolinyl-urea derivatives. *DE Pat. 3 623 503.*

Sauer, G., Heindl, J. and Wachtel, H. (1988a) Electrophilic substitution of lithiated ergolines. *Tetrahedron Lett.,* **29,**6425–6428.

Sauer, G., Schröter, B. and Künzer, H. (1988b) Striking influence of the reaction conditions on the stereoselectivity in electrophilic substitution of a 10-lithioergolinyl-urea. *Tetrahedron Lett.,* **29,**6429–6432.

Sauer, G., Brumby, T., Wachtel, H., Turner, J. and Löschmann, P.A. (1990) 13-Brom und 13, 14-Dibrom-Ergoline, ihre Herstellung und Verwendung in Arzneimitteln. *EPPat. 0 418 990.*

Sauer, G., Brumby, T. and Künzer, M. (1991). Preparation of 2-hydroxymethylergolines. *DE Pat. 4 020 341.*

Seifert, K. and Johne, S. (1979) Synthese von Zuckerestern der Lysergsäure und 9, 10-Dihydrolysergsäure. *J. Prakt. Chem.,* **321,**171–174.

Seifert, K. and Johne, S. (1980) Verfahren zur Herstellung von 2-Alkoxyergolinen. *DD Pat. 149 667.*

Seifert, K., Härtling, S. and Johne, S. (1983) Regiocontrolled electrochemical cyanation of ergolines. *Tetrahedron Lett.,* **24,**2841–2842.

Seifert, K., Härtling, S. and Johne, S. (1986) Preparation and characterization of ergoline thiazolidinones and an ergoline imidazolidinone. *Arch. Pharm. (Weinheim,Ger.),* **319,** 266–270.

Seifert, K., Phuong, N.M. and Vincent, B.R. (1992) Electrochemical oxidation of ergolines. *Helv. Chim. Acta,* **75,**288–293.

Semonský, M. (1970) Mutterkornalkaloide und ihre Analoga. *Pharmazie,* **32,**899–907.

Schneider, H.R., Stadler, P.A., Stütz, P., Troxler, F. and Seres, J. (1977) Synthesis and properties of bromocriptine. *Experientia,* **33,**1412–1413.

Schreier, E. (1976) Radiolabelled peptide ergot alkaloids. *Helv. Chim. Acta,* **59,**585–606.

Stadler, P.A. and Stütz, P. (1975) The ergot alkaloids. In R.H.F.Manske, (ed.), *The Alkaloids,* Vol. XV, Academic Press New York-San Francisco-London, Chap. 1, pp. 1–44.

Stadler, P.A., Stürmer, E., Weber, H.P. and Loosli, H.R. (1981) 2-Aza-dihydroergotamin. *Eur. J. Med. Chem.,* **16,**349–354.

Stanovnik, B., Tišler, M., Jurgec, M. and Ručman, R. (1981) Bromination of α-ergocryptine and other ergot alkaloids with 3-bromo-6-chloro-2-methylimidazo [1, 2-b] pyridazine-bromine complex as a new brominating agent. *Heterocycles,* **16,**741–745.

Stoll, A. and Hofmann, A. (1937) Racemische Lysersäure und ihre Auflösung in die optischen Antipoden. *Hoppe-Seylers Z. Physiol. Chem.,* **250,**7.

Stoll, A. and Hofmann, A. (1943) Die Optisch aktiven Hydrazide der Lysergsäure und der Isolysergsäure. *Helv. Chim. Acta,* **26,** 922–928.

Stoll, A. and Hofmann, A. (1965) The ergot alkaloids. In R.H.F.Manske (ed.), *TheAlkaloids,* Vol. VIII, Academic Press, New York, Chap. 21, pp. 725–783.

Stoll, A., Petrzilka, Th. and Becker, B. (1950) Beitrag zur Kenntnis des Polypeptidteils von Mutterkornalkaloiden. (Spaltung der Mutterkornalkaloide mit Hydrazin.). *Helv.Chim. Acta,* **33,** 57–67.

Stütz, P. and Stadler, P.A. (1973) A novel approach to cyclic β-carbonyl-enamines. $\Delta^{7,8}$-Lysergic acid derivatives via the Polonovski reaction. *Tetrahedron Lett.,* **51,** 5095–5098.

Ščigelová, ML, Křen, V. and Nilsson, K.G.I. (1994) Synthesis of α-mannosylated ergot alkaloids employing α-mannosidase. *Biotechnol. Lett.,* **16,** 683–688.

Šmidrkal, J. and Semonský, M. (1982a) Alkylation of ergoline derivatives at position N [1]. *Collect. Czech. Chem. Commun.,* **47,** 622–624.

Šmidrkal, J. and Semonský, M. (1982b) D-Carboxymethyl-8β-carboxy-6-methylergoline and some 1, 8-disubstituted ergolines derived from it. *Collect. Czech. Chem. Commun.,* **47,** 625–629.

Taimr, J. and Křepelka, J. (1987) 1-Tosylderivatives of ergoline and their synthesis. *CSPat. 262 200* (in Czech).

Taimr, J., Beneš, J. and Křepelka, J. (1987a) 2-Trifluoracetylderivatives of ergoline and their synthesis. *CS Pat. 262 581* (in Czech).

Taimr, J., Křepelka, J. and Řežábek, K. (1987b) 5β, 10α, -l-Formyl-8α-formylamino-2, 3dihydro-6-methylergoline and the process for its preparation. *CS Pat. 267 284* (in Czech).

Taimr, J., Křepelka, J. and Valchář, M. (1987c) 13-Acetyl-2-bromo- a 2-acetyl-13-bromoderivatives of ergoline and the process for their preparation. *CS Pat. 262 283* (in Czech).

Temperilli, A. and Bernardi, L. (1980a) Secoergolinderivate. *DE Pat. 3 018 543.*

Temperilli, A. and Bernardi, L. (1980b) Dérivés de 6, 7 secoergoline, leur procédé de preparation et leur utilisation pharmaceutique. *FR Pat. 80 11002.*

Temperilli, A., Mantegani, S., Arcari, G. and Caravaggi, A.M. (1980) 5(10→9) *Abeo*ergoline derivatives, their preparation and therapeutic compositions containing them. *EP Pat. 0 016 411.*

Temperilli, A., Brambilla, E., Gobbini, M. and Cervini, M.A. (1987a) The synthesis of ergoline *tert*-butylderivatives. *SU Pat. 1 547 708* (in Russian).

Temperilli, A., Eccel, R., Brambilla, E. and Salvati, P. (1987b) New tetracyclic indole derivatives. *EP Pat. 0 254 527.*

Timms, G.H. and Tupper, D.E. (1985) Ergoline derivatives and their use as pharmaceuticals. *EP Pat. 0 180 463.*

Timms, G.H., Tupper, D.E. and Morgan, S.E. (1989) Synthesis of novel 8- and 10-substituted clavine derivatives. *J. Chem. Soc., Perkin Trans. 1,* **1989,** 817–822.

Toda, T. and Oshino, N. (1981) Biotransformation of lisuride in the hemoglobin-free perfused rat liver and in the whole animal. *Drug Metab. Dispos.,* **9,** 108–113.

Troxler, F. and Hofmann, A. (1957a) Substitutionen am Ringsystem der Lysergsäure II. Alkylierung. *Helv. Chim. Acta,* **40,** 1721–1732.

Troxler, F. and Hofmann, A. (1957b) Substitutionen am Ringsystem der Lysergsäure I. Substitutionen am Indol-Stickstoff. *Helv. Chim. Acta,* **40,** 1706–1720.

Troxler, F. and Hofmann, A. (1959) Oxidation von Lysergsaure Derivaten in 2,3Stellung. *Helv. Chim. Acta,***42,**793–802.

Tupper, D.E., Pullar, I.A., Clemens, J.A., Fairhurst, J., Risius, F.C., Timms, G.H. and Wedley, S. (1993) Synthesis and dopamine antagonist activity of 2-thioether derivatives of the ergoline ring system. *J. Med. Chem.,***36,**912–918.

Voges, R. (1988) Tritiated compounds for *in vivo* investigations, part II. Low-dosed drugs: CQP 201–403, a case study. In R.R.Mucino (ed), *Synthesis and Applicationsof Isotopically Labelled Compounds 1985, Proceedings of the Third InternationalSymposium,*Elsevier, Amsterdam, 33–40.

Voges, R., von Wartburg, B.R. and Loosli, H.R. (1985) Tritiated compounds for *in vivo* investigations: CAMP and ^{3}H-NMR-spectroscopy for synthesis planning and process control. In T.A.Baillie, J.R.Jones (eds), *Synthesis and Applications of IsotopicallyLabelled Compounds, Proceedings of the Second International Symposium,*Elsevier, Amsterdam, 371–376.

Watts, S.W., Gackenheimer, S.L., Gehlert, D.R. and Cohen, M.L. (1994) Autoradiographic comparison of (1–125) LSD-labeled 5-HT$_2$A receptor distribution in rat and Guinea-pig brain. *Neurochem. Int.,***24,**565–574.

Wheeler, W.J. (1986) Wittig methylenation of 9, 10-didehydro-6-methylergolin-8-one, a novel synthesis of lysergene and its subsequent conversion to agroclavine. *Tetrahedron Lett.,***27,**3469–3470.

Wheeler, W.J. (1987) The synthesis of 8-β-[(methylsulfinyl-[^{18}O]-methyl]-6-propylergoline. *J. Labelled Compd. Radiopharm.,***24,**1123–1129.

Wheeler, W.J. (1988) Preparation of ^{14}C-labeled 8, 9-didehydro-6, 8-dimethyl-2-methylthioergoline mesylate, a dopamine antagonist potentially useful in the treatment of schizophrenia. *J. Labelled Compd. Radiopharm.,***25,**667–674.

Wheeler, W.J., Kau, D.L.K. and Bach, N.J. (1990) The synthesis of [^{2}H], [^{3}H] and [^{14}C]-labeled 8β-[(methylthio)methyl]-6-propylergoline mesylate (pergolide mesylate), a potent, long-acting dopamine agonist. *J. Labelled Compd. Radiopharm.,***28,**273–295.

Wyss, P.A., Rosenthaler, J., Nüesch, E. and Aellig, W.H. (1991) Pharmacokinetic investigation of oral and IV dihydroergotamine in healthy subjects. *Eur. J. Clin.Pharmacol.,***41,**597–602.

9.
BIOTRANSFORMATIONS OF ERGOT ALKALOIDS

VLADIMÍR KŘEN

Laboratory of Biotransformation, Institute of

Microbiology,Academy of Sciences of the Czech Republic, Vídeňská

1083,CZ-142 20 Prague 4, Czech Republic

9.1.
INTRODUCTION

From all naturally occurring ergot alkaloids only two are used in therapy: ergotamine and ergometrine. The rest of the medicinally important ergot compounds underwent some chemical changes, e.g. halogenation, alkylation, 9, 10-double bond hydrogenation, etc.

Although total synthesis of various ergot alkaloids has been demonstrated, it is ruled out for economical reasons. On the other hand, chemical modifications of natural compounds possessing ergot skeleton are very important. Bioconversions can fulfil some of these tasks employing their advantages. Bio-reactions often occur at chemically "nonactivated" positions. Biological systems exhibit regional selectivity upon polyfunctional molecules. This is of a great advantage over many chemical reagents that cannot distinguish among multiple similar functional groups. A high degree of stereoselectivity (both substrates and products) is a typical feature of bioconversions. This area is rapidly moving toward a more refined approach to choosing enzymes or organisms for their predictable type-reactions (Křen, 1991).

Much of the work on alkaloid transformations was motivated by the need to produce more effective drugs from naturally available compounds. Main effort has been devoted to specific oxidations of ergot alkaloids to produce desired amount of substrates for semisynthetic preparations. Many enzymatic reactions were employed for obtaining of new ergot derivatives, as e.g., their glycosides.

For biotransformation of ergot alkaloids various systems were used. Enzymes (Taylor and Shough, 1967; Shough and Taylor, 1969; Křen *et al.,* 1992), crude cell homogenates (Taylor *et al.,* 1966; Wilson *et al.,* 1971; Gröger, 1963), microbial prokaryotic (Béliveau and Ramstad, 1966; Yamatodani *et al.,* 1962; Davis, 1982) and eukaryotic cells (Béliveau and Ramstad, 1966; Brack *et at.,* 1962; Tyler *et al.,* 1965; Sieben *et al.,* 1984; Křen *et at.,* 1989) including the alkaloid producing strains of *Claviceps,* plant suspension cells (Ščigelová *et al.,* 1995; Křen *et al.,* 1996a; Křen *et al.,* 1996b) and also such unusual systems as plant seedlings (Taylor *et al.*, 1966). The biotransformation methods involved both free and immobilised cells (Křen *et al.*, 1989).

Although for many biotransformations the alkaloid producing *Claviceps* strains were used, some alkaloids fed to the organisms can be regarded as "xenobiotics". It is especially true in the case when added alkaloids are not naturally produced by the respective *Claviceps* strain used (e.g., feeding lysergols to *C. purpurea*). Similar situation occurs when the converted alkaloid is normally produced as a minor component and by its high concentration the alkaloid can enter alternative metabolic pathways affording thus new compounds (e.g., feeding high concentrations of chanoclavine into *C. fusiformis*). Part of the studies on ergot alkaloid bioconversions was stimulated also by problems of metabolism of ergot drugs in mammals.

Biotransformations of ergot alkaloids, as few other groups of natural products, offer rich variety of methodologies used and the results. Not only many important basic findings was obtained during the decades of this research but some of them can be exploited industrially.

9.2.

BIOCONVERSIONS OF CLAVINE ALKALOIDS

Clavine alkaloids as the simplest ergot alkaloids are valuable starting material for many semisynthetic derivatives. The biotransformation studies in early sixties (Tyler *et al.*, 1965) were stimulated by not entirely known biogenesis of ergot skeleton. Later, mostly oxidative bioreactions were investigated aiming at the production of lysergol, elymoclavine and lysergic acid on an industrial scale.

9.2.1.
Agroclavine

Agroclavine (**1**) can be produced in high yields by selected strains *C. fusiformis* and *C. purpurea*. However, till recently it was not considered to be a useful drug. Therefore, main effort in agroclavine bioconversions was targeted to its oxidations to elymoclavine (**6**) that is an important substrate for semisynthetic ergot alkaloid-based drugs. Chemical oxidation of **1** to elymoclavine is infeasible.

Enzyme systems from outside genus *Claviceps* transform agroclavine mostly to 8-hydroxyderivatives (Figure 1). These conversions are mediated mainly by peroxidases. The 8-oxidation of 8, 9-ergolenes is accompanied by the shift of double bond to 9, 10-position. Intermediates of this reaction are in some cases 10-hydroxy- or 8, 9-epoxy-derivatives.

More than 100 species of filamentous fungi and other microorganisms oxidise agroclavine to setoclavine (**2**) and isosetoclavine (**3**) (Béliveau and Ramstad, 1966, Yamatodani *et al.*, 1962) (Table 1). *Psilocybe semperviva* converts agroclavine with certain degree of stereoselectivity to setoclavine (Brack *et al.*,

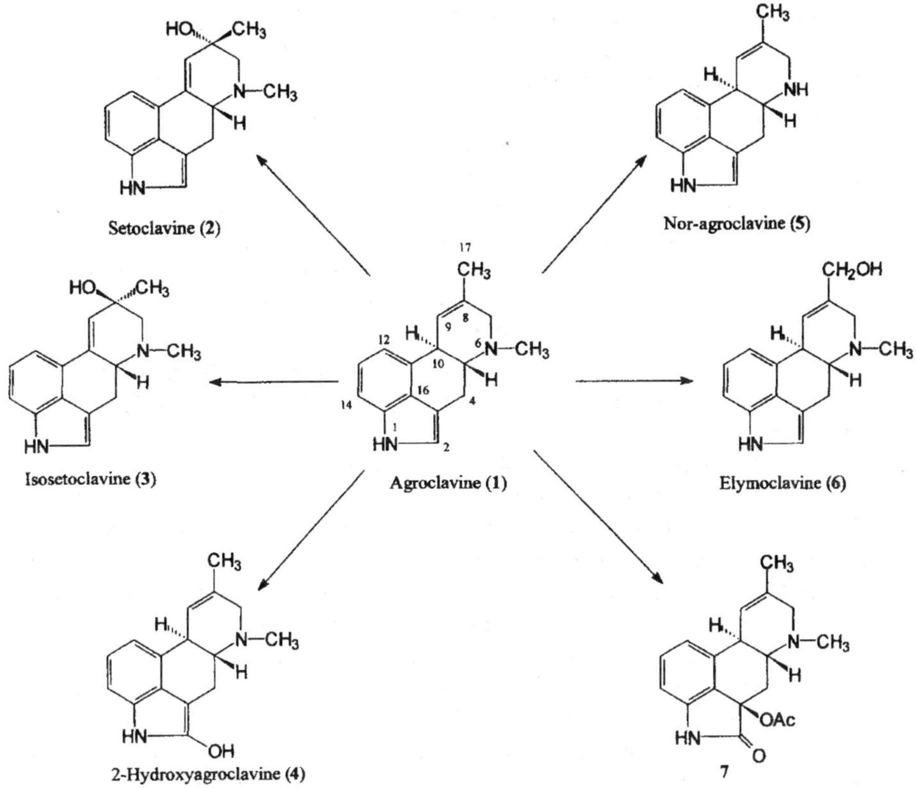

Figure 1 Typical bioconversions of agroclavine

1962). The same reaction is performed also by some prokaryotic microorganisms as, e.g., *Streptomycetes, Nocardias* (Béliveau and Ramstad, 1966; Yamatodani *et al.,* 1962) and *Pseudomonas aeruginosa* (Davis, 1982).

Peroxidase rich homogenates from tomato fruits, potato sprouts, horse radish and morning glory *(Convolvulaceae)* seedlings catalyse the oxidation of agroclavine in presence of H_2O_2 to setoclavine and isosetoclavine (Taylor and Shough, 1967; Shough and Taylor, 1969; Taylor *et al.,* 1966). Shough and Taylor (1969) found that 10-hydroxyagroclavine and 10-hydroxy-8,9-epoxyagroclavine were intermediates of this reaction. The later compound was recently isolated as a product of agroclavine biotransformation (besides both setoclavines) by horse radish peroxidase. This biotransformation gave the same product under both aerobic and anaerobic conditions (Křen and Kawuloková, *unpublished results*).

Series of plant cell cultures exhibiting high peroxidase activity were tested for the agroclavine biotransformation (Ščigelová *et al.,* 1995). HPLC analyses revealed that some cultures gave rise to a mixture of setoclavine and isosetoclavine in nearly equimolar ratio *(Ajuga reptans, Atropa belladonna,Papaver somniferum)*, whereas the other ones showed substantial degree of stereoselectivity *(Armoracia rusticana, Duboisia myoporoides, Euphorbia* **Table 1***(Continued)* *calyptrata*) producing isosetoclavine in high excess (Table 2). Previous experiments with plant homogenates did not reveal any stereospecifity of those systems (Taylor *et at.,* 1966).

Agroclavine can be also *N*-6 demethylated by peroxidase leading to the formation of nor-agroclavine **(5)** and formaldehyde (Chan Lin *et al.,* 1967b). Also mammalian tissue homogenates (rat liver, guinea pig adrenal) produce from agroclavine nor-agroclavine, elymoclavine **(6)** and small amount of both setoclavines **(2,3)** (Wilson *et al.,* 1991).

Abe (1967 and Abe *et al.,* 1963) found as a metabolite of agroclavine by *Corticium sasakii* besides setoclavines also 2-hydroxyagroclavine **(4)**. The later compound must, however, exist in its keto-form.

Some peroxidase system are able to oxidise also the pyrrole part of ergoline skeleton while degrading it. So called "degradation products" of ergot alkaloids **(8, 9)** (Figures 2, 3) are obviously products of peroxidase attack to the previously produced agroclavine and elymoclavine (Flieger *et al.,* 1991). These compounds are products of enzymes not directly involved in the alkaloid biosynthesis. The same compound **(9)** was obtained as the oxidation product of agroclavine by halogenperoxidase from *Streptomyces aureofaciens* (Figure 3) (Křen *et al.,* 1997). The main product in this case was, however, 2, 3-dihydro-6, 8dimethyl-3β-propionyloxy-8-ergolen-2-one (10), where the 3-propionyloxy group was introduced stereoselectively (3β). When acetate instead of propionate buffer was used in this reaction, analogous product 2, 3-dihydro6, 8-dimethyl-3β-acetoxy-8-ergolen-2-one (7) was produced without higher oxidation product as in the previous case (Křen *et al.,* 1997).

Some strains of *C. purpurea* are able, besides the main conversion product elymoclavine, to convert agroclavine to setoclavines and to festuclavine and pyroclavine (Agurell and Ramstad, 1962).

The most desired and the most economically important agroclavine conversion is its oxidation to elymoclavine or lysergol. Although minute amount of elymoclavine was found as agroclavine conversion product in rat liver homogenate system (Wilson *et al.,* 1971) and by *Penicillium roqueforti*(Abe, 1966), no organism, except for the *Claviceps* strains is able to perform this reaction at a reasonable rate.

Hsu and Anderson (1971) found agroclavine 17-hydroxylase activity in *C. purpurea* PRL 1980 cytosol, Kim *et al.* (1981) localised this activity to the microsomal fraction of the same strain. Sieben *et al.* (1984) conducted biotransformation of agroclavine derivatives with *C. fusiformis* SD-58 to obtain corresponding derivatives of elymoclavine. The strain was able to transform 1-alkyl-, 1-benzyl-, 1-hydroxymethyl-, 2-halo and 2, 3-dihydroagroclavine and 6-ethyl-6-noragroclavine to the corresponding elymoclavine derivatives. It was shown that the substrate specificity of the agroclavine 17-hydroxylase is high

Table 1 Agroclavine bioconversions

Reaction	Product	Conversion system (culture)	Reference
8-Hydroxylation	Setoclavine	Absidia spinosa *ATCC-6648, Aspergillus carbonarius PCC-104, Bispora effusa CBS, Cladosporium fulvus ATCC-10391, Fusarium solani PCC-143, Epicoccum sp., Giberella zeae Ull, Helminthosporium carbonum Ull, Mucor angulisporus CBS, Streptomyces annulatus PCC-A-111, S. griseus PCC, Nocardia rubra PCC-252	Béliveau and Ramstad (1966)
		Psilocybe semperviva	Brack et al. (1962)
		Penicillium viridicatum	Tyler et al. (1965)
8-Hydroxylation	Setoclavine + Isosetoclavine	Corticium sasakii	Abe et al. (1963)
		Pseudomonas aeruginosa	Davis (1982)
		Horse radish peroxidase	Taylor and Shough (1967), Shough and Taylor (1969)
		Tomato homogenate, Potato sprouts homogenate, Morning glory seedlings homogenate	Taylor et al. (1966)
2-Hydroxylation	2-Hydroxyagro-clavine	C. purpurea (more strains)	Agurell and Ramstad (1962)
		Corticium sasakii	Yamatodani et al. (1963)
2-Oxidation, 3-hydroxyacylation	2-Keto-3-O-acylagroclavines	Haloperoxidase from Streptomyces aureofaciens	Křen et al. (1997)
10-Hydroxylation	10-Hydroxyagro-clavine	Horse radish peroxidase	Shough and Taylor (1969), Chan Lin et al. (1967b)

8,9-Epoxidation	10-Hydroxy-8,9-epoxyagroclavine	Horse radish peroxidase	Shough and Taylor (1969), Chan Lin et al. (1967b)
17-Hydroxylation	Elymoclavine	*Penicillium roqueforti*	Abe et al. (1966)
		Rat liver homogenate	Wilson et al. (1971)
		Microsomal fraction from	Agurell and Ramstad (1962),
		C. purpurea PEPTY 695/S	Kim et al. (1981)
		C. purpurea	Sieben et al. (1984),
		C. fusiformis	Křen et al. (1989),
			Tyler et al. (1965)
		C. paspali	Tyler et al. (1965)
		Claviceps sp. strains	Sieben et al. (1984)
		KK-2, Se-134, 47A, SD-58	
N-6 Demethylation	Noragroclavine	*Streptomyces roseochromogenes,*	Yamatodani et al. (1962)
		S. punipalus, S. purpurascens	
		Rat liver homogenate	Wilson et al. (1971)
		Horse radish peroxidase (+ O$_2$)	Chan Lin and Ramstad (1967)
N-6-Demethylation + 8-Hydroxylation	Norsetoagroclavine	Horse radish peroxidase	Shough and Taylor (1969)
8,9-Hydrogenation	Festuclavine + Pyroclavine	*C. purpurea*	Agurell and Ramstad (1962)

Reaction	Product	Conversion system (culture)	Reference
17-Hydroxylation 1-Alkyl- 1-Benzyl- 1-Hydroxymethyl- 2-Bromo- 2,3-Dihydro- 6-Ethyl-6-nor-	Corresponding elymoclavines	C. fusiformis SD-58	Sieben et al. (1984)

* Abbreviations of culture collections:

ATCC=American Type Culture Collection, Washington, D.C., U.S.A.
CBS=Centralbureau voor Schimmelculturen, The Netherlands.
IMUR=Institute de Micologia Universitade Recife, Recife, Brazil.
LSHTM=London School of Tropical Medicine, Great Britain.
NRRL=Northern Regional Research Laboratory, Peoria, Illinois, U.S.A.
PCC=Purdue Culture Collection, Lafayette, Indiana, U.S.A.
U11=Collection of Dr. A.J.Ullstrup, Purdue University, Lafayette, Indiana, U.S.A.

Table 2 Plant cell cultures used in agroclavine biotransforniation experiments. Yields of setoclavine and isosetoclavine were calculated using molar extinction coefficients (Ščigelová *et al.,* 1995)

Plant cell culture	Setoclavine (%)	Isosetoclavine (%)
Ajuga reptans	5	5
Armoracia rusticana	10	80
Atropa belladonna	10	12
Duboisia myoporoides	5	90
Euphorbia calyptrata	traces	85
Papaver somniferum	20	30
Solanum aviculare	20	70

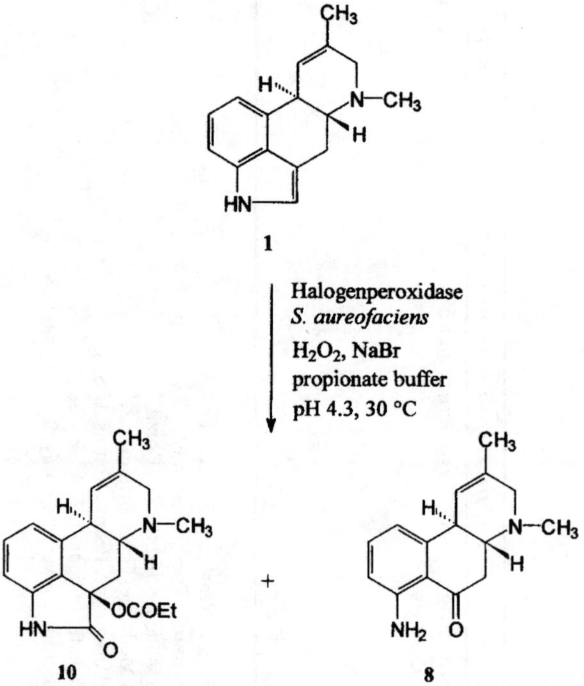

Figure 2 Biotransformation of agroclavine by halogenperoxidase from *Streptomycesaureofaciens*

with respect to the 8,9-double bond and to the tertiary state of N-6, whereas the specificity is low for variations in the pyrrole partial structure (N-l, C-2, C-3). The N-1 alkylated agroclavines are hydroxylated faster probably due to better penetration of the substances into the cells because of more lipophilic attributes. Noragroclavine and lysergine were oxidised by the above system (Eich and Sieben, 1985) to the corresponding 8α-hydroxyderivatives, i.e. norsetoclavine and setoclavine. The system is highly stereospecific (to 8α-oxidation) in contrary to horse radish peroxidase that gives rise to a mixture of 8α and 8β isomers.

For industrial bioconversion of agroclavine to elymoclavine, the high production strains *C. fusiformis* or selected strains *C. paspali* are the most suitable ones (Křen *et al.,* 1989). Use of immobilised and permeabilised *C. fusiformis* cells for this bioconversion was successfully tested (Křen *et al.,* 1989).

9.2.2.
Elymoclavine

Most of the elymoclavine (6) bioconversions were focused to its oxidation to lysergic acid (14) or paspalic acid (15) (Figure 3). Similarly as in agroclavine biooxidations, the C-17 oxidation activity is confined to the *Claviceps* genus. Other biosystems as fungi (Béliveau and Ramstad, 1966; Abe, 1966; Abe *et al.,* 1963; Tyler *et al.,* 1965; Sebek, 1983), bacteria (Béliveau and Ramstad, 1966; Yamatodani *et al.,* 1963; Abe, 1966) and plant preparations (Taylor *et al.,* 1966; Chan Lin *et al.,* 1967a, b; Gröger, 1963) introduce hydroxy-group into position C-8 by peroxidase reaction—analogously to agroclavine—giving rise to penniclavine (11) and isopenniclavine (12). Production of complicated mixture of

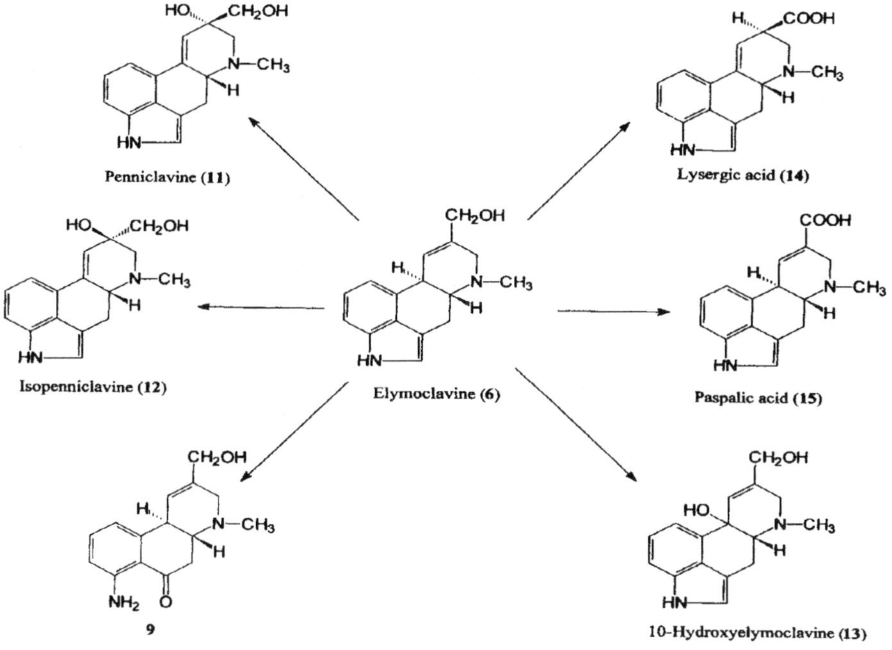

Figure 3 Typical bioconversions of elymoclavine

penniclavine, isopenniclavine, lysergol (20) (Béliveau and Ramstad, 1966) and 10-hydroxyderivatives (Chan Lin *et al.,* 1967a, b) was frequently observed (Table 3).

Elymoclavine transformation was tested also with plant cell cultures as in the case of agroclavine (Table 2) (Ščigelová *et al.,* 1995). Mixture of penniclavine (11) and isopenniclavine (12) was obtained with *Atropa belladonna* and *Papaver somniferum* cultures. *Euphorbia calyptrata* culture revealed a strict stereoselectivity, yielding almost exclusively 11. Penniclavine and isopenniclavine mixture produced by *A. belladonna* was accompanied with a polar substance not detected in other systems. The compound was isolated and identified as 10-hydroxyelymoclavine (13). This compound was suggested previously to be a tentative intermediate of peroxidase conversion of elymoclavine to penniclavine and isopenniclavine by plant homogenates (Chan Lin *et al.,* 1967a).

Biological reduction of elymoclavine leading to agroclavine was accomplished by several fungi (Yamatodani *et al.,* 1963; Abe, 1966; Abe *et al.,* 1963; Sebek, 1983). Oxidation of elymoclavine to lysergic acid and to its derivatives (18, 19) is practicable only by selected strains *C. purpurea* and *C. paspali* (Sebek, 1983; Maier *et al.,* 1988; Kim *et al.,* 1983; Mothes *et al.,* 1962; Philippi and Eich, 1984). Enzyme system responsible for this reaction is localised in microsomal fraction (Maier *et al.,* 1988; Kim *et al.,* 1983) and belongs obviously to the cytochrome P-450 family. Derivatives of elymoclavine are analogously converted by *C. paspali* to respective lysergic acid derivatives (Philippi and Eich, 1984). Biotransformations of elymoclavine leading to its glycosides are described separately.

9.2.3.
Lysergol, Lysergene, and Lysergine

Lysergol (21), lysergine (24), and their isomers occur as minor alkaloids in the *Claviceps* cultures. Their biogenetic relations to other alkaloids remain still unclear. Most of lysergolns bioconversio were aimed at discovering their biogenetic relations to other alkaloids.

Lysergol and lysergine are oxidised by *C. fusiformis* by similar mechanism as elymoclavine to penniclavine and setoclavine, respectively (Eich and Sieben, 1985). However, lysergol is not converted by elymoclavine 17-oxygenase from *C. purpurea* to paspalic acid (Maier *et al.,* 1988). Lysergols were also enzymatically glycosylated—*vide infra.*

Lysergene (23) is the only ergot alkaloid containing an exo-methylene group. Owing to its conjugation with the $\Delta^{9,10}$-double bond, this molecule is rather susceptible to oxidation or addition. The compound can be prepared easily and in good yield from elymoclavine (Křen *et al.,* 1996b) that makes 23 a favourite starting material for various semisynthetic ergot alkaloids. Its biotransformation using plant cell culture of *Euphorbia calyptrata* gave a unique spirooxadimer that can be formally derived from a putative intermediate, 8, 17-epoxylysergene **(26)** produced probably by action of peroxidase. **Table 3***(Continued)* Addition of this compound to **23** (presumably by radical mechanism) could then lead to the dimer **27** (Křen *et al.*, 1996b).

Lysergene was also converted by various strains *C. purpurea* to lysergol, isolysergol, penniclavine and isopenniclavine (Chan Lin *et al.*, 1967a; Agurell and Ramstad, 1962),

9.2.4.
Chanoclavine

Chanoclavine-I **(28)** is a common precursor of most ergot alkaloids in the *Claviceps* genus and its conversion to agroclavine and elymoclavine by enzymatic system of various *Claviceps* strains has been clearly proved (Ogunlana *et al.*, 1970; Erge *et al.*, 1973; Sajdl and Řeháček, 1975). Ogunlana *et al.* (1969) reported cyclisation of chanoclavine-I to elymoclavine as a sole product by pigeon-liver acetone-powder (+ATP+Mg^{2+}). This is probably the only report referring to this reaction in a system of *non-Claviceps* origin.

The conversion of exogenous chanoclavine by intact mycelia of *C. fusiformis* strain W1 gave chanoclavine-I aldehyde, elymoclavine and agroclavine. However, comparing with analogous agroclavine biotransformation in the same system, the conversion proceeded slowly due to, presumably low, transport rate of chanoclavine into cells. More polar alkaloids like chanoclavine and elymoclavine enter cells in a lower rate than less polar agroclavine. Beside these products also mono- and difructoside of chanoclavine were identified (Flieger *et al.*, 1989). These glycosides are formed by the action of *C. fusiformis* β-fructofuranosidase using as a β-fructofuranosyl donor sucrose (Figure 8).

Suspension plant cell cultures of *Euphorbia calyptrata*, *Atropa belladona,Armoracia rusticana*, and *Solanum aviculare* were tested for biotransformation of chanoclavine **(28)**. All tested cultures produced similar spectrum of biotransformation products from **28** (Figure 6). *Euphorbia calyptrata* gave the highest yields and, therefore, it was chosen for the preparatory purposes (Křen *et al.*, 1996a). Observed oxidative biotransformation of chanoclavine mimics the synthetic process in the reversed order. It also probably demonstrates typical biooxidative pattern of secoclavines by peroxidases.

8,9-Dihydrochanoclavine and isochanoclavine are converted by *Claviceps* sp. SD-58 to festuclavine and pyroclavine, respectively (Johne *et al.*, 1972).

Table 3 Elymoclavine bioconversions

Reaction	Product	Conversion system (culture)	Reference
8-Hydroxylation	Penniclavine	Aspergillus fumigatus, Mucor corticolus *CBS, Rhizopus arrbizus CBS, Streptomyces rimosus NRLL-2234, S. scabies ATCC-3352, Streptomyces sp. PCC	Béliveau and Ramstad (1966)
		Penicillium viridicatum	Tyler et al. (1965)
		Euphorbia calyptrata plant cell culture	Ščigelová et al. (1995)
		Ipomoea (leaves)	Gröger (1963)
8-Hydroxylation	Isopenniclavine	Absidia spinosa ATCC-6648, Fusarium graminearum PCC-140	Béliveau and Ramstad (1966)
8-Hydroxylation	Penniclavine + Isopenniclavine	Colletatrislium graminicolum, Cunningbamela echinulata CBS, Giberela zeae U11, Helmithosporium victorie, Rhizopus circinans CBS, R. nigricans PCC-199	Béliveau and Ramstad (1966)
		Streptomyces lipanii, Fusarium niveum, Corticium sasakii	Yamatodani et al. (1963), Abe (1966)
		Atropa belladonna, Papaver somniferum plant cell cultures	Ščigelová et al. (1995)
		Tomato homogenate, Potato sprouts homogenate, Morning glory seedlings homogenate	Taylor et al. (1966)
10-Hydroxylation	10-Hydroxy-elymoclavine	Horse radish peroxidase	Chan Lin et al., 1967a,b
		Horse radish peroxidase	Chan Lin et al., 1967a,b
		Atropa belladonna plant cell culture	Ščigelová et al. (1995)
8,9-Epoxidation	10-Hydroxy-8,9-epoxyelymoclavine	Horse radish peroxidase	Chan Lin et al. 1967a,b

Table 3 *(Continued)*

Reaction	Product	Conversion system (culture)	Reference
17-Oxidation	Paspalic acid	Claviceps spp.	Sebek (1983)
		Claviceps sp. PCCE1 (purpurea?) mycelial fraction	Kim et al. (1983)
17-Oxidation	Lysergic acid	C. purpurea PEPTY 695/S microsomal fraction	Maier et al. (1988)
17-OH-Reduction	Agroclavine	C. paspali SO 70/5/2	Philippi and Eich (1984)
		Aspergillus fumigatus, Corticium sasakii	Abe (1966), Abe et al. (1963), Sebek (1983)
8,9-Double bond isomerisation	Lysergol	Beauveria bassiana PCC-122, Fusarium lini ATCC-9593, F. lycopersici PCC-141, F. roseum PCC-142, Mucor adventitius CBS, Monascus pitosus IMUR-165, Rhizopus chinensis CBS, Streptomyces parvus NRRL-B-1456, S. griseus (3 strains)	Béliveau and Ramstad (1966)
Isomerisation	Lysergol + isolysergol	Claviceps spp.	Agurell and Ramstad (1962)
Isomerisation + hydroxylation	Lysergol + penniclavine + isopenniclavine	Aspergillus chevalieri NRRL-78, A. fumigatus (3 strains), A. niger ATCC-6277, A. quadrilineatus PCC-115, Botryosporium sp. PCC-284, Ceratocystis ulmi, Fusarium graminearum PCC-144, Mucor corticolus CBS, M. flavus PCC-256, M. globosus CBS, Melanospora destruens LSHTM	Béliveau and Ramstad (1966)

17-Oxidation	Ergine	BB-168, *Sclerotinia sclerotiorum*, *Nocardia convoluta* PCC-109, *Streptomyces rimosus* NRRL-2234, *S. rubrireticuli* NRRL-B-1484, *S. scabies* ATCC-3352	Mothes *et al.* (1962)
17-Oxidation + derivatization	Ergotamine	*C. paspali* LI 189+	Mothes *et al.* (1962),
17-Oxidation		*C. purpurea* (sclerotia)	Winkler and Mothes (1962)
Elymoclavine derivatives	Corresponding lysergic acid α-hydroxyethyl-amides	*C. paspali*	Philippi and Eich (1984)
1-Alkyl-			
1-Benzyl-			
1-Hydroxymethyl-			
2-Bromo-			
2,3-Dihydro-			
6-Ethyl-6-nor-			

* For the abbreviations of culture collections□see footnote of Table 1.

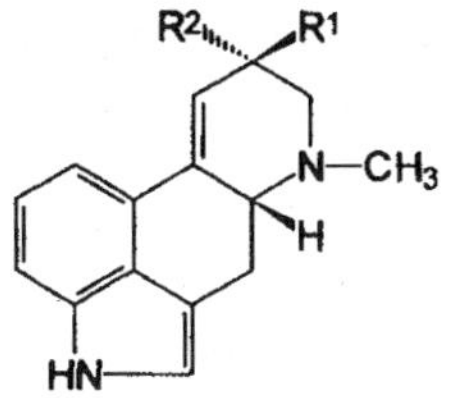

Figure 4 9, 10-Ergolenes, lysergic acid derivatives

(14) Lysergic acid R^1=COOH, R^2=H

(16) *iso*-Lysergic acid R^1=H, R^2=H

(17) Ergine R^1=CONH$_2$, R^2=H

(18) Ergomome R^1=H, R^2=CONH$_2$

(19) Ergometrine R^1=CONHCH(CH$_3$)CH$_2$OH, R^2=H

(20) Lysergic acid α-hydroxyethylamide R^1=CONHCH(CH$_3$)OH, R^2=H

(21) Lysergol R^1=CH$_2$OH, R^2=H

(22) *iso*-Lysergol R^1=H, R^2=CH$_2$OH

(23) Lysergene $R^{1,2}$= =CH$_2$

(24) Lysergine R^1=CH$_3$, R^2=H

(25) 8-Hydroxyergine R^1=CONH$_2$, R^2=OH

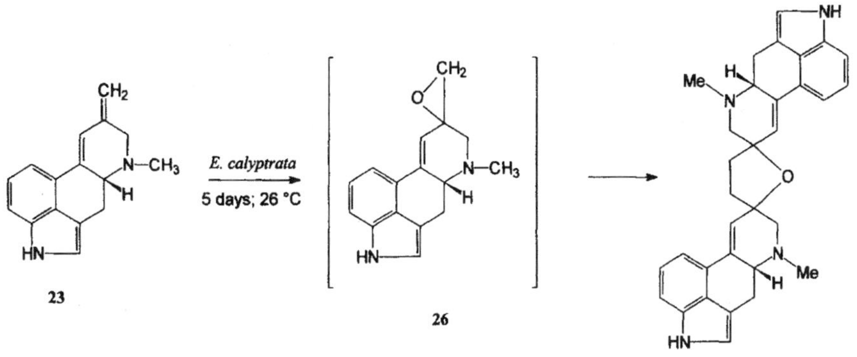

Figure 5 Dimerisation of lysergene by plant cell culture of *Euphorbia calyptrata*

9.2.5.
Semisynthetic Clavine Alkaloids

Recent interest in lergotrile (**33**) (Figure 7) stems from its putative dopaminergic activity and inhibition of prolactin secretion. Toxicity in clinical trials prompted the exploration of microbial transformation for producing less toxic derivatives.

Davis *et al.* (1979) screened nearly 40 organisms for their ability to produce metabolites of lergotrile. Five microorganisms (*Cunninghamelaechinulata* UI 3655, *Streptomyces rimosus* ATCC 23955, *S. platensis* NRRL 2364, *S. spectabilis* UI-C632, *S. flocculus* ATCC 25435) biotransformed lergotrile **(33)** to nor-lergotril **(34)** by *N*-demethylation. *S. platensis* exhibited complete conversion, and preparative-scale incubation was accomplished with an isolated yield of 50%. Additional organisms have been screened for their ability to produce hydroxylated metabolites of lergotrile **(35, 36)** that has been found in humans and in mammals but these efforts have been unsuccessful (Smith and Rosazza, 1982). Microbial *N*-demethylation is important because chemical demethylation of the compounds like lergotrile is rather difficult (Smith and Rosazza, 1982). In guinea pig liver, lergotrile is demethylated, and hydroxylated on C-13 and on nitrile group (Parli and Smith, 1975).

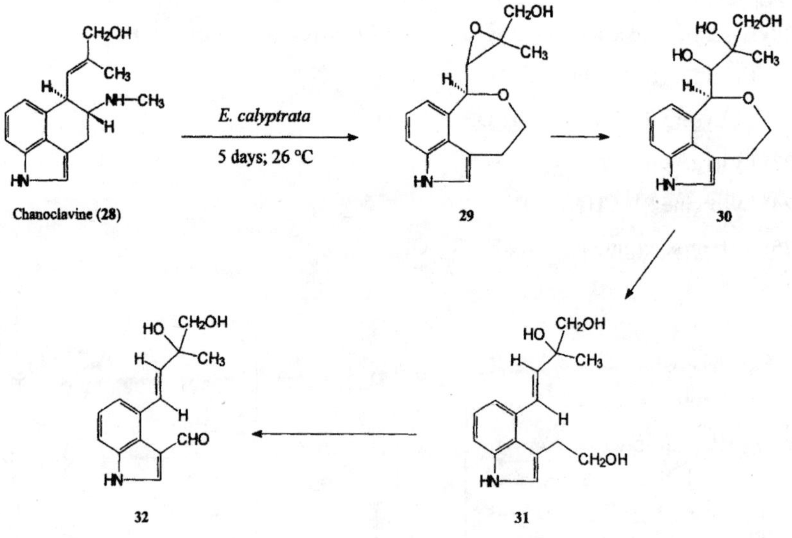

Figure 6 Oxidative sequential biotransformation of chanoclavine by plant cell culture of *Euphorbia calyptrata*

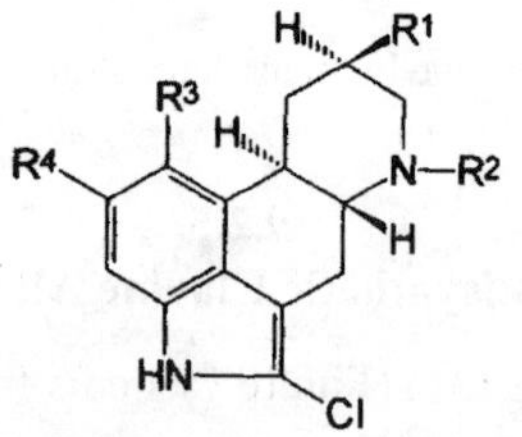

Figure 7 Semisynthetic ergot drugs and their metabolites
(33) Lergotrile $R^1=CH_2CN$, $R^2=CH_3$, $R^3=R^4-H$

(34) Nor-lergotrile $R^1=CH_2CN$, $R^2=R^3=R^4=H$
(35) 12-Hydroxylergotrile $R^1=CH_2CN$, $R^2=CH_3$, $R^3=OH$, $R^4=H$
(36) 13-Hydroxylergotrile $R^1=CH_2CN$, $R^2=CH_3$, $R^3=H$, $R^4=OH$

9.2.6.
Ergot Alkaloid Glycosylations

Ergot alkaloid fructosylation

Glycosides of ergot alkaloids were isolated as naturally occurring products (Floss et al., 1967) and recently large series of them was prepared by chemical and enzymatic methods (Křen, 1997). Their promising physiological effects stimulate future research in this field. Most of the work with the enzymatic ergot alkaloid glycosylations has been performed with clavine alkaloids.

Fructosylation of elymoclavine by C. purpurea was described (Křen et al., 1990). This reaction is mediated by transfructosylating activity of β-fructofuranosidase in sucrose containing media. Beside mono- (38) and difructoside (Figure 8), probably higher fructosides (tri- and tetra-) (39) are formed (Figure 9) (Havlíček et al., 1994).

The reaction is strongly dependent on pH (optimum 6.5) and substrate concentration (optimum 75 g sucrose/l). Similar reaction occurs with chanoclavine (Flieger et al., 1989) and also with other clavines bearing primary OH group (lysergols). This bioconversion can be effectively performed in a culture of C. purpurea with alkaloid production selectively blocked by 5-fluorotryptophan (Křen et al., 1993). The bioconversions of ergot alkaloids by Claviceps strains are often complicated by production of alkaloids de novo that might compete with added "xeno" alkaloids and make the mixture after conversion rather complex. Strains C. purpurea with glycosylation activity produce normally high amount of elymoclavine that is glycosylated at a high rate and thus competes with added lysergoles. 5-Fluorotryptophane blocks the alkaloid production at the first reaction—dimethylallyltryptophan synthase and consequent steps remain active (Pažoutová et al., 1990). The growth and differentiation of the culture was not impaired, so the bioconversion proceeds in "physiologically" normal culture and without any endogenous alkaloid production.

Glycosylating Claviceps strains (mostly C. fusiformis) produce most of elymoclavine in the form of β-fructofuranosides that complicates isolation of elymoclavine. Hydrolysis of fructosides by HC1 is not suitable for a large scale process due to aggression of the acid solution and losses of elymoclavine. A more elegant method is a bioconversion employing high invertase activity of Saccharomyces cerevisiae. At the end of the production cultivation, a suspension of baker's yeast is added to the fermentation broth (without the product isolation). The hydrolysis is completed within 1 hour (37°C) (Křen et al., 1990).

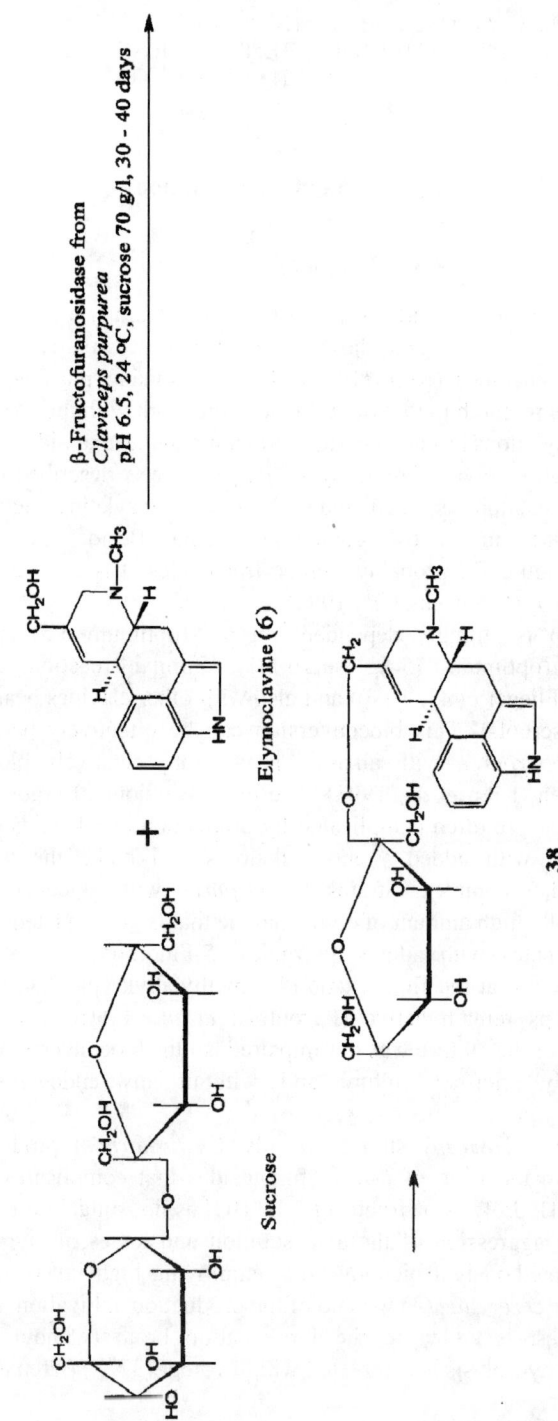

Figure 8β-Fructosylation of elymoclavine by *Claviceps fusiformis*

For high submerged production of elymoclavine it is advantageous to screen for the strains having high transfructosylation activity. Most of the elymoclavine produced is fructosylated (extracellulary) and due to higher polarity it does not re-enter the cell avoiding thus the feedback inhibition of agroclavine oxygenase (cytochrome P-450). Fructosylated elymoclavine can be eventually hydrolysed to its aglycone by the above method.

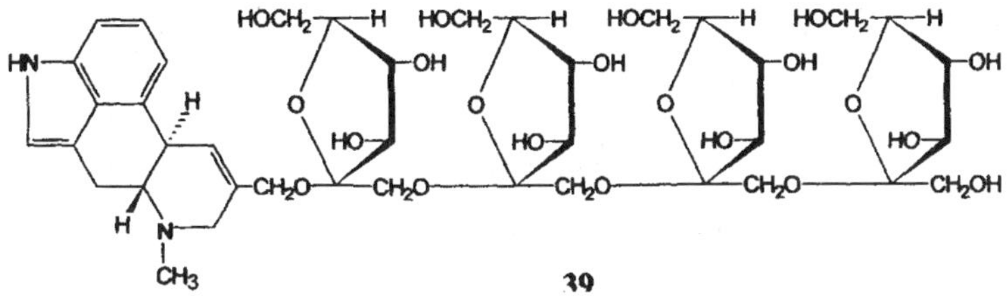

Figure 9 Elymoclavine tetrafructoside

Ergot alkaloid glycosylation using glycosidases

Interesting physiological effects of alkaloid fructosides (Křen, 1997) stimulated preparation of other glycosides, as e.g., galactosides, glucosides, *N*-glucosaminides and complex alkaloid glycosides.

β-Galactosylation of elymoclavine **(6)**, chanoclavine **(28)**, lysergol **(21)**, 9, 10dihydrolysergol, and ergometrine **(19)** was accomplished by *β*-galactosidase from *Aspergillus oryzae* using *p*-nitrophenyl-*β*-D-galactopyranoside or lactose as a glycosyl donors. Transglycosylation yields ranged from 13 to 40% (Křen *et al.,* 1992). This enzymatic method enabled for the first time to glycosylate ergometrine bearing in the molecule amidic bond.

Aminosugar-bearing alkaloids had been expected to have immunomodulatory activities and also this glycosylation would create basis for further extension of carbohydrate chain (introduction of LacNAc or sialyl residue—*vide infra*).

This task was accomplished by tranglycosylation using *β*-hexosaminidase from *A. oryzae.* Representatives of each class of ergot alkaloids, e.g. clavines— elymoclavine **(6)**, secoclavines—chanoclavine **(28)** and lysergic acid derivatives —ergometrine **(19)** were chosen to demonstrate the wide applicability of this method. As a donor *p*-nitrophenyl-*β*-*N*-acetylglucosaminide or -galactosaminide were used, the yields ranged from 5 to 15% (Křen *et al.,* 1994a).

Enzymatic mannosylation of the alkaloids by α-mannosidase from *Canavalia ensiformis* (Jack beans) was accomplished by two different strategies— transglycosylation using *p*-nitrophenyl-α-D-mannopyranoside or reversed glycosylation using high concentration of mannose (Ščigelová *et al.,* 1994). In the case of chanoclavine **(28)** higher yield of respective α-mannoside **(30)** was obtained in shorter time by using of transglycosylation concept (Figure 10). Lower yields in reversed glycosylation are, however, compensated by

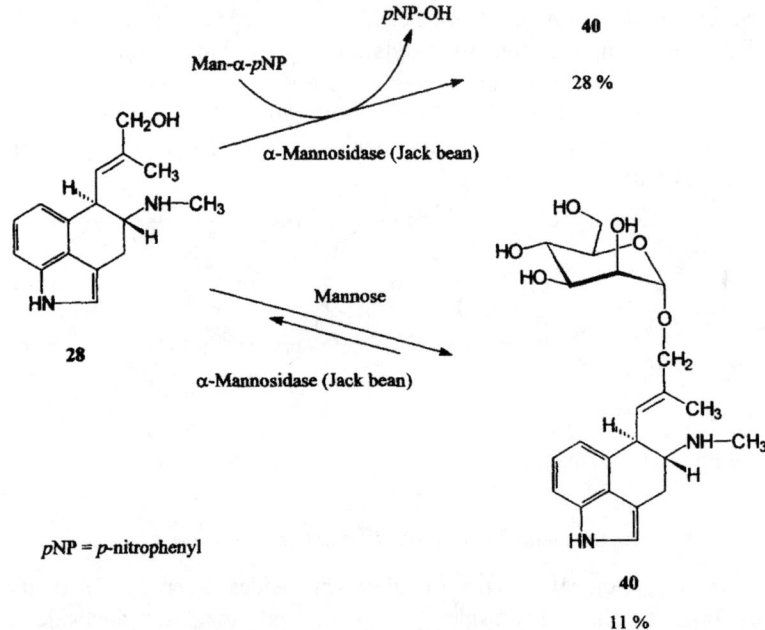

Figure 10 Mannosylation of chanoclavine with α-mannosidase from jack bean: transglycosylation using *p*-nitrophenyl-α-mannoside and reversed mannosylation using high concentration of free mannose.

considerably cheaper mannosyl donor (mannose). Unreacted aglycone can be nearly quantitatively recuperated.

Series of other glycosides of ergot alkaloids was prepared by enzymatic transglycosylations using activated *p*-nitrophenylglycosides as donors. *β*-Glucosides of, e.g., elymoclavine and dihydrolysergol, were prepared by *β*-glucosidase from *Aspergillus oryzae* (Křen *et al.,* 1996c), α-glucosides of e.g., elymoclavine and chanoclavine were prepared by α-glucosidase from *Bacillus stearothermophilus* or by α-glucosidase from rice, α-galactosides of the same alkaloids were prepared by α-galactosidase from *Coffea arabica* (green coffee beans) or that one from *A. niger.*

Ergot alkaloid gly cosy lations by glycosyltransferases—complex glycosides

Complex alkaloid glycosides bearing e.g., lactosyl (Lac), lactosaminidyl (LacNAc) or sialyl (Neu5Ac) moieties were required for immunomodulation tests *(vide infra).* Because of paucity of starting material (alkaloid monoglycosides) and because of the need of regioselective glycosylation only enzymatic reactions were practicable.

For preparation of D-galactopyranosyl (1→4)-2-acetamido-2-deoxy-β-Dglucopyranosyl-(1→O)-elymoclavine (32) the extension of previously prepared 2-acetamido-2-deoxy-β-D-glucopyranosyl-(1→O)-elymoclavine (31) by the use of bovine β-1, 4-galactosyltransferase was chosen. Uridine 5′-diphosphogalactose (UDP-Gal) served as a substrate (Křen *et at.,* 1994b).

β-Lactosyl elymoclavine (32) was prepared from respective β-glucoside by the use of bovine β-1, 4-galactosyltranferase in presence of α-lactalbumine. Analogously, β-Lac and β-LacNAc derivatives of other ergot alkaloids, e.g., 9, 10-dihydrolysergol were prepared (Křen *et al.,* 1994b).

Attachment of 5-*N*-acetylneuraminic acid to β-LacNAc-elymoclavine (32) yielding 33 was accomplished by the use of α-2, 6 sialyltransferase from rat liver (Figure 11) (Křen *et al.,* 1994b).

9.3.

BIOCONVERSIONS OF LYSERGIC ACID DERIVATIVES

9.3.1.
Hydrolysis of Lysergic Acid Amides

This group includes simple lysergic acid amides and peptide ergot alkaloids. Most of these substances are directly used in pharmaceutical preparations. However, reports referring to their biotransformation are scarce.

One of the most desired transformations is a hydrolysis of above substances resulting in free lysergic acid, the substrate for many semisynthetic preparations. Chemical hydrolysis of its derivatives gives low yields (50–65%). Fermentative production of lysergic acid is somehow complicated. Enzymatic hydrolysis of peptide alkaloids is still impracticable. Common proteolytic enzymes (papain, subtilisin, chymotrypsin, termolysin) do not attack peptidic bond in ergokryptine (34) probably due to the steric reasons (Křen—unpublished results).

Amici *et al.* (1964) reports the hydrolysis of ergine (17) and erginine (18) by a strain *C. purpurea* yielding 80–90% lysergic acid. The bioreaction is aerobic, cultivation medium was supplemented by ergine dissolved in organic solvent up to final conc. 1000 mg/1.

Ergotoxine producing strain *C. purpurea* is able to split lysergyl-amino acidmethyl esters into corresponding lysergyl-amino acids as well as into lysergic acid and alanine and valine (Maier *et al.,* 1974). This reaction proceeds inside the cells, so the transport is involved. Ergotoxine producing strains are more active than the nonproducing strains. Also lysergyl-oligopeptides (*d*-lysergyl-L-valyl-L-leucine-OMe, *d*-lysergyl-L-valyl-L-valine-OMe and *d*-lysergyl-L-valyl-L-proline-OMe) are split into their components after feeding to the mycelium of *C. purpurea* Pepty 695 (ergotoxine producing) (Maier *et al.,* 1978).

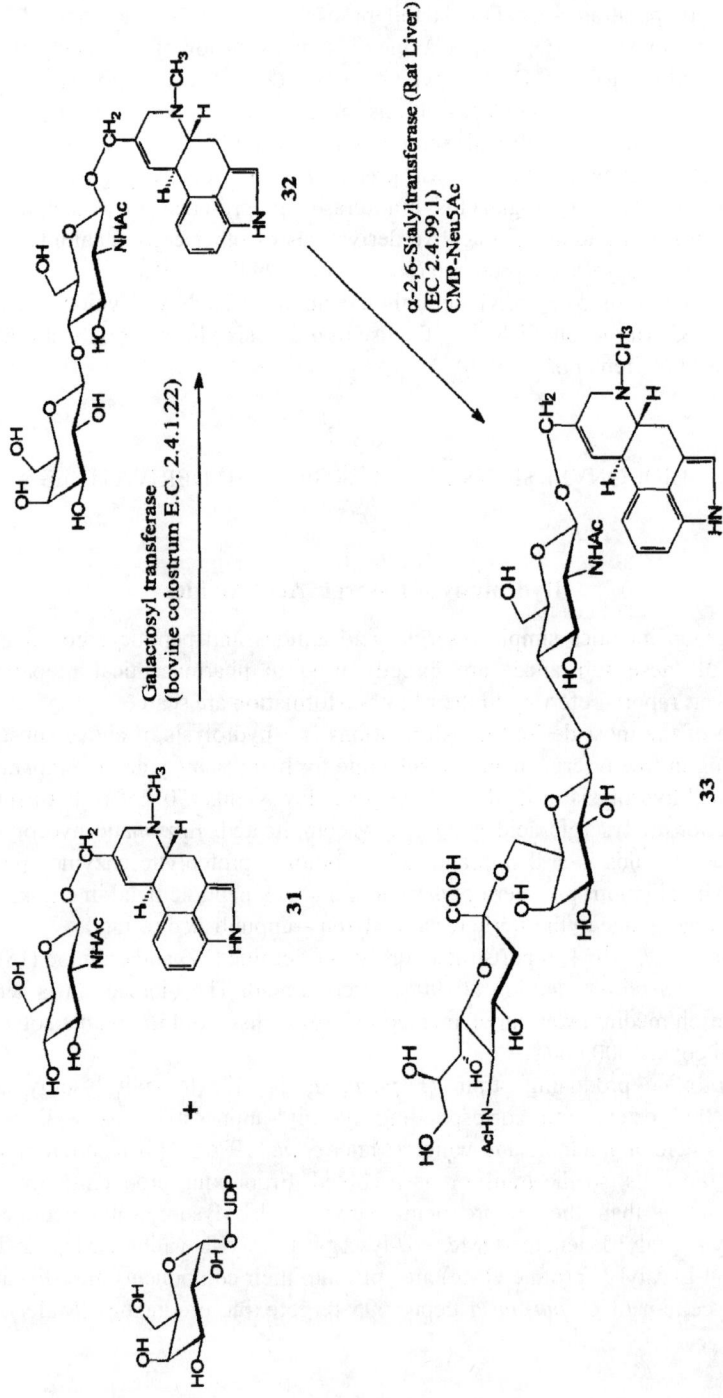

Figure 11 Preparation of complex ergot alkaloid glycosides using glycosyltransferases

9.3.2.
Oxidation of Lysergic Acid Derivatives

Ergotamine **(36)** and ergometrine **(19)** are partially isomerised on C-8 (but not hydroxylated) to ergotaminine and ergometrinine by *Psilocybe semperviva* and partially decomposed (Brack *et al.,* 1962).

Lysergic acid derivatives can be oxidised, analogously to most clavine alkaloids, on C-8 due to action of peroxidases. 8α-Hydroxy-α-ergokryptine **(35)** has been found in α-ergokryptine **(34)** producing *C. purpurea* strain (Cvak *et al.,* 1997). Oxidation of exogenous ergine **(17)** to 8-hydroxyergine **(25)** by *C. paspali* MG-6 was accomplished by Bumbová-Linhartová *et al.* (1991).

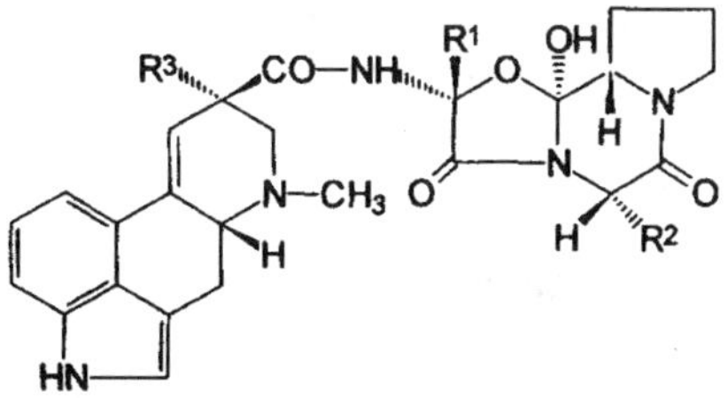

Figure 12 Peptide ergot alkaloids and their biotransformation products
(34) α-Ergokryptine R^1=CH(CH$_3$)$_2$, R^2=CH$_2$CH(CH$_3$)$_2$, R^3=H
(35) 8α-Hydroxy-α-ergokryptine R^1=CH(CH$_3$)$_2$, R2=CH$_2$CH(CH$_3$)$_2$, R^3=O
(36) Ergotamine R^1=CH$_3$, R^2=CH$_2$C$_6$H$_5$, R^3=H
(36a) 8α-Hydroxyergotamine R^1=CH$_3$, R^2=CH$_2$C$_6$H$_5$, R^3=OH

9.3.3.
Biotransformations of LSD and its Homologues

Lysergic acid amides, most notably lysergic acid diethylamide (LSD) **(37),** are of interest because of their hallucinogenic activity. Microbial metabolites of LSD might serve as more active medical preparations. Another biotransformation products correspond to LSD metabolites found in humans. Both these facts led Ishii *et al.* (1979a,b; 1980) to an extensive study of microbial bioconversions of the LSD and its derivatives.

Initial studies with LSD showed that many cultures were capable of attacking the *N*-6 and amide *N*-alkyl substituents (Ishii *et al.,* 1979a,b; 1980). *Streptomyces lavendulae* IFM 1031 demethylated only the *N*-6 position yielding nor-LSD. Conversely, *Streptomyces roseochromogenes* IFM 1081 attacked only the *N*-amide alkyl group to yield lysergic acid ethylamide **(38),** lysergic acid ethylvinylamide **(39),** and lysergic acid ethyl 2-hydroxyethylamide **(40)** (Figure 13). Other *Streptomycetes* and *Cunninghamella* strains produced all four metabolites. The high degree of substrate stereoselectivity in *Streptomyces roseochromogenes* was proved by the fact that this organism could not metabolise *iso*-LSD while, in contrary, *S. lavendulae* yielded *iso*-nor-LSD.

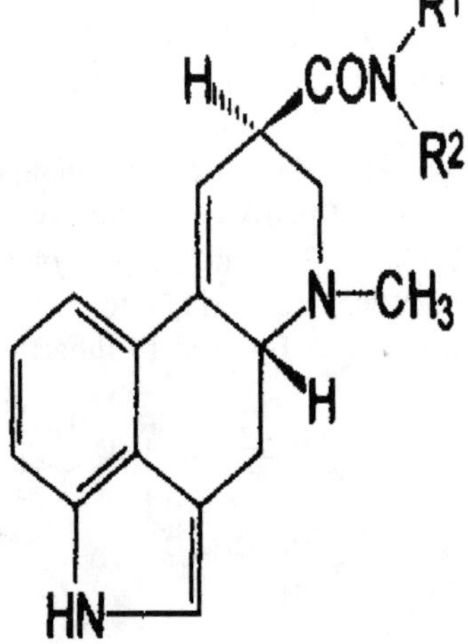

Figure 13 LSD and its metabolites
(**37**) Lysergic acid diethylamide (LSD) $R^1=R^2=CH_2CH_3$
(**38**) Lysergic acid ethylamide $R1=H$, $R^2=CH_2CH_3$
(**39**) Lysergic acid ethylvinylamide $R^1=CH_2CH_3$, $R^2=CH=CH_2$
(**40**) Lysergic acid ethyl 2-hydroxyethylamide $R^1=CH_2CH_3$, $R^2=CH_2CH_2OH$

C-17-Amide dealkylations were examined in greater detail using a series of lower and higher alkyl homologues of LSD (Figure 14). The following results were observed: Lysergic acid dimethylamide (**41**) was only dealkylated to the monomethylamide (**42**), lysergic acid diethylamide (**37**) was also dealkylated to yield monoethyiamide (**38**), and other metabolites like ethylvinylamide (**39**) and ethylhydroxyethylamide (**40**). Neither lysergic acid di-*n*-propylamide (**43**) nor lysergic acid di-*n*-butylamide (**47**) were dealkylated, but rather yielded two epimeric alcohols resulting from ω-1 hydroxylation, i.e. (**44**) and (**46**) from (**43**), and (**48**) and (**50**) from (**47**), as well as the further oxidation products, ketones (**45**) and (**49**), respectively. Based on these results, the authors proposed that the chain length regulates the site of oxygenation, with ω-1 hydroxylation occurring, if possible. The dimethyl-derivative (**41**) has no ω-1 position, thus C-α-(methyl) hydroxylation yields the carbinolamide with resulting *N*-demethylation.

These studies resulted in a proposed active site for the hydroxylase (*N*-dealkylase) of *S. roseocbromogenes* that accounts for the general mode of metabolism of the homologues. Authors also use the diagram to explain the stereochemical control of the system, based on the observation that one epimeric alcohol predominates in the hydroxylation of (**43**) or (**47**). This argument is based on a favoured binding of one alkyl group over the other, implying non-equivalence of two alkyl groups (Ishii *et al.*, 1979b; Davis, 1984). Microsomes

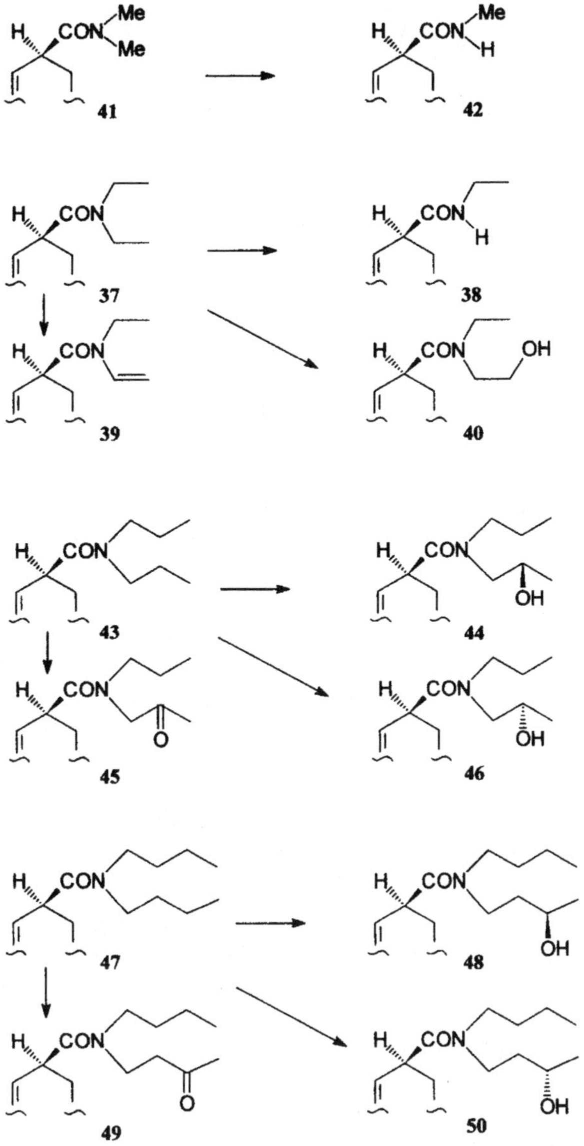

Figure 14 Metabolism of lysergic acid dialkylamide homologues by *Streptomycesroseochromogenes*

from mammalian liver (+NADPH+O_2) oxidise LSD to 2-hydroxyLSD (Axelrod *et al.*, 1956).

9.4.

BIOCONVERSION OF ERGOT ALKALOIDS AS A TOOL FOR STUDY OF THEIR METABOLISM IN MAMMALS

Elucidation of mammalian metabolic pathways is important in attempting to rationalise detoxification, and to evaluate potentially active metabolites. In case of some ergot alkaloids with strong hallucinogenic activity (e.g. LSD) it is desirable to detect its metabolites for use in anti-doping screening and forensic chemistry. However, such studies have been hampered by a lack of minor metabolite availability. For this reason, Ishii *et al.* (1979a, b, 1980) examined a series of microorganisms and animals for parallel routes of metabolism of LSD and of the related compounds. They assumed that the metabolism of these xenobiotics would proceed in a similar manner in both mammals and microorganisms (Ishii *et al.,* 1980). Both nor-LSD and lysergic acid ethylamide produced by the microbial conversion of LSD are known metabolites of LSD in mammals, and the authors were able to use the lysergic acid ethylvinyl amide generated in the microbial studies to determine its presence in mammals.

This work has allowed a high degree of predictability regarding the *N*-(amide) -alkyl   oxidation   of   lysergic   acid   amide   derivatives   using *Streptomycesroseochromogenes* (see above). Analysis of a series of homologues from lysergic acid dimethylamide through lysergic acid dibutylamide resulted in a proposal for the enzyme active site, as well as rationalisation of the resulting metabolites. This study also enabled preparation of the minor metabolites to be made for further investigation.

Lergotrile **(33)** (see above) was hydroxylated at various positions in mammal systems (C-13, C-12, hydrolysis of nitrile group) and demethylated at *N*-6 (Smith and Rozassa, 1982; Axelrod *et al.,* 1956). To find a parallel for this reaction and to prepare some standards of the metabolites more than 30 microorganisms have been screened for their hydroxylation ability (Davis *et al.,* 1979). Only *N*-6 demethylation activity has been found in *Streptomycesplatensis.*

Microbial   systems   parallel   mammalian   metabolism   with   an   other dopaminergic ergoline, e.g., pergolide **(53a)** (Figure 16). Metabolism in mammals centres on the methyl sulfide moiety, which is sequentially oxidised to the   sulfoxide   **(53b)**   and   the   sulfone   **(53c)**—*vide   infra.*   Similarly, *Aspergillusalliaceus* UI-315 catalysed the same sequential oxidative transformation. In contrast, *Helminthosporium* sp. NRRL 4761 stops at the sulfoxide stage and also catalyses reduction of the sulfoxide back to pergolide (Davis, 1984). No stereoselectivity in sulfoxide formation was observed, in contrast to the high degree of product stereoselectivity often observed in this microbial-type reactions (Davis, 1984; Auret *et al.,* 1981).

9.4.1.
Metabolism of the Therapeutically Used Ergot Alkaloids in Human Organism

Ergot alkaloids undergo many metabolic changes in human organism. The study of their metabolites is important both from pharmacodynamic and pharmacokinetic views. Liver is probably the most active organ in ergot alkaloid detoxification. Oxidative and conjugation systems in liver catalyse many reaction analogous to those discussed in previous parts (e.g., cytochrome P-450, oxygenases, peroxidases, glucuronyltransferases, etc.). Typical degradative reactions are *N*-demethylation, alkylamide dealkylations, hydroxylations (oxidative), epimerization at C-8, oxidation at C-2 followed by oxidative ring B opening and *N*-6 oxidation. The complex knowledge of the "weak" positions in alkaloid structures enables predicting of possible metabolic transformations of respective ergot alkaloid derivative.

Nicergoline

Nicergoline **(50)** used as cerebrovascular dilatans with strong α-adrenolytic activity (see the chapter 14 in this book) is metabolised in human organism into two products (Arcamone *et al.,* 1971; Arcamone *et al.,* 1972; Gabor *et al.,* 1995). The hydrolysis yields 5-bromonicotinic acid and 1-methyl-10α-methoxydihydrolysergol **(51)** which is further demethylated via 1-hydrohymethylderivative **51a** into l0α-methoxydihydrolysergol **(52)** (Banno *et al.,* 1991) (Figure 15). The metabolite **51** can be, due to its free hydroxyl group, a subject of further *β*-glucuronidation.

Pergolide

Pergolide **(53a)** is one of the strongest dopamine agonists and prolactin inhibitors. It is metabolised into four products resulting from oxidation and *N*-6 dealkylation (Clemens *et al.,* 1993; Kerr *et al.,* 1981). Pergolide sulfoxide **(53b)** and pergolide sulfone **(53c)** are formed by the oxidation of **53a**.

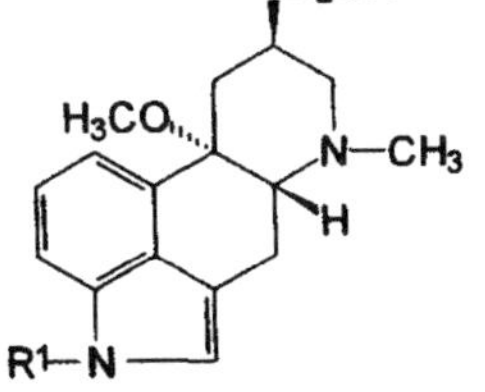

50 R^1 = CH$_3$, R^2 =

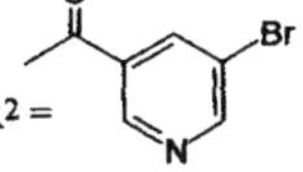

51 R^1 = CH$_3$, R^2 = H

51a R^1 = CH$_2$OH, R^2 = H

52 R^1 = H, R^2 = H

Figure 15 Nicergoline **(50)** and its metabolites

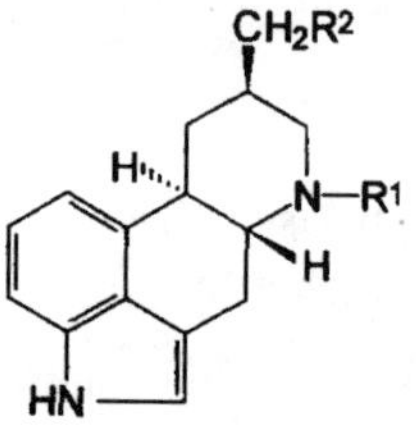

	R^1	R^2
53a	C$_3$H$_7$	SCH$_3$
53b	C$_3$H$_7$	SOCH$_3$
53c	C$_3$H$_7$	SO$_2$CH$_3$
53d	H	SCH$_3$
53e	H	SOCH$_3$

Figure 16 Pergolide (53a) and its metabolites

Dealkylation of **53a** yields in despropylpergolide **(53d)** and analogously metabolite **53b** is dealkylated into despropylpergolidesulfoxid **(53e)** (Figure 16).

2-Bromo-β-ergokryptine

2-Bromo-*β*-ergokryptine **(54a)** is very strong dopaminergic agonist and inhibits secretion of pituitary hormones, e.g., prolactine, somatotropic hormone (growth hormone) and ACTH (adrenocorticotropic hormone). It is, therefore, used for the treatment of disorders associated with the pathological overproduction of these hormones, as, e.g., hyperprolactinemia, acromegaly and Parkinson's disease. Metabolism of **54a** was studied in rats, and from their bile 18 metabolites were isolated (Maurer *et al.*, 1982) (Figure 17). From their structure following metabolic changes can be expected:

— epimerisation at C-8 of lysergyl moiety is quite common in its derivatives and can be also spontaneous. Therefore, occurrence of metabolites **54b** (bromokryptinine), 2-bromoisolysergic acid **(54g)** and its amide **(54c)** need not to be a result of the enzymatic processes.
— oxidation in the position 8′ of proline fragment of the peptide part is non-stereoselective and yields a mixture of four isomers at C-8 and C-8′ **(54j-m)**.
— further oxidation of 8′-hydroxylated isomers leads either to introduction of second OH-group at the position 9′ and followed by their glucuronidation **(54q-s)**. Non-conjugated 8′, 9′ dihydroxyderivative can undergo rearrangement leading to opening of proline ring **(54e)**.

Occurrence of the above metabolites was also monitored in human urine. As the main metabolites 2-bromolysergic acid and its 8-epimer and the respective amides were found **(54c, d, g, h)** (Maurer *et al.*, 1983). It is interesting that oxidative attack did not start at ergoline ring, especially in pos. 2 or 3. Bromination probably improves the drug stability.

Terguride and lisuride

Terguride (55) and lisuride (56) are structurally very similar ergoline derivatives with urea side chain (Figure 18). Lisuride is selective D_2-agonist and   together with its transdihydroderivative terguride, having partial agonistic activity towards CNS-dopamine receptor, they are both used as prolactin inhibiting drugs. They are extensively metabolised in mammal organisms. The studies of their biotransformation both *in vivo* and *in vitro* (Gieschen *et al.*, 1994, Hümpel *et al.*, 1989) revealed that the main metabolic changes are *N*-dealkylations at urea moiety leading to, e.g., deethylterguride (55b) and deethyllysuride (56b). Monodealkylation is catalysed by human cyt P450 2D6 (Rauschenbach *et al.*, 1997). Further dealkylation yielded dideethylderivatives (55c, 56c) (Hümpel *et al.*, 1989). These reactions resemble to those of LSD metabolism—*vide supra*.

Analogous metabolic changes, e.g., side chain dealkylation undergoes semisynthetic alkaloid CQA 206–291 (*N,N*-diethyl-*N'*-(1-ethyl-6-methylergoline-8α-*yl*)-sulfamide that is effective against Parkinson's disease. Human liver microsomes rich in cytochrome P-450 deethylated this compound exclusively to *N*-monodeethylated metabolite (Ball *et al.*, 1992).

Proterguride

Proterguride (57a) is a new ergoline derivative with a strong agonistic activity towards dopamine receptors reducing thus prolactine levels. *In vitro* studies revealed similar metabolic changes analogous to terguride, e.g., *N*-dealkylation yielding 57b (Figure 19). Further oxidative changes lead to *N*-6 oxide (57c) and oxidative degradation of indole ring starting with 2-oxoderivative (57d)*via* 3-hydroxy-2-oxoderivative (57e). 57e is further oxidised to 2-oxoderivative which is unique by the full aromatisation of C-ring. Another oxidative degradation of 57e yields despyrrole derivative (57f) (Krause *et al.*, 1993). This pathway, reminding kynurenine pathway of tryptophan degradation, has its parallel in the oxo-derivatives of agroclavine and elymoclavine (8, 9) (Figures 2, 3) obtained mostly by peroxidase action—*vide supra*.

Cabergoline

Cabergoline (58a) is another modern prolactine inhibitor with prolonged effect (Figure 20). This drug is metabolised in human organism similarly as   the above compounds undergoing, primarily dealkylation (58b) (minor metabolite). Dealkylated metabolites are finally hydrolysed giving dihydrolysergic acid derivative (58c) (major metabolite) found in urine (Cocchiara and Benedetti, 1992).

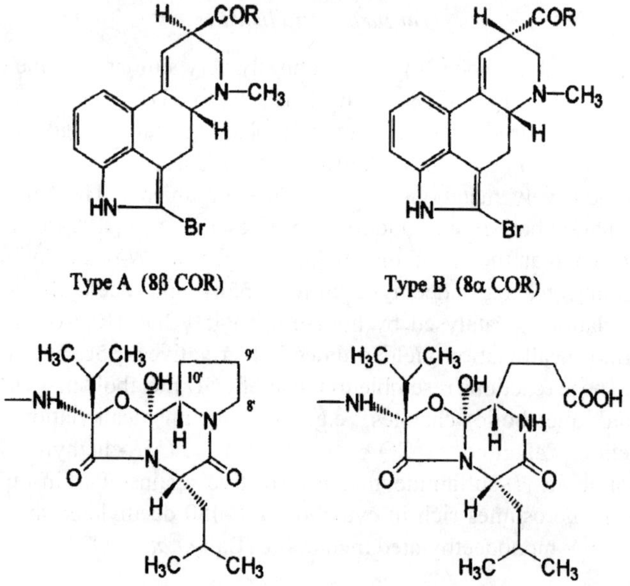

Type A (8β COR)   Type B (8α COR)

R: X   Y

Metabolite	Type	R	Other substituents
54a	A	X	
54b	B	X	
54c	B	NH₂	
54d	A	NH₂	
54e	B	Y	
54f	A	Y	
54g	B	OH	
54h	A	OH	
54i	A	X	8'-oxo, 10'-hydroxy
54j	A	X	8'α-OH
54k	A	X	8'β-OH
54l	B	X	8'α-OH
54m	B	X	8'β-OH
54n	B	X	8',9'-didehydro
54o	A	X	8'β-O-glucuronide
54p	B	X	8'β-O-glucuronide
54q	A	X	8'α-OH, 9'β-O-glucuronide
54r	B	X	8'α-OH, 9'β-O-glucuronide
54s	B	X	8'β-O-glucuronide, 9'β OH

Figure 17 Bromokryptine (**54a**) and its metabolites

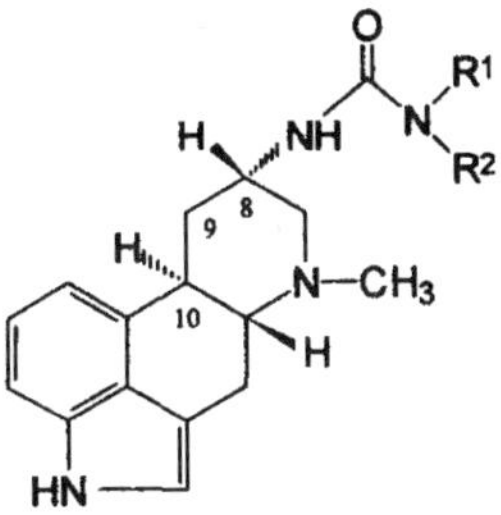

55a R^1 = Et, R^2 = Et
55b R^1 = H, R^2 = Et
55c R^1 = H, R^2 = H
56a 8,9-didehydro, R^1 = Et, R^2 = Et
56b 8,9-didehydro, R^1 = H, R^2 = Et
56c 8,9-didehydro, R^1 = H, R^2 = H

Figure 18 Terguride **(55a)** and lisuride **(56a)** metabolites

Dihydroergotamine

Dihydroergotamine **(59a)** has been used for a long time for a treatment of migraine and orthostatic hypotension. This compound is oxidised in human organism analogously as bromokryptine at 8′-position of the proline part of cyclopeptide moiety to the mixture of hydroxyderivatives **(59b)**. Also dihydrolysergic acid and its amide (dihydroergine) are produced in large quantities (Maurer and Frick, 1984) (Figure 21).

In vitro transformations produced further oxidative products, e.g., dihydroxyderivative **59c** and derivative **59d** arisen by oxidative opening of the proline ring. Glucuronidation of dihydroxyderivatives could be also expected as in the case of bromokryptine. Oxidative attack to the pyrrole moiety results in its opening. Metabolites **59e, f** are obvious intermediates of the next oxidative product found, e.g., in the case of proterguride metabolism **(57g)**.

Other peptide ergot alkaloids and their dihydroderivatives are metabolised analogously as bromokryptine or dihydroergotamine. Human cytochrome *P450s* 3A exhibits high affinity to the cyclopeptide moiety of ergopeptines catalysing its hydroxylation (Peyronneau *et al.,* 1994). Metabolism of ergopeptines is inhibited by erythromycin and oleandromycin and stimulated by co-administration of dexamethasone that is *P450S* 3A inducer (Peyronneau *et al.,* 1994).

6-Methyl-8β-(2, 4-dioxo-1-imidazolidinylmethyl) ergoline

This compound belongs to the intensively studied α_1-adrenoreceptors and S_2-receptors blockers. It is probably first ergoline derivative influencing the cardiovascular system without substantial side effects (Salvati *et al.,* 1989) observed in its analogues. Metabolic studies with it demonstrated that this compound is excreted almost unchanged. Its *N*-6 oxide is probably reduced in the organism back to the original drug. Small quantities of *N*-6 demethyl derivative are formed as well.

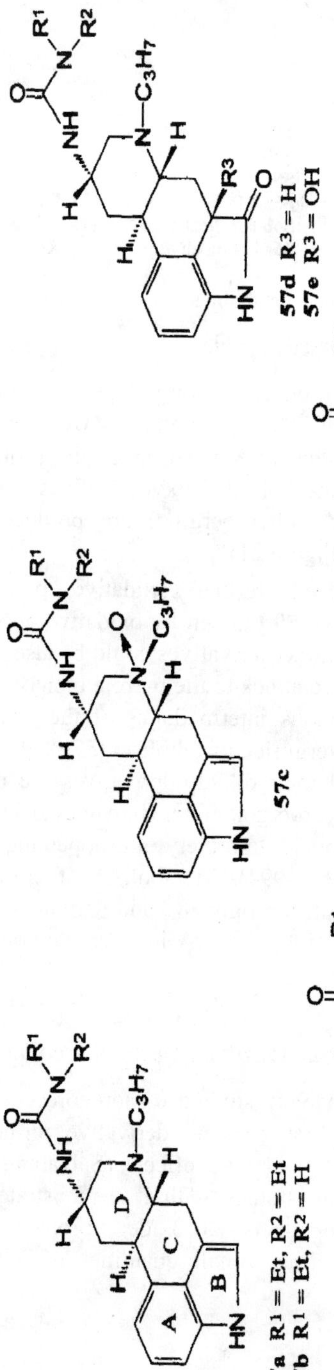

Figure 19 Proterguride (57a) and its metabolites

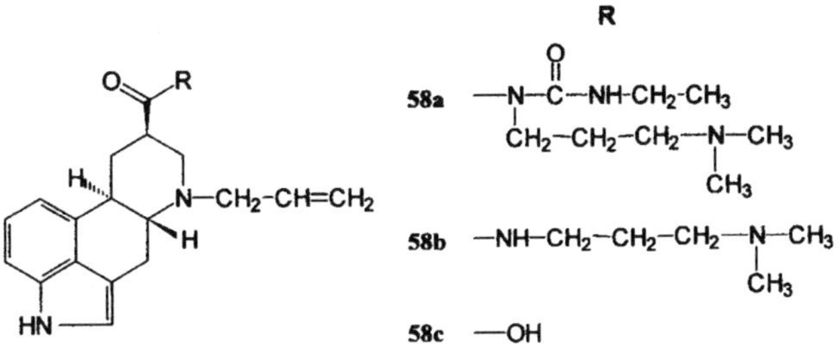

Figure 20 Cabergoline (**58a**) and its metabolites

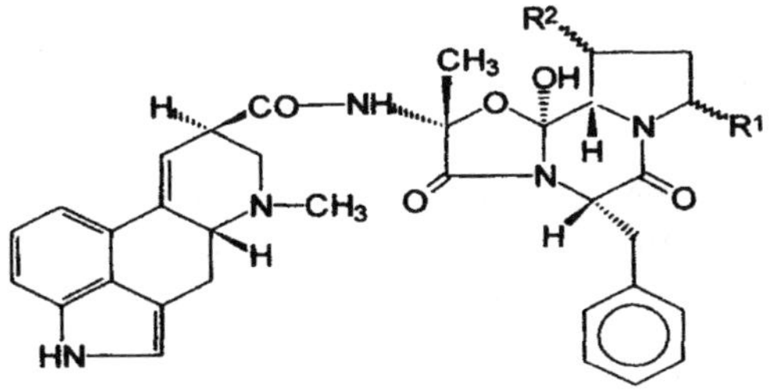

59a R1= H,   R2 = H
59b R1 = OH, R2 = H
59c R1 = OH, R2 = OH

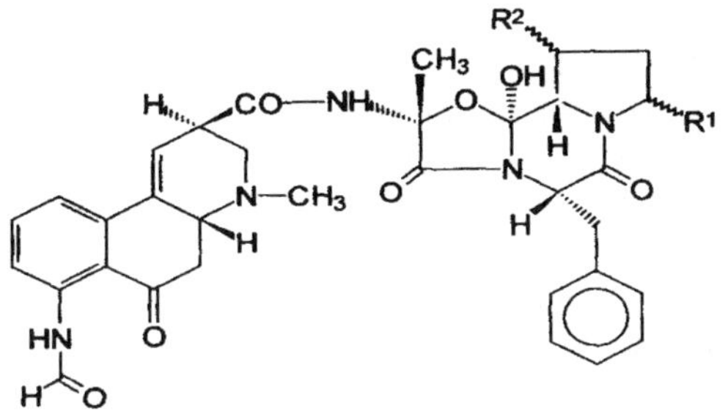

59e R1 = OH, R2 = H
59f R1 = OH, R2 = OH

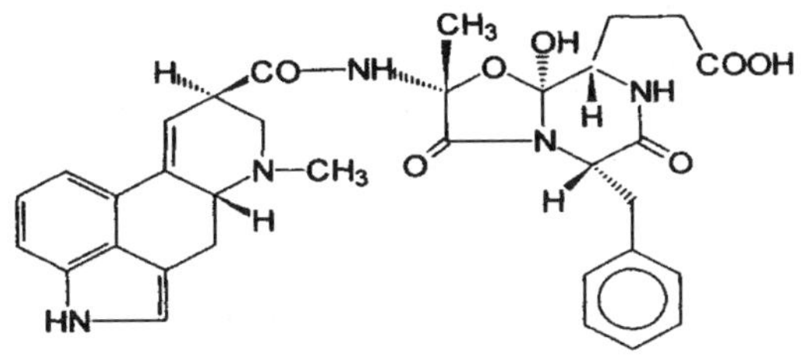

59d

Figure 21 Dihydroergotamine (**59a**) and its metabolites

9.5.
PROSPECTS TO THE FUTURE

Production of most desired ergot substances, e.g. elymoclavine, paspalic acid and lysergol, used mainly for semisynthetic ergot preparatives, is rather complicated. Bioconversion of their precursors may improve the effectivity of their preparation.

The most important ergot alkaloid bioconversion from industrial point of view is ergoline oxidation at C-17. The precursors in the lower oxidation degree, i.e., agroclavine and chanoclavine, can be produced in considerably high amounts. Elymoclavine acts as a feedback inhibitor of its production in *Claviceps,* so its yield in the batch fermentation is limited. Another highly desired reaction is the bioconversion of clavine alkaloids to paspalic acid. Both these biooxidations seem to be performed only by organisms within *Claviceps* genus. Although some work on these problems has already been done, both processes should be amended and adopted for large scale application.

In some cases, lysergic acid is produced by alkaline hydrolysis of peptide ergot alkaloids or lysergic acid derivatives. Harsh conditions of this chemical reaction cause a drop in the yields due to decomposition of ergoline skeleton. This could be another challenge for a further search for bioconversion methods.

Immobilised cells are applicable for most bioconversions. Mixed cultures, either free or immobilised, could simplify performance of these bioreactions. However, meeting different nutritional requirements of mixed cultures is a problem that must be overcome.

Ergot alkaloids may be also industrial pollutant. Their biodegradation has been, therefore, addressed as well. Various yeasts were tested for reduction of ergot alkaloid in industrial wastes and some of them *(Hansenula anomala,Pichia kluyveri, Candida utilis)* degraded up to 70–80% of the alkaloid content. The best degraders were isolated directly from the wastes containing ergot alkaloids (Recek *et al.,* 1984).

Finally, thorough and detailed study of various products of alkaloid biotransfromations can help to predict their most plausible bioconversion in animal organisms. From the large set of data on various ergot alkaloid biotransformations it can be concluded that the most sensitive parts of ergoline skeleton are positions 1, 2 and 3 (analogy to kynurenine degradation), *N*-6 (dealkylation, oxidation) and 8, 9, and 10—depending on the double bond position. Contrary to that, position C-4 is extremely resistant to any substitution (till now no chemical or biological modification of this position has been biotransformations it can be concluded that the most sensitive parts of ergoline skeleton are positions 1, 2 and 3 (analogy to kynurenine degradation), *N*-6 (dealkylation, oxidation) and 8, 9, and 10—depending on the double bond position. Contrary to that, position C-4 is extremely resistant to any substitution (till now no chemical or biological modification of this position has been reported), and also position 7 is very stable. The above general conclusions can be very helpful for estimation of chemical and/or metabolic reactivity of various ergot alkaloids.

ACKNOWLEDGEMENT

Support by a grant No. 203/96/1267 from the Grant Agency of the Czech Republic is gratefully acknowledged.

REFERENCES

Abe, M. (1966) Biogenetic interrelations of ergot alkaloids. *Abh. Deut. Akad. Wiss.Berlin, Kl. Chem. Geol. Biol.,*393–404; *Chem. Abstr.,* (1967), **66,**113082j.

Abe, M., Yamatodani, S., Yamano, T., Kozu, Y. and Yamada, S. (1963) Biosynthetic interrelation between agroclavine and elymoclavine. *Agric. Biol. Chem.,***27,**659–662.

Agurell, S. and Ramstad, E. (1962) Biogenetic interrelations of ergot alkaloids. *Arch.Biochem. Biophys.,***98,**457–470.

Amici, A.M., Minghetti, A., Spalla, C. and Tonolo, A. (1964) Noveau precédé de preparation de l'acide lysergique. *French Pat.*No. 1362876, C 12 d, 27.4.1964.

Arcamone, F., Glaesser, A.H., Minghetti, A. and Nicoletta, V. (1971) Metabolism of ergoline derivatives. *Boll. Chim. Farm.,*110, 704–711.

Arcamone, F., Glaesser, A.G., Grafnetterova, J., Minghetti, A. and Nicoletta, V. (1972) Metabolism of ergoline derivatives. Metabolism of nicergoline [8β-(5-bromonicotinoyloxymethyl)-1, 6-dimethyl-10α-methoxyergoline] in man and in animals. *Biochem. Pharmacol.,***21,**2205–2213.

Auret, B.J., Boyd, D.R., Breen, F. and Grene, R.M.E. and Robinson, P.M. (1981) Stereoselective enzyme-catalyzed oxidation-reduction of thioacetals-thioacetal sulfoxides by fungi. *J. Chem. Soc. Perkin Trans.1,*930–933.

Axelrod, J., Brady, R.O., Witkop, B. and Evans, E.W. (1956) Metabolism of lysergic acid diethylamide. *Nature,***178,**143–144.

Ball, S.E., Maurer, G., Zollinger, M., Ladona, M. and Vickers, A.E.M. (1992) Characterization of the cytochrome P-450 gene family responsible for the *N*-dealkylation of the ergot alkaloid CQA 206–291 in humans. *Drug Metab. Dispos.,***20,** 56–63.

Banno, K., Horimoto, S. and Mabuchi, M. (1991) Assay of nicergoline and three metabolites in human plasma and urine by high-performance liquid chromatographyatmospheric pressure ionization mass spectrometry. *J. Chromatogr.,* **568,**375–384.

Béliveau, J. and Ramstad, E. (1966) 8-Hydroxylation of agroclavine and elymoclavine by fungi. *Lloydia,*29,234–238.

Brack, A., Brunner, R. and Kobel, H. (1962) 33. Mikrobiologische Hydroxylierung an Mutterkornalkaloiden vom Clavin-Typus mit dem mexikanischen Rauschpilz *Psilocybe semperviva* Heim et Cailleux. *Helv. Chim. Acta.,***45,**278–281.

Bumbová-Linhartová, R., Flieger, M., Sedmera, P. and Zima, J. (1991) New aspects of submerged fermentation of *Claviceps paspali. Appl. Microbiol. Biotechnol.,***34,** 703–706.

Chan Lin, W.-N., Ramstad, E. and Taylor, E.H. (1967a) Enzymology of ergot alkaloid biosynthesis. Part III. 10-Hydroxyelymoclavine, an intermediate in the peroxidase conversion of eiymoclavine to penniclavine and isopenniclavine. *Lloydia,*30,202–208.

Chan Lin, W.-N., Ramstad, E., Shough, H.R. and Taylor, E.H. (1967b) Enzymology of ergot alkaloid biogenesis. V. Multiple functions of peroxidase in the conversion of clavines. *Lloydia,*30,284P.

Clemens, J.A., Okimura, T. and Smalstig, E.B. (1993) Dopamine agonist activities of pergolide, its metabolites, and bromocriptine as measured by prolactin inhibition, compulsive turning, and stereotypic behavior. *Arzneim.-Forsch./Drug. Res.,*43, 281–286.

Cocchiara, G. and Benedetti, M.S. (1992) Excretion balance and urinary metabolic pattern of [^{3}H]cabergoline in man. *Drug Metab. Drug Interact.,*10,203–211.

Cvak, L., Jegorov, A., Pakhomova, S., Kratochvil, B., Sedmera, P., Havlíček, V. and Minář, J. (1997) *Phytochemistry,*44,365–369.

Davis, P.J. (1982) Microbial transformation of alkaloids. In J.P.Rosazza (ed.), *Microbialtransformation of bioactive compounds,*Vol. 11, CRC Press, Inc., Boca Raton, Florida, p. 67.

Davis, P.J. (1984) Natural and semisynthetic alkaloids. In *Biotransformations,Biotechnology,Vol. 6a,*K.Kieslich (ed.) Verlag Chemie, Basel, p. 207–215.

Davis, P.J., Glade, J.C., Clark, A.M. and Smith, R.V. (1979) *N*-Demethylation of lergotrile by *Streptomyces platensis. Appl. Environ. Microbiol.,*38,891–893.

Eich, E. and Sieben, R. (1985) 8α-Hydroxylierung von 6-Nor-agroclavine und Lysergine durch *Claviceps fusiformis. Planta Medica,*282–283.

Erge, D., Maier, W. and Gröger, D. (1973) Enzymic conversion of chanoclavine I. *Biochem. Physiol. Pflanzen,*164,234–247.

Flieger, M., Sedmera, P., Novák, J., Cvak, L., Zapletal, J. and Stuchlík, J. (1991) Degradation products of ergot alkaloids. *J. Nat. Prod.,*54,390–395.

Flieger, M., Křen, V., Zelenkova, N.F., Sedmera, P., Novák, J. and Sajdl, P. (1989) Ergot alkaloids glycosides from saprophytic cultures of *Claviceps.* II. Chanoclavine fructosides. *J. Nat. Prod.,*53,171–175.

Floss, H.G., Günther, H., Mothes, U. and Becker, I. (1967) Isolierung von Elymoclavin-*O*-β-D-fructosid aus Kulturen des Mutterkornpilzes. *Z. Naturforschg.,*22b,399–402.

Gabor, F., Hamilton, G. and Pittner, F. (1995) Drug-protein conjugates: Haptenation of 1-methyl-10α-methoxydihydrolysergol and 5-bromonicotinic acid to albumin for the production of epitope-specific monoclonal antibodies against nicergoline. *J. Pharm. Sci.,*84,1120–1125.

Gieschen, H., Hildebrand, M. and Salomon, B. (1994) Metabolism of two dopaminergic ergot derivatives in genetically engineered V79 cells expressing CYP450 enzymes. In *Cytochrome P450 Biochemistry, Biophysics and Molecular Biology.*8th Int. Conf. Paris. Lechner, M.C. (ed.), John Libbey Eurotext, Paris, p. 467.

Gröger, D. (1963) Über die Umwandlung von Elymoclavine in *Ipomoea-Blättern. Planta Medica,*11,444–449.

Havlíček, V., Flieger, M., Křen, V. and Ryska, M. (1994) Fast atom bombardment mass spectrometry of elymoclavine glycosides. *Biol. Mass Spectrom.,*23,57–60.

Hsu, J.C. and Anderson, J.A. (1971) Agroclavine hydroxylase of *Claviceps purpurea. Biochim. Biophys. Acta,*230,518–525.

Hümpel, M.D., Sostarek, H., Gieschen, H. and Labitzky, C. (1989) Studies on the biotransformation of lonazolac, bromerguride, lisuride and terguride in laboratory animals and their hepatocytes. *Xenobiotica,* **19,**361–377.

Ishii, H., Hayashi, M., Niwaguchi, T. and Nakahara, Y. (1979a) Studies on lysergic acid diethylamide and related compounds. VII. Microbial transformation of lysergic acid diethylamide and related compounds. *Chem. Pharm. Bull.,* **27,**1570–1575.

Ishii, H., Hayashi, M., Niwaguchi, T. and Nakahara, Y. (1979b) Studies on lysergic acid diethylamide and related compounds. IX. Microbial transformation of amides related to lysergic acid diethylamide by *Streptomyces roseochromogenes. Chem.Pharm. Bull.,* **27,** 3029–3038.

Ishii, H., Niwaguchi, T., Nakahara, Y. and Hayashi, M. (1980) Studies on lysergic acid diethylamide and related compounds. Part 8. Structural identification of new matabolites of lysergic acid diethylamide obtained by microbial transformation using *Streptomyces roseochromogenes. J. Chem. Soc. Perkin Trans. 1,*902–905.

Johne, S., Gröger, D., Zier, P. and Voigt, R. (1972) Metabolism of [U-^{3}H]labeled dihydrochanoclavine in submerged cultures of *Pennisetum* ergot. *Pharmazie,* **27,** 801–802.

Kim, I.-S., Kim, S.-U. and Anderson, J.A. (1981) Microsomal agroclavine hydroxylase of *Claviceps* species. *Photochemistry,* **20,**2311–2314.

Kim, S.-U., Cho, Y.-J., Floss, H.G. and Anderson, J.A. (1983) Conversion of elymoclavine to paspalic acid by a paniculate fraction from an ergotamine-producing strain of *Claviceps* sp. *Planta Medica,* **48,**145–148.

Kerr, K.M., Smith, R.V. and Davis, P.J. (1981) High-performance liquid chromatographic determination of pergolide and its metabolite, pergolide sulfoxide, in microbial extracts. *J. Chrom.,* **219,**317–320.

Krause, W., Düsterberg, B., Jakobs, U. and Hoyer, G.-A. (1993) Biotransformation of proterguride in the perfused rat liver. *Drug Metab. Dispos.,* **21,**203–208.

Křen, V. (1991) Bioconversions of ergot alkaloids. *Adv. Biochem. Eng.,* **44,**124–142.

Křen, V. (1997) Enzymatic and chemical glycosylations of ergot alkaloids and biological aspects of new compounds. *Topics Curr. Chem.,* **186,**45–65.

Křen, V., Břemek, J., Flieger, M., Kozová, J., Malinka, Z. and Řeháček Z. (1989) Bioconversion of agroclavine by free and immobilized *Claviceps fusiformis* cells. *Enzyme Microbiol. Technol.,* **11,**685–691.

Křen, V., Flieger, M. and Sajdl, P. (1990) Glycosylation of ergot alkaloids by free and immobilized *Claviceps purpurea* cells. *Appl. Microbiol. Biotechnol.,* **32,**645–650.

Křen, V., Sedmera, P., Havlíček, V. and Fišerová, A. (1992) Enzymatic galactosylation of ergot alkaloids. *Tetrahedron Lett.,* **33,**7233–7236.

Křen, V., Svatoš, A., Vaisar, T., Havlíček, V., Sedmera, P., Pažoutová, S. and Šaman, D. (1993) Fructosylation of ergot alkaloids by submerged culture *Claviceps purpurea* inhibited in alkaloid production. *J. Chem. Res.,*89.

Křen, V., Ščigelová, M., Přikrylová, V., Havlíček, V. and Sedmera P. (1994a) Enzymatic preparation of β-N-acetylhexosaminides of ergot alkaloids. *Biocatalysis,*10, 181–193.

Křen, V., Augé, C, Sedmera, P. and Havlíček V. (1994b) β-Glucosyl and β-galactosyl transfer catalysed by β-1, 4-galactosyltransferase in preparation of glycosylated alkaloids. *J. Chem. Soc. Perkin Trans. 1,*2481–2484.

Křen, V., Sedmera, P., Přikrylová, V., Polášek, M., Minghetti, A., Crespi-Perellino, N. and Macek, T. (1996a) Biotransformation of chanoclavine by *Euphorbia calyptrata* cell culture. *J. Nat. Prod.,* **59,**481–484.

Křen, V., Sedmera, P., Polášek, M., Minghetti, A. and Crespi-Perellino, N. (1996b) Dimerisation of lysergene by *Euphorbia calyptrata* cell cultures. *J. Nat. Prod.,***59**, 609–611.

Křen, V., Fišerová, A., Augé, C., Sedmera, P., Havlíček, V. and Šíma, P. (1996c) Ergot alkaloid glycosides with immunomodulatory activities. *Bioorg. Med. Chem.,***4**, 869–876.

Křen, V., Kawuloková, L., Sedmera, P., Polášek, M., Lindhorst, T.K. and Karl-Heinz van Pée, K.-H. (1997) Biotransformation of ergot alkaloids by haloperoxidase from *Streptomyces aureofaciens:* Stereoselective acetoxylation and propionoxylation. *Liebigs Ann.,* in press.

Maier, W., Schumann, B. and Gröger, D. (1988) Microsomal oxygenases involved in ergoline alkaloid biosynthesis of various *Claviceps* strains. *J. Basic. Microbiol.,***28**, 83–93.

Maier, W., Erge, D. and Gröger, D. (1974) Metabolism of lysergyl derivatives in *Claviceps purpurea* (Fr.) Tulasne. *Biochem. Physiol. Pflanzen,***165**,479–485.

Maier, W., Baumert, A. and Gröger, D. (1978) Synthesis of lysergyl peptides and their metabolism in an ergotoxine producing *Claviceps-strain. Biochem. Physiol. Pflanzen,* **172**,15–26.

Maurer, G. and Frick, W. (1984) Elucidation of the structure and receptor binding studies of the major primary metabolite of dihydroergotamine in man. *Eur. J. Clin.Pharmacol.,* **26**,463–470.

Maurer, G., Schreier, E., Delaborde, S., Loosli, H.R., Nufer, R. and Shukla, A.P. (1982) Fate and disposition of bromocriptine in animals and man. I: Structure elucidation of the metabolites. *Eur. J. Drug Metab. Pharmacokin.,***7**,281–292.

Maurer, G., Schreier, E., Delaborde, S., Loosli, H.R., Nufer, R. and Shukla, A.P. (1983) Fate and disposition of bromocriptine in animals and man. II: Absorption, elimination and metabolism. *Eur. J. Drug Metab. Pharmacokin.,***8**,51–62.

Mothes, K., Winkler, K., Gröger, D., Floss, H.G., Mothes, U. and Weygand, F. (1962) Über die Umwandlung von Elymoclavine in Lysergsäurederivate durch Mutterkornpilze *(Claviceps). Tetrahedron Lett.,*933–937.

Ogunlana, E.O., Wilson, B.J., Tyler, V.E. and Ramstad, E. (1970) Biosynthesis of alkaloids. Enzymatic closure of ring D of the ergolene nucleus. *Chem. Commun.,***1970**, 775–776 .

Ogunlana, E.O., Ramstad, E., Tyler, V.E. and Wilson, B.J. (1969) Enzymatic ring closure in the ergoline biosynthesis. *Lloydia,***32**,524P.

Pažoutová, S., Křen, V. and Sajdl, P. (1990) The inhibition of clavine biosynthesis by 5-F-Trp; an useful tool for the study of regulatory and biosynthesis relationships in *Claviceps. Appl. Microbiol. Biotechnol.,***33**,330–334.

Parli, C.J. and Smith, B. (1975) Metabolic fate in G. pig of the prolactine secretion inhibitor 2-chloro-6-methylergolene-8-β-acetonitrile (Lergotrile). *Fed. Proc.,***34**,813.

Peyroneau, M.A., Dalaforge, M., Riviere, R., Renaud, J.-P. and Mansuy, D. (1994) High affinity of ergopeptides for cytochromes P450 3A. Importance of their peptide moiety for P450 recognition and hydroxylation of bromocriptine. *Eur. J. Biochem.,***223**, 947–956.

Philippi, U. and Eich E. (1984) Microbiological C-17-oxidation of clavine alkaloids, 2: Conversion of partial synthetic elymoclavines to the corresponding lysergic acids/ isolysergic acids by *Claviceps paspuli. Planta Medica,***50**,456–457.

Rauschenbach, R., Gieschen, H., Salomon, B., Kraus, C., Kühne, G. and Hildebrand, M. (1997) Development of a V79 cell line expressing human cytochrome P450 2D6 and its application as a metabolic screening tool. *Environ. Toxicol. Pharmacol.,***3**,31–39.

Recek, M., Kučan, E., Čadež, J. and Raspor, P. (1994) Can be ergot alkaloids eliminated from waste water?*Proc. IUMS Congresses 6.,*p. 537, July 3–8, 1994, Prague, Czech Republic.

Salvati, P., Bondiolotti, G.P., Caccia, C., Vaghi, F. and Bianchi, G. (1989) Mechanism of the antihypertensive effect of FCE 22716, a new ergoline derivative, in the spontaneously hypertensive rat. *Pharmacology,***38**,78–92.

Sajdl, P. and Řeháček, Z. (1975) Cyclization of chanoclavine-I by cell-free preparations from saprophytic *Claviceps* strain. *Folia Microbiol.,***20**,365–367.

Ščigelová, M., Křen, V. and Nilsson, K.G.I. (1994) Synthesis of α-mannosylated ergot alkaloids employing α-mannosidase. *Biotecbnol. Lett.,***16**,683–688.

Ščigelová, M., Macek, T., Minghetti, A., Macková, M., Sedmera, P., Přikrylová, V. and Křen, V. (1995) Biotransformation of ergot alkaloids by plant cell cultures with high peroxidase activity. *Biotecbnol. Lett.,***17**,1213–1218.

Sebek, O.K. (1983) Fungal transformations as a useful method for the synthesis of organic compounds. *Mycologia,***75**,383–394.

Shough, H.R. and Taylor, E.H. (1969) Enzymology of ergot alkaloid biosynthesis. IV. Additional studies on the oxidation of agroclavine by horseradish peroxidase. *Lloydia,* **32**,315–326.

Sieben, R., Philippi, U. and Eich, E. (1984) Microbiological C-17-oxidation of clavine alkaloids. I. Substrate specificity of agroclavine hydroxylase of *Claviceps fusiformis. J. Nat. Prod.,***47**,433–438.

Smith, R.V. and Rosazza, J.P. (1982) Microbial transformation as means of preparing mammalian drug metabolites. In J.P.Rosazza, (ed.), *Microbial transformation ofbioactive compounds,*CRC Press, Inc., Boca Raton, Florida, pp. 1–43.

Taylor, E.H. and Shough, H.R. (1967) Enzymology of ergot alkaloid biosynthesis. II. The oxidation of agroclavine by horseradish peroxidase. *Lloydia,***30**,197–201.

Taylor, E.H., Goldner, K.J., Pong, S.F. and Shough, H.R. (1966) Conversion of $\Delta^{8,9}$-ergolines to $\Delta^{9,10}$-8-hydroxyergolines in plant homogenates. *Lloydia,***29**,239–244.

Tyler, V.E. Jr., Erge, D. and Gröger, D. (1965) Biological conversions of agroclavine and elymoclavine. *Planta Medica,***13**,315–325.

Wilson, B.J., Ramstad, E., Jansson, I. and Orrenius, S. (1971) Conversion of agroclavine by mammalian cytochrome P-450. *Biochim. Biophys. Acta,***252**,348–356.

Winkler, K. and Mothes, K. (1962) Significance of the sclerotial condition of *Clavicepspurpurea* for alkaloid synthesis. *Planta Medica,***10**,208–220.

Yamatodani, S., Kozu, Y., Yamada, S. and Abe, M. (1962) XLVIII. Microbial conversions of clavine type alkaloids, agroclavine, and elymoclavine. *Takeda Kenkyusho Nempo (Ann. Repts. Takeda Res. Lab.),***21**,88–94; *Chem. Abstr.,*(1963) **59**,3099c.

10.
ANALYTICAL CHEMISTRY OF ERGOT ALKALOIDS

ALEXANDR JEGOROV

Galena a.s., Research Unit, Branišovská 31, CZ 370 05České Budějovice, Czech Republic

The development of drugs based on ergot alkaloids (EA) was probably the primary driving force leading to the first attempts at the standardization of ergot preparations in the second half of the 19th century. Elemental analysis, titration, melting point, optical rotation, some color reactions, various physiological experiments and precipitation tests were the only guides used for this purpose practically till the thirties of the 20th century (Evers, 1927). Standardization of ergot preparations, and particularly the monitoring of some degradation processes were the main objectives of the analytical chemistry of EA between the wars (for a review see Swanson *et al.*, 1932). The first colorimetric methods introduced by Evers (1927) and van Urk (1929) can be considered a milestone in the analytical chemistry of EA. Since these procedures are limited to the determination of the total content of EA only, regardless of their activity, bioassays dominated the field until the fifties, when the first quantitative chromatographic methods appeared (Fuchs, 1953; Gyenes and Bayer, 1961). The development of planar chromatographic techniques after the World War II, hand in hand with the general improvement of separation techniques, led to the isolation and description of the majority of EA in the fifties and sixties, as well as to the discovery of the majority of reactions responsible for their instability. High performance liquid chromatography was introduced to the analysis of EA in the early seventies. Almost simultaneously, immunological methods were developed as a convenient alternative to the chromatographic methods (Van Vunakis *et al.*, 1971; TauntonRigby *et al.*, 1973; Castro *et al.*, 1973; Loeffler and Pierce, 1973). Sophisticated TLC and HPLC methods enabling the distinguishing and quantitative analysis of all common EA have been described in the eighties; with some minor modifications, and with much better chromatographic material quality, they are still in use. These procedures are nowadays supplemented also by modern coupled chromatographic-mass spectrometric methods.

10.1.
COLORIMETRY, SPECTROPHOTOMETRY, AND
FLUORIMETRY

The first colorimetric test based on the reaction of EA with concentrated sulphuric acid was described by Tanret as early as 1875. Many slight modifications of this procedure have appeared since then and a careful survey of the literature would indicate that this reaction has been "rediscovered" several times. The first semiquantitative analytical method for the determination of EA developed by Evers (1927) uses the same principle. Two years later, van Urk (1929) described the reaction of EA with *p*-dimethylaminobenzaldehyde and its use for quantitative analysis. This method has undergone many modifications (for a review see Gyenes and Bayer, 1961; Michelon and Kelleher, 1963; Lászlóne and Gábor, 1967), but its principle has remained practically unchanged. Spectrophotometric assay with *p*-dimethylaminobenzaldehyde reagent has been employed as an official assay for the determination of EA according to the European Pharmacopeia (1997), and as a fast in-process quantification of EA produced by submerged fermentation. This reagent is also a key component of the Ehrlich reagent used for visualisation of EA on TLC plates.

An exposure to the light was necessary for the full color development according to the procedure by van Urk (1929). Irradiation could be eliminated by adding a small amount of ferric chloride to the reagent (Allport and Cocking, 1932) and/or these two factors could be combined. The third principal modification was introduced by Sprince (1960) who had noticed that an additional spraying of paper chromatograms developed with the *p*-dimethylaminobenzaldehyde reagent with sodium nitrite considerably reduced the time necessary for the development as well as the tendency of the spots to fade. Similarly as with other indoles, two molecules of an ergot derivative condens with one molecule of *p*-dimethylaminobenzaldehyde in the acidic media to produce an intense blue color. 2-Substitued-derivatives create red color, but 2,3-disubstitued analogues do not react at all (Pöhm, 1953). Comparison of various Spectrophotometric methods has been published by Michelon and Kelleher (1963). Depending on the particular modification of the method, slightly different UV spectra of the product(s) could be obtained resulting in a different slope of the calibration curves. Since the reaction conditions must be carefully followed to avoid the formation of the colored byproducts, a Spectrophotometric determination of EA with ninhydrin has been proposed recently as an alternative method (Zakhari *et al.*, 1991).

A number of EA exhibit a strong fluorescence upon irradiation. This feature is extensively used for the detection of EA after their separation by HPLC or TLC. Without any preceding separation, the fluorescence of ergot derivatives can also be used for the development of some fluorimetric assays (Gyenes and Bayer, 1961; Hooper *et al.*, 1974). The limitation of all direct methods of EA determination including colorimetry, spectrophotometry or fluorimetry is the lack of specificity regarding related alkaloids or their own degradation products. Consequently, the majority of these methods has now been abandoned, but the specific features or reactions of EA are further employed for their detection by more sophisticated HPLC and TLC methods.

In addition to various methods employing the absorbance or fluorescence of EA in the UV-vis or near infrared region, reflectance spectroscopy was examined as a possible alternative to HPLC analysis for ergovaline in tall fescue (Roberts *et al.*, 1997). In contrast to the above mentioned methods permitting EA determination only in rather simple matrices, NIR spectroscopy is a nondestructive technique making it possible to rapidly process a large number of whole-tissue samples. In spite of the precision and detection limit achieved (hundreds of μg of ergovaline per kilogram of tall fescue), the narrow range of samples examined so far will have to be substantially expanded to obtain more general evaluation of this method in the future. However, with regard to obvious problems with the validation of this procedure, it cannot be expected that NIR spectroscopy might replace current methods such as HPLC or immunoassays.

10.2.
PLANAR CHROMATOGRAPHY

Interesting circular paper chromatographic methods were described by Berg (1951, 1952), but they were soon replaced by procedures based on descending paper chromatography enabling the exact localization and extraction of individual chromatographic spots. Buffers and solvent mixtures such as butanol-acetic acid-water (organic layer) or aqueous phase of the same mixture for the paper impregnated with silicon were used in the original methods (Tyler and Schwarting, 1952). Since the impregnation of paper with formamide provided better and more reproducible resolution for the majority of alkaloids, this procedure has later become an analytical standard (Tyler and Schwarting, 1952; Stoll and Rüegger, 1954; Macek *et al.*, 1954, 1956; Horák and Kudrnáč, 1956; Klošek, 1956a; Pöhm, 1958). Since some alkaloids were still not resolved, their mutual ratio was determined by the amino acid and keto acid analysis of their acid or alkaline hydrolysates, respectively (Stoll and Rüegger, 1954; Pöhm, 1958). Extraction of individual spots in combination with the reaction with *p*-dimethylaminobenzaldehyde enabled the development of the first quantitative colorimetric or photometric assays for the determination of individual EA (Klošek, 1956b; Pöhm, 1958). A paper chromatographic method has been described also for LSD (Faed and McLeod, 1973) and clavine alkaloids using benzene-chloroform-ethanol or butanolwater mixtures (Kornhauser *et al.*, 1970). More recently, paper chromatography and thin layer chromatography were used to analyze the EA produced by the toxic plant *Ipomoea muelleri* (Marderosian *et al.*, 1974). The time necessary for one analysis, poor resolution as well as low reproducibility caused paper-chromatographic methods to be quickly and almost completely replaced by thin-layer chromatography (TLC) on silica gel in the sixties; now they have a merely historical significance.

The thin layer chromatography of EA was first described about 1958, and since then it has become one of the most important methods for the quantitative and qualitative analysis of this group of compounds. It is worth mentioning that TLC is also the official method for the analysis of related alkaloids described in the current European Pharmacopeia (1997). Among several tens of the literature references, the majority of separations were carried out on common silica gel. The use of aluminium oxide or even chalk was also described (for a review see Reichelt and Kudrnáč, 1973). Separation on cellulose plates was reported as an improvement of the older paper chromatographic techniques (Teichert et al., 1960; Fowler et al., 1972; Prošek et al., 1976a, b) but, generally, none of these materials seems to offer any advantage in resolution when compared with a separation on silica gel. Some representative methods for various classes of EA are summarized in Table 1. Among many published methods, the following papers can be particularly recommended for clavines: Klavehn and Rochelmeyer (1961), Marderosian (1974), Wilkinson et al. (1987); for ergopeptines: Fowler et al. (1972), Reichelt and Kudrnáč (1973), Bianchi et al. (1982), and Crespi-Perellino et al. (1993).

All natural EA are colorless, the slightly yellow or even greenish color in some preparations being caused by some degradation processes on exposure to air and/or light. Hence, some appropriate reaction must be used to visualize the TLC spots to enable their examination by naked eye. In the simplest way the spots can be developed by iodine vapours to produce a transient yellow color. Van Urk's color reaction, or its numerous modifications (including also Ehrlich's reagent: p-dimethylaminobenzaldehyde-conc. HC1), are undoubtedly the most popular methods, described in various pharmacopeias. Some limitations of this method have already been mentioned in the part on colorimetry. People enganged in environmental analysis and new compound hunters would surely appreciate the subtle color differences accompanying slight structural changes, e.g., epimerization at C8 or the presence of benzyl group in the peptide moiety. Similarly, various colors ranging from blue, azure, green, pink to yellow, were used to indicate the structural modifications and presence of some functional groups in the clavine series (Yamatodani, 1960; Klaven and Rochelmeyer, 1961; Křen et al., 1986, 1996b). However, the proper interpretation of these differences demands a previous experience. The main advantage of this method is the stability of the majority of colors that enables one to store the original plates for years as documentation material. Besides the native fluorescence of some EA derivatives, the reactions with perchloric acid-$FeCl_3$ reagent or with the Dragendorff reagent can be alternatively used to visualize these spots (Keipert and Voigt, 1972). EA can also be visualized by charring with 50% sulphuric acid at 110°C for l0 min (Scott and Kennedy, 1976). Alternatively, a procedure with 5–10% sulphuric acid in ethanol provides a better estimation of individual EA and more specific indication of unrelated impurities (Křen, private communication). The π-complex formation of aromatic rings in EA with some π-acceptors, e.g., 7, 7, 8, 8-tetracyanoquinodimethane, 2, 4, 7-trinitrofluorenone or 2, 4, 5, 7-tetranitrofluorenone was also tested (Rücker and Taha, 1977), but neither of these nor other reagents described in the literature have found a widespread use.

After the separation, EA can be quantitatively analyzed either directly *"in situ"*, or indirectly, by scraping off the spots, eluting the active substances and by some subsequent qualitative or quantitative detection. The original procedures used extraction of individual spots and photometric or fluorimetric determination of the alkaloids (Dal Cortivo *et al.,* 1966; Keipert and Voigt, 1972). Although this method could be used also with the current commercially available TLC plates (Wilkinson *et al.,* 1987), it is very cumbersome, particularly in the case of multicomponent analysis. Hence, it is used now for the off-line identification of unknown components rather than for quantitative analysis. Ergot alkaloid mixtures can be quantitatively analysed by TLC—densitometry, e.g., after the reaction with the van Urk reagent (Genest, 1965). To avoid any inhomogeneity caused by spraying, *p*-dimethylaminobenzaldehyde can also be directly added to the solvent system and the development is accomplished in HCl vapours. Alternatively, EA might be quantitatively analysed with higher sensitivity also by TLC—fluorodensitometry (Niwaguchi and Inoue, 1971; Eich and Schunack, 1975; Prošek *et al.,* 1976a, b; Nelson and Foltz, 1992a). However, it should be noted, that some alkaloids or their semisynthetic derivatives do not exhibit fluorescence (Gröger and Erge, 1963).

10.3.
GAS CHROMATOGRAPHY

Some low molecular weight clavine alkaloids and simple lysergic acid amides lacking hydrophilic functional groups could be directly separated on a gas chromatographic column (Radecka and Nigam, 1966; Agurel and Ohlsson, 1971; Nichols *et al.,* 1983; Japp *et al.,* 1987; Clark, 1989; Flieger *et al.,* 1991). Several derivatization reactions have been tested in order to overcome the low volatility of the alkaloids and to reduce the peak tailing. Silylation of clavines was found to produce single derivatives (Barrow and Quigley, 1975), but the silylation at the *N*1-atom is incomplete even when using a high excess of the silylation reagent and thus cannot be used for quantitative analysis (Agurel and Ohlsson, 1971; Křen and Sedmera, 1996a). Trifluoroacetylation of clavines was found to be unsuitable due to the formation of multiple derivatives and laborious reaction conditions (Barrow and Quigley, 1975).

The continued use of lysergic acid diethylamide (LSD), whose simple production, low cost and wide availability cause the increase of its popularity among drug addicts, has stimulated also efforts to develop effective analytical methods for its determination in body fluids. Among other methods currently used (fluorimetry, TLC, HPLC and immunoassays), gas chromatography represents a method of choice (for a review see Nelson and Foltz, 1992a). Several derivatization strategies have been successfully applied using either silylation with bis(trimethylsilyl)trifluoroacetamide (BSTFA) (Paul *et al.,* 1990; Bukowski and Eaton, 1993), trifluoroacetylation at *N*1 with trifluoroacetylimidazole in the presence of 1, 4-dimethylpiperazine (Lim *et al.,* 1988), or treatment with a mixture of trimethylsilylimidazole, bis(trimethylsilyl)acetamine, and trimethylchlorsilane (3:3:2, v/v/v) (Nakahara *et al.,* 1996). Derivatives can be analysed by conventional capillary columns, but serious attention has to be paid to the fact that the silyl-derivative is moisture sensitive and LSD exhibits a strong tendency to undergo irreversible adsorption on the glass during the chromatographic process (Paul *et al.,* 1990; Nelson and Foltz, 1992a; White *et al.,* 1997). Nowadays, a confirmation of LSD and its isomer lysergic acid *N*-methyl, *N*-*n*-propylamide (LAMPA), which is also a controlled drug, is usually performed by a combination of some chromatographic and mass spectrometric techniques that make it possible to distinguish these compounds (Nichols *et al.,* 1983; Paul *et al.,* 1990; Nelson *et al.,* 1992a,b; White *et al.,* 1997).

Because of their thermal instability, low vapor pressure, and potential occurrence of isomerization and/or decomposition products direct measurement of ergopeptine molecules by gas chromatography had been found unsuitable. In the first attempt, whole blood sample containing an ergot alkaloid was partly hydrolyzed with 6 N HC1 and characteristic decomposition products were analysed by GC (de Zeeuv *et al.,* 1978). However, partial hydrolysis leads to irreproducible formation of several products and thus cannot be used for quantitative analysis. Nevertheless, total hydrolysis followed by the GC/MS analysis was used for the indirect determination of some ergopeptines containing unusual amino acids (Jegorov *et al.,* 1997). Without the preceding hydrolysis, several direct methods based on the quantitative thermal decomposition of ergopeptines in the injection port of a gas chromatograph and the GC analysis of the decomposition products have been developed (Szepesi and Gazdag, 1976; van Mansvelt *et al.,* 1978; Plomp *et al.,* 1978; Larsen *et al.,* 1979; Feng *et al.,* 1992). The injector temperature is the key factor influencing the decomposition of ergopeptines. Initially, ergopeptines are split into lysergic acid amide (or corresponding dihydrolysergic acid amide) and a fragment derived from the tripeptide moiety. Several subsequent reactions can occur and the fragmentation is similar to that observed under the electron impact conditions in the mass spectra. In general, the reproducibility and accuracy of these methods is slightly lower than that of some HPLC methods, but they can be advantageously used in toxicological screening (van Mansvelt *et al.,* 1978; Uboh *et al.,* 1995). It should be noted that the annotation of these methods as "GC methods for ergopeptide alkaloids" in the literature, cited also by many authors, is slightly misleading as all these methods are in fact based on mass fragmentography (MF/GC).

Table 1 Selected multicomponent TLC methods for the determination of EA

Alkaloid class	Method*	Reference
Clavines and simple lysergic acid amides	silica gel (AcOEt : EtOH : DMF) silicic acid	Klavehn and Rochelmeyer, 1961 Gröger and Erge, 1963
	(AcOEt : EtOH : DMF) Al_2O_3 GF ($CHCl_3$: EtOH) silica gel HR' (acetone : piperidine) silica gel ($CHCl_3$: MeOH : NH_4OH) 2D methods on a silica gel : (MeOH : $CHCl_3$ x DEA : $CHCl_3$) (MeOH : $CHCl_3$ x benzene : DMF)	Genest, 1965 Genest, 1965 Scott and Kennedy, 1976 Marderosian et al., 1974 Wilkinson et al., 1987
Ergopeptines	silica gel G (benzene : DMF) silica gel G (AcOEt : DMF : EtOH) silica gel, Al_2O_3, cellulose[†] silica gel H ($CHCl_3$: MeOH) silica gel impregnated with DMF (DIPE : THF : toluene : DEA) silica gel G (AcOEt : DMF : EtOH) silica gel 60 ($CHCl_3$: isopropanol) silica gel (CH_2Cl_2 : isopropanol)	McLaughlin et al., 1964 McLaughlin et al., 1964 Fowler et al., 1972[†] Vanhaelen and Vanhaelen-Fastré, 1972 Reichelt and Kudrnáč, 1973 Eich and Schunack, 1975 Bianchi et al., 1982 Crespi-Perellino et al., 1993
Dihydroergopeptines	cellulose (AcOEt : heptane : DEA)	Prošek et al., 1976a, b
Peptide lactams	silica gel ($CHCl_3$: toluene : acetone : EtOH) (CH_2Cl_2 : DIPE : MeOH : benzene)	Trtík and Kudrnáč, 1982 Kudrnáč and Kakáč, 1982
LSD	silica gel (trichloroethane : MeOH) silica gel G (MeOH : $CHCl_3$: hexane) silica gel G (acetone : $CHCl_3$) silica gel G (MeOH : $CHCl_3$)	Dal Cortivo et al., 1966 Niwaguchi and Inoue, 1971 Clark, 1989
Lisuride maleate	silica gel ($CHCl_3$: MeOH)	Amin, 1987
Nicergoline	silica gel (CH_2Cl_2 : acetone : water)	Banno et al, 1991b

* Abbreviations: AcOEt—ethyl acetate, EtOH—ethanol, MeOH—methanol, DEA—diethylamine, DMF—dimethylformamide, DIPE—diisopropyl ether.
† Fourteen alkaloids in eighteen solvent systems.

10.4.
LIQUID CHROMATOGRAPHY

First attempts to separate EA by liquid chromatography were made on unmodified silica gel (NP-HPLC) with mixtures such as diisopropyl ether-acetonitrile, chloroform-methanol, chloroform-acetonitrile or diethyl ether-methanol (Heacock et al., 1973; Wittwer and Kluckhohn, 1973). Separation of EA on silica gel provides an excellent resolution of individual EA groups in the elution order ergopeptinines>ergopeptines>ergometrine and clavines, which is very similar to that on TLC plates, and is also employed in the industrial purification of EA. Individual series of ergopeptines are resolved partially, i.e., ergotamines, ergoxines, and ergotoxines, but their individual members usually coelute (Szepesy et al., 1978, 1980; Žorž et al., 1985). Since analytical methods based on "normal-phase" chromatography suffered from serious disadvantages (low resolution, peak tailing, strong adsorption of polar constituents), methods using unmodified silica were soon abandoned and replaced by liquid chromatography on various modified phases. Besides the use of silica in the "normal phase" mode, an interesting separation of lysergic acid, lysergamide, LSD and iso-LSD in the "reversed-phase" mode using methanol—0.2 N aqueous ammonium nitrate (3:2, v/v) is worth mentioning (Jane, 1975).

Separation of EA by reversed-phase liquid chromatography (RP-HPLC) was developed in the seventies and eighties. Nowadays all, or almost all, routine analyses are carried out on C-8 or C-18 columns. Similarly, some other modified silicas, e.g., phenyl (Otero et al., 1993; Cvak et al., 1994) or cyanopropyl (Sochor et al., 1995) can be used in the reversed-phase mode. Ergot alkaloids are eluted according to their increasing hydrophobicity, i.e., in the order: clavines<lysergic acid amides<ergopeptines<ergopeptinines, and for the particular series of ergotoxines in the order: ergocornine< β-ergokryptineα-ergokryptine<ergocristine<ergogaline. Ergopeptinines (8-epimers) are always eluted later than the corresponding ergopeptines. Études on HPLC methods were described by many authors on columns from various producers and some examples of problems which could be solved by HPLC are summarized in Table 2. An additional information can be found also in three excellent reviews summarizing various aspects of EA analytical chemistry (Nelson and Foltz, 1992a; Garner et al., 1993; Flieger et al., 1997). In all HPLC methods, methanol-water or acetonitrile-water adjusted usually to slightly alkaline pH (ammonium or triethylamine buffers) were used as eluents. At alkaline pH, EA are separated as free bases providing usually much better selectivity compared with chromatography in "ion-pair" mode at low pH. Since ergopeptines are hydrophobic, water should be avoided for dissolution of samples due to their limited solubility and possible adsorption in injector loops (Otero et al., 1993). Most frequently, the detection of EA was accomplished by measuring the UV absorbance (usual detector setting to 310 nm for ergopeptines and 280 nm for dihydroergopeptines). A higher sensitivity and selectivity was achieved by

fluorescence monitoring, e.g., with ergopeptines (Edlund, 1981; Cieri, 1987; Rottinghaus et al., 1991; Cox et al., 1992; Hill et al., 1993; Miles et al., 1996; Shelby and Flieger, 1997), dihydroergopeptines (Žorž et al., 1983; Zecca et al., 1983; Humbert et al., 1987), lysergic acid amides or simple ergolene derivatives (Cox et al., 1992; Nelson and Foltz, 1992a; de Groot et al., 1993; Wolthers et al., 1993; Brooks et al., 1997). The detector wavelength combination: excitation at 310 nm and emission at 415 nm, is recommended for natural ergopeptines (Shelby and Flieger, 1997). However, it is important to note that the fluorescence intensity and position of maximum in the emission spectrum of alkaloids largely depends on the pH and solvent used (Hooper et al., 1974).

Table 2 Selected problems solved by RP-HPLC methods

Alkaloid class	Matrix	Reference
Clavines and simple lysergic acid amides	ergometrine in pharm. preparations	Sondack, 1978; Tokunaga et al., 1983
	lysergic acid and simple amides	Gill and Key, 1985; Flieger et al., 1993
	clavine glycosides	Flieger et al., 1989, 1990; Křen et al., 1990
	ergometrine in plasma	Edlund, 1981; Cox et al., 1992
	ergometrine pharmacokinetics	de Groot et al., 1993
	clavines produced by Penicillium	Zelenkova et al., 1996
Ergopeptines	ergotamine preparations	Bethke et al., 1976; Erni et al., 1976; Cieri, 1987
	various systems and columns	Dolinar, 1977
	EA in wheat, flour, bread, pancakes	Scott and Lawrence, 1980, 1982; Scott et al., 1992; Fajardo et al., 1995
	ergotamine in plasma	Edlund, 1981; Ibraheem et al., 1982, 1983
	α-, β-ergokryptine distinguishing	Herényi and Görög, 1982
	stability of alkaloids in sclerotia	Young et al., 1983
	eight ergopeptines and their 8-epimers	Magg and Ballschmiter, 1985
	in vitro ruminal digestion of ergovaline	Moyer et al., 1993
	EA in blood	Moubarak et al., 1996
	EA in tall fescue	Rottinghaus et al., 1991; Hill et al., 1993; Miles et al., 1996; Shelby and Flieger, 1997
Dihydroergopeptines	dihydroergotoxines in plasma	Žorž et al., 1983; Zecca et al., 1983
	dihydroergotoxine components	Hartmann et al., 1978; Chervet and Plas, 1984; Papp, 1990
	dihydroergotamine	Humbert et al., 1987; Niazi et al., 1988

The main area of use of HPLC methods is the control of impurities in bulk pharmaceutical substances (according to current general rules, all impurities above 0.1 rel % are analyzed and declared), and control of final dosage forms (stabilities, accelerated stress tests). Particular attention is paid to the stability of

Alkaloid class	Matrix	Reference
Semisynthetic drugs	LSD in biological fluids	Christie et al., 1976; Harzer, 1982; Webb et al., 1996
	LSD review	Nelson and Foltz, 1992a
	nicergoline impurities	Flieger et al., 1984b
	bromokryptine in plasma	Phelan et al., 1990
	nicergoline in plasma	Kohlenberg-Müller et al., 1991
	lisuride in plasma	Wolthers et al., 1993
	bromokryptine in tablet formulation	Foda and Elshafie, 1996
	sergolexole and LY215840	Brooks et al., 1997

essential drugs during the shipment to the tropics, e.g., of ergometrine and methylergometrine maleates (Hogerzeil et al., 1992; Hogerzeil and Walker, 1996). Increasing attention is devoted also to the determination of the metabolic fate of EA. Studies of the metabolism of bromokryptine (Maurer et al., 1982, 1983) and dihydroergotamine (Maurer and Frick, 1984) revealed that the primary metabolic pathway is the hydroxylation of the C-8' proline atom in the peptidic part. Splitting of the peptide bond between the lysergic acid and the tripeptide moiety was also detected. In addition, the metabolism of ergot-related drugs has been studied with some semisynthetic derivatives: lisuride (Toda and Oshino, 1981), pergolide (Rubin et al., 1981; Bowsher et al., 1992), methysergide (Bredberg and Paalzov, 1990), proterguride (Krause et al., 1993), nicergoline (Kohlenberg-Müller et al., 1991; Banno et al., 1991b; Sioufy et al., 1992), CQA 206–291 (Ball et al., 1992), cabergoline (Rains et al., 1995), sergolexole (Brooks et al., 1997), and LSD (Nelson and Foltz, 1992a,b; White et al., 1997). These studies revealed that the ergolene (ergoline) moiety can also be modified at several positions including $N(1)$-dealkylation (methysergide, nicergoline), $N(6)$-dealkylation (lisuride, cabergoline, pergolide, LSD), N-dealkylation in the urea or sulfonamide moiety (lisuride, CQA 206–291, proterguride, cabergoline), hydroxylation at the C-13 or C-15 ergolene atoms (lisuride, LSD), and formation of 2-oxo derivatives (proterguride, LSD). Glucuronides are supposed to be among the structures being excreted (Toda and Oshino, 1981; Maurer et al., 1982, 1983; Nelson and Foltz, 1992a). However, in a number of cases, only the principal metabolites were identified and the structures of others remain to be elucidated (e.g., Rubin et al., 1981). Isolation of several natural ergopeptines or clavine derivatives hydroxylated at the C-8 atom (Pakhomova et al., 1996a; Cvak et al., 1997) indicates that, e.g., also metabolites of this type can be expected. This simple list of sites of possible metabolic attack clearly illustrates that the metabolism of EA is still far from being completely understood, offering thus opportu-nities for future research (see also the chapter 9 on EA biotransformation in this book).

It is worth mentioning that in modern history, practically since 1993, the majority of new EA was isolated by RP-HPLC (Flieger *et al.*, 1993; Cvak *et al.*, 1994; Szántay *et al.*, 1994; Cvak *et al.*, 1997). Analysis of EA in various agricultural products dominates among other topics (Scott and Lawrence, 1980, 1982; Rottinghaus *et al.*, 1991; Hill *et al.*, 1993; Shelby and Flieger, 1997). By the end of the eighties, the basic development of HPLC methods was practically finished, and individual methods no longer appear in the literature without some special use or without coupling to some identification technique (LC/MS).

In the shadow of the use of C-8 or C-18 columns, some interesting methods based on other chromatographic sorbents also have been described. Even if these methods are not used routinely, they have an undeniable importance in EA research. Some disadvantages of silica gel could be at least partly overcome also by its modification with aminopropylsilane (NH_2 columns). Mixtures of chloroform, dichloromethane or diethyl ether with aliphatic alcohols (methanol, ethanol or 2-propanol) are typically used as eluents (Wurst *et al.*, 1978, 1979; Flieger *et al.*, 1981, 1982, 1984a, b). The same column type was also used for the supercritical fluid chromatographic (SFC) analysis of some clavine alkaloids using carbon dioxide-methanol (9:1, v/v) as an eluent (Berry *et al.*, 1986). The merits inherent in the use of amino-modified columns are several. Firstly, as expected, the elution of EA corresponds to "normal phase" chromatography and is thus complementary to "reversed-phase" chromatography on C-18 columns. Secondly, the method very sensi-tively responds to even subtle stereochemical changes in the structure, e.g., epimerization at proline or other part of the peptide moiety, conversion to -inines, etc., and makes it possible, e.g., to resolve α-ergopeptines (containing Leu) and β-ergopeptines (He). This approach provided an almost complete resolution of the clavine and ergopeptine series (Wurst *et al.*, 1978, 1979) and also contributed to the isolation of some new ergopeptines (Flieger *et al.*, 1981, 1984a).

10.5.
ELECTROANALYTICAL METHODS

In contrast to various chromatographic techniques, much less attention has been paid to the development of electroanalytical methods for the determination of EA. However, these alkaloids possess several features which could be used to develop such a method. Titration of EA in nonaqueous solvents with potentiometrical endpoint determination (glass/calomel electrode system) belongs to the oldest, but in various pharmacopeias still surviving, techniques (Gyenes and Bayer, 1961). Although it is not an electroanalytical procedure, the separation of EA by counter-current distribution is also based on the difference of their dissociation constants (Galeffi and Miranda delle Monache, 1974).

Electrophoretic analysis of EA has been tested on various thin layers (Agurell, 1965; Kornhauser and Perpar, 1965, 1967). For example, silica gel G plates sprayed with an acetic acid-pyridine buffer (pH 5.6) were used for the separation at 1500 volts (Agurell, 1965). With the exception of ergopeptines, which showed some tailing tendency, EA appeared as distinct spots and the resolution was comparable to that obtained by TLC. Although the basic principle of separation by electrophoresis is different from TLC, which might have some value for the separation of complex mixtures, the method has not found its followers most probably due to the fact that the same information can be obtained more simply by a combination of several TLC methods.

Much later, the same principle was used to separate ergot alkaloids by capillary electrophoresis. Fanali *et al.* (1992) have demonstrated the resolution of some EA enantiomers and epimers using capillary electrophoresis with the background electrolyte supplemented with γ-cyclodextrin. Although this method is very interesting, it should be noted that the analytical problem is slightly artificial, since ergot alkaloid enantiomers are encountered neither in nature nor in synthetic mixtures, due to the fact that they are synthesized from different precursors. The use of EA derivatives themselves as chiral selectors in capillary electrophoresis is much better idea in terms of practical application. These compounds were developed as a simple modification of some known EA-based drugs by L. Cvak and applied as chiral selectors both in capillary electrophoresis (Ingelse *et al.*, 1996a, b) and HPLC (Sinibaldi *et al.*, 1994; Padiglioni *et al.*, 1996; Messina *et al.*, 1996). The same HPLC column can also be used for the resolution of some semisynthetic ergot derivatives themselves (Flieger *et al.*, 1994).

Analyses of real samples by capillary electrophoresis were demonstrated by Ma *et al.* (1993). They developed a high performance capillary electrophoresis method for the determination of ergovaline in the seeds of *Festucaarundinacea* (tall fescue) infected with the endophytic fungus *Acremoniumcoenophialum*. Micrograms of ergovaline per kilogram of seeds can be detected and quantified. Capillary electrophoresis has also been successfully applied for the screening and quantitation of LSD in seized illicit substances (Walker *et al.*, 1996). Only recently, the potential advantages of capillary electrophoresis have been demonstrated on the simultaneous determination of caffeine and ergotamine in pharmaceutical dosage forms (Aboul-Enein and Bakr, 1997).

Ergot alkaloids are easily oxidized in the indole moiety by a two-electron transfer process affording a deep purple highly conjugated dimer as the principal product (Dankházi *et al.*, 1993). This feature can be used, e.g., for the electrochemical detection of EA after their HPLC separation (Pianezzola *et al.*, 1992). Based on this principle, several differential pulse voltametric methods have been developed as well aimed at the analyses of a single compound (ergotamine, ergometrine, ergocristine) in a drug formulation without any preceding separation from the excipients (Belal and Anderson, 1986; Wang and Ozsoz, 1990; Inczeffy *et al.*, 1993). The presence of bromonicotinic acid in nicergoline, a semisynthetic EA derivative, made it also possible to develop a polarographic method for its determination based on its electrochemical reduction (Sturm *et al.*, 1992).

10.6.
IMMUNOASSAYS

The main advantages of immunoassays are the possibility of analyzing very complex mixtures such as plasma and other biological fluids without extensive purification and, in comparison with chromatographic techniques, usually also with much higher sensitivity. Accordingly, the majority of immunoassays have been developed for studies of the pharmacokinetic profiles and bioequivalence studies of various dosage forms (Table 3). It is noteworthy that these methods were sometimes published separately from the subsequent pharmacokinetic studies. Thus, in order to obtain an additional information about the method performance, coefficients of variation, detection limits and usual concentrations of EA, it is advisable to read reports of these studies, e.g., for dihydroergotoxines (Kleimola *et al.,* 1977), dihydroergopeptines and bromokryptine (Kanto, 1983), α-dihydroergokryptine (Grognet *et al.,* 1991), lisuride (Krause *et al.,* 1991 a), dihydroergotamine (Wyss *et al.,* 1991; Lau *et al.,* 1994; Humbert *et al.,* 1996), dihydroergocristine (Grognet *et al.,* 1992; Coppi *et al.,* 1992), bromokryptine (Rabey *et al.,* 1990), or cabergoline (Persiani *et al.,* 1994; Andreoti *et al.,* 1995). However, it should be noted that once the therapeutic dose has been established, the routine monitoring of the EA plasma levels is not necessary in clinical practice.

The possibility of a routine analysis of a large number of samples is very useful. However, the limiting factor of all immunological methods is the cross reactivity with molecules having similar epitopes. Such structurally related molecules that might be naturally present in the matrix (nonspecific binding) could originate from related structures (e.g., from related alkaloids in the fermentation broths) and in most cases also from some metabolites of the parent compound. In contrast to the chromatographic methods, which provide information on the concentration of any resolved compound or metabolite, depending on the specificity of the antibody cumulative value or specific value is the result of all immunological methods. This value always represents the sum of concentrations of all recognized compounds multiplied by their relative response factors. Accordingly, the overestimation of immunoassays with respect to a target analyte ranges typically from a few rel. % to several multiples and is strongly dependent on the antibody type. With regard to the fact that some metabolites of EA-derivatives still retain their pharmacological activity, e.g., the metabolites of ergotamine (Tfelt-Hansen and Johnson, 1993), dihydroergotamine (Aeling, 1984; Müller-Schweinitzer, 1984), and/or pergolide (Clemens *et al.,* 1993), the development of a specific assay for the determination of the parent compound combined with the determination of the sum of its concentration together with its metabolites by a polyclonal antibody has been used in recent studies and is also recommended for the evaluation of bioequivalence of different dosage forms and for the comparison of various routes of administration (Lau *et al.,* 1994; Ezan *et al*, 1996; Humbert *et al.,* 1996).

Table 3 Selected areas of use of immunoassays

Topic	Ergot alkaloid	Reference
Ergot fermentation	ergotamine, ergocristine lysergic acid derivatives	Arens and Zenk, 1980 Lehtola et al., 1982
Drug monitoring and pharmacokinetic studies	LSD	Van Vunakis et al., 1971; Taunton-Ringby et al., 1973; Castro et al., 1973; Loeffler and Pierce, 1973; Lopatin and Voss, 1974; Ratclife et al., 1977; Peel and Boynton, 1980; Stead et al., 1986; Altunkaya and Smith, 1990; Webb et al., 1996; Cassells et al., 1996; Francis and Craston, 1996
	ergopeptines and dihydroergopeptines	Rosenthaler and Munzer, 1976; Koskinen and Kleimola, 1976; Kleimola, 1978; Ala-Hurula et al., 1979a, b; Collignon and Pradelles, 1984; Rosenthaler et al., 1984; Ezan et al., 1996
	bromokryptine	Rosenthaler et al., 1983; Bevan et al., 1986; Valente et al., 1996
	nicergoline	Bizollon et al., 1982; Gabor et al., 1995; Chen et al., 1996
	terguride	Krause et al., 1991b
	lisuride	Hümpel et al., 1981
	pergolide	Bowsher et al., 1992
	cabergoline	Rains et al., 1995
Environmental analysis	ergopeptines-grain, flour	Shelby and Kelley, 1990, 1992
	ergometrine-tall fescue	Shelby and Kelley, 1991
	ergoline alk.-tall fescue	Hill and Agee, 1994; Hill, 1997
	ergopeptines-tall fescue	Shelby, 1996

The key step determining the specificity of an antibody is the manner of alkaloid coupling to the carrier protein used for immunization. The part of the molecule that is most distal to the site of conjugation to the carrier is that one recognized by an antibody. EA offer several possibilities of conjugation. Mannich condensation, which was originally used for conjugation of lysergic acid amide to human serum albumin (HSA) through the nitrogen atom of the indole group (Taunton-Rigby *et al.,* 1973), was later used also by other authors to conjugate, e.g., ergotamine and ergocristine to bovine serum albumin (BSA) (Arens and Zenk, 1980), *N*-demethylnicergoline to BSA (Bizollon *et al.,* 1982), dihydroergocornine, dihydroergokryptine, or dihydroergocristine to BSA (Collignon and Pradelles, 1984), dihydroergokryptine to egg albumin (Collignon and Pradelles, 1984), or ergotamine to BSA or egg albumin (Shelby and Kelley, 1990; Shelby, 1996). This manner of preparation affords an antibody specific for the peptide portion of the molecule. For example, if the conjugate of some member of the ergotoxine family was used for the haptenation, the cross reactivity of all members of the ergotoxine series was detected even in the case of some modification at the lysergic acid moiety (Valente *et al.,* 1996). However, metabolic hydroxylation of the cyclopeptide moiety diminished the cross reactivity (Valente *et al.,* 1996). There is a good evidence that both alkyl groups of the first and the second amino acid of the cyclol moiety are involved in the recognition by an antibody. Examples illustrating this fact are the absence of cross reactivity of ergovaline against ergotamine antibodies, low reactivity of ergotamine against ergotoxine antibodies, and differences within the ergotoxine series (Shelby, 1996; Valente *et al.,* 1996).

Alternatively, EA having a carboxylic group can be simply conjugated via an amide bond with the free amino group of a carrier using carbodiimides as coupling reagents. This method was used for binding the *N*-6-carboxymethylderivative of dihydroergotamine to BSA (Rosenthaler and Munzer, 1976), lysergic acid to polylysine (Lopatin and Voss, 1974; Shelby and Kelley, 1991), bromolysergic acid to BSA (Valente *et al.,* 1996), lysergic acid to a hydrazide derivative of agarose (Loeffler and Hinds, 1975), and/or 2-(10α-methoxy6-methylergoline-8β-methoxy)-acetic acid to BSA (Chen *et al.,* 1996). A two-step modification of this procedure consists in the esterification of a free OH group of an ergot derivative with succinic or glutaric anhydride to produce hemisuccinate or hemiglutarate esters. The free carboxyl group is subsequently conjugated to a carrier by the carbodiimide method. This technique was used to conjugate, e.g., ergometrine or lysergol to HSA (Shelby and Kelley, 1991; Hill *et al.,* 1994; Hill and Agee, 1994). Such antibodies exhibit affinity for various compounds with an intact ergoline skeleton (i.e., both ergopeptines and clavines), but they have little affinity for compounds modified in this part (bromination, hydrogenation). Hence, immunoassay based on these antibodies can be used for monitoring the total content of alkaloids in fermentation broths or in agricultural products. It is worth mentioning that also EA-producing endophytic fungi can be directly detected by an immunoassay (Miles *et al.,* 1996).

Purified specific antibodies have also been used to develop immunoaffinity chromatographic methods for on-line purification of EA from biological samples. Extracts from urine, bile, serum, and tissue are directly injected on the column with the covalently bound antibodies with little or no sample preparation (Loeffler and Hinds, 1975; Francis and Craston, 1996). The specific binding of an analyte to an antibody renders it possible to remove most of the interferring substances and permits its trace enrichment so that the subsequent analysis by conventional techniques (usually GC/MS or HPLC/MS) can provide its unequivocal identification in forensic samples. An alternative to covalently bound antibody columns is the use of a system in which a small amount of an antibody is held on a column by non-covalent interaction (e.g., by using G-protein), and, after each purification procedure, washed off the column together with an analyte. Then the chromatographic column is equilibrated with the fresh portion of an antibody (Rule and Henion, 1992; Webb *et al.,* 1996). A combination with column-switching techniques could be used which would permit automated sample extraction with an on-line mass spectral identification of drugs directly from urine.

10.7.
COMBINED MASS SPECTROMETRIC TECHNIQUES

The result of one-dimensional techniques such as GC, HPLC or TLC is a peak or a spot whose elution time or position is characteristic for a given compound and the area under the curve or the peak height is directly proportional to its concentration. However, these techniques do not provide any proof of whether a given peak or spot represents actually the target compound or whether it is an interference eluting at the same position. With the more complex matrices and with an increasing number of peaks the correct assignment becomes more and more ambiguous. Determination of LSD, whose doses as low as 25 µg can cause central nervous system disturbances, is the typical example (Clark, 1989). Hence, concentration of LSD and its metabolites in plasma are likely to be very low. At this concentration level, both gas chromatography and high-performance liquid chromatography with fluorescence detection exhibit some interferences that might prevent conclusive drug identification. Furthermore, the analyte volatility, its termal instability and tendency to incur adsorptive losses during gas chromatographic analysis, all contribute to the difficulty of developing a method for unambiguous confirmation of LSD in body fluids (Lim *et al.,* 1988). Similarly, since the therapeutic dose of nearly all EA is very low, also a number of other analytical methods are operating on the edge of the limit of detection.

Whereas the UV-absorbance or fluorescence are features common for many compounds, monitoring of a particular molecular ion or a characteristic fragment provides much more selective detection. This allows a sensitive determination of a component in a rather complex matrix without previous separation; a combination of the retention time with the molecular mass information then provides a strong evidence of the target structure. Tandem mass spectrometry techniques (MS/MS) can distinguish nearly unambiguously between individual structures even in the case of several isobaric compounds (having the same molecular weight) (Halada *et al.,* 1997). The MS/MS instrumentation itself can be considered as a separation technique. This approach of MS/MS is to utilize one stage of the mass separation to isolate the compound of interest from the matrix and a second stage of the mass separation for analysis. Thus, without any preceding separation, EA can be identified directly in complex mixtures (Plattner *et al.*, 1983; Yates *et al.*, 1985; Belesky *et al.*, 1988). Contamination of dairy food by ergot sclerotia can be indirectly proved using GC/MS trace analysis of ricinoleate, a special and prominent feature of oil-rich ergot tissue (Mantle, 1996). Selective ion monitoring makes it possible to reach very low limits of detection that are particularly useful in pharmacokinetic studies (Haering *et al.*, 1985; Sanders *et al.*, 1986; Häring *et al.*, 1988). It should be mentioned that mass spectrometry together with NMR spectroscopy and X-ray crystallography belong also to the most powerful identification techniques.

As the GC/MS interface is relatively simple, gas chromatographic/mass spectrometric methods were the first coupled techniques used for the analyses of LSD and mass fragmentographic studies of ergopeptines. The combination of liquid chromatography/mass spectrometry (LC/MS) has been introduced in the seventies. The major drawback of this method is certainly the flow rate typically used with the high performance liquid chromatographic columns interfering with the high vacuum necessary for the performance of mass spectrometers. Consequently, the routine application of LC/MS has been waiting about 20 years for the development of a suitable LC/MS interface (Hopfgartner *et al.*, 1993). At present, several such devices are available, and in a few cases they have already been used for EA analyses (Table 4). However, it should be noted that regardless of the development of hyphenated methods, the importance of pre-purification steps does not decrease, and in some cases the chromatographic or mass spectrometric analysis represents only a short final step.

10.8.
STANDARD DEFINITION AND PHASE ANALYSIS

Continuously increasing demands on the quality of pharmaceuticals tighten up also the criteria for validation of analytical methods. No wonder, therefore, that this development includes also the quality of the analytical standards. The presence of various polymorphs in drugs can dramatically influence their pharmacokinetic profile (differences in dissolution kinetics). Similarly, the presence of a solvate in the analytical standard can significantly affect the slope of the calibration curve. Since various polymorphs and solvates differ in the IR spectra, the use of an incorrect standard can also cause a misidentification (compare, e.g., Harris and Kane, 1991; with Neville *et al.*, 1992). Differences in the X-ray powder diffractograms enable one to develop methods for the phase analysis of the mixtures of various crystalline forms occurring either in the bulk substances or originating from the production of pills. For example, such method was developed for the analyses of mixtures of "low" and "high" melting nicergoline polymorphic forms (Hušák *et al.*, 1994b).

Table 4 Analytical techniques coupled with mass spectrometry

Method*	Topic	Reference
GC/MS (EI, CI, CID)	LSD	Nichols *et al.*, 1983; Paul *et al.*, 1987, 1990; Francom *et al.*, 1988; Lim *et al.*, 1988; Papac and Foltz, 1990; Nelson and Foltz, 1992a, b; Bukowski and Eaton, 1993; Nakahara *et al.*, 1996
GC/MF/MS (CI, EI)	ergopeptines	Van Mansvelt *et al.*, 1978; Feng *et al.*, 1992; Uboh *et al.*, 1995
	dihydroergopeptines	Plomp *et al.*, 1978
	bromokryptine	Larsen *et al.*, 1979
SFC/EI/MS	*C. purpurea* extracts	Berry *et al.*, 1986
HPLC/EI,CI/MS	*C. purpurea* extracts	Eckers *et al.*, 1982
HPLC/CI/MS	bromokryptine	Schellenberg *et al.*, 1987
HPLC/APCI/MS	nicergoline metabolites	Banno *et al.*, 1991b; Banno and Horimoto, 1991
HPLC/IS,ESI/MS	LSD	Duffin *et al.*, 1992; Hopfgartner *et al.*, 1993; Webb *et al.*, 1996; Cai and Henion, 1996b; White *et al.*, 1997
HPLC/FAB/MS	ergopeptines, tall fescue	Miles *et al.*, 1996
HPLC/ESI/MS		Shelby *et al.*, 1997
Capillary electrophoresis/MS	LSD metabolism	Cai and Henion 1996a
Direct MS/MS	*C. purpurea* extracts	Plattner *et al.*, 1983
	ergopeptines, tall fescue	Yates *et al.*, 1985
	ergotamine in plasma	Haering *et al.*, 1985
	bromokryptine, plasma	Häring *et al.*, 1988
	ergovaline, tall fescue	Belesky *et al.*, 1988
TLC/SIMS	nicergoline	Banno *et al.*, 1991a

* GC-gas chromatography, HPLC-high performance liquid chromatography, SFC-supercritical fluid chromatography, TLC-thin layer chromatography, MS-mass spectrometry, MF-mass fragmentography, CI-chemical ionization, EI-electron ionization, ECI-electron capture ionization, IS-ionspray, FAB-fast atom bombardment, ESI-electrospray, SIMS-secondary ion mass spectrometry.

The first precedent example of inconsistence of two standards of LSD was published by Neville *et al.* (1992). However, according to our experience more than 50% of various EA derivatives examined within the last few years exhibit two or even more crystalline forms. In some cases, the solvates are so unstable that they decompose instantaneously upon isolation from the mother liquor. However, some solvates are considerably stable even under reduced pressure. In order to provide some data about the occurrence of polymorphism among ergot alkaloid derivatives, some examples are summarized in Table 5.

Table 5 Examples of solvation or existence of several polymorphic forms of EA

Ergot derivative	Crystalline form	Reference
Ergometrine	several forms, solvates?	Grant and Smith, 1936; Stoll and Hofmann, 1943a; Hušák et al., 1998
LSD o-iodobenzoate	monohydrate	Baker et al., 1972
Ergopeptine bases	two forms, one hydrate?	Stoll and Hofmann, 1943b
Ergocristine	acetone solvate	Pakhomova et al., 1997
Ergogaline	monohydrate	Cvak et al., 1994
Ergotamine tartrate	methanol and ethanol solvates	Pakhomova et al., 1995, 1996b
Hydroxyergotamine	methanol solvate	Pakhomova et al., 1996a
8α-Hydroxy-α-ergokryptine	methanol solvate monohydrate	Cvak et al., 1997
Dihydroergocristine	bis(dioxane) solvate	Čejka et al., 1997a
Dihydroergotamine mesylate	monohydrate	Hebert, 1979
Dihydroergocristine mesylate	monohydrate	Čejka et al., 1995
Dihydro-α-ergokryptine mesylate	monohydrate ethanol solvate	Čejka et al., 1997b
Dihydro-β-ergokryptine mesylate	monohydrate methanol solvate	
Bromokryptine methanesulfonate	isopropanol solvate	Camerman et al., 1979
	unsolvated	Hušák et al., 1997
Nicergoline	two unsolvated forms	Fabregas and Beneyto, 1981; Hušák et al., 1994a, b
Terguride	several crystalline forms	Kratochvíl et al., 1993, 1994

Ergot derivative	Crystalline form	Reference
Terguride maleate	monohydrate	Hušák *et al.*, 1993
8β-[(Benzoyloxycarbonyl) amino-methyl]- 6-methyl-10α-ergoline	monohydrate	Seratoni *et al.*, 1980
2-(10-Methoxy-1,6- dimethyl-8β-ergolinyl) ethyl 3,5-dimethyl- 1*H*-2-pyrrolecarboxylate	toluene hemisolvate	Foresti *et al.*, 1988

10.9.
FUTURE TRENDS

Several distinct trends in the analytical development can be observed with respect to various areas of the research and use of EA. The most important area is undoubtedly the industrial production of EA and their derivatives and the subsequent production of dosage forms. As already mentioned, the majority of official assays for EA or their impurities, as described in various pharmacopeias, are still based on spectrophotometry and thin layer chromatography. Although TLC is usually cheaper and more suitable for the routine analyses of a large number of samples, its resolving power and accuracy for trace components are limited. Accordingly, the use of some HPLC method is frequently requested as an additional specification for the determination of impurities in connection with the bulk pharmaceuticals trade. It can thus be expected that these methods will be soon implemented also in pharmacopeias. Similarly, HPLC methods, and particularly LC/MS methods, have found an inreplaceable position in the analyses of final drug forms, in studies of pharmacokinetics and metabolism of EA derivatives.

Another dynamically developing area is immunoassays. Due to the fact that these methods make it possible to analyse a large number of samples without extensive purification and without an involvement of highly qualified personnel, they have a great chance to replace some chromatographic methods still applied in pharmacokinetic studies and environmental analyses. However, an important factor underlying the development of immunoassays is the determination of cross reactivities with various metabolites. Since our knowledge in this field is very limited even in the case of the most widespread LSD (see Nelson and Foltz, 1992b), it can be expected that the necessity of validation of immunoassays will stimulate also the research of the ergot alkaloid metabolism.

REFERENCES

Aboul-Enein, H.Y. and Bakr, S.A. (1997) Simultaneous determination of caffeine and ergotamine in pharmaceutical dosage formulations by capillary electrophoresis. *J. Liquid Chromatogr. Rel. Technol.*,**20**,47–45.

Aellig, W.H. (1984) Investigation of the venoconstrictor effect of 8'-hydroxydihydroergotamine, the main metabolite of dihydroergotamine, in man. *Eur. J. Clin.Pharmacol.*,**26**,239–242.

Agurell, S. (1965) Thin-layer chromatographic and thin-layer electrophoretic analysis of ergot alkaloids. Relations between structure, R_M value and electrophoretic mobility in the clavine series. *Acta Pharm. Suecica*,**2**,357–374.

Agurell, S. and Ohlsson, A. (1971) Gas chromatography of ergot alkaloids. *J. Chromatogr.*,**61**,339–342.

Ala-Hurula, V., Myllylä, V.V., Arvela, P., Heikkila, J., Kärki, N. and Hokkanen, E. (1979a) Systemic availability of ergotamine tartrate after oral, rectal and intramuscular administration. *Eur. J. Clin. Pharmacol.*,**15**,51–55.

Ala-Hurula, V., Myllylä, V.V., Arvela, P., Kärki, N. and Hokkanen, E. (1979b) Systemic availability of ergotamine tartrate after three successive doses and during continuous medication. *Eur. J. Clin. Pharmacol.*,**16**,355–360.

Allport, N.L. and Cocking, T.T. (1932) Colorimetric assay of ergot. *Quart. J. Pharm.Pharmacol.*,**5**,341–346.

Altunkaya, D. and Smith, R.N. (1990) Evaluation of a commercial radioimmunoassay kit for the detection of lysergide (LSD) in serum, whole blood, urine and stomach contents. *Forensic. Sci. Int.*,**47**,113–121.

Amin, M. (1987) Quantitative determination and stability of lisuride hydrogen maleate in pharmaceutical preparations using thin-layer chromatography. *Analyst,***112**,1663–1665.

Andreotti, C., Pianezzola, E., Persiani, S., Pacciarini, M.A., Strolin Benedetti, M. and Pontiroli, A.E. (1995) Pharmacokinetics, pharmacodynamic, and tolerability of cabergoline, a prolactin-lowering drug, after administration of increasing oral doses (0. 5, 1.0, and 1.5 miligrams) in healthy male volunteers. *J. Clin. Endocrinol. Metab.*,**80**, 841–845.

Arens, H. and Zenk, M.H. (1980) Radioimmuntest für die Bestimmung der Peptidealkaloide Ergotamin und Ergocristin in *Claviceps purpurea*. *Planta Med.*,**38**, 214–226.

Baker, R.W., Chothia, C., Pauling, P. and Weber, H.P. (1972) Molecular structure of LSD. *Nature,***178**,614–615.

Ball, S.E., Maurer, G., Zollinger, M., Ladona, M. and Vickers, A.E.M. (1992) Characterization of the cytochrome P-450 gene family responsible for the *N*-dealkylation of the ergot alkaloid CQA 206–291 in humans. *Drug Metab. Disp.*,**20**, 56–63.

Banno, K. and Horimoto, S. (1991) Separation and quantitation of nicergoline and related substances by high-performance liquid chromatography/atmospheric pressure ionization mass spectrometry. *Chromatographia,***31**,50–54.

Banno, K., Matsuoka, M. and Takahashi, R. (1991a) Quantitative analysis by thinlayer chromatography secondary ion mass spectrometry. *Chromatographia,***32**,179–181.

Banno, K., Horimoto, S. and Mabuch, M. (1991b) Assay of nicergoline and three metabolites in human plasma and urine by high-performance liquid chromatographyatmospheric pressure ionization mass spectrometry. *J. Chromatogr.,* **568,**375–384.

Barrow, K.D. and Quigley, F.R. (1975) Ergot alkaloids. II. Determination of agroclavine by gas-liquid chromatography. *J. Chromatogr.,* **105,**393–395.

Belal, F. and Anderson, J.L. (1986) Flow injection determination of ergonovine maleate with amperometric detection at the Kel-F-graphite composite electrode. *Talanta,* **33,** 448–450.

Belesky, D.P., Stuedemann, J.A., Planner, R.D. and Wilkinson, S.R. (1988) Ergopeptine alkaloids in grazed tall fescue. *Agron. J.,* **80,**209–212.

Berg, A.M. (1951) De papierchromatographie der moederkoornalkaloiden. *Pharm.Weekbl.,* **86,**900–901.

Berg, A.M. (1952) Papierchromatographisch onderyoek van ergotoxine en ergotinine. *Pharm. Weekbl.,* **87,**282–283.

Berry, A.J., Games, D.E. and Perkins, J.R. (1986) Supercritical fluid chromatographic and supercritical fluid chromatographic-mass spectrometric studies of some polar compounds. *J. Chromatogr.,* **363,**147–158.

Bethke, H., Delz, B. and Stich, K. (1976) Determination of the content and purity of ergotamine preparations by means of high-pessure liquid chromatography. *J. Chromatogr.,* **123,**193–203.

Bevan, J.S., Baldwin, D. and Burke, C.W. (1986) Sensitive and specific bromocriptine radioimmunoassay with iodine label: Measurement of bromocriptine in human plasma. *Ann. Clin. Biochem.,* **23,**686–693.

Bianchi, M.L., Crespi-Perellino, N., Gioia, B. and Minghetti, A. (1982) Production by *Claviceps purpurea* of two new peptide ergot alkaloids belonging to a new series containing α-aminobutyric acid. *J. Nat. Prod.,* **45,**191–196.

Bizollon, Ch. A., Rocher, J.P. and Chevalier, P. (1982) Radioimmunoassay of nicergoline in biological material. *Eur. J. Nucl. Med.,* **7,**318–321.

Bowsher, R.R., Apathy, J.M., Compton, J.A., Wolen, R.L., Carlson, K.H. and Desante, K.A. (1992) Sensitive, specific radioimmunoassay for quantifying pergolide in plasma. *Clin. Chem.,* **38,**1975–1980.

Bredberg, U. and Paalzow, L. (1990) Pharmacokinetic of methysergide and its metabolite methylergometrine in the rat. *Drug. Metab. Disp.,* **18,**338–343.

Brooks, S.A., Lachno, D.R. and Obermeyer, B.D. (1997) Automated high-performance liquid chromatographic method for the analysis of two novel ergoline compounds in human plasma. *J. Chromatogr. B,* **691,**383–388.

Bukowski, N. and Eaton, A.N. (1993) The confirmation and quantitation of LSD in urine using gas chromatography/mass spectrometry. *Rapid Commun. MassSpectrom.,* **7,** 106–108.

Cai, J. and Henion, J. (1996a) Elucidation of LSD *in vitro* metabolism by liquid chromatography and capillary electrophoresis coupled with tandem mass spectrometry. *J. Anal. Toxicol.,* **20,**27–37.

Cai, J. and Henion, J. (1996b) On-line immunoaffinity extraction-coupled column capillary liquid chromatography/tandem mass spectrometry: trace analysis of LSD analogs and metabolites in human urine. *Anal. Chem.,* **68,**72–78.

Camerman, N., Chan, L.Y.Y. and Camerman, A. (1979) Stereochemical characteristic of dopamine agonists: Molecular structure of bromocriptine and structural comparisons with apomorphine. *Mol. Pharmacol.,* **16,**729–736.

Cassells, N.P., Craston, D.H., Hand, C.W. and Baldwin, D. (1996) Development and validation of nonisotopic immunoassay for the detection of LSD in human urine. *J. Anal. Toxicol.,***20,**409–415.

Castro, A., Grettie, D.P., Bartos, F. and Bartos, D. (1973) LSD radioimmunoassay. *Res.Commun. Chem. Pathol. Pharmacol.,***6,**879–886.

Čejka, J., Ondráček, J., Hušák, M., Kratochvíl, B., Jegorov, A. and Stuchlík, J. (1995) Absolute crystal structure determination of ergot alkaloid—dihydroergocristine methanesulfonate monohydrate. *Coll. Czech. Chem. Commun.,***60,**1333–1342.

Čejka, J., Kratochvil, B., Jegorov, A. and Cvak, L. (1997a) Crystal structure of dihydroergocristine bis(dioxane) solvate $(C_{35}H_{41}N_5O_5)$ $(C_4H_8O_2)_2$. *Z. Kristallogr.,***212,** 111–112.

Čejka, J., Kratochvíl, B., Jegorov, A. and Cvak, L. (1997b) Crystal structures of dihydro-α-ergokryptine and dihydro-β-ergokryptine methanesulfonates. *Z. Kristallogr.,* in press.

Chen, P., Tian, Z., Digenis, G.A. and Tai, H.-H. (1996) Enzyme immunoassay of two nicergoline metabolites, 10α-methoxy-9, 10-dihydrolysergol (MDL) and 1-methyl10α-methoxy-9, 10-dihydrolysergol *(MMDL)* . *Res. Commun. Mol. Pathol. Pharmacol.,***92,** 315–328.

Chervet, J.P. and Plas, D. (1984) Fast separation of some ergot alkaloids by high-performance liquid chromatography. *J. Chromatogr.,***295,**282–290.

Christie, J., White, M.W. and Wiles, J.M. (1976) A chromatographic method for the detection of LSD in biological fluids. *J. Chromatogr.,***120,**496–501.

Cieri, U.K. (1987) Determination of ergotamine tartrate in tablets by liquid chromatography with fluorescence detection. *J. Assoc. Off. Anal. Chem.,***70,**538–540.

Clark, C.C. (1989) The differentiation of lysergic acid diethylamide (LSD) from N-methyl, N-propyl and N-butyl amides of lysergic acid. *J. Forensic Sci.,***34,**532–546.

Clemens, J.A., Okimura, T. and Smalstig, E.B. (1993) Dopamine agonist activities of pergolide, its metabolites, and bromocriptine as measured by prolactin inhibition, compulsive turning, and stereotypic behaviour. *Arzneim.-Forsch.,***43 (I),**281–286.

Collignon, F. and Pradelles, P. (1984) Highly sensitive and specific radioimmunoassays for dihydroergotoxine components in plasma. *Eur. J. Nucl. Med.,***9,**23–27.

Coppi, G., Zanotti, A. and Mailland, F. (1992) Untersuchungen zur Pharmakokinetik von Dihydroergocristin bei freiwilligen Probanden nach oraler Verabreichung von drei Formulierungen. *Arzneim.-Forsch.,***42 (II),**1397–1399.

Cox, S.K., van Manen, R.T. and Oliver, J.W. (1992) Determination of ergonovine in bovine plasma samples. *J. Agric. Food Chem.,***40,**2164–2166.

Crespi-Perellino, N., Malyszko, J., Ballabio, M., Gioia, B. and Minghetti, A. (1993) Identification of ergobine, a new natural ergot alkaloid. *J. Nat. Prod.,***56,**489–493.

Cvak, L., Jegorov, A., Sedmera, P., Havlíček, V., Ondráček, J., Hušák, M., Pakhomova, S., Kratochvíl, B. and Granzin, J. (1994) Ergogaline, a new ergot alkaloid, produced by *Claviceps purpures:* Isolation, identification, crystal structure and molecular conformation. *J. Chem. Soc. Perkin Trans 2,*1861–1865.

Cvak, L., Jegorov, A., Pakhomova, S., Kratochvíl, B., Sedmera, P., Havlíček, V. and Minář, J. (1997) 8α-Hydroxy-α-ergokryptine, an ergot alkaloid. *Phytochemistry,***44,** 365–369.

Dal Cortivo, L.A., Dihrberg, A. and Newman, B. (1966) Identification and estimation of lysergic acid diethylamide by thin layer chromatography and fluorometry. *Anal.Chem.,***38,**1959–1960.

Dankháyi, T., Fekete, É., Paál, K. and Farsang, G. (1993) Electrochemical oxidation of lysergic acid-type ergot alkaloids in acetonitrile. Part 1. Stoichiometry of the anodic oxidation electrode reaction. *Anal. Chim. Acta,* **282,** 289–296.

De Groot, A.N.J.A., Vree, T.B., Hekster, Y.A., Baars, A.M., van den Biggelaar-Martea, M. and van Dongen, P.W.J. (1993) High-performance liquid chromatography of ergometrine and preliminary pharmacokinetics in plasma of men. *J. Chromatogr.,* **613,** 158–161.

De Zeeuw, R.A., van Mansvelt, F.J.W. and Greving, J.E. (1978) Detection of ergotamine in blood by means of gas chromatography-mass spectrometry. *Adv. Mass Spectrom.,* **7B,** 1555–1560.

Dolinar, J. (1977) Separation of ergot alkaloids by reversed-phase liquid chromatography. *Chromatographia,* **10,** 364–367.

Duffin, K.L., Wachs, T. and Henion, J.D. (1992) Atmospheric pressure ion-sampling system for liquid chromatography/mass spectrometry analyses on a benchtop mass spectrometer. *Anal. Chem.,* **64,** 61–68.

Eckers, C., Games, D.E., Mallen, D.N.B. and Swann, B.P. (1982) Studies of ergot alkaloids using high-performance liquid chromatography-mass spectrometry and *B/E* linked scans. *Biomed. Mass Spectrom.,* **9,** 162–173.

Edlund, P.O. (1981) Determination of ergot alkaloids in plasma by high-performance liquid chromatography and fluorescence detection. *J. Chromatogr.,* **226,** 107–115.

Eich, E. and Schunack, W. (1975) Die directe Auswertung von Dünnschichtchromatogrammen durch Remissions- und fluoreszenzmessungen. *Planta. Med.,* **27,** 58–64.

Erni, F., Frei, R.W. and Linder, W. (1976) A low-cost gradient system for high-performance liquid chromatography. Quantitation of complex pharmaceutical raw materials. *J. Chromatogr.,* **125,** 265–274.

European Pharmacopeia, 3rd Ed. (1997) Council of Europe, Strasbourg.

Evers, N. (1927) A colour test for ergot alkaloids. *Pharm. J. Pharmacist,* **25,** 721–723.

Ezan, E., Ardouin, T., Delhotal Landes, B., Flouvat, B., Hanslik, T., Legeai, J.M. and Grognet, J.-M. (1996) Bioequivalence study of α-dihydroergocryptine: utility of metabolite evaluation. *Int. J. Clin. Pharmacol. Ther.,* **34,** 32–37.

Fabregas, J.L. and Beneyto, J.E. (1981) Las formas cristalinas de la nicergolina. *Il Farmaco -Ed. Pr.,* **36,** 256–261.

Faed, E.M. and McLeod, W.R. (1973) A urine screening test for lysergide (LSD-25). *J. Chromatogr. Set.,* **11,** 4–6.

Fajardo, J.E., Dexter, J.E., Roscoe, M.M. and Nowicki, T.W. (1995) Retention of ergot alkaloids in wheat during processing. *Cereal Chem.,* **72,** 291–298.

Fanali, S., Flieger, M., Steinerová, N. and Nardi, A. (1992) Use of cyclodextrins for the enantioselective separation of ergot alkaloids by capillary zone electrophoresis. *Electrophoresis,* **13,** 39–43.

Feng, N., Minder, E.I., Grampp, T. and Vonderschmitt, D.J. (1992) Identification and quantification of ergotamine in human plasma by gas chromatography-mass spectroscopy. *J. Chromatogr.,* **575,** 289–294.

Flieger, M., Wurst, M., Stuchlík, J. and Řeháček, Z. (1981) Isolation and separation of new natural lactam alkaloids of ergot by high-performance liquid chromatography. *J. Chromatogr.,* **207,** 139–144.

Flieger, M., Sedmera, P., Vokoun, J., Řičicová, A. and Řeháček, Z. (1982) Separation of four isomers of lysergic acid α-hydroxyethylamide by liquid chromatography and thei spectroscopic identification. *J. Chromatogr.,***236,**453–459.

Flieger, M., Sedmera, P., Vokoun, J., Řeháček, Z., Stuchlík, J., Malinka, Z., Cvak, L. and Harazim, P. (1984a) New alkaloids from a saprophytic culture of *Clavicepspurpurea. J. Nat. Prod.,***47,**970–976.

Flieger, M., Sedmera, P., Vokoun, J., Řeháček, Z., Stuchlík, J. and Černý, A. (1984b) Liquid chromatography of semisynthetic ergot preparations. I. Nicergoline. *J. Chromatogr.,***284,**219–225.

Flieger, M., Zelenkova, N.F., Sedmera, P., Křen, V., Novák, J., Rylko, V., Sajdl, P. and Řeháček, Z. (1989) Ergot alkaloid glycosides from saprophytic cultures of *Claviceps.* I. Elymoclavine fructosides. *J. Nat. Prod.,***52,**506–510.

Flieger, M., Křen, V., Zelenkova, N.F., Sedmera, P., Novák, J. and Sajdl, P. (1990) Ergot alkaloid glycosides from saprophytic cultures of *Claviceps.* II. Chanoclavine fructosides. *J. Nat. Prod.,***53,**171–175.

Flieger, M., Sedmera, P., Novák, J., Cvak, L., Zapletal, J. and Stuchlík, J. (1991) Degradation products of ergot alkaloids. *J. Nat. Prod.,***54,**390–395.

Flieger, M., Sedmera, P., Havlíček, V., Cvak, L. and Stuchlík, J. (1993) 10-Hydroxy-*cis*- and 10-hydroxy-*trans*-paspalic acid amide: New alkaloids from *Claviceps paspali. J. Nat. Prod.,***56,**810–814.

Flieger, M., Sinibaldi, M., Cvak, L. and Castellani, L. (1994) Direct resolution of ergot alkaloid enantiomers on a novel chiral silica-based stationary phase. *Chirality,***6,** 549–554.

Flieger, M., Wurst, M. and Shelby, R. (1997) Ergot alkaloids—sources, structures and analytical methods. *Folia Microbiol.,***42,**3–30.

Foda, N.H. and Elshafie, F. (1996) Quantitative analysis of bromocriptine mesylate in tablet formulations by HPLC. *J. Liq. Chromatogr.,***19,**3201–3209.

Foresti, E., Sabatino, P., Di Sanseverino, L.R., Fusco, R., Tosi, C. and Tonati, R. (1988) Structure and molecular orbital study of ergoline derivatives. 1-(6-Methyl-8*β*-ergolinylmethyl) imidazolidine-2, 4-dione (I) and 2-(10-Methoxy-1, 6-dimethyl-8*β*-ergolinyl) ethyl 3, 5-dimethyl-1*H*-2-pyrrolecarboxylate toluene hemisolvate (II) and comparison with nicergoline (III). *Acta Cryst.,***B44,**307–315.

Fowler, R., Gomm, P.J. and Patterson, D.A. (1972) Thin-layer chromatography of lysergide and other ergot alkaloids. *J. Chromatogr.,***72,**351–357.

Francis, J.M. and Craston, D.H. (1996) Development of a stand alone affinity clean-up for lysergic acid diethylamide in urine. *Analyst,***121,**177–182.

Francom, P., Lim, H.K., Andrenyak, D., Jones, R.T. and Foltz, R.L. (1988) Determination of LSD in urine by capillary column gas chromatography and electron impact mass spectrometry. *J. Anal. Toxicol.,***12,**1–8.

Fuchs, L. (1953) Die Standardisierung von Secale cornutum und Secale-Präparaten. *Scientia Pharm.,***21,**20–30.

Gabor, F., Hamilton, G. and Pittner, F. (1995) Drug-protein conjugates: Haptenation of 1-methyl-10α-methoxydihydrolysergol and 5-bromonicotinic acid to albumin for the production of epitope-specific monoclonal antibodies against nicergoline. *J. Pharm. Sci.,***84,**1120–1125.

Galeffi, C. and Miranda delle Monache, E. (1974) Separation of dihydroergotoxine (hydrergine) into dihydroergocornine, dihydroergocryptine and dihydroergocristine by counter-current distribution. *J. Chromatogr.,***88,**413–415.

Garner, G.B., Rottinghaus, G.E., Cornell, C.N. and Testereci, H. (1993) Chemistry of compounds associated with endophyte/grass interaction: ergovaline- and ergopeptine-related alkaloids. *Agriculture, Ecosystems Environment,* **44,**65–80.

Genest, K. (1965) A direct densitometric method on thin-lasyer plates for the determination of lysergic acid amide, isolysergic acid amide and clavine alkaloids in morning glory seeds. *J. Chromatogr.,* **19,**531–539.

Gill, R. and Key, J.A. (1985) High performance liquid chromatography system for the separation of ergot alkaloids with applicability to the analysis of illicit lysergide (LSD). *J. Chromatogr.,* **346,**423–427.

Grant, L.R. and Smith, S. (1936) Dimorphism of ergometrine. *Nature,* 137, 154.

Grognet, J.M., Istin, M., Zanotti, A., Mailland, F. and Coppi, G. (1991) Pharmacokinetics of α-dihydroergocryptine in rats after intravenous and oral administration. *Arzneim.-Forsch.,* **41(II),**689–691.

Grognet, J.M., Riviére, R., Istin, M., Zanotti, A. and Coppi, G. (1992) Pharmakokinetik von Dihydroergocristin nach intravenöser und oralen Applikation bei Ratten. *Arzneim.-Forsch.,* **42(II),**1394–1396.

Gröger, D. and Erge, D. (1963) Dünnschichtchromatographie von Mutterkornalkaloiden. *Pharmazie,* **18,**346–349.

Gyenes, I. and Bayer, J. (1961) Über verschiedene Verfahren zur quantitativen Bestimmung der Mutterkornalkaloide. *Pharmazie,* **16,**211–217.

Haering, N., Settlage, J.A., Sanders, S.W. and Schuberth, R. (1985) Measurement of ergotamine in human plasma by triple sector quadrupole mass spectrometry with negative ion chemical ionization. *Biol. Mass Spectrom.,* **12,**197–199.

Halada, P., Jegorov, A., Cvak, L., Ryska, M. and Havlíček, V. (1997) Amino acid analysis in ergopeptines. *J. Mass. Spectrom.,* in press.

Häring, N., Salama, Z. and Jaeger, H. (1988) Triple stage quadrupole mass spectrometric determination of bromocriptine in human plasma with negative ion chemical ionization. *Arzneim.-Forsch.,* **38(II),**1529–1532.

Harris, H.A. and Kane, T. (1991) A method for identification of lysergic acid diethylamide (LSD) using a microscope sampling device with fourier transform infrared (FT/IR) spectroscopy. *J. Forensic Sci.,* **36,**1186–1191.

Hartmann, V., Rödiger, M., Abieidinger, W. and Bethke, H. (1978) Dihydroergotoxine: Separation and determination of four components by high performance liquid chromatography. *J. Pharm. Sci.,* **67,**98–103.

Harzer, K. (1982) Detection of LSD in body fluids with high-performance liquid chromatography. *J. Chromatogr.,* **249,**205–208.

Heacock, R.A., Langille, K.R., MacNeil, J.D. and Frei, R.W. (1973) A preliminary investigation of the high-speed liquid chromatography of some ergot alkaloids. *J. Chromatogr.,* **77,**425–430.

Hebert, H. (1979) The crystal structure and absolute configuration of (-)-dihydroergotamine methanesulfonate monohydrate. *Acta Cryst.,* **B35,**2978–2984.

Herényi, B. and Görög, S. (1982) Ready separation of ergocornine, α and β-ergocryptine by high-performance liquid chromatography. *J. Chromatogr.,* **238,**250–252.

Hill, N.S., Rottinghaus, G.E., Agee, C.S. and Schultz, L.M. (1993) Simplified sample preparation for HPLC analysis of ergovaline in tall fescue. *Crop. Sci.,* **33,**331–333.

Hill, N.S. and Agee, C.S. (1994) Detection of ergoline alkaloids in endophyte-infected tall fescue by immunoassay. *Crop. Sci.,* **34,**530–534.

Hill, N.S., Thompson, F.N., Dawe, D.L. and Stuedemann, J.A. (1994) Antibody binding of circulating ergot alkaloids in cattle graying tall fescue. *Am. J. Vet. Res.*, **55**, 419–424.

Hill, N.S. (1997) Affinity of anti-lysergol and anti-ergonovine monoclonal antibodies to ergot alkaloid derivatives. *Crop. Sci.*, **37**, 535–537.

Hogerzeil, H.V., Battersby, A., Srdanovic, V. and Stjernstrom, N.E. (1992) Stability of essential drugs during shipment to the tropic. *Br. Med. J.*, **304**, 210–212.

Hogerzeil, H.V. and Walker, G.J.A. (1996) Instability of (methyl)ergometrine in tropical climates—an overview. *Eur. J. Obstetric Gynecol. Reproductive Biol.*, **69**, 25–29.

Hooper, W.D., Sutherland, J.M., Eadie, M.J. and Tyrer, J.H. (1974) Fluorimetric assay of ergotamine. *Anal. Chim. Acta*, **69**, 11–17.

Hopfgartner, G., Wachs, T., Bean, K. and Henion, J. (1993) High-flow ion spray liquid chromatography/mass spectrometry. *Anal. Chem.*, **65**, 439–446.

Horák, P. and Kudrnáč, S. (1956) Zkušenosti s chromatografickým dělením námelových alkaloidů na impregnovaných papírech. *Cesk. Farm.*, 595–596.

Humbert, H., Denouel, J., Chervet, J.P., Lavene, D. and Kiechel, J.R. (1987) Determination of sub-nanogram amounts of dihydroergotamine in plasma and urine using liquid chromatography and fluorimetric detection with off-line and on-line solid-phase drug enrichment. *J. Chromatogr.*, **417**, 319–329.

Humbert, H., Cabiac, M.-D., Dubray, C. and Lavéne, D. (1996) Human pharmacokinetics of dihydroergotamine administred by nasal spray. *Clin. Pharmacol. Tber.*, **60**, 265–275.

Hümpel, M., Nieuweboer, B., Hasan, S.H. and Wendt, H. (1981) Radioimmunoassay of plasma Hsuride in man following intravenous and oral administration of lisuride hydrogen maleate; effect on plasma prolactin level. *Eur. J. Clin. Pharmacol.*, **20**, 47–51.

Hušák, M., Kratochvíl, B., Sedmera, P., Stuchlík, J. and Jegorov, A. (1993) A conformation study of the semisynthetic ergot alkaloid—terguride. *Coll. Czech. Chem.Commun.*, **58**, 2944–2954.

Hušák, M., Kratochvíl, B., Ondráček, J., Maixner, J., Jegorov, A. and Stuchlík, J. (1994a) The crystal and absolute structure of "low melting" nicergoline (form II). *Z.Krist-allogr.*, **209**, 260–262.

Hušák, M., Had, J., Kratochvíl, B., Cvak, L., Stuchlík, J. and Jegorov, A. (1994b) X-ray absolute structure of nicergoline (Form *I*). Quantitative analysis of the nicergoline phase mixture Form *I*/Form *II*. *Coll. Czech. Chem. Commun.*, **59**, 1624–1636.

Hušák, M., Kratochvíl, B. and Jegorov, A., (1997a) Crystal structure of 2-bromo-α-ergokryptine methanesulfonate, $(C_{32}H_{41}BrN_5O_5)^+(CH3_SO3)^-$. *Z. Kristallogr.*, **212**, 39–40.

Hušák, M., Kratochvíl, B., Jegorov, A. and Cvak, L. (1998) Ergometrine ethylacetate solvate (2:1). *Z. Kristallogr.*, **213**, 195–196.

Ibraheem, J.J., Paalzov, L. and Tfelt-Hansen, P. (1982) Kinetics of ergotamine after intravenous and intramuscular administration to migraine sufferers. *Eur. J. Clin.Pharmacol.*, **23**, 235–240.

Ibraheem, J.J., Paalzov, L. and Tfelt-Hansen, P. (1983) Low bioavailability of ergotamine tartrate after oral and rectal administration in migraine sufferers. *Br. J.Clin. Pharmacol.*, **16**, 695–699.

Inczeffy, J., Somodi, Z.B., Pap-Sziklay, Z. and Farsang, G. (1993) The study of the differential pulse voltametric behaviour of ergot alkaloids and their determination by DC amperometric detection in a FIA system. *J. Pharm. Biomed. Anal.*, **11**, 191–196.

Ingelse, B.A., Reijenga, J.C., Claessens, H.A., Everaerts, F. and Flieger, M. (1996a) Ergot alkaloids as novel chiral selectors in capillary electrophoresis. *J. High Res.Chromatogr.,***19,**225–226.

Ingelse, B.A., Flieger, M., Claessens, H.A. and Everaerts, P.M. (1996b) Ergot alkaloids as chiral selectors in capillary electrophoresis. Determination of the separation mechanism. *J. Chromatogr. A,***755,**251–259.

Jane, I. (1975) The separation of a wide range of drugs of abuse by high-pressure liquid chromatography. *J. Chromatogr.,***111,**227–233.

Japp, M., Gill, R. and Osselton, M. (1987) The separation of lysergide (LSD) from related ergot alkaloids and its identification in forensic science casework samples. *J. Forensic Sci.,***32,**933–940.

Jegorov, A., Šimek, P., Heydová, A., Cvak, L. and Minář, J. (1997) Free and bonded homoisoleucine in sclerotia of the parasitic fungus *Clauiceps purpurea. AminoAcids,* **12,**9–19.

Kanto, J. (1983) Clinical pharmacokinetics of ergotamine, dihydroergotamine, ergotoxine, bromocriptine, methysergide, ang lergotrile. *Int. J. Clin. Pharmacol. Ther.Toxicol.,***21,**135–142.

Keipert, S. and Voigt, R. (1972) Quantitative Dünnschichtchromatographie von Mutterkornalkaloiden. *J. Chromatogr.,***64,**327–340.

Klavehn, M. and Rochelmeyer, H. (1961) Zur Analytik der Mutterkornalkaloide. 4. Mitteilung: Trennung und Nachweis der Alkaloide aus der Clavinreihe. *Deut.Apot.- Ztg.,***17,**477–481.

Kleimola, T., Mäntylä, R. and Kanto, J. (1977) Renal excretion of dihydrogenated alkaloids of ergotoxine in man. *Acta Pharmacol. Toxicol.,***40,**541–544.

Kleimola, T.T. (1978) Quantitative determination of ergot alkaloids in biological fluids by radioimmunoassay. *Br. J. Clin. Pharmacol.,***6,**255–260.

Klošek, J. (1956a) Beitrag zur papierchromatographischen Bestimmung der Mutterkornalkaloide. *Microchim. Acta,*1375–1388.

Klošek, J. (1956b) Eine Metode zur quantitativen Bestimmung der wasserunlöslichen Mutterkornalkaloide auf papierchromatographischem Wege. *Microchim. Acta,* 1662–1671.

Kohlenberg-Müller, K., Meier, D.H., Kunz, K., Wauschkuhn, C.H. and Schaffler, K. (1991) Vergleichende Untersuchungen zur Bioverfügbarkeit von Nicergolin aus zwei unterschiedlichen Darreichungsformen im Steady State. *Arzneim.-Forsch.,***41(II),** 728–731.

Kornhauser, A. and Perpar, M. (1965) Die Trennung der Mutterkorn-Alkaloide undFarbstoffe mit Hilfe der Papier- und Dünnschicht-Elektrophorese. *Archiv Pharm.,* **298,**321–325.

Kornhauser, A. and Perpar, M. (1967) Kvantitativno določanje alkaloidov rženega rožička v prisotnosti barvil the droge. *Farm. Vest. (Ljubljana),***18,**1–8.

Kornhauser, A., Perpar, M. and Gašperut, L. (1970) Beitrag zur Analytik der ClavinAlkaloide. *Archiv. Pharm.,***303,**882–885.

Koskinen, E.H. and Kleimola, T. (1976) Radioimmunoassay for ergot alkaloids in biological fluids. *Acta Physiol. Scand. Supp.,***440,**122.

Kratochvíl, B., Ondráček, J., Novotný, J., Hšáak, M., Jegorov, A. and Stuchlík, J. (1993) The crystal and molecular structure of terguride monohydrate. *Z. Kristallogr.,***206,** 77–86.

Kratochvíl, B., Novotný, J., Hušák, M., Had, J., Stuchlík, J. and Jegorov, A. (1994) X-ray structural study of terguride solvates—terguride methanol solvate. *Coll. Czech.Chem. Commun.,***59,**149–158.

Krause, W., Mager, T., Kühne, G., Duka, T. and Voet, B. (1991a) The pharmacodynamics of lisuride in healthy volunteers afrer intravenous, intramuscular, and subcutaneous injection. *Eur. J. Clin. Pharmacol.,***40,**339–403.

Krause, W., Kühne, G. and Seifert, W. (1991b) Pharmacokinetics of ^{3}H-terguride in eldery volunteers. *Arzneim.-Forsch.,***41(I),**373–377.

Krause, W., Düsterberg, B., Jakobs, U. and Hoyer, G.-A. (1993) Biotransformation of proterguride in the perfused rat liver. *Drug. Metab. Disp.,***21,**203–208.

Křen, V., Pažoutová, S., Řeháček, Z. and Rylko, V. (1986) High-production mutant *Claviceps purpurea* 59 accumulating secoclavines. *FEMS Microbiol. Lett.,***37,**31–34.

Křen, V., Flieger, M. and Sajdl, P. (1990) Glycosylation of ergot alkaloids by free and immobilized cells of *Claviceps purpurea. Appl. Microbiol. Biotechnol.,***32,**645–650.

Křen, V. and Sedmera, P. (1996a) *N*-1-Trimethylsilyl derivatives of ergot alkaloids. *Coll.Czech. Chem. Commun.,***61,**1248–1253.

Křen, V., Sedmera, P., Přikrylová, V., Mingheti, A., Crespi-Prerellino, N., Polášek, M. and Macek, T. (1996b) Biotransformation of Chanoclavine by *Euphorbia calyptrata* cell cultures. *J. Nat. Prod.,***59,**481–484.

Kudrnáč, S. and Kakáč, B. (1982) Analytické studie námelových alkaloidů a jejich derivátů. V. Chromatografie necyklolových alkaloidů. *Czech. Farm.,***31,**44–47.

Larsen, N.-E., Ohman, R., Larsson, M. and Hvidberg, E.F. (1979) Determination of bromocriptine in plasma: Comparison of gas chromatography, mass fragmentography and liquid chromatography. *J. Chromatogr.,***174,**341–349.

Lászlóné, V. and Gábor, V. (1967) Adatok az anyarozsalkaloidák meghatározásához. *Acta Pharm. Hung.,***37,**67–70.

Lau, D.T.-V., Yu, Z., Aun, R.L., Hassell, A.E. and Tse, F.L.S. (1994) Pharmacokinetics of intranasally-administred dihydroergotamine in the rat. *Pharm. Res.,***11,**1530–1534.

Lehtola, T., Huhtikangas, A., Lundell, J. and Vallanen, T. (1982) Radioimmunoassay of lysergic acid derivatives from *Claviceps paspali* fermentation cultures. *Planta Med.,***45,** 161.

Lim, H.K., Andrenyak, D., Francom, P. and Foltz, R.L. (1988) Quantification of LSD and *N*-demethyl-LSD in urine by gas chromatography/resonance electron capture ionization mass spectrometry. *Anal. Chem.,***60,**1420–1425.

Loeffler, L.J. and Pierce, J.V. (1973) Radioimmunoassay for lysergide (LSD) in illicit drugs and biological fluids. *J. Pharm. Sci.,***62,**1817–1820.

Loeffler, L.J. and Hinds, C.J. (1975) Isolation of lysergide (LSD) with agarose-bound antibodies to lysergic acid. *J. Pharm. Sci.,***64,**1890–1892.

Lopatin, D.E. and Voss, E.W., Jr. (1974) Anti-lysergyl antibody: Measurement of binding parameters in IgG fractions. *Immunochemistry,***11,**285–293.

Ma, Y., Meyer, K.G., Aizal, D. and Agena, E.A. (1993) Isolation and quantification of ergovaline from *Festuca arundinacea* (tall fescue) infected with the fungus *Acremonium coenophialum* by high-performance capillary electrophoresis. *J. Chromatogr. A.,***652,**535–538.

Macek, K., Černý, A. and Semonský, M. (1954) Mutterkornalkaloide I. Die Bewertung von Ergotaminpräpreparaten mittels der Papierchromatographie. *Pharmazie,***9,** 388–390.

Macek, K., Hacaperková, J. and Kakáč, B. (1956) Systematische Analyse von Alkaloiden mittels Papierchromatographie. *Pharmazie,* **11,**533–538.

Magg, H. and Ballschmiter, K. (1985) Retention indices of some ergopeptines in reversed-phase high-performance liquid chromatography. *J. Chromatogr.,* **331,**245–251.

Mantle, P.G. (1996) Detection of ergot *(Claviceps purpurea)* in a dairy feed component by gas-chromatography and mass-spectrometry. *J. Dairy Sci.,* **79,**1988–1991.

Marderosian, A.D., Cho, E. and Chao, J.M. (1974) The isolation and identification of the ergoline alkaloids from *Ipomea muelleri. Planta Med.,* **25,**6–16.

Maurer, G., Schreier, E., Delaborde, E., Loosli, H.R., Nufer, R. and Shukla, A.P. (1982) Fate and disposition of bromocriptine in animals and man. I: Structure elucidation of the metabolites. *Eur. J. Drug Metabol. Pharmacokin.,* **7,**281–292.

Maurer, G., Schreier, E., Delaborde, E., Loosli, H.R., Nufer, R. and Shukla, A.P. (1983) Fate and disposition of bromocriptine in animals and man. II: Absorption, elimination and metabolism. *Eur. J. Drug Metabol. Pharmacokin.,* **8,**51–62.

Maurer, G. and Frick, W. (1984) Elucidation of the structure and receptor binding studies of the major primary, metabolite of dihydroergotamine in man. *Eur. J. Clin.Pharmacol.,* **26,**463–470.

McLaughlin. J.L., Goyan, J.E. and Paul, A.G. (1964) Thin-layer chromatography of ergot alkaloids. *J. Pharm. Set.,* **53,**306–310.

Messina, A., Girelli, A.M., Flieger, M., Sinibaldi, M., Sedmera, P. and Cvak, L. (1996) Direct resolution of optically active isomers on chiral packings containing ergoline skeletons. 5. Enantioseparation of amino acid derivatives. *Anal. Chem.,* **68,**1191–1196.

Michelon, L.E. and Kelleher, W.J. (1963) The spectrophotometris determination of ergot alkaloids. A modified procedure employing paradimethylaminobenzaldehyde. *Lloydia,* **26,**192–201.

Miles, C.O., Lane, G.A., di Menna, M.E., Garthwaite, I., Piper, E.L., Ball, O.J.-P., Latch, G.C.M., Allen, J.M., Hunt, M.B., Busch, L.P., Min, F.K., Fletcher, I. and Harris, P.S. (1996) High levels of ergonovine and lysergic acid amide in toxic *Achnatheruminebrians* accompany infection by an *Acremonium*-like endophytic fungus. *J. Agric.Food Chem.,* **44,**1285–1290.

Moubarak, A.S., Piper, E.L., Johnson, Z.B. and Flieger, M. (1996) HPLC method for detection of ergotamine, ergosine, and ergine after intravenous injection of a single dose. *J. Agric. Food Chem.,* **44,**146–148.

Moyer, J.L., Hill, N.S., Martin, S.A. and Agee, C.S. (1993) Degradation of ergoline alkaloids during *in vitro* ruminal digestion of tall fescue forage. *Crop. Sci.,* **33,**264–266.

Müller-Schweinitzer, E. (1984) Pharmacological action of the main metabolites of dihydroergotamine. *Eur. J. Clin. Pharmacol.,* **26,**699–705.

Nakahara, Y., Kikura, R., Takahashi, K., Foltz, R.L. and Mieczkowski, T. (1996) Detection of LSD and metabolite in rat hair and human hair. *J. Anal. Toxicol.,* **20,** 323–329.

Nelson, C.C. and Foltz, R.L. (1992a) Chromatographic and mass spectrometric methods for determination of lysergic acid diethylamide (LSD) and metabolites in body fluids. *J. Chromatogr.,* **580,**97–109.

Nelson, C.C. and Foltz, R.L. (1992b) Determination of lysergic acid diethylamide (LSD), iso-LSD, and *N*-demethyl-LSD in body fluids by gas chromatography/tandem mass spectrometry. *Anal. Chem.,* **64,**1578–1585.

Neville, G.A., Beckstead, H.D., Black, D.B., Dawson, B.A. and Ethier, J.-C. (1992) USP lysergic acid diethylamide tartrate (Lot I) authentic substance recharacterized for authentication of a house supply of lysergide (LSD) tartrate. *Can. J. Appl. Spect.,***37,** 149–157.

Niazy, E.M., Molokhia, A.M. and Elgorashi, A.S. (1988) Quick and simple determination of dihydroergotamine by high-performance liquid chromatography. *Anal. Lett.,***21,** 1833–1843.

Nichols, H.S., Anderson, W.H. and Stafford, D.T. (1983) Capillary GC separation of LSD and LAMPA. *J. High Res. Chromatogr., Chromatogr. Commun.,***6,**101–103.

Niwaguchi, T. and Inoue, T. (1971) Studies on quantitative *in situ* fluorometry of lysergic acid diethylamide (LSD) on thin-layer chromatograms. *J. Chromatogr.,***59,**127–133.

Otero, G.C.F., Lucangioli, S.E. and Carducci, C.N. (1993) Adsorption of drugs in high-performance liquid chromatography injector loops. *J. Chromatogr. A,***654,**87–91.

Padiglioni, P., Polcaro, C.M., Marchese, S., Sinibaldi, M. and Flieger, M. (1996) Enantiomeric separation of halogen-substituted 2-aryloxypropionic acids by high-performance liquid chromatography on a terguride-based chiral stationary phase. *J. Chromatogr. A,***756,**119–127.

Pakhomova, S., Ondráček, J., Hušák, M., Kratochvíl, B., Jegorov, A. and Stuchlík, J. (1995) Ergotamine tartrate bis(ethanol) solvate. *Acta Cryst.,***C51,**308–311.

Pakhomova, S., Ondráček, J., Kratochvil, B., Jegorov, A. and Cvak, L. (1996a) The structure of ergot alkaloid hydroxyergotamine. *Z. Kristallogr.,***211,**39–42.

Pakhomova, S., Ondráček, J., Kratochvíl, B. and Jegorov, A. (1996b) Crystal structure of 12′-hydroxy-2′-methyl-3′, 6′, 18′-trioxo-5′α-(phenylmethyl) ergotamanium tartrate bis (methanol) solvate, $C_{37}H_{46}N_5O_{10}$. *Z. Kristallogr.,***211,**555–556.

Pakhomova, S., Ondráček, J., Hušák, M., Kratochvil, B., Jegorov, A., Cvak, L., Sedmera, P. and Havlíček, V. (1997) Conformation of ergopeptam and ergopeptine alkaloids (ergocristam and ergocristine). *Z. Kristallogr.,***212,**593–600.

Papac, D.I. and Foltz, R.L. (1990) Measurement of lysergic acid diethylamide (LSD) in human plasma by gas chromatography/negative ion chemical ionization mass spectrometry. *J. Anal. Toxicol.,***14,**189–190.

Papp, E. (1990) Fast separation of dihydro ergot alkaloids on an octylsilica column. *J. Chromatogr.,***502,**241–242.

Paul, B.D., Mitchell, J.M., Mell, L.D., Jr., Sroka, R. and Irving, J. (1987) Gas chromatography—electron impact mass fragmentometric identification and quantitation of lysergic acid diethylamide (LSD) in urine. *Clin. Chem.,***36,**971.

Paul, B.D., Mitchell, J.M., Burbage, R., Moy, M. and Sroka, R. (1990) Gas chromatographic-electron-impact mass fragmentographic determination of lysergic acid diethylamide in urine. *J. Chromatogr.,***529,**103–112.

Peel, H.W. and Boynton, A.L. (1980) Analysis of LSD in urine using radioimmunoassay-excretion and storage effects. *Can. Soc. Forensic Sci. J.,***13,**23–28.

Persiani, S., Sassolas, G., Piscitelli, G., Bizollon, C.A., Poggesi, I., Pianezzola, E., Edwards, D.M.F. and Strolin Benedetti, M. (1994) Pharmacodynamics and relative bioavailability of cabergoline tablets vs solution in healthy volunteers. *J. Pharm. Sci.,* **83,**1421–1424.

Phelan, D.G., Greig, N.H., Rapoport, S.I. and Soncrant, T.T. (1990) High-performance liquid chromatographic assay of bromocriptine in rat plasma and brain. *J. Chromatogr.,* **533,**264–270.

Pianezzola, E., Bellotti, V., La Croix, R. and Benedetti, M.S. (1992) Determination of cabergoline in plasma and urine by high performance liquid chromatography with electrochemical detection. *J. Chromatogr.,***574,**170–174.

Piattner, R.D., Yates, S.G. and Porter, J.K. (1983) Quadrupole mass spectrometry of ergot cyclol alkaloids. *J. Agric. Food Chem.,***31,**785–789.

Plomp, T.A., Leferink, J.G. and Maes, R.A.A. (1978) Quantitative analysis of dyhydroergotoxine alkaloids by gas chromatography and gas chromatography-mass spectrometry. *J. Chropmatogr.,***151,** 121–132.

Pöhm, M. (1953) Die Umsetzung von Lysergsäure mit p-Dimethylaminobenzaldehyd. *Archiv Pharm.,***286,**509–511.

Pöhm, M. (1958) Ein Verfahren zur Trennung und Bestimmung der Mutterkornalkaloide mittels Papierchromatographie. *Archiv Pharm.,***291,**468–480.

Prošek, M., Kučan, E., Katić, M. and Bano, M. (1976a) Quantitative fluorodensitometric determination of ergot alkaloids. *Chromatographia,***9,**273–276.

Prošek, M., Kučan, E., Katić, M. and Bano, M. (1976b) Quantitative fluorodensitometric determination of ergot alkaloids. II. Determination of hydrogenated ergot alkaloids of the ergotoxine group. *Chromatographia,***9,**325–327.

Rabey, J.M., Oberman, Z., Scharf, M., Isakov, M., Bar, M. and Graff, E. (1990) Bromocriptine blood levels after the concominant administration of levodopa, amantadine and biperiden in Parkinson's disease. *Acta Neurol. Scand.,***81,**411–415.

Radecka, C. and Nigam, I.C. (1966) Detection of trace amounts of lysergic acid diethylamide in sugar cubes. *J. Pharm. Sci.,***55,**861–864.

Rains, C.P., Bryson, H.M. and Fitton, A. (1995) Cabergoline. A review of its pharmacological properties and therapeutic potential in the treatment of hyperprolactinaemia and inhibition of lactation. *Drugs,***49,**255–279.

Ratcliffe, W.A., Fletcher, S.M., Moffat, A.C., Ratcliffe, J.G., Harland, W.A. and Levitt, T.E. (1977) Radioimmunoassay of lysergic acid diethylamide (LSD) in serum and urine by using antisera of different specificities. *Clin. Chem.,***23,**169–174.

Reichelt, J. and Kudrnáč, S. (1973) Analytical studies on ergot alkaloids and their derivatives. I. Separation of ergot alkaloids of the ergotoxine and ergotamine groups by thin-layer chromatography. *J. Chromatogr.,***87,**433–436.

Roberts, C.A., Joost, R.E. and Rottinghaus (1997) Quantification of ergovaline in tall fescue by near infrared reflectance spectroscopy. *Crop. Sci.,***37,**281–284.

Rosenthaler, J. and Munzer, H. (1976) 9, 10-Dihydroergotamine: Production of antibodies and radioimmunoassay. *Experientia,***32,**234–236.

Rosenthaler, J., Munzer, M. and Voges, R. (1983) Immunoassay of bromocriptine and specificity of antibody: criteria for choice of antiserum and marker compound. In: Reid, E. and Leppard, J.P., Eds., Drug metabolite isolation and determination. Plenum Press, New York, p. 215–223.

Rosenthaler, J., Munger, H., Voges, R., Andres, H., Gull, P. and Bolliger, G. (1984) Immunoassay of ergotamine and dihydroergotamine using a common ^{3}H-labelled ligand as tracer for specific antibody and means to overcome experienced pitfalls. *Int. J. Nucl. Med. Biol.,***11,**85–89.

Rottinghaus, G.E., Garner, G.B., Cornell, C.N. and Ellis, J.L. (1991) HPLC method for quantitating ergovaline in endphyte-infested tall fescue: Seasonal variation of ergovaline levels in stems with leaf sheats, leaf blades, and seed heads. *J. Agric.Food Chem.,***39,** 112–115.

Rubin, A., Lemberger, L. and Dhahir, P. (1981) Physiologic disposition of pergolide. *Clin. Pharmacol. Ther.,***30,**258–265.

Rule, G.S. and Henion, J.D. (1992) Determination of drugs from urine by on line immunoaffinity chromatography-high-performance liquid chromatography-mass spectrometry. *J. Chromatogr.,***582**,103–112.

Rücker, G. and Taha, A. (1977) The use of π-acceptors for detection of alkaloids on thin layers. *J. Chromatogr.,***132**,165–167.

Sanders, S.W., Haering, N., Mosberg, J. and Jaeger, H. (1986) Pharmacokinetics of ergotamine in healthy volunteers following oral and rectal dosing. *Eur. J. Clin.Pharmacol.,***30**,331–334.

Schellenberg, K.H., Under, M., Groeppelin, A. and Erni, F. (1987) Experience with routine application of liquid chromatography-mass spectrometry in the pharmaceutical industry. *J. Chromatogr.,***394**,239–251.

Scott, P.M. and Kennedy, B.P.C. (1976) Analysis of blue cheese for roquefortine and other alkaloids from *Penicillium roqueforti. Agric. Food Chem.,***1976**,865–868.

Scott, P.M. and Lawrence, G.A. (1980) Analysis of ergot alkaloids in flour. *Agric. FoodChem.,***28**,1258–1261.

Scott, P.M. and Lawrence, G.A. (1982) Losses of ergot alkaloids during making of bread and pancakes. *J. Agric. Food Chem.,***30**,445–450.

Scott, P.M., Lombaert, G.A., Pellaers, P., Bacler, S. and Lappi, J. (1992) Ergot alkaloids in grain foods sold in Canada. *J. AOAC Int.,***75**,773–779.

Serantoni, E.F., Di Sanseverino, L.R. and Sabatino, P. (1980) An antiserotonic drug metabolite: 8β-[(Benzoyloxycarbonyl)aminomethyl]-6-methyl-10α-ergoline monohydrate. *Acta Cryst.,***B36**,2471–2473.

Shelby, R.A. and Kelley, V.C. (1990) An immunoassay for ergotamine and related alkaloids. *J. Agric. Food. Chem.,***38**,1130–1134.

Shelby, R.A. and Kelley, V.C. (1991) Detection of ergot alkaloids in tall fescue by competitive immunoassay with a monoclonal antibody. *Food. Agric. Immunol.,***3**, 169–177.

Shelby, R.A. and Kelley, V.C. (1992) Detection of ergot alkaloids from *Claviceps* species in agricultural products by competitive ELISA using a monoclonal antibody. *J. Agric.Food Chem.,***40**,1090–1092.

Shelby, R.A. (1996) Detecting ergot alkaloids by immunoassay. *ACS Symp. Ser.,***621**, 231–242.

Shelby, R.A. and Flieger, M. (1997) Improved method of analysis for ergovaline in tall fescue by high-performance liquid chromatography. *J. Agric. Food Chem.,***45**, 1797–1800.

Shelby, R.A., Olšovská, J., Havlíček, V. and Flieger, M. (1997) Analysis of ergot alkaloids in edophyte-infected tall fescue by liquid chromatography. *J. Agric. Food Chem.,***445**, 4674–4679.

Sinibaldi, M., Flieger, M., Cvak, L., Messina, A. and Pichini, A. (1994) Direct resolution of optically active isomers on chiral packings containing ergoline skeletons. II. Enantioseparation of carboxylic acids. *J. Cbromatogr. A,***666**,471–478.

Sioufy, A., Sandrenan, N. and Godbillon, J. (1992) Determination of 10α-methoxy-9, 10dihydrolysergol, a nicergoline metabolite, in human urine by high performance liquid chromatography. *Biomedical Chromatogr.,***6**,9–11.

Sochor, J., Klimeš, J. and Šurmová, M. (1995) High-performance liquid chromatographic determination of terguride in solid dosage forms and plasma. *J. Cbromatogr. B,***663**, 309–313.

Sondack, D.L. (1978) High-performance liquid chromatographic analysis of ergonovine maleate formulations. *J. Chromatogr.,* **166,** 615–618.

Sprince, H. (1960) A modified Ehrlich benzaldehyde reagent for detection of indoles on paper chromatograms. *J. Chromatogr.,* **3,** 97–98.

Stead, A.H., Watton, J., Goddard, C.P., Patel, A.C. and Moffat, A.C. (1986) The development and evaluation of a ^{125}I radioimmunoassay for the measurement of LSD in body fluids. *Forensic Sci. Int.,* **32,** 49–60.

Stoll, A. and Hofmann, A. (1943a) Partialsyntese von Alkaloiden von Typus des Ergobasins. *Helv. Chim. Acta,* **26,** 944–965.

Stoll, A. and Hofmann, A. (1943b) Die Alkaloide der Ergotoxingruppe: Ergocristin, Ergokryptin und Ergocornin. *Helv. Chim. Acta,* **26,** 1570–1601.

Stoll, A. and Rüegger, A. (1954) Zur papierchromatographischen Trennung der Mutterkornalkaloide. *Helv. Chim. Acta,* **37,** 1725–1732.

Sturm, J.C., Nunez-Vergara, L.J. and Squella, J.A. (1992) Electrochemical reduction of nicergoline and its analytical determination in dosage forms. *Talanta,* **39,** 1149–1154.

Swanson, E.E., Powell, C.E., Stevens, A.N. and Stuart, E.H. (1932) VIII. The standardization and stabilization of ergot preparations. *Am. Pharm. Assoc.,* **21,** 229–239 continued at 320–324.

Szántay, C., Jr., Bihari, M., Brlik, J., Csehi, A., Kassai, A. and Aranyi, A. (1994) Structural elucidation of two novel ergot alkaloid impurities in α-ergokryptine and bromokryptine. *Acta Pharm. Hung.,* **64,** 105–108.

Szepesi, G. and Gazdag, M. (1976) Determination of dihydroergotoxine alkaloids by gasliquid chromatography. *J. Chromatogr.,* **122,** 479–485.

Szepesi, L., Fehér I., Szepesi, G. and Gazdag, M. (1978) High-performance liquid chromatography of ergot alkaloids. *J. Chromatogr.,* **149,** 271–280.

Szepesi, G., Gazdag, M. and Terdy, L. (1980) Separation of ergotoxine alkaloids by high-performance liquid chromatography on silica. *J. Chromatogr.,* **191,** 101–108.

Tanret, Ch. (1875) Sur la presence d'un nouvel alcaloide, l'ergotinine, dans le seigle ergoté. *Compt. Rend. Acad. Sci. Fr.,* **81,** 896–897.

Taunton-Rigby, A., Sher, S.E. and Kelley, P.R. (1973) Lysergic acid diethylamide: Radioimmunoassay. *Science,* **181,** 165–166.

Teichert, K., Mutschler, E. and Rochelmeyer, H. (1960) Beiträge zur analytischen Chromatographie. *Deut. Apotb.-Ztg.,* **11,** 283–286.

Tfelt-Hansen, P. and Johnsòn, E.S. (1993) Ergotamine. In: J. Olesen, P.TfeltHansen, and K.M.A. Welch Eds. *The Headaches,* Raven Press, Ltd., New York, pp. 313–322.

Toda, T. and Oshino, N. (1981) Biotransformation of lisuride in the hemoglobin-free perfused rat liver and in the whole animal. *Drug. Metab. Disp.,* **9,** 108–113.

Tokunaga, H., Kimura, T. and Kawamura, J. (1983) Determination of ergometrine maleate and methylergometrine maleate in pharmaceutical preparations by high-performance liquid chromatography . *Chem. Pharm. Bull.,* **31,** 3988–3993.

Trtík, B. and Kudrnáč, S. (1982) Analytické studie námelových alkaloidů a jejich derivátů. VI. Stanovení laktamových alkaloidů. *Czech. Farm.,* **31,** 48–50.

Tyler, V.E., Jr. and Schwarting, A.E. (1952) The separation of the ergot alkaloids by paper partition chromatography. *J. Am. Pharm. Assoc., Sc. Ed.,* **41,** 354–355.

Uboh, C.E., Rudy, J.A., Railing, F.A., Enright, J.M., Shoemaker, J.M., Kahler, M.C., Shellenberger, J.M., Kemecsei, Z., Das, D.N., Soma, L.R. and Leonard, J.M. (1995) Postmortem tissue samples: An alternative to urine and blood for drug analysis in racehorses. *J. Anal. Toxicol.,* **19,** 307–315.

Valente, D., Ezan, E., Crémion, C., Delaforge, M., Benech, H., Pradelles, P. and Grognet, J.-M. (1996) Enzyme immunoassays for bromocriptine and its metabolites. *J. Immunoassay,* **17,**297–320.

Vanhaelen, M. and Vanhaelen-Fastré, R. (1972) Dosage photodensitométrique *in situ* des alkaloides dans *Secale cornutum. J. Chromatogr.,* **72,**139–144.

Van Mansvelt, F.J.V., Greving, J.E. and De Zeeuw, R.A. (1978) Identification of ergot-peptide alkaloids, based on gas-liquid chromatography of the peptide moiety. *J. Chromatogr.,* **151,**113–120.

Van Urk, H.W. (1929) Een nieuwe gevoelige reactie op de moederkoornalkaloiden ergotamine, ergotoxine en ergotinine en de toepassig voor het onderzoek en de colorimetrische bepaling in moederkoornpreparaten. *Pharm. Weekbl,* **66,**473–481.

Van Vunakis, H., Farrow, J.T., Gjika, H.B. and Levine, L. (1971) Specificity of the antibody receptor site to D-lysergamide: Model of a physiological receptor for lysergic acid diethylamide. *Proc. Nat. Acad. Sci. USA,* **66,**1483–1487.

Walker, J.A., Marche, H.L., Newby, N. and Bechtold, E.J. (1996) A free capillary electrophoresis method for the quantitation of common illicit drug samples. *J. Forensic Sci.,* **41,**824–829.

Wang, J. and Ozsoz, M. (1990) Hydrophobic uptake and volatametry of ergot alkaloids at lipid-coated electrodes. *Electroanalysis,* **2,**595–599.

Webb, K.S., Baker, P.B., Cassells, N.P., Francis, J.M., Johnson, D.E., Lancaster, S.L., Minty, P.S., Reed, G.D. and White, S.A. (1996) The analysis of lysergide (LSD): The development of novel enzyme immunoassay and immunoaffinity extraction procedures together with an HPLC-MS confirmation procedure. *J. Forensic Sci.,* **41,** 938–946.

White, S.A., Cattterick, T., Harrison, M.E., Johnston, D.E., Reed, G.D. and Webb, K.S. (1997) Determination of lysergide in urine by high-performance liquid chromatography combined with electrospray ionisation mass spectrometry. *J. Chromatogr. B,* **689,** 335–340.

Wilkinson, R.E., Hardcastle, W.E. and McCormic, C.S. (1987) Seed ergot alkaloid contents of *Ipomea hederifolia, I. quamocit, I coccinea* and *I. wrightii. J. Sci. FoodAgric.,* **39,** 335–339.

Wittwer, J.D., Jr. and Kluckhohn, J.H. (1973) Liquid chromatographic analysis of LSD. *J. Chromatogr. Sci.,* **11,**1–6.

Wolthers, E.G., Kamerbeek, W.D.J.V., van Beusekom, C.M., Elshof, F., de Ruyter Buitenhuis, A.W., Brunt, E.P.R. and Lakke, J.P.W.F. (1993) Quantitative determination of the dopamine agonist lisuride in plasma using high-pergormance liquid chromatography with fluorescence detection. *J. Chromatogr.,* **622,**33–38.

Wurst, M., Flieger, M. and Řeháček, Z. (1978) Analysis of ergot alkaloids by high-performance liquid chromatography. I. Clavines and simple derivatives of lysergic acid. *J. Chromatogr.,* **150,**477–483.

Wurst, M., Flieger, M. and Řeháček, Z. (1979) Analysis of ergot alkaloids by high-performance liquid chromatography. II. Cyclol alkaloids (ergopeptines). *J. Chromatogr.,* **174,**401–407.

Wyss, P.A., Rosenthaler, J., Nüesch, E. and Aellig, W.H. (1991) Pharmacokinetic investigation of oral and IV dihydroergotamine in healthy subjects. *Eur. J. Clin.Pharmacol.,* **41,**597–602.

Yamatodani, S. (1960) Researches on ergot fungus. XLI. Studies on the color reaction of ergot alkaloids. *Ann. Repts. Takeda Research Labs.,* **19,**8–14.

Yates, S.G., Plattner, R.D. and Garner, G.B. (1985) Detection of ergopeptine alkaloids in endophyte infected, toxic Ky-31 tall fescue by mass spectrometry/mass spectrometry. *J. Agric. Food Chem.,***33,**719–722.

Young, J.C., Chen, Z.-J. and Marquardt, R.R. (1983) Reduction in alkaloid content of ergot sclerotia by chemical and physical treatment. *J. Agric. Food Chem.,***31,**413–415.

Zakhari, N.A., Hassan, S.M. and El-Shabrawy, Y. (1991) Spectrophotometric determination of ergot alkaloids with ninhydrin. *Acta Pharm. Nord.,***3,**151–154.

Zecca, L., Bonini, L. and Bareggi, S.R. (1983) Determination of dihydroergocristine and dihydroergotamine in plasma by high-pergormance liquid chromatography with fluorescence detection. *J. Chromatogr.,***272,**401–405.

Zelenkova, N.F., Arinbasarov, M.U. and Kozlovskii, A.G. (1996) Identification and determination of clavine alkaloids by liquid and thin-layer chromatography. *Zh. Anal. Chim.,***51,**679–683.

Žorž, M., Marušič, A., Smerkloj, R. and Prošek, M. (1983) Quantitative determination of low concentrations of DHETX m.s. in human plasma by high performance liquid chromatography with fluorescence detection. *J. High Res. Chrom., Chrom.Commun.,***6,** 306–309.

Žorž, M., Čulig, J., Kopitar, Z., Milivojević, D., Marušič, A. and Bano, M. (1985) HPLC method for determination of ergot alkaloids and some derivatives in human plasma. *Human Toxicol.,***4,**601–607.

11.
PARASITIC PRODUCTION OF ERGOT ALKALOIDS

ÉVA NÉMETH

University of Horticulture and Food Industry, Villányi str. 29–

43H-1114 Budapest, Hungary

11.1.
MAIN CHARACTERISTICS OF THE ONTHOGENESIS OF CLAVICEPS PURPUREA (FR.) TUL.

For the optimal agricultural production of ergot, knowledge of the principal developmental stages of its onthogenesis is necessary. High yields of appropriate quality in a field production can be achieved only by considering the natural vegetation circle of the fungus, its biological properties, parasitic behaviour and ecological requirements.

As an ascomycete, onthogenesis of *Claviceps purpurea* (Fr.) Tul. has two main phases, the generative as well as the vegetative developmental phases. The generative period begins by the germination of the sclerotium *(Secalecornutum)* that is the overwintering stage of ergot. In the early spring, the sclerotium develops handled stromas, with perithecia containing the ascus. By hyfae conjugation and meiotic cell divisions ascospores are formed, which infect early flowering gramineous species. Although the spores are able to infect different plant organs—especially in juvenile stadium—(Lewis, 1968), the main and practically important way of infection should be the female propagative organ, the pistil. Detailed mechanism of the infection, sphacelium and sclerotium formation are presented in the Chapter 2.

11.2.
GENOTYPES AND BREEDING FOR FIELD PRODUCTION

Naturally occuring ergot as a plant disease have usually small alkaloid content (1–2‰) composed of ergotamine, ergometrine and ergotoxines or even clavines (Bojor, 1968; Ruokola, 1961).

In parallel with development of cultivation techniques, genetical improvement was carried out to optimise the biological background of the production and to develop special ergot genotypes having high alkaloid yield and desired composition. By the intensive breeding, alkaloid content could reach about 9–11‰ (Ruokola, 1972, Šnejd, 1974). Improved genotypes are registered mainly as strains of *Claviceps purpurea* and often protected by patents (see also the Chapter 12).

Selection is the main method of the ergot breeding for the field production. It can start both from vegetative as well as generative cells of *Clavicepspurpurea*. Single sclerotium selection was started by a collection of wild growing materials in different areas (Stadler, 1982). Today, the origin is the breeding field, which must have a proper isolation distance from any other ergot growing field. Strains for the selection work are propagated on small plots (about $2m^2$) of rye, infected and harvested by hand. The ripen sclerotia are tested for their alkaloid content and compositional purity. For the analysis only one half of the sclerotium is used, the other half is left for the propagation. It was demonstrated, that cultures originating from the surface of the ergot did not produce significantly different yields or alkaloid contents, than those originating from the internal part of the sclerotia (Ruokola, 1972). Although the sclerotia are not always homogenous, relatively small number of the heterogeneous types does not damage the method efficacy (Békésy, 1967). The advantageous individual pieces are coming for propagation. Examination can be carried out during the winter-period, such as next year selection circle may be adjusted to the vegetation of rye.

The selected sclerotia are propagated in *in vitro* saprophytic culture on a solid medium at room temperature for some weeks. Form, growth intensity and colour of cultures are in most cases characteristic for the genotype or the strain, which may be used as an additional information in selection. Repeated inoculation and continuous maintenance in saprophytic culture may result in decrease of production capacity of the strain or in loss of sclerotia viability (Békésy, 1967). Further propagation of the selected cultures is carried out in suspension cultures which provides the inoculation material for rye in the next breeding circle.

Trials for single conidium selection are also known (Ruokola, 1972). Starting material for these cultures is also the half sclerotium. On the agar slants the culture produces large number of conidia which are then transferred into sucrose solution and plated on the Petri dishes. Then the individual colonia are selected. Inoculation onto rye is necessary in each phase.

A more uncertain and laborious method having, however, eventually a high efficacy in the strain improvment for the field production, is a selection by generative propagation. Sclerotia are made to germinate in a soil with the meiotic process taking place in the perithecia. Ascospores are then used for the inoculation of the saprophytic culture (Ruminska, 1973).

For the development of new advantageous genotypes, mutation process might also be used, e.g., by irradiation of sclerotia before inoculation on the nutritive medium. Whilst the selection of individual sclerotia or conidia is rather reliable method for continuous, stepwise strain improvement, in the last two cases rare, neocombinant genotypes may provide new sources of genetical variation.

11.3.
ENVIRONMENTAL REQUIREMENTS

11.3.1.
Abiotic Factors

The optimal ecological conditions for the development of ergot vary in the course of its onthogenesis. For vernalization of the scerotia winter temperatures around 0°C (Chapter 3). Germination starts only when the average daily temperature reaches 9–10°C in the springtime. However, optimal temperature for this process lays as high as 18–22°C.

During the time of infection by conidia the optimal weather conditions are characterised by moderately warm temperatures (about 18–20°C), high air humidity (near saturation) or drizzling rain (Golenia *et al.*, 1961; Ruokola, 1961). Low temperatures up to 12°C—especially during the night 3–4 weeks prior to flowering may have an advantageous effect on the infection efficacy because of the stronger degeneration or abnormal development of anthers (McLaren and Wehner, 1992; Watkins and Littlefeld, 1976). At the time of the mycelia development similar ecological conditions are considered to be optimal. In warm and windy weather the risk of drying out of the conidia increases, the flowering period of the rye becomes shorter decreasing the possibility for a successful secondary infection.

The best sclerotia growth is assured by dry, constantly moderate weather conditions. In hot weather, the growing period of sclerotia is shortened causing less developed scerotia of lower alkaloid content. According to the experiments of Kiniczky (1992) under controlled conditions, the alkaloid level of the sclerotia at 22°C was 5.1% , while the rising the temperature to 25°C decreased the alkaloid content to 2.7% . Higher temperature (up to 30°C) had a significant effect also on the individual sclerotium mass, but their number in the ear was not influenced.

In general, ergot may be grown in the areas, where the annual amount of precipitation is over 500 mm and the conditions are considered as advantageous also for the rye. It is desirable to protect the plot against strong and drying wind so that it has a humid climate and regular early morning mist (Kybal and Strnadová, 1968).

11.3.2.
Biotic Factors

Host plants are rather important factor for the production. *Claviceps purpurea* (Fr.) Tul. has been found as an obligate parasite on about 70 species belonging to the plant families *Poaceae, Cyperaceae* and *Juncaceae* (Bojor, 1968). According to the latest publication the host number might reach 400 species (see Chapters 2 and 3). The most important genera from this point of view are *Agropyron,*

Avena, Brachypodium, Bromus, Calamagrostis,Dactylis, Festuca, Hordeum, Lolium, Phleum, Poa, Triticum, and *Secale,* having the highest economical significance (Ubrizsy, 1965). Susceptibility of the hosts may be different (Lewis, 1968). The form and size of the sclerotium is determined by the plant host (flower organs, seed morphology) (Blažek and Starý, 1962). These features might be also strain-dependent (Frey and Brack, 1968). Although the rye (*Secale cereale* L.) is the most important host plant giving higher sclerotium production than any other grass species, the alkaloid content of the wild ergot types is often higher on the other species (Bojor, 1968).

Sensitivity for an ergot infection of rye cultivars or genotypes shows rather big differences which can be explained by several reasons (Singh *et al.,* 1992). Long lasting flowering caused by light stand density or cool weather increases the possibility for the effective infection. However, metabolic differences may also play a considerable role, e.g. the varying stimulative effect of pollen extracts (see Chapter 2). According to Caspar *et al.* (1990) hybrid rye varieties showed specificity of response to some particular alkaloid-producing ergot strain.

Also some agrotechnological characteristics are of special importance for the respective variety usage. Ripening time (length of vegetation period), stem height and stability and long ears are the main ones (Kybal and Strnadova, 1968). For the companies producing mainly or exclusively ergot the production of more rye varieties of different ripening time seems to be advantageous because it assures a continuous work in the fields. For the companies producing also other crops, mainly early cultivars are advisable to avoid an accumulation of the work during cereal harvesting. In Hungary, for a long period tetraploid rye cultivars were considered as the most advantageous because of ergot amount and alkaloid yield (e.g. cultivar 'Sopronhorpácsi Tetra'). Its unsuitable high growth could be diminished by growth regulation treatment (Vásárhelyi *et al.,* 1980). In the eighties, a new variety was registered especially for ergot production: 'Kisvárdai 101'. In Europe, among others the Petkuser and Dankowskie variety groups were widely used for this purpose.

Although in the seventies perennial rye *(Secale)* species seemed to be promising and productive host plants, their usage was not distributed (Snejd, 1974).

At the end of the seventies and beginning of eighties breeding of Fl hybrid cereal varieties (rye, triticale) attracted most interest. Several male sterile lines were developed and observations showed that both hybrids and especially the cytoplasmic male sterile genotypes exhibited increased sensitivity to the ergot infection (Cauderon *et al.,* 1984; Mielke, 1993). Later examinations proved differences according to the nuclear genetic factors among the ergot susceptibility of male sterile genotypes (Rai and Thakur, 1995). Geiger and Bausback (1979) reported the first experiments where provocative infection of male sterile lines of rye was carried out aiming at the development of a new method for parasitic ergot production. About ten years later, male sterile host cultivars have been introduced into the agrotechnology. These materials, e.g.

'Hyclaro' by Rentschler company became immediately popular in consequence of their high production potential especially in the Czech Republic, Germany, France and Hungary. However, still up to now fertile rye varieties are used, mainly because of the considerable lower price of their propagation material and higher alkaloid content in the sclerotia.

In the case of male sterile genotypes, beside high sterility percentage (above 95%) the above mentioned technical properties, especially stem stability are of prominent importance.

11.4.
AGROTECHNOLOGY OF ERGOT CULTIVATION

11.4.1.
Historical Aspects

Although the first idea for growing ergot as an agricultural crop emerged already in the second half of the last century, until the third decade of this century the alkaloids of ergot could only be used by collecting the sclerotia from spontaneously infected cereal populations (Kerekes, 1969). The effectiveness and the alkaloid yield was extremely low. The first large scale technology for parasitic production of ergot was developed and patented by Hungarian specialist, Miklós Békésy in 1934. The basic method went through several improvements and technological development by the same specialist as well as by foreign researchers (Grzybowska et al., 1970; Kiniczky, 1992).

Parasitic ergot production used to be an important source for raw material supply in several countries, such as: Canada, Czech Republic, Finland, Germany, Hungary, India, Poland, Romania, Slovenia and Switzerland (Aho, 1953; Gaspar et al., 1990; Heeger, 1956; Németh, 1993; Kumar et al., 1993; Pageau and Wauthy, 1995; Ruminska, 1973). Although in some of these countries production of ergot becomes more and more based on saprophytic culture, field cultivation remains an adequate alternative method. Its agriculture exhibits advantages both from the point of view of economics as well as of the possibility of producing all kind of ergot alkaloids. Nevertheless, the demand for ergot sclerotia originating from field production seems to be not stable.

Although a great number of fungi in the genus Claviceps are registered as parasites of different cereals thorough the world, pharmaceutical exploitation and field cultivation is known only in the case of C. purpurea. The cultivation practices can be divided into two parallel parts: the so called "pricking method" (technology on the normal, male fertile rye varieties and "on-pricking" method, (technology on genetically male sterile host varieties). The latter kind of technology was developed in the late eighties, and it will be discussed separately.

11.4.2.
Sowing and Care of Host Plant

Male Fertile Varieties

Appropriate selection of previous crop and preparation of soil does not require any specialities compared to normal rye cultivation. When growing different chemotypes of ergot (producing different alkaloid spectra), an isolation distance of minimum 500 m must be kept between the populations. Although rye allows a monoculture for some years, when grown for ergot production, the sclerotia fallen out in the previous years may cause a significant contamination in a consequence of generative propagation of the fungus. Thus, either the same chemical form of ergot (alkaloid spectrum)

Figure 1 Male fertile rye field for ergot production

should be produced in the same locality or a crop rotation is advisable. Sowing time in Central Europe is October, the same as in the case of cereal production. The sowing is carried out in stripes, leaving 70 cm wide free areas for machine movement (Figure 1). The sown strips includes usually ten rows at a row distance of 12 cm.

Seed rate may vary according to the rye cultivar and vegetation year (germination capacity, thousand seed mass), however, assuring 4.5 to 5 millions germs per hectare (10000 m^2) is essential.

Male Sterile Varieties

Male sterile genotypes usually develop a higher number of shoots, than fertile ones. Strong branching results in an uneven height of the plant stand. Even though it is an undesired phenomenon in the fertile varieties because of the infection technology, it makes no problems in the male sterile genotype production. The optimal sowing time is about 2–3 weeks earlier than that of the fertile varieties. The most usual sowing occurs in continuous rows at a distance of 12–15 cm. The seed rate is lower than in the cereal technology the number of living germs is 3.5 millions (about 90 kg) per hectare.

Care of the rye stand does not differ considerably from the grain producing populations. Herbicide application can be carried out according to the normal technology. It has a big importance in assuring an even growth, optimal condition and clean stand thoroughout the vegetation. Similarly, nutrition supply and insecticide application should be arranged for assuring the described optimal stage. In phytotron experiments, it was found that the main nutrients have different roles to the development of ergot. Nitrogen has a more significant, positive effect on the number of sclerotia and the level of alkaloids, phosphorus influences the number of sclerotia, while potassium supports mainly the individual sclerotium mass (Bernáth, 1985). There exist experimental data about the advantageous effect of nitrogen on the ergot yield also in triticale (Naylor and Munro, 1992). Biomass production of the host and that of the ergot are in tight linear correlation (Bernáth, 1985).

In the case of fungicide application before infection preference should be given to active agents whose harmlessness on ergot has been experimentally proved (triadimefon, benomyl). During the vegetation of C. purpurea no fungicide application is allowed.

Ergot sclerotia are rarely damaged by insects or parasites. Cladosporiumherbarum (Pers.) Link, as well as Fusarium spp. might cause eventual losses, however, it occurs especially in weakened stands and they do not need any special treatment (Békésy, 1967).

11.4.3.
Infection

Male Fertile Rye Technology (Pricking Technology)

By the infection procedure, conidia of C. purpurea ought to be inoculated into the ovary. According to the spontaneous infection mechanism, inoculation of conidia into the ovary means the basic criteria for parasitic ergot production. Although the most obvious way would be infection through the stigma—like the natural propagation process—in normal fertile rye genotypes it can not assure a stable and high infection rate, sclerotium and alkaloid mass. In the eve of ergot field cultivation, trials were carried out on flowering (Lewis, 1945). During the flowering a huge amount of pollen is present (Scoles and Evans, 1979) which competes severely with ergot conidia. After fertilisation of rye ovules the probability of effective infection quickly decreases. This feature was proved first of all in cytoplasmic male sterile cereals. In the experiments of Kiniczky (1992), provocative infection and pollination was carried out in different time intervals. Infection with ergot conidia before fertilisation of the ovary resulted in 3.4 sclerotia/ear. Even, in case of the treatment by conidia and pollen at the same time, no decrease of yield was found (3.7 sclerotium/ear). However, infection of rye by 48 hours after fertilisation was almost ineffective (0.1 sclerotia/ear). In the male sterile wheat, the number of sclerotia reached 11 pcs/ear when inoculated 1 day prior anthesis, it was reduced to 6, 2 in plants inoculated at the anthesis and further 53% decrease was observed in the case of inoculation 1 day after anthesis (Watkins and Littlefield, 1976). Similar experience was made with the male sterile barley (Watkins and Littlefield, 1976) and with *C. sorghi* and *C. africana* in the rye (Musabyimana *et al.*, 1995).

Experimental and practical results demonstrated that *optimal time* for infection was the phenological phase, when the ovary was fully developed,

Figure 2 Main developmental phases of rye. * optimal phase for infection of male fertile rye varieties. ** optimal phase for infection of cytoplasmic male sterile rye varieties

however flowers still were not open (Bekesy, 1967; Watkins and Littlefield, 1976). According to the accepted scale for the developmental phases of cereals (Large, 1954), the biologically optimal stage for ergot infection is the stage O, when the ear and the stem reached their final size but the stamina cannot be seen, yet. Still, in the practice often an earlier **N** stage (ears developed, however among the upper leaves yet) is used, because of organisational purposes and because of the fact that in this stage the homogeneity of plant height is much better (Figure 2). In Central Europe the mentioned phenological stage is usualy reached in the beginning of May. Duration of the appropriate phases enables an artificial infection usually within a two weeks period, (depending on weather conditions) which is one of the advantages of the fertile technology.

Infection is the most effective in the morning hours. Formerly, it was suggested because of the flowering dynamic (Lewis, 1945). Today, it is also accepted the influence of the morning dew and moderate temperatures preventing the conidia from sudden drying (Hornok, 1990).

For the preparation of *infecting material* several methods are known. One of the most widespread method had been the surface culture of ergot spores on an agar-agar nutritive medium supplemented by 10% malt extract in Kolbe flasks (Bekesy and Garay, 1960). Cereal grains (rye, wheat) are also used as solid medium for culturing the spores mixed with the same amount of water and having grown for 5–6 weeks after inoculation by a heavily sporulating stock culture (Lewis, 1945). Considerable disadvantage of these methods is a short viability of the cultures. Preservation of the conidia in sucrose solution by deep freezing is, therefore, used.

The best known and widespread methods include large scale fermentation process (Gláz, 1955), resuspension of the biomass in an osmotically stable medium and an immediate freezing of the suspension at −40 to −70°C followed by a storage at −20°C.

Infection of male fertile rye populations is usualy carried out by the frozen conidium preparations. Transportation from the place of fermentation/storage to the final utilisation place requires a freezing chain. Before usage, the material should be melted at lower temperatures and diluted with water. The melted material must not be stored longer than 12 hours or frozen again that causes considerable losses of the viability (Hornok, 1990).

Optimal concentration of the infecting material varies in a large scale, from 5 to 300 thousand spores/ml (Békésy, 1967). Under conditions in Hungary a concentration of 30000 spores/ml proved to be optimal that means 3×10^{12} spores per hectare using the usual amount of water for the treatment (100–150l/ha). Altogether, the number of conidia seems to be much more important than the concentration of suspension (Lewis, 1945).

Principle of the *infecting technology* is sticking the ears by needles—without damaging the spikes—and injecting the solution into it. Originaly, this work was done by hand operating injecting panels or multiple apparatus (Lewis, 1945). Today the infection process is carried out by special equipment. They operate by metal roller pairs which consist of a pinned roller and a rubber one (Figure 3). In some older constructions the roller pairs were placed horizontally, however, the way and the position of ears was hardly to be optimised. In modern constructions they are placed vertically, turning into the opposite directions, the rye ears passing between them. The number of roller pairs varies between 5–12, the length of them reaches about 25–30 cm, assuring pricking of ears within this height level (Németh, 1993). The infecting material (spore-suspension) is sprayed onto the pricked ears through spraying nozzles positioned above the roller row (Figure 4). The constructions may have characteristics depending on the local conditions (Figure 5). Beside mounted equipment, self-propelled ones are also known and used. For the latter type the classical model is the Swiss "Goldhamster".

For optimal production, an optimal number of pricks should be achieved. Bernáth (1985) demonstrated that in the case of a single pricking it may be unsuccessful by 36–45% probability. He measured an optimal number of sclerotia (5–6 pcs/ear) by making 5–15 pricks on each ear. In the Hungarian production practice it is accepted that 70–80% of the ears in the rye stand needs to be pricked, and 7–10 pricks are to be placed on each single ear. The infection effectivity can be enhanced by about 45% by another motion. It is carried out in opposite direction thorough the rows so that the other side of ears also gets the necessary inoculation (Kiniczky and Nagy, 1976). As a further method also known for increasing the productivity and the purity of the final product is the treatment of rye by chemicals of gametocide effect. It is made together with the infection, by a dosage of 8–12l/ha (Balassa et al., 1987). By appropriate gametocide application a further 20–25% yield increase and a purity above 90% can be achieved.

Male Sterile Rye Technology (Cultivation Without Pricking)

Sterile technology based on the male sterile rye and triticale genotypes became videly used in the last decade. In this case the *biologically optimalstage* of the host plant for infection is different from the described phenological phases of male fertile taxa. As the amount of pollens in the rye stand is minimal, the best way of infection is to get the ergot conidia onto the stigma, because no considerable competition with pollen can be expected. Flowering of a single spike generally lasts 10–12 days and the spikes of the fertile varieties are closed right after pollination (Lewis, 1945). According to the results of Kiniczky (1992), the maximal sensitivity of the stigma is in the first three days of flowering (spike opening). From the 4th day, the sensitivity is decreasing rapidly, and on the 8–10th day the stigma is getting perished   and cannot be infected. Within each ear, flowering of the spikes has a definite sequence: beginning from the middle, going to upwards, thereafter downwards. Duration of the flowering process depends on weather conditions, especially on the air temperature. As it was shown in the phytotron chambers, at 30–35°C flowering of the rye population takes place in 5–6 days, whereas at 20–25°C more than 8 days (Kiniczky, 1992). It is of a big importance for the infecting process: in the case of hot and dry pre-summer weather (in the Central Europe end of May) optimal infection of the stand means a high cumulation of work.

Infecting material in the sterile technology may be the deep-frozen conidium culture, fermented industrially and stored in unit bottles.

The characteristic flowering mechanism of the male sterile spikes allows also the utilisation of other infecting materials. The best known second possibility is the solid spores preparation. The method elaborated by Kubec et al. (1974) was published as the first one for solving this problem. The concentration

Figure 3 Rollers of the infecting machine (left: pinned roller; right: rubber roller)

preservation of viability could be improved by the Hungarian method, when the homogenisation of the fermented biomass is carried out by colloidal hydrophilic silicium-dioxide (Siloxide) of high hygroscopy and dispersity grade, granulated and dried by fluidisation under 30°C (Kiniczky et al., 1982).

The required quantity of this formulation lays between 400–1000 g/ha. The homogenous spreading out of spores requires an appropriate formulation by carrier material(s). The carrier should be inert with good adhesion capacity and low specific volume, furthermore the size of carrier grains needs to suit to those of conidia (3–13 μm) (Ubrizsy, 1965).

Infection technology on male sterile cereals does not require any special field equipment. Universal machines constructed and used for plant protective purposes are used for this process. The optimal dosage per hectare is about 2–3 times higher, than in the fertile technology. Repeated application during the

Figure 4 Infecting machine at work in male fertile rye stand

Figure 5 Infecting adapter of the "pricking technology"

second half of flowering results in a significant increase of sclerotium yield. The amount of spraying water is 200–350 I/ha in order to reach an entire, homogenous covering of the stand. In the case of solid infecting material use, plant protecting dusting machines are applied. The optimal dosage of the above described solid ergot spores preparation varies between 6–10 kg/ha, depending on the carrier and the spore concentration, as well as on the technical parameters of the dusting machine (Kiniczky *et al.*, 1989). Dusting is effective only in the case of suitable weather conditions.

The *secondary infection* plays a significant role in the effectiveness and yield of the production. According to Kybal and Strnadova (1968) this may cause a 30–70% yield increase. The conidia of ergot are carried out mainly by insects collecting honeydew on the infected ears, and transferred over to the ears of the side shoots. The amount of it is, therefore, higher in the male sterile genotypes as a consequence of the lower seed rate and their biological characteristics. The insect activity is highly influenced by the weather.

11.4.4.

Harvesting

Fertile Rye Technology (Pricking Technology)

The first ripe sclerotia appear 4–6 weeks after infection, depending mainly on the growing area, weather circumstances and the vegetation length of the rye cultivar. Also the ergot strain has a slight effect on the growing period. In the fertile technology, on the average 2–3 sclerotia/ear and by application of gametocides 3–4 sclerotia/ear develop (Jurkevich and Misenin, 1976). The average sclerotium weight often reaches 80–100 mg/piece. They are seated among the glumellae of the spike, and fall off easily. Besides, the ripening process within a stand is not homogenous in a consequence of the development, localisation of ears as well as of the secondary infection. These facts may cause sever losses in the yield, especially when promoted by wind.

That is why, a *continuous gathering* of sclerotia seems to be necessary in these stands. Most usual, harvesting is made in two or three steps. During the first motions only the ripen scerotia are gathered and the unripe ones as well as the rye grains remain on their place. For this procedure the optimal equipment consists of brush cylinder pairs placed horizontally which lift out the sclerotia gently from the ears by turning movement. Released sclerotia fall into a metal trough that should be emptied at the end of the rows. The equipment is mounted onto the same basic machine as the infecting adapter. The "Goldhamster" equipment represents the basic and widest known type. Higher temperatures and sunshine causes further drying and loosening of the sclerotia in the spikes, thus the harvesting in early morning decreases yield loss.

The *final gathering* is accomplished by a cereal combine harvester, when both the remained sclerotia as well as the ripen rye seeds are harvested (in the Central Europe end of June-beginning July). While the pre-harvested material consists more than 90% of ergot, and only the ear debris is to be separated thereafter, the finally harvested product contains only 30–40% ergot that requires a vast effort to be separated. In the case of gametocide application, much higher purity (above 90%) can be achieved. This is one of the main advantages of this technology.

After harvesting the material must be *dried* so that the water content of sclerotia is reduced to 7–9% (Hornok, 1990). Drying is a very important procedure in the preservation of the alkaloid content. It should be done immediately after harvesting. Storage and heating of the fresh material in the container of the harvester, even for few hours may result in severe losses in the alkaloid level. Drying can be carried out in several type of dryers. Today, almost exclusively artificial dryers are used. Control of drying temperature is essential and it must not exceed 40°C, especially at the beginning of the procedure or in the products containing 35–40% water (Ruminska, 1973). In some cases, when the harvested material is clean and dry enough—water content is around 10% (final harvest)—higher temperatures up to 80°C may be shortly (10–20 minutes) used.

Cleaning of the harvested material involves different type of work in the case of the first and the last harvesting. Rests of ears or other plant parts can be separated easily by different sieving machines. Cleaning of the combine material goes on in several steps. Larger parts, such as stem, leaf, ear pieces, dust and other contamination can be separated by sieves and ventilating screens. Rye grains are removed from ergot sclerotia also by sieving. The up-to-date equipment universally used for separation of cereals or other seeds are able to clean the material to about 85–95% purity. Afterwards, some specific methods should be employed.

Flotation in saline is an effective method for both of small and large scale. Optimal concentration is 15% NaCl (Grzybowska *et al.,* 1970). In this solution rye seeds sink while sclerotia flotate. Alternatively, a fine cleaning of the rye-sclerotium mixture can be carried out by photoelectric separator on the basis of colour differences.

By the described technology an average ergot yield of 300–350 kg/ha, with gametocides up to 500 kg/ha can be calculated (Ruokola, 1961; Hornok, 1990).

Male Sterile Rye Technology (Cultivation Without Pricking)

Ergot production on the male sterile varieties is of considerable advantage for harvesting. In these populations, the number of sclerotia per ear reaches 15–30 pieces, (depending on the infection effectiveness and condition of rye). The mass of individual sclerotia is however much lower (20–40 mg) than in the case of the fertile rye populations. Therefore, falling out from the ears is minimised. Ripening is more even thus there is no need for pre-gathering of the early ripen sclerotia. Harvesting is made by combine harvester at the time of full ripeness both of the ergot and the rye. This period is somewhat later than in the male fertile technology—in the Central Europe the first half of July.

Drying and cleaning of the harvested material uses the same technology, as above. Separation requires moderate expenditure compared to the fertile varieties, because of much higher purity of the harvested material (above 90%, frequently 95%).

Sclerotia yield is usually much higher than in the case of the production on the male fertile rye. As it was proved by Bernáth (1985), the sclerotium number has a more important role in the total production than the individual sclerotium mass. General yields achieved by this technology are about 1000kg per hectare but in good years and using advanced technology doubling of the crop can be achieved (Németh, 1993).

The alkaloid content of the individual ergot sclerotia is in tight correlation with their number in the ear. The sclerotium mass decreases with their number according to inverse correlation (r=−0.97). However, the alkaloid content grows with the size of sclerotia according to a positive correlation (r=0.88) (Bernáth, 1985; Kiniczky, 1992). Therefore, smaller sclerotia can be considered as the main drawback of the alkaloid production on male sterile cereals. Thus, 0.6 to 1. 0 g/m^2 alkaloid yields can be achieved by a usual ergot strain, which means about 8–10 kg alkaloids per one hectare. Nevertheless, it was shown, that extremely high yield potential of these male sterile materials assures also a considerably higher alkaloid yield compared to the male fertile populations' capacity (4–5 kg/ha for the same strain).

11.8.
CONCLUSIONS

For economic ergot production the most important factors seems to be: optimal climatic conditions, advantageous genotype and condition of host plant, well organised technology, infecting material of high quality. The price of the propagation material for male sterile genotypes definitely arises the costs. In the agrotechnology, care of the rye stand, infection and harvesting mean rather stable expenses that may be, however, considerable increased if intensive drying and much separation are needed. As for the income for the producing company both the sclerotium yield per hectare and the unit price of the sclerotia is of a crucial importance. While the first one depends on the rye variety and the applied agrotechnology, the price is determined by bilateral contracts between the producer farming company and the pharmaceutical company. Generally, it depends on the alkaloid accumulation level of the drug and the market relations.

REFERENCES

Aho, E. (1953) On the artificial production of ergot in Finland. *FarmaseuttinenAikakaus Lehti*(Pharmaceutical Bulletin), **54**,115–129.

Balassa, J., Horváth, P., Reichert, P., Waller, I., Léber, F. and Varga, I. (1987) *Hungarian Pat.,*195.387.

Bekesy, M. (1967) Az anyarozstermesztés problémái. (Problems of ergot production.) *Herba Hungarica,***6**,95–125.

Békésy, M. and Garay, A. (1960) *Az anyarozs. Claviceps purpurea (Fr.) Tul. (Ergot.Clavicepspurpurea (Fr.) Tul),*Akadémiai Kiadó, Budapest.

Bernáth, J. (1985) *Speciális növényi anyagok produkció-biológiája (Productionbiology of special materials in plants).*DSc Thesis, Hungarian Academy of Sciences, Budapest.

Blažek, Z. and Starý, F. (1962) Gewichtsvariabilität der Mutterkornsklerotien verschiedener Grasarten. *Pharmazie,***35**,461–465.

Bojor, O. (1968) Über die Auslese einiger Alkaloide inhaltenden Stämme von *Clavicepspurpurea* Tul. aus der wildwachsenden Flora Rumäniens. *Herba Hungarica,***7**, 137–144.

Cauderon, Y., Cauderon, A., Gay, G. and Roussel, J. (1984) Alioplasmic lines and nucleocytoplasmic interactions in triticale. *Proc. of the 3rd Eucarpia meeting of theCereal Section on Triticale,*July, 1984.

Frey, H.P. and Brack, H. (1968) Mutterkorn-Stämme (*Claviceps purpurea* Tul.) mit Nadelförmigen Sklerotien auf Roggen. *Herba Hungarica,***7**,149–153.

Caspar, I., Ignatescu, I., Madej, L. and Raibuh, A. (1990) Possibilities of using male-sterile forms of rye in producing sclerotia of *Claviceps purpurea* (Fr.) Tul.*Ana. Inst. Cerc. Planta Techn.,***58**,41–59.

Geiger, H.H. and Bausback, G.A. (1979) Untersuchungen über die Eignung pollensterilen Roggens zur parasitischen Mutterkornerzeugung. *Z. Pflanzenzücbtg.,***83**,163–175.

Gláz, T. (1955) Researches about the viability and preservation of the conidia of *Claviceps purpurea* Tul. grown in submerged culture. *Acta Microbiol. Acad. Set.Hung.,***2**,316–325.

Golenia, A., Pawelczyk, E., Speichert, H. and Tyborska, B. (1961) Rezultaty porównawczej hodowli roznych ras grzyba *Claviceps purpurea* w ciagu kilku lat. *Biol. Inst.Roslin Leczn.,***7**,33–38.

Grzybowska, T., Lozykowska, A. and Zalaski, K. (1970) Oddzielanie sporuszu od zyta w roztworze soli kuchennej. *Herba Polonica,***16**,149–154.

Heeger, E.F. (1956) *Handbuch der Arznei- und Gewürzpflanzenbaues.Drogengewinnung,*Deutscher Bauernverlag, Leipzig, p. 85.

Hornok, L. (1990) *Cultivation and processing of medicinal plants,*Akadémiai Kiadó, Budapest, pp. 111–118.

Jurkevich, I.D. and Misenin, I.D. (1976) *Lekarstvennie rastenie i ih primenenie (Medicinal plants and their utilization),*Isd. Nauka, Minsk, pp. 34–36.

Kerekes, J. (1969) *Gyógynövénytermesztés (Medicinal plant production),*Mezôgaz-dasági Kiadó, Budapest, pp. 99–108.

Kiniczky, M. (1992) Az anyarozs (*Claviceps purpurea* Fr. Tul.) produkciójának optimalizálása hímsteril gazdanövényen (Optimalization of the production of ergot on male sterile host). PhD Thesis, Research Institute of Medicinal Plants, Budakalász.

Kiniczky, M. and Nagy, J. (1976) Eljárás anyarozs termesztésére, valamint fertôzôgép az eljárás foganatosítására (Method for ergot production and infection mashine for its implement). *Hungarian Pat.,*175.469.

Kiniczky, M., Tétényi, P., Kiss, I., Zámbó, I. and Balassa, J. (1982) Eljárás anyarozs spóra és élesztô tartósítására (Method for conservation of ergot spores and yeast). *Hungarian Pat.,*181.706.

Kiniczky, M., Tétényi, P., Vásárhelyi Gy., Baranyi, I., Bernáth, J., Lôrincz Cs., Szarvady, B. and Kováts, L. (1989) Eljárás anyarozs termesztésére (Method for ergot production). *Hungarian Pat.,*198.366.

Kubec, K., Culik, K., Hilbert, O., Kybal, J. and Severa, Z. (1974) *Czechosl. Pat.*155.440

Kumar, N., Khader, J.B.M.M.A., Randgaswami, P., Irulappan, I., Abdul-Khader, J.B.M.M. and Khader, A (1993) *Introduction to species, plantation crops, medicinal andaromatic plants.*Rajalakshimi Publ., Tamil Nadu, p. 278.

Kybal, J. and Strnadová, K. (1968) Den Ertrag von Mutterkorn bei feldmässiger Kultur bestimmende Faktoren. *Herba Hungarica,*7,123–135.

Large, E.G. (1954) Growth stages in cereals. Illustration of the Feekes scale. *PlantPatbology,*4,128–129.

Lewis, R.W. (1945) The field inoculation of rye with *Claviceps purpurea. Phytopathology,*35,353–360.

Lewis, R.W. (1968) Ergot, parasitism and differentiation. *Herba Hungartea,*7,49–51.

McLaren, N.W. and Wehner, F.C. (1992) Pre-flowering low temperature predisposition of sorghum to sugary disease *(Claviceps africana). J. Phytopathol.,*135,328–334.

Mielke, H. (1993) Untersuchungen zur Bekämpfung des Mutterkorns. *Nachrichtenblatt des Deutschen Pflanzenschutzdienstes,*45,97–102.

Musabyimana, T., Sehene, C. and Bandyopadhyay, R. (1995) Ergot resistance in sorghum in relation to flowering, inoculation technique and disease development. *Plant Pathl.,* 44,109–115.

Németh, É. (1993) *Clavicepspurpurea.* In Bernáth, J. (ed.) *Vadon termô és termesztettgyógynövények (Wild growing and cultivated medicinal plants),*Mezôgazda Kiadó, Budapest, pp. 197–204.

Naylor, R.E.L. and Munro, L.M. (1992) Effects of nitrogen and fungicide application on the incidence of ergot *(Claviceps purpurea)* in triticale. *Tests of Agrochem. Cultiv.,*13, 28–29.

Pageau, D. and Wauthy, J.M. (1995) Seeding date and seeding rate effects on ergot development on barley. *Can. J. Plant Sci.,*75,511–513.

Rai, K.N. and Thakur, R.P. (1995) Ergot reaction of pearl millet hybrids affected by fertility restoration and genetic resistence of parental lines. *Euphytica,*83,225–231.

Ruminska, A. (1973) *Rosliny lecznicze (Medicinal plants),*Panstwowe Wydaw. Naukowe, Warszawa, pp. 364–376.

Ruokola, A. (1961) Possibilities of ergot production in Finland. *Maatalous jaKoetoimiuta (Agricultural Experiments),*14,248–259.

Ruokola, A. (1972) Breeding of ergot in Finland. *Ann. Agricult. Fenniae,*11,361–370.

Scoles, G.J. and Evans, L.E. (1979) The effect of temperatutes on pollen fertility and anther dehiscence of cytoplasmic male sterile rye. *Can. J. Plant Sci.,*59,627–633.

Singh, H.P., Singh, H.P. and Singh, K.P. (1992) Varietal susceptibility of rye *(Secalecereale)* for ergot *(Claviceps purpurea* Tul.) infection. *New Botanist,*19, 225–227.

Snejd, J. (1974) Perennierender Kulturroggen. III. Neue Möglichkeiten für die Züchtung und Erzeugung des Mutterkornpilzes (*Claviceps purpurea* Tul.). *Z. Pflanzenzüchtg.,***72,** 346–351.

Stadler, P.A. (1982) Neue Ergebnisse der Mutterkornalkaloid-Forschung (New results of ergot alkaloid research). *Planta Med.,***46,**131–144.

Ubrizsy, G. (1965) *Növénykórtan II (Plant pathology),*Akadémiai Kiadó, Budapest, p. 223.

Vásárhelyi, Gy., Tétényi, P.-né, Tétényi, P. and Lassányi, Zs. (1980) Eljárás anyarozs gazdanövény vegyszeres kezelésére (Method for chemical treatment of ergot host plant). *Hungarian Pat.,*175–924.

Watkins, J.E. and Littlefield, L.J. (1976) Relationship of anthesis in waldron wheat to infection by *Claviceps purpurea. Trans. Br. Mycol. Soc.,***66,**362–363.

12.
SAPROPHYTIC CULTIVATION OF
CLAVICEPS

ZDENĚK MALINKA

Galena a.s., 747 70 Opava 9, Czech Republic

12.1.
INTRODUCTION

12.1.1.
History

Biological, non-parasitic production of ergot alkaloids is carried out by saprophytic cultivation of production strains of different species of the genus *Claviceps*. The saprophytic cultivations of *Claviceps* spp. were experimentally performed as early as in the last century (Bové, 1970). Mycelial saprophytic cultures in nutrient media were reported since the 1920s (Bonns, 1922; McCrea, 1931; Schweizer, 1941; De Tempe, 1945). These experiments provided the basis of cultivation of the fungi *Claviceps* under artificial nutritional conditions but did not yet serve for alkaloid production or were not reproducible (McCrea, 1933).

Successful work oriented at the directed use of the saprophytic cultivation for alkaloid manufacture depended on the isolation of clavine alkaloids from saprophytic cultures (Abe, 1951; Abe *et al.,* 1951, 1952, 1953). The processes developed in a number of laboratories aimed at the industrial production of therapeutically applicable alkaloids or their precursors (Stoll *et al.,* 1953; Stoll *et al.,* 1954a; Rochelmeyer, 1959; Rutschman and Kobel, 1963a, b; Rutschman *et al.,* 1963). (For the history of *Claviceps* fermentation see Chapter 1 of this book.)

Fermentation makes possible to produce ergopeptines, paspalic acid, simple derivatives of lysergic acid and clavine alkaloids. Ergopeptines can be used for therapeutical purposes directly or after semisynthetic modification. Simple derivatives of lysergic acid, paspalic acid as well as clavines serve as a basal structure for the subsequent semisynthetic production of pharmaceutically utilizable alkaloids. From the simple derivatives of lysergic acid only ergometrine is used in therapy. (For details see Chapter 13.)

12.2.
PRODUCTION MICROORGANISMS

12.2.1.
Sources

Fermentation production of ergot alkaloids based on saprophytic cultivation of production strains selected from different species of the genus *Claviceps* represents the most important way of biological production of the alkaloids. However, also other filamentous fungi, able to produce ergot alkaloids, can serve as a source of production strains. (For more details see Chapter 18 of this book.) The patent literature mentiones, besides *Claviceps,* only fungi of genera *Aspergillus* (Siegle and Brunner, 1963), *Hypomyces* (Yamatoya and Yamamoto, 1983) and *Penicillium* (Kozlovsky *et al.,* 1979). In the *Claviceps* fungi, selection of strains has been described for the species *C. purpurea,C. paspali* and *C. fusiformis.* The use of tissue cultures of plants of the *Convolvulaceae* family represents so far only theoretical possibility.

Saprophytic cultures can easily be obtained from sclerotia of the appropriate species of *Claviceps.* After pre-soaking with ethanol or propanol, the sclerotium surface is sterilized by a suitable agent—resorcinol, mercury dichloride or Lugol solution (Desai *et al.,* 1982b; Mantle, 1969; Strnadová *et al.,* 1986). After washing under sterile conditions, the plectenchymatic tissue of the sclerotium is cut and the slices are transferred on the surface of an agar growth medium. Another method of inoculum preparation by mechanical decomposition of a sclerotium was described by Řičicová and Řeháček (1968). Preparation of a saprophytic culture from the honeydew of a host plant invaded by *Claviceps* sp. was also described (Janardhan and Husain, 1984). A surface saprophytic mycelium starts growing on the medium and different asexual spores are generated on the hyphal tips. They are mostly classified as conidia. These spores are used for the further transfer and culture propagation, monosporic isolation (Kybal *et al.,* 1956; Nečásek, 1954) and the following stabilization of the culture.

Culture isolation from ascospores is another method. A fungal sclerotium, after a cold storage period, forms under suitable conditions fruiting bodies. After ripening they release sexual ascospores, which germinate on the surface of an agar medium and form the saprophytic mycelium (Vásárhelyi *et al.,* 1980b). Monosporic isolation can be performed both directly with the ascospores or with asexual spores formed during the further saprophytic cultiva tion of a mycelium grown up from an ascospore.

12.2.2.
Breeding and Selection of Production Strains

Classical methods of selection pressure, mutagenesis and recombination or their mutual combinations, can be applied to breeding production strains for fermentative alkaloid production. With the *Claviceps* fungi these methods are to a certain extent complicated by an incomplete information about the cell nucleus for a number of potential sources. The production strains are often highly heterogeneous and include both heterokaryotic and homokaryotic ones (Amici *et al.*, 1967c; Didek-Brumec *et al.*, 1991a; Mantle and Nisbet, 1976; Olasz *et al.*, 1982; Spalla *et al.*, 1969; Strnadová and Kybal, 1974). (See also Chapter 5 on the genetics of *Claviceps*.)

Nutrient components are mostly used for selection pressure. Such principle of selection in growth media, simulating the composition of phloem juice of a host plant, is described by Strnadova *et al.* (1986). Another example is, *e.g.*, acquisition of new strains with modified production qualities through regeneration of protoplasts (Schumann *et al.*, 1982, 1987).

The principles of mutagenesis of ergot alkaloid producers are the same as those used in bacteria and fungi. According to the literature, physical mutagenes are often used as mutational agents—UV light (Nordmann and Bärwald, 1981; Strnadová, 1964a, b), X-rays and gamma irradiation (Zalai *et al.*, 1990). Chemical agents include derivatives of N-nitrosoguanidine and nitrosourea, ethyl methane sulfonate (Keller, 1983) or their combinations (Řeháček *et al.*, 1978a), and nitrous acid (Strnadová and Kybal, 1976).

The simplest way is to expose to the chemical or physical mutagens a suspension of fungal spores. This technique facilitates the subsequent simple monosporic isolation, cultivation and selection of isolates originating from a single cell. Problems arise when asporogenic fungal strains are to be selected. In this case a suspension of hyphae or hyphal fragments can be directly exposed to a mutagenic agent, but it brings difficulties with culture heterogeneity in the subsequent transfer and selection. In this case it is advantageous to perform the mutagenesis on protoplasts (Křen *et al.*, 1988c). Protoplasts can be also prepared from spores of sporulating strains. When protoplasts are used, the mutation frequency is much higher (Baumert *et al.*, 1979b; Keller, 1983; Olasz *et al.*, 1982; Zalai *et al.*, 1990).

Mutants were also prepared with the ergot alkaloid biosynthetic pathway blocked on different levels (Maier *et al.*, 1980a; Pertot *et al.*, 1990). When supplemented by a modified precursor these strains can be employed for effective mutational biosynthesis (Erge *et al.*, 1981; Maier *et al.*, 1980b).

With *Claviceps* spp., breeding using DNA recombination can be done in two basic ways—meiotic recombination and fusion of protoplasts. In former method a corresponding strain is cultivated parasitically to form a sclerotium which, after its germination, then serves as the source of sexual ascospores (Tudzynski *et al.*, 1982; Vásárhelyi *et al.*, 1980b). Protoplast fusion methods have been therefore elaborated for the common species *C. purpurea,C. paspali* and *C. fusiformis.* A problem of genetic markers had to be solved since the markers of auxotrophy or resistance against fungicides in most cases negatively influence the alkaloid production level of progeny strains (DidekBrumec *et al.,* 1991b). To eliminate these disadvantages, methods were developed resulting in nearly 30% increase of production compared to parent strains (Didek-Brumec *et al.,* 1992, 1993). Interspecies hybrids that have been prepared by fusion of protoplasts from *C. purpurea+C. paspali* (Spalla and Marnati, 1981) and *C. purpurea+C. fusiformis* (Nagy *et al.,* 1994) represent further possibilities of selection of production strains. A question remains to what extent these hybrids will be stable during manifold transfers.

Application of the above methods is followed by selection of isolates having higher production capability or altered in some other way. The testing of all isolates in submerged cultivation on a shaker or by stationary surface cultivation imposes a high material requirements. Correlations were therefore studied between morfological and physiological characteristics and alkaloid production (Srikrai and Robbers, 1979). Selection methods based on pigmentation (Borowski *et al.,* 1976; Kobel and Sanglier, 1976; Molnár *et al.,* 1964; Udvardy Nagy, E. *et al.,* 1964; Wack *et al.,* 1973), specific colour reaction (Zalai *et al.,* 1990), fluorescence (Gaberc-Porekar *et al.,* 1981, 1983), enzyme activity profiles (Schmauder and Gröger, 1983) or antimicrobial effects of clavine alkaloids (Homolka *et al.,* 1985) were worked out.

In the following steps an intimate selection can be performed on shakers or in laboratory fermentors for submerged or stationary cultivation. For the most promising isolates these works are organically interconnected with optimization of a medium and production conditions.

Developments of molecular biology and genetics of the genus *Claviceps* open new perspectives in obtaining of suitable production organisms (for more details see Chapter 4 in this book).

12.2.3.
Maintenance Improvement

Selected high-yielding strains of *Claviceps* spp., similarly as those of other microorganisms, degenerate (Kobel, 1969). Also, problems of transfer of original cultures in a fermentation technological process are connected with this fact. Different producers of clavine alkaloids born to be transferred 6–9 times without decrease of producing capability (Malinka *et al.,* 1988).

It is necessary to maintain the optimal qualities of a selected production strain by two parallel ways—conservation and dynamic ones. The conservation way consists in keeping of stock cultures of the production strain under conditions of maximal possible elimination of biological effects given by transfer of cultures, ageing and other external influences. The dynamic way comprises systematically performed selection in the frame of the maintenance improvement, which consists in continual testing of monosporic isolates (in sporogenic lines) or at least hyphal isolates (in asporogenic lines) made from stock cultures of the production strain, and positive choice of a culture with optimal producing qualities for the subsequent work. This activity can be a part of optimization of other factors having an influence on a level and parameters of production of the final product. Here, there is also possible to apply the before-cited procedures of rational selection and apply selection pressure methods.

12.2.4.
Long-Term Preservation of Production Strains

For long-term preservation of the production strains of *Claviceps* common methods used for other fungi can be applied as reviewed *e.g.* by Kirsop and Snell (1984) and by Hunter-Cervera and Belt (1996). Besides preservation of sporulated cultures on rye grains placed in a refrigerator or a deep-freezer it is also possible to keep frozen dried suspensions or gelatine disks. First of all non-sporulating strains, being more sensitive to different conservation procedures, can be preserved as cultures on agar plates under a mineral oil or, for a single use, as a suspension of mycelium at—18°C (Křen *et al.*, 1988c). Keeping of lyofilized cultures and cultures frozen in liquid nitrogen are probably the most universal methods, though technically more complicated. Břemek (1981) compared these methods with different *Claviceps* strains and found that the both are suitable. During lyofilization diverse protective media are applied, *e.g.* serum, milk, peptone, sugars, sodium glutamate or combinations of previous (Chomátová *et al.*, 1985; Ustyuzhanina *et al.*, 1991). Procedures of lyofilization of non-sporulating strains are described by Křen *et al.* (1988c) and Pertot *et al.* (1977). A modification of the lyofilization process for preservation of cultures from regenerated protoplast was worked out by Baumert *et al.* (1979b).

As a theoretical alternative seems a method according that strains, which produce ergot alkaloids saprophytically, are preserved as sclerotia formed on an infected, proper host plant, *e.g.* rye for strains of the species *Clavicepspurpurea*. Viability of the sclerotia when stored in refrigerator is several years. Questions of contingent changes of strain production characteristics due to alternation of saprophytic and parasitic phases were treated by Breuel and Braun (1981), and Breuel *et al.* (1982). During surface stationary production of peptide alkaloids it was possible to keep production strain in the form of dried mycelium at 4°C for 3 years without any influence to production capability (Kybal, Malinka, unpublished results).

As a source of production strains serve internationally established culture collections (*e.g.* ATCC, CBS, CCM, NRRL) or collections in certain institutes (*e.g.* MZKIBK—Cimerman *et al.*, 1992). However, industrially usable strains are mostly patented; in the collections they are stored according to the Budapest Convention and not commonly accessible.

12.3.
FERMENTATION TECHNOLOGY

All species of ergot alkaloid producers from the *Pyrenomycetes* class as well as an overwhelming majority of other fungal producers are parasites of different plants and fungi. Principle of saprophytic cultivation is growth of a production fungi on a synthetic medium. The saprophytic cultivation makes possible better optimization of a production level, elimination of biosynthesis of accompanying undesirable matters and regulation of ergot alkaloids production through rational outside interventions. On the other side it is much more exacting on the technological equipment. All industrially adopted processes have the same basic aim—the maximal production of a matter with the minimum of undesirable compounds, got in the shortest time with minimized costs of medium, equipment and labour.

For ergot alkaloid manufacture different fermentation technologies can be employed. In principle, they can be produced by (i) stationary cultivation, when microorganisms are growing on the surface of a cultivation medium, both liquid (Abe, 1951; Kybal and Vlček, 1976; Malinka, 1988) and solid (Trejo-Hernández *et al.*, 1992; Trejo-Hernández and Lonsane, 1993), or (ii) submerged cultivation with agitation of a suspension of microorganisms (Abe *et al.*, 1951; Amici *et al.*, 1966; Arcamone *et al.*, 1960; Bianchi *et al.*, 1976; Kobel and Sanglier, 1986). Semicontinuous and continuous cultivations (Kopp and Rehm, 1984; Křen *et al.*, 1986b) as well as those using immobilized microorganisms (Komel *et al.*, 1985; Kopp and Rehm, 1983; Křen *et al.*, 1989a) represent specific modifications of the submerged cultivation. General reviews of biosynthesis and production of ergot alkaloids were published by a number of authors (Esser and Düvell, 1984; Kobel and Sanglier, 1986; Křen *et al.*, 1994; Mantle, 1975; Řeháček, 1983a, b; 1984, 1991; Řeháček and Sajdl, 1990; Robbers, 1984; Sočič and Gaberc-Porekar, 1992; Udvardy Nagy, 1980). (For special cultivation procedures see the Chapter 7.)

In all types of fermentation it is necessary to use optimized media. Generally, the cultivation media should fulfill the same requirements as those for saprophytic cultivation of other fungi, *i.e.* they have to contain sources of energy, carbon, nitrogen, phosphorus and with advantage also certain trace elements and some complex matters. Price of the medium should always be taken into account and optimal variants be chosen from the point of view of costs per an unit operation. (For media components and physiology of production see Chapter 6.)

During the production phase, according to a kind of fermentation and elaboration of a given fermentation process, it is desirable to follow utilization of individual nutrients, activities of particular enzymes and a course of the proper synthesis of the alkaloids. It is also necessary to control and regulate basic physical and physico-chemical parameters of the culture—pH, dissolved oxygen concentration, dissolved carbon dioxide concentration, concentration of carbon dioxide in outlet, temperature, pressure, agitator speed or other physical characteristics typical for a given kind of fermentation and a type of fermentor used. During the pre-inoculation and inoculation phases, depending on requirements of their optimal course, only some fermentation parameters should be controlled and measured. Values of these parameters and their course during cultivation cannot be generalized for they are very often specific for the each production strain and the sort of the end product. To industrial production of the ergot alkaloids there are related pertinent regulations and requirements of the state and international institutions and offices (Priesmeyer, 1997). Basic pharmacopoieal demands on fermentation processes are presented by Anonymous (1997).

Specific problems of fermentation production of the ergot alkaloids consist in microbial contamination. Compared to an overwhelming majority of other secondary metabolites production processes, fermentation of the ergot alkaloids is marked by two negative factors—slow growth of mycelium on rich media and none or weak antibiotic activity of the produced alkaloids. In spite of reported antibacterial effect of clavine alkaloids (Eich and Eichberg, 1982; Eich et al., 1995) probability of their pronounced exercise in autoprotection against contamination is only small. For these reasons an effect of broad spectrum antibiotics on the production strain of C. fusiformis W1 was investigated (Břemek et al., 1986b; Křen et al., 1986a); chloramphenicol was found as the most suitable antibiotic. Its use can be advantageous also in semicontinuous and continuous processes (Křen et al., 1985; Křen et al., 1986b). Kopp (1987) having been working with immobilized cells of Claviceps used streptomycin. The positive effect of streptomycin and the negative one of oxytetracycline and nystatin on production of alkaloids by not closer specified strain of Claviceps described Slokoska et al. (1992).

12.3.1.
Stationary Surface Cultivation

The stationary cultivation is commonly used for stock and starting cultures (growth on the surface of an agar medium) without respect what a kind of cultivation process will be used in the production phase. On the production scale there are in particular described processes where fungal mycelium capable to produce ergot alkaloids was growing on the surface of a liquid medium (Adams, 1962; Kobel et al., 1962; Kybal et al., 1960; Molnár et al., 1964; Rochelmeyer, 1965; Stoll et al., 1953; Strnadová et al., 1981, 1986). Necessity of gaining the surface as large as possible under aseptic condition and difficulties of automation represent the main problems of that kind of cultivation. An equipment for the stationary cultivation representing a simple stationary fermentors (Figure 1) was developed (Vlček and Kybal, 1974; Kybal and Vlček, 1976) with plastic bags filled by the inoculated medium. The bags are manufactured by cross welding of a sterile polyethylene hose of a proper width. After that the bags are filled up by

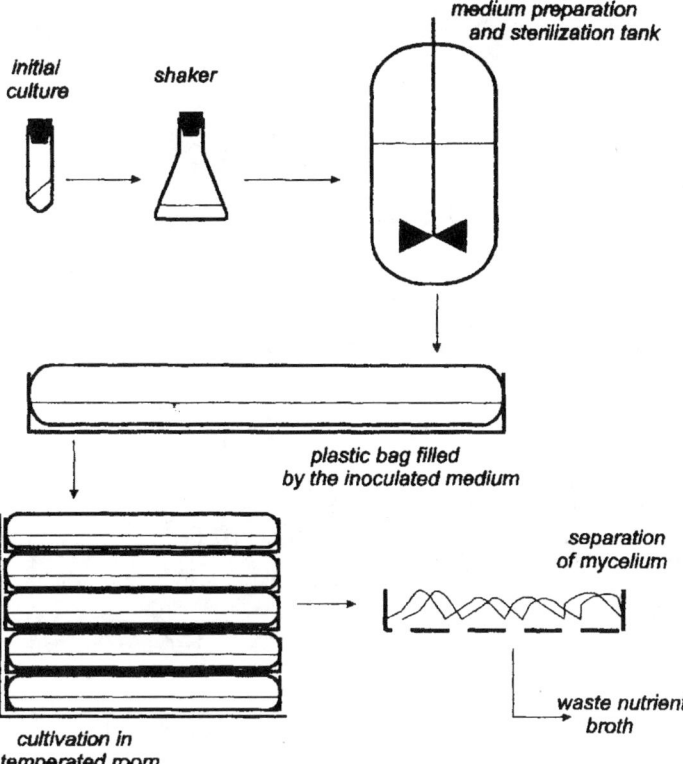

Figure 1 Flowsheet of stationary surface cultivation in plastic bags

the inoculated medium and equipped by manifolds for controlled aeration (Malinka, 1982). During the following stationary fermentation, which is performed in a tempered room, the mycelium is growing on the surface of the medium. The water soluble alkaloids are excreted into the medium whilst the hydrophobic ones remain in the mycelium as intracellular metabolites. Manipulation with the filled cultivation bags can be performed by a high lift truck, commonly used in stores (Figure 2). Large cultivation area is an advantage of the plastic bag cultivation while the fact that each bag during long-term cultivation behaves as a separated fermentor made problems with product standardization. Even if manipulation with the bags is mechanized the load of workers during bag filling and harvesting of a mycelium produced is increased. More exacting cleaning of the used polyethylene foil prior to recycling represents a non-negligible aspect, as well. Into the industrial scale this method was introduced for production of ergocornine, and α- and β-ergokryptine.

The stationary surface cultivation in the plastic bags can also be adopted for production of physiologically active asexual spores of production strains of

Figure 2 Stationary cultivation equipment

Claviceps purpurea (Fr.) Tul., which are used as an infection agent at field parasitic cultivation of ergot (Harazim *et al.,* 1984; Valík and Malinka, 1992). This methods is employed for cyclosporin A production (Mat'ha, 1993; Mat'ha *et al.,* 1993), for cultivation of entomopathogenic fungi with the aim of production of spores used in manufacturing of bioinsecticides, and for cultivation of the mould *Trichoderma harzianum* producing mycofungicide (Kybal and Nesrsta, 1994; Nesrsta, 1989).

Kybal and Strnadová (1982) described also other, technically more complicated equipments for stationary surface cultivation to prepare inoculum for field parasitic cultivation of ergot.

12.3.2.
Submerged Cultivation

The submerged cultivation is used for manufacturing of different microbial products and of ergot alkaloids as well. By the means of laboratory scale submerged cultivation most of knowledge of biogenesis of the ergot alkaloids and physiology of their producers was gained. Contemporary expertise makes possible to control effectively individual production stages, influence biosynthesis of alkaloids and to a considerable extent eliminate unfavorable factors typical for selected high producing strains, such as *e.g.* loss of sporulating

Figure 3 Cultivation of shaker culture

capability, production of glucans and minimal adaptation to variable cultivation conditions.

A basis of the submerged fermentation on laboratory scale as well as the primary step in an overwhelming majority of industrial scale processes is a shaker culture (Figure 3). In industrial scale production the aim of this cultivation is to obtain a sufficient amount of inoculum for the next cultivation step. Medium composition is subordinated to the aim of reaching fast germination of spores of production microorganism and fast growth of mycelial hyphae, or, as the case may be in specific processes, fast sporulation, and obtaining a mixture of hyphae and asexual spores.

For inoculation there is usually used a suspension of spores and/or hyphae of the aerial mycelium from the surface of primal cultures growing on agar solid media, or hyphal fragments when non-sporulating strains are worked with. Lyofilized cultures or microorganisms kept in liquid nitrogen can be also used as an inoculating material.

Preparation of the shaker cultures is usually made on rotary shakers. If for some production strains less mechanical stress of hyphae is more suitable, reciprocal shakers can be used. The shaker culture can be replaced by a culture from a laboratory fermentor.

The following steps are always run in fermentors. Volume of the end production step is decisive for the number of previous cultivation steps for

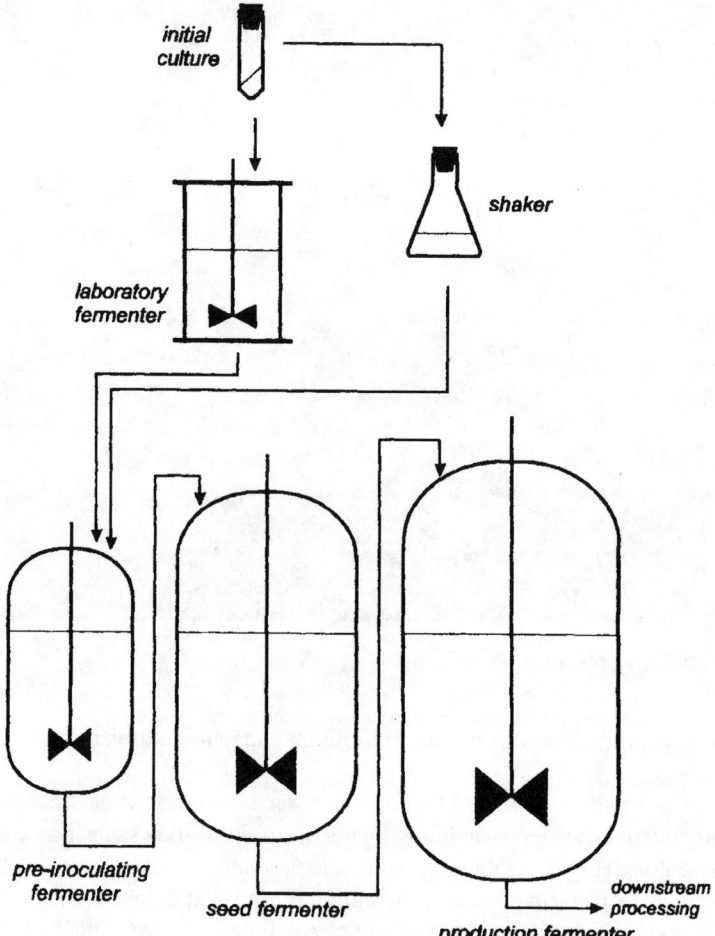

initial culture

shaker

laboratory fermenter

pre-inoculating fermenter

seed fermenter

production fermenter

downstream processing

Figure 4 Flowsheet of submerged cultivation

propagation of necessary amount of inoculum. During the inoculum preparation it can be also advantageously manipulated to evoke an optimal state of the culture for biosynthesis of ergot alkaloids in the production step (Sočič *et al.,* 1985, 1986). Most often there are three cultivation steps, *viz.* cultivation in pre-inoculating tanks, seed tanks and production fermentors (Figure 4). Increasing the number of the cultivation steps is usually undesirable for with increasing transfers production capability of the culture is diminished. This fact limits also the possibility of recirculation of a part of the cultivation medium from the production step for inoculation of the following cultivation.

Similarly as at other filamentous fungi, during *Claviceps* cultivation fermentors have to be used enabling to work with viscous media. Non-newtonian character of liquid flow becomes obvious only during the course of cultivation, on the one hand due to growth of hyphal filaments, on the other hand because of production of glucans.

12.3.3.
Alternative Fermentation Processes

Ergot alkaloids can be manufactured also by alternative fermentation processes, *e.g.* by those using nontraditional substrates or immobilized cells of *Claviceps* spp. or their subunits. Semicontinuous or continuous cultivations represent other alternatives of the saprophytic cultivation.

Stationary solid state cultivation on the surface of solid substrates soaked by a liquid medium has been reported by Trejo Hernández *et al.* (1992) and Trejo Hernández and Lonsane (1993). In these studies, growth and production of alkaloids were investigated with different species of the genus *Claviceps* on sugar-cane pith bagasse and significant dependence of the production level and the spectrum of synthesized alkaloids on composition of the medium used for solid substrate impregnation was found. However, there is discussible a possibility of application of this process on the industrial scale.

Other alternative types of cultivation are described in Chapter 6.

12.4.
MANUFACTURE OF CLAVINE ALKALOIDS

After transfer to a saprophytic culture a number of parasitic strains of ergot, both wild or improved, is able to synthesize only clavine alkaloids and lose the ability to perform subsequent biosynthetic steps. For this reason a vast number of production strains exists, most frequently of the species *Clavicepspurpurea, e.g.* CP 7/274 CCM F-632 (Řeháček *et al.*, 1978b), 88-EP/1988 (Křen *et al.*, 1988c), IBP 182 ZIMET 43673 (Schumann *et al.*, 1984), IBP 180 ZIMET PA 138 (Baumert *et al.*, 1979a), Pepty 695/e (Erge *et al.*, 1984), Pepty 695/ch-I (Gröger *et al.*, 1991; Maier *et al.*, 1988a,b), 59 CC 5/86 (Řeháček *et al.*, 1986a), SL 096 CCM F-733 (Flieger *et al.*, 1989b), EK 10 (Pažoutová *et al.*, 1990), AA218 (Harris and Horwell, 1992), CBS 164.59 (Kopp, 1987) and *C. fusiformis, e.g.* W1 (Křen *et al.*, 1985), F 27 (Křen *et al.*, 1985), MNG 00211 (Trinn *et al.*, 1983), NCAIM 001107 (Trinn *et al.*, 1990), CF 13 (Rozman *et al.*, 1985, 1987).

The species *Claviceps paspali* has been reported only sporadically, *e.g.* strain DSM 2838, a producer of festuclavine (Wilke and Weber, 1985a) and strains Li 342 (ATCC 34500) (Erge *et al.*, 1972) and Li 342/SE 60 (Gröger, 1965), producers of chanoclavine-I; the same holds for not closely identified strains of the genus *Claviceps, e.g.* DSM 2837, which produces chanoclavine (Wilke and Weber, 1985b), IBFM-F-401, a producer of elymoclavine (Kozlovsky *et al.*, 1978), and the strains 47A and 231, from whose cultures norsetoclavine was isolated for the first time (Ramstad *et al.*, 1967).

The patent literature reports, besides the genus *Claviceps,* also strains *Hypomyces aurantus* IFO 773, which produce ergocornine, agroclavine, elymoclavine and chanoclavine (Yamatoya and Yamamoto, 1983), and *Penicillium corylophillum* IBFM-F-152 (Kozlovsky *et al.,* 1979), which produces epoxyagroclavine I.

A number of clavine alkaloids, in addition to the mentioned fungal species, was isolated also from *Claviceps gigantea, Claviceps* spp. originating from different host plants, fungi of the *Penicillium, Aspergillus, Rhizopus* and other genera, and from seeds of plants of the *Convolvulaceae* family. A review was compiled by Flieger *et al.* (1997). Apart from the genus *Claviceps,* nothing is known about other fungal species employed in the selection of industrially applicable production strains, except for the two above cases. It is, however, possible that some isolates of *Claviceps* sp., mentioned in connection with the isolation of certain clavine alkaloids, served as a starting material in the selection of production strains (Stoll *et al.,* 1954b).

12.4.1.
Production of $\Delta^{8,9}$-Ergolenes

Industrial processes for manufacture of agroclavine and elymoclavine are best elaborated among the methods for acquiring of clavine alkaloids. Due to the direct biosynthetic succession of these alkaloids they are usually produced in mixtures (Adams, 1962; Baumert *et al.,* 1979b; Břemek *et al.,* 1986c, 1989; Erge *et al.,* 1984; Řeháček *et al.,* 1978a, b, c; 1984a; 1986b; Řeháček and Rylko, 1985; Takeda Pharm. Ind. 1956; Trinn *et al.,* 1983; Wack *et al.,* 1966; Windisch and Bronn, 1960; Yamatoya and Yamamoto, 1983).

As a substrate for subsequent chemical operations elymoclavine is superior to agroclavine and efforts were therefore made to develop processes leading to elymoclavine production with maximal possible elimination of agroclavine. Generally, this goal can be achieved by selecting proper production strains and by optimizing cultivation conditions. The direct biosynthetic succession of the two alkaloids makes it possible to use also bioconversion processes.

Production strains for elymoclavine manufacture and general cultivation conditions have been described in a number of patents (Kozlovsky *et al.,* 1978; Křen *et al.,* 1985, 1988c; Řeháček *et al.,* 1984b; Schumann *et al.,* 1984; Trinn *et al.,* 1983, 1990). In addition to the patent literature many works refer to results of investigation of individual aspects of physiology and biochemistry of clavine alkaloid synthesis. These works were done either directly with production strains, *e.g. C. purpurea* 129 (later classified as *C. fusiformis*) producing 4500–7000 mg L^{-1} and isolates selected from it (Desai and Řeháček, 1982; Křen and Řeháček, 1984; Křen *et al.,* 1984, 1987; Pažoutová *et al.,* 1977, 1980, 1981; Pažoutová and Řeháček, 1978, 1981a, b, 1984; Řeháček *et al.,* 1977; Sajdl *et al.,* 1978; Voříšek *et al.,* 1981) or with strains of different provenance, *e.g. Clavices* sp. SD-58 (ATCC 26019) (later classified as *C. fusiformis;* Desai *et al.,* 1982a, 1983, 1986;

Eich and Sieben., 1985; Kozikowski *et al.*, 1993; Křen *et al.*, 1987; Otsuka *et al.*, 1980; Patel and Desai, 1985; Robbers *et al.*, 1972, 1978, 1982; Robertson *et al.*, 1973; *Rylkoetal.*, 1986, 1988a; Schmauder *et al.*, 1981a, b, 1986; Vaidya and Desai, 1981a, b, 1982, 1983a, b), *C. purpurea* 59 *(C. fusiformis)* (Pažoutová *et al.* 1986, 1987a, b, 1988, 1989, 1990; Pažoutová and Sajdl, 1988; Sajdl *et al.*, 1988b), *Claviceps* sp. CP II (Krustev *et al.*, 1984; Slokoska *et al.*, 1981, 1985, 1988) and *Claviceps* sp. PRL 1980, ATCC 26245 (Kim *et al.*, 1981; Taber, 1964).

Bioconversion of agroclavine to elymoclavine can be done by both free and immobilized cells of suitable production strains (Břemek *et al.*, 1986a; Křen *et al.*, 1989a) with efficiency of up to 97% (Malinka and Břemek, 1989). In addition to strains producing clavine alkaloids, also those synthesizing simple derivatives of lysergic acid can be used for the conversion. In this case a preferential bioconversion of agroclavine to elymoclavine can be brought about by a simple modification of cultivation conditions (Flieger *et al.*, 1989a; Harazim *et al.*, 1989). Flieger *et al.*, 1989b described also a process of purification of clavine alkaloids combined with a conversion to elymoclavine and lysergic acid α-hydroxyethylamide; these products can be easily separated and used for semisynthesis. Other strains able to convert agroclavine to elymoclavine are *C. fusiformis* SD-58, and *Claviceps* sp. KK-2, Se-134 and 47A (Sieben *et al.*, 1984). An exhaustive review on the bioconversion of ergot alkaloids was worked out by Křen (1991) and a review can be also found in Chapter 10.

During the cultivation of commonly used strains, growth of a culture and production of clavine alkaloids are accompanied by the concurrent biosynthesis of glucans. These compounds unfavourably influence medium rheology, complicate proper mixing and aeration, slow down oxygen transfer and make the cultivation medium foam. Processes elaborated to eliminate glucan production employ a special composition of an inoculation medium, two-stage preparation of inoculum and a special composition of a production medium (Břemek *et al.*, 1986c). In this way, physiological conditions are reached which decrease or eliminate the synthesis of glucans and, at the same time, have a positive effect on alkaloid biosynthesis. Production of up to 4600 mg L^{-1} of total alkaloids (out of which 2300 mg L^{-1} is due to elymoclavine) in shaker cultures has been reported; in fermentors the production reaches 2836 mg L^{-1} (2322 mg L^{-1} of elymoclavine). The concentration of undesirable of glucans can be decreased by the addition of 0.4–0.5 g L^{-1} of sodium phenobarbitale from the original 38.4–42.2 g L^{-1} to 0–11 g L^{-1} (Řeháček and Rylko 1985). An addition of barbiturates into a medium, influencing cytochrome P-450, was also described by Trinn *et al.* (1983) but without relationship to glucan suppression. Processes with feedback inhibition of glucan-synthesizing enzymes induced by addition of glucans into cultivation media during inoculation were proposed (Kybal, personal communication). The use of specific production strains with lowered or eliminated glucan production, such as *Claviceps purpurea* CP 7/5/35 CC-2/1985 (Řeháček *et al.*, 1984a), seems to be economically optimal. This strain produces a mixture of 10–30% of elymoclavine, 65–90% of agroclavine and 1–5% of chanoclavine—I. However, the employment of the strain *Claviceps purpurea* 88-EP/1988 (Křen *et al.*, 1988c) is more advantageous since the strain produces nearly 2500 mg L^{-1} of elymoclavine and this alkaloid represents almost 90% of total alkaloids. The use of two inoculation stages for clavine producers is suitable not only for elimination of glucan production but also for reaching an optimum physiological state for maximal biosynthesis of the alkaloids. The production of alkaloids by the strain *C. fusiformis* W1 is decreased by 11.4–57.8% (Malinka *et al.*, 1986) when a single-stage inoculum is used.

An ihibition effect of phosphate ions upon biosynthesis of alkaloids plays an important role and a positive effect of phosphate deficiency was described already by Windisch and Bronn (1960). The problem consists in the fact that phosphate is necessary for biomass growth; it is therefore necessary to find an optimal ratio between biomass growth (and proliferation of cells able to produce alkaloids) and the alkaloid synthesis rate. Most of the processes described here employ a low content of phosphate in production media combined with the use of a dense inoculum. Some production strains are marked by a higher resistance of alkaloid biosynthesis to phosphates (Řeháček *et al.*, 1984a). This problem was solved in a particular way in the patent of Břemek *et al.* (1989) by using gradually utilized hexaamidotriphosphazene as a phosphate source and at the same time as a supplementary source of nitrogen.

Křen *et al.* (1989b) described a process of production of fructosides of elymoclavine, namely elymoclavine-*O*-β-D-fructofuranoside (Floss *et al.*, 1967) and elymoclavine-*O*-β-D-fructofuranosyl-(2→1)-*O*-β-D-fructofuranoside (Flieger *et al.*, 1989d). The efficiency of glycosylation fluctuated between 10 and 62% . The strain *C. purpurea* 88-EP-47 was selected for the preparation of fructosides of elymoclavine (Křen *et al.*, 1989c); during fermentation this strain produces fructosides in a concentration of 920 mg L^{-1} while the concentration of the total alkaloids reached 2800 mg L^{-1}. Due to the high glycosylation activity the strain could be used for production of fructosides of alkaloids added to cultures (Křen *et al.*, 1989b). Glycosides of ergot alkaloids exhibit interesting physiological effects and can be also used as substrates for the preparation of semisynthetic derivatives.

12.4.2.
Production of 6,7-secoergolenes

Besides agroclavine and elymoclavine, also other clavine alkaloids can be used for the preparation of certain semisynthetic derivatives. Although no derivatives prepared by modification of the clavine molecule are used in therapy, some preparation procedures yielding such clavine alkaloids are protected by patents. Among 6, 7-secoergolenes, *i.e.*, alkaloids with an open ring D of the ergoline structure, chanoclavine-I and chanoclavine-I aldehyde have been patented.

Production processes employing specific production strains are also protected by patents. Thus the production of chanoclavine-I or a mixture of chanoclavine-I and chanoclavine-I aldehyde has been described because all chanoclavine-I aldehyde represents a suitable substrate for subsequent semisynthesis. Wilke and Weber (1985b) described a method of chanoclavine manufacture with the strain *C. purpurea* DSM 2837 giving 390 mg L^{-1} of the alkaloid. Baumert *et al.* (1979a) reported on the use of the strain *C. purpurea* IBP 180, ZIMET PA 138, in which the total production of alkaloids was 500–600 mg L^{-1} and this amount comprised 80% of chanoclavine-I and 20% of chanoclavine-I aldehyde. Maier *et al.* (1980a, b) and Baumert and Gröger (1982) described another strain, denoted Pepty 695/ch, which produced chanoclavine-I and chanoclavine-I aldehyde; these secoergolenes were produced in a concentration of 300–350 mg L^{-1} in a ratio of 3:1 (Erge *et al.*, 1984). A substantially higher production was mentioned by Řeháček *et al.* (1986a) for the strain *C. purpurea* 59 CC5/86 selected from the parent strain *C. purpurea* 129 (Pažoutová *et al.*, 1987a) which produced as much as 3000–6000 g L^{-1} of total alkaloids, composed of 40–60% chanoclavine-I, 20–30% chanoclavine-I aldehyde, 10–15% elymoclavine and 5–10% agroclavine. Besides the patent literature, Gröger (1965) described the strain *C. paspali* Li 342/SE 60 producing 400 mg L^{-1} of alkaloids, 40% of which was chanoclavine. Chanoclavine-I was isolated not only from fungi of the genus *Claviceps* (Abe *et al.*, 1959; Agurell and Ramstad, 1965; Hofmann *et al.*, 1957; Stauffacher and Tscherter, 1964), but also from other fungi—*Penicillium concavo-rugulosum* (Abe *et al.*, 1969), *Aspergillus fumigatus* (Yamano *et al.*, 1962) and *Hypornycesaurantius* (Yamatoya and Yamamoto, 1983).

12.4.3.
Production of Ergolines

Processes for the production of festuclavine and epoxyagroclavine I from the group of ergolines are described in patents. Festuclavine was isolated from cultures of *Aspergillus fumigatus* and from sclerotia of *Claviceps gigantea* (Agurell and Ramstad, 1965). In the patent of Wilke and Weber (1985a) a method of production of festuclavine is described using the production strain *C. paspali* 2338. During a 7–9-day cultivation the concentration of festuclavine reached 2280 mg L^{-1}. Epoxyagroclavine I has so far been found only as a metabolite of *Penicillium corylophilum* (Kozlovsky *et al.*, 1982) and process of its production is patented (Kozlovsky *et al.*, 1979).

12.4.4.
Production of $\Delta^{9,10}$-Ergolenes

Lysergol as well as isolysergol from the group of $\Delta^{9,10}$-ergolenes can be used as suitable substrates for the production of semisynthetic derivatives. Lysergol, together with lysergene and lysergine were isolated from the saprophytic fungi

Claviceps spp. originating from ergot parasitizing on *Elymus mollis* (Abe *et al.,* 1961). Isolysergol was isolated from the saprophytic cultures of *Claviceps* sp. 47 A derived from ergot parasitizing on *Pennisetum typhoideum* (Agurell, 1966).

12.5.
LYSERGIC ACID, ITS SIMPLE DERIVATIVES AND PASPALIC ACID

This group of ergot alkaloids encompasses both compounds directly applicable in therapy (ergometrine) and compounds, which can be employed for the production of semisynthetic alkaloids (lysergic acid, ergine, lysergic acid α-hydroxyethylamide and their isomers, paspalic acid).

Lysergic acid and paspalic acid were isolated in 1964 from cultures of *C.paspali* (Kobel *et al.,* 1964) and in 1966 from cultures of *C. purpurea* (Castagnoli and Mantle, 1966). Ergine (Arcamone *et al.,* 1961; Kobel *et al.,* 1964) and lysergic acid α-hydroxyethylamide (Arcamone *et al.,* 1960; Flieger *et al.,* 1982) were also isolated from the cultures of *C. paspali*. Ergometrine (ergonovine) was isolated from both *C. purpurea* (Stoll, 1952) and *C. paspali* (Kobel *et al.,* 1964).

Analogously to clavine alkaloids, a number of ergolene-production strains has been isolated. Strains of the genus *Claviceps* for direct biosynthesis of lysergic acid, paspalic acid and lysergic acid α-hydroxyethylamide were obviously selected only from the species *Claviceps paspali* that grows on grasses of the genus *Paspalum* in diverse parts of the world, *e.g.* strains *C. paspali* F-140 (ATCC 13895), F-S 13/1 (ATCC 13892), F-237 (ATCC 13893), F-240 (ATCC 13894) (Chain *et al.,* 1960), NRRL 3027, NRRL 3166 (Rutschmann and Kobel, 1963b), NRRL 3080, NRRL 3167 (Kobel and Schreier, 1966; Rutschmann *et al.,* 1963), ATCC 14988 (Tyler, 1963), C-60 and its derivatives (Mary *et al.,* 1965), FA CCM F-731 (Řičicová *et al.,* 1982b), CP 2505, YU 6 (Harazim *et al.,* 1986), CCM 8061 (Flieger *et al.,* 1989a), CCM 8063 (Harazim *et al.,* 1989), CCM 8176 (Satke *et al.,* 1994).

The sporogenic strain *C. paspali* MG-6 played for ergolenes a similar role as the strain *Claviceps fusiformis* SD-58 for clavine alkaloids—the basic knowledge of physiology and biochemistry of formation of simple derivatives of lysergic acid was gained using this organism (Bumbová-Linhartová *et al.,* 1991; Linhartová *et al.,* 1988; Řeháček and Malik, 1971; Řeháček *et al.,* 1971; Rylko *et al.,* 1988d). The same holds for the strains *C. paspali* 31 (Rosazza *et al.,* 1967) and L-52, identical with the strain ATCC 13892 (Sočič *et al.,* 1986). Mantle (1969) described the production of a mixture of lysergic and paspalic acids by saprophytic strains, not selected by mutagenesis, isolated from ergotoxine containing sclerotia of *C. purpurea*. Philippi and Eich (1984) demonstrated the bioconversion of elymoclavine to lysergic acid by the strain *C. paspali* SO 70/5/ 2, Maier *et al.* (1988b) reported on an analogous bioconversion using a microsomal fraction of the ergopeptine producer *C. purpurea* Pepty 695/S.

Besides the genus *Claviceps* formation of lysergic acid and its derivatives has been reported in a number of strains of different species of the genus *Aspergillus (A. clavatus, A. repens, A. umbrosus,A. fumigatus, A. caespitosus, A. nidulans, A. ustus, A. flavipes, A. versicolor,A. sydowi, A. humicola, A. terreus, A. niveus, A. carneus, A. niger, A. phoenicus)* (Siegle and Brunner, 1963).

12.5.1.
Production of Simple Derivatives of Lysergic Acid

Production of amides of lysergic acid is described more often than the production of the acid itself. These amides are isomers of lysergic acid α-hydroxyethylamide; lysergic acid can be prepared from them by bioconversion. Amici *et al.* (1963) described bioconversion with 95% efficiency in cultures of *Claviceps purpurea* without further specification. Some production strains are very sensitive to surplus iron ions (Chain *et al.*, 1960; Řičicová *et al.*, 1982b) or they require ions of iron and zinc, and sometimes also other inorganic ions, in defined proportions (Mary *et al.*, 1965; Rutschmann and Kobel, 1963b). The process reported by Chain *et al.* (1960) needs so-called virulentation of the strain in a rye embryo to get sufficient production. Concentrations between 450 and 1600 mg L^{-1} are reached during the submerged cultivation. Iron ions did not interfere with the process described by Tyler (1963). The procedure according to Řičicová *et al.* (1982a) employed the asporogenic production strain *C. paspali* FA CCM F-731; that brought problems with the preparation of a standard inoculum for the production phase. The strain produced over 2000 mg L^{-1} of alkaloids from which 80% was lysergic acid α-hydroxyethylamide. Řičicová *et al.* (1981, 1986) also reported on the strain *C. paspali* F 2056 that produced nearly 2000 mg L^{-1} of alkaloids with the same proportion of lysergic acid α-hydroxyethylamide. Production of max. 2000 mg L^{-1} of simple derivatives of lysergic acid was described using the strain *C. paspali* ATCC 13892 and optimized cultivation conditions (Pertot *et al.*, 1984). Rutschmann and Kobel (1963b) reported that the strain *C. paspali* NRRL 3027 had formed over 1000 mg L^{-1} of alkaloids. The concentration of total alkaloids in a shaker culture of the strain NRRL 3166 reached as much as 2210 mg L^{-1} from which 80% was formed by amides of lysergic and isolysergic acids; in a fermentor the concentration was 1820 mg L^{-1}, with 87% of amides. Harazim *et al.* (1986) dealt with the optimization of the inoculation phase of lysergic acid α-hydroxyethylamide production. The asporogenic strain *C. paspali* CP 2505 and the sporogenic one YU 6, selected from natural material of a different geographic origin, were found to have the same requirements for optimal media composition. Pertot *et al.* (1990) reported the strain *C. paspali* L-52 which produced as much as 2647 mg L^{-1} of a mixture of ergometrine, lysergic acid amide and lysergic acid α-hydroxyethylamide, and from it selected a mutant CP 2 with a totally blocked synthesis of ergometrine and with the production of as much as 1552 mg L^{-1} of lysergic acid derivatives. Flieger *et al.* (1989a) and

Harazim *et al.* (1989) described the production strains of *C. paspali* mentioned earlier in connection with biotransformation of agroclavine to elymoclavine. The strain CCM 8061 (Flieger *et al.,* 1989a) produced 1220 mg L^{-1}of simple derivatives of lysergic acid; at the same time it showed a high activity of bioconversion (almost 95%) of clavine alkaloids to simple derivatives. Also, by adding clavines the actual biosynthesis of lysergic acid derivatives was increased by 33.5% . The strain could also be used in the immobilized form for semicontinuous production of lysergic acid derivatives by *de novo* biosynthesis and/or by clavine conversion. The strain CCM 8063 (Harazim *et al.,* 1989) is characteristic by the production of lysergic acid α-hydroxyethylamide in concentrations of up to 2200 mg L^{-1}. When clavine alkaloids were added to the medium nearly, 5070 mg L^{-1} of lysergic acid α-hydroxyethylamide was produced as a consequence of their concurrent conversion. A semicontinuous process was also described using cells of this strain entrapped in alginate. Procedures reported for these two strains were later worked out to produce simple derivatives of lysergic acid, first of all its α-hydroxyethylamide, ergometrine and partially also ergine, by means of aggressive bioconversion of clavine alkaloids (Flieger *et al.,* 1989b); induction of lysergic acid derivatives took place at the same time. In a batch cultivation, the concentration of lysergic acid derivatives reached almost 5400 mg L^{-1} while in a large scale industrial fermentor the concentration was 2920 mg L^{-1}. The concentration of total alkaloids in cultures of the strain *C. paspali* CCM 8062 after clavine conversion reached 2130 mg L^{-1}; out of this amount 78% was ergometrine, 11% ergine and 11% lysergic acid α-hydroxyethylamide. A mixed cultivation of the strain *C. purpurea* CCM F-733 (producer of clavine alkaloids) and *C. paspali* CCM 8061 yielded 4890 mg L^{-1} of alkaloids during a fortnight cultivation; alkaloids suitable for semisynthesis made up 97.3% (lysergic acid α-hydroxyethylamide 73%, elymoclavine 24.3%). A mixture of lysergic acid α-hydroxyethylamide and ergine was also produced by the strain *C. paspali* MG-6. Derivatives of lysergic and paspalic acids—8-hydroxyergine and 8-hydroxyerginine (Flieger *et al.,* 1989c), and 10-hydroxy-*cis*- and 10-hydroxy-*trans*-paspalic acid amide (Flieger *et al.,* 1993)—were isolated from the culture medium of this strain in the post-production phase. The bioconversion of elymoclavine to ergine by the strain *C. paspali* LI 189+was described by Mothes *et al.* (1962). Matošić *et al.* (1988a, b) used an immobilized strain of *C. paspali* which produced a mixture of lysergic acid α-hydroxyethylamide and ergometrine. He also tried to increase the production by means of surfactants. During a 60-day cultivation with six medium replacements the total production of alkaloids reached 8290 mg L^{-1}.

12.5.2.
Production of Ergometrine

Ergometrine was isolated from both *Claviceps purpurea* (Stoll, 1952) and *C. paspali* (Kobel *et al.,* 1964) cultures. There is a number of described strains of the both species that produce ergometrine: *C. paspali* CCM 8062 (Flieger *et al.*, 1989b), NRRL 3081, NRRL 3082 (Rutschmann and Kobel, 1963a), ATCC 13892 (Gaberc-Porekar *et al.*, 1987), *C. paspali* without additional marking, isolated from *Paspalum commersonii* (Janardhan and Husain, 1984), *C. purpurea* IMET PA 130 (ZIMET 43769), IMET PA 135 (2IMET 43695) (Borowski *et al.*, 1976; Volzke *et al.*, 1985), NCAIM 001106 (Zalai *et al.*, 1990), OKI 22/1963 (Molnár *et al.*, 1964; Udvardy-Nagy, I. *et al.*, 1964), OKI 620 125 (Molnár and Tétényi, 1962), Pepty 695 (Baumert and Gröger 1982, Erge *et al.*, 1972), PRL 1578 (ATCC 14934) (Taber and Vining, 1958). Ergometrine is formed biosynthetically via the intermediate lysergylalanine, the common precursor of ergoptinyle (Řeháček and Sajdl, 1990). *Claviceps purpurea* which, unlike *C. paspali,* is able to synthesize ergopeptine alkaloids, normally produces ergometrine together with a certain amount of ergopeptines. Both components are easily separable and most of ergopeptines find application in therapy.

Stoll *et al.* (1953) described a surface cultivation of *Claviceps purpurea* during which low amounts of ergometrine and ergotamine are formed in strict dependence on the concentration of iron and zinc ions, similarly as in lysergic acid amide production (Rutschmann and Kobel, 1963b). Windisch and Bronn (1960) reported on cultivations in which production of clavines, ergometrine and ergopeptines was induced by anaerobic conditions elicited by respiration inhibition. The process could hardly be implemented on the industrial scale, because of a very low production and other factors. Later, more efficient processes were developed having with the aid of better production strains. Molnár *et al.* (1964) described the production of a mixture of ergometrine and ergotoxine, rich in ergocristine, by submerged as well as surface cultivation yielding a minimum concentration of alkaloids 300 mg L^{-1}. A patent of Molnár and Tétényi (1962) described a production of a mixture of ergometrine, ergokryptine and ergocornine during both stationary and submerged cultivations. In the surface cultivation, the concentration of alkaloids in the mycelium was 0. 6% , 30% of which was ergometrine, in the submerged one the total alkaloid production was 480 mg L^{-1}. The process according to Rutschmann and Kobel (1963a) made use of the production strains *C. paspali* NRRL 3081 and 3082; they formed higher concentrations of a product without ergopeptines. Gaberc-Porekar *et al.* (1987) published data on the asporogenic strain *C. paspali* ATCC 13892 that produced 1200 mg L^{-1} of alkaloids consisting from 50–60% by ergometrine and from 25–30% by lysergic acid α-hydroxyethylamide. Using a mutagenic effect of gamma irradiation, they selected from this strain a daughter one able to form conidia. In the process of Zalai *et al.* (1990) the production strain *C. purpurea* NCAIM 001106 was selected by mutagenesis of protoplasts. The produced mixture of alkaloids contained 1100 mg L^{-1} of ergometrine, 450 mg L^{-1} of ergocornine and 600 mg L^{-1} of ergokryptine. Borowski *et al.* (1976) described submerged cultivations of the strain *C. purpurea* IMET PA 130 where the total alkaloid concentrations reached 2430–2460 mg L^{-1}, out of which the ergotoxine group ergopeptines comprised 1150-1300 mg L^{-1} and ergometrine

340–550 mg L^{-1}. The procedures were further elaborated by Volzke *et al.* (1985) who used the strains IMET PA 130 (ZIMET 43769) and IMET PA 135 (ZIMET 43695); by changing the limitation and/or nutrient sources they were able to change the proportions of ergometrine and individual alkaloids of the ergotoxine group. The total concentration of alkaloids was as high as 4000 mg L^{-1}; under different cultivation regimes ergometrine was produced in concentrations between 420 and 800 mg L^{-1}. The maximal proportion of ergometrine was reached when urea was used together with partial limitation by the phosphorus source, oxygen saturation was kept at 48–86% and pH under 7. The process described by Flieger *et al.* (1989b), in which the strain *C. paspali* CCM 8062 produced 1660 mg L^{-1} of ergometrine and small amounts of ergine and lysergic acid amide when clavine alkaloids were added as precursors, has been mentioned earlier. The paragraph concerning clavine alkaloids also report on the production of ergometrine mixed with clavine alkaloids by immobilized cells of the strain *C. purpurea* CBS 164.59 (Kopp, 1987).

12.5.3.
Production of Paspalic Acid

Paspalic acid is another suitable substrate for preparation of semisynthetic derivatives. The patent literature contains description of its production by strains *C. paspali* NRRL 3080, and NRRL 3167 (Kobel and Schreier, 1966, Rutschmann *et al.,* 1963) and *C. paspali* CCM 8176 (Satke *et al.,* 1994). The strain NRRL 3167 formed 3330 mg L^{-1} of total alkaloids, out of which paspalic acid represented 89% . The strain CCM 8176, in dependence on sugar and organic acid components used, produced as much as 7927 mg L^{-1} of total alkaloids. Paspalic acid formed 54.4% (4257 mg L^{-1}) and the rest was formed by isopaspalic, lysergic and isolysergic acids. A cell-free extract of the strain *C. purpurea* PCCE1 was able to convert elymoclavine to paspalic acid with a 95% efficiency (Kim and Anderson, 1982; Kim *et al.,* 1983).

12.6.
PRODUCTION OF ERGOPEPTINES

12.6.1.
Production of Ergotamine Group Alkaloids

So far, ergotamine is the only natural alkaloid from the ergotamine group of ergopeptines which has found a therapeutical use. It was detected only in the strain *Claviceps purpurea* (Flieger *et al.,* 1997; Stoll, 1952). The first isolation from saprophytic mycelia was shown by Kybal and Starý (1958). The fermentative production of ergotamine was performed with many strains—*C. purpurea* IBP 74, IMET PA 135 (Baumert *et al.,* 1979b, c), JAP 471 (Erge *et al.,*

1984; Schmauder and Gröger, 1986), JAP 471/1 (Maier *et al.,* 1983), I.M.I. 104437 (ATCC 15383) (Amici *et al.,* 1964), 275 F.I. (Amici *et al.,* 1966, 1967a; Crespi-Perellino *et al.,* 1981; Floss *et al.,* 1971b), F.I. 32/17 (ATCC 20102) (Amici *et al.,* 1968; Keller *et al.,* 1980) and its derived strain 1029 (Keller *et al.,* 1988; Lohmeyr and Sander, 1993); L-4 (ATCC 20103) (Komel *et al.,* 1985), CP II (Sarkisova and Smirnova, 1984), 312-A (Sarkisova, 1990; Ustyuzhanina *et al.,* 1991).

In addition to the process mentioned earlier, which produces small amounts of ergotamines (Windisch and Bronn, 1960), other processes were successively developed with higher industrial utility. Kybal *et al.* (1960) described both surface and submerged cultivation of non-specified strains *Claviceps purpurea*. In the surface cultivation the yield of ergotamine was 0.14% in dry biomass, while in the submerged one 0.07% . Amici *et al.* (1964) working with the strain I.M.I. 104437, obtained as much as 1300 mg L^{-1} of ergotamine. In their experiments with the strain 275 F.I., which produced 1–150 mg L^{-1} of alkaloids, Amici *et al.,* found correlation between the production capability of alkaloids and lipids (Amici *et al.,* 1967a). Procedures described by Amici *et al.* (1968) with the strain F.I. 32/17 served to increase the production of ergotamine and α-ergokryptine in shaker cultures up to 2000 mg L^{-1} with an approximately equal proportion of the two components; in a fermentor the production reached 1200 mg L^{-1}. Baumert *et al.* (1979b) described, *e.g.,* a procedure of selection of the production strain IBP 47, IMET PA 135; Baumert *et al.* (1979c) developed cultivation processes for this strain. The total alkaloid production reached 900–1500 mg L^{-1}; the total alkaloid mass was composed of 75–80% ergotamine, 10–15% chanoclavine, 5–6% ergometrine, 5% ergokryptine, a maximum of 4% ergosine and traces of other clavines. The strain JAP 471 gave about 800 mg L^{-1} of alkaloids out of which 70% was ergotamine and 30% was clavine alkaloids (Erge *et al.,* 1984). In the submersion mycelium of the strain *C. purpurea* II, the content of alkaloids reached 0.4% of dry mass (Sarkisova and Smirnova, 1984). The strain L-4 (ATCC 20103) produced about 1500 mg L^{-1} of ergotamine (Komel *et al.,* 1985).

Long-term production of a mixture of ergotamine and ergokryptine by immobilized cells was studied by Dierkes *et al.* (1993) in semicontinuous and continuous systems. When cells of the strain *C. purpurea* 1029/N5 entrapped in alginate were cultivated in a 500 mL bubble column reactor for 30 days, alkaloid productivity was 17–40 mg L^{-1} per day.

Ergotamine producing strains were also used in different studies as model organisms for research on various aspects of *Claviceps* biology and alkaloid biosyntesis, *e.g.* the original parasitic ergotamine strain *C. purpurea* Pla-4 (Majer *et al.,* 1967; Řeháček and Kozová, 1975), or the strain PCCE1 (Quigley and Floss, 1981).

Ergosine, another representative of the ergotamine group, exerts very similar pharmacological effects as ergotamine. In spite of the fact that it has not yet been used in therapy, processes of its production are described in the patent literature.

Amici *et al.* (1969) described a concurrent production of ergocornine and ergosine by the strain *C. purpurea* F.I. 43/14, ATCC 20106, when the production of total alkaloids was 950–1100 mg L^{-1} and ergosine content 40–45% . Gröger *et al.* (1977) and Maier *et al.* (1981) employed the strains *C. purpurea* MUT 168 and MUT 168/2 for both surface and submerged cultivation with a production of 300–350 mg L^{-1} of alkaloids, containing 90% of ergosine and ergosinine together with 10% of clavine alkaloids, or 80% of ergosine and 20% of chanoclavine-I, respectively. Baumert *et al.* (1979b, 1980) described the selection of the production strain IBP 179, IMET PA 136 and its submerged cultivation. The concentration of total alkaloids was in this case 900–1300 mg L $^{-1}$, with 80–90% of ergosine and ergosinine. The ergosine strain *C. purpurea* MUT 170 (Baumert and Gröger, 1982; Schmauder and Gröger, 1986) produced a mixture of ergosine and clavine alkaloids in amounts of about 700 mg L^{-1} (Erge *et al.,* 1984).

Dihydroergopeptines (dihydroergotamine, dihydroergocristine *etc.*) which are produced from common ergopeptines by chemical methods have significant therapeutic use. The only dihydroergopeptine found in nature is dihydroergosine (Mantle and Waight, 1968) isolated from *Claviceps africana* (formerly *Sphacelia sorghi*); its biosynthetic precursors are dihydroelymoclavine and dihydrolysergic acid (Barrow *et al.,* 1974). These findings opened the possibility of fermentative production of dihydroergopeptines by common strains when these precursors were used.

12.6.2.
Production of Alkaloids of the Ergotoxine Series

Ergocristine, ergocornine, α-ergokryptine and β-ergokryptine from alkaloids of this group are used in therapy. As drugs they are used both separately (*e.g.* ergocristine) and in mixtures (ergocornine, α- and β-ergokryptine), with the native molecule or hydrogenated. All these alkaloids were isolated from the species *C. purpurea* (Schlientz *et al.,* 1968; Stoll, 1952). Besides the above mentioned production strain *Hypomyces aurantus* (Yamatoya and Yamamoto, 1983) and the only one described production strain *Clavicepspaspali* (Wilke and Weber, 1984), all the strains mentioned in the literature originated from the species *C. purpurea:* CCM F-508 (Strnadová and Kybal, 1976), CCM F-725 (Strnadová *et al.,* 1981), CCM 8043 (Strnadová *et al.,* 1986), IBP 84, ZIMET 43768 (Schumann *et al.,* 1986), IMET PA 130, ZIMET 43769 (Ludwigs *et al.,* 1985; Volzke *et al.,* 1985), DH 82, ZIMET 43695 (Erge *et al.,* 1982), F.I. 101a (Amici *et al.,* 1967b), F.I. 43/14, ATCC 20106 (Amici *et al.,* 1969), F.I. S40, ATCC 20103 (Minghetti *et al.,* 1967), F.I. 7374 (Bianchi *et al.,* 1974), Exy 20, Ech K 420 (Kobel and Sanglier, 1976) Ecc 93 (Keller *et al.,* 1988), MNG 022, MNG 0083, MNG 00186 (Udvardy-Nagy *et al.,* 1981; Wack *et al.,* 1981), OKI 88/ 1972 (Richter Gedeon V.G., 1973), 231 F.I., ATCC 20106 (Bianchi *et al.,* 1976, Crespi-Perellino *et al.,* 1987, 1992, 1993), 563 E (Miličić *et al.,* 1984), L-16 (Puc *et al.,* 1987), L-17 (Didek-Brumec *et al.,* 1991a, b; Gaberc-Porekar *et al.,* 1990; Miličič *et al.,* 1987, 1989; Sočič *et al.,* 1985), L-18 (Didek-Brumec *et al.,* 1988), Pepty 695 (Maier *et al.,* 1971; Schmauder and Gröger, 1986), Pepty 695/S (Erge *et al.,* 1984; Maier *et al.,* 1980b, 1988b), 1029 (Lohmeyer and Sander, 1993; Lohmeyer *et al.,* 1990).

The original procedures elaborated for fermentation production of ergotoxine alkaloids were not introduced into practice both for practical reasons, as, *e.g.,* in the patent of Windisch and Bronn (1960), and for economic ones, given by the very low productivity. For example, surface cultivation was developed producing 0.18 g of total ergotoxine alkaloids per l00 g of dry mass with the ergocristine/ergocornine/ergokryptine ratio of 3:1:2 (Kybal *et al.,* 1960). At that time no technological process was available for the industrial application of this cultivation but later the process of cultivation in plastic bags was developed (Kybal and Vlček, 1976; Vlček and Kybal, 1974) and the high-producing strains *C. purpurea* CCM F-725 (Strnadová *et al.,* 1981) and CCM 8043 (Strnadová *et al.,* 1986) were selected. On a rich medium the strain *C. purpurea* CCM F-725 formed mycelia containing 1.5% of alkaloids per dry mass. This product contained ergocornine, α-ergokryptine and β-ergokryptine in a 6:5:1 ratio, small amounts of ergometrine and traces of ergosine, ergocristine, ergotamine and ergoxine. Later the strain *C. purpurea* CCM 8043 was selected which produced as much as 3.5% alkaloids per mycelia dry mass; the mixture of alkaloids contained α-ergokryptine, ergocornine, β-ergokryptine, ergometrine and traces of ergosine. From the end of the 1970s pharmacopoeias requirements became more strict as regards the mutual proportion of α- and β-ergokryptine in ergotoxine substances and drugs. The mutual ratio of biologically synthesized alkaloids of the ergotine group can be influenced by the addition of amino acids that form the peptidic moiety of the ergopeptine structure (Kobel and Sanglier 1978). Kybal *et al.* (1979) described a surface cultivation giving a controlled proportion of ergocornine, α-ergokryptine and β-ergokryptine. Threonine, the biosynthetic precursor of isoleucine, was also used besides the amino acids forming the peptidic part of ergopeptines. Experiments with additions of threonine, leucine and isoleucine into media provided 0.51–0.86% alkaloids per dry mass, with the ratio of ergokryptines to ergocornine 1.5–3:1 and α-ergokryptine to β-ergokryptine 1:5–100:1. When precursors were employed, three new alkaloids were isolated—5′-epi-β-ergokryptine from the ergopeptine group, and β-ergokryptame and β, β-ergoanname (Flieger *et al.,* 1984). The procedure was further optimized by using the economically more favourable threonine; as a result, production strains D3–18 and D2–B1 were able to produce mycelia containing the precise proportion of ergocornine and ergokryptine components in ergotoxine preparations required by the Pharmacopoeia (Malinka *et al.,* 1987). A process for the controlled biosynthesis of ergocornine, α-ergokryptine and β-ergokryptine was also elaborated for submerged cultivation (Udvardy-Nagy *et al.,* 1981). Production strains MNG 0022, MNG 0083 a MNG 00186 provided 80–200 mg L^{-1} of ergocornine, 15–150 mg L^{-1} of α-ergokryptine, nearly 80–100 mg L^{-1} of β-ergokryptine and 150–180 mg L^{-1} of a mixture of ergocorninine and ergokryptinines. A broader spectrum of compounds was used as precursors— besides threonine and isoleucine also homoserine, homocysteine, methionine and α-ketobutyric acid. Wack *et al.* (1981) reported on the use of valine and isoleucine as precursors in the cultivation of strain MNG 00186 also. The precursor addition enhanced the original production of 150 mg L^{-1} of ergocornine, 40 mg L^{-1} of α-ergokryptine and 90 mg L^{-1} of β-ergokryptine to 320 mg L^{-1} of ergocornine, 60 mg L^{-1} of α-ergokryptine and 160 mg L^{-1} of β-ergokryptine. Increased production of α-ergokryptine using leucine as a precursor in cultures of *C. purpurea* strains IMET PA 130 or ZIMET PA 43769 was described by Ludwigs *et al.* (1985). When 2–5 g L^{-1} of L-leucine was added to the medium, the concentration of ergotoxine alkaloids reached 900–1200 mg L^{-1} or 1400–2500 mg L$^{-1,}$ with 65–85% of α-ergokryptine. Puc *et al.* (1987) described the use of valine as a precursor with strain L-16. Depending on the amount of valine added into a submerged culture of the strain producing 1800 mg L^{-1} of total alkaloids, with a proportion of ergocornine to ergokryptines 1:2, this proportion was changed up to 4.5:1. Another method of production control, in addition to leucine precursoring, is described in the patent of Volzke *et al.* (1985). The production of ergocornine by *C. purpurea* strains IMET PA 130 and ZIMET 43769 can be supported by partial phosphate limitation and by continuous addition of ammonium ions; production of α-ergokryptine can be increased by simultaneous addition of urea or ammonium salts and phosphate. In addition to these procedures with directed precursoring, a number of patents describes the production of ergocornine or ergokryptines without precursors. The production strain *C. purpurea* IBP 84, ZIMET 43768 used for production of a mixture of α-ergokryptine and ergosine formed 700–1400 mg L^{-1} of total alkaloids with 80% of α-ergokryptine and 20% of ergosine (Schumann *et al.,* 1986). According to the patent of Amici *et al.* (1967b), the production of ergokryptine in cultures of the strain *C. purpurea* F.I. 101a reached 1100–1500 mg L$^{-1.}$ Production of a mixture of ergokryptine and ergotamine was mentioned earlier (Amici *et al.,* 1968), and so was the production of ergokryptinine, ergokryptine and other alkaloids by a fungi of the genus *Hypomyces* (Yamatoya and Yamamoto, 1983). Wilke and Weber (1984) reported the production of 525 mg L^{-1} of α-ergokryptine with the asporogenic strain *C. paspali* DSM 2836. Patent of Richter Gedeon V.G. (1973) described the production of a mixture of ergocornine and α-ergokryptine; when the strain *C. purpurea* OKI 88/1972 was employed, 1046–1246 mg L^{-1} of total alkaloids were produced from which a mixture of ergocornine and ergokryptine represented 646 mg L^{-1} and the concentration of ergometrine was 202–310 mg L^{-1}. A mixture of ergocornine and ergokryptine was also the main component of the 2000 mg L^{-1} alkaloids which were produced by the sporogenic strain L-17 bred by combined mutagenesis and selection from an originally parasitic strain (Didek-Brumec *et al.,* 1991a,b). The production of a mixture of ergocornine and ergosine was reported by Amici *et al.* (1969). The strain *C. purpurea* F.I. 43/14 ATCC 20106 formed in different cultivation media 950–110 mg L^{-1} of a mixture containing 40–45% of ergosine and 55–60% of ergocornine. Special strains for β-ergokryptine production are referred to by Bianchi *et al.* (1974, 1976). The strains *C. purpurea* 231 F.I. and F.I. 7374 produced the total amount 1200 mg L^{-1} of ergopeptines with 30% of ergokryptine. The patent of Kobel and Sanglier (1976) described the production of ergocornine and ergokryptine by the strain *C. purpurea* Exy 20, and the production of ergocristine by the strain *C. purpurea* Ech K 420; a so called pre-culture was used in the process. The cultivation production of total ergopeptines was 770 mg L^{-1} from which ergokryptine and ergokryptinine comprised 203 mg L^{-1}, ergocornine and ergocorninine 200 mg L $^{-1}$, ergocristine and ergocristinine 206 mg L^{-1} and other alkaloids 170 mg L^{-1}. The fermentative production of ergocristine was described in patent of Minghetti *et al.* (1967). The process employed the production strain *C. purpurea* F.I. S40 (ATCC 20103) which, when cultivated in a fermentor, gave 920 mg L^{-1} of ergocristine. Another process described by Erge *et al.* (1982) employed the strain *C. purpurea* DH 82, ZIMET 43695; in a fermentor the production of alkaloids reached 600-1000 mg L^{-1} of which ergocristine represented 400–550 mg $L^{-1.}$ Didek-Brumec *et al.* (1988) referred to the asporogenic strain L-18 that formed 2000 mg L^{-1} of ergocristine. The production about l000–1200 mg L^{-1} of total alkaloids by the strain Pepty 695/S, which contained 50–60% of ergotoxines composed of a mixture of ergocornine, ergokryptine and a 20% of ergometrine, was reported by Maier *et al.* (1980b, 1988b) and Erge *et al.* (1984). The original parent strain Pepty 695 showed the total alkaloids production of about 400–450 mg L^{-1}, with 50% of ergotoxines (ergocornine to ergokryptine ratio 3:1) and 15–20% of ergometrine (Floss *et al.,* 1971 a). Gaberc-Porekar *et al.* (1990) used the strain *C. purpurea* L-17 to produce 2400 mg L^{-1} of total ergotoxine alkaloids, mostly ergocornine and ergokryptine. In the same study, devoted to carbohydrate metabolism, the hexose monophosphate shunt metabolizing glucose during the vegetative phase of fermentation was shown to be replaced by glycolysis during the period of increasing production of alkaloids. This strain served for further research on biochemistry and physiology of high-producing strains, *e.g.* the correlation between the intermediary metabolism and secondary metabolite synthesis (Gaberc-Porekar *et al.,* 1992a). In the case of ergopeptines, the further direction of production processes development—immobilization of producers and possible continualization—is only at its beginning (Lohmeyer and Sander, 1993).

12.6.3.
Derivatives and Analoga of Ergopeptines

The fungus *Claviceps purpurea* is able to incorporate the amino acids, present in the medium, into the peptidic moiety of ergopeptines and to perform similar reaction also with their precursors. This fact was used in a controlled fermentation with the directed application of precursors, as mentioned in previous paragraphs. *C. purpurea* is also able to incorporate a number of other different amino acids and their analoga into the peptidic moiety (Beacco *et al.,* 1978). Thus ergobutine (from the group of ergoxines) and ergobutyrine (from the group of ergotoxines) were isolated from the saprophytic cultures of the strain *C. purpurea* 231 F.I. (Bianchi *et al.,* 1982). 5'-Epi-*β*-ergokryptine from the ergotoxine group, *β*-ergokryptame from the ergotaxame group and *β, β*-ergoanname from the ergoanname group, isolated from saprophytic surface cultures of the strain *C. purpurea* D-3–18 on addition of different stereomers of isoleucine and threonine as precursors (Flieger *et al.,* 1984), were mentioned earlier. Addition of L-norvaline as a precursor into cultures of *C. purpurea* 231 F.I. yielded unnatural ergopeptines—ergorine, ergonorine and ergonornorine (Crespi-Perellino *et al.,* 1992). These capabilities of *C. purpurea* were used in the development of processes for preparation of ergopeptines analoga. The procedure according to Beacco *et al.* (1977) employed the specially selected mutant strains *C. purpurea* ATCC 15383, ATCC 20103 and ATCC 20019, dependent upon different nonhydroxylated amino acids—leucine, phenylalanine, halogenated phenylalanine, thienylalanine, pyrazolylalanine, furylalanine, pyridylalanine, *etc.* A number of derivatives of ergopeptine with the adrenolytic effect (blockade of α-receptors), *e.g.* 5'-debenzyl-5'-*p*-chlorobenzyl-dihydroergocristine or 5'-debenzyl-5'-*p*-fluorobenzyl-ergotamine, was obtained. The process proposed by Baumert *et al.* (1981) employed the strains *C. purpurea* IBP 179 and MUT 168, which produced ergosine. Addition of 3–6 g L^{-1} of the proline analogue—the anticancer substance 1, 3-thiazolidine-4-carboxylic acid-resulted in the synthesis of 1' *β*-methyl-5'α-isobutyl-9'-thiaergopeptine (Thiaergosine). The biosynthesis of similar compounds has been described in the patent of Kobel *et al.* (1982). Addition of appropriate precursors to cultures of the strain *C. purpurea* NRRL 12043, which produces ergotamine and ergotaminine, and to those of the strain NRRL 12044, that produces ergocristine and ergocristinine, yielded a number of substances. These derivatives of peptidic alkaloids exert a spectrum of physiological and therapeutical effects (dopaminergic stimulation, prolactin inhibition, vasoconstriction activity, *etc.*). 9'-Thia-ergocristine and 9'-thia-ergotamine can be mentioned as representatives of such substances.

12.7.
CONTROL AND MODELLING OF ERGOT ALKALOID FERMENTATION

Processes in which final yields of products were influenced by precursor addition or by limitation and dosing of individual nutrients were mentioned earlier. Tryptophan, the building unit of the ergoline nucleus (Floss, 1976), can also be used to increase production of many alkaloids. Detailed research into the topic was done by, *e.g.,* Gaberc-Porekar *et al.* (1992b). A comparative study with a number of strains producing different ergot alkaloids was performed by Erge *et al.* (1984). More details are given in Chapter 7.

Bianchi *et al.* (1981) and Crespi-Perellino *et al.* (1994) carried out certain generalization of results with controlled precursor addition promoting ergopeptine production. An amino acid at position 3 of the ergopeptine molecule is specific, amino acids at positions 1 and 2 can be changed. Amino acids with a lipophilic side chain can be introduced into the ergopeptine molecule depending on the number of C atoms in the side chain.

Some model procedures for a more complex control of fermentative production of ergot alkaloids were elaborated based on, *e.g.* mathematical models of clavine and ergopeptine alkaloid production in batch cultivation. A model based on the concentration of extracellular and intraceliular phosphate was published for clavine alkaloids (Votruba and Pažoutová, 1981). A mathematical simulation of different technological alternatives of clavine production was done on this basis (Pažoutová *et al.*, 1981b). Apart from this model, also a hypothesis was published on gene expression in *Claviceps* biosynthetic pathways (Pažoutová and Sajdl, 1988). A regulation model of the gene expression for alkaloid biosynthesis was proposed according to which the tryptophan-induced synthesis is mediated by an activator binding tryptophan and stimulating the transcription of pertinent genes. The kinetics of clavine alkaloids production was also investigated (Flieger *et al.*, 1988). Based on these results, processes were developed in which elimination of feed-back inhibition by fermentation products lead to a higher production of clavine alkaloids and ergometrine (Flieger *et al.*, 1987).

In the case of ergopeptine alkaloids, batch submerged cultivation was modelled on the basis of the predicted concentrations of biomass, alkaloids and sucrose. Good agreement was achieved between the calculated and found values of the former two parameters (Grm *et al.*, 1980). Using a previously found correlation between growth and alkaloid biosynthesis on the one hand (Miličić *et al.*, 1987) and the effects of cultivation conditions on morphology and alkaloid synthesis on the other (Miličić *et al.*, 1989), Miličić *et al.* (1993) elaborated a more general model. Models of microorganism "life span", "microbial growth" and "alkaloid synthesis" were elaborated on the basis of the specific growth rates and morphological analysis of proliferation.

Preliminary studies, whose results could be used for model building were performed with producers of simple derivatives of lysergic acid. BumbováLinhartová *et al.* (1991) divided the production process of these derivatives into three phases—production, post-production and degradation ones —and set up their characteristics.

Other procedures leading to increased effectivity of the production processes are also described in the patent literature. Rochelmayer (1965) stimulated alkaloid biosynthesis by adding parts of the Thallophyta, especially fungi and bacteria, into the medium. A similar principle was adopted in the patent of Fiedler *et al.* (1989) where elicitors were used to enhance the activity of biosynthetic enzymes in microorganisms and higher plants. Another way increasing of alkaloid biosynthesis by more than 100%, described by Rylko *et al.* (1988a, b, c), made use of suitable inducers of cytochrome P-450. A positive effect of substances modifying cell lipids of production strains has also been demonstrated (Křen *et al.*, 1988a,b). These substances increased alkaloid production by almost 74%; the same effect has been shown with high-producing strains (Sajdl *et al.*, 1988a). The relationship between morphology of saprophytic cells and production capability has not been explicitly elucidated yet (Esser and Tudzynski, 1978; Didek-Brumec *et al.*, 1991 a). The production of alkaloids is supported by such cultivation conditions that cause mycelial differentiation to sclerotium cells (Kybal, 1981; Wichmann and Voigt, 1962) and are connected with specific manifestations of the primary metabolism (Kleinerová, 1975; Kybal *et al.*, 1978, 1981; Zalai and Jaksa, 1981). These findings have been complemented by Lösecke *et al.* (1980, 1981, 1982) by the data on the relationship between the ultrastructure of cells from submerged culture and alkaloid production.

12.8.
PRODUCTION OF INOCULATION MATERIAL FOR PARASITIC ERGOT PRODUCTION

The infection material for inoculation has been mentioned by Németh in Chapter 11 "Parasitic production of ergot alkaloids". Asexual sporesconidia—are exclusively used as a source of the primary infection in the parasitic production of ergot. When the infection inoculation material is to be prepared, the saprophytic cultivation aims at obtaining the maximum amount of vital infectious spores. Nutritional sources and the cultivation process itself are adapted to support growth and differentiation of hyphae to obtain massive conidiation.

Cultivation processes are generally identical with those for the production of alkaloids. It is possible to employ cultivation on solid substrates as well as stationary or submerged cultivation in liquid media. Grains, which were reported as the solid-state medium of choice since the 1940s, has been mentioned by Chapter 11 (see also Kybal, 1955; Sastry *et al.*, 1970b). The grains supplemented by nutrients (Kybal, 1963) was still used in the 1980s as an optimum substrate for production of high-quality inoculation material and a reference standard for comparison with other inoculation materials. This material, or conidia from surface agar cultures, were used also in experimental parasitic cultivations (Corbett *et al.*, 1974; Kybal and Strnadova., 1968; Singh *et al.*, 1992) while inoculation material from submerged cultivations has been used less frequently (Košir *et al.*, 1981). Cultivations on the surface of liquid media are performed with the plastic bags (Harazim *et al.*, 1984; Kybal and Vlček, 1976; Strnadová *et al.*, 1986) or other suitable equipment (Kybal and Strnadová, 1982). Submerged cultivations, depending on the properties of the parasitic production strains, are multi-stage. The produced infection material can be conserved in a sucrose solution (Kubec *et al.*, 1974), mixed with an inert filler, granulated and dried (Kiniczky *et al.*, 1982; Kybal *et al.*, 1990) or frozen in an osmotically stabilized medium (Yásárhelyi *et al.*, 1980a). In experiments done by Czech authors, optimum results were achieved with an inoculation material dried together with SiO_2 (Valík and Malinka, 1992).

Quality evaluation of the inoculation material can be done by vital staining of conidia, but methods based on germination ability (Švecová, 1985) and determination of the unit infection dose appears more optimal. An optimal number of conidia and a procedure of infection of a host spike should be experimentally determined for each kind of inoculation material and a type of host (Sastry *et al.*, 1970a, c).

12.9.

PRODUCTION OF OTHER SUBSTANCES BY *CLAVICEPS*

Fungi of the genus *Claviceps* have been shown to produce not only ergot alkaloids but also other substances. Tryptophan is used as a starting material for biosynthesis of ergot alkaloids. The use of the *Claviceps* fungi is mentioned in the patent of Enatsu and Terui (1967) describing L-tryptophan production. In the process reported by Dinelli *et al.* (1972), enzyme complexes isolated among others from *Claviceps* are employed for the production of L-tryptophan from indole and serine. The patent of Lapis *et al.* (1978) describes the manufacture of antitumor basic proteins with molecular weight of 1, 8–3, 5 kDa from the mycelium of *C. purpurea* and *C. fusiformis*.

The rice leaf binding component, produced in aerobic cultures of the strain *Claviceps purpurea* ATCC 9605 or of a number of other microorganisms (Oishi *et al.*, 1984), can be used to increase the rice crop. The active component increases the yield and shortens the production period. Also other metabolites of *Claviceps* can find application in agriculture practice. Patent of Dowd *et al.* (1988) describes the use of tremorgenic mycotoxins as insecticides against corn earworm and fall armyworm. Gubaňski and Lowkis (1964) have demonstrated the inhibition of the tobacco mosaic virus by a substance isolated from *C. purpurea*.

Detoxification of methyl-*N*-methylanthranilate to methyl-anthranilate using a number of microorganisms, among others *Claviceps* spp., is described in the patent of Page *et al.* (1989). The detoxification was carried out in a 4–8-day fermentation process.

There are two patents describing production of lipids. The patent of Fukuda (1986) referred to a method of isolation of lipids from lipid-producing algae and fungi, including *Claviceps*. In the patent of Sarkisova *et al.* (1987) the strain *C. purpurea* 312A produced lipids with composition similar to that of cotton-seed oil. After a 10-day cultivation in a liquid medium, the biomass contained 27–46% of lipids.

Production of carbohydrates is also mentioned in two patents. Glucans, undesirable in ergot alkaloid fermentation but applicable in a number of other industrial branches including the pharmaceutical industry, are produced according to the patent of Johal and Cash (1989) by different filamentous fungi, including *Claviceps*. Glucans are remarkable for their pharmacological, especially immunomodulatory, effects and their future therapeutic use can be envisaged. The patent of Senda *et al.* (1989) describes the production of another type of carbohydrates—oligoinulosaccharides—using β-fructofuranosidase from different microorganisms (also from *Claviceps purpurea*). These oligosaccharides can be employed, *e.g.,* in the food industry.

12.10.
DOWN-STREAM PROCESSES

Procedures for the subsequent processing of fermentation products depend on their properties; they are different for products of stationary cultivations and of submerged ones, and also for water soluble (clavines, lysergic acid and its derivatives, ergometrine) or insoluble (ergopeptines) substances.

The concentration of ergopeptines in a medium during their surface production is negligible; the mycelium is processed in this case. Isolation procedures are very similar to those used for the isolation of alkaloids from ergot sclerotia grown during field parasitic cultivation. When clavines, lysergic acid or its derivatives are produced by surface fermentation it is advantageous to process both the mycelia and the medium.

In the case of submerged cultivation the whole volume of a medium with the mycelium is processed. Due to the mechanical stress and the resulting injury to the hyphae a non-negligible amount of hydrophobic alkaloids can be contained in the medium.

To decide what isolation process should be used it is necessary to take into account also other substances produced by the fungi. Besides the already mentioned glucans, which complicate manipulation with the end product of fermentation and make the isolation more expensive, these effects can be exerted also by lipids and pigments.

Isolation processes used in industry are detailed in Chapter 13.

ACKNOWLEDGEMENT

The author thanks to Ms. Jiřina Kaštovská and Ms. Marcela Kazimírská for their technical help.

REFERENCES

Anonymous (1997) PA/P Exp. 7/T (97) 15 ANP Products of fermentation. *Pharmeuropa,* **9,**86–88,

Abe, M. (1951) Researches of ergot fungus. *Annual Reports Takeda Research Labs.,***10,** 73–239. (Parts III to XVII) (In Japan).

Abe, M., Ohmono, S., Ohashi, T. and Tabuchi, T. (1969) Isolation of chanoclavine—(I) and two new interconvertible alkaloids, rugulovasine A and B, from the cultures of *Penicillium concavo-rugulosum. Agric. Biol. Chem.,***33,**469–471.

Abe, M., Yamano, T., Kozu, Y. and Kusumoto, M. (1951) Production of ergot alkaloids in submerged cultures. *J. Agr. Chem. Soc. Japan,***24,**416–422.

Abe, M., Yamano, T., Kozu, Y. and Kusumoto, M. (1952) A new water-soluble ergot alkaloid, elymoclavine. *J. Agr. Chem. Soc. Japan,***25,**458–459.

Abe, M., Yamano, T., Kozu, Y. and Kusumoto, M. (1953) Isolation of a mutant productive of agroclavine rather excellently even in submerged culture. *J. Agr.Chem. Soc. Japan,***27,**18–23.

Abe, M., Yamano, T., Yamatodani, S., Kozu, Y., Kusumoto, M., Komatsu, H. and Yamada, S. (1959) On the new peptide-type ergot alkaloids, ergosecaline and ergosecalinine. *Bull. Agr. Chem. Soc. Jap.,***23,**246–248.

Abe, M., Yamatodani, S., Yamano, T. and Kusumoto, M. (1961) Isolation of lysergol, lysergene and lysergine from the saprophytic cultures of ergot fungi. *Agric. Biol.Chem.,***25,**594–595.

Adams, R.A. (1962) Process for the production of ergot alkaloids. *US pat. 3 117 917.*

Agurell, S. (1966) Isolysergol from saprophytic cultures of ergot. *Acta Pharmacol.Suecica,***3,**7–10.

Agurell, S. and Ramstad, E. (1965) A new ergot-alkaloid from Mexican maize ergot. *Acta Pharmacol. Suecica,***2,**231–238.

Amici, A.M., Minghetti, A., Scotti, T., Spalla, C. and Tognoli, L. (1966) Production of ergotamine by a strain of *Claviceps purpurea* (Fr.) Tul.*Experientia,***22,**415–418.

Amici, A.M., Minghetti, A., Scotti, T., Spalla, C. and Tognoli, L. (1967a) Ergotamine production in submerged culture and physiology of *Claviceps purpurea. Appl.Microbiol.,* **15,**597–602.

Amici, A.M., Minghetti, A., Scotti, T., Spalla, C. and Tognoli, L. (1967b) Precédé de fermentation pour la préparation de l'ergocryptine. *FR pat. 1 531 205.*

Amici, A.M., Minghetti, A., Scotti, T. and Spalla, S. (1968) Ergotamine and ergocryptine. *GB pat. 1 158 380.*

Amici, A.M., Minghetti, A. and Spalla, C. (1969) Mikrobiologisches Verfahren zur gleichzeitigen Herstellung von Ergocornin und Ergosin. *CH pat. 518 361.*

Amici, A.M., Minghetti, A., Tonolo, A. and Spalla, C. (1963) Preparation of lysergic acid. *GB pat. 975 880.*

Amici, A.M., Scotti, T., Spalla, C. and Tognoli, L. (1967c) Heterokaryosis and alkaloid production in *Claviceps purpurea. Appl. Microbiol.,* **15,**611–615.

Amici, A.M., Spalla, C., Scotti, T. and Minghetti, A. (1964) Verfahren zur biosynthetischen Herstellung einer Mischung von Ergotamin und Ergotaminin in Submerszüchtung unter aeroben Bedingungen. *CH pat. 447 191.*

Arcamone, F., Bonino, C., Chain, E.B., Ferretti, A., Pennella, P., Tonolo, A. and Vero, L. (1960) Production of lysergic acid derivatives by a strain of *Claviceps paspali* Stevens et Hall in submerged culture. *Nature* (London), **187,**238–239.

Arcamone, F., Chain, E.B., Ferretti, A., Minghetti, A., Pennella, P., Tonolo, A. and Vero, L. (1961) Production of a new lysergic acid derivative in submerged culture by a strain of *Claviceps paspali* Stevens & Hall. *Proc. R. Soc.,***155B,**26–54.

Barrow, K.D., Mantle, P.O. and Quigley, F.R. (1974) Biosynthesis of dihydroergot alkaloids. *Tetrahedron Lett.,***16,**1557–1560.

Baumert, A., Erge, D. and Gröger, D. (1981) Verfahren zur Herstellung von Ergopeptinen mit modifiziertem Prolinteil. *DD pat. 200 571.*

Baumert, A., Erge, D., Gröger, D., Maier, W., Schmauder, H.-P. and Schumann, B. (1979a) Verfahren zur Herstellung von Chanoclavin—I. *DD pat. 230 019.*

Baumert, A., Erge, D., Gröger, D., Maier, W., Schmauder, H.-P. and Schumann, B. (1980) Verfahren zur Herstellung von Ergosin. *EP pat. appl. 022 973.*

Baumert, A., Erge, D., Gröger, D., Maier, W., Schmauder, H.-P., Schumann, B., Breuel, K. and Höhne L. (1979b) Verfahren zur Herstellung von Mutterkornalkaloiden. *DDpat. 234 171.*

Baumert, A., Erge, D., Gröger D.,Maier, W., Schmauder, H.-P., Schumann, B., Breuel, K. and Höhne, L. (1979c) Verfahren zur Herstellung von Ergotamin. *DD pat.234 172.*

Baumert, A. and Gröger, D. (1982) Proteolytische Enzymaktivitäten in submers kultivierten Stämmen von *Claviceps purpurea. Biochem. Physiol. Pflanzen,***177,**18–28.

Beacco, E., Bianchi, M.L., Minghetti, A. and Spalla, C. (1977) Ergot alkaloids. *GB pat.1 584 464.*

Beacco, E., Bianchi, M.L., Minghetti, A. and Spalla, C. (1978) Directed biosynthesis of analogues of ergot peptide alkaloids with *Claviceps purpurea. Experientia,***34,** 1291–1293.

Bianchi, M.L., Cattaneo, P.A., Crespi-Perellino, N., Guicciardi, A., Minghetti, A. and Spalla, C. (1981) Mechanisms of control in the qualitative biosynthesis of peptidic ergot alkaloids. In: *Abstract book, FEMS Symp. Overprod. Microbial Products.*Hradec Králové, Czechoslovakia, August 9–14, 1981, p. 71–72.

Bianchi, M.L., Crespi-Perellino, N., Giola, B. and Minghetti, A. (1982) Production by *Claviceps purpurea* of two new peptide ergot alkaloids belonging to a new series containing α-amino-butyric acid. *J. Nat. Prod.,***45,**191–196.

Bianchi, M., Minghetti, A. and Spalla, C. (1974) Process for production of β-ergocryptine. *IT pat. 1 059 516* (In Italian).

Bianchi, M., Minghetti, A. and Spalla, C. (1976) Production of beta-ergocryptine by a strain of *Claviceps purpurea* (Fr.) Tul. in submerged culture. *Experientia,***32,**145–146.

Bonns, W.W. (1922) A preliminary study of *Claviceps purpurea* in culture. *Am. J. Bot.,***9,** 339–353.

Borowski, E., Braun, K., Breuel, K., Dauth, Ch., Erge, D., Grawert, W., Gröger, D., Höhne, L., Knothe, E., Müller, M., Nordmann, G., Schirutschke, R. and Volzke, K.-D. (1976) Process for fermentative preparation of ergoline derivatives*CS pat. 206 196* (In Czech).

Bové, F.J. (1970) *The Story of Ergot,*S. Karger, Basel, New York.

Břemek, J. (1981) Conservation of *Claviceps purpurea* (Fr.) Tul. producing strains. Diss., J.E.P. University, Brno. (In Czech).

Břemek, J., Kouřil, M., Alberti, M., Bartoš, V., Kiss, G., Kühnel, E., Lovecká, H., Malinka, Z. and Příhoda, J. (1989) Process for fermentative production of clavine ergot alkaloids. *CZ pat. 272 992* (In Czech).

Břemek, J., Křen, V., Řeháček, Z., Sajdl, P., Spáčil, J., Malinka, Z., Kozová, J., Krajíček, A., Flieger, M. and Pilát, P. (1986a) Process for biological conversion of agroclavine into elymoclavine by suspensed and immobilized cultures of *Claviceps* fungi. *CSpat. 257 311* (In Czech).

Břemek, J., Malinka, Z. and Harazim, P. (1986b) Influence of chlortetracycline on clavine ergot alkaloids production by species *Claviceps fusiformis* In: *AbstractBook, 17.*

*Congr. Czechoslov. Microbiol. Soc.,*České Budějovice, Czechoslovakia, 23–25 September, 1986, A-14 (In Czech).

Břemek, J., Řeháček, Z., Pilát, P., Malinka, Z., Pažoutová, S., Chomátová, S., Spáčil, J., Rylko, V., Bárta, M., Krajíček, A., Kozová, J., Sajdl, P. and Homolka, L. (1986c) Process for preparation of clavine ergot alkaloids with decreased contents of glucanes by submerged cultivation of saprophytic *Claviceps* cultures. *CS pat. 261 951* (In Czech).

Breuel, K. and Braun, K. (1981) Regulationsphänomene bei der PeptidalkaloidBiosynthese von *Claviceps purpurea. Die Pharmazie,***36,**211.

Breuel, K., Braun, K., Dauth, C. and Gröger, D. (1982) Qualitative changes in peptide alkaloid biosynthesis of *Claviceps purpurea* by environmental conditions. *PlantaMedica,***44,**121–122.

Bumbová-Linhartová, R., Flieger, M., Sedmera, P. and Zima, J. (1991) New aspects of submerged fermentation of *Claviceps paspali. Appl. Microbiol. Biotecbnol.,***34,** 703–706.

Castagnoli, N. Jr. and Mantle, P.G. (1966) Occurence of D-lysergic acid and 6-methylergol-8-ene-8-carboxylic acid in cultures of *Claviceps purpurea. Nature,***211,** 859–860.

Chain, E.B., Bonino, C. and Tonolo, A. (1960) Process for the production of alkaloid derivatives of lysergic acid. *US pat. 3 038 840.*

Chomátová, S., Ryšánová, H., Břemek, J., Řeháček, Z., Pilát, P., Homolka, L., Bárta, M., Malinka, Z. and Kozová, J. (1985) Process for long-term preservation of saprophytic spores of *Claviceps* fungus. *CS pat. 251 000* (In Czech).

Cimerman, A., Gunde-Cimerman, N. and Meze-Blatnik, J. (1992) MZKIBK—the fungal culture collection at Boris Kidrič Institute of Chemistry, Ljubljana, Slovenia. *Abstract Book, Biotechnology in Central European Initiative Countries,*Graz, Austria, April 13–15, 1992, P86.

Corbett, K., Dickerson, A.G. and Mantle, P.G. (1974) Metabolic studies on *Clavicepspurpurea* during parasitic development on rye. *J. Gen. Microbiol.,***84,**39–58.

Crespi-Perellino, N., Ballabio, M., Gioia, B. and Minghetti, A. (1987) Two unusual ergopeptines produced by a saprophytic culture of *Claviceps purpurea. J. Nat.Prod.,***50,** 1065–1074.

Crespi-Perellino, N., Guicciardi, A., Minghetti, A. and Spalla, C. (1981) Incorporation of α-aminobutyric acid into ergostine by *Claviceps purpurea. Experientia,***37,**217–218.

Crespi-Perellino, N., Malyszko, J., Ballabio, M., Gioia, B. and Minghetti, A. (1992) Directed biosynthesis of unnatural ergot peptide alkaloids. *J. Nat. Prod.,***55,**424–427.

Crespi-Perellino, N., Malyszko, J., Ballabio, M., Gioia, B. and Minghetti, A. (1993) Identification of ergobine, a new natural peptide ergot alkaloid. *J. Nat. Prod.,***56,** 489–493.

Crespi-Perellino, N., Malyszko, J. and Minghetti, A. (1994) Considerations on the biosynthesis of ergopeptines. *In:Abstract Book, IUMS Congresses 94,*Prague, Czech Republic, July 3rd-8th, 1994, p. 452.

Desai, J.D., Desai, A.J. and Patel, H.C. (1983) Effect of biotin on alkaloid production during submerged cultivation of *Claviceps* sp. strain SD-58. *Appl. Environ. Microbiol.,* **45,**1694–1696.

Desai, J.D., Desai, A. and Shah, S.R. (1982a) Activities of catabolic pathways and alkaloid biogenesis during submerged cultivation of *Claviceps* sp. SD-58. *FoliaMicrobiol.,***27,**245–249.

Desai, J.D., Desai, A.J. and Vaidya, H.C. (1982b) A new method for isolation of saprophytic cultures of *Claviceps fusiformis* from sclerotia. *Folia Microbiol.,***27**, 182–185.

Desai, J.D., Patel, H.C. and Desai, A.J. (1986) Effect of Tween series surfactants on alkaloid production by submerged cultures of *Claviceps* species. *J. Ferment. Tech.,***64**, 499–501.

Desai, J.D. and Řeháček, Z. (1982) Clavine-alkaloid production & cell-lipid accumulation during the submerged cultivation of *Claviceps purpurea* (Fr.) Tul.*IndianJ. Exp. Biol.,* **20**,181–183.

De Tempe, J. (1945) Synthesis of alkaloids by *Claviceps purpurea* (Fr.) Tul. in saprophytic culture. Diss. Amsterdam (In Dutch).

Didek-Brumec, M., Gaberc-Porekar, V., Alačević, M., Druškovič, B. and Sočič, H. (1991a) Characterization of sectored colonies of a high-yielding *Claviceps purpurea* strain. *J. Basic Microbiol.,***31**,27–35.

Didek-Brumec, M., Gaberc-Porekar, V., Alačević, M., Miličič, S. and Sočič, H. (1991b) Activation of ergot alkaloid biosynthesis in prototrophic isolates by *Clavicepspurpurea* protoplast fusion. *J. Biotecbnol.,***20**,271–278.

Didek-Brumec, M., Gaberc-Porekar, V., Alačević, M. and Sočič, H. (1992) Strain improvement of *Claviceps purpurea* by genetic recombination. *Abstract Book, Biotechnologyin Central European Initiative Countries,*Graz, Austria, April 13–15, 1992, P123.

Didek-Brumec, M., Gaberc-Porekar, V., Alačević, M. and Sočič, H. (1993) Strain improvement of *Claviceps purpurea* by protoplast fusion without introducing auxotrophic markers. *Appl. Microbiol. Biotechnol.,***38**,746–749.

Didek-Brumec, M., Jezernik, K., Puc, A. and Sočič, H. (1988) Ultrastructural characteristics of *Claviceps purpurea* seed cultures. *J. Basic Microbiol.,***28**,589–598.

Dierkes, W., Lohmeyer, M. and Rehm, H.-J. (1993) Long-term production of ergot peptides by immobilized *Claviceps purpurea* in semicontinuous and continuous culture. *Appl. Environ. Microbiol.,***59**,2029–2033.

Dinelli, D., Morisi, F. and Cecere, F. (1972) A process for the enzymatic production of L-tryptophan. *GB pat. 1 386 674.*

Dowd, P.P., Cole, R.J. and Vesonder, R.F. (1988) Control of insects by fungal tremorgenic mycotoxins. *US pat. 4 973 601.*

Eich, E. and Eichberg, D. (1982) Zur antibakteriellen Wirkung von Clavinalkaloiden und deren partialsynthetischen Derivaten. *Planta Medica,***48**,146–147.

Eich, E., Eichberg, D., Schwarz, G., Clas, F. and Loos, M. (1995) Antimicrobial activity of clavines. *Arzneim.-Forsch./DrugRes.,***35 (II),**1760–1763.

Eich, E. and Sieben, R. (1985) 8 α-Hydroxylierung von 6-Nor-Agroclavin und Lysergin durch *Claviceps fusiformis* . *Planta Medica,***47**,282–283.

Enatsu, T. and Terui, G. (1967) Fermentation process for producing 1-tryptophane. *USpat. 3 296 090.*

Erge, D., Maier, W. and Gröger, D. (1981) Mutational biosynthesis in *Clavicepspurpurea* (Fr.) Tul. *Abstract Book, FEMS Symp. Overprod. Microbial Prod.,*Hradec Králové, Czechoslovakia, August 9–14, 1981, 221.

Erge, D., Schumann, B. and Gröger, D. (1984) Influence of tryptophan and related compounds on ergot alkaloid formation in *Claviceps purpurea* (Fr.) Tul.*Z. Allgem.Mikrobiol.,***24**,667–678.

Erge, D., Schumann, B., Schmauder, H.-P., Gröger, D., Maier, W., Baumert, A., Braun, K., Breuel, K., Höhne, L. and Ludwigs, J. (1982) Verfahren zur Herstellung von Ergocristin. *DD pat. 212 749.*

Erge, D., Wenzel, A. and Gröger, D. (1972) Physiology of alkaloidsynthesis in species of *Claviceps. Biochem. Physiol. Pflanzen,*163, 288. (In German)

Esser, K. and Düvell, A. (1984) Biotechnological exploitation of ergot fungus *(Claviceps purpurea). Process Biochemistry,*19,142–149.

Esser, K. and Tudzynski, P. (1978) Genetics of the ergot fungus *Claviceps purpurea.* I. Proof of a monoecious life cycle and segregation patterns for mycelial morphology and alkaloid production. *Theor. Appl. Genet.,*53,145–149.

Fiedler, F., Zenk, M., Gundlach, H., Weber, A. and Kennecke, M. (1989) Verfahren zur Steigerung von Enzym-Aktivitäten und der Syntheseleistung von Mikroorganismen und höheren Pflanzen. *EP pat. 0325 933.*

Flieger, M., Linhartová, R., Řeháček, Z., Sajdl, P., Stuchlík, J., Malinka, Z., Harazim, P., Cvak, L. and Břemek, J. (1989a) Industrial strain of microorganism *Claviceps paspali* Stevens et Hall. CCM 8061. *CZ pat. 276 430* (In Czech).

Flieger, M., Linhartová, R., Řeháček, Z., Sajdl, P., Stuchlík, J., Malinka, Z., Harazim, P., Cvak, L. and Břemek, J. (1989b) Process for production of lysergic acid simple derivatives. *CZ pat. 281 196* (In Czech).

Flieger, M., Linhartová, R., Sedmera, P., Zima, J., Sajdl, P., Stuchlík, J. and Cvak, L. (1989c) New alkaloids of *Claviceps paspali. J. Nat. Prod.,*52, 1003–1007.

Flieger, M., Sedmera, P., Havlíček, V., Cvak, L. and Stuchlík, J. (1993) 10-hydroxy-*cis*- and 10-hydroxy-*trans*-paspalic acid amide: New alkaloids from *Claviceps paspali. J. Nat. Prod.,*56,810–814.

Flieger, M., Sedmera, P., Vokoun, J., Řeháček, Z., Stuchlík, J., Malinka, Z., Cvak, L. and Harazim, P. (1984) New alkaloids from a saprophytic culture of *Claviceps purpurea. J. Nat. Prod.,*47,970–976.

Flieger, M., Sedmera, P., Vokoun, J., Řičicová, A. and Řeháček, Z. (1982) Separation of four isomers of lysergic acid α-hydroxy-ethylamide by liquid chromatography and their spectroscopic identification. *J. Chromatogr.,*236,453–459.

Flieger, M., Votruba, J., Křen, V., Pažoutová, S., Rylko, V., Sajdl, P. and Řeháček, Z. (1988) Physiological control and process kinetics of clavine alkaloid production by *Claviceps purpurea. Appl. Microbiol. Biotechnol.,*29,181–185.

Flieger, M., Votruba, J., Sajdl, P., Stuchlík, J., Cvak, L., Harazim, P. and Malinka, Z. (1987) Process for submerged fermentation production of ergot alkaloids by directed physiological conditions. *CS pat. 268 211* (In Czech).

Flieger, M., Wurst, M. and Shelby, R. (1997) Ergot alkaloids-sources, structures and analytical methods. *Folia Microbiol.,*42,3–30.

Flieger, M., Zelenkova, N.F., Sedmera, P., Křen, V., Novák, J., Rylko, V., Sajdl, P. and Řeháček, Z. (1989d) Ergot alkaloid glycosides from saprophytic cultures of *Claviceps.* l. Elymoclavine. *J. Nat. Prod.,*52,506–510.

Floss, H.G. (1976) Biosynthesis of ergot alkaloids and related compounds. *Tetrahedron,* 32,873–912.

Floss, H.G., Günther, H., Mothes, U. and Becker, I. (1967) Isolierung von Elymoclavine-O-*β*-D-fruktosid aus Kulturen des Mutterkornpilzes. *Z. Naturforsch.,*22b,399–402.

Floss, H.G., Basmadjian, G.P., Tcheng, M., Gröger, D. and Erge, D. (1971a) Biosynthesis of peptide-type ergot alkaloids. Ergocornine and ergocryptine. *Lloydia,*34,446–448.

Floss, H.G., Basmadjian, G.P., Tcheng, M., Spalla, C. and Minghetti, A. (1971b) Biosynthesis of peptide-type ergot alkaloids. Ergotamine. *Lloydia,* **34**,442–445.

Fukuda, H. (1986) Process for secretive fermentation of lipids by fungi or algae. *EPpat. 207 475.*

Gaberc-Porekar, V., Didek-Brumec, M. and Sočič, H. (1981) Direct selection of active *Claviceps* colonies on agar plates. *Abstract Book, FEMS Symp. Overprod. MicrobialProd.,*Hradec Králové, Czechoslovakia, August 9–14, 1981, 220.

Gaberc-Porekar, V., Didek-Brumec, M. and Sočič, H. (1983) Direct selection of active *Claviceps* colonies on agar plates. *Z. Allgem. Mikrobiol.,* **23**,95–98.

Gaberc-Porekar, V., Didek-Brumec, M. and Sočič, H. (1990) Carbohydrate metabolism during submerged production of ergot alkaloids. *Appl. Microbiol. Biotechnol.,* **34**, 83–86.

Gaberc-Porekar, V., Didek-Brumec, M. and Sočič, H. (1992a) Correlation studies between tricarboxylic acid cycle and metabolic pathways of lipids and ergot alkaloids in a high-yielding *Claviceps purpurea* strain. *Abstract Book, Biotechnology in Central European Initiative Countries,*Graz, Austria, 13–15 April 1992, P124.

Gaberc-Porekar, V., Didek-Brumec, M. and Sočič, H. (1992b) Metabolic block of tryptophan synthesis in Claviceps purpurea cultures. *Abstract Book, Biotechnologyin Central European Initiative Countries,*Graz, Austria, 13–15 April 1992, P125.

Gaberc-Porekar, V., Sočič, H., Pertot, E. and Miličić, S. (1987) Metabolic changes in a conidia-induced *Claviceps paspali* strain during submerged fermentation. *Can. J.Microbiol.,* **33**,602–606.

Grm, B., Mele, M. and Kremser, M. (1980) Model of Growth and Ergot Alkaloid Production by *Claviceps purpurea. Biotechnol. Bioeng.,* **22**,255–270.

Gröger, D. (1965) Über die Bildung von Lysergsäurederivaten in Submerskultur von *Claviceps paspali.* 3. Mitteilung: Izolation von Chanoclavin—(I). *Pharmazie,* **30**, 523–524.

Gröger, D., Gröger, L., D'Amico, D., He, M.-X. and Floss, H.G. (1991) Steric course of the N-methylation in the biosynthesis of ergot alkaloids by *Claviceps purpurea. J.Basic Microbiol.,* **31**,121–125.

Gröger, D., Schmauder, H.-P., Johne, S., Maier, W. and Erge, D. (1977) Verfahren zur Gewinnung von Ergosin. *DD pat. 129 801.*

Gubański, M. and Lowkis, L. (1964) Effect of the ergot inhibitor *(Claviceps purpurea)* on the tobacco mosaic virus (TMV). *Acta Microbiol. Polonica,* **13**,169–175.

Harazim, P., Malinka, Z. and Břemek, J. (1986) Influence of inoculation mediums composition with different carbon source on fermentative production of lysergic acid alpha-hydroxyethylamide. *Abstract Book, 17th Congr. Czecboslov. Microbiol.Soc.,*České Budějovice, Czechoslovakia, September 23–25, 1986, p. A-16 (In Czech).

Harazim, P., Malinka, Z., Stuchlík, J., Cvak, L., Břemek, J., Flieger, M, Linhartová, R., Řeháček, Z. and Sajdl, P. (1989) Industrial strain of microorganism *Claviceps paspali* Stevens et Hall. CCM 8063. *CZ pat. 276 032* (In Czech).

Harazim, P., Valík, J., Malinka, Z. and Kybal, J. (1984) Process for fermentative production of ergot inoculating material. *CS pat. 243 009* (In Czech).

Harris, J.R. and Horwell, D.C. (1992) Conversion of agroclavine to lysergol. *SyntheticCommunications,* **22**,995–999.

Hofmann, A., Brunner, R., Kobel, H. and Brack, A. (1957) New alkaloid from saprophytic culture of ergot fungus in *Pennisetum typhoideum* Rich. *Helv. Chim.Acta,* **40,**1358–1373 (In German).

Homolka, L., Pilát, P., Řeháček, Z., Bárta, M., Chomátová, S., Malinka, Z. and Břemek, J. (1985) Process for evaluation of producing ability of microbial eukaryote isolates, producing ergot alkaloids. *CS pat. 259 670* (In Czech).

Hunter-Cervera, J.C. and Belt, A. (1996) *Maintaining Cultures for Biotechnology andIndustry.*Academic Press, New York.

Janardhan, K.K. and Husain, A. (1984) A new strain of *Claviceps paspali* Stevens et Hall producing ergometrine in submerged culture. *Proc. Indian natn. Sci. Acad.,***B50,** 438–440.

Johal, S.S. and Cash, H.S. (1989) Recovery of polysaccharides by employing a divalent cation with a water miscible organic solvent. *US pat. 4 960 697.*

Keller, U. (1983) Highly efficient mutagenesis of *Claviceps purpurea* by using protoplasts. *Appl. Environment. Microbiol.,***46,**580–584.

Keller, U., Han, M. and Stoffler-Meilicke, M. (1988) D-lysergic acid activation and cell-free synthesis of D-lysergyl peptides in enzyme fractions from the ergot fungus *Claviceps purpurea. Biochemistry,***27,**6164–6170.

Keller, U., Zocher, R. and Kleinkauf, H. (1980) Biosynthesis of ergotamine in protoplasts of *Claviceps purpurea. J. Gen. Microbiol.,***118,**485–494.

Kim, S.-U. and Anderson, J.A. (1982) Conversion of elymoclavine to paspalic acid by a paniculate fraction from an ergotamine-producing strain of *Claviceps* sp. *PlantaMedica,* **47,**141.

Kim, S.-U., Cho, Y.-J., Floss, H.G. and Anderson, J.A. (1983) Conversion of elymoclavine to paspalic acid by a paniculate fraction from an ergotamine-producing strain of *Claviceps* sp. *Planta Medica,***48,**145–148.

Kim, I.-S., Kim, S.-U. and Anderson, J.A. (1981) Microsomal agroclavine hydroxylase of *Claviceps* species. *Phytochemistry,***20,**2311–2314.

Kiniczky, M., Tétényi, P., Kiss, J., Zámbó, I. and Balossa, J. (1982) Process for conservation of ergot and yeast spores. *HU pat. 181 706* (In Hungarian).

Kirsop, B. and Snell, I.I. (eds.) (1984) *Maintenance of Microorganisms. A Manual ofLaboratory Methods.*Academic Press, London.

Kleinerová, E. (1975) Metabolism of ergot *Claviceps purpurea* (Fr.) Tul. and relationship with alkaloid biosynthesis. *Dissertation,*Charles Univ., Prague (In Czech).

Kobel, H. (1969) Degenerationsprobleme bei Produktionsstämmen von *Claviceps. Pathologia Microbiol.,***34,**249–251.

Kobel, H., Brunner, R. and Brack, A. (1962) Vergleich der Alkaloidbildung von *Clavicepspurpurea* in Parasitischer und Saprophytischer Kultur. *Experientia,***18,** 140–142.

Kobel, H. and Sanglier, J.-J. (1976) Precédé de fabrication d'alcaloides du groupe de l'ergotoxine par fermentation combinée. *FR pat. 2 307 870.*

Kobel, H. and Sanglier, J.-J. (1978) Formation of ergotoxine alkaloids by fermentation and attempts to control their biosynthesis. In: R.Hütter, T.Leisinger, J.Nüesch, W. Wehrli, (eds.) *Antibiotics and other secondary metabolites. FEMS Symposium No. 5.*Academic Press, London,pp. 233–242.

Kobel, H. and Sanglier, J.-J. (1986) Ergot Alkaloids. In H.J.Rehm and G.Reed, (eds.). *Biotechnology,*Vol. 4, VCH, Weinheim, Germany, pp. 569–609.

Kobel, H., Sanglier, J.-J., Tscherter, H. and Boilinger, G. (1982) Ergot peptide alkaloids, their preparation and pharmaceutical compositions containing them. *GB pat.2109 795.*

Kobel, H. and Schreier, E. (1966) Verfahren für Gewinnung von Lyserg Säure Derivaten. *CH pat. 482 831.*

Kobel, H., Schreier, E. and Rutschmann, J. (1964) Mutterkorn Alkaloiden 60. 6-Methyl $\Delta^{8,9}$-ergolen-8-carbosäure ein neues Ergolinderivat aus Kulturen eines Stammes von *Claviceps paspali* Stevens et Hall. *Helv. Chim. Acta,* **47,** 1052–1064.

Komel, R., Rozman, D., Puc, A. and Sočič, H. (1985) Effect of immobilization on the stability of *Claviceps purpurea* protoplasts. *Appl. Microbiol. Biotechnol.,* **23,** 106–109.

Kopp, B. (1987) Long-term alkaloid production by immobilized cells of *Clavicepspurpurea.* In: *Methods in Enzymology,* **136,** 317–329.

Kopp, B. and Rehm, H.J. (1983) Alkaloid production by immobilized mycelia of *Claviceps purpurea. Eur.J. Appl. Microbiol. Biotechnol.,* **18,** 257–263.

Kopp, B. and Rehm, H.J. (1984) Semicontinuous cultivation of immobilized *Clavicepspurpurea. Appl. Microbiol. Biotechnol.,* **19,** 141–145.

Košir, B., Smole, P. and Povšič, Z. (1981) Influence of some factors on yield and quality of ergot sclerotiums. *Farm. Vestn. (Ljubljana)* **32,** 21–25 (In Slovenian).

Kozikowski, A.P., Chen, C., Wu, J.-P., Shibuya, M., Kim, C.-G. and Floss, H.G. (1993) Probing ergot alkaloid biosynthesis: Intermediates in the formation of ring C. *J. Am.Chem. Soc.,* **115,** 2482–2488.

Kozlovsky, A.G., Arinbasarov, M.U., Solovieva, T.F., Angelov, T.J., Slokoska, L.S., Angelova, M.B. and Veličkov, I.G. (1978) Process for production of elymoclavine. *SUpat. 735 010* (In Russian).

Kozlovsky, A.G., Solovieva, T.F., Adanin, V.M. and Skryabin, G.K. (1979) Process for preparing epoxyagroclavine. *SU pat. 955 693* (In Russian).

Kozlovsky, A.G., Solovieva, T.F., Sakharovsky, V.G. and Adanin, V.M. (1982) Ergot alkaloids agroclavine-I and epoxyagroclavine-I-metabolites of *Penicilliumcorylophilum. Prikl. Biokhim. Mikrobiol.,* **18,** 535–541 (In Russian).

Křen, V. (1991) Bioconversions of ergot alkaloids. In A.Fiechter (ed.), *Advances inBiochemical Engineering/Biotechnology,* Springer, Germany, pp. 123–144.

Křen, V., Břemek, J., Flieger, M., Kozová, J., Malinka, Z. and Řeháček, Z. (1989a) Bioconversion of agroclavine by free and immobilized *Claviceps fusiformis* cells. *Enzyme Microb. Technol.,* **11,** 685–691.

Křen, V., Chomátová, S., Břemek, J., Pilát, P. and Řeháček, Z. (1986a) Effect of some broad-spectrum antibiotics on the high-production strain *Claviceps fusiformis* W I. *Biotechnol. Lett.,* **8,** 327–332.

Křen, V., Harazim, P. and Malinka, Z. (1994) *Claviceps purpurea* (ergot): culture and bioproduction of ergot alkaloids. In Y.P.S.Bajaj, (ed.), *Biotechnology in Agricultureand Forestry 28, Medicinal and Aromatic Plants VII,* Springer-Verlag, Berlin, Germany, pp. 139–156.

Křen, V., Kozová, J. and Řeháček, Z. (1986b) Production of ergot alkaloids by *Clavicepsfusiformis* immobilized cells in semicontinuous and continuons process. In: *Abstract Book, 17. Congres Czechoslov. Microbiol. Soc.,* České Budějovice, Czechoslovakia, 23–25. September, 1986, p. A-17 (In Czech).

Křen, V., Mehta, P., Rylko, V., Flieger, M., Kozová, J., Sajdl, P. and Řeháček, Z. (1987) Substrate regulation of elymoclavine formation by some saccharides. *Zentralbl.Mikrobiol.,* **142,** 71–85.

Křen, V., Pažoutová, S., Řezanka, T. and Sajdl, P. (1988a) Chlorophenoxyacids influence ergot alkaloid production in *Claviceps* cultures by modification of cell-membrane lipid composition. *Abstract Book, 2nd Internal. Symp. Overprod.Microbial Products,*České Budějovice, Czechoslovakia, July 3–8, 1988, 156.

Křen, V., Pažoutová, S., Rylko, V., Sajdl, P., Wurst, M. and Řeháček, Z. (1984). Extracellular metabolism of sucrose in a submerged culture of *Claviceps purpurea:* Formation of monosaccharides and clavine alkaloids. *Appl. Environ. Microbiol.,***48**, 826–829.

Křen, V., Pažoutová, S. and Sajdl, P. (1988b) Chlorophenoxyacids alters lipid composition and protoplast membrane fluidity in *Claviceps* cells. *Abstract Book, IUB 14thInternal. Congress of Biochemistry,*Prague, Czechoslovakia, July 10–15, 1988, TU, p. 97.

Křen, V. and Řeháček, Z. (1984) Feedforward regulation of phosphofructokinase in a submerged culture of *Claviceps purpurea* producing clavine alkaloids. *Speculat. inSci. and Technol.,***7,**223–226.

Křen, V., Řeháček, Z., Kozová, J., Sajdl, P. and Ludvík, J. (1985) Process for production of elymoclavine alkaloid by semicontinuous cultivation of fungus *Claviceps* immobilized cells. *CS pat. 250 474* (In Czech).

Křen, V., Sajdl, P., Flieger, M. and Svatoš, A. (1989b) Method for preparing ergot alkaloid fructosides. *CZ pat. 280 415* (In Czech).

Křen, V., Sajdl, P., Malinka, Z. and Břemek, J. (1989c) Industrial strain of *Clavicepspurpurea* (Fr.) Tul. 88-EP-47/1989. *CS pat. 275 000* (In Czech).

Křen, V., Sajdl, P., Řeháček, Z., Malinka, Z. and Břemek, J. (1988c) Industrial strain of *Claviceps purpurea* (Fr.) Tul. 88-EP/1988 for elymoclavine preparation. *CZ pat.272 562* (In Czech).

Krustev, H.I., Slokoska, L.S. and Grigorov, I.V. (1984) Electron microscopic investigations of *Claviceps* sp. CP II, producer of clavine alkaloids. *Dokl. Bolg. Akad. Nauk,***37,**1233–1236 (In Bulgarian).

Kubec, K., Culik, K., Hilbert, O., Kybal, J. and Severa, Z. (1974) Process for preparation of durable storable material for preparation of inoculum by field ergot production. *CS pat. 155 440* (In Czech).

Kybal, J. (1955) Process for preparation of durable rye inoculating material of ergot. *CS pat. 86 302* (In Czech).

Kybal, J. (1963) Process for *Claviceps purpurea* (Fr.) Tul. cultivation. *CS pat. 114 933* (In Czech).

Kybal, J. (1981) Vegetative Differenzierung von *Claviceps purpurea* in Saprophytischer Kultur als Voraussetzung der Alkaloidsynthese. *Pharmazie,***36,**212.

Kybal, J., Horák, P., Brejcha, V. and Kudrnáč, S. (1956) Monokonidienisolation und deren Bedeutung bei der Selektion von Rassen und Stämmen des Mutterkorns. *Abh. Deutsch. Akad. Wiss. Berlin. f. Chem. Geol.-Biol.,***7,**236–242.

Kybal, J., Malinka, Z., Harazim, P., Stuchlík, J., Spáčil, J. and Svoboda, E. (1979) Process for fermentative production of ergot alkaloids with directed ratio of ergocornine, α-ergokryptine and β-ergokryptine. *CS pat. 202 613* (In Czech).

Kybal, J. and Nesrsta, M. (1994) Ecological plant protection-reneval of equilibrium between pests and their natural antagonists. In: *Abstract Book, IUMS Congresses,*Prague, Czech Republic, July 3–8, 1994, p. 464.

Kybal, J., Nesrsta, M., Strnadová, K., Břemek, J., Valík, J. and Považská, H. (1990) Process for production of storable preparations with high content of live spores of filamentous fungi. *CS pat. 276 530* (In Czech).

Kybal, J., Protiva, J., Strnadová, K., Starý, F. and Čekan, Z. (1960) Process for artificial cultivation of ergot fungus. *CS pat. 104 613* (In Czech).

Kybal, J. and Starý, F. (1958) Beitrag zur biologischen Problematik der Saprophytischen Kultur von *Claviceps purpurea* (Fries) Tul. *Planta Medica,***6,**404–409.

Kybal, J. and Strnadová, K. (1968) Den Ertrag von Mutterkorn bei Feldmässiger Kultur bestimmende Faktoren. *Herba Hungar.***7,**123–135.

Kybal, J. and Strnadová, K. (1982) Process for production of large amount of physiologicaly high active conidies of *Claviceps purpurea* (Fr.) Tul. *CS pat. 227 392* (In Czech).

Kybal, J., Svoboda, E. and Kleinerová, E. (1978) Utilization of malic acid and its importance for the biosynthesis of peptide ergot alkaloids. *Abstract Book, 14th Ann.Meet, of Czechoslov. Soc. Microbiol.,*Prague, Czechoslovakia, October 17–19, 1978.

Kybal, J., Svoboda, E., Strnadová, K. and Kejzlar, M. (1981) Role of organic acid metabolism in the biosynthesis of peptide ergot alkaloids. *Folia Microbiol.,***26,** 112–119.

Kybal, J. and Vlček, V. (1976) A simple device for stationary cultivation of microorganisms. *Biotechnol. Bioeng.,***18,**1713–1718.

Lapis, K., Kopper, L., Szende, B., Patthy, A., Molnar, G. and Tyihak, E. (1978) Antitumour protein prepared from *Claviceps* strain. *HU pat. appl. T 14 568* (In Hungarian).

Linhartová, R., Pažoutová, S., Sajdl, P. and Řeháček, Z. (1988) Production and degradation of simple lysergic acid derivatives in submerged cultures of *Clavicepspaspali* MG-6. *Abstract Book, 2nd Internal. Symp. Overprod. Microbial Products,*České Budějovice, Czechoslovakia, July 3–8, 1988.

Lohmeyer, M., Dierkes, W. and Rehm, H.-J. (1990) Influence of inorganic phosphate and immobilization on *Claviceps purpurea. Appl. Microbiol. Biotechnol.,***33,**196–201.

Lohmeyer, M. and Sander, B. (1993) Cultivation of immobilized cells of *Clavicepspurpurea* in bioreactors. *J. Ferment. Bioeng.,***76,**376–381.

Lösecke, W., Neumann, D. and Gröger, D. (1981) Zusammenhänge zwischen Alkaloidbildung und Ultrastruktur bei *Claviceps purpurea* in Submerskultur. *Pharmazie,***36,**211.

Lösecke, W., Neumann, D., Schmauder, H.-P. and Gröger, D. (1980) Ultrastructure of submerged non-alkaloid producing strains of *Claviceps purpurea* (Fr.) Tul. *Biochem. Physiol. Pflanzen,***175,**552–561.

Lösecke, W., Neumann, D., Schmauder, H.-P. and Gröger, D. (1982) Changes in cytoplasmic ultrastructure during submerged cultivation of a peptide alkaloids-producing strain of *Claviceps purpurea* (Fr.) Tul. *Z.Allgemeine Mikrobiol.,***22,**49–61.

Ludwigs, J., Volzke, K., Knothe, E., Schmauder, H.-P., Gröger, D. and Schumann, B. (1985) Verfahren zur Herstellung von alpha—Ergokryptin. *DD pat. 288 836.*

Maier, W., Erge, D. and Gröger, D. (1971) Zur Biosynthese von Ergotoxinalkaloiden in *Claviceps purpurea. Biochem. Physiol. Pflanzen,***161,**559–569.

Maier, W., Erge, D. and Gröger, D. (1980a) Mutational biosynthesis in a strain of *Claviceps purpurea. Planta Medica,***40,**104–108.

Maier, W., Erge, D. and Gröger, D. (1983) Further studies on cell-free biosynthesis of ergotamine in *Claviceps purpurea*. *FEMS Microbiol. Lett.*, **20**,233–236.

Maier, W., Erge, D., Schmidt, J. and Gröger, D. (1980b) A blocked mutant of *Clavicepspurpurea* accumulating chanoclavine-I-aldehyde. *Experientia*, **36**,1353–1354.

Maier, W., Erge, D., Schumann, B. and Gröger, D. (1981) Incorporation of L-[U-^{14}C] leucine into ergosine by cell-free extracts of *Claviceps purpurea* (Fr.) Tul. *Biochem. Biophys. Res. Commun.*, **99**,155–162.

Maier, W., Schumann, B. and Gröger, D. (1988a) Effect of chlorsulfuron, a potent inhibitor of acetohydroxyacid synthase, on metabolism of *Claviceps purpurea*. *Z.Naturforsch.*, **43c**,403–407.

Maier, W., Schumann, B. and Gröger, D. (1988b) Microsomal oxygenases involved in ergoline alkaloid biosynthesis of various *Claviceps* strains. *J. Basic Microbiol.*, **28**, 83–93.

Majer, J., Kybal, J. and Komersová, I. (1967) Ergot alkaloids. I. Biosynthesis of the peptide side chain. *Folia Microbiol.*, **12**,489–491.

Malinka, Z. (1982) Konstanter Luftdurchsatz für den biologischem Prozess. *Messen +Steuern*, **44**,37.

Malinka, Z. (1988) Physiological state of the culture by fermentation production of the ergot alkaloids In: *Industrial biocbemicals-Interbiotech '88*,Bratislava, June 29-July 1, 1988, 61.

Malinka, Z. and Břemek, J. (1989) Biotransformation of the clavine alkaloids by the submerged cultures of *Claviceps.Abstract Book, 19th Meeting FEBS*,Rome, Italy, July 2–7, 1989, FR251.

Malinka, Z., Břemek, J., Harazim, P. and Pilát, P. (1986) Clavine ergot alkaloids production by *Claviceps* culture in dependence on inoculating material. *AbstractBook, 17th Congr. Czechoslov. Microbiol. Soc.*,České Budějovice, Czechoslovakia, September 23–25, 1986, A-15. (In Czech).

Malinka, Z., Břemek, J. and Pilát, P. (1988) Production of clavine alkaloids at maintenance and transfers of *Claviceps* cultures. In: *Abstract Book, 2nd International Symp. Overprod. Microbial Products*,České Budějovice, Czechoslovakia, July 3–8, 1988, 158.

Malinka, Z., Harazim, P., Břemek, J. and Valík, J. (1987) Influence of aminoacids in ergot alkaloid biosynthesis. *Bulletin Čs. spol. biochem. při ČSAV*, **15**,107 (In Czech).

Mantle, P.G. (1969) Development of alkaloid production *in vitro* by a strain of *Claviceps purpurea* from *Spartia townsendii*. *Trans. Br. Mycol. Soc.*, **52**,381–392.

Mantle, P.G. (1975) Industrial exploitation of ergot fungi. In: J.E.Smith, D.R.Berry, (eds.) *The Filamentous Fungi 1. Industrial Mycology*,Arnold, London, pp. 281–300.

Mantle, P.G. and Nisbet, L.J. (1976) Differentiation of *Claviceps purpurea* in axenic culture. *J. Gen. Microbiol.*, **93**,321–324.

Mantle, P.G. and Waight, E.S. (1968) Dihydroergosine: A new naturally occurring alkaloid from the sclerotia of *Sphacelia sorghi* (McRae). *Nature*, **218**,581–582.

Mary, N.Y., Kelleher, W.J. and Schwarting, A.E. (1965) Production of lysergic acid derivatives in submerged culture. III. Strain selection on defined media. *Lloydia*, **28**, 218–229.

Mat'ha, V. (ed.) (1993) *The Story of the Czech Cyclosporin A*.Galena, Opava-Komárov, Czech Republic.

Mat'ha, V., Jegorov, A., Weiser, J., Harazim, P., Malinka, Z. and Stuchlík, J. (1993) Production of cyclosporins by *Tolypocladium terricola* in stationary cultivation. *Microbios,***75,**83–90.

Matošić, S., Mehak, M., Ercegović, L. and Brajković, N. (1988a) Improvement of ergot alkaloids biosynthesis by means of immobilized cells of *Claviceps paspali*. *AbstractBook, 2nd Internat. Symp. Overprod. Microbial Products,*České Budějovice, Czechoslovakia, July 3–8, 1988, 63.

Matošić, S., Mehak, M., Ercegović, L. and Brajković, N. (1988b) Effect of surfactans on biosynthesis of ergot alkaloids by means of immobilized mycelium of *Clavicepspaspali. Abstract Book, 8th Internat. Biotechnology Symp.,*Paris, France, July 17–22, 1988, 309.

McCrea, A. (1931) The reactions of *Claviceps purpurea* to variations of environment. *Am. J. Bot.,***18,**50–78.

McCrea, A. (1933) Ergot preparation and process of obtaining same. *U.S. pat. 2 056 360*

Miličić, S., Kremser, M., Gaberc-Porekar, V., Didek-Brumec, M. and Sočič, H. (1987) Correlation between growth and ergot alkaloid biosynthesis in *Claviceps purpurea* batch fermentation. *Appl. Microbiol. Biotechnol.,***27,**117–120.

Miličić, S., Kremser, M., Gaberc-Porekar, V., Didek-Brumec, M. and Sočič, H. (1989) The effect of aeration and agitation on *Claviceps purpurea* dimorphism and alkaloid synthesis during submerged fermentation. *Appl. Microbiol. Biotechnol.,***31,**134–137.

Miličić, S., Kremser, M., Povšič, Z. and Sočič, H. (1984) Growth and sporulation of the fungus *Claviceps purpurea* in batch fermentation. *Appl. Microbiol. Biotechnol.,***20,** 356–359.

Miličić, S., Velušček, J., Kremser, M. and Sočič, H. (1993) Mathematical modeling of growth and alkaloid production in *Claviceps purpurea* batch fermentation. *Biotechnol. Bioeng.,***41,**503–511.

Minghetti, A., Spalla, C. and Tognoli, L. (1967) Fermentatives Verfahren zur Herstellung von Ergocristin. *DE pat. 1 806 984.*

Molnár, G. and Tétényi, P. (1962) Process of ergometrine, ergokryptine and ergocornine production by saprophytic cultivation. *HU pat. 150 631* (In Hungarian).

Molnár, G., Tétényi, P., Udvardy Nagy, E., Wack, G. and Wolf, L. (1964) Verfahren zur biosynthetischen Herstellung eines Mutterkornalkaloidgemisches. *CH pat. 455 820.*

Mothes, K., Winkler, K., Gröger, D., Floss, H.G., Mothes, U. and Weygand, F. (1962) Über die Umwandlung von Elymoclavin in Lysergsäurederivate durch Mutterkornpilze *(Claviceps). Tetrahedron Lett.,*21, 933–937.

Nagy, A., Zalai, K., Manczinger, L. and Ferenczy, L. (1994) Interspecific protoplast fusion between *Claviceps fusiformis* and *Claviceps purpurea* auxotrophic mutants. *Abstract Book, 7th IUMS Congresses,*Prague, Czech Rep., July 3–8, 1994, 454.

Nečásek, J. (1954) Monosporic isolation of fungi. *Preslia,*26, 105–109 (In Czech).

Nesrsta, M. (1989) Effectioness of the industrial production of mycopesticides. In. *Abstract Book, Biopesticides, theory and practice,*September 25–28, 1989, České Budějovice, Czechoslovakia, 22.

Nordmann, G. and Bärwald, G. (1981) Untersuchungen zur Mutabilität von *Clavicepspurpurea* unter besonderer Berücksichtigung der Leistungssteigerung saprophytischer Alkaloidbildner. *Pharmazie,*36, 211.

Oishi, K., Sugiyama, S. and Yokota, T. (1984) Preparation of plant-growth controlling component by microorganism. *JP pat. 61 067 491* (In Japanese).

Olasz, K., Gaál, T. and Zalai, K. (1982) Improvement of *Claviceps purpurea* by mutagenic treatment of protoplast. *Acta Biochim. Biophys. Acad. Sci. Hungar.,* **17,** 126.

Otsuka, H., Quigley, F.R., Gröger, D., Anderson, J.A. and Floss, H.G. (1980) *In vivo* and *in vitro* evidence for N-methylation as the second pathway-specific step in ergoline biosynthesis. *Planta Medica,* **40,** 109–119.

Page, G.V., Scire, B. and Farbood, M. (1989) A method for the preparation of "natural" methyl anthranilate. *WO pat appl. 89/00203.*

Patel, H.C. and Desai, J.D. (1985) Phosphate transport in *Claviceps* sp. strain SD-58. *Experientia,* **41,** 96–97.

Pažoutová, S., Flieger, M., Rylko, V., Křen, V. and Sajdl, P. (1987a) Effect of cultivation temperature, clomiphene and nystatin on the oxidation and cyclization of chanoclavine in submerged cultures of the mutant strain *Claviceps purpurea* 59. *Curr. Microbiol.,* **15,** 97–101.

Pažoutová, S., Flieger, M., Sajdl, P. and Řeháček, Z. (1981a) The relationship between intensity of oxidative metabolism and predominance of agroclavine or elymoclavine in submerged *Claviceps purpurea* cultures. *J. Nat. Prod.,* **44,** 225–235.

Pažoutová, S., Křen, V., Řezanka, T., Amler, E., Flieger, M., Rylko, V. and Sajdl, P. (1989) Effect of triadimefon on lipids, sterols and membrane fluidity in submerged cultures of *Claviceps purpurea*. *Pestic. Biochem. Physiol.,* **34,** 211–217.

Pažoutová, S., Křen, V., Řezanka, T. and Sajdl, P. (1988) Effect of clomiphene on fatty acids, sterols and membrane fluidity in clavine producing *Claviceps purpurea* strains. *Biochem. Biophys. Res. Commun.,* **152,** 190–196.

Pažoutová, S., Křen, V., Rylko, V. and Řeháček, Z. (1986) Direction of clavine alkaloid biosynthesis and conversion in submerged cultures of *Claviceps* strains. *AbstractBook, 17th Congr. Czechoslov. Microbiol. Soc.,* České Budějovice, Czechoslovakia, September 23–25, 1986, A-9 (In Czech).

Pažoutová, S., Křen, V., Rylko, V. and Sajdl, P. (1987b) Composition of lipids and membrane enzymes in *Claviceps* fungi. *Bull Čs. spol. biochem.,* **15,** 109 (In Czech).

Pažoutová, S., Křen, V. and Sajdl, P. (1990) The inhibition of ciavine biosynthesis by 5-fluorotryptophan; a useful tool for the study of regulatory and biosynthetic relationships in *Claviceps*. *Appl. Microbiol. Biotechnol.,* **33,** 330–334.

Pažoutová, S., Pokorný, V. and Řeháček, Z. (1977) The relationship between conidiation and alkaloid production in saprophytic strain of *Claviceps purpurea*. *Can. J. Microbiol.,* **23,** 1182–1187.

Pažoutová, S. and Řeháček, Z. (1978) Regulation of biosynthesis of clavine alkaloids by phosphate and citrate. *Abstract Book, 14th Annual Meet. Czechoslov. Soc. Microbiol.,* Prague, Czechoslov., October 17–19, 1978, 30–31.

Pažoutová, S. and Řeháček, Z. (1981a) Correlation of sugar catabolism changes with alkaloid synthesis in *Claviceps purpurea* 129. *Abstract Book, FEMS Symp. Overprod. Microbial Products,* Hradec Králové, Czechoslovakia, August 9–14, 1981, 145.

Pažoutová, S. and Řeháček, Z. (1981b) The role of citrate on the oxidative metabolism of submerged cultures of *Claviceps purpurea* 129. *Arch. Microbiol.,* **129,** 251–253.

Pažoutová, S. and Řeháček, Z. (1984) Phosphate regulation of phosphatases in submerged cultures of *Claviceps purpurea* 129 producing clavine alkaloids. *Appl. Microbiol. Biotechnol.,* **20,** 389–392.

Pažoutová, S., Řeháček, Z. and Voříšek, J. (1980) Induction of sclerotialike mycelium in axenic cultures of *Claviceps purpurea* producing clavine alkaloids. *Can.J. Microbiol.,* **26,**363–370.

Pažoutová, S. and Sajdl, P. (1988) Hypothesis on gene expression of alkaloid biosynthetic pathway in *Claviceps.* In: *Abstract Book, 2nd Internal. Symp. Overprod.Microbial Products,*České Budějovice, Czechoslovakia, July 3–8, 1988, p. 129.

Pažoutová, S., Votruba, J. and Řeháček, Z. (1981b) A mathematical model of growth and alkaloid production in the submerged culture of *Claviceps purpurea. Biotechnol. Bioeng.,* **23,**2837–2849.

Pertot, E., Čadež, J., Miličić, S. and Sočič, H. (1984) The effect of citric acid concentration and pH on the submerged production of lysergic acid derivatives. *Appl. Microbiol. Biotechnol.,* **20,**29–32.

Pertot, E., Gaberc-Porekar, V. and Sočič, H. (1990) Isolation and characterization of an alkaloid-blocked mutant of *Claviceps paspali. J. Basic Microbiol.,* **30,**51–56.

Pertot, E., Puc, A. and Kremser, M. (1977) Lyophilization of nonsporulating strains of the fungus *Claviceps. European J. Appl. Microbiol.,* **4,**289–294.

Philippi, U. and Eich, E. (1984) Microbiological C-17-oxidation of clavine alkaloids, 2: Conversion of partial synthetic elymoclavines to the corresponding lysergic acids/ isolysergic acids by *Claviceps paspali. Planta Medica,* **50,**456–457.

Priesmeyer, K. (1997) Fermentation: A regulatory perspective. *Pharmaceut. Technol.,* **21,** 40–54.

Puc, A., Miličić, S., Kremser, M. and Sočič, H. (1987) Regulation of ergotoxine biosynthesis in *Claviceps purpurea* submerged fermentation. *Appl. Microbiol.Biotechnol.,* **25,**449–452.

Quigley, F.R. and Floss, H.G. (1981) Mechanism of amino acid α-hydroxylation and formation of the lysergyl moiety in ergotamine biosynthesis. *J. Org. Chem.,* **46,** 464–466.

Ramstad, E., Chan Lin, W.-N., Shough, H.R., Goldner, K.J., Parikh, R.P. and Taylor, E.H. (1967) Norsetoclavine, a new clavine-type alkaloid from *Pennisetum* ergot. *Lloydia,* **30,**441–444.

Řeháček, Z. (1983a) New trends in ergot alkaloid biosynthesis. *Process Biochem.,* **18,** 22–29, 33.

Řeháček, Z. (1983b) Physiological aspects of ergot alkaloid formation. *Prikl. Biokhim.Mikrobiol.,* **19,**267–276 (In Russian).

Řeháček, Z. (1984) Biotechnology of ergot alkaloids. *Trends in Biotechnology,* **2,** 166–182.

Řeháček, Z. (1991) Physiological controls and regulation of ergot alkaloid formation. *Folia Microbiol.,* **36,**323–342.

Řeháček, Z., Desai, J.D., Sajdl, P. and Pažoutová, S. (1977) The cellular role of nitrogen in the biosynthesis of alkaloids by submerged culture of *Claviceps purpurea* (Fr.)Tul. *Can. J. Microbiol.,* **23,**596–600.

Řeháček, Z. and Kozová, J. (1975) Production of alkaloids and differentiation in submerged culture of *Claviceps purpurea* (Fr.) Tul. *Folia Microbiol.,* **20,**112–117.

Řeháček, Z. and Malik, K.A. (1971) Cell-pool tryptophan phases in ergot alkaloid fermentation. *Folia Microbiol.,* **16,**359–363.

Řeháček, Z., Pažoutová, S., Kozová, J. and Sajdl, P. (1978a) Industrial mutant of *Claviceps purpurea* CP 7/5 CCM F-630 with production of alkaloids agroclavine and elymoclavine. *CS pat. 204 301* (In Czech).

Řeháček, Z., Pažoutová, S., Kozová, J. and Sajdl, P. (1978b) Industrial mutant of *Claviceps purpurea* CP7/274 CCM F-632. *CS pat. 204 303* (In Czech).

Řeháček, Z., Pažoutová, S., Kozová, J., Sajdl, P., Křen, V. and Flieger, M. (1984a) Producing strain of *Claviceps purpurea* CP7/5/35 CC-2/85. *CS pat. 246 704* (In Czech).

Řeháček, Z., Pažoutová, S., Křen, V., Rylko, V., Kozová, J. and Sajdl, P. (1986a) Producing strain of *Claviceps purpurea* (Fr.) Tul. 59 CC 5/86. *CS pat. 252 603* (In Czech).

Řeháček, Z., Pažoutová, S., Rylko, V., Kozová, J., Sajdl, P., Chomátová, S., Pilát, P., Bárta, M., Homolka, L., Břemek, J., Malinka, Z., Krajíček, A. and Spáčil, J. (1986b) Industrial strain of *Claviceps fusiformis* Loveless F 27CC-4/85. *CS pat. 253 097* (In Czech).

Řeháček, Z. and Rylko, V. (1985) Process for suppression of extracellular glucanes and conidias in submerged cultures of *Claviceps* producing alkaloids agroclavine and elymoclavine. *CS pat. 249 461* (In Czech).

Řeháček, Z. and Sajdl, P. (1990) *Ergot Alkaloids. Chemistry, Biological Effects,Biotechnology,* Academia, Praha, Czechoslovakia.

Řeháček, Z., Sajdl, P., Kozová, J., Křen, V. and Rylko, V. (1984b) Industrial mutant of *Claviceps fusiformis* Wl CC-1/85. *CS pat. 248 612* (In Czech).

Řeháček, Z., Sajdl, P., Kozová, J., Malik, K.A. and Řičicová, A. (1971) Correlation of certain alterations in metabolic activity with alkaloid production by submerged *Claviceps. Appl. Microbiol,.* **22,**949–956.

Řeháček, Z., Spáčil, J., Pažoutová, S., Sajdl, P., Kozová, J., Flieger, M., Krajíček, A. and Malinka, Z. (1978c) Process for production of alkaloids agroclavine and elymoclavine by submerged fermentation. *CS pat. 199 986* (In Czech).

Richter Gedeon Vegyészeti Gyár R.T. (1973) Fermentation process for preparing ergot alkaloids . *GB pat. 1 401 406.*

Řičicová, A., Flieger, M. and Řeháček, Z. (1981) Production of simple derivatives of lysergic acid by the strain *Claviceps paspali* F 2056. *Abstract Book, FEMS Symp.Overprod. Microbial Products,* Hradec Králové, Czechoslovakia, August 9–14, 1981.

Řičicová, A., Flieger, M. and Řeháček, Z. (1982a) Quantitative changes of the alkaloid complex in a submerged culture of *Claviceps paspali. Folia Microbiol.,* **27,**433–445.

Řičicová, A., Flieger, M., Sedmera, P., Vokoun, J., Řeháček, Z., Stuchlík, J. and Krumphanzl, V. (1982b) Process for the production of lysergic acid derivatives by submerged fermentation. *CS pat. 224 532* (In Czech).

Řičicová, A., Pokorný, V. and Řeháček, Z. (1986) Characteristics of the saprophytic reference strain FA of *Claviceps paspali. Folia Microbiol.,* **31,**32–43.

Řičicová, A. and Řeháček, Z. (1968) Preparation of inoculum from sclerotia of the ascomycete *Claviceps purpurea. Folia Microbiol.,* **13,**156–157.

Robbers, J.E. (1984) The fermentative production of ergot alkaloids. In: A.Mizuki, A.L.Van Wezel, (eds.) *Advances in Biotechnological Process,* **Vol. 3,** Alan R. Liss, New York, pp.197–239.

Robbers, J.E., Eggert, W.W. and Floss, H.G. (1978) Physiological studies on ergot: Time factor influence on the inhibitory effect of phosphate and the induction effect of tryptophan on alkaloid production. *Lloydia,* **41,**120–129.

Robbers, J.E., Robertson, L.W., Hornemann, K.M., Jindra, A. and Floss, H.G. (1972) Physiological studies on ergot: Further studies on the induction of alkaloid synthesis by tryptophan and its inhibition by phosphate. *J. Bacteriol.,***112**,791–796.

Robbers, J.E., Srikrai, S., Floss, H.G. and Schlossberger, H.G. (1982) Physiological studies on ergot. The induction of ergot alkaloid biosynthesis by the tryptophan bioisosteres, β-1- and β-2-naphthylalanine. *J. Nat. Prod.,***45**,178–181.

Robertson, L.W., Robbers, J.E. and Floss, H.G. (1973) Some characteristics of tryptophan uptake in *Claviceps* species. *J. Bacteriol.,***114**,208–219.

Rochelmeyer, H. (1959). Verfahren zur Extraktion der Alkaloide aus dem Mycel und den Sklerotien alkaloidhaltiger *Clavicipiteen. DE pat. 1125 940.*

Rochelmeyer, H. (1965) Verfahren zur Gewinnung von Mutterkornalkaloiden in saprophytischer Kultur. *DE pat. 1492 109.*

Rosazza, J.P., Kelleher, W.J. and Schwarting, A.E. (1967) Production of lysergic acid derivatives in submerged culture. IV. Inorganic nutrition studies with *Clavicepspaspali. Appl. Microbiol.,***15**,1270–1283.

Rozman, D., Komel, R. and Pertot, E. (1987) Soybean peptone and its fractions as nutrients in fermentations with immobilized *Claviceps fusiformis* cells. *Vestn. Slov.Kem. Drus.,***34**,457–463.

Rozman, D., Pertot, E., Belič, I. and Komel, R. (1985) Soybean peptones as nutrients in the fermentative production of clavine ergot alkaloids with *Claviceps fusiformis. Biotechnol. Lett.,***7**,563–566.

Rutschmann, J. and Kobel, H. (1963a) Neues mikrobiologisches Verfahren zur Gewinnung von Ergobasin. *CH pat. 433 357.*

Rutschmann, J. and Kobel, H. (1963b) Process for the production of lysergic acid derivatives. *GB pat. 1041 246.*

Rutschmann, J., Kobel, H. and Schreier, E. (1963) Verfahren für Herstellung von 6-Methyl- $\Delta^{8,9}$-ergolen-8-Carboxyl Säure. *CH pat. 468 465.*

Rylko, V., Flieger, M., Sajdl, P. and Řeháček, Z. (1988a) Induction of cytochrome P-450 in *Claviceps* cultures stimulates alkaloid formation and overcomes its inhibition by phosphate. *Abstract Book, 2nd Internal. Symp. Overprod. Microbial Products,*České Budějovice, Czechoslovakia, July 3–8, 1988.

Rylko, V., Flieger, M., Sajdl, P. and Řeháček, Z. (1988b) Effects of inducers and inhibitors on ergot cytochrome P-450 and alkaloid production by *Claviceps* spp. *Abstract Book, IUB 14th Internal. Congr. Biochemistry,*Prague, Czechoslovakia, July 10–15, **IV**, 94.

Rylko, V., Flieger, M., Sajdl, P., Řeháček, Z., Malinka, Z., Harazim, P. and Stuchlík, J. (1988c) Process for production of ergot alkaloids by *Claviceps* with induced increase of productive ability. *CS pat. appl. 4725–88.*

Rylko, V., Linhartová, R., Sajdl, P. and Řeháček, Z. (1988d) Formation of conidia in a saprophytic strain *Claviceps paspali* producing simple lysergic acid derivatives. *Folia Microbiol.,***33**,425–429.

Rylko, V., Šípal, Z., Sajdl, P. and Řeháček, Z. (1986) Development of the cytochrome P-450 contents in submerged culture *Claviceps* sp. SD-58. *Abstract Book, 17th Congr.Czechoslov. Soc. Microbiol.,*České Budějovice, Czechoslovakia, September 23–25, 1986, A-l8 (In Czech).

Sajdl, P., Kozová, J. and Řeháček, Z. (1978) Synthesis of clavin alkaloids and basic metabolism of a submerged culture of *Claviceps purpurea* 129. *Abstract Book, 14thAnn. Meet. Czechoslov. Soc. Microbiol.,*Prague, October 17–19, 1978, 30.

Sajdl, P., Křen, V., Pažoutová, S., Řezanka, T., Malinka, Z. and Harazim, P. (1988a) Process for fermentative production of ergot alkaloids by modification of cellular lipids. *CZ pat. 273 373* (In Czech).

Sajdl, P., Pažoutová, S., Křen, V., Rylko, V. and Řeháček, Z. (1988b) Relationships between *Claviceps* membranes and alkaloid biosynthesis. *Abstract Book, 2ndInternal. Symp. Overprod. Microbial Products,*České Budějovice, Czechoslov., July 3–8, 1988, 62.

Sarkisova, M.A. (1990) Strain of *Claviceps purpurea* fungus as producer of peptide ergoalkaloids. *SU pat. 1 342 011* (In Russian).

Sarkisova, M.A. and Smirnova, V.I. (1984) Physiological and biochemical tests of ergot, *Claviceps purpurea* (Fr.) Tul. in saprophytic culture. *Mikol. Filopatol.,*18,393–396 (In Russian).

Sarkisova, M.A., Zhadskaya, E.S., Rubinczik, M.A., Konczalovskaya, M.E., Sokolova, I.A. and Radchenko, L.M. (1987) Process for production of lipids with the same composition as cotton oil. *SU pat. 1 493 671* (In Russian).

Sastry, K.S.M., Gupta, J.H., Thakur, R.N. and Pandotra, V.R. (1970a) Effect of different concentrations of inoculum of ergot fungus (*Claviceps purpurea* (Fr.) Tul.) in relation to the yield of ergot sclerotia. *Proc. Indian Acad. Sci.,*71B,33–35.

Sastry, K.S.M., Pandotra, V.R., Thakur, R.N., Gupta, J.H., Singh, K.P. and Husain, A. (1970b) Studies on parasitic cultivation of ergot (*Claviceps purpurea* (Fr.) Tul.) in India. *Proc. Indian Acad. Sci.,*72B,99–114.

Sastry, K.S.M., Thakur, R.N., Pandotra, V.R., Singh, K.P. and Gupta, J.H. (1970c) Studies on number of inoculations on rye spikes (*Secale cereale* L.) by needle board puncture method in relation to the yield of ergot sclerotia. *Proc. Indian Acad. Sci.,*71B, 28–32.

Satke, J., Mat'ha, V., Jegorov, A., Cvak, L., Stuchlík, J. and Kadlec, Z. (1994) New industrial producing strain of microorganism Claviceps paspali. *CZ pat. appl.2456–94* (In Czech).

Schlientz, W., Brunner, R., Ruegger, A., Berde, B., Sturmer, E. and Hofmann, A. (1968) β-Ergokryptin, ein neues Alkaloid der Ergotoxin-gruppe. *Pharm. Acta Helv.,*43, 497–509.

Schmauder, H.-P. and Gröger, D. (1983) Selection und Eigenschaften potentiell alkaloidbilder *Claviceps-Stämme. Acta Biotechnol.,*3,379–382.

Schmauder, H.-P. and Gröger, D. (1986) Biosynthesis of phenylalanine and tyrosine in *Claviceps. Planta Medica,*49,395–397.

Schmauder, H.-P., Maier, W. and Gröger, D. (1981a) Einfluss von Amitrol und substituierten Indolderivaten auf Wachstumm Alkaloidbildung und Enzymaktivitäten von *Claviceps. Pharmazie,*36,211.

Schmauder, H.-P., Maier, W. and Gröger, D. (1981b) Einfluss von Amitrol auf *Clavicepsfusiformis,* Stamm SD 58. *Z. Allgemeine Mikrobiol.,*21,689–692.

Schumann, B., Erge, D., Gröger, D., Maier, W., Schmauder, H.-P. and Baumert, A. (1984) Verfahren zur Herstellung von Elymoclavin. *DD pat. 222 634.*

Schumann, B., Erge, D., Maier, W. and Gröger, D. (1982) A new strain of *Clavicepspurpurea* accumulating tetracyclic clavine alkaloids. *Planta Medica,*45, 11–14.

Schumann, B., Gröger, D., Schmauder, H.-P., Lehmann, H.-R. and Maier, W. (1986) Verfahren zur Herstellung von α-Ergokryptin und Ergosin. *DD pat. 280 977.*

Schumann, B., Maier, W. and Gröger, D. (1987) Characterization of some *Claviceps* strains derived from regenerated protoplasts. *Z. Naturforsch.,***42c,**381–386.

Schweizer, C. (1941) Über die Kultur von *Claviceps purpurea* (Tul.) auf kaltsterilisierten Nährboden. *Phytopath. Ztg.,***13,**317–350.

Senda, T., Okura, T., Yoshihiro, Y. and Miyazawa, F. (1989) Production of inulooligosaccharide. *JP pat. 02182 195* (In Japanese).

Sieben, R., Philippi, U. and Eich, E. (1984) Microbiological C-17-oxidation of clavine alkaloids, I. Substrate specificity of agroclavine hydroxylase of *Claviceps fusiformis. J. Nat. Prod.,***47,**433–438.

Siegle, H. and Brunner, R. (1963) Process for the production of lysergic acid and its derivatives by fermentation. *FR pat. 1 350 280* (In French).

Singh, H.P., Singh, H.N. and Singh, K.P. (1992) Varietal susceptibility of rye (*Secalecereale* L.) for ergot (*Claviceps purpurea* Fr.) Tul. infection. *New Botanist,***19,** 225–227.

Slokoska, L., Angelova, M., Pashova, S. and Genova, L. (1992) Effect of some antibiotics on the biosynthesis of clavine alkaloids from *Claviceps* sp. *Dokl. Bolg. Akad. Nauk.,***45,** 105–107.

Slokoska, L., Grigorov, I., Angelov, T., Angelova, M., Kozlovsky, A., Arinbasarov, M. and Solovyeva, T. (1981) Studies on the alkaloid-producing strain *Claviceps* sp. CP II. I. Relation between the different development stages and the alkaloid accumulation. *Acta Microbiol. Bulgar.,***8,**44–50 (In Russian).

Slokoska, L.S., Grigorov, I., Angelova, M.B., Nikolova, N.S. and Pashova, S.B. (1985) Studies into the connection between the alkaloid biosynthesis and the lipid metabolism in strain *Claviceps* sp. CP II. *Dokl. Bolg. Akad. Nauk.,***38,**1025–1028.

Slokoska, L., Grigorov, I., Angelova, M., Pashova, S., Stefanov, V. and Nikolova, N. (1988) Effect of laser irradiation on strain *Claviceps* sp. *Abstract Book, 2nd Internal.Symp. Overprod. Microbial Products,*České Budějovice, Czechoslov., July 3– 8, 1988, 111.

Sočič, H. and Gaberc-Porekar, V. (1992) Biosynthesis and physiology of ergot alkaloids. In: O.K.Arora, R.P.Elander and K.G.Mukerji, (eds.), *Handbook of AppliedMycology,*vol. 4. *Fungal Biotechnology,*Dekker, New York.

Sočič, H., Gaberc-Porekar, V. and Didek-Brumec, M. (1985) Biochemical characterization of the inoculum of *Claviceps purpurea* for submerged production of ergot alkaloids. *Appl. Microbiol. Biotecbnol.,***21,**91–95.

Sočič, H., Gaberc-Porekar, V., Pertot, E., Puc, A. and Miličić, S. (1986) Developmental studies of *Claviceps paspali* seed cultures for the submerged production of lysergic acid derivatives. *J. Basic Microbiol.,***26,**533–539.

Spalla, C., Amici, A.M., Scotti, T. and Tognoli, L. (1969) Heterokaryosis of alkaloid-producing strains of *Claviceps purpurea* in saprophytic and parasitic conditions. In D.Perlman, (ed.), *Fermentation Advances,*Academic Press, New York, pp. 611–628.

Spalla, C. and Marnati, M.P. (1981) Aspects of the interspecific fusion of protoplasts of alkaloid producing strain of *Claviceps purpurea* and *Claviceps paspali. AbstractBook, FEMS Symp. Overprod. Microb. Prod.,*Hradec Králové, Czechoslovakia, August 9–14, 1981.

Srikrai, S. and Robbers, J.E. (1979) Mutation studies on the ergot fungus, *Claviceps* species, strain SD 58. *Proceedings,***42,**689.

Stauffacher, D. and Tscherter, H. (1964) Isomere des Chanoclavins aus *Clavicepspurpurea* (Fr.) Tul. *(Secale cornutum). Helv. Chim. Acta,***47,**2186–2194.

Stoll, A. (1952) Recent investigations on ergot alkaloids. *Fortschr. Chem. Org.,* **9,** 114–174.

Stoll, A., Brack, A., Hofmann, A. and Kobel, H. (1953) Process for production of ergotamine, ergotaminine and ergobasine. *CH pat. 321 323.*

Stoll, A., Brack, A., Hofmann, A. and Kobel, H. (1954a) Process for production of ergotamine, ergotaminine and ergobasine by saprophytic cultivation of ergot fungus (*Claviceps purpurea* (Fr.) Tul.) *in vitro* and isolation of so produced alkaloids. *CH pat. 330 722.*

Stoll, A., Brack, A., Kobel, H., Hofmann, A. and Brunner, R. (1954b) Die Alkaloide eines Mutterkornpilzes von *Pennisetum typhoideum* Rich und deren Bildung in saprophytishem Kultur. *Helvetica Chimica Acta,* **37,** 1815–1827.

Strnadova, K. (1964a) UV-mutanten bei *Claviceps purpurea. Planta Medica,* **12,** 521–527.

Strnadová, K. (1964b) Methoden zur Isolierung und Ermittlung auxotropher Mutanten bei *Claviceps purpurea* (Fr.) Tul. *Z. f. Pflanzenzucht.,* **51,** 167–171.

Strnadová, K. and Kybal, J. (1974) Ergot alkaloids. V. Homokaryosis of the sclerotia of *Claviceps purpurea* (Fr.) Tul. *Folia Microbiol.,* **19,** 272–280.

Strnadová, K. and Kybal, J. (1976) Improvement of microorganisms of the genus *Claviceps purpurea* (Fr.) Tul. *CS pat. 176 803* (In Czech).

Strnadová, K., Kybal, J., Svoboda, E. and Spáčil, J. (1981) Strain of microorganism *Claviceps purpurea* (Fr.)Tul. CCM F-725. *CS pat. 222 391* (In Czech).

Strnadová, K., Kybal, J., Svoboda, E., Spáčil, J., Krajíček, A., Malinka, Z. and Harazim, P. (1986) Process of improvement of genus *Claviceps* microorganisms and new industrial strain of *Claviceps purpurea* (Fr.) Tul. CCM 8043. *CS pat. 267 573* (In Czech).

Švecová, M. (1985) Evaluation of properties *Claviceps purpurea* (Fr.) Tul. conidias, which are used for rye infection by field production of ergot. Dissertation, Charles University, Prague (In Czech).

Taber, W.A. (1964) Sequential formation and accumulation of primary and secondary shunt metabolic products in *Claviceps purpurea. Appl. Microbiol.,* **12,** 321–326.

Taber, W.A. and Vining, L.C. (1958) The influence of certain factors on the *in vitro* production of ergot alkaloids by *Claviceps purpurea* (Fr.) Tul. *Can. J. Microbiol.,* **4,** 611–626.

Takeda Pharmaceutical Industries (1956) Water-soluble ergot alkaloid, elymoclavine. *GB pat. 757 696.*

Trejo Hernández, M.R. and Lonsane, B.K. (1993) Spectra of ergot alkaloids produced by *Claviceps purpurea* 1029c in solid-state fermentation system: Influence of the composition of liquid medium used for impregnating sugar-cane pith bagasse. *Process Biochem.,* **28,** 23–27.

Trejo Hernández, M.R., Raimbault, M., Roussos, S. and Lonsane, B.K. (1992) Potential of solid state fermentation for production or ergot alkaloids. *Letters in AppliedMicrobiology,* **15,** 156–159.

Trinn, M., Kordik, G., Nagy, E.U., Vida, Z. and Zsóka, E. (1983) Bei Gärung zur Clavinmutterkornalkaloiderzeugung, insbesondere Elymoclavinerzeugung, fähiger *Claviceps fusiformis*—Variantenstamm sowie Verfahren zu seiner Ilerstellung und seine Verwendung zur Herstellung von Clavinmutterkornalkaloiden. *DE pat.3 429 573.*

Trinn, M., Kordik, G., Nagy, E.U., Vida, Z. and Zsoka, E. (1984) New *Clavicepsfusiformis* strain, method for obtaining said strain and fermentation process for producing clavine alkaloids using said strain. *GB pat. 2148 278.*

Trinn, M., Manczinger, L., Polestyukné, N.Á., Kordik, G., Pécsné, R.Á., Zalai, K., Beszedics, G., Ferenczy, L., Nagy, L., Robicsek, K. and Szegedi, M. (1990) Process for the elymoclavine production and preparation of new strain. *HU pat. 209 325* (In Hungarian).

Tudzynski, P., Esser, K. and Gröschel, H. (1982) Genetics of the ergot fungus. II. Exchange of genetic material via meiotic recombination. *Theor. Appl. Genet.,***61,** 97–100.

Tyler, V.E. Jr. (1963) Production of ergot alkaloids. *US pat. 3 224 945.*

Udvardy Nagy, E. (1980) Consideration on the development of an ergot alkaloid fermentation process. *Process Biochem.,***15,**5–8.

Udvardy Nagy, E., Budai, M., Fekete, G., Görög, S., Herényi, B., Wack, G. and Zalai, K. (1981) Method of regulation of quantity and ratio of alkaloids content. *CS pat.236 660* (In Czech).

Udvardy Nagy, E., Wack, G. and Prócs, T. (1964) Process for preparation of producing strains of *Claviceps purpurea* (Fr.) Tul. cultures for ergot alkaloids production. *ATpat. 269 363* (In German).

Udvardy-Nagy, I., Wack, G. and Wolf, L. (1964) Process for biosynthesis of ergoline compounds. *HU pat. 152 238* (In Hungarian).

Ustyuzhanina, S.V., Sarkisova, M.A. and Gorin, S.E. (1991) Preservation of *Clavicepspurpurea,* an organism producing peptide ergoalkaloids, by L-drying. *Antibiotiki iChimioter.,***36,**6–7 (In Russian).

Vaidya, H.C. and Desai, J.D. (1981a) Some metabolic studies in *Claviceps* species strain SD 58: Effect of phosphate. *Abstract Book, FEMS. Symp. Overprod. MicrobialProducts,*Hradec Králové, Czechoslov., August 9–14, 1981, 142.

Vaidya, H.C. and Desai, J.D. (1981b) Cell differentiation & alkaloid production in *Claviceps* sp. strain SD 58. *Indian J. Exp. Biol.,***19,**829–831.

Vaidya, H.C. and Desai, J.D. (1982) Effect of phosphate on growth, carbohydrate catabolism & alkaloid biogenesis in *Claviceps* sp. strain SD 58. *Indian J. Exp. Biol.,***20,** 475–478.

Vaidya, H.C. and Desai, J.D. (1983a) Alkaloid production by *Claviceps* sp. SD-58; Involvement of phosphatase isozymes. *Folia Microbiol.,***28,**12–16.

Vaidya, H.C. and Desai, J.D. (1983b) Alkaloid production in *Claviceps* sp. strain SD-58: Physiology of phosphate effect. *Indian J. Biochem. Biopbys.,***20,**222–225.

Valík, J. and Malinka, Z. (1992) New inoculate material for parasitical production of ergot. *Abstract Book, Biotechnology in Central European Initiative Countries,*Graz, Austria, 13–15 April, 1992, P81.

Vásárhelyi, G., Tétényi, P., Tétényi, P. and Lassányi, Zs. (1980a) Proces of ergot cultivation. *HU pat. 175 924* (In Hungarian).

Vásárhelyi, G., Tétényi, P., Zambo, I., Kinicky, M., Balassa-Barkanyi, I., Kiss, J., Pecs-Razso, A. and Zsoka-Somkuti, E. (1980b) Strain of *Claviceps purpurea,* virulent on rye, for ergot alkaloids, especially ergocristine production and process for preparation of it. *DE pat. 3 006 989* (In German).

Vlček, V. and Kybal, J. (1974) Equipment for cultivation of microorganisms. *CS pat. 172552* (In Czech).

Volzke, K., Knothe, E., Langer, J., Gross, E., Decker, K.-L., and Schmauder, H.-P. (1985) Verfahren zur gesteurten Production von Ergolinderivaten durch Fermentation. *DD pat. 289 562.*

Voříšek, J., Sajdl, P., and Řeháček, Z. (1981) Fine structural localization of alkaloid and lipid synthesis in endoplasmic reticulum of submerged *Claviceps purpurea*. *Abstract Book, FEMS Symp. Overprod. Microbial Products,*Hradec Králové, Czechoslov., August 9–14, 1981, 146.

Votruba, J. and Pažoutová, S. (1981) Modelling of the optimal conditions for the maximum alkaloid synthesis. In *Abstract Book, FEMS Symp. Overprod. MicrobialProducts,*Hradec Králové, Czechoslovakia, August 9–14, 1981, p. 104.

Wack, G., Kiss, J., Nagy, L., Udvardy Nagy, E., Zalai, K., and Zsoka, E. (1981) Precede pour la préparation d'alcaloides de l'ergot de seigle, en particulier de l'ergocornine et de la *β*-ergocryptine. *FR pat. 2 475 573.*

Wack, G., Nagy, L., Székély, D., Szolnoky, J., Udvardy Nagy, E., and Zsóka, T. (1973) Fermentation process for the preparation of ergot alkaloids. *US pat. 3 884 762.*

Wack, G., Perényi, I., Udvardy Nagy, E., and Zsoka, E. (1966) Process for Production of Ergot Alkaloids. *GB pat. 1170 600.*

Wichmann, D. and Voigt, R. (1962) Zur Alkaloidbildung des Roggenmutterkorns in saprophytischer Kultur. *Pharmazie,***17,**411–418.

Wilke, D. and Weber, A. (1984) Verfahren zur Herstellung von α-Ergokryptin. *DE pat.3 420 953.*

Wilke, D. and Weber, A. (1985a) Process for production of festuclavine. *CS pat. 271 314* (In Czech).

Wilke, D. and Weber, A. (1985b) Process for production of chanoclavine. *CS pat.272 207* (In Czech).

Windisch, S. and Bronn, W. (1960) Ergot alkaloids. *US pat. 2 936 266.*

Yamano, T., Kishino, K., Yamatodani, S., and Abe, M. (1962) Researches on ergot fungus. Part XLIX. Investigation on ergot alkaloids occured in cultures of *Aspergillusfumigatus* Fres. *Ann. Rep. Takeda Res. Lab.,***21,**95–101.

Yamatoya, S. and Yamamoto, I. (1983) Preparation of alkaloid. *JP pat. 59140 892* (In Japanese).

Zalai, K. and Jaksa, I. (1981) Contribution to metabolism research of *Clavicepspurpurea*. *Pharmazie,***36,**212 (In German).

Zalai, K., Kordik, G., Manczinger, L., Pécsné, R.Á., Beszedics, G., Polestyukné, N.Á., Ferenczy, L., Olasz, K., Szegedi, M., and Trinn, M. (1990) Process for production of ergolen compounds, mainly ergometrine, and process of colour selection. *HU pat.209 324* (In Hungarian).

13.
INDUSTRIAL PRODUCTION OF ERGOT ALKALOIDS

LADISLAV CVAK

Galena a.s.Opava, 747 70 Czech Republic

13.1.
INTRODUCTION

This chapter contains information about all the therapeutically used ergot alkaloids and their manufacture. Not all such information can be found in the literature and supported by references. The technology used for manufacture can be traced in the patent literature but not all the patented processes are actually used in the production and, on the other hand, not all the technologies used have been patented. So at least a part of this information is based on personal communication only or is deduced from some indirect cues—for instance the profile of impurities. Further information, not usually published is the amount of individual manufactured products. These estimations are based on a long-term experience in ergot alkaloid business. Even if all such estimations can be inaccurate, I believe that in a book published in a series "Industrial profiles" they cannot be omitted.

The history of industrial production of ergot alkaloids began in 1918, when Arthur Stoll patented the isolation of ergotamine tartrate (Stoll, 1918), which the Sandoz company introduced on the market in 1921. Until the end of World War II, Sandoz remained virtually the only real industrial ergot alkaloid producer. The first competitors appeared in the fifties. Sandoz is still the world leading ergot alkaloid producer (lately under the name Novartis). The company sells the whole production in its own pharmaceutical products. Other major producers sell most of their products as "bulk pharmaceutical chemicals": Boehringer Ingelheim (Germany), Galena (Czech Republic), Gedeon Richter (Hungary), Lek (Slovenia), Poli (Italy). Besides these producers manufacturing a broad spectrum of ergot alkaloids, two other companies influenced ergot research, manufacture and business—see Chapter 1. History of ergot research. Farmitalia (Italy, now a part of Pharmacia-Upjohn) developed and produces nicergoline (Sermion) and cabergoline (Dostinex), and Eli Lilly developed and produces pergolide mesylate (Permax). Some others, usually locally active producers, exist in India, Finland and Poland and other companies produce some products from purchased intermediates: Rhone Poulenc (France), Indena and Linea Nuova (both Italy) producing nicergoline and Sanofi and Piere Fabre (both France) manufacturing dihydroergotamine and dihydroergocristine. Schering AG (Germany) and Maruko Seiyaku (Japan) were, or are, active in ergot alkaloid research.

A distinct trend can be seen in the use of ergot alkaloids in the last few decades. While the therapeutic use of classical ergot alkaloids (ergotamine, dihydroergotamine, dihydroergotoxine) has been stable for many years and their production has been increasing only a little, the therapeutic use and consumption of new, semisynthetic derivatives is growing quickly (nicergoline, pergolide). The annual world production of ergot alkaloids can be estimated at 5000–8000 kg of all ergopeptines and 10000–15000 kg of lysergic acid, used for the manufacture of semisynthetic derivatives, mainly nicergoline. The larger part of this production comes from fermentations (about 60%), the rest comes from the field ergot. The estimation of individual product volumes is given in part 4 of this chapter.

13.2.
SOURCES OF ERGOT ALKALOIDS

13.2.1.
Field Ergot

Collected wild ergot was the only source of ergot alkaloids throughout the history, and ergot from artificial cultivation has remained an important source for alkaloid production. Two world leading alkaloid manufacturers, Boehringer Ingelheim and Galena, are the main producers of ergot.

Wild ergot was poorly suited for the isolation of alkaloids because of its great variability in alkaloid content and spectrum. In fact, it was the success of the artificial cultivation of ergot which created a basis for large-scale production of ergot alkaloids (Well, 1910; Hecke, 1922, 1923). Enormous effort was devoted to the selection of strains producing a defined spectrum of alkaloids. Later, similar effort was aimed at the economical parameters: yield of ergot and alkaloid content. While in the forties the average yield of ergot was 400 kg/ha (Stoll and Brack, 1944), today the yields of leading producers are over 1000 kg/ha. Similar development has taken place in the content of ergot alkaloids. Producer strains used by the leading manufacturers produce above 1% of alkaloids. In regard of the alkaloids produced, there are strains producing all the desired ergopeptines as separate single alkaloids, or producing an optimal mixture of alkaloids (e.g. a mixture of ergotoxine alkaloids for manufacture of dihydroergotoxine).

A special problem is the content of undesirable minor ergopeptines potential impurities in the final products. In spite of all the effort devoted to minimising their formation, they always persist in the ergot and the purification processes used by individual producers can remove them to a different extend. This can be demonstrated by the isolation of many novel alkaloids in the laboratories of leading ergot alkaloid manufacturers (Krajíček *et al.,* 1979; Szantay *et al.,* 1994; Cvak *et al.,* 1994, 1996, 1997). Besides ergopeptines, each ergot contains some simple lysergic acid derivatives, mainly ergine (lysergic acid amide) and ergometrine. These are not usually taken as undesirable because they can be easily removed during ergopeptine purification and, moreover, can be used for lysergic acid manufacture. Ergometrine, when present in a higher concentration (sometimes up to 0.1% of the total alkaloid content of about 1%), can be isolated as a by-product.

All the aspects of parasitic ergot production are described in detail in Chapter 11—Parasitic production of ergot.

Ergot Extraction

Extraction of ergot is described mainly in older literature devoted to first isolations of new alkaloids (for example Stoll, 1945), or in the patent literature. When analysing the patent literature, one has to be careful. Many patented procedures are so complicated that they can hardly be used for industrial production. While a two-stage process is usually used in lab-scale extraction of alkaloids—defatting by a nonpolar solvent being followed by alkaloid extraction by a more polar solvent, a one-stage direct extraction is used on industrial scale.

Patented procedures use both organic solvent and water extraction. Although the solubility of ergopeptines in diluted aqueous acids is satisfactory, ergot swells in such solvents and the problems connected with this fact have never been overcome. Only organic solvents are therefore used for industrial-scale extraction. Methylenechloride, trichloroethylene, ethyl acetate, acetone, methylisobutyl ketone and mixtures of toluene with methanol or ethanol and ether with ethanol are or were used. Percolation technology is used to reach satisfactory yield, using a battery of percolators or some type of a continual extractor (usually carousel-type extractor). Extraction of at least 95% of alkaloids present in the ergot is usually accepted as economically satisfactory.

Primarily obtained extracts are usually subjected to liquid-liquid extraction using aqueous diluted acids. Alkaloids are transferred into the water phase, whereas fats remain in the organic reffinate. Further processing of aqueous extracts depends on the experience of individual producers. In any case, the product of ergot extraction is a crude concentrate of alkaloids containing all the alkaloids present in ergot (sometimes excluding the water-soluble ergometrine) and only a low amount of other ballast components. A very important factor which is necessary to take into account is the epimerisation of lysergic acid derivatives—ergopeptines into isolysergic acid derivatives—ergopeptinines—see Figure 1. Individual processes differ in the rate of epimerisation and each crude alkaloid concentrate contains higher or lower amounts of ergopeptinines or the sole product of extraction is the respective ergopeptinine.

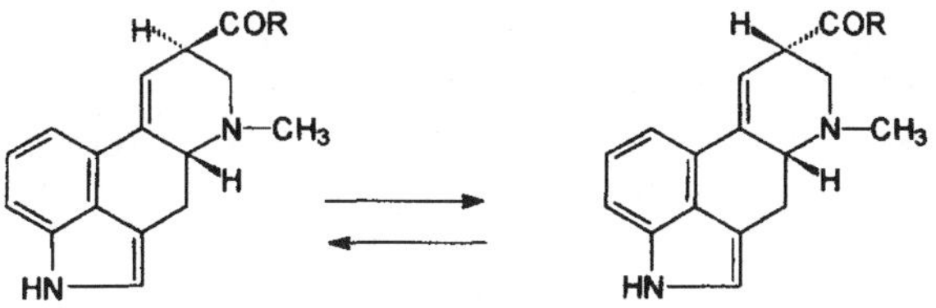

| Lysergic acid | R = OH | Isolysergic acid |
| Ergopeptine | R = aminocyclol | Ergopeptinine |

Figure 1 Epimerisation of lysergic acid derivatives

Purification of Ergot Alkaloids

The processes used for purification of individual alkaloids depend on the quality of the starting crude concentrate and the required quality of the product. There are so many processes developed by individual producers that only their general features can be mentioned here.

The goal of the purification process is the complete removal of both ballast components and minor undesirable ergopeptines or other alkaloids. While the complete elimination of ballast components is not so difficult, the complete elimination of minor alkaloids was successful only in some cases and practically each purified product (ergopeptine or dihydroergopeptine) contains some minor ergopeptines. In the past, many processes for separation of individual ergopeptines were developed using crystallisation, liquid-liquid extraction (the Craig process) or preparative-scale chromatography. Such processes are usually no longer used, because better strains producing individual alkaloids were developed.

Two main separation operations are used for ergopeptine purification: 1. Crystallisation of alkaloids, both bases and their salts, from different solvents. 2. Preparative-scale chromatography on silica or alumina. Also epimerisation of ergopeptinine into ergopeptine is always a part of the purification process. The basic procedure for epimerisation of ergotaminine was described by Stoll (1945). The procedure was later developed for the epimerisation of all the ergopeptinines and it was repeatedly improved to reach higher yield and better quality of the product (for example Terdy *et al.,* 1981; Schinutschke *et al.,* 1979).

13.2.2.
Fermentation

Fermentation of ergot alkaloids is the subject of Chapter 12 and so only *thestate of the art* in the industrial-scale production of ergot alkaloids will be mentioned here.

Only submerged (deep) fermentation is used for ergot alkaloid production. The fermenter size depends on the quantity of the product required: ergot alkaloids are medium-size products and medium size fermenters are therefore used for their production—10 to 50 m^3. Inoculation of such fermenters must be done in multiple stages—3 or 4. The duration of the production stage is 12 to 21 days. Fermentation is used for the production of both ergopeptines and simple ergoline compounds used for partial synthesis of therapeutically used derivatives. Because these two cases differ in downstream processing, they will be discussed separately.

Ergopeptines

Processes for ergopeptine fermentation were developed by Sandoz (e.g., Kobel and Sanglier, 1976, 1978), Farmitalia (Amici *et al.,* 1966, 1969), Gedeon Richter (Udvardy *et al.,* 1982), Lek and Poli. The production of ergopeptines presented in the literature is below 1 g/1 but the top production is now between 1 and 2 grams per liter of fermentation broth. The solubility of all ergopeptines in water at a pH value suitable for the fermentation process is low and this is the reason why most of produced alkaloids remain in the biomass (mycelium)—the liquid phase usually contains less than 5% of all the alkaloids of the fermentation broth.

Two types of downstream processes are used for alkaloid extraction. In a two-stage process the mycelium is filtered off and the alkaloids are isolated from the mycelium only. The filtrate is usually processed in a waste-water-treatment plant. The extraction of alkaloids from the mycelium is a process similar to ergot extraction, water-miscible organic solvents being usually used for this operation. In a one stage-process (direct extraction) the whole fermentation broth is subjected to extraction with a water-immiscible solvent (ethyl acetate, butyl acetate). The two-stage process is less effective but it does not require a special centrifugal extractor which is used for the direct extraction.

The processes for purification of ergopeptines are the same as those used for the isolation of crude alkaloid concentrates from ergot and they are therefore not discussed here.

Simple Ergolines

The need for simple ergoline derivatives was initiated by the progress in synthetic chemistry which enabled both the synthesis of natural alkaloids from their ergoline precursors (ergometrine, ergopeptines) and the synthesis of new semisynthetic derivatives providing pharmacological and therapeutical benefits (methylergometrine, methysergide, nicergoline). A cheap source of lysergic acid or some other ergoline precursor was a prerequisite for such syntheses. The first suitable product available by fermentation was elymoclavine—Figure 7 (Abe *et al.,* 1952), to be followed by lysergic acid hydroxyethylamide—Figure 6 (Arcamone *et al.,* 1961) and by paspalic acid—Figure 5 (Kobel *et al.,* 1964).

Also ergometrine—Figure 11—is now available by submerged fermentation (Rutschmann and Kobel, 1963). Lysergic acid hydroxyethylamide and paspalic acid are now the most important simple ergoline products obtained by fermentation. They are converted into lysergic acid which is the starting material for chemical syntheses. Fermentation processes used for their production can produce broth containing up to 5 grams of alkaloids per litter.

All the above mentioned simple ergoline products are relatively well soluble in water and are therefore present mostly in the liquid phase of the fermentation broth. The mycelium is usually discharged after filtration and only the filtrate is used for alkaloid isolation. Two different processes can be used for this purpose: liquid-liquid extraction into an organic solvent (with the exception of paspalic acid which cannot be extracted into any organic solvent) and sorption on an ion exchanger. The latter is the preferable method of isolation of simple ergoline products from fermentation broths.

13.2.3.
Higher Plants

The occurrence of ergot alkaloids in higher plants is discussed in Chapter 18. Of practical importance is the industrial isolation of lysergol from the Kaladana seeds. Kaladana is the aboriginal name for a plant, botanically classified as *Ipomoea (Ipomoea hederacea, Ipomoea parasitica, CalonictionIpomoea)* and growing wildly in the for-Himalaya area of India. Its seeds contain up to 0.5% of lysergol and only a low amount of other alkaloids. Patents belonging to the Italian company Simes (later Farmex) describe the isolation of lysergol from these seeds and the process for nicergoline manufacture from lysergol (Simes, 1971; Mora, 1979; Bernardelli, 1987). Production of nicergoline from this source is not very important and its competitivity is questionable. It depends on the crop of wildly growing Kaladana seeds and the reliability of such a source is low.

13.2.4.
Organic Synthesis

Considerable effort was devoted to the total synthesis of ergoline compounds. Information about this area can be found in a review (Ninomyia and Kiguchi, 1988). Although many interesting approaches were developed, a process producing ergot alkaloids more effectively than is their isolation from natural material was never found.

Partial synthesis of more complex alkaloids from simple ergoline precursors brought more success. The first total synthesis of the peptidic part of ergopeptines (the cyclol part) was achieved by Sandoz researchers (Hofmann et al., 1961). This synthesis (Figure 2) was extended to all the natural ergopeptines, their dihydroderivatives and other non-natural analogues and derivatives—(i.e. Stadler et al., 1963, 1969; Stadler and Hofmann, 1969; Hofmann et al., 1963;

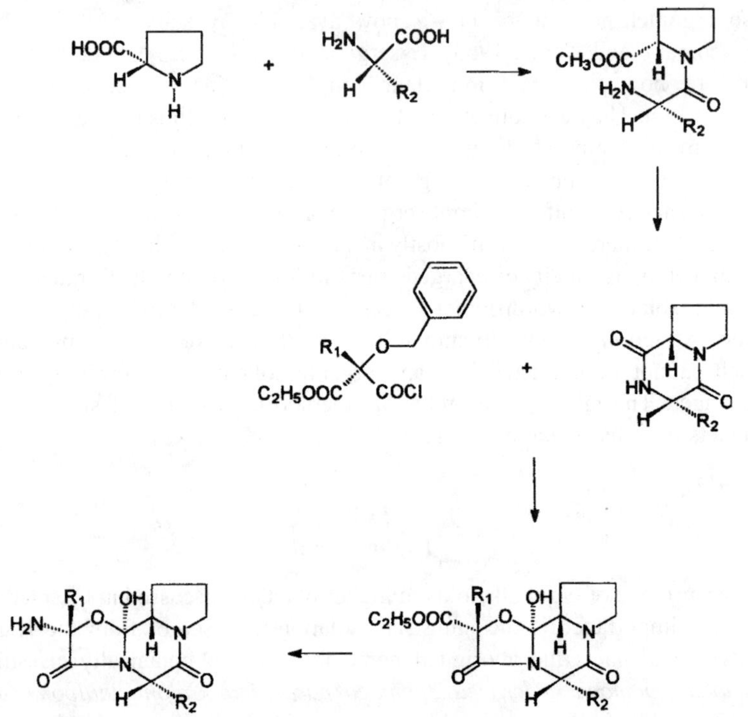

Figure 2 Synthesis of cyclol moiety of natural ergopeptines: R_1=methyl, ethyl or isopropyl; R_2=benzyl, isopropyl, isobutyl or *sec*-butyl

Stütz *et al.*, 1969; Guttmann and Huguenin, 1970). Alternative syntheses of the cyclol moiety were described by Stadler (1978 and 1980) and Losse (1982). The synthesis of the cyclol part of natural ergopeptines was developed by Sandoz up to the industrial scale and it is used for the manufacture of ergopeptines and dihydroergopeptines. The source of the ergoline part is a mixture of lysergic, isolysergic and paspalic acids of fermentation origin.

Chemical modification of the ergoline skeleton is described in Chapter 8 —"Chemical modification of ergot alkaloids". From the point of view of industrial production, the first such modification was hydrogenation of ergopeptines to dihydroergopeptines, first described by Stoll and Hofmann (1943a). Later, many modifications of hydrogenation were patented using different catalysts (PtO_2, palladium, Raney-nickel) and claiming some special conditions. Lysergic acid derivatives are hydrogenated easily (at atmospheric pressure) and stereoselectively, forming dihydroderivatives with *trans* connection of C and D rings. Isolysergic acid derivatives have to be hydrogenated at a higher pressure and a mixture of *trans* (ergoline-I) and *cis*

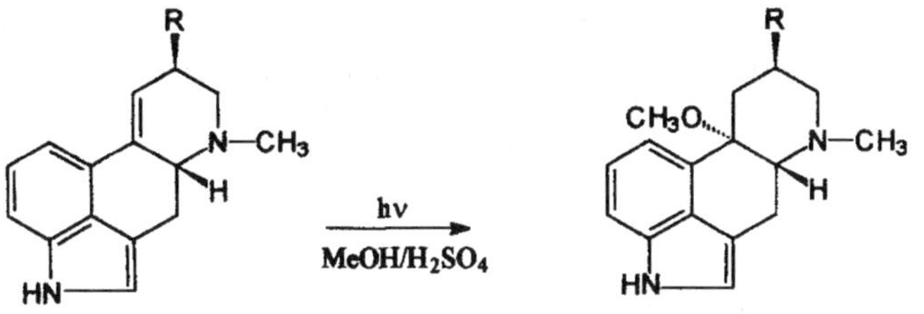

| Methyl lysergate | R = COOCH$_3$ | Methyl 10α-methoxydihydrolysergate |
| Lysergol | R = CH$_2$OH | 10α-Methoxydihydrolysergol |

Figure 3 Photochemical methoxylation of methyl lysergate and lysergol

(ergoline-II) dihydroderivatives is obtained. The ratio of ergoline-I to ergoline-II can be modified by reaction conditions (Sauer *et al.*, 1986).

Successful therapeutic use of some semisynthetic ergolines initiated the search for new synthetic methods giving higher yield and better product quality. Looking for new, patentable processes was another goal. Many procedures for bromination of ergoline compounds were developed aiming at the synthesis of bromokryptine (Troxler and Hofmann, 1957; Ručinan *et al.*, 1977; Stanovnik *et al.*, 1981; Börner *et al.*, 1983; Megyeri *et al.*, 1986; Cvak *et al.*, 1988). Investigation of a new process for the manufacture of nicergoline brought new procedures for indole nitrogen alkylation (Troxler and Hofmann, 1957a; Ručman, 1978; Šmidrkal and Semonský, 1982; Cvak *et al.*, 1983; Gervais, 1986; Marzoni and Garbrecht, 1987).

Very interesting is the photochemically initiated addition of methanol to the ergolene skeleton (methyl lysergate or lysergol), which is the key step in nicergoline synthesis (Figure 3). The original work of Hellberg (1957) on the water addition to ergopeptines, in which their acidic aqueous solutions were irradiated by UV light (10-hydroxy derivatives called lumi-derivatives were produced), was extended to an industrial-scale method. The photomethoxylation is a stereoselective process (more than 90% of 10α-methoxy derivative) giving a quantum yield of 0.48 (Cvak, 1985). It is one of the rare industrial applications of photochemistry (Bernardi *et al.*, 1966; Stres and Ručman, 1981; Bombardelli and Mustich, 1985).

Another frequently used industrial synthesis is the coupling of lysergic or dihydrolysergic acids with amines, which is the key step of the syntheses of ergometrine and methylergometrine and ergopeptines and dihydroergopeptines. Many coupling reagents were suggested for this purpose (Pioch, 1956; Garbrecht, 1959; Frey, 1961; Hofmann and Troxler, 1962; Černý and Semonský, 1962; Patelli and Bernardi, 1964; Stuchlík *et al.*, 1985), but only a few are really used on the industrial scale.

Some other chemical modifications of ergot alkaloids are used for production of particular semisynthetic, therapeutically used derivatives. They are mentioned in part 4 of this chapter.

13.3.
INTERMEDIATES FOR INDUSTRIAL PARTIAL
SYNTHESES OF ERGOT ALKALOIDS

13.3.1.
Lysergic Acid

Lysergic acid is the basic and universal intermediate for the syntheses of all the therapeutically used ergot alkaloids. It is produced in the chiral form with configuration 5R and 8R (designations d-lysergic acid or D-lysergic acid are also used). The annual world production of lysergic acid can be estimated at 10–15 tons. Most of this quantity is used for nicergoline manufacture, the rest for ergometrine, methylergometrine and methysergide. Novartis company uses lysergic acid for the syntheses of ergopeptines.

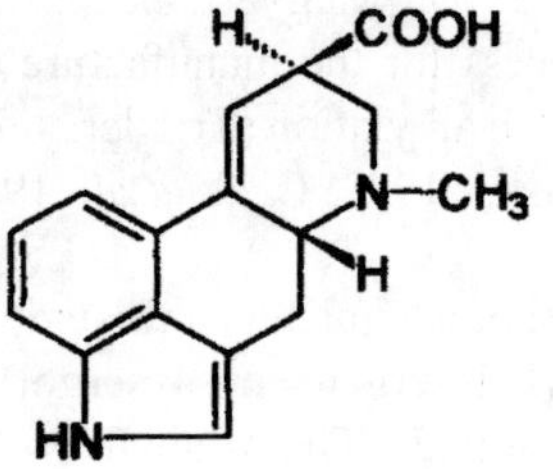

Figure 4 d-Lysergic acid

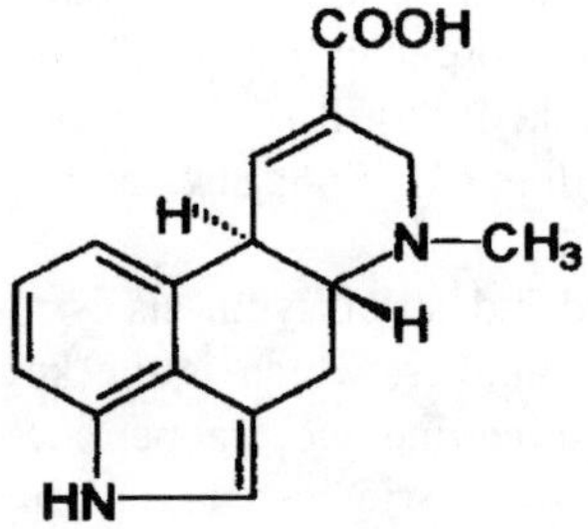

Figure 5 Paspalic acid

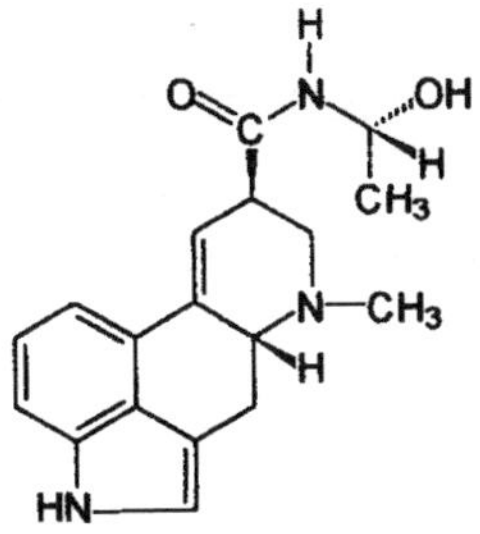

Figure 6 Lysergic acid hydroxyethylamide

There are two methods for lysergic acid manufacture. The first one is hydrolysis of ergopeptines isolated from ergot or of fermentation origin, the second one is the direct fermentation of one of its simple precursor—paspalic acid (Figure 5) or lysergic acid hydroxyethylamide (Figure 6). The former process is based on works of Jacobs and Craig (1934, 1934a, 1935, 1935a, 1936) on alkaline hydrolysis of ergopeptines. Many patents appeared later, specifying reaction conditions or isolation and purification of the product (i.e. Ručman, 1976; Cvak et al., 1978).

The majority of lysergic acid is produced fermentatively. Because there exists no strain producing lysergic acid as the main secondary metabolite, it is manufactured indirectly via its available precursors. While paspalic acid is converted into lysergic acid very easily (Troxler, 1968), the lysergic acid hydroxyethylamide is easily hydrolysed only to ergine and erginine, which must be hydrolysed to lysergic acid by alkaline hydrolysis similarly as ergopeptines.

13.3.2.
Dihydrolysergic Acid

Dihydrolysergic acid can be used only for the manufacture of dihyhro ergopeptines, metergoline, pergolide, terguride and cabergoline. Its world production is very limited. It can be obtained by the hydrolysis of dihydroergopeptines (often wastes from their purification) or by hydrogenation of lysergic or paspalic acids.

13.3.3.
Lysergol

As mentioned above, lysergol (Figure 8), isolated from the Kaladana seeds is used for the manufacture of nicergoline. There are two other processes for lysergol production. Methyl lysergate can be reduced to lysergol by lithium aluminium hydride (Stoll et al., 1949) or sodium borohydride (Beran et al., 1969). The latter process uses elymoclavine (Figure 7) available by fermentation. Eich (1975) described the isomerisation of elymoclavine to lysergol.

13.3.4.
Dihydrolysergol

Dihydrolysergol (Figure 9) is the intermediate for the production of pergolide. It is produced by the hydrogenation of lysergol or elymoclavine. Production from dihydrolysergic acid *via* reduction of its methyl ester is also possible.

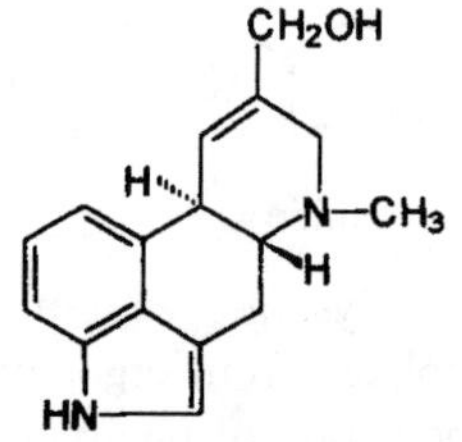

Figure 7 Elymoclavine

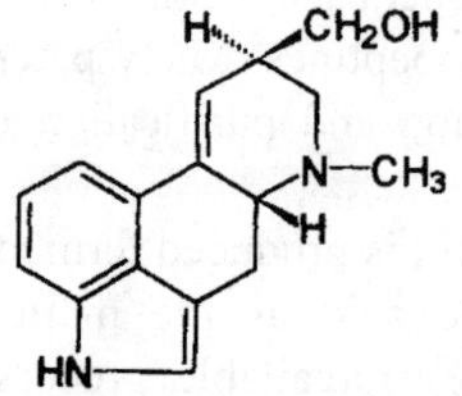

Figure 8 Lysergol

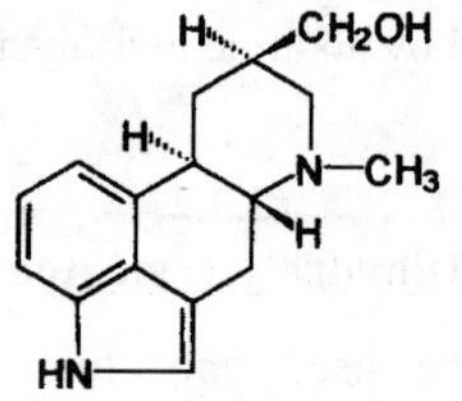

Figure 9 Dihydrolysergol

13.3.5.
Other Intermediates

There are some other intermediates used for the manufacture of other therapeutically used alkaloids. Lisuride can be prepared from erginine (Sauer and Haffer, 1981; Bulej *et al.*, 1990), which is obtained by partial hydrolysis of ergopeptines or lysergic acid hydroxyethylamide. Dihydroergine, produced by partial hydrolysis of dihydroergopeptines, can be used for the manufacture of metergoline.

13.4.
THERAPEUTICALLY USED ERGOT ALKALOIDS AND THEIR PRODUCTION

All the therapeutically used ergot alkaloids and their derivatives are described in the following part of the chapter. The main qualitative requirements of actual world leading pharmacopoeias (Eur. Ph. 1997, USP 23 and JP XIII), in which the substances have been incorporated and the names of pharmaceutical specialities with ergot alkaloids (Negwer, 1994) are presented here.

Ergot alkaloids are rather complicated molecules. As a consequence, many chemical names of ergot alkaloids, both correct and faulty, can be found in the literature. Only some examples of different types of nomenclature are presented here, namely the nomenclature according to Chemical Abstracts, where the trivial names ergoline for the tetra-cyclic system and ergotaman for the seven-cyclic ergopeptine system are used, and the nomenclature according to the IUPAC rules for heterocyclic compounds.

Pharmacology, toxicology and metabolism of therapeutically used ergot alkaloids were reviewed in monograph of Berde and Schild (1978) and therefore only references to newly developed products or to some new findings and reviews of older products are presented here.

13.4.1
Ergotamine

Chemical names: 2'-Methyl-5'-benzyl-ergopeptine;
12'-hydroxy-2'-methyl-5'-(phenylmethyl)-ergotaman-3',6', 18-trione;
(6aR, 9R)-N-[(2R, 5S, 10aS, 10bS)-5-benzyl-10b-hydroxy2-methyl-3, 6-dioxo-octahydro-8H-oxazolo [3, 2-a] pyrolo[2, 1-c]-pyrazin-2-yl]-7-methyl-4,   6,   6a,   7,   8,   9-hexahydroindolo-[4, 3-fg]quinoline-9-carboxamide

Structural formula: See Figure 10
Empirical formula: base $C_{33}H_{35}N_5O_5$
                         tartrate $(C_{33}H_{35}N_5O_5)_2 \cdot C_4H_6O_6$
Molecular weight: base   581.7
                         tartrate   1313.4
CAS No.   base   113–15–5
                         tartrate   379–79–3

Specifications and their requirements:
Eur. Ph. 1997 Ergotamine tartrate assay (titration): 98.0–101.0% in dry substance

		total impurities: not more than 1.5% (TLC)
USP23	Ergotamine tartrate	only one impurity more than 0.5% (TLC) assay (titration): 97.0–100.5% in dry substance
		total impurities: not more than 2.0% (TLC)
JP XIII	Ergotamine tartrate	only one impurity more than 1.0% (TLC) assay (titration): not less than 98.0% in dry substance
		total impurities: not more than 2.0% (TLC)

Typical impurities: ergotaminine
aci-ergotamine
material isolated from both field ergot and fermentation broths contains usually some minor ergopeptines (ergosine, ergostine, ergocristine, α-ergokryptine or 8-hydroxyergotamine)

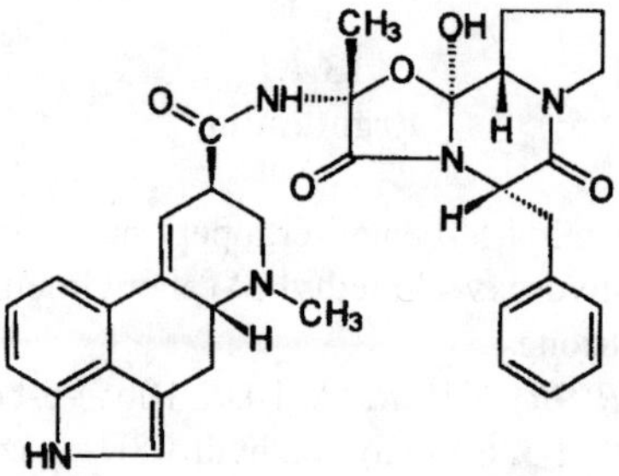

Figure 10 Ergotamine

Dosage forms: Avetol, Bedergot, Cornutamin, Enxak, Ercal, Ergam, Ergane, Ergate, Ergocito, Ergofeina, Ergogene, Ergogyn, Ergo-Kranit mono, Ergomar "Fisons", Ergomigrin, Ergomine-S, Ergo-sanol SL, Ergostat, Ergostin, Ergota "Kanto", Ergotan, Ergotartrat, Ergoton-A, Ergotrat AWD), Etin, Exmigra, Exmigrex, Femergin, Fermergin, Gynecorn, Gynergen, Gynofort, Ingagen, Lagen, Lingrän, Lingraine, Lingrene, Masekal, Migretamine, Migrexa, Migtamin, NeoErgotin, Neo-Secopan, Pannon, Rigetamin, Ryegostin, Secagyn, Secanorm, Secotamin, Secupan, Synergan, Vigrame, Wigrettes

Therapeutic use: uterotonic, antimigrenic, vasoconstrictor, hemostatic

Introduction: 1921
World production: 1000–1500 kg per year
Bulk substance
manufactures: Boehringer Ingelheim (Germany)—isolation from ergot
                       Galena (Czech Rep.)—isolation from ergot
                       Lek (Slovenia)—fermentation
                       Novartis (Switzerland)—synthesis
                       Poli (Italy)—fermentation
Manufacture: 1. Isolation from field ergot
                       2. Isolation from fermentation
                       broth
                       3. Synthesis from d-lysergic
                       acid and synthetic peptidic
                       moiety
References: Kreilgard 1976 (anal.), Holger 1994 (therap.use)

13.4.2.
Ergometrine

Other names: Ergobasine
                       Ergonovine
Chemical names: d-Lysergic acid-L-(+)-1-(hydroxymethyl)ethylamide;
                       9,   10-Didehydro-N-[(S)-2-hydroxy-1-methylethyl]-6-
                       methylergoline-8β(S)-carboxamide;
                       (6aR,   9R)-N-[(S)-2-hydroxy-1-methylethyl]-7-methyl-4,
                       6, 6a, 7,8, 9-hexahydroindolo [4, 3-fg]quinoline-9-
                       carboxamide
Structural formula: See Figure 11
Empirical formula: base $C_{19}H_{23}N_3O_2$
                       maleate $C_{19}H_{23}N_3O_2 \cdot C_4H_4O_4$
                       tartrate $(C_{19}H_{23}N_3O_2)_2 \cdot C_4H_6O_6$
Molecular weight: base   325.4
                       maleate   441.5
                       tartrate   800.8
CAS No.:   base   60–79–7
                       maleate   129–51–1
                       tartrate   129–50–0

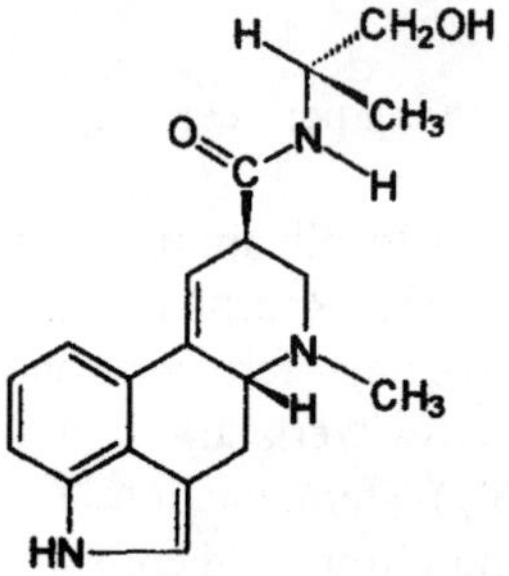

Figure 11 Ergometrine

Specifications and their requirements:

Eur. Ph. 1997	Ergometrine maleate	assay (titration): 98.0–101.0% in dry substance no impurity above 1.0% (TLC) only one impurity above 0.5% (TLC)
USP23	Ergonovine maleate	assay (spectrophotometric): 97.0–103.0% total impurities: not more than 2.0% (TLC)
JP XIII	Ergometrine maleate	assay (spectrophotometric): not less than 98.0% total impurities: not more than 2.0% (TLC)

Typical impurities: ergometrinine
other impurities are specific for individual producers and depend on their manufacturing process

Dosage forms: Arconovina, Basergin, Cornocentin, Cryovinal, Ergofar, Ergomal, Ergomar Nordson, Ergomed "Promed", Ergomet, Ergometine, Ergometron, Ergomine, Ergostabil, ErgotonB, Ergotrate Maleate, Ermalate, Ermeton, Ermetrin, Hemogen, Margonovine, Metriclavin, Metrisanol, Neofemergen, Novergo, Panergal, Secalysat-EM, Secometrin, Takimetrin, Uteron

Therapeutic use: uterotonic, oxytocic
Introduction: 1936
World production: 100–200 kg per year
Bulk substance
manufactures: Boehringer Ingelheim (Germany)
Galena (Czech Rep.)

Lek (Slovenia)
Lonza (Switzerland)
Novartis (Switzerland)

Manufacture: 1. Isolation from field ergot as a minor by-product
2. Isolation from fermentation broth
3. Synthesis from d-lysergic acid and L-(+)-2-aminopropanol using different coupling reagents

References: Rutschmann and Kobel 1967 (fermentation), Stoll and Hofmann 1948 (synth.)
Reif 1982 (anal.)

13.4.3.
Dihydroergotamine

Chemical names: 2'-Methyl-5'-benzyl-dihydroergopeptine;
9, 10-Dihydro-12'-hydroxy-2'-methyl-5'-(phenylmethyl) ergotaman-3', 6', 18-trion;
(6aR, 9R, 10aR)-N-[(2R, 5S, 10aS)-5-benzyl-10b-hydroxy-2-methyl-3, 6-dioxo-octahydro-8H-oxazolo[3, 2-a]-pyrolo[2, 1-c] pyrazin-2-yl]-7-methyl-4, 6, 6a, 7, 8, 9, 10, 10aoctahydro-indolo [4, 3-fg]quinoline-9-carboxamide

Structural formula: See Figure 12

Empirical formula: base $C_{33}H_{37}N_5O5$
mesylate $C_{33}H_{37}N_5O_5 \cdot CH_4O_3S$
tartrate $(C_{33}H_{37}N_5O_5)_2 \cdot C_4H_6O_6$

Molecular weight: base   583.7
mesylate 679.8
tartrate   1317.5

CAS No.:   base   511–12–6
mesylate 6190–39–2
tartrate   5989–77–5

Specifications and their requirements:
Eur. Ph. 1997 Dihydroergotamine mesilate assay (spectrophotometric): 97. 0–97.0–103.0% in dry substance
no impurity above 0.5% and only two impurities above 0.2% (TLC)

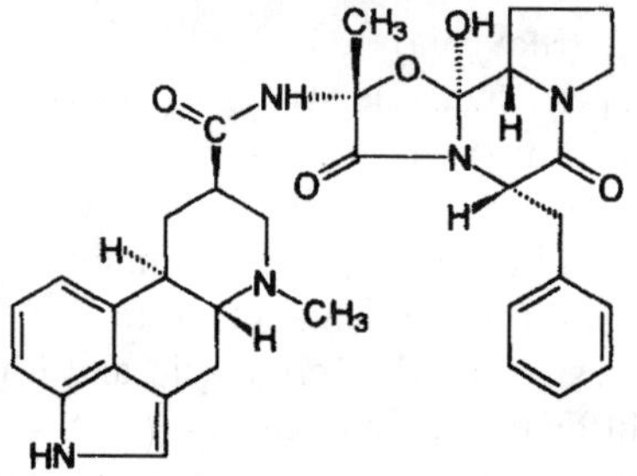

Figure 12 Dihydroergotamine

	Dihydroergotamine tartrate	assay (spectrophotometric): 97.0–103.0% in dry substance no impurity above 0.5% and only two impurities above 0.2% (TLC)
USP23	Dihydroergotamine mesylate	assay (spectrophotometric): 97.0–103.0% in dry substance total impurities: not more than 2.0% (TLC)
JP XIII	Dihydroergotamine mesilate	assay (titration): not less than 97.0% in dry substance no impurity above 0.5% and only two impurities above 0.2% (TLC)

Main impurities: ergotamine, egotaminine, aci-dihydroergotamine material produced *via* extraction from ergot or fermentation broth usually contains some minor dihydroergopeptines (dihydroergosine, dihydroergostine, dihydroergocristine, dihydro-α-ergokryptine, dihydro-8-hydroxy-ergotamine)

Dosage forms: Adhaegon, Agit, Angionorm, Biosupren, Bobinium, Clavigrenin, Cornhidral, Cozetamin, Dergiflux, Dergolyoc, Dergot, Dergotamine, Detemes, DETMS, DHE 45, DHE-MS, DHE-Puren, DHE-Ratiopharm, DE-Ergotamin, DHE-Tablinen, DHE-Tamin, Diaperos "Materia", Diergo-Spray, Di-ergotan, Di-got, Digotamin, Dihydergot, Dihydroergotamin-Sandoz, Dihy-ergot, Dihytam, Dihytamin, Diidergot, Dirgotarl, Disecotamin, Ditamin, Divegal, D-Tamin, Eldoral Dumex, Elmarine Genepharm, Endophleban, Ergomimet, Ergont, Ergospaon, Ergotex, Ergotonin, Ergott, Ergovasan, Esikmin, For You, Hidergot, Hidrotate, Hydro-Tamin, Hyporal, Ikaran, Itomet, Kidira, Kodamaine, Kouflem, Migergon D, Migretil, Migrifen, Mitagot, Morena,

Neomigran, Orsta-norm, Ortanorm, Panergot, Pefanicol, Pervone "Sanofi-Greece", Phlebit, Rayosu, Rebriden, Restal Tokyo "Tanabe", Seglor, Tamik, Tariyonal, Tonopres, Vasogin, Verladyn, Verteblan, Youdergot, Yougovasin

Therapeutic use: antimigrenic, sympatholytic, vasoconstric
Introduction: 1946
World production: 1500–2000 kg per year
Bulk substance
manufactures: Boehringer Ingelheim (Germany)
Galena (Czech Rep.)
Lek (Slovenia)
Novartis (Switzerland)
Piere Fabre (France)

Poli (Italy)
Sanofi (France)
Manufacture: 1. Hydrogenation of ergotamine isolated from field ergot or fermentation broth
2. Synhesis from dihydrolysergic acid and synthetic peptidic part
References: Marttin 1997 (pharmacokinetics)

13.4.4.
Dihydroergotoxine

Other names: Ergoloid
Codergocrine
Chemical name: Dihydroergotoxine is a mixture of Dihydroergocristine, Dihydoergocornine, Dihydro-α-ergokryptine and Dihydroβ-ergokryptine
Structural formula: See Figure 13

Empirical formula: and Molecular weight

Dihydroergocristine base	$C_{35}H_{41}N_5O_5$	611.7
Dihydroergocristine mesylate	$C_{35}H_{41}N_5O_5 \cdot CH_4O_3S$	707.9
Dihydroergocornine base	$C_{31}H_{41}N_5O_5$	563.7
Dihydroergocornine mesylate	$C_{31}H_{41}N_5O_5 \cdot CH_4O_3S$	659.8
Dihydro-α-ergokryptine base	$C_{32}H_{43}N_5O_5$	577.7
Dihydro-α-ergokryptine mesylate	$C_{32}H_{43}N_5O_5 \cdot CH_4O_3S$	673.8
Dihydro-β-ergokryptine base	$C_{32}H_{43}N_5O_5$	577.7
Dihydro-β-ergokryptine mesylate	$C_{32}H_{43}N_5O_5 CH_4O_3S$	673.8

CAS No.: Dihydroergotoxine mesylate   8067-24-1
Specifications and their requirements:

Eur. Ph. 1997   Not implemented
BP93   Co-dergocrine mesylate   Assay (HPLC): 97.0–
                                                       103.0% in dry substance
                                                       30.0–36.5%   of
                                                       dihydroergocristine
                                                       mesylate

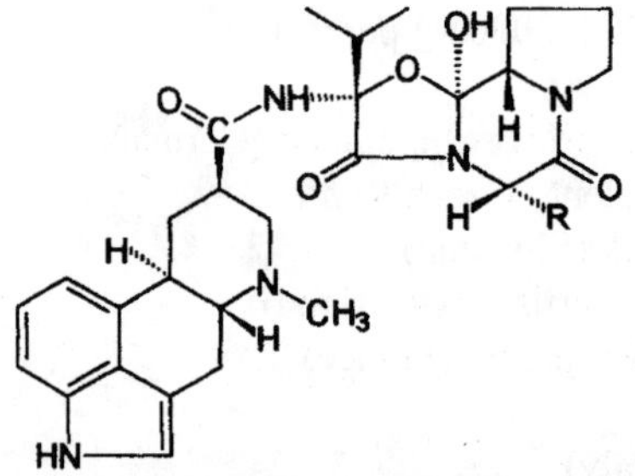

Figure 13 Dihydroergotoxine: R=benzyl for dihydroergocristine, isopropyl for dihydroergocornine, isobutyl for dihydro-α-ergokryptine and *sec-butyl* for dihydro-βergokryptine

		30.0–36.5% of dihydroergocornine mesylate
		30.0–36.5% of dihydroergokryptine mesylate
		ratio of α and β dihydroergokryptine is not less than
		1.5:1.0 and not more than 2.5: 1.0 (HPLC)
USP23	Ergoloid mesylates	Assay (HPLC): 97.0–103.0% in dry substance
		30.3–36.3% of dihydroergocristine mesylate
		30.3–36.3% of dihydroergocornine mesylate
		30.3–36.3% of dihydroergokryptine mesylate

	ratio of α and β dihydroergokryptine is not less than 1.5 :1.0 and not more than 2.5 :1.0 (HPLC)

JP XIII — Not implemented, draft presented in JP Forum Vol. 5 No. 3 (1996)

Dosage forms: Alizon, Alkergot, Apolamin, Aramexe, Artedil, Artergin, Astergina, Baroxin "Toa Eiyo", Bordesin, Brentol, Capergyl, Carlom, CCK 179, Cervitonic, Circanol, Clavor, Coax, *Co-dergocrine mesylate,* Coplexina, Coristin, Cortagon, Cursif, Dacoren, DCCK, Deapril-ST, Defluina N, Demanda, Derginal, D-Ergotox, DH-Ergotoxin, DH-Tox-Tablinen, Dihydren, Dilaten, Doctergin, Dorehydrin, Dulcion, Ecuor, Elmesatt, Enirant "Gepepharm", Epos, Ercalon, Erginemin, Ergoceps, Ergocomb, Ergodesit, Ergodilat, Ergodina, Ergodose, Ergogine, Ergohydrin, Ergokod, Ergokrinol, *Ergoloid mesylates,* Ergomax, Ergomed Kwizda, Ergomolt, Ergoplex, Ergoplus, Ergotox v. ct, Ergoxyl, Erlagine, Fermaxin, Fluzal, Geroplus, H.E.A., Hidergo, Hidrosan, HY 71, Hyderan, Hyderaparl, Hydergin(e), Hydervek, Hydro-Cebral, Hydro-Ergot, Hydrolid G, Hydro-Toxin, Hydroxina, Hydroxium, Hynestim, Hyperloid, Ibergal, Indolysin, Inorter, Iresolamin, Iristan, Ischelium Kerasex, Kylistop, Larvin, Latergal, Lysergin, Medixepin, Memoxy, Milepsin, Minerizine, Necabiol, Nehydrin N, Niloric, Nor-madergin, Normanomin, Novofluën, Nulin Velka, Optamine, Orphol, Pallotrinate, Pérénan, Phenyramon, Primarocin, Progeril Midy-Milano; Sanofi-Basel, Redergin, Redergot, Redicor, Regotand, Relark, Samyrel, Santamin, Scamin, Secamin Lab, Secatoxin, Segol, Senart, Simactil, Siokarex, Sponsin, Stofilan, Theo-Nar, Theragrin-S, Toterjin, Tredilat Tri-Ergone, Trifargina, Trigot, Trihydrogen Goldline, Tusedon, TY-0032, Ulatil, Vasergot, Vasolax, Vimotadine, Youginin, Zenium, Zidrol, Zinvalon, Zodalin

Therapeutic use: cerebral and peripheral vasodilator

World production: 1000–1500 kg per year

Bulk substance manufacturers: Boehringer Ingelheim (Germany)
Galena (Czech Rep.)
Gedeon Richter (Hungary)
Lek (Slovenia)

Novartis (Switzerland)

Poli (Italy)

Manufacture: 1. Isolation of individual alkaloids or their mixtures from field ergot or fermentation broths, their hydrogenation, preparation of salts with methane sulphonic acid and adjustment to required ratio of individual components.

2. Synthesis of individual dihydroergotoxines from dihydrolysergic acid and coresponding synthetic peptidic parts, preparation of salts with methane sulphonic acid and adjustment to required ratio of individual components.

References: Schoenleber *et al.,* 1978 (anal.), Baer and Jenike, 1991 (therap. use), Wadworth and Chrisp, 1992 (pharmacology and use in geriatry), Ammon *et al.,* 1995 (clin.)

13.4.5.
Dihydroergocristine

Chemical names: 2'-Isopropyl-5'-benzyl-dihydroergopeptine;

9, 10-Dihydro-12'-hydroxy-2'-(1-methylethyl)-5'-(phenyl methyl)ergotamane-3', 6', 18-trione;

(6a*R*, 9*R*, 10a*R*)-N-[(2*R*, 5*S*, 10a*R*, 10b*S*)-5-benzyl-10b-hydroxy2-(1-methyl-ethyl)-3, 6-dioxo-octahydro-8H-oxazolo[3, 2-a]pyrolo[2, 1-c] pyrazin-2-yl]-7-methyl-4, 6, 6a, 7, 8, 9, 10,10α-octahydroindolo [4, 3-fg] quinoline-9-carboxamide

Structural formula: See Figure 14

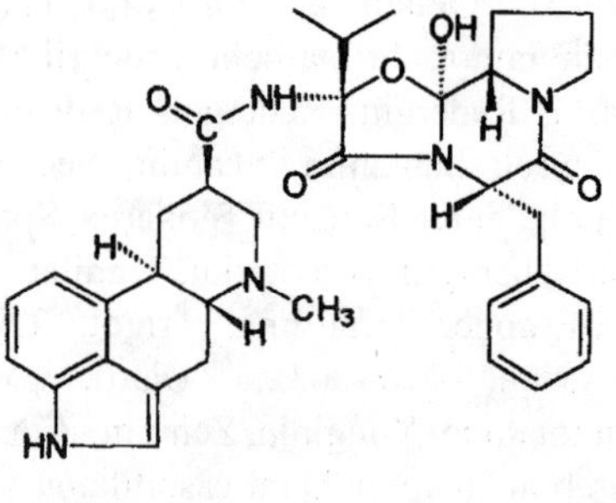

Figure 14 Dihydroergocristine

Empirical formula: base $C_{35}H_{41}N_5O_5$

mesylate $C_{35}H_{41}N_5O_5 \cdot CH_4O_3S$

$C_{35}H_{41}N_5O_5$

Molecular weight: base 611.7

mesylate 707.9

CAS No.:	base	17479–19–5
	mesylate	24730–10–7

Specifications and their requirements: Substance is included only in Czechoslovak pharmacopoeia.

Producers declare its quality by HPLC analysis.

Typical impurities: Ergocristine, ergocristinine, aci-dihydroergocristine, dihydroergine

Material manufactured from isolated ergocristine (from ergot or fermentation broths) usually contains some other dihydroergopeptines (dihydroergotamine, dihydroergocornine, dihydro-α-ergokryptine and others), dihydroergocristam or dihydroergometrine

Dosage forms: Angiodil, DCS 90, Decme Italmex; Spitzner; Zyma, Decril, Defluina Simes, Diertina, Diertine, Diertix, Difluid, Dirac, Enirant, Ergo Foletti, Ergocris, Ergodavur, Fluiben, Gral, Hydrofungin, Insibrin, Iskemil, Iskevert, Nehydrin, Nor-mosedon, Unergol, Vigoton

Therapeutic use: sympatholytic, peripheric vasodilator

World production: 1000–1500 kg per year

Bulk substance manufacturers:
Boehringer Ingelheim (Germany)
Galeana (Czech Rep.)
Gedeon Richter (Hungary)
Lek (Slovenia)
Piere Fabre (France)
Poli (Italy)

Manufacture: 1. Isolation ergocristine from ergot or fermentation broth, its hydrogenation and salt formation with methane sulphonic acid
2. Synthesis from dihydrolysergic acid and synthetic pepetidic part

References: Mailand, 1992 (pharm. and clin. review), Malacco and Di Cesare, 1992 (therap. use), Franciosi and Zavattini, 1994 (use in geriatry)

13.4.6.
Dihydro-α-ergokryptine

Chemical names: 2'-Isopropyl-5'-isobutyl-dihydroergopeptine;
9, 10-Dihydro-12'-hydroxy-2'-(1-methylethyl)-5'-(2-methyl propyl)ergotamane- 3', 6', 18-trione;

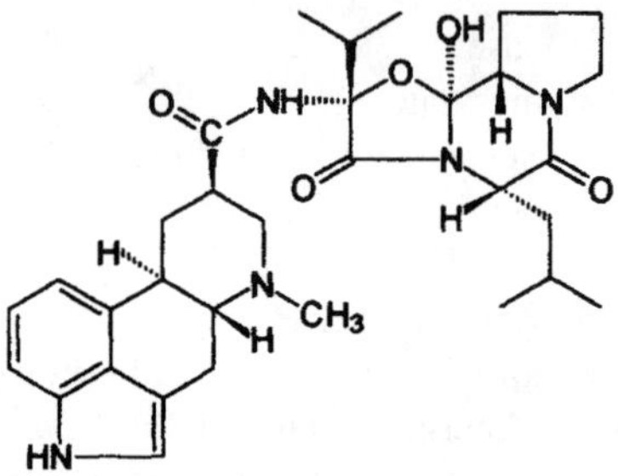

Figure 15 Dihydro-α-ergokryptine

(6a*R*, 9*R*, 10a*R*)-N-[(2*R*, 5*S*, 10a*S*, 10b*S*)-10b-hydroxy2-(1 methylethyl)-5-(2-methylpropyl)-3, 6-dioxo-octahydro8H-oxazolo[3, 2-a]pyrrolo[2, 1-c] pyrazin-2-yl]-7-methyl-4, 6,6a, 7, 8, 9, 10, 10a-octahydro-indolo[4, 3-fg]quinoline-9-carboxamide

Structural formula:	See Figure 15	
Empirical formula:	base	$C_{32}H_{43}N_5O_5$
	mesylate	$C_{32}H_{43}N_5O_5 \cdot CH_4O_3S$
Molecular weight:	base	577.7
	mesylate	673.9
CAS No.:	base	25447–66–9
	mesylate	29261–93–6

Specifications and their requirements: The substance is not monographed in any pharmacopoeia.

Producers declare its purity by HPLC analysis.

Dosage forms:	Daverium, Myrol, Vasobral
Therapeutic use:	antiparkinsonian, prolactine inhibitor, cerebral vasodilator
Introduction:	1989
World production:	400–600 kg per year
Bulk substance manufacturers:	Galena (Czech Rep.), Poli (Italy)
Manufacture:	Hydrogenation of α-ergokryptine isolated from ergot or fermentation broth and salt formation with methane sulphonic acid.
References:	Poli, 1990; Coppi, 1991; Scarzela *et al.*, 1992 (all pharmacol. and therap.)

13.4.7.
Bromokryptine

Chemical names: 2-Bromo-α-ergokryptine;

2-Bromo-2'-isopropyl-5'-isobutyl-ergopeptine;
2-Bromo-12'-hydroxy-2'-(1-methylethyl)-5'-(2-
methylpropyl)-ergotamane-3', 6', 18-trione;

(6aR, 9R)-*N*-[(2*R*, 5*S*, 10a*S*, 10b*S*)-10b-hydroxy2-(1-
methyl   ethyl)-5-(2-methylpropyl)-3,   6-dioxo-
octahydro-8Hoxazolo[3, 2-a]pyrolo[2, 1-c]pyrazin-2-
yl]-5-bromo-7methyl-4, 6, 6a, 7, 8, 9-hexahydro-indolo[4,
3-fg]quinoline-9carboxamide

Structural formula:	See Figure 16	
Empirical formula:	base	$C_{32}H_{40}BrN_5O_5$
	mesylate	$C_{32}H_{40}BrN_5O_5 \cdot CH_4O_3S$
Molecular weight:	base	656.6
	mesylate	750.7
CAS No.:	base	25614–03–3
	mesylate	22260–51–1

Specifications and their requirements:		
Eur. Ph. 1997	Bromocriptine mesilate	Assay (titration): 98.0–101.0% in dry substance No impurity above 0.4% (TLC) One impurity above 0. 2% (TLC)
USP23	Bromocriptine mesylate	Assay (titration): 98.0–102.0% in dry substance Total impurities (TLC): not more than 1.0% No impurity above 0.5% (TLC)
JP XIII	Bromocriptine mesilate	Assay (titration): Not less than 98.0% in dry substance No impurity above 0.5% (TLC) One impurity above 0. 25% (TLC)
Typical impurities:	Bromokryptinine, α-ergokryptine, 2-chloro-α-ergokryptine Material manufactured from α-ergokryptine isolated from ergot or fermentation broth usually contains 2-bromoderivatives of some	

other ergopeptines (2-bromo-β-ergokryptine, 2-bromoergocristine, 2-bromo-ergogaline or others)

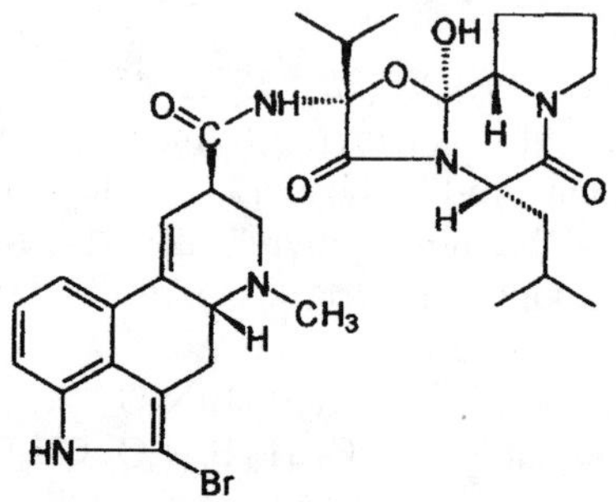

Figure 16 Bromokryptine

	Material of some producers can contain dibromoderivatives (2, 12-dibromo-α-ergokryptine and/or 2, 13-dibromoα-ergokryptine)
Dosage forms:	Bromocorn, Bromopar, CB 154, Criten, Deparo, Elkrip, Erenant, Ergolactin, Grifocriptina, Kirim, Lactismine, Maylaktin, Morolack, NSC-169774, Padoparine, Palolactin, Antipark, Atlodel, Axialit, Bagren, Bromed, Bromergon, Parilac, Parlodel, Parlomin, Parodel, Parukizone, Practin, Pravidel, Prigost, Proctinal, Prospeline, Serocryptin, Serono-Bagren, Sintiacrin, Sulpac, Syntocriptine, Umprel, Upnol B
Therapeutical use:	dopamine agonist, antiparkinsonian, prolactine inhibitor, treatment of acromegaly
Introduction:	1975
World production:	1000 kg per year
Bulk substance manufacturers:	Galena (Czech Rep.)
	Gedeon Richter (Hungary)
	Lek (Slovenia)
	Novartis (Switzerland)
	Poli (Italy)
Manufacture:	Bromination of α-ergokryptine by different brominating agents (N-bromo-succinimide, pyrolidone hydrotribromide, N-bromosacharine and other N-bromo-

derivatives, trimethylsilylbromide/ dimethylsulphoxide, bromine/hydrobromide, bromine/bortrifluoride etherate and others) Starting material, α-ergokryptine is isolated from ergot or fermentation broth or is synthetised from lysergic acid.

References: Flückiger and Troxler, 1973; Ručman *et at.,* 1977; Stanovnik *et al.,* 1981; Börner *et al.,* 1983; Megyeri *et al.,* 1986; Cvak *et al.,* 1988 (all manufacture), Giron-Forest and Schoenleber 1979 (anal.), Vigouret *et al.,* 1978, Lieberman and Goldstein 1985, Weil 1986 (all pharmacol. and therap.)

13.4.8.
Nicergoline

Chemical names: 10α-Methoxy-1, 6-dimethylergoline-8β-methanol 5-bromonicotinate(ester);
10α-Methoxy-1-methyl-dihydrolysergol 5-bromonicotinate;
5-Bromopyridine-3-carboxylate of [(6a*R*, 9*R*, 10a*S*)-10a-methoxy-4, 7-dimethyl-4, 6, 6a, 7, 8, 9, 10, 10a-octahydro-indolo[4, 3-fg]quinoline-9-yl]methyl

Structural formula: See Figure 17
Empirical formula: $C_{24}H_{26}BrN_3O_3$
Molecular weight: 484.4
CAS No.:   27848–84–6
Specification and their requirements: Substance is monographed only in French Pharmacopoeia, 10th edition, January 1995
Assay (titration):   98.0–101.0% in dry substance
Purity (HPLC):   No impurity above 1.0%
Not more than 2 impurities above 0.5%
Not more than 4 impurities above 0.2%
Typical impurities: chloronicergoline
1-demethylnicergoline
10α-methoxy-1-methyl-dihydrolysergol
5-bromonicotinic acid

Dosage forms: Adavin, Cergodum, Circo-Maren, Dasovas,
                      Dilasenil, Dospan, Duracebrol, Ergobel, Ergotop,
                      F.I.6714, Fisifax, Mariol, Memoq, Nardil Gödecke,
                      Nargoline, Nicergolent, Nicergolyn, Nicerhexal,
                      Nicerium, 19561 R.P., Sermion, Sincleron, Specia,
                      287; Varson, Vasospan Exa, Vetergol
Therapeutic use: cerebral vasodilator
Introduction: 1978
World production: 7000–10000 kg per year
Manufacturers: Farmitalia (Pharmacia-Upjohn)
                      Galena (Czech Rep.)
                      Indena , Linea Nuova (both Italy)
                      Rhone Poulenc (France)
Manufacture: 1. From lysergic acid
                      2. From lysergol
References: Bernardi *et al.,* 1966; Arcari *et al.,* 1972; Ručman
                      and Jurgec, 1977; Ručman, 1978; Stres and
                      Ručman, 1981; Černý *et al.,* 1983; Cvak *et al.,*
                      1983, 1985; Bombardeli and Mustich 1985 and
                      1985a; Gervais 1986 (all manufacture);

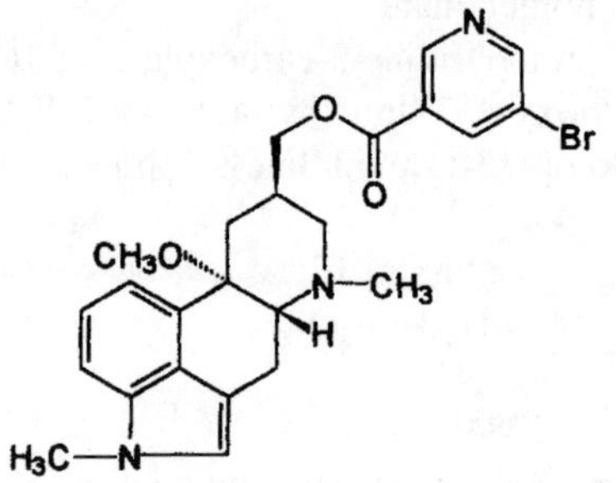

Figure 17 Nicergoline

Bernardi, 1979 (pharmacol. review); Banno, 1989 (analytics and stability)

13.4.9.
Metergoline

Chemical names: 1-Methyl-*N*-carbonybenzyloxy-dihydrolysergamine; [(1,
                      6-Dimethylergolin-8β-yl)-methyl]-carbamic   acid
                      phenylmethyl ester;
Structural formula: See Figure 18
Empirical formula: $C_{25}H_{29}N_3O_2$

Molecular weight:	403.5
CAS No.:	17692–51–2
Specifications:	Substance is not monographed in any pharmacopeia; manufacturers have their own specifications
Dosage forms:	Al-Migren, Contralac, Liserdol
Therapeutic use:	serotonine antagonist, antimigrenic, prolactine inhibitor
Introduction:	1987
World production:	20–50 kg per year
Manufacturers:	Farmitalia (Pharmacia-Upjohn), Galena (Czech Rep.), Poli (Italy)
Manufacture:	1. From 1-methyl-dihydroergine *via* reduction with lithium aluminiumhydride to 1-methyldihydrolysergamine and its reaction with benzylchloroformate. 1-Methyl-dihydroergine is available from dihydroergotamine or another dihydroergopeptine. 2. From dihydrolysergole
References:	Bernardi *et al.,* 1964; Camerino *et al.,* 1966 (manufacture)

13.4.10.
Methylergometrine

Other names:	Methylergobrevine Methylergobasine Methylergonovine
Chemical names:	d-Lysergic acid L-(+)-1-(hydroxymethyl) propylamide;

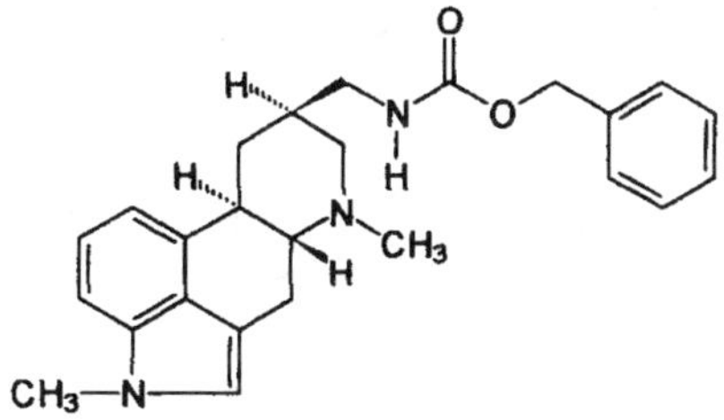

Figure 18 Metergoline

9, 10-Didehydro-*N*-[(*S*)-1-(hydroxymethyl)propyl]-6-methylergoline-8*β(R)*-carboxamide;
(6a*R,* 9*R*)-N-[*(S)*-1-(hydroxymethyl)propyl]-7-methyl-4, 6, 6a7, 8, 9-hexahydro-indolo[4, 3-fg]quinoline-9-carboxamide

Structural formula: See Figure 19

Empirical formula: base $C_{20}H_{25}N_3O_2$

                    maleate $C_{20}H_{25}N_3O_2 \cdot C_4H_4O_4$

                    tartrate $(C_{20}H_{25}N_3O_2)_2 \cdot C_4H_6O_6$

Molecular weight: base   339.4

                    maleate   455.5

                    tartrate   828.9

CAS No.:   base   113–42–8

                    maleate   7432–61–8

                    tartrate   6209–37–6

Specifications:

USP23   Methylergonovine   Assay
                    Maleate   (spectrophotometric):   97.
                                          0–103.0% in dry substance
                                          Total impurities (TLC): not
                                          more than 2.0%

JP XIII   Methylergometrine   Assay
                    Maleate   (spectrophotometric):   95.
                                          0–105.0% in dry substance
                                          Purity (TLC): No impurity
                                          above 1.0%

Typical impurities: methylergometrinine
                    d-Lysergic acid D-(—)-1-(hydroxymethyl)propylamide
                    other impurities are specific for individual producers and
                    depends on their manufacturing processes

Dosage forms: Basofortina, Demergin, Derganin, Elamidon, Elpan S,
                    Emifarol, Enovine, Erezin, Ergobacin, Ergopartin,
                    Ergotyl, Ergovit-Amp., Levospan, Mergot, Metenarin,
                    Methecrine, Methergen, Methergin, Metiler, Mitrabagin-
                    C, Mitrosystal, Mitrotan, Myomergin, NSC-186067,
                    Obstet, Partergin,

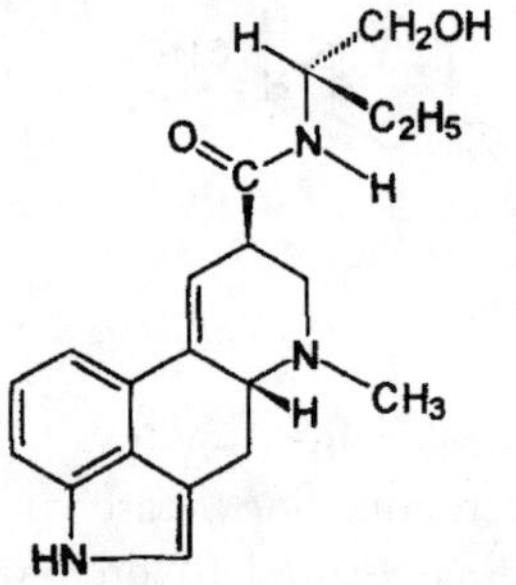

Figure 19 Methylergometrine

Ryegonovin, Santargot, Secotyl, Spametrin-M, TakimetrinM, Telpalin, Unidergin, Utergine, Uterin

Therapeutic use:	uterotonic, oxytocic
Introduction:	1946
World production:	80–150 kg per year
Bulk substance manufacturers:	Galena (Czech Rep.), Lek (Slovenia), Novartis (Switzerland)
Manufacture:	Synthesis from d-lysergic acid and L-(+)-2-aminobutanol using diffrent coupling reagents
References:	Stoll and Hofmann, 1941 and 1943 (manuf.)

13.4.11.
Methysergide

Chemical names:	1-Methyl-d-lysergic acid-L-(+)-1-(hydroxymethyl) propylamide;	
	9, 10-Didehydro-N-[(S)-1-(hydroxymethyl)propyl]-1, 6dimethylergoline-8$\beta(R)$-carboxamide;	
	(6aR, 9R)-N-[(S)-1-(hydroxymethyl)propyl]-4, 7-dimethyl-4, 6, 6a, 7, 8, 9-hexahydroindolo[4, 3-fg] quinoline-9-carboxamide	
Structural formula:	See Figure 20	
Empirical formula:	base	$C_{21}H_{27}N_3O_2$
	maleate	$C_{21}H_{27}N_3O_2 \cdot C_4H_4O4$
Molecular weight:	base	353.4
	maleate	469.5
CAS No.:	base	361–37–5
	maleate	129–49–7
Specifications:	USP23	
Typical impurities:	methylergometrine (1-demethyl-methysergide) iso-methysergide	
Dosage form:	Deseril, Sansert	
Therapeutic use:	serotonine antagonist, antimigrenic	

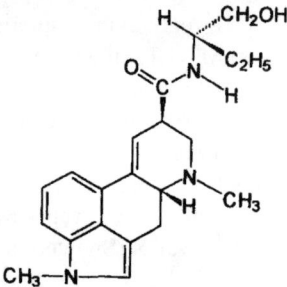

Figure 20 Methysergide

Introduction: 1960
World production: 30–50 kg per year
Bulk substance manufacturers: Novartis (Switzerland)
Manufacture: 1. Synthesis from lysergic acid *via* 1-methyl lysergic acid
    2. Methylation of methylergometrine
References: No new reference

13.4.12.
Lisuride

Chemical names: 3-(9, 10-Didehydro-6-methylergolin-8α-yl)-1, 1-diethylurea; *N*-(6-methyl-8-isoergolenyl)-*N'*, *N'*-diethylurea;
Structural formula: See Figure 21
Empirical formula: base $C_{20}H_{26}N_4O$
    maleate $C_{20}H_{26}N_4O \cdot C_4H_4O_4$
Molecular weight: base   338.4
    maleate   454.5
CAS No.:   base   18016–80–3
    maleate   19875–60–6
Specification: Substance is monographed only in Czechoslovak Pharmacopoeia
Dosage forms: Apodel, Cuvalit, Dispergol, Dopergin, Eunal, Lisenil, Lysenyl, Prolactam, Revanil
Therapeutic use: serotonine antagonist, antimigrenic, prolactine inhibitor, antiparkinsonic
Introduction: 1971 (Czechoslovakia), 1987 other countries
World production: 20–30 kg per year
Bulk substance manufacturers: Galena (Czech Rep.)

Manufacture: 1. Synthesis from lysergic acid

                      2. Synthesis from erginine

References: Zikán and Semonský, 1960; Sauer and Haffer, 1981; Bulej *et al.,* 1990 (all manufacture); Calve *et al.,* 1983 (pharmacol. and therap.)

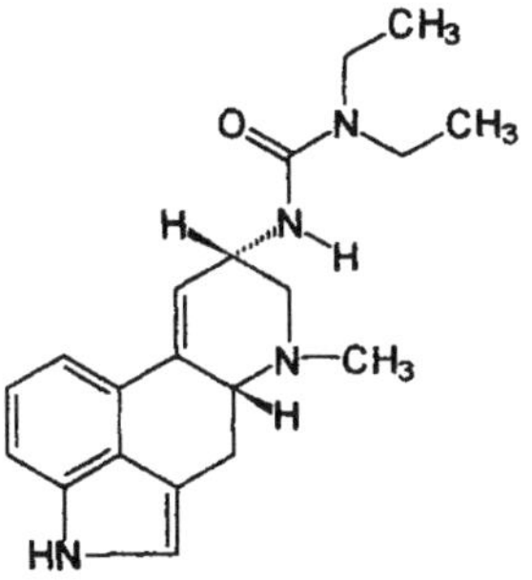

Figure 21 Lisuridc

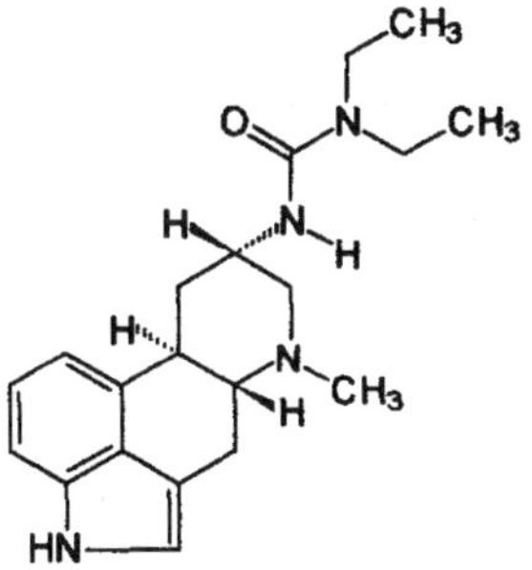

Figure 22 Terguride

13.4.13.
Terguride

Other name: Transdihydrolisuride

Chemical names: 3-(6-Methylergolin-8α-yl)-1,   1-diethylurea;   *N*-(6-Methyl-8-isoergolinyl)*N', N'*-diethylurea;

Structural formula: See Figure 22

Empirical formula: base $C_{20}H_{28}N_4O$

                       maleate $C_{20}H_{28}N_4O \cdot C_4H_4O_4$

Molecular weight: base   340.4

                       maleate   456.5

CAS No.:   base   37686–84–3

maleate
Specification: Substance is not monographed in any pharmacopeia
Dosage forms: Mysalfon
Therapeutic use: dopamine agonist, prolactine inhibitor, antiparkinsonian
Introduction: 1986 (Czechoslovakia), clinical trials in other countries
World production: 10kg per year
Bulk substance manufacturers: Galena (Czech Rep.)
Manufacture: 1. Hydrogenation of lisuride
                      2. Reduction of lisuride by lithium in liquid
                          ammonia
References: Zikán *et al.,* 1972; Sauer, 1980; Sauer *et al.,* 1986 (all
                      manufacture); Kratochvil *et al.,* 1993 (properties), Calve
                      et al. , 1983; Golda and Cvak, 1994 (both pharmacol. and
                      therap.)

13.4.14.
Pergolide

Chemical names: 8β-[(Methylthio)methyl)-6-propylergoline; Methyl](6a*R*,
                      9*R*, 10a*R*)-7-propyl-4, 6, 6a, 7, 8, 9, 10, 10a-
                      octahydroindolo[4, 3-fg]quinoline]-9-methylsulphide
Structural formula: See Figure 23
Empirical formula: base $C_{19}H_{26}N_2S$
                      mesylate $C_{19}H_{26}N_2S \cdot CH_4O_3S$

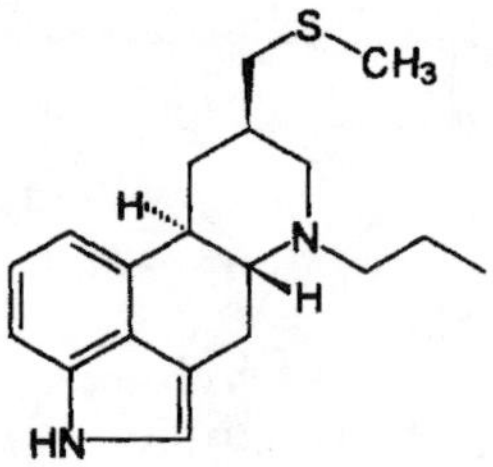

Figure 23 Pergolide

Molecular weight: base   314.5
                      mesylate   410.6
CAS No.:   base   66104–22–1
                      mesylate   66104–23–2
Specification: Substance   is   not   monographed   in   any
                      pharmacopoeia
Dosage forms: Celance, Parkotil, Permax, Pharken

Therapeutic use:	dopamine agonist, prolactine inhibitor, antiparkinsoniac
Introduction:	1989
World production:	50 kg per year
Bulk substance manufacturers:	Eli Lilly (USA) and Galena (Czech Rep.)
Manufacture:	Synthesis from dihydrolysergol
References:	Kornfeld and Bach, 1979; Misner, 1993; Misner *et al.*, 1996, 1997; Kennedy, 1997 (all manufacture); Sprankle and Jensen, 1992; Kerr *et al.*, 1981; Bowsher *et al.*, 1992 (all anal.); Owen, 1981 (pharmacol.)

13.4.15.
Cabergoline

Chemical name:	1-[(6-Allylergolin-8β-yl)carbonyl)-1-[3-(dimethylamino) propyl]-3-ethylurea;
Structural formula:	See Figure 24

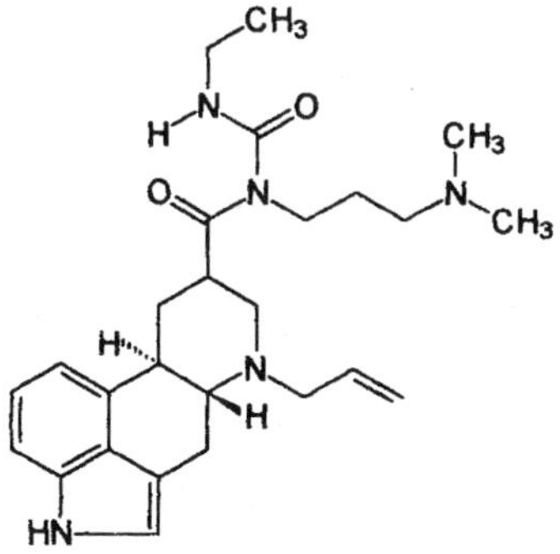

Figure 24 Cabergoline

Empirical formula:	base	$C_{26}H_{37}N_5O_2$
	phosphate	$C_{26}H_{37}N_5O_2 \cdot (H_3PO_4)_2$
Molecular weight:	base	451.6
	phosphate	647.7
CAS No.:	base	81409–90–7
	phosphate	85329–89–1
Specification:	Substance is not monographed in any pharmacopoeia	
Dosage forms:	Dostinex, Galastop, Cabaser	

Therapeutic use:	dopamine agonist, prolactine inhibitor, antiparkinsonian
Introduction:	1993
World production:	20–30 kg per year
Bulk substance manufacturer:	Pharmacia-Upjohn
Manufacture:	Synthesis from dihydrolysergic acid
References:	Bernardi *et al.,* 1982; Salvati *et al.,* 1985; Brambilla *et al.,* 1989 (all manufacture); Lera *et al.,*1993; Rabey *et al.,* 1994 (both therap. use)

REFERENCES

Abe, M., Yamano, T., Kozu, Y. and Kusomoto, M. (1952) A preliminary report on a new water-soluble ergot alkaloid, "Elymoclavine". *J. Agr. Chem. Soc. Japan,***25**:458.

Amici, A., Minghetti, A., Scotti, T., Spalla, C. and Tognoli, L. (1966) Production of ergotamine by a strain of *Claviceps purpurea* (Fr.) Tul. *Experientia,***22**,415–418.

Amici, A., Minghetti, A., Scotti, T., Spalla, C. and Tognoli, L. (1969) Production of peptide ergot alkaloids in submerged culture by three isolates of *Claviceps paspali. Appl. Microbiol.,***18**,464–468.

Ammon, R., Sharma, R., Gambert, S.R. and Lal Gupta, K. (1995) Hydergine revisited: A statistical analysis of studies showing efficacy in the treatment of cognitively impaired elderly. *Age,***18**,5–9.

Arcamone, F., Chain, E.B., Ferretti, A., Minghetti, A., Pennella, P., Tonolo, A. and Vero, L. (1961) Production of a new lysergic acid derivative in submerged cultures by a strain of *Claviceps paspali* Stivens & Hall. *Proc. of the Royal Soc., B,***155**,26–54.

Arcari, G., Bernardi, L., Bosisio, G., Coda, S., Fregnan, G.B. and Glaesser, A.H. (1972) 10-Methoxyergoline derivatives as α-adrenergic blocking agents. *Experientia,***28**, 819–820.

Börner, H., Haffer, G. and Sauer, G. (1983) Verfahren zur Herstellung von 2-Brom-8-ergolinyl-Verbindungen. *DE pat. 33 40 025.*

Baer, L. and Jenike, M.A. (1991) Hydergine in Alzheimer's disease. *J. Geriatric PsychiatryNeurol.,***4**,122–128.

Banno, K., Matsuoka, M., Matsuo, M., Kato, J., Shimizu, R. and Kinumaki, A. (1989) Nicergoline: Physicochemical properties and stability studies of nicergoline. *Iyakuhin Kenkyu,***20**,621–638.

Beran, M., Semonský, M. and Řežábek, K. (1969) Ergot alkaloids XXXV. Synthesis of D-6-methyl-8β-hydroxyethylergolene. *Collect. Czech. Chem. Commun.,***34**, 2819–2823.

Berde, B. and Schild, O. (1978) *Handbook of experimental pharmacology: Ergotalkaloids and related compounds.*Springer- Verlag, Berlin, Heidelberg, New York.

Bernardelli, G. (1987) Precede pour la fabrication d'esters du 1-méthyl-10α-méthoxylumilysergol. *FR pat. 2 616 788.*

Bernardi, L., Camerino, B., Patelli, B. and Radealli, S. (1964) Derivati della ergolina. Nota I. Derivati della D-6-metil-8β-aminometil-10α-ergolina. *Gazz. Chim. Ital.,***94**, 936–946.

Bernardi, L., Bosisio, G. and Goffredo, O. (1966) Lumilysergol derivatives. *US pat.32 28 943.*

Bernardi, L. (1979) From Ergot alkaloids to nicergoline. Review of nicergoline pharmacology. *Arzneim. Forsch.,* **29,** 1203–1316.

Bernardi, L., Temperilli, A. and Brambilla, E. (1982) Ergoline derivatives. *GB pat. 2 103 603.*

Bombardelli, E. and Mustich, G. (1985) Process for the preparation of N1-methyl10alpha-methoxylumilysergol and esters thereof, and intermediates for their preparation. *Eur. Pat. Appl. 156 645.*

Bombardelli, E. and Mustich, J. (1985) A process for preparing lysergol derivatives. *Eur. Pat. Appl. 171 988.*

Bowsher, R.R., Apathy, J.M., Compton, J.A., Wolen, R.L., Carlson, K.H. and DeSante, K.A. (1992) Sensitive, specific radioimmunoassay for quantifying pergolide in plasma. *Clin. Chem.,* **38,** 1975–1980.

Brambilla, E., di Salle, E., Briatico, G., Mantegani, S. and Temperilli, A. (1989) Synthesis and nidation inhibitory activity of a new class of ergoline derivatives. *Eur. J. Med.Chem.,* **24,** 421–426.

Bulej, P., Cvak, L., Stuchlík, J., Markovič, L. and Beneš, J. (1990) Process for manufacture of N-(D-6-methyl-8α-ergolenyl)-N', N'-diethylurea. *CS pat. 278 725.*

Calve, D., Horowski, R., McDonald, R. and Wuttke, W. (eds.) (1983) *Lisuride and otherdopamine agonists.*Raven Press, New York.

Camerino, B., Patelli, B. and Glaesser, A. (1966) Derivatives of 6-methyl and 1, 6dimethylergoline I. *US pat. 32 38 211.*

Coppi, G. (1991) Dihydro-alpha-ergokryptine, a new anti-parkinson drug: A pharmacological and clinical review. *Arch. Gerontol. Geriatr.,*Suppl. 2, 185.

Cvak, L., Stuchlík, J., Bořecký, M. and Krajíček, A. (1978) Process for purification of lysergic acid. *CS pat. 222 404* (in Czech).

Cvak, L., Stuchlík J., Černý, A., Křepelka, J. and Spáčil J. (1983) Manufacture of 1-alkylderivatives of dihydrolysergol. *CS pat. 234 498* (in Czech).

Cvak, L. (1985) Unpublished results.

Cvak, L., Stuchlík, J., Roder, L., Markovič, L., Krajič, A. and Spáčil, J. (1985) Process for manufacture of N-1 alkylated derivatives of dihydrolysergol. *CS pat. 247 570.*

Cvak, L., Stuchlík, J., Flieger, M., Sedmera, P., Zapletal, J., Beneš, K., Opálka, M., Roder, L., Krajíček, A. and Spáčil, J. (1988) Production of 2-bromo-alpha-ergokryptine and its acid addition salts. *CA pat. 1 294 956.*

Cvak, L., Jegorov, A., Sedmera, P., Havlíček, V., Ondráček, J., Hušák, M., Pakhomova, S., Kratochvil, B. and Granzin, J. (1994) Ergogaline, a new ergot alkaloid, produced by *Claviceps purpurea:* Isolation, identification, crystal structure and molecular conformation. *J. Chem. Soc. Perkin Trans.* **2,** 1861–1865.

Cvak, L., Minář, J., Pakhomova, S., Ondráček, J., Kratochvíl, B., Sedmera, P., Havlíček, V. and Jegorov, A. (1996) Ergoladinine, an ergot alkaloid. *Phytochemistry,* **42,** 231–233.

Cvak, L., Jegorov, A., Pakhomova, S., Kratochvil, B., Sedmera, P., Havlíček, V. and Minář, J. (1997) 8α-Hydroxy-α-ergokryptine, an ergot alkaloid. *Phytocbemistry,* **44,** 365–369.

Černý, A. and Semonský, M. (1962) Mutterkornalkaloide XIX. Über die Verwendung von N, N'-Carbonyldiimidazol zur Synthese der Lysergsäure-, dihydrolysergsäure-und 1-Methyldihydrolysergsaureamide. *Collect. Czech. Chem. Commun.,* **27,** 1585–1592.

Černý, A., Křepelka, J., Stuchlík, J., Cvak, L. and Spáčil J. (1982) Process for manufacture of D-1, 6-dimethyl-8β-(5-bromonicotinoyl)oxyrnethyl-10α-methoxyergolin. *CS pat.229 086* (in Czech).

Eich, E. (1975) Partial Synthese neuer Ergolinderivat aus Clavinalkaloiden. *Pharmazie,***30,** 516–520.

Eur. Ph. 1997=*European Pharmacopeia 3rd Edition 1997.*Concil of Europe, Strasbourg Cedex, 1996.

Flückiger, E. and Troxler, F. (1973) 2-Bromo-α-ergokryptine as lactation inhibitor. *USpat. 37 52 888.*

Franciosi, A. and Zavattini, G. (1994) Dihydroergocristine in the treatment of elderly patients with cognitive deterioration: A double-blind, placebo-controled, doseresponse study. *Curr. Ther. Res.,***55,**1391–1401.

Frey, A. (1961) Nouveaux halogenures d'acides de la serie lysergique et didydrolysergique. *FR. pat. 1 308 758.*

Garbrecht, W.L. (1959) Synthesis of amides of lysergic acid. *J. Org. Chem.,24:*368–372.

Gervais, Ch. (1986) Procédé de preparation des derives N-méthyles du lysergol et du méthoxy-10alpha lumilysergol. *Eur. Pat. Appl. 209 456.*

Giron-Forest, D.A. and Schoenleber, W.D. (1979) Bromocriptine methanesulphonate. In K.Florey (ed.) *Analytical profiles of drug substances,***8,**Academic Press, New york, pp. 47–81.

Golda, V. and Cvak, L. (1994) Terguride but not bromocriptine alleviated glucose tolerance abnormalities and hyperlipidaemia in obese and lean genetically hypertensive Koletsky rats. *Physiol. Res.,***43,**299–305.

Guttmann, S. and Huguenin, R. (1970) Verfahren zur Herstellung neuer heterocyclischer Verbindungen. *DE. pat. 2 029 447.*

Hecke, L. (1922) *Schweiz Apoth. Ztg.,***60,**45.

Hecke, L. (1923) *Wien. Landw. Ztg.,***73,**1–2.

Hellberg, H. (1957) On the photo-transformation of ergot alkaloids. *Acta Chem.Scand.,* **11,**219–227.

Hofmann, A., Frey, A.J. and Ott, H. (1961) Die Totalsynthese des Ergotamins. *Experientia,***17:**206–207.

Hofmann, A. and Troxler, F. (1962) Nouveaux derives de I'uree appartenant a la serie de I'acide lysergique ou dihydrolysergique et leur preparation. *FR. pat. 1 303 288.*

Hofmann, A., Ott, H., Griot, R., Stadler, P.A. and Frey, A.J. (1963) Die Synthese und Stereochemie des Ergotamins. *Helv. Chim. Acta,46:*2306–2328.

Hofmann, A. (1964) *Die Mutterkoralkaloide.*Ferdinand Enke, Verlag, Stuttgard.

Holger, W. (1994) Ergotamin. *Deutsche Apot. Ztg.,*134, 35–38.

Jacobs, W.A. and Craig, L.C. (1934) The ergot alkaloids. II. The degradation of ergotinine with alkali. Lysergic acid. *J. Biol. Chem.,***104,**547–551.

Jacobs, W.A. and Craig, L.C. (1934a) The ergot alkaloids. III. Lysergic acid. *J. Biol.Chem.,* **106,**393–399.

Jacobs, W.A. and Craig, L.C. (1935) Ergot alkaloids. V. Hydrolysis of ergotinine. *J. Biol.Chem.,***110,**521–530.

Jacobs, W.A. and Craig, L.C. (1935) Ergot akaloids. VI. Lysergic acid. *J. Biol. Chem.,* **111,**455–465.

Jacobs, W.A. and Craig, L.C. (1936) Ergot alkaloids. IX. Structure of lysergic acid. *J. Biol. Chem.,***113,**767–778.

JP XIII=*The Japanese Pharmacopeia 13th Edition.*The Society of Japanese Pharmacopeia, Tokyo1966.

Kennedy, J.H. (1997) HPLC purification of pergolide using Silica gel. *Organic ProcessResearch and Development,***1**,68–71.

Kerr, K.M., Smith, R.V. and Davis, P.J. (1981) High-performance liquid chromatographic determination of pergolide and its metabolite, pergolide sulfoxide, in microbial extracts. *J. Chrom.,***219**,317–320.

Kobel, H., Schreier, E. and Rutschmann, J. (1964) 6-Methyl-$\Delta^{8,9}$-ergolen-carbonsäure, ein neues Ergolinderivat aus Kulturen eines Stammes von *Claviceps paspali* Stevens et Hall. *Helv. Chim. Acta,***47**,1052–1064.

Kobel, H. and Sanglier, J.J. (1976) Process for production of ergotoxine group alkaloids by combined fermentation. *FR pat. 2 307 87.*

Kobel, H. and Sanglier, J.J. (1978) Formation of ergotoxine alkaloids by fermentation and attempts to control their biosynthesis. In R.Hutter, T.Leisinger, J.Nuesch and W.Wehrli (eds.) *Antibiotics and other secondary metabolites,*pp. 233–242, Academia PressNew York.

Kornfeld, E.C. and Bach, N.J. (1979) 6-N-Propyl-8-methoxymethyl or methylmercaptomethylergolines and related compounds. *US pat. 4 166182*

Krajíček, A., Trtík, B., Spáčil, J., Sedmera, P., Vokoun, J. and Řeháček, Z. (1979) 8-Hydroxyergotamine, a new ergot alkaloid. *Collect. Czech. Chem. Commun.,***44**, 2255–2260.

Kratochvíl, B., Ondráček, J., Novotný, J., Hušák, M., Jegorov, A. and Stuchlík, J. (1993) The crystal and molecular structure of terguride monohydrate. *Z.Krystallogr.,***206**, 77–86.

Kreilgard, B. (1976) Ergotamine Tartrate. In K.Florey (ed.), *Analytical profiles ofdrug substances,***6**,Academic Press, New York, pp. 113–159.

Lera, G., Vaamonde, J., Rodriquez, M. and Obeso, J.A. (1993) Cabergolin in Parkinson's disease: Long-term folow-up. *Neurology,***43**,2587–2588.

Lieberman, A.N. and Goldstein, M. (1985) Bromocriptine in Parkinson disease. *Pharmacol. Rev.,***37**,217–227.

Losse, G. and Strobel, J. (1984) Improved synthetic routes to the cyclol system of ergot peptide alkaloids. *J. Pract. Chem.,***326**,765–778.

Mailand, F. (1992) Dihydroergocristin. Aktueller Stand von Forschung und Entwicklung. *Arzneimittel-Forsch.,***42**,1379–1422.

Malacco, E. and Di Cesare, F. (1992) Effects of dihydroergocristine treatment on carbohydrate tolerance and cognitive function in patients with non-insulin-dependent diabetes. *Current Ther. Res.,***51**,515–523.

Marttin, E., Romeijn, S.G., Verhoef, J.C. and Merkus, F.W.H.M. (1997) Nasal absorption of dihydroergotamine from liquid and powder formulations in rabbits. *J. Pharm. Sci.,* **86**,802–807.

Marzoni, G. and Garbrecht, W.L. (1987) N-1 Alkylation of dihydrolysergic acid. *Synthesis,*651–653.

Marzoni, G., Garbrecht, W.L., Fludzinski, P. and Cohen, M.L. (1987) 6-Methylergoline-8-carboxylic acid esters as serotonin antagonists: N-1 substituent effects on $5HT_2$ receptor affinity. *J. Med. Chem.,***30**,1823–1826.

Megyeri, G., Keve, T., Galambos, J., Kovacs, L., Stefko, B., Bogsch, E. and Trischler, F. (1986) Verfahren zur Herstellung von 2-Brom-α-ergokryptin. *DE pat. 36 19 617.*

Minghetti, A., Spalla, C. and Tognoli, L. (1971) Fermentative process for producing ergocristine. *US pat. 3 567 582.*

Misner, J.W. (1993) One-pot process for preparing pergolide. *Eur. Pat. Appl 571 202.*

Misner, J.W., Kennedy, J.H. and Biggs, W.S. (1996) Pergolide: Process design challenges of a potent drug. *Chemtech.,*November, 28–33.

Misner, J.W., Kennedy, J.H. and Biggs, W.S. (1997) Integration of highly selective demethylation of quarternized ergoline into one-pot synthesis of pergolide. *OrganicProcess Research and Development,***1,**77–80.

Mora, E.G. (1979) A process for preparing lysergol derivatives. *Eur. Pat. Appl. 004 664.*

Negwer, M. (1994) *Organic chemical drugs and their synonyms.*Akademia Verlag, Berlin.

Ninomiya, I. and Kiguchi, T. (1990) Ergot alkaloids. In A.Brossi (ed.), *The alkaloids.Chemistry and pharmacology,*Academic Press, Inc.New York.

Owen, R.T. (1981) Pergolide mesylate. *Drugs Put.,***6,**231–233.

Patelli, B. and Bernardi, L. (1964) Process for the preparation of lysergic acid amides. *US pat. 3 141 887.*

Pioch, R.P. (1956) Preparation of lysergic acid amides. *US pat. 2 736 728.*

Poli, S. (1990) Use of alpha-dihydroergocryptin for treatment of Parkinson's syndrome, depression and cephalalgia. *DE pat. 3 525 390.*

Rabey, J.M., Nissipeanu, P., Inzelberg, R. and Korczyn, A.D. (1994) Beneficial effect of Cabergoline, new long lasting D-2 agonist in the treatment of Parkinson disease. *Clin. Neuropharmacol.,***17,**286–293.

Reif, V.D. (1982) Ergonovine maleate. In K.Florey (ed.), *Analytical profiles of drugsubstances,***11,**Academic Press, New york, pp. 273–312.

Ručman, R. (1976) Verfahren zur Herstellung von d-Lysergsaure. *DE pat. 2 610 859.*

Ručman, R. and Jurgec, M. (1977) Verfahren zur Herstellung von l0α-Methoxy-dihydrolysergol-5-bromnicotin-säureester. *DE pat. 27 52 533.*

Ručman, R., Korsič, J. and Kotar, M. (1977) Verfahren zur Herstelleng von 2-Brom-α-Ergokryptin. *DE pat. 27 52 532.*

Ručman, R. (1978) N-Substituirte 9, 10-dihydrolysergsäurester sowie ein Verfahren zu deren Hersteilung. *Eur. pat. 000 533.*

Rutschmann, J. and Kobel H. (1967) Neues mikrobiologisches Verfahren zur Gewinnung von Ergobasin. *Swiss pat. 433 357.*

Salvati, P., Caravaggi, A.M., Temperilli, A., Bosisio, G., Sapini O. and di Salle, E. (1985) Dimethylaminoalkyl -3-(ergoline-8β-carbonyl)-ureas. *US pat. 4 526 892.*

Sauer, G. (1980) Verfahren zur Herstellung von 8α-substituirten 6-Methylergolinen. *DEpat. 3 001 752*

Sauer, G. and Haffer, G. (1981) Process for the preparation of ergoline derivatives. *DEpat. 3 135 305.*

Sauer, G., Haffer, G. and Wachtel, H. (1986) Reduction 8α-substituted 9, 10-didehydro ergolines. *Synthesis,*1007–1010.

Scarzella, L., Bono, G. and Bergamasco, B. (1992) Dihydroergocryptine in the management of senile psychoorganic syndrome. *Int. J. Clin. Pharm. Res.,***12,**37–46.

Schinutschke, R., Wolf, I., Neumann, B. and Braun, K. (1979) Verfahren zur Gewinnung von epimerenfreiem Ergotoxin. *DD pat. 161 251*

Schoenleber, W.D., Jacobs, A.L. and Brewer, G.A. (1978) Dihydroergotoxine methanesulfonate. In K.Florey (ed.), *Analytical profiles of drug substances,***7,**Academic Press, New York, pp. 81–147.

Simes (1971) Precédé pour l'extraction de lysergol et d'alkaloides ergoliques d'une planta du gentre Ipomoea. *BE Pat. 778 087.*

Šmidrkal, J. and Semonský, M. (1982) Alkylation of ergoline derivatives at position N1. *Collect. Czech. Chem. Commun.,* **47,** 622–624.

Smith, S. and Timmis, G.M. (1932) Alkaloids of ergot III. Ergine, a new base obtained by the degradation of ergotoxine and ergotinine. *J. Chem. Soc.,* 763–766.

Sprankle, D.J. and Jensen, C. (1992) Pergolide mesylate. In H.G. Brittain (ed.) *Analyticalprofiles of drug substances,* **21,** Academic Press, New York, pp.

Stadler, P.A., Frey, A.J. and Hofmann, A. (1963) Herstellimg der optisch aktiven Methyl-benzyloxymalonsäure-halbester und Bestimmung ihrer absoluten Konfiguration. *Helv. Chim. Acta,* **47,** 2300–2305.

Stadler, P.A. and Hofmann, A. (1969) Verfahren zur Herstellung von heterocyclischen Verbindungen. *Swiss pat. 512 490.*

Stadler, P.A., Guttmann, S., Hauth, H., Huguenin, R.L., Sandrin, E., Wersin, G., Willems, H. and Hofmann, A. (1969) Die Synthese der Alkaloide der Ergotoxin Gruppe. *Helv. Chim. Acta,* **52,** 1549–1564.

Stadler, P.A. (1978) Verfahren zur Herstellung von 2-Amino-3,6-dioxo-octahydro-8H-oxazolo[3, 2-a]pyrolo[2, 1-c]pyrazin Derivaten. *DE pat. 2 800 064.*

Stadler, P.A. (1980) Recent Advances in Ergot Research. *Kem. Ind.,* **29,** 207–216.

Stanovnik, B., Tišler, M., Jurgec, M. and Ručman, R. (1981) Bromination of a-ergo-kryptine and other ergot alkaloids with 3-bromo-6-chloro-2-methylimidazolo[1, 2-b]-pyridazine-bromine complex as a new brominating agent. *Heterocycles,* **16,** 741–745.

Stoll, A. (1918) Verfahren zur Isolierung eines hochwertigen Präparates aus Secale cornutum. *Swiss pat. 79 819.*

Stoll, A. and Burckhard, E. (1937) Ergocristin und Ergocristinin, ein neues Alkaloid-paar aus Mutterkorn. *Hoppe Seller's Z. Phisiol. Chem.,* **250,** 1–6.

Stoll, A. and Hofmann, A. (1938) Partialsynthese des Ergobasins, eines natürlichen Mutterkornalkaloids sowie seines optischen Antipoden. *Z. Physiol. Chem.,* **251,** 155–163.

Stoll, A. and Hofmann, A. (1943) Partialsynthese von Alkaloiden vom Typus des Ergobasins. *Helv. Chim. Acta,* **26,** 944–965.

Stoll, A. and Hofmann, A. (1943a) Die Dihydroderivate der natürlichen linksdrehenden Mutterkornalkaloide. *Helv. Chim. Acta,* **26,** 2070–2081.

Stoll, A. and Brack, A. (1944) *Pharm. Acta Helv.,* **19,** 118–123.

Stoll, A. (1945) Über Ergotamin. *Helv. Chim. Acta,* **28,** 1283–1308.

Stoll, A. and Hofmann, A. (1948) Optically active salts of the lysergic acid and isolysergic acid derivatives and a process for their preparation and isolation. *US pat. 24 47 214.*

Stoll, A., Hofmann, A. and Schlientz, W. (1949) Die stereoisomeren Lysergole und Dihydrolysergole. *Helv. Chim. Acta,* **32,** 1947–1956.

Stoll, A. and Hofmann, A. (1955) Amide der stereoisomeren Lysergsäuren und Dihydrolysergsäuren. *Helv. Chim. Acta,* **38,** 421–433.

Stres, J. and Ručman, R. (1981) Study of photochemical methoxylation of lysegic acid. *Hem. Ind.,* **35,** 41–43 (In Slovenian).

Stuchlík, J., Cvak, L., Kejzlarová, J., Schreiberová, M., Krajíček, A. and Spáčil, J. (1985) Method for manufacture of amides of lysergic acid. *CS pat. 246 643* (in Czech).

Stütz, P., Stadler, P.A. and Guttmann, S. (1969) Neues Verfahren zur Herstellung von Mutterkornpeptidalkaloiden. *Swiss pat. 530 374.*

Szantay Jr., C., Bihari, M., Brlik, J., Csehi, A., Kassai, A. and Aranyi, A. (1994) Structural elucidation of two novel ergot alkaloid impurities in α-ergokryptine and bromokryptine. *Acta Pharm. Hung.,***64,**105–108.

Terdy, L., Kiss, J., Trompler, A., Zambo, I., Foldesi, Z., Dancsi, L., Kassai, A., Gazdag, M. (1981) Nouveau procédé de production d'alkaloides de l'ergot de seigle. *FR pat. 2477155.*

Troxler, F. and Hofmann, A. (1957) Substitutionen am Ringsystem der Lysergsaure II. Alkylierung. *Helv. Chim. Acta,***40,**1721–1732.

Troxler, F. and Hofmann, A. (1957a) Substitutionen am Ringsystem der Lysergsaure III. Halogenierung. *Helv. Chim. Acta,***40,**2160–2170.

Troxler, F. (1968) Beitrage zur Chemie der 6-Methyl-8-ergolen-8-carbonsäure. *Helv.Chim. Acta,***51:**1372–1381.

Udvardy, E., Budai, M., Fekete, G., Goeroeg, S., Herenyi, B., Wack, G. and Zalai, K. (1982) Process for ergot alkaloids production. *DE Pat. 3 104 215.*

USP 23=*The United States Pharmacopeia 23.*United States Pharmacopeial Convention, INC, Rockville, MD, 1994.

Vigouret, J.M., Burki, H.R., Jaton, A.L., Zuger, P.E. and Loew, D.M. (1978) Neurochemical and neuropharmacological investigations with four ergot derivatives: Bromocriptine, Dihydroergotoxine, CF 25–397 and CM 29–712. *Pharmacology,***16,** Suppl 1, 156–173.

Wadworth, A.N. and Chrisp, P. (1992) Co-Dergocrine Mesylate. A review of its pharmacodynamic and pharmacokinetic properties and therapeutic use in age-related cognitive decline. *Drugs & Aging,***2,**153–173.

Weil, C. (1986) The safety of bromocriptine in long-term use: A review of the literature. *Curr. Med. Res. Opin.,***10,**25–51.

Well, R. (1910) Verfahren zur Züchtung des Mutterkornpilzes. *DE pat. 267 560.*

Zikán, V. and Semonský, M. (1960) Mutterkornalkaloide XVI. Einige N-(D-6-Methyl-isoergolenyl-8)-, N-(D-6-Methylergolenyl-8)- und N-(D-6-Methylergolin-I-yl-8)-N′-substituirte Harnstoffe. *Collect. Czech. Chem. Commun.,***25,**1922–1928.

Zikán, V., Semonský, M., Řeábek, E., Aušková, M. and Šeda, M. (1972) Some N-(D-6-methyl-8-isoergolin-I-yl) and N-(D-6-methyl-8-isoergolin-II-yl)-N′-substitued ureas. *Collect. Czech. Chem. Commun.,***37,**2600–2605.

14.

ERGOT ALKALOIDS AND THEIR DERIVATIVES ASLIGANDS FOR SEROTONINERGIC, DOPAMINERGIC,AND ADRENERGIC RECEPTORS

HEINZ PERTZ and ECKART EICH

Institut für Pharmazie II, Freie Universität Berlin,Königin-Luise-Strasse 2, 14195 Berlin (Dahlem),Germany

14.1

INTRODUCTION

Among compounds from natural sources ergolines are of paramount importance as ligands for serotonin (5-hydroxytryptamine, 5-HT) receptors, dopamine receptors, and adrenoceptors. The tetracyclic structure of the ergolines contains the essential features of the monoamine neurotransmitters 5-HT, dopamine, and noradrenaline, and it is not surprising that many naturally occurring and (semi) synthetic ergolines have been shown to act as agonists, partial agonists or antagonists at receptors for these neurotransmitters. It is difficult to explain the complexity of the pharmacological profile of the ergolines without encountering the issue of receptor heterogeneity. The extent of the multiplicity of 5-HT receptors, dopamine receptors and adrenoceptors became fully apparent in the early 1990s, since at least 14 distinct subtypes of 5-HT receptors (Hoyer *et al.,* 1994; Martin and Humphrey, 1994; Boess and Martin, 1994), 5 subtypes of dopamine receptors (Sibley and Monsma, 1992; Strange, 1993; Seeman and Van Tol, 1994), and at least 10 subtypes of adrenoceptors (Bylund *et al.,* 1994; Hieble *et al.,* 1995a, b) could be identified on the basis of structural, transductional and operational information obtained from molecular biological, second messenger and radioligand binding as well as functional studies.

Although a number of structurally diverse classes of ligands demonstrate high affinity for serotoninergic, dopaminergic and adrenergic receptors, among the ergolines only few show specificity with regard to the different monoamine receptor systems and selectivity among subtypes of each of these major groups. Nevertheless, a number of ergolines has emerged as real targets for the treatment of vascular and neurological diseases and other disorders (Table 1). Moreover, it is entirely possible that any subtype-selective drugs that will be developed on the basis of the molecular biological advances of the past decade may not be as effective clinically as those that are currently available but less selective.

Table 1 Current therapeutical applications for selected natural and semisynthetic ergolines in clinical relevant doses

Compounds	Therapeutical application	Receptors mostly involved	Quality of action
Ergotamine, Dihydro-ergotamine	Migraine (acute)	$5\text{-HT}_{1B/1D}$	Partial agonism
Methysergide	Migraine (prophylactic)	5-HT_{2B}	Antagonism
Ergometrine, Me-Ergometrine	Postpartum haemorrhage	5-HT_{2A}	Partial agonism
Bromokryptine	Parkinsonism	D_2-like	(Partial) agonism
	Suppression of the secretion of prolactin	D_2-like	(Partial) agonism
Lisuride	Parkinsonism	D_2-like	(Partial) agonism
	Suppression of the secretion of prolactin	D_2-like	(Partial) agonism
Pergolide	Parkinsonism	D_1-like and D_2-like	(Partial) agonism
Cabergoline	Suppression of the on secretion of prolactin	D_2-like	(Partial) agonism

14.2.
ERGOLINES AS LIGANDS FOR 5-HT RECEPTORS

There is a continued interest in the biological actions of ergot alkaloids and their semisynthetic derivatives at 5-HT receptors which can be divided into 4 main classes, termed 5-HT_1, 5-HT_2, 5-HT_3, and 5-HT_4. The amino acid sequence of 3 additional types of receptors, denoted as 5-ht_5, 5-ht_6, and 5-ht_7 has been identified but their functional role is not yet clear; therefore they are abbreviated using lower case letters. With the exception of the 5-HT_3 receptor, which is a ligand-gated ion-channel, all receptors mentioned belong to the superfamily of G-protein-coupled receptors (proteins characterized by 7 transmembrane domains). The class of 5-HT1 receptors comprises 5-HT_{1A}, 5-HT_{1B}, $5\text{-HT}_{1D\alpha}$, and $5\text{-HT}_{1D\beta}$ receptors and two less well characterized subtypes (5-HT_{1E} and 5-HT_{1F}). The nomenclature of $5\text{-HT}_{1B/1D}$ receptors has recently been simplified: rat 5-HT_{1B} and human $5\text{-HT}_{1D\beta}$ receptors are now termed $r5\text{-HT}_{1B}$ and $h5\text{-HT}_{1B}$, respectively, whereas human $5\text{-HT}_{1D\alpha}$ receptors are now termed $h5\text{-HT}_{1D}$ (Hartig et al., 1996) (vide infra). The class of 5-HT_2 receptors includes 5-HT_{2A} (formerly "5-HT_2"), 5-HT_{2B}, and 5-HT_{2c} (formerly 5-HT_{1c}) receptors. Whereas 5-HT_1 receptors are negatively coupled to adenylyl cyclase and 5-HT_4, 5-ht_6, and 5-ht_7

receptors are positively coupled to adenylyl cyclase, 5-HT$_2$ receptors stimulate phospholipase C.

Among naturally occurring ergolines, ergopeptines such as ergotamine and simple lysergic acid amides such as ergometrine show high affinity for different 5-HT receptor subtypes (e.g., 5-HT$_{1A}$, 5-HT$_{1B}$, 5-HT$_{1D}$, 5-HT$_{2A}$, 5-HT$_{2C}$), whereas clavines show moderate affinity for rat 5-HT^{2A} receptors and high affinity (e.g., lysergol) for human 5-HT$_{1D}$ receptors and 5-HT$_{1F}$ receptors (for review, see Eich and Pertz, 1994; Pertz, 1996). The main disadvantage of naturally occurring ergot alkaloids is their lack of selectivity for each of the individual 5-HT receptor subtypes. During the last two decades various structural modifications of the ergoline skeleton have been reported and led to the discovery of highly potent and even more selective serotoninergic ligands. This chapter will focus on naturally occurring and semisynthetic ergolines which in concert with nonergoline derivatives represent the wide range of valuable tools to characterize 5-HT receptors (Table 2). Since the most important **Table 2**(Continued) effects of ergolines with regard to 5-HT are due to their action on the central nervous system (CNS) and the cardiovascular (CV) system, we will emphasize newer developments in the pharmacology of ergolines acting at 5-HT receptors in the CNS and the CV system.

14.2.1.
Ergolines are Nonselective Ligands with High Affinity for 5-HT$_{1A}$Receptors

Most drugs with partial agonist properties at 5-HT$_{1A}$ receptors are used for CNS applications. 5-HT$_{1A}$ receptors in the CNS are localized on the cell bodies and dendrites of 5-HT neurones in the raphe nuclei and function as somatodendritic autoreceptors which mediate the inhibition of cell firing. Clinical interest in 5-HT$_{1A}$ receptor partial agonists and silent antagonists with sufficient brain penetration is related to the putative involvement of 5-HT$_{1A}$ receptors in anxiety and depression (Traber and Glaser, 1987; Fletcher et al., 1993). Furthermore, the administration of 5-HT$_{1A}$ receptor agonists results in a decrease of arterial blood pressure due to the inhibition of central sympathetic neurones (McCall and Clement, 1994). In peripheral tissues presynaptic 5-HT$_{1A}$ heteroreceptors are possibly involved in the modulation of the gastointestinal motility. It has been shown that the activation of presynaptic 5-HT$_{1A}$ receptors by e.g., lysergic acid diethylamide (LSD) can inhibit the electrically stimulated [^{3}H]acetylcholine release from cholinergic neurones of guinea-pig ileum (Pfeuffer-Friederich and Kilbinger, 1985; Fozard and Kilbinger, 1985). On the whole, ergolines (e.g., lisuride, dihydroergotamine, LSD, methylergometrine, ergotamine, ergometrine, metergoline, methysergide, and bromokryptine) show high affinity for 5-HT$_{1A}$ receptors but have not been used as principle agents for the determination of 5-HT$_{1A}$ receptor-mediated activity due to their poor pharmacological selectivity (Hoyer, 1989).

Table 2 Pharmacological characterization of serotonin (5-HT) receptors by means of ergolines in functional and radioligand binding studies

Receptor name	5-HT$_{1A}$	5-HT$_{1B/1D}$	5-HT$_{1E}$	5-HT$_{1F}$
Compounds	Ergotamine	Ergometrine		
	DHE	DHE		
	Ergometrine	Ergometrine		
	Me-ergometrine	Me-ergometrine		Me-ergometrine
	Metergoline	Metergoline[a]		
	Methysergide	Methysergide		Methysergide
	LSD	LSD		
	Lisuride	Lysergol		Lysergol
	Bromokryptine			
Quality of action	(partial) agonism	partial agonism	–	agonism
Radioligand	–	–	–	2-[^{125}I]LSD

Receptor name	5-HT$_{2A}$	5-HT$_{2B}$	5-HT$_{2C}$	5-HT$_3$
Compounds	LY53857	LY53857	LY53857	–
	Mesulergine	Mesulergine	Mesulergine	
	LSD[b]		LSD[b]	
	Methysergide	Methysergide	Methysergide	
	Me-ergometrine[c]	Me-ergometrine		
	Ergometrine[c]			
	Metergoline		Metergoline	
	Ergotamine[d]	Ergotamine[c]	Ergotamine[c]	
	DHE[d]	DHE[d]	DHE[d]	
	Sergolexole	Lisuride		
	Amesergide			
	LY215840			
Quality of action	antagonism	antagonism	antagonism	–
Radioligand	–	–	[^{3}H]Mesulergine	–

Receptor name	5-HT$_4$	5-ht$_{5A/5B}$	5-ht$_6$	5-ht$_7$
Compounds	–	Ergotamine	Lisuride	Lisuride
			Metergoline	Metergoline
		LSD	2-Br-LSD	LSD
			Pergolide	Pergolide
			Lergotrile	Bromocriptine
			DHE	Mesulergine
		Methysergide		Methysergide
				LY215840
Quality of action	–	?	?	antagonism
Radioligand	–	2-[^{125}I]LSD	2-[^{125}I]LSD	2-[^{125}I]LSD

Data shown are derived from a variety of published reports cited in the text. [a]Metergoline shows partial agonist activity (Miller *et al.*, 1992) or antagonist activity (Hamel and Bouchard, 1991; Bax *et al.*, 1992) at 5-HT$_{1D}$ receptors. [b]LSD acts as a partial agonist at 5-

HT^{2A} and 5-HT$_{2C}$ receptors (Kaumann, 1989; Glennon, 1990; Pierce and Peroutka, 1990; SandersBush et al., 1988). [c]Ergometrine and methylergometrine are partial agonists at 5-HT$_{2A}$ receptors (Hollingsworth et al., 1988; Milhahn et al., 1993). [d]Ergotamine and dihydroergotamine (DHE) possess high affinity for 5-HT$_{2A}$ receptors in mammalian brain membranes (Hoyer, 1989). [e]Ergotamine and dihydroergotamine show partial agonist activity at 5-HT$_{2B}$ receptors (Glusa and Roos, 1996) and full agonist activity at 5-HT$_{2C}$ receptors (Brown et al., 1991)

14.2.2.
Ergolines are Partial Agonists of High Potency at 5-HT$_{1B/1D}$Receptors

According to molecular biological studies, 5-HT$_{1B}$ and 5-HT$_{1D}$ receptors form a subfamily of related receptors (Hartig et al., 1992, 1996). Two human 5-HT$_{1D}$ receptors (called 5-HT$_{1D\alpha}$ and 5-HT$_{1D\beta}$) within this subfamily have been cloned, which, due to their operational characteristics, resemble the 5-HT$_1$-like receptor of the functionally-based receptor classification of Bradley and colleagues (Bradley et al., 1986). The human 5-HT$_{1D\beta}$ receptor (h5-HT$_{1B}$) is a species homologue of the rat 5-HT$_{1B}$ receptor (r5-HT$_{1B}$). Both are presynaptic autoreceptors which are localized on the axon terminals of 5-HT neurones mediating inhibition of 5-HT release (Göthert et al., 1996). In addition, 5-HT$_{1B/1D\beta}$ receptors occur as presynaptic 5-HT heteroreceptors on sympathetic nerve terminals in blood vessels (Göthert et al., 1996). Whereas the rat 5-HT$_{1B}$ receptor and the human 5-HT$_{1D\beta}$ receptor display striking differences in their pharmacological binding properties, the pharmacological profiles of human 5-HT$_{1D\alpha}$ and 5-HT$_{1D\beta}$ receptors are quite similar (Weinshank et al., 1992). 5-HT$_{1D\alpha}$ receptors have been identified as presynaptic inhibitory 5-HT heteroreceptors on sympathetic axon terminals in human atrial appendages (Molderings et al., 1996). On the other hand, 5-HT$_1$-like receptors which mediate constrictor effects of 5-HT in blood vessels of various species including man, may correspond to the 5-HT$_{1D\beta}$ subtype (Martin, 1994).

Recent evidence has implicated 5-HT$_{1D}$ (5-HT$_1$-like) receptors in neurological and cardiovascular diseases such as the acute migraine attack or cerebral and coronary vasospasm. 5-HT$_{1D}$ (5-HT$_1$-like) receptor agonists such as the nonergoline derivative sumatriptan and the ergolines ergotamine and dihydroergotamine belong to the most effective drugs in aborting migraine attacks (Moskowitz, 1992; Ferrari and Saxena, 1993). The acute migraine attack is characterized by a pathological dilatation of extracerebral intracranial arteries which evokes an increase in vascular pulsations followed by a stimulation of perivascular sensory afferents of the Vth cranial nerve to cause the typical symptoms such as pain, nausea, vomiting and photophobia. In addition, a neurogenic inflammatory response, which mediates plasma protein extravasation in blood vessels of dura mater and is induced via the release of vasoactive neuropeptides from perivascular nerve terminals, may play an important role in the pathogenesis of migraine attacks. The effectiveness of antimigraine drugs such as sumatriptan, ergotamine and dihydroergotamine is based both on the 5-HT_{1D} receptor-mediated contraction of pathologically dilated intracranial arteries and the inhibition of neuropeptide release via activation of prejunctional 5-HT_{1D} receptors, thereby blocking the development of neurogenic inflammation. The 5-$HT_{1D\beta}$ receptor is presumably responsible as the mediator of cranial vasoconstriction, since mRNA for this subtype has been found in human cerebral arteries (Hamel et al., 1993). Furthermore, cardiac side effects such as coronary vasospasm, myocardial infarction, and possibly stroke which may complicate the treatment of migraine with the 5-HT_{1D} receptor agonists sumatriptan, ergotamine, and dihydroergotamine can be explained by the stimulatory effect of these drugs on 5-$HT_{1D\beta}$ receptors in the coronary vasculature (Kaumann et al., 1993, 1994). On the other hand, it has been suggested that the 5-HT_{1D} receptor which modulates neuropeptide release in migraine, may be of the 5-$HT_{1D\alpha}$ subtype, since mRNA for this subtype has been found in human trigeminal ganglia (Rebeck et al., 1994).

A large number of studies has shown conclusively that ergolines possess high affinity for 5-HT_{1B} and 5-HT_{1D} receptors (Figure 1). For example, it has been demonstrated that dihydroergotamine displays high affinity for the rat 5-HT_{1B} receptor (Hamblin et al., 1987), and ergolines such as lysergol, ergotamine, dihydroergotamine, LSD, metergoline, and methysergide are ligands with high affinities for human 5-$HT_{1D\alpha}$ and 5-$HT_{1D\beta}$ receptors (Weinshank et al., 1992; Hamblin and Metcalf, 1991; Jin et al., 1992; Hamblin et al., 1992; Oksenberg et al., 1992; Miller et al., 1992; Peroutka, 1994; Levy et al., 1992a; Demchyshyn et al., 1992). Among the ergolines mentioned, it was the natural clavine alkaloid lysergol that exhibited the highest affinity among 20 nonergoline and ergoline-based 5-HT receptor ligands tested for both human 5-HT_{1D} receptor subtypes (Weinshank et al., 1992). Within a series of pharmacological agents, which were analyzed for their ability to discriminate effectively between the two closely related subtypes, the nonergoline 5-HT_{2A} receptor antagonist ketanserin showed 120-fold selectivity and ergolines such as ergotamine, dihydroergotamine, metergoline, and methysergide showed 10 to 26-fold selectivity for the 5-$HT_{1D\alpha}$ receptor relative to the 5-$HT_{1D\beta}$ subtype (Peroutka, 1994).

It is beyond any doubts that ergolines are among the most potent partial agonists at vascular 5-HT_1-like receptors. Ergotamine and dihydroergotamine, for example, acted as powerful partial agonists at 5-HT_1-like receptors in human basilar artery, of which a contractile response via α-adrenoceptors could be excluded at least for ergotamine (Müller-Schweinitzer, 1983, 1992). Furthermore, ergotamine was about 100-fold more potent than the antimigraine drug sumatriptan as a constrictor of human coronary artery, which is characterized by the coexistence of 5-HT_1-like receptors, S-HT_{2A} receptors, α-adrenoceptors and other (as yet unknown) receptors (Bax et al., 1993) In this connection it is worth pointing out that the carotid vascular effects of ergotamine and dihydroergotamine in the pig were only partially blocked by the 5-HT_1-like

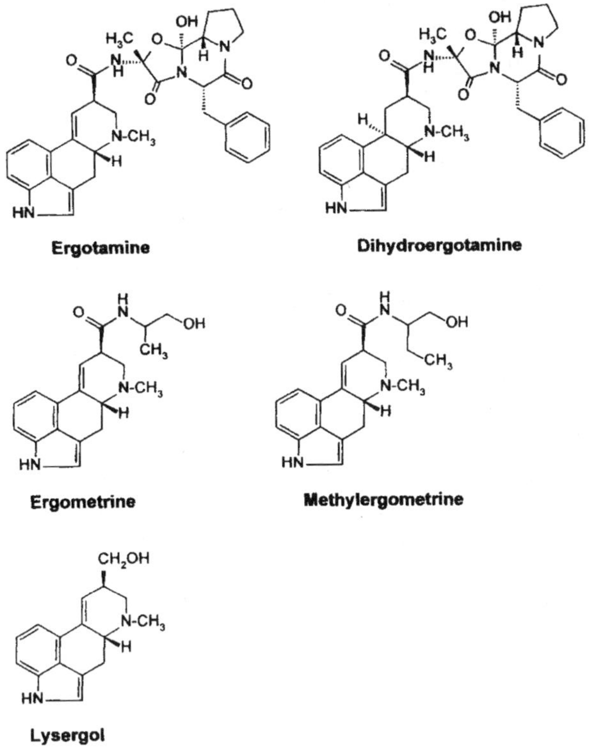

Ergotamine

Dihydroergotamine

Ergometrine

Methylergometrine

Lysergol

Figure 1 Structures of ergolines with partial agonist activity at 5-HT$_{1B/1D}$ receptors. Metergoline and methysergide are further partial agonists of which the structures are shown in Figure 2

receptor antagonist methiothepin, suggesting that the vasoconstrictor response to these ergolines is not exclusively mediated via 5-HT$_1$-like receptors but additionally *via* an unknown receptor or mechanism (Den Boer *et al.*, 1991). Such a phenomenon has also been reported in rabbit saphenous vein (MacLennan and Martin, 1990) and guinea-pig iliac artery (Pertz, 1993), where contractions to ergometrine, methylergometrine, and methysergide resulted in biphasic concentration-response curves, of which only the first phase was mediated via 5-HT$_1$-like receptors and the second phase mediated via unknown receptors. It is also worth pointing out that ergometrine, although exhibiting only weak activity at α$_1$-adrenoceptors (Müller-Schweinitzer and Weidmann, 1978), provokes coronary artery spasm in patients with Prinzmetal's angina and is used in the diagnosis of this disease (Prinzmetal *et al.*, 1959; Maseri *et al.*, 1977; Heupler *et al.*, 1978). Affinity and efficacy of ergometrine and methylergometrine for 5-HT$_1$-like receptors were higher than those of methysergide (MacLennan and

Martin, 1990; Pertz, 1993). In human and canine blood vessels powerful but extremely slow development of the contractile response to ergotamine and ergometrine has been observed (Mikkelsen *et al.*, 1981; Bax *et al.*, 1993; Brazenor and Angus, 1981). Therefore, undesirable cardiac side effects in the treatment of migraine may limit the therapeutical benefit of the highly efficient vasoconstrictor ergotamine.

14.2.3.
Ergolines Show Low Affinity for 5-HT$_{1E}$ Receptors and High Affinity for 5-HT$_{1F}$ Receptors

The most distinguishing feature between the 5-HT$_{1D}$ receptor and the 5-HT$_{1E}$ receptor (Levy *et al.*, 1992b; McAllister *et al.*, 1992; Zgombick *et al.*, 1992) is the low affinity for 5-carboxamidotryptamine (5-CT) and ergotamine (Beer *et al.*, 1993). The 5-HT$_{1E}$ receptor has been detected only in the brain where it possibly plays a role of an autoreceptor. There are no functional correlates and no selective ligands for this subtype. Another less well characterized receptor within the 5-HT$_1$ receptor group is the 5-HT$_{1F}$ receptor, which has the highest homology to the the 5-HT$_{1E}$ receptor. 5-HT$_{1F}$ receptors has been detected in the brain, uterus, and mesentery (Adham *et al.*, 1993a). The high affinity of the antimigraine drug sumatriptan for this subtype indicates a possible role of 5-HT$_{1F}$ and/or 5-HT$_{1D}$ receptors in migraine. This idea was supported by the agonist profile of the ergolines lysergol, methylergometrine, and methysergide, which showed similar affinity and efficacy for 5-HT$_{1F}$ receptors than did sumatriptan (Adham *et al.*, 1993b).

14.2.4.
Ergolines are Potent Antagonists/Partial Agonists at S-HT$_{2A}$Receptors

It has been suggested that the classical hallucinogenic agent LSD may exert its psychotic effect by acting as partial agonist at 5-HT$_{2A/2C}$ receptors (Glennon, 1990; Pierce and Peroutka, 1990). The ability of many 5-HT$_{2A}$ receptor antagonists to block both 5-HT$_{2A}$ and 5-HT$_{2C}$ receptors seems to be a key factor for the therapeutical benefit in the treatment of psychiatric disorders (Clarke, 1992). Therapeutic indications for S-HT$_{2A/2C}$ receptor antagonists include schizophrenia, depression, and anxiety (Peroutka, 1995). The antipsychotic therapy has been improved due to the recent development of so-called atypical antipsychotic drugs with combined D$_2$/5-HT$_{2A}$ receptor blocking properties which avoid extrapyramidal side effects (Meltzer and Nash, 1991). Recently it has been shown that 5-HT$_{2A}$ receptor antagonism alone may be sufficient for antipsychotic activity, since selective 5-HT$_{2A}$ receptor blockade increases dopamine release in the prefrontal cortex, thereby providing an improvement of the negative symptoms of schizophrenia (Schmidt *et al.*, 1995). In the CV system, the ability of S-HT$_{2A}$ receptor antagonists to inhibit 5-HT-induced platelet aggregation which is mediated via platelet 5-HT$_{2A}$ receptor stimulation, and to block the direct contractile effects of platelet-released 5-HT on vascular smooth muscle (Vanhoutte, 1990), makes them candidates for the treatment of ischemic heart disease and other vascular occlusive disorders (for review, see Audia and Cohen, 1990). An important application for certain simple lysergic acid amides is related to their powerful contractile effect on uterine smooth muscle. Ergometrine and methylergometrine are used frequently in the treatment of postpartum haemorrhage. Functional studies have suggested that the stimulatory effect of the partial agonist ergometrine in rat uterus is mediated via S-HT$_{2A}$ receptors due to the blockade by methysergide and the nonergoline 5-HT$_{2A}$ receptor antagonist ICI 169,369, respectively (Hollingsworth et al., 1988).

A variety of ergolines (e.g., metergoline, methysergide, dihydroergotamine, LSD, lisuride, α-dihydrergokryptine, ergometrine, ergotamine, methylergometrine, and β-dihydroergokryptine) displays high affinities for 5-HT$_{2A}$ receptors in radioligand binding studies (Hoyer, 1989). The interaction of ergolines, however, with vascular 5-HT receptors which may be of the 5-HT$_1$-like type, the 5-HT$_{2A}$ type or a mixture of both (Saxena and Villalón, 1990), is complex, since ergolines may act as partial agonists at 5-HT$_1$-like receptors (vide supra) and as antagonists or partial agonists at 5-HT$_{2A}$ receptors (Figure 2). To characterize a given response as being of the 5-HT$_{2A}$ type, the investigation of the antagonism by both the nonergoline 5-HT$_{2A}$ antagonist ketanserin and the ergoline-based 5-HT$_{2A}$ antagonist methysergide is of great value. Ketanserin shows high affinity for S-HT$_{2A}$ receptors, low affinity for 5-HT$_1$ receptors (Hoyer, 1989) but appreciable affinity for α$_1$-adrenoceptors (Van Nueten et al., 1981). In contrast, the potent nonselective 5-HT$_{2A}$ receptor antagonist methysergide has negligible affinity for α$_1$-adrenoceptors (Bradley et al., 1986).

The disadvantage of "classical" 5-HT$_{2A}$ receptor antagonists as tools for receptor classification is based on their lack of specificity and selectivity. For example, nonergoline 5-HT$_{2A}$ receptor antagonists (e.g., ketanserin and mianserin) interact with α$_1$-adrenoceptors and histamine H$_1$ receptors, whereas ergoline 5-HT$_{2A}$ receptor antagonists interact with dopamine receptors (e.g., LSD, metergoline, and mesulergine) and/or fail to discriminate between 5-HT$_1$ receptors and 5-HT$_2$ receptors (e.g., LSD, metergoline, and methysergide) (Closse et al., 1984; Hoyer, 1989). Thus the search for more specific and selective S-HT$_{2A}$ receptor antagonists is of special interest.

Among the ergolines, derivatives with high antagonist activity for 5-HT$_{2A}$ receptors and negligible α1 adrenergic, histaminergic and dopaminergic blocking properties, include the isopropyldihydrolysergic acid esters LY53857 and sergolexole as well as the amides amesergide and LY215840, which have been developed by Lilly Research Laboratories (Garbrecht, 1971; Marzoni et al., 1987; Garbrecht et al., 1988; Misner et al., 1990). It has been shown that LY53857, sergolexole, and amesergide potently block 5-HT$_{2A}$ receptors on both blood vessels and platelets, thereby inhibiting 5-HT-stimulated platelet

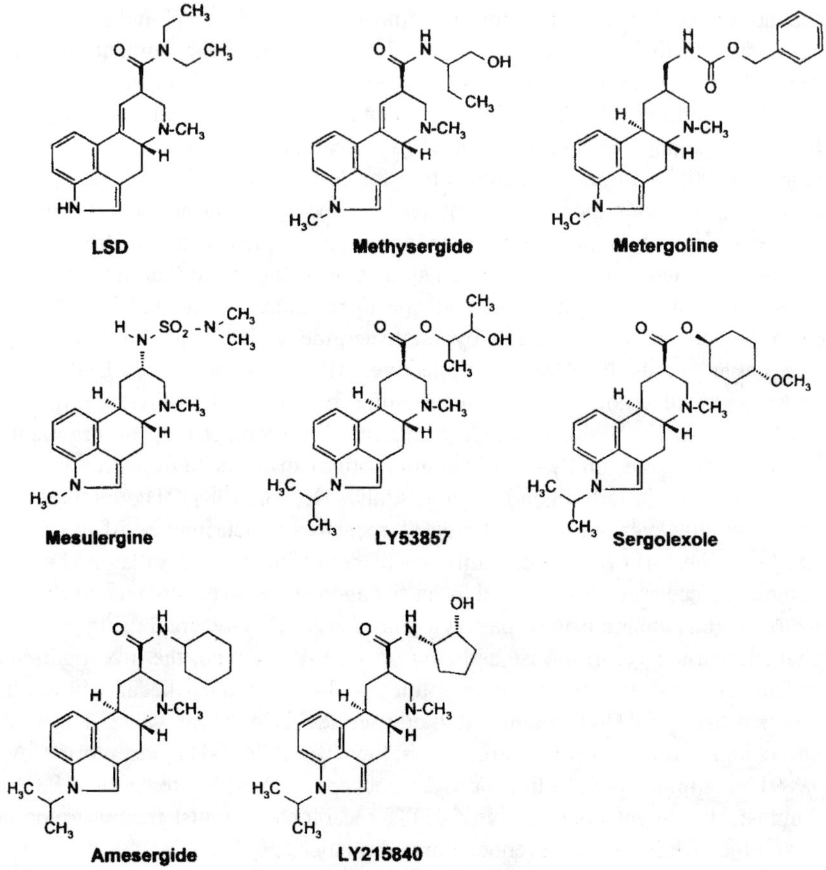

Figure 2 Structures of ergolines that show potent antagonist activity at receptors. LSD is a partial agonist at 5-HT$_{2A/2C}$ receptors. Additionally, mesulergine, methysergide, and LY53857 are potent antagonists at 5-HT$_{2B/2C}$ receptors, and metergoline is a potent antagonist at 5-HT$_{2C}$ receptors. Ergotamine and dihydroergotamine (see Figure 1) are agonists at 5-HT$_{2B/2C}$ receptors

aggregation (McBride *et al.,* 1990). In addition, LY53857 and amesergide, respectively, inhibit 5-HT-amplified ADP-induced aggregation in rabbit platelets (Wilson *et al.,* 1991; Cohen *et al.,* 1994), and LY215840 in both rabbit and human platelets (Cohen *et al.,* 1992). This additional mechanism may contribute to the potential effectiveness of such 5-HT$_{2A}$ receptor antagonists in the treatment of vascular disorders. On historical grounds LY53857 justifies special mention as a prototype in this series of ergoline 5-HT$_{2A}$ receptor antagonists (Cohen *et al.,* 1983). The compound represents a mixture of 4 diastereomers, all of which individually display nearly equal affinity for the 5-HT$_{2A}$ receptor (Cohen *et al.,* 1985). Although LY53857 antagonizes central as well as peripheral S-HT$_{2A}$

receptors, it does not lower the blood pressure in the spontaneously hypertensive rat (Cohen *et al.*, 1983). Similarly, no marked effect on blood pressure has been found with sergolexole (Cohen *et al.*, 1988), amesergide (Foreman *et al.*, 1992), and LY215840 (Cohen *et al.*, 1992). Thus S-HT$_{2A}$ receptor bockade *per se* appears to be not sufficient to cause a reduction in blood pressure. Sergolexole has recently been shown to be ineffective for migraine prophylaxis (Chappell *et al.*, 1991). Since sergolexole equipotently blocks 5-HT$_{2A}$ and 5-HT$_{2C}$ receptors (Tfelt-Hansen and Pedersen, 1992), this would speak against the previously postulated theory that 5-HT$_{2C}$ receptors are involved in the initiation of migraine (Fozard and Gray, 1989; Fozard, 1992) *(vide infra)*. Amesergide which proved to be 10–100 times more potent than LY53857 and sergolexole, respectively, in augmenting sexual responses of male rats might be useful in the treatment of sexual dysfunctions (Foreman *et al.*, 1989, 1992). Since amesergide shows nearly equal affinity than LY53857 and sergolexole, respectively, for 5-HT$_{2A}$ receptors (Nelson *et al.*, 1993), the amplification of male rat sexual behaviour caused by amesergide may be attributed to its interaction with 5-HT$_{2C}$ receptors.

The ergoline 5-HT$_{2A}$ receptor antagonists LY53857, sergolexole, amesergide, and LY215840 share the structural property to be substituted with an isopropyl group at the indole nitrogen (N1) (see Figure 2). It has recently been shown that ergolines with an N1-isopropyl group have higher affinity for the rat versus the human, monkey, and pig 5-HT$_{2A}$ receptor, whereas the corresponding N1-unsubstituted ergolines have higher affinity for the human, monkey, and pig versus the rat 5-HT$_{2A}$ receptor (Nelson *et al.*, 1993; Johnson *et al.*, 1993). The findings are consistent with the higher affinities of mesulergine and methysergide, which both are N1-methylergolines, for the rat versus the human and pig 5-HT$_{2A}$ receptor (Pazos *et al.*, 1984a). The affinity profile of further ergoline-based compounds confirms this pattern: the rat 5-HT$_{2A}$ receptor prefers Nl-methylergolines such as metergoline and nicergoline, whereas the human 5-HT$_{2A}$ receptor prefers N1-unsubstituted ergolines such as ergotamine, dihydroergotamine, ergometrine, LSD, lisuride, and pergolide (Hagen *et al.*, 1994). Mutational studies have shown that a single amino acid difference at position 242 in TMH 5 of the 5-HT$_{2A}$ receptor protein accounts for species variability. Point mutation of the Ser242 in the human 5-HT$_{2A}$ receptor to alanine resulted in an affinity for mesulergine that closely resembled that at the rat 5-HT$_{2A}$ receptor (Kao *et al.*, 1992). Similarly, the change from alanine to serine in the rat 5-HT$_{2A}$ receptor explained all of the affinity differences seen for a large number of N1-alkylated ergolines and their N1-unsubstituted analogues in different species (Johnson *et al.*, 1994). Hence it has been concluded that the amino acid 242 seems to be in close proximity to the N1-position of the indole nucleus of the ergolines and may serve as an important contact point in the 5-HT$_{2A}$ receptor by allowing a favourable hydrogen-bonding interaction of N1-unsubstituted ergolines with Ser242 of the human 5-HT$_{2A}$ receptor and a favourable Van der Waals interaction of N1-alkylated ergolines with the Ala242 of the rat 5-HT$_{2A}$ receptor (Johnson *et al.*, 1994). It has additionally been shown by alignment of the TMH 5 of the cloned 5-HT2A and 5-HT$_{2C}$ receptors that the locus which corresponds to Ser/Ala in the S-HT$_{2A}$ receptor is characterized by an alanine in the 5-HT$_{2C}$ receptor of both humans and rats. As a consequence, N1-alkylated ergolines such as mesulergine displayed higher selectivity for human 5-HT$_{2C}$ versus 5-HT$_{2A}$ receptors, whereas N1-unsubstituted ergolines such as LSD, lisuride, and ergometrine displayed higher affinity for human 5-HT$_{2A}$ versus 5-HT$_{2C}$ receptors (Almaula *et al.*, 1996). Thus, N1-unsubstituted ergolines and those with suitable substituents (methyl or isopropyl) are useful tools not only for unmasking species differences among 5-HT$_{2A}$ receptors but for determining subtype selectivity between human 5-HT2A and 5-HT$_{2C}$ receptors. Additional loci within the 5-HT$_{2A}$ receptor necessary for high affinity receptor binding have been detected with the use of several ergolines: the conserved aspartic acid residue at position 155 which has been found to be essential for LSD binding (Wang *et al.*, 1993) and the conserved phenylalanine residue at position 340 of which the phenyl moiety may allow a specific aromatic-aromatic interaction (e.g., π-π or hydrophobic) with the aromatic ring of the ergoline nucleus of non-peptide ergolines such as mesulergine, metergoline, methysergide, lisuride, LY53857, ergometrine, lergotrile, and amesergide (Choudhary *et al.*, 1995). By means of the potent 5-HT2A receptor antagonism of 1-isopropylelymoclavine, the parent drug of a series of ergoline reverse esters, and some other simple clavines such as 1-isopropylagroclavine and 1-isopropylfestuclavine of which the tetracyclic skeleton represents more or less the complete molecule, it has recently been demonstrated in the rat that the ergoline nucleus plays a crucial role in determining 5-HT$_{2A}$ receptor affinity and not the substituent at position 8 (Pertz *et al.*, 1995).

Insurmountable antagonists of 5-HT at vascular 5-HT$_{2A}$ receptors such as the ergolines LSD, methysergide, and LY53857 have been reported to bind to an allosteric site of this receptor, thereby inducing a conformational change of the receptor protein which is responsible for the depression of the 5-HT maximum response (Kaumann, 1989). The model of the allosteric 5-HT$_{2A}$ receptor system was supported by the pharmacological properties of 9, 10didehydro-6-methyl-8β-ergolinylmethyl *R, S*-2-methylbutyrate, a derivative of the naturally occurring clavine lysergol, which was able to reverse the depressent effect of the insurmountable 5-HT$_{2A}$ receptor antagonist methysergide (Pertz and Eich, 1992).

14.2.5.
The Complexity of the Interaction of the Ergolines with 5-HT$_{2B}$and 5-HT$_{2C}$Receptors

Based on their pharmacological profile, 5-HT$_{2B}$ receptors are closely related to 5-HT$_{2A}$ and even more to 5-HT$_{2C}$ receptors (Bonhaus *et al.*, 1995). The 5-HT$_{2B}$ receptor is the receptor that mediates the contractile response to 5-HT in the rat stomach fundus. Although this tissue has been used as a bioassay for 40 years (Vane, 1957), the classification of the fundal contractile receptor within the 5-HT receptor family has proven difficult. Its exact characterization as a 5-HT_{2B} receptor could be established only after the successful cloning of this subtype in the early 1990s (Foguet *et al.*, 1992; Kursar *et al.*, 1992). Ergolines such as methysergide, metergoline, mesulergine, 2-Br-LSD, LY53857, and amesergide possessed high affinity for the cloned rat 5-HT_{2B} receptor in radioligand binding studies (Wainscott *et al.*, 1993). On the other hand, mesulergine, LY53857, and methysergide displayed complex behaviour as antagonists of 5-HT at the cloned rat 5-HT_{2B} receptor and in rat stomach fundus. Mesulergine acted as a potent and surmountable antagonist, while LY53857 and methysergide showed potent but insurmountable antagonism of the effects of 5-HT (Wainscott *et al.*, 1993; Baxter *et al.*, 1995).

Interestingly, the human stomach does not contain a contractile 5-HT_{2B} receptor. Thus, after the successful cloning of the human 5-HT_{2B} receptor (Schmuck *et al.*, 1994) the question arose what function could be ascribed to this receptor type in humans. There are some facts that speak at that time for an involvement of 5-HT_{2B} receptors in the onset of migraine attacks, although 5-HT_{2C} receptor activation has previously been suggested to be a key step in the initiation of migraine. It has been shown that *m*-chlorophenylpiperazine (*m*-CPP), originally characterized as a potent 5-HT_{2C} receptor agonist, acts as an inducer of migraine-like headache (Fozard and Grey, 1989; Fozard, 1992). Since *m*-CPP also stimulates 5-HT_{2B} receptors in concentrations which induce headache, it has been suggested that drugs that prevent migraine may do so by blocking 5-HT_{2B} receptors. Indeed, the most consistently effective drugs in migraine prophylaxis, lisuride, methysergide, pizotifen, and propranolol have in common antagonist effects at 5-HT_{2B} receptors (Kalkman, 1994; Roos and Glusa, 1998). Also of interest is that methylergometrine, the major metabolite and active principle of methysergide in man (Müller-Schweinitzer and Tapparelli, 1986; Bredberg *et al.*, 1986) has found to be a potent antagonist at 5-HT_{2B} receptors (Fozard and Kalkman, 1994). A further argument that supports the idea of an involvement of 5-HT_{2B} receptors in the initiation of migraine is the finding that 5-HT_{2B} receptors which are present on endothelial cells, including those lining the cerebral blood vessels (Ullmer *et al.*, 1995), mediate vascular relaxation by the release of nitric oxide (NO). Clinical evidence points to a key role for NO in the initiation of migraine (Olesen *et al.*, 1994). In agreement with the findings in the rat stomach fundus, mesulergine acted as a potent and surmountable antagonist at endothelial 5-HT_{2B} receptors in rat jugular vein, while methysergide and LY53857 produced insurmountable antagonism in this tissue (Bodelsson *et al.*, 1993). In contrast, ergopeptines such as ergotamine and dihydroergotamine which proved to be highly efficient as anti-migraine drugs were potent agonists at endothelial 5-HT_{2B} receptors in porcine pulmonary artery (Glusa and Roos, 1996). Therefore, the relevance of 5-HT_{2B} receptor antagonists as efficient drugs in migraine seems to be a point of controversy, although the agonist activity of ergotamine and dihydroergotamine does not rule out a role for 5-HT_{2B} receptors in the initiating event of migraine. The therapeutic benefit of ergotamine and dihydroergotamine in the acute migraine attack and *not* in migraine prevention is based on the potent agonist activity of these drugs at 5-HT$_1$-like 5-HT$_{1B/1D}$ receptors *(vide supra)*.

Based on similarities between the former "5-HT$_{1C}$ receptor" and the receptor according to structural, transductional and operational criteria, the 5-HT$_{1C}$ receptor has been suggested to be a member of the 5-HT$_2$ class and is now termed 5-HT$_{2C}$ (Hoyer *et al.*, 1994). 5-HT$_{2C}$ receptors have definitely been detected only in the CNS, where they are localized with high density in the choroid plexus (Hoyer, 1988). [^{3}H]5-HT and [^{3}H]mesulergine have been used as high affinity radioligands for labelling 5-HT$_{2C}$ receptors (Pazos *et al.*, 1984b). The lack of selective agonists and antagonists at the 5-HT$_{2C}$ receptor has hampered elucidation of its pharmacological effects for a long time. Ergolines such as mesulergine, methysergide, and LY53857, originally characterized as 5-HT$_{2A}$ receptor antagonists, and the nonselective ergot derivative metergoline, show high affinity for the 5-HT$_{2C}$ receptor (Hoyer, 1989). LSD has found to be a partial agonist at this site (Sanders-Bush *et al.*, 1988). In the CNS nonselective 5-HT$_{2C/2A}$ receptor antagonists display an anxiolytic profile, whereas selective 5-HT$_{2A}$ receptor antagonists fail to produce anxiolysis. Consequently, it has been suggested that anxiolysis is mediated via the blockade of 5-HT$_{2C}$ receptors (Kenneth, 1992). Further evidence for the involvement of 5-HT$_{2C}$ receptors in anxiety has been provided by means of LY53857, of which a 5-fold selectivity for 5-HT$_{2C}$ versus S-HT$_{2A}$ receptors may be responsible for its marked anxiolytic effect (Kenneth *et al.*, 1994).

The pharmacological profile of ergotamine and dihydroergotamine at 5-HT$_{2C}$ receptors resembles that of these drugs at 5-HT$_{2B}$ receptors. Both drugs behaved as powerful 5-HT$_{2C}$ receptor agonists in piglet choroid plexus (Brown *et al.*, 1991). It is worth mentioning in this connection that the ability of ergotamine and dihydroergotamine to produce headache when taken in excess, may result from sufficient brain penetration at high doses followed by the activation of cerebral 5-HT$_{2C}$ receptors (Brown *et al.*, 1992). On the other hand, headache as an adverse reaction seen with bromokryptine (antiparkinsonian drug) and dihydroergotoxine (used in the treatment of senile dementia), is due to vasodilatation rather than to 5-HT$_{2C}$ receptor stimulation (Brown *et al.*, 1992).

14.2.6.
No Role for Ergolines at 5-HT$_3$ and 5-HT$_4$ Receptors

With the exception of LY53857 which possesses moderate affinity for 5-HT$_3$ receptors (Kennett *et al.*, 1994), only few informations exist about the interaction of ergolines with these sites. Similarly, ergolines play no role as ligands for 5-HT$_4$ receptors. It has been shown that ergolines such as metergoline, mesulergine, and methysergide possess low affinity for 5-HT$_4$ receptors (Dumuis *et al.*, 1988).

14.2.7.
Ergolines as Useful Tools for the Characterization of 5-ht$_5$, 5-ht$_6$, and 5-ht$_7$Receptors

Two G-protein coupled 5-HT receptors from both mouse and rat brain, designated 5-ht$_{5A}$ and 5-ht$_{5B}$, have recently been cloned, of which the amino acid sequence and the pharmacological profile is sufficiently distinct from those of the well characterized receptors 5-HT1—5-HT$_4$. Due to its high affinity for both 5-ht$_{5A}$ and 5-ht$_{5B}$ receptors, 2-[^{125}I]LSD has been used as radioligand for these sites. 5-ht$_{5A}$ and 5-ht$_{5B}$ receptors display high affinity for ergotamine and methysergide (Matthes *et al.*, 1993). The functional role of 5-ht$_5$ receptors remains to be established.

The recently cloned rat 5-ht$_6$ receptor is exclusively localized in the CNS (especially in the corpus striatum and various limbic and cortical systems). Competition for 2-[^{125}I]LSD binding by a number of drugs revealed high affinity for 5-ht$_6$ receptors not only for ergolines such as lisuride, dihydroergotamine, 2-Br-LSD, pergolide, metergoline, and lergotrile but for tricyclic antipsychotic and antidepressant drugs such as clozapine, amoxipine, and amytriptyline. This suggests a possible role for 5-ht$_6$ receptors in several neuropsychiatric disorders that involve serotoninergic systems (Monsma *et al.*, 1993).

5-ht$_7$ receptors showed high affinity for ergolines (e.g., lisuride, metergoline, pergolide, mesulergine, bromokryptine, and methysergide) and antipsychotic/antidepressant drugs (e.g., clozapine, loxapine, and amitriptyline). In this regard 5-ht$_7$ receptors resemble 5-ht$_6$ receptors, although their transmembrane regions exhibit homology of only 44% (Shen *et al.*, 1993). Second messenger coupling, pharmacological profile, and tissue distribution suggest a possible role for 5-ht$_7$ receptors in relaxation of smooth muscle systems (Bard *et al.*, 1993). For example, the precontracted guinea-pig ileum can be relaxed by 5-HT receptor agonists, including 5-CT and 5-HT, due to activation of 5-ht$_7$ receptors. The relaxant effect of 5-CT has most potently been blocked by ergolines, including LSD, mesulergine, and methysergide, followed by nonergolines, including spiperone and clozapine (Carter *et al.*, 1995). Among structurally related ergolines such as LY53857, sergolexole, amesergide, and LY215840 which have originally been characterized as potent S-HT$_{2A}$ receptor antagonists *(vide supra)*, LY215840 has recently been identified as a high-affinity 5-ht$_7$ receptor antagonist of 5-HT-induced relaxation in canine coronary artery (Cushing *et al.*, 1996).

14.3.
ERGOLINES AS LIGANDS FOR DOPAMINE RECEPTORS

Dopamine receptors are targets for antipsychotic drugs, antiparkinsonian drugs, and agents that affect the activity of the hypothalamic-pituitary system, particularly the release of prolactin from the pituitary gland. Antipsychotic agents (neuroleptics) specifically block dopamine receptors, whereas antiparkinsonian and prolactin-lowering agents stimulate dopamine receptors. Due to their agonist activity at dopamine receptors, a number of ergolines are widely used as antiparkinsonian drugs (e.g., bromokryptine, lisuride, and pergolide) and as inhibitors of prolactin release (e.g., bromokryptine, cabergoline, and lisuride) (Figure 3).

The knowledge of the existence of multiple dopamine receptors and their localization in different tissues is important for understanding how dopaminergic agents achieve their therapeutic effects and how adverse reactions may arise. Biochemical and pharmacological studies led to the identification of two native dopamine receptors, designated D1 and D_2 (Kebabian and Calne, 1979). In the early 1990s, molecular biology techniques have corrected the oversimplified picture of two dopamine receptors and defined five different isoforms D_1—D_5 which may be divided into D_1-like (D_1, D_5) and D_2-like (D_2, D_3, D_4) subfamilies on the basis of their structural and pharmacological properties (Sibley and Monsma, 1992; Strange, 1993). The D_1-like and D_2-like subfamilies of cloned receptor isoforms correspond to the D_1 and D_2 receptors of the former receptor classification. D_1 and D_5 (both positively coupled to adenylyl cyclase) and D_2 (negatively coupled to adenylyl cyclase) belong to the superfamily of G-protein coupled receptors. Within the neostriatum, the brain region where dopamine is important for control of motor function, the principal receptor subtypes are D_1 and D_2. In addition D_2 receptors are localized at high levels in the pituitary gland. The preferential localization of D_3 receptors is in limbic regions of the brain. This suggests an important role for D_3 receptors in the control of aspects of behaviour, emotion, motivation, and cognition (Strange, 1991). The D_4 receptor appears to be distributed at lower levels in the brain and at higher levels in the CV system (Van Tol et al., 1991). Therefore, the D_4 receptor may be considered as a peripheral D_2-like receptor (O'Malley et al., 1992).

Ergolines presumably exert their antiparkinsonian effect via stimulation of neostriatal D_1 and D_2 receptors (Strange, 1993), whereas their inhibitory effect on prolactin release from the anterior pituitary may be mediated via stimulation of D_2 or D_4 receptors (Sokoloff et al., 1993). Parkinson's disease is caused by a loss of dopaminergic neurones innervating the striatum (Strange, 1992). The treatment of Parkinson's disease with mixed D_1/D_2 receptor agonists which lack D_3, D_4, and D_5 receptor affinity may lead to a facilitation of motor function by maintaining the balance between D1 receptor activation via a direct neostriatal pathway and D_2 receptor inhibition via an indirect neostriatal pathway in favour of the direct pathway (Strange, 1993). The improved efficiency of D_2-like receptor agonists such as bromokryptine and lisuride, when used in association with L-DOPA as a prodrug of dopamine which stimulates both D_1 and D_2 receptors in the neostriatum, provides some support for the need of D_1 and D_2 receptor occupancy in the treatment of Parkinson's disease (Agnoli et al., 1985).

Ergolines with dopaminergic activity that have been most widely studied include bromokryptine, lisuride, lergotrile, pergolide, and cabergoline. Prior to molecular cloning of dopamine receptor subtypes, the dopaminergic profile of these drugs had been established on the basis of the existence of two native dopamine receptors, D_1 (i.e. D_1-like) and D_2 (i.e. D_2-like) *(videsupra)*. For example, bromokryptine seemed to be relatively selective at D_2 (i.e. D_2-like) receptors, whereas lisuride and lergotrile acted as a D_2 (i.e. D_2-like) receptor agonists with antagonist activity at D_1 (i.e. Di-like) receptors (Kebabian and Calne, 1979; Cote *et al.,* 1985). It should be mentioned that lisuride lacks specificity for dopamine receptors due to its high affinity for 5-HT_{1A}, 5-HT_{1D}, 5-HT_{2A}, 5-HT_{2B}, and 5-HT_{2C} receptors (Hoyer, 1989; Roos and Glusa, 1998). On the other hand, pergolide showed increased specificity for dopamine receptors and behaved as an agonist at both D_1 (i.e. D_1-like) and D_2 (i.e. D_2-like) receptors (Fuller and Clemens, 1991). Due to its mixed D_1-like and D_2-like receptor agonist activity, pergolide appears to be more suitable to treat parkinsonism than pure D_2-like receptor agonists *(videsupra)*. Cabergoline, another ergoline with enhanced specificity for dopamine receptors, showed marked D_2 (i.e. D_2-like) receptor agonist activity and long duration of action (Pontiroli *et al.,* 1987). Due to its poor brain penetration, the clinical application of cabergoline was restricted to the inhibition of prolactin release from the anterior pituitary, which is not protected by the blood-brain barrier. Unfortunately, only few informations are available about the interaction of ergolines with the cloned receptor isoforms D_1-D_5. Bromokryptine has been shown to possess comparably high affinity for both D_2 and D_3 receptors with marginal affinity for D_1, D_4, and D_5 receptors, while pergolide seems to be an equipotent ligand for D_1, D_2, and D_3 receptors (Seeman and Van Tol, 1994). It is worth mentioning that the interaction of bromokryptine and pergolide with D_3 receptors may be responsible for unwanted psychic side effects (e.g., confusion, hallucination) observed in the treatment of Parkinson's disease with these drugs.

A phenomenon not directly related to subtypes deals with the partial D_2-like receptor agonism of ergolines such as CF 25–397 and terguride (*trans*-dihydrolisuride), respectively. Partial agonists have the advantage of being recognized as full agonists by dopamine receptors when neostriatal dopaminergic neurones are relatively low or absent as in Parkinson's disease. On the other hand, D_2-like receptors of non-striatal systems mediating nausea and emesis may remain untouched due to the antagonist properties of partial agonists. Thus, the use of partial agonists may lead to a better balance between therapeutic actions and unwanted side effects (Carlsson, 1993). Further examples for partial dopamine receptor agonists within the family of ergolines are SDZ208911 and SDZ208912. Whereas SDZ208912 is a potent D_2-like receptor antagonist with only marginal intrinsic activity, SDZ208911 exhibits higher intrinsic activity and less D_2-like receptor blockade. Based on their reduced parkinsonian side effects, SDZ208911 and SDZ208912 may be useful drugs in the treatment of schizophrenia (Coward *et al.,* 1990). It has recently been shown that non-addictive dopamine receptor agonists such as bromokryptine, lisuride, and pergolide may ameliorate some of the symptoms of psychostimulant withdrawal.

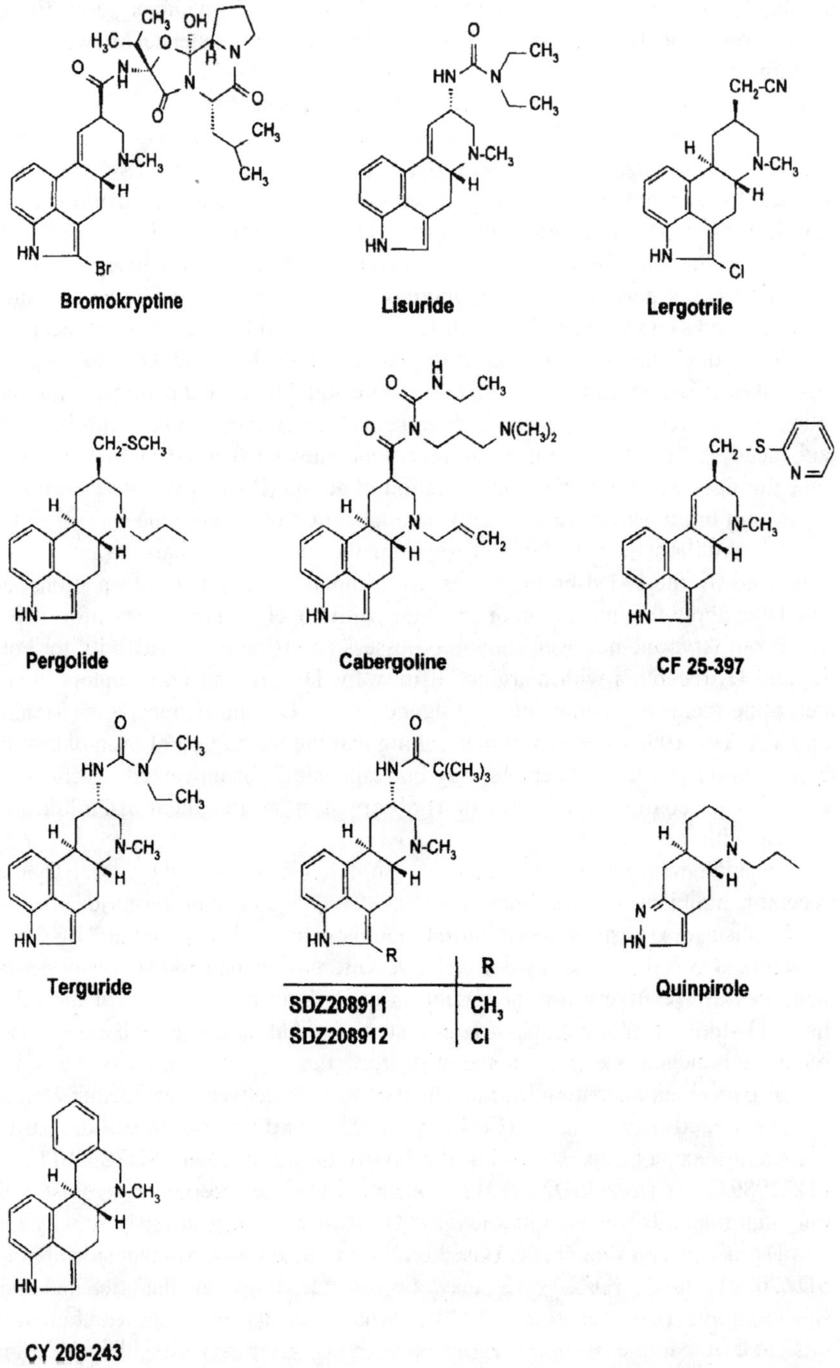

Figure 3 Structures of ergolines that show (partial) agonist activity at D_2-like receptors. CY 208–243 is an agonist at D_1-like receptors, and pergolide shows mixed D_1-like and D_2-like receptor agonist activity

It should be noted that partial agonists such as terguride, SDZ208911, and SDZ208912 are of special importance as candidates for the treatment of psychostimulant dependence due to their normalizing effect on dopamine neurotransmission during the various phases of psychostimulant addiction. The therapeutical benefit of partial agonists can be ascribed to their antagonist activity under conditions of dopamine hyperactivity following the exposure to psychostimulants and to their agonist activity during psychostimulant withdrawal which is characterized by a low dopamine tone (Pulvirenti and Koob, 1994).

Early studies with lergotrile and pergolide suggested that the indole NH group of the ergoline nucleus is bioisosteric with the *m*-hydroxy group of dopamine, and that the rigid pyrrolethylamine portion of the molecule might be the pharmacophoric constituent of the ergolines (Bach *et al.,* 1980). Potent D_2-like receptor agonism of linear tricyclic analogues of ergolines confirmed the hypothesis. Tricyclic ergoline partial structures such as LY141865 were comparable in potency with the highly active ergoline pergolide (Bach *et al.,* 1980). It could be demonstrated that dopamine receptor agonist activity of racemic LY141865 is a property of its *R, R*-(—)enantiomer quinpirole (Titus *et al.,* 1983) which showed high affinity for D_2 receptors and somewhat lower affinity for D_3 and D_4 receptors (Seeman and Van Tol, 1994). On the other hand, enhanced D_1-like receptor selectivity could be induced by fusion of a benzene ring across the 8,9-bond of the ergoline skeleton. Benzo-fused pentacyclic ergolines ("benzergolines") such as CY 208–243 represent the first structural class of potent and selective non-catechol D_1-like receptor agonists which allow efficient penetration into the CNS (Seller *et al.,* 1991, 1993).

14.4.
ERGOLINES AS LIGANDS FOR ADRENOCEPTORS

Since the end of the 1940s structure-activity relationship studies by means of natural and synthetic ligands have led to the detection of an increasing number of distinct adrenoceptor subtypes. Drugs interacting with these subtypes have proven useful in a variety of diseases such as hypertension, angina pectoris, congestive heart failure, cardiac arythmia, asthma, depression, prostatic hypertrophy, and glaucoma (Bylund *et al.,* 1994).

From a historical point of view ergot alkaloids are closely linked to the classification of adrenoceptors into two major subtypes (α and β). The discrimination between α- and β-adrenoceptors was based on the insensitivity of the latter to ergot alkaloids or β-haloalkylamines (Nickerson, 1949). On the basis of structural, transductional, and operational criteria it became apparent that the existence of two subtypes of α-adrenoceptors, the α_1-adrenoceptor, sensitive to blockade by prazosin, and the α_2-adrenoceptor, sensitive to blockade by yohimbine or rauwolscine, makes it more appropriate to classify adrenoceptors into three major subtypes: the α_1-adrenoceptors, α_2-adrenoceptors, and β-adrenoceptors (Bylund, 1988).

The interaction of ergolines with these three major subtypes (α_1, α_2, β) appears to be highly complex even if newer developments considering the existence of further subtypes are neglected *(vide infra)*. Therefore, we will only mention some general aspects of the effects of ergolines on adrenoceptors, particularly since this subject has been excellently reviewed years ago (MüllerSchweinitzer, 1978). In radioligand binding studies ergopeptines such as ergotamine, dihydroergotamine, dihydroergotoxine, α-dihydroergokryptine, and dihydroergocristine generally displayed higher affinity for α_2- than for α_1-adrenoceptors (Closse *et al.,* 1984). Functional studies showed that dihydroergotamine, dihydroergotoxine, and dihydroergocristine acted as partial agonists at α_2-adrenoceptors and antagonists at α_1-adrenoceptors in the peripheral vascular system and in vas deferens (Roquebert and Grenié, 1986; Roquebert *et al.,* 1983, 1984, 1985). Slightly higher affinity for α_2-adrenoceptors than for α_1-adrenoceptors has been obtained *in vivo* and *in vitro* for ergotamine (Megens *et al.,* 1986). Moreover, a combined 5-HT$_{1D}$/α_2-receptor agonist activity has been reported for ergotamine in dog saphenous vein (Müller-Schweinitzer, 1992). In contrast simple lysergic acid amides such as ergometrine, methylergometrine, and methysergide are only weakly active at α_1-adrenoceptors (Müller-Schweinitzer, 1978). However, this is not always the case. For example, LSD exhibits a highly complex pharmacology which is the result of its interference not only with 5-HT receptors and dopamine receptors but also with α_1-, α_2-, and β-adrenoceptors (Closse *et al.,* 1984; MaronaLewicka and Nichols, 1995; Dolphin *et al.,* 1978). In addition, ergolines such as bromokryptine, lisuride, lergotrile, and pergolide, which play an important role in the treatment of Parkinson's disease and as inhibitors of prolactin secretion due to their dopaminergic activity *(vide supra),* also display high affinity for α_1- and α_2-adrenoceptors (Closse *et al.,* 1984; Ruffolo *et al.,* 1987). This may be taken as a further evidence for the lack of ergolines to interact specifically with monoamine receptor systems.

With the rapid development of additional pharmacological tools which displayed improved selectivity for either α_1, α_2- or β-adrenoceptors in functional and radioligand binding studies, and the advent of molecular biological techniques by cloning of distinct adrenoceptor subtypes being in accord with their native correlates, the existence of additional subtypes became apparent. At present, the family of adrenoceptors comprises three α_1-adrenoceptor subtypes (α_{1A}, α_{1B}, α_{1D}), four α_2-adrenoceptor subtypes (α_{2A}, α_{2B}, α_{2C}, α_{2D}), and three β-adrenoceptor subtypes (β_1, β_2, β_3).

The present study makes it clear that ergolines as ligands of low specificity and selectivity generally play a limited role in the characterization of so many closely related subtypes of adrenoceptors. A worth-mentioning exception is BAM-1303 (8β[(2-phenylimidazol-l-yl)methyl]-6-methylergoline) which represents a useful pharmacological tool in this field due to its ability to discriminate between the closely related α_{2A}- and α_{2D}-receptors on the one hand, and α_{2B}- and α_{2C}-receptors on the other hand (Simmoneux *et al.*, 1991). The relatively high affinity of methysergide, originally characterized as a 5-HT$_{1D}$ receptor partial agonist and as a 5-HT$_{2A/2B/2C}$ receptor antagonist, for α_{2B}-adrenoceptors should be mentioned as a further evidence for the extremely complex activity profile of the ergolines (Brown *et al.*, 1990).

14.5.
CONCLUSION

Despite their low specificity and selectivity, ergot alkaloids and their derivatives are highly efficient tools for the characterization of serotonin (5-HT) receptors, dopamine receptors, and adrenoceptors, where they display complex behaviour as agonists, partial agonists or antagonists. Analysis of the interaction at these sites shows that ergolines are "dirty" drugs of which the therapeutical benefit as well as the unwanted side effects can be related to the involvement of a variety of different receptor subtypes. Predominant targets for therapeutically used ergolines are 5-HT$_{1D}$ receptors in the treatment of the acute migraine attack (ergotamine, dihydroergotamine), 5-HT$_{2B}$ receptors in migraine prophylaxis (methysergide), 5-HT2A receptors in the control of postpartum bleeding (ergometrine, methylergometrine), D$_2$-like receptors in the inhibition of prolactin release from the anterior pituitary (e.g., cabergoline), and D$_2$-like (e.g., bromokryptine) or D$_1$-like and D$_2$-like receptors (e.g., pergolide) in the treatment of parkinsonism. The vasoactive properties of certain ergolines (e.g., dihydroergotamine, dihydroergotoxine) involve partial α_2-adrenoceptor activation and α_1-adrenoceptor blockade in the peripheral vascular system. Their interaction with further subtypes of adrenoceptors remains to be established.

REFERENCES

Adham, N., Kao, H.-T., Schechter, L.E., Bard, J., Olsen, M., Urquhart, D., Durkin, M., Hartig, P.R., Weinshank, R.L. and Branchek, T.A. (1993a) Cloning of another human serotonin receptor (5-HT$_{1F}$): a fifth 5-HT$_1$ receptor subtype coupled to the inhibition of adenylyl cyclase. *Proc. Natl. Acad. Sci. USA,***90,**408–412.

Adham, N., Borden, L.A., Schechter, L.E., Gustafson, E.L., Cochran, T.L., Vaysse, P.J.-J., Weinshank, R.L. and Branchek, T.A. (1993b) Cell-specific coupling of the cloned human 5-HT$_{1F}$ receptor to multiple signal transduction pathways. *NaunynSchmiedeberg's Arch. Pharmacol.,***348,**566–575.

Agnoli, A., Ruggieri, S., Baldassarre, M., Stocchi, F., Denaro, A. and Falaschi, P. (1985) Dopaminergic ergots in parkinsonism. In D.B.Calne, R.Horowski, R.J.McDonald, W.Wuttke (eds.), *Lisuride and Other Dopamine Agonists,*Raven Press, New York, pp. 407–417.

Almaula, N., Ehersole, B.J., Ballcsteros, J.A., Weinstein, H. and Sealfon, S.C. (1996) Contribution of a helix 5 locus to selectivity of hallucinogenic and nonhallucinogenic ligands for the human 5-hydroxytryptamine$_{2A}$ and 5-hydroxytryptamine$_{2C}$ receptors:

direct and indirect effects on ligand affinity mediated by the same locus. *Mol. Pharmacol.,***50,**34–42.

Audia, J.E. and Cohen, M.L. (1991) Serotonin modulators and cardiovascular/gastrointestinal diseases. *Ann. Rep. Med. Chem.,***26,**103–112.

Bach, N.J., Kornfeld, E.C., Jones, N.D., Chancy, M.O., Dorman, D.E., Paschal, J.W., Clemens, J.A. and Smalstig, E.B. (1980) Bicyclic and tricyclic ergoline partial structures. Rigid 3-(2-aminocthyl)-pyrroles as dopamine antagonists. *J. Med. Chem.,***23,**481–491.

Bard, J.A,. Zgombick, J., Adham, N., Vaysse, P., Branchek, T.A. and Weinshank, R.L. (1993) Cloning of a novel human serotonin receptor (5-HT$_7$) positively linked to adenylyl cyclase. *J. Biol. Chem.,***268,**23422–23426.

Bax, W.A., Van Heuven-Nolsen, D., Bos, E., Simoons, M.L. and Saxena, P.R. (1992) 5-Hydroxytryptamine-induced contractions of the human isolated saphenous vein: involvement of 5-HT$_2$ and 5-HT$_{1D}$ -like receptors, and a comparison with grafted veins. *Naunyn-Schmiedeberg's Arch. Pharmacol.,***345,**500–508.

Bax, W.A., Renzenbrink, G.J., Van Heuven-Nolsen, D., Thijssen, E.J.M., Bos, E. and Saxena, P.R. (1993) 5-HT receptors mediating contractions of the isolated human coronary artery. *Eur. J. Pharmacol.,***239,**203–210.

Baxter, G.S., Murphy, O.E. and Blackburn, T.P. (1994) Further characterization of 5-hydroxytryptamine receptors (putative 5-HT$_{2B}$) in rat stomach fundus longitudinal muscle. *Br. J. Pharmacol.,***112,**323–331.

Beer, M.S., Middlemiss, D.N. and McAllister, G. (1993) 5-HT$_1$-like receptors: six down and still counting. *Trends Pharmacol. Sci.,***14,**228–231.

Bodelsson, M., Törnebrandt, K. and Arneklo-Nobin, B. (1993) Endothelial relaxing 5-hydroxytryptamine receptors in the rat jugular vein: similarity with the 5-hydroxytryptamine$_{1C}$ receptor. *J. Pharmacol. Exp. Ther.,***264,**709–716.

Boess, F.G. and Martin, I.L. (1994) Molecular biology of 5-HT receptors. *Neuropharmacology,***33,**275–317.

Bonhaus, D.W., Bach, C., DeSouza, A., Salazar, F.H.R., Matsuoka, B.D., Zuppan, P., Chan, H.W. and Eglen, R.M. (1995) The pharmacology and distribution of human 5-hydroxytryptamine$_{2B}$ (5-HT$_{2B}$) receptor gene products: comparison with 5-HT2A and 5-HT$_{2c}$ receptors. *Br. J. Pharmacol.,***115,**622–628.

Bradley, P.B., Engel, G., Feniuk, W., Fozard, J.R., Humphrey, P.P.A., Middlemiss, D.N., Mylecharane, E.J., Richardson, B.P. and Saxena, P.R. (1986) Proposals for the classification and nomenclature of functional receptors for 5-hydroxytryptamine. *Neuropbarmacology,***25,**563–576.

Brazenor, R.M. and Angus, J.A. (1981) Ergometrine contracts isolated canine coronary arteries by a serotonergic mechanism: no role for alpha adrenoceptors. *J. Pharmacol. Exp. Ther.,***218,** 530–536.

Bredberg, U., Eyjolfsdottir, G.S., Paalzow, L., Tfelt-Hansen, P. and Tfelt-Hansen, V. (1986) Pharmacokinetics of methysergide and its metabolite methylergometrine in man. *Eur. J. Clin. Pharmacol.,***30,**75–77.

Brown, C.M., MacKinnon, A.C., McGrath, J.C., Spedding, M. and Kilpatrick, A.T. (1990) α_2-Adrenoceptor subtypes and imidazoline-like binding sites in the rat brain. *Br. J.Pharmacol.,***99,**803–809.

Brown, A.M., Patch, T.L. and Kaumann, A.J. (1991) The antimigraine drugs ergotamine and dihydroergotamine are potent 5-HT$_{1C}$ receptor agonists in piglet choroid plexus. *Br. J. Pharmacol.,***104,**45–48.

Brown, A.M., Patch, T.L. and Kaumann, A.J. (1992) Ergot alkaloids as 5-HT$_{1C}$ receptor agonists: Relevance to headache. In J.Olesen, P.R.Saxena. (eds.), *5-Hydroxytryptamine Mechanisms in Primary Headaches,*Raven Press, New York, pp. 247–251.

Bylund, D.B. (1988) Subtypes of α_2-adrenoceptors: pharmacological and molecular biological evidence converge. *Trends Pharmcol. Sci.,***9**,356–361.

Bylund, D.B., Eikenberg, D.C., Hieble, J.P., Langer, S.Z., Lefkowitz, R.J., Minneman, K.P., Molinoff, P.B., Ruffolo, R.R., Jr. and Trendelenburg, U. (1994) Nomenclature of adrenoceptors. *Pharmacol. Rev.,***46**,121–136.

Carlsson, A. (1993) Thirty years of dopamine research. In H.Narabayashi, T.Nagatsu, N.Yanagisawa, Y.Mizuno (eds.), *Advances in Neurology,*vol. 60, Raven Press, New York, pp. 1–10.

Carter, D., Champney, M., Hwang, B. and Eglen, R.M. (1995) Characterization of a postjunctional 5-HT receptor mediating relaxation of guinea-pig isolated ileum. *Eur.J. Pbarmacol.,***280**,243–250.

Chappell, A.S., Bay, J.M. and Botzum, G.D. (1991) Sergolexole maleate and placebo for migraine prophylaxis. *Cephalalgia,***11**(suppl. 11), 170–171.

Choudhary, M.S., Sachs, N., Uluer, A., Glennon, R.A., Westkaemper, R.B. and Roth, B.L. (1995) Differential ergoline and ergopeptine binding to 5-hydroxytryptamine$_{2A}$ receptors: ergolines require an aromatic residue at position 340 for high affinity binding. *Mol. Pharmacol.,***47**,450–457.

Clarke, D.E. (1992) A synopsis of the pharmacology of clinically used drugs at 5-HT receptors and uptake sites. In J. Olesen, P.R.Saxena (eds.), *5-HydroxytryptamineMechanisms in Primary Headaches,*Raven Press, New York, pp. 118–128.

Closse, A., Frick, W., Dravid, A., Bolliger, G., Hauser, D., Sauter, A. and Tobler, H.-J. (1984) Classification of drugs according to receptor binding profiles. *NaunynSchmiedeberg's Arch. Pharmacol.,***327**,95–101.

Cohen, M.L., Fuller, R.W. and Kurz, K.D. (1983) LY53857, a selective and potent serotonergic (5-HT$_2$) receptor antagonist, does not lower blood pressure in the spontaneously hypertensive rat. *J. Pbarmacol. Exp. Ther.,***227**,327–332.

Cohen, M.L., Kurz, K.D., Mason, N.R., Fuller, R.W., Marzoni, G.P. and Garbrecht, W L. (1985) Pharmacological activity of the isomers of LY53857, potent and selective 5-HT$_2$ receptor antagonists. *J. Pbarmacol. Exp. Ther.,***235**,319–323.

Cohen, M.L., Fuller, R.W., Kurz, K.D., Parli, C.J., Mason, N.R., Meyers, D.B. Smallwood, J.K. and Toomey, R.E. (1988) Preclinical pharmacology of a new serotonergic receptor antagonist, LY281067. *J. Pharmacol. Exp. Ther.,***244**,106–112.

Cohen, M.L., Robertson, D.W., Bloomquist, W.E. and Wilson, H.C. (1992) LY215840, a potent 5-hydroxytryptamine (5-HT)$_2$ receptor antagonist, blocks vascular and platelet 5-HT$_2$ receptors and delays occlusion in a rabbit model of thrombosis. *J. Pharmacol. Exp. Ther.,***261**,202–208.

Cohen, M.L., Kurz, K.D., Fuller, R.W. and Calligaro, D.O. (1994) Comparative 5-HT$_2$receptor antagonist activity of amesergide and its active metabolite 4-hydroxyamesergide in rats and rabbits. *J. Pharm. Pharmacol.,***46**,226–229.

Cote, T.E., Eskay, R.L., Frey, E.A., Grewe, C.W., Munemura, M., Tsuruta, K., Brown, E.M. and Kebabian, J.W. (1985) Actions of lisuride on adrenoceptors and dopamine receptors. In D.B.Calne, R.Horowski, R.J.McDonald, W.Wuttke (eds.), *Lisurideand Other Dopamine Agonists,*Raven Press, New York, pp. 45–53.

Coward, D.M., Dixon, A.K., Urwyler, S., White, T.G., Enz, A., Karobath, M. and Shearman, G. (1990) Partial dopamine-agonistic and atypical neuroleptic properties of the amino-ergolines SDZ 208–911 and SDZ 208–912. *J. Pharmacol. Exp. Ther.,***252**, 279–285.

Gushing, D.J., Zgombick, J.M., Nelson, D.A. and Cohen, M.L. (1996) LY215840, a high-affinity 5-HT receptor ligand, blocks serotonin-induced relaxation in canine coronary artery. *J. Pharmacol. Exp. Ther.,***277,**1560–1566.

Demchyshyn, L., Sunahara, R.K., Miller, K., Teitler, M., Hoffman, B.J., Kennedy, J.L., Seeman, P., Van Tol, H.H.M. and Niznik, H.B. (1992) A human serotonin 1D receptor variant (5HT1D*β*) encoded by an intronless gene on chromosome 6. *Proc.Natl. Acad. Sci. USA,***89,**5522–5526.

Den Boer, M.O., Heiligers, J.P.C. and Saxena, P.R. (1991) Carotid vascular effects of ergotamine and dihydroergotamine in the pig: no exclusive mediation via 5-HT$_1$-like receptors. *Br. J. Pharmacol.,***104,**183–189.

Dolphin, A., Enjalbert, A., Tassin, J.-P., Lucas, M. and Bockaert, J. (1978) Direct interaction of LSD with central "beta"-adrenergic receptors. *Life Sci.,***22,**345–352.

Dumuis, A., Bouhelal, R., Sebben, M., Croy, R. and Bockaert, J. (1988) A nonclassical 5-hydroxytryptamine receptor positively coupled with adenylate cyclase in the central nervous system. *Mol. Pharmacol.,***34,**880–887.

Eich, E. and Pertz, H. (1994) Ergot alkaloids as lead structures for differential receptor systems. *Pharmazie,***49,**867–877.

Eerrari, M.D. and Saxena, P.R. (1993) Clinical and experimental effects of sumatriptan in humans. *Trends Pharmacol. Sci.,***14,**129–133.

Fletcher, A., Cliffe, I.A. and Dourish, C.T. (1993) Silent 5-HT$_{1A}$ receptor antagonists: utility as research tools and therapeutic agents. *Trends Pharmacol. Sci.,***14,**441–448.

Foguet, M., Hoyer, D., Pardo, L.A., Parekh, A., Kluxen, F.W., Kalkman, H.O., Summer, W. and Lübbert, H. (1992) Cloning and functional characterization of the rat stomach fundus serotonin receptor. *EMBO J.,***11,**3481–3487.

Foreman, M.M., Hall, J.L. and Love, R.L. (1989) The role of the 5-HT$_2$ receptor in the regulation of sexual performance of male rats. *Life Sci.,***45,**1263–1270.

Foreman, M.M., Fuller, R.W., Nelson, D.L., Calligaro, D.O., Kurz, K.D., Misner, J.W., Garbrecht, W.L. and Parli, C.J. (1992) Preclinical studies on LY237733, a potent and selective serotonergic antagonist. *J. Pharmacol. Exp. Ther.,***260,**51–57.

Fozard, J.R. (1992) 5-HT$_{1C}$ receptor agonism as an initiating event in migraine. In J.Olesen, P.R.Saxena. (eds.), *5-Hydroxytryptamine Mechanisms in Primary Headaches,*Raven Press, New York, pp. 200–212.

Fozard, J.R. and Gray, J.A. (1989) 5-HT$_{1C}$ receptor activation: a key step in the initiation of migraine?*Trends Pharmacol. Sci.,***10,**307–309.

Fozard, J.R. and Kalkman, H.O. (1994) 5-Hydroxytryptamine (5-HT) and the initiation of migraine: new perspectives. *Naunyn-Schmiedeberg's Arch. Pharmacol.,***350,**225–229.

Fozard, J.R. and Kilbinger, H. (1985) 8-OH-D DPAT inhibits transmitter release from guinea pig enteric cholinergic neurons by activating 5-HT$_{1A}$ receptors. *Br. J.Pharmacol.,***86,**601 P.

Fuller, R.W. and Clemens, J.A. (1991) Pergolide: a dopamine agonist at both D! and D$_2$ receptors. *Life Sci.,***49,**925–930.

Garbrecht, W.L. (1971) Hydroxyesters of hexa- and octahydroindoloquinolines. U.S. Patent No 3,580,916.

Garbrecht, W.L., Marzoni, G., Whitten, K.R. and Cohen, M.L. (1988) (8β)-Ergoline-8-carboxylic acid cycloalkyl esters as serotonin antagonists: structure-activity study. *J. Med. Chem.*,**31**,444–448.

Glennon, R.A. (1990) Do classical hallucinogens act as 5-HT_2 agonists or antagonists? *Neuropharmacology*,**3**,509–517

Glusa, E. and Roos, A. (1996) Endothelial 5-HT receptors mediate relaxation of porcine pulmonary arteries in response to ergotamine and dihydroergotamine. *Br.J. Pharmacol.*,**119**,330–334.

Göthert, M., Fink, K., Frölich, D., Likungu, J., Molderings, G., Schlicker, E. and Zentner, J. (1996) Presynaptic 5-HT auto- and heteroreceptors in the human central and peripheral nervous system. *Behav. Brain Res.*,**73**,89–92.

Hagen, J.D., Pierce, P.A. and Peroutka, S.J. (1994) Differential binding of ergot compounds to human versus rat 5-HT_2 cortical receptors. *Biol Signals*,**3**,223–229.

Hamblin, M.W. and Metcalf, M.A. (1991) Primary structure and functional characterization of a human 5-HT1D-type serotonin receptor. *Mol. Pharmacol.*,**40**, 143–148.

Hamblin, M.W., Ariani, K., Adriaenssens, P.I. and Ciaranello, R.D. (1987) [³H] Dihydroergotamine as a high affinity, slowly dissociating radioligand for 5-HT_{1B} binding sites in rat brain membranes: evidence for guanine nucleotide regulation of agonist affinity states. *J. Pharmacol. Exp. Ther.*,**243**,989–1001.

Hamblin, M.W., Metcalf, M.A., McGuffin, R.W. and Karpells, S. (1992) Molecular cloning and functional characterization of a human 5-HT_{1B} serotonin receptor: a homologue of the rat 5-HT_{1B} serotonin receptor with 5-HT_{1D}-like pharmacological specificity. *Biochem. Biophys. Res. Commun.*,**184**,752–759.

Hamel, E. and Bouchard, D. (1991) Contractile 5-HT_1 receptors in human isolated pial arterioles: correlation with 5-HT_{1D} binding sites. *Br. J. Pharmacol.*,**102**,227–233.

Hamel, E., Fan, E., Linville, D., Ting, V., Villemure, J.-G. and Chia, L.-S. (1993) Expression of mRNA for the serotonin 5-hydroxytryptamine$_{1D\beta}$ receptor subtype in human and bovine cerebral arteries. *Mol. Pharmacol.*,**44**,242–246.

Hartig, P.R., Branchek, T.A. and Weinshank, R.L. (1992) A subfamily of 5-HT_{1D} receptor genes. *Trends Pharmacol. Set.*,**13**,152–159.

Hartig, P.R., Hoyer, D., Humphrey, P.P.A. and Martin, G.R. (1996) Alignment of receptor nomenclature with the human genome: classification of 5-HT_{1B} and 5-HT_{1D} receptor subtypes. *Trends Pharmacol. Set.*,**17**,103–105.

Heupler, F.A., Proudfit, W.L., Razavi, M., Shirey, E.K., Greenstreet, R. and Sheldon, W.C. (1978) Ergonovine maleate provocative test for coronary arterial spasm. *Am.J. Cardiol.*,**41**,631–640.

Hieble, J.P., Bylund, D.B., Clarke, D.E., Eikenburg, D.C., Langer, S.Z., Lefkowitz, R.J., Minneman, K.P. and Ruffolo, R.R., Jr. (1995) Recommendation for nomenclature of α_1-adrenoceptors: consensus update. *Pharmacol. Rev.*,**47**,267–270.

Hieble, J.P., Bondinell, W.E. and Ruffolo, R.R., Jr. (1995) α- and β-Adrenoceptors: from the gene to the clinic. 1. Molecular biology and adrenoceptor subclassification. *J. Med. Chem.*,**38**,3415–3444.

Hollingsworth, M., Edwards, D. and Miller, M. (1988) Ergometrine—a partial agonist at 5-HT receptors in the uterus isolated from the oestrogen-primed rat. *Eur.J. Pharmacol.*, **158**,79–84.

Hoyer, D. (1988) Molecular pharmacology and biology of 5-HT_{1c} receptors. *TrendsPharmacol. Set.*,**9**,89–94.

Hoyer, D. (1989) 5-Hydroxytryptamine receptors and effector coupling mechanisms in peripheral tissues. In J.R.Fozard (ed.), *The Peripheral Actions of 5-Hydroxytryptamine,*Oxford University Press, Oxford, pp. 72–99.

Hoyer, D., Clarke, D.E., Fozard, J.R., Hartig, P.R., Martin, G.R., Mylecharane, E.J., Saxena, P.R. and Humphrey, P.P.A. (1994) Classification of receptors for 5-hydroxytryptamine (serotonin). *Pharmacol. Rev.,***46,**157–203.

Jin, H., Oksenberg, D., Ashkenazi, A., Peroutka, S.J., Duncan, A.M.V., Rozmahel, R., Yang, Y., Mengod, G., Palacios, J.M. and O'Dowd, B.F. (1992) Characterization of the human 5-hydroxytryptamine$_{1B}$ receptor. *J. Biol. Chem.,***267,**5735–5738.

Johnson, M.P., Audia, J.E., Nissen, J.S. and Nelson, D.L. (1993) N(1)-substituted ergolines and tryptamines show species differences for the agonist-labeled 5-HT$_2$ receptor. *Eur. J. Pharmacol.,***239,**111–118.

Johnson, M.P., Loncharich, R.J., Baez, M. and Nelson, D.L. (1994) Species variations in transmembrane region V of the 5-hydroxytryptamine type 2A receptor alter the structure-activity relationship of certain ergolines and tryptamines. *Mol. Pharmacol.,* **45,**277–286.

Kalkman, H.O. (1994) Is migraine prophylactic activity caused by 5-HT$_{2B}$ or 5-HT$_{2C}$ receptor blockade?*Life Set.,***54,**641–644.

Kao, H.-T., Adham, N., Olsen, M.A., Weinshank, R.L., Branchek, T.A. and Hartig, P.R. (1992) Site-directed mutagenesis of a single residue changes the binding properties of the serotonin 5-HT$_2$ receptor from a human to a rat pharmacology. *FEBS Lett.,***307,** 324–328.

Kaumann, A.J. (1989) The allosteric 5-HT$_2$ receptor system. In J.R.Fozard (ed.), *ThePeripheral Actions of 5-Hydroxytryptamine,*Oxford University Press, Oxford, pp. 45–71.

Kaumann, A.J., Parsons, A.A. and Brown, A.M. (1993) Human arterial constrictor serotonin receptors. *Cardiovasc. Res.,***21,**2094–2103.

Kaumann, A.J., Frenken, M., Posival, H. and Brown, A.M. (1994) Variable participation of 5-HT$_1$-like receptors and 5-HT$_2$ receptors in serotonin-induced contraction of human isolated coronary arteries. *Circulation,***90,**1141–1153.

Kebabian, J.W. and Calne, D.B. (1979) Multiple receptors for dopamine. *Nature (London),***277,**93–96.

Kennett, G.A. (1992) 5-HT$_{1C}$ receptor antagonists have anxiolytic-like actions in the rat social interaction test. *Psychopharmacology,***107,**397–402.

Kennett, G.A., Pittaway, K. and Blackburn, T.P. (1994) Evidence that 5-HT$_{2c}$ receptor antagonists are anxiolytic in the rat Geller-Seifter model of anxiety. *Psychopharmacology,***114,**90–96.

Kursar, J.D., Nelson, D.L., Wainscott, D.B., Cohen, M.L. and Baez, M. (1992) Molecular cloning, functional expression, and pharmacological characterization of a novel serotonin receptor (5-hydroxytryptamine$_{2F}$) from rat stomach fundus. *Mol. Pharmacol.,* **42,**549–557.

Levy, P.O., Gudermann, T., Perez-Reyes, E., Birnbaumer, M., Kaumann, A.J. and Birnbaumer, L. (1992) Molecular cloning of a human serotonin receptor (S12) with a pharmacological profile resembling that of the 5-HT$_{1D}$ subtype. *J. Biol. Chem.,***267,** 7553–7562.

MacLennan, S.J. and Martin, G.R. (1990) Actions of non-peptide ergot alkaloids at 5-HT$_1$-like and 5-HT$_2$ receptors mediating vascular smooth muscle contraction. *Naunyn-Schmiedebergs Arch. Pharmacol.,***342,**120–129.

Marona-Lewicka, D. and Nichols, D.E. (1995) Complex stimulus properties of LSD: a drug discrimination study with α_2-adrenoceptor agonists and antagonists . *Psychopharmacology,***120,**384–391.

Martin, G.R. and Humphrey, P.P.A. (1994) Receptors for 5-hydroxytryptamine: current perspectives on classification and nomenclature. *Neuropharmacology,***33,**261–273.

Martin, G.R. (1994) Vascular receptors for 5-hydroxytryptamine: distribution, function and classification. *Pharmacol. Ther.,***62,**283–324.

Marzoni, G., Garbrecht, W.L., Fludzinski, P. and Cohen, M.L. (1987) 6-Methylergoline-8-carboxylic acid esters as serotonin antagonists: N'-substituent effects on $5HT_2$ receptor affinity. *J. Med. Chem.,***30,**1823–1826.

Maseri, A., Pesola, A., Marzilli, M., Serveri, S., Parodi, O., L'Abbate, A., Ballestra, A.M., Maltinti, G., De Nes, D.M. and Biagini, A. (1977) Coronary vasospasm in angina pectoris. *Lancet,*713–717.

Matthes, H., Boschert, U., Amlaiky, N., Grailhe, R., Plassat, J.-L., Muscatelli, F., Mattei, M.-G. and Hen, R. (1993) Mouse 5-hydroxytryptamine$_{5A}$ and 5-hydroxytryptamine$_{5B}$ receptors define a new family of serotonin receptors: cloning, functional expression, and chromosomal localization. *Mol. Pharmacol.,***43,**313–319.

McAllister, G., Charlesworth, A., Snodin, C., Beer, M.S., Noble, A.J., Middiemiss, D.N., Iversen, L.L. and Whiting, P. (1992) Molecular cloning of a serotonin receptor from human brain (5HT1E): a fifth SHT1-like subtype. *Proc. Natl. Acad. Set. USA,***89,** 5517–5521.

McBride, P.A., Mann, J.J., Nimchinsky, E. and Cohen, M.L. (1990) Inhibition of serotonin-amplified human platelet aggregation by ketanserin, ritanserin, and the ergoline $5\text{-}HT_2$ receptor antagonists LY53857, sergolexole, and LY237733. *Life Set.,***47,** 2089–2095.

McCall, R.B. and Clement, M.E. (1994) Role of serotonin$_{1A}$ and serotonin$_2$ receptors in the central regulation of the cardiovascular system. *Pharmacol. Rev.,***46,**231–243.

Megens, A.A.H.P., Leysen, J.E., Awouters, F.H.L. and Niemegeers, C.J.E. (1986) Further validation of in vivo and in vitro pharmacological procedures for assessing the α_2/α_1-selectivity of test compounds: (1) α-adrenoceptor antagonists. *Eur. J. Pharmacol.,***129,** 49–55.

Meltzer, H.Y. and Nash, J.F. (1991) Effects of antipsychotic drugs on serotonin receptors. *Pharmacol. Rev.,***43,**587–604.

Mikkelsen, E., Pedersen, O.L., Østergaard, J.F. and Pedersen, S.E. (1981) Effects of ergotamine on isolated human vessels. *Arch. Int. Pharmacodyn.,***252,**241–252.

Milhahn, H.-C., Pertz, H. and Eich, E. (1993) Differential effects of low molecular ergolines at $5\text{-}HT_2$ receptors of rat tail artery. *Arch. Pharm. (Weinheim),***326,**P76.

Miller, K.J., King, A., Demchyshyn, L., Niznik, H. and Teitler, M. (1992) Agonist activity of sumatriptan and metergoline at the human $5\text{-}HT_{1D\beta}$ receptor.- further evidence for a role of the $5\text{-}HT_{1D}$ receptor in the action of sumatriptan. *Eur. J. Pharmacol. Mol.Pharmacol. Sect.,***227,**99–102.

Misner, J.W., Garbrecht, W.L., Marzoni, G., Whitten, K.R. and Cohen, M.L. (1990) (80)6-Methylergoline amide derivatives as serotonin antagonists: N^1-substituent effects on vascular $5HT_2$ receptor activity. *J. Med. Chem.,***33,**652–656.

Molderings, G.J., Frölich, D., Likungu, J. and Göthert, M. (1996) Inhibition of noradrenaline release via presynaptic $5\text{-}HT_{1D\alpha}$ receptors in human atrium. *NaunynSchmiedebergs Arch. Pharmacol.,***353,**272–280.

Monsma, F.J., Jr., Shen, Y., Ward, R.P., Hamblin, M.W. and Sibley, D.R. (1993) Cloning and expression of a novel serotonin receptor with high affinity for tricyclic psychotropic drugs. *Mol. Pharmacol.,***43**,320–327.

Moskowitz, M.A. (1992) Neurogenic versus vascular mechanisms of sumatriptan and ergot alkaloids in migraine. *Trends Pharmacol. Sci.,***13**,307–311.

Müller-Schweinitzer, E. (1983) Vascular effects of ergot alkaloids: a study on human basilar arteries. *Gen. Pharmacol.,***14**,95–102.

Müller-Schweinitzer, E. (1992) Ergot alkaloids in migraine: Is the effect via 5-HT receptors? In J.Olesen, P.R.Saxena (eds.), *5-Hydroxytryptamine Mechanisms inPrimary Headaches,*Raven Press, New York, pp. 297–304.

Müller-Schweinitzer, E. and Weidmann, H. (1978) Basic pharmacological properties. In B.Berde, H.O.Schild (eds.), *Ergot Alkaloids and Related Compounds*(Handbook of experimental pharmacology, vol. 49), Springer, Berlin, pp. 87–232.

Müller-Schweinitzer, E. and Tapparelli, C. (1986) Methylergometrine, an active metabolite of methysergide. *Cephalalgia,***6**,35–41.

Nelson, D.L., Lucaites, V.L., Audia, J.E., Nissen, J.S. and Wainscott, D.B. (1993) Species differences in the pharmacology of the 5-hydroxytryptamine$_2$ receptor: structurally specific differentiation by ergolines and tryptamines. *J. Pharmacol. Exp. Ther.,***265**, 1272–1279.

Nickerson, M. (1949) The pharmacology of adrenergic blockade. *Pharmacol. Rev.,***1**, 27–101.

Oksenberg, D., Marsters, S.A., O'Dowd, B.F., Jin, H., Havlik, S., Peroutka, S.J. and Ashkenazi, A. (1992) A single amino-acid difference confers major pharmacological variation between human and rodent 5-HT$_{1B}$ receptors. *Nature (London),***360**,161–163.

Olesen, J., Thomsen, L.L. and Iversen, H. (1994) Nitric oxide is a key molecule in migraine and other vascular headaches. *Trends Pharmacol. Sci.,***15**,149–153.

O'Malley, K.L., Harmon, S., Tang, L. and Todd, R.D. (1992) The rat dopamine D$_4$ receptor: sequence, gene structure and demonstration of expression in the cardiovascular system. *New Biologist,***4**,137–146.

Pazos, A., Hoyer, D. and Palacios, J.M. (1984a) Mesulergine, a selective serotonin-2 ligand in the rat cortex, does not label these receptors in porcine and human cortex: evidence for species differences in brain serotonin-2 receptors. *Eur. J.Pharmacol.,***106**, 531–538.

Pazos, A., Hoyer, D. and Palacios, J.M. (1984b) The binding of serotonergic ligands to the porcine choroid plexus: characterization of a new type of serotonin recognition site. *Eur. J. Pharmacol.,***106**,539–546.

Peroutka, S.J. (1994) Pharmacological differentiation of human 5-HT$_{1B}$ and 5-HT$_{1D}$ receptors. *Biol. Signals,***3**,217–222.

Peroutka, S.J. (1995) 5-HT receptors: past, present and future. *Trends Neurosci.,***18**, 68–69.

Pertz, H. (1993) 5-Hydroxytryptamine (5-HT) contracts the guinea-pig isolated iliac artery via 5-HT$_1$-like and 5-HT$_2$ receptors. *Naunyn-Schmiedeberg's Arch. Pharmacol.,* **348**,558–565.

Pertz, H. (1996) Naturally occurring clavines: antagonism/partial agonism at 5-HT$_{2A}$ receptors and antagonism at α$_1$-adrenoceptors in blood vessels. *Planta Med.,***62**, 387–392.

Pertz, H. and Eich, E. (1992) O-Acylated lysergol and dihydrolysergol-I derivatives as competitive antagonists of 5-HT at 5-HT$_2$ receptors of rat tail artery. Allosteric modulation instead of pseudoirreversible inhibition. *Naunyn-Schmiedeberg's Arch.Pharmacol.*,**345**,394–401.

Pertz, H., Milhahn, H.-C. and Eich, E. (1995) The parent drug of ergoline reverse esters, 1-isopropylelymoclavine, is a potent 5-HT$_{2A}$ antagonist in rat tail artery. *Naunyn-Schmiedeberg's Arch. Pharmacol.*,**351**,R143.

Pfeuffer-Friederich, I. and Kilbinger, H. (1985) The effects of LSD in the guinea pig ileum. Inhibition of acetylcholine release and stimulation of smooth muscle. *Naunyn-Schmiedeberg's Arch. Pharmacol.*,**331**,311–315.

Pierce, P.A. and Peroutka, S.J. (1990) Antagonist properties of *d*-LSD at 5-hydroxytryptamine$_2$ receptors. *Neuropharmacology,***3**,503–508.

Pontiroli, A.E., Cammelli, L., Baroldi, P. and Pozza, G. (1987) Inhibition of basal and metoclopramide-induced prolactin release by cabergoline, an extremely long-acting dopaminergic drug. *J. Clin. Endocrinol. Metab.*,**65**,1057–1059.

Prinzmetal, M., Kannamer, R., Merliss, R., Wada, T. and Bor, N. (1959) Angina pectoris I. A variant form of angina pectoris. Preliminary report. *Am. J. Med.*,**27**,375–388.

Pulvirenti, L. and Koob, G.F. (1994) Dopamine receptor agonists, partial agonists and psychostimulant addiction. *Trends Pharmacol. Sci.*,**15**,374–379.

Rebeck, G.W., Maynard, K.I., Hyman, B.T. and Moskowitz, M.A. (1994) Selective 5-HT$_{1D\alpha}$ serotonin receptor gene expression in trigeminal ganglia: implications for antimigraine drug development. *Proc. Natl. Acad. Sci. USA,***91**,3666–3669.

Roos, A. and Glusa, E. (1998) J-HT-induced relaxation of porcine pulmonary arteries is mediated through endothelial 5-HT$_{2B}$ receptors. *Naunyn-Schmiedeberg's Arch.Pharmacol.*,**357**,R30.

Roquebert, J. and Grenié, B. (1986) α_2-Adrenergic agonist and α_1-adrenergic antagonist activity of ergotamine and dihydroergotamine in rats. *Arch. Int. Pharmacodyn.*,**284,** 30–37.

Roquebert, J. and Demichel, P. (1985) α-Adrenergic agonist and antagonist activity of dihydroergotoxine in rats. *J. Pharm. Pharmacol.*,**37**,415–420.

Roquebert, J., Demichel, P., Gomond, P. and Malek, A. (1983) Activity of dihydroergocristine at pre- and postsynaptic α-adrenoceptors in the isolated rat vas deferens. *J. Pharmacol.*,**14**,151–159.

Roquebert, J., Malek, A., Gomond, P. and Demichel, P. (1984) Effects of dihydroergocristine on blood pressure and activity at peripheral α-adrenoceptors in pithed rats. *Eur. J. Pharmacol.*,**97**,21–27.

Ruffolo, R.R., Jr., Cohen, M.L., Messick, K. and Horng, J.S. (1987) Alpha-2-adrenoceptor-mediated effects of pergolide. *Pharmacology,***35**,148–155.

Sanders-Bush, E., Burris, K.D. and Knoth, K. (1988) Lysergic acid diethylamide and 2, 5-dimethoxy-4-methylamphetamine are partial agonists at serotonin receptors linked to phosphoinositide hydrolysis. *J. Pharmacol. Exp. Ther.*,**246**,924–928.

Saxena, P.R. and Villalón, C.M. (1991) Cardiovascular effects of serotonin agonists and antagonists. *J. Cardiovasc. Pharmacol.*,**15**,S17-S34.

Schmidt, C.J., Sorensen, S.M., Kehne, J.H., Carr, A.A. and Palfreyman, M.G. (1995) The role of 5-HT$_{2A}$ receptors in antipsychotic activity. *Life Sci.*,**56**,2209–2222.

Schmuck, K., Ullmer, C., Engels, P. and Lübbert, H. (1994) Cloning and functional characterization of the human 5-HT$_{2B}$ serotonin receptor. *FEBS Lett.*,**342**,85–90.

Seeman, P. and Van Tol, H.H.M. (1994) Dopamine receptor pharmacology. *TrendsPharmacol. Sci.*,**15**,264–270.

Seiler, M.P., Hagenbach, A., Wüthrich, H.-J. and Markstein, R. (1991) *trans*-Hexahydroindolo[4, 3-*ab*]phenanthridines ("Benzergolines"), the first structural class of potent and selective dopamine D_1 receptor agonists lacking a catechol group. *J. Med. Chem.,***34,**303–307.

Seiler, M.P., Floersheim, P., Markstein, R. and Widmer, A. (1993) Structure-activity relationships in the *trans*-Hexahydroindolo[4, 3-*ab*]phenanthridine ("Benzergoline") series. 2. Resolution, absolute configuration, and dopaminergic activity of the selective D_1 agonist CY 208–243 and its implification for an "extended rotamer-based dopamine receptor model". *J. Med. Chem.,***36,**977–984.

Shen, Y., Monsma, F.J., Jr., Metcalf, M.A., Jose, P.A., Hamblin, M.W. and Sibley, D.R. (1993) Molecular cloning and expression of a 5-hydroxytryptamine$_7$ serotonin receptor subtype. *J. Biol. Chem.,***268,**18200–18204.

Sibley, D.R. and Monsma, F.J., Jr. (1992) Molecular biology of dopamine receptors. *Trends Pharmacol. Sci.,***13,**61–69.

Simonneaux, V., Ebadi, M. and Bylund, D.B. (1991) Identification and characterization of α_{2D}-adrenergic receptors in bovine pineal gland. *Mol. Pharmacol.,***40,**235–241.

Sokoloff, P., Martres, M.-P. and Schwartz, J.-C. (1993) La famille des récepteurs de la dopamine. *Medicine/Sciences,***9,**12–20.

Strange, P.G. (1991) Interesting times for dopamine receptors. *Trends Neurosci.,***14,** 43–45.

Strange, P.G. (1992) *Brain Biochemistry and Brain Disorders,*Oxford University Press, Oxford.

Strange, P.G. (1993) New insights into dopamine receptors in the central nervous system. *Neurochem. Int.,***22,**223–236.

Tfelt-Hansen, P. and Pedersen, H.R. (1992) Migraine prophylaxis with 5-HT$_2$ partial agonists and antagonists. In J. Olesen, P.R.Saxena. (eds.), *5-HydroxytryptamineMechanisms in Primary Headaches,*Raven Press, New York, pp. 305–310.

Titus, R.D., Kornfeld, E.G., Jones, N.D., Clemens, J.A., Smalstig, E.B., Fuller, R.W., Hahn, R.A., Hynes, M.D., Mason, N.R., Wong, D.T. and Foreman, M.M. (1983) Resolution and absolute configuration of an ergoline-related dopamine agonist, trans-4, 4a, 5, 6, 7, 8, 8a, 9-octahydro-5-propyl-*H* (or 2*H*)-pyrazolo[3, 4-g]quinoline. *J. Med. Chem.,***26,**1112–1116.

Traber, J. and Glaser, T. (1987) 5-HT$_{1A}$ receptor-related anxiolytics. *Trends Pharmacol.Set.,***8,**432–437.

Ullmer, C., Schmuck, K., Kalkman, H.O. and Lübbert, H. (1995) Expression of serotonin receptor mRNAs in blood vessels. *FEBS Lett.,***370,**215–221.

Vane, J.R. (1957) A sensitive method for the assay of 5-hydroxytryptamine. *Br. J.Pharmacol. Chemother.,***12,**344–349.

Vanhoutte, P.M. (1990) Vascular effects of serotonin and ischemia. *J. Cardiovasc.Pharmacol.,***16,**S15-S19.

Van Nueten, J.M., Janssen, P.A.J., van Beek, J., Xhonneux, R., Verbeuren, T.J. and Vanhoutte, P.M. (1981) Vascular effects of ketanserin (R41468), a novel antagonist of 5-HT$_2$ serotonergic receptors. *J. Pharmacol. Exp. Ther.,***218,**217–230.

Van Tol, H.H.M., Bunzow, J.R., Guan, H.C., Sunahara, R.K., Seeman, P., Niznik, H.B. and Civelli, O. (1991) Cloning of the gene for a human dopamine D_4 receptor with high affinity for the antipsychotic clozapine. *Nature (London),***350,**610–614.

Wainscott, D.B., Gohen, M.L., Schenck, K.W., Audia, J.E., Nissen, J.S., Baez, M., Kursar, J.D., Lucaites, V.L. and Nelson, D.L. (1993) Pharmacological characteristics of the newly cloned rat 5-hydroxytryptamine$_{2F}$ receptor. *Mol. Pharmacol.,***43,**419–426.

Wang, C.-D., Gallaher, T.K. and Shih, J.C. (1993) Site-directed mutagenesis of the serotonin 5-hydroxytryptamine$_2$ receptor: identification of amino acids necessary for ligand binding and receptor activation . *Mol. Pharmacol.,***43,**931–940.

Weinshank, R.L., Zgombick, J.M., Macchi, M.J., Branchek, T.A. and Hartig, P.R. (1992) Human serotonin 1D receptor is encoded by a subfamily of two distinct genes: 5-HT$_{1Da}$ and 5-HT$_{1D\beta}$. *Proc. Natl. Acad. Sci. USA,***89,**3630–3634.

Wilson, H.C., Goffman, W., Killam, A.L. and Cohen, M.L. (1991) LY53857, a 5HT$_2$ receptor antagonist, delays occlusion and inhibits platelet aggregation in a rabbit model of carotid artery occlusion. *Thromb. Haemost.,***66,**355–360.

Zgombick, J.M., Schechter, L.E., Macchi, M., Hartig, P.R., Branchek, T.A. and Weinshank, R.L. (1992) Human gene S31 encodes the pharmacologically defined 5-hydroxytryptamine$_{1E}$ receptor. *Mol. Pharmacol.,***42,**180–184.

15.
ANTIMICROBIAL AND ANTITUMOR EFFECTS OFERGOT ALKALOIDS AND THEIR DERIVATIVES

ECKART EICH and HEINZ PERTZ

Institut für Pharmazie II, Freie Universität Berlin,Koenigin-Luise-Str. 2+4, D-14195 Berlin (Dahlem),Germany

15.1.
INTRODUCTION

Ergot alkaloids and their derivatives are able to inhibit the growth of certain hormone-dependent tumors *via* the inhibition of prolactin release from the anterior pituitary gland (Cassady and Floss, 1977). The therapy of the prolactinoma, an adenoma of this gland, and other prolactin-producing tumors by e.g. bromokryptine or lisuride is of clinical relevance (Thorner *et al.,* 1980). The inhibition of prolactin release is mediated by stimulation of dopamine D_2-like receptors. However, there is a completely different mode of action which is responsible for antitumor properties of certain ergolines. The latter was detected in connection with the discovery of the antimicrobial activity of clavine-type ergot alkaloids.

15.2.
ANTIMICROBIAL EFFECTS

Investigations on the transformation of different ergot alkaloids by microorganisms in the early eighties led to the observation that the addition of agroclavine to cultures of *Streptomyces purpurascens* caused a pronounced growth inhibiting effect (Schwarz, 1978; Schwarz and Eich, 1983). The biogenetically closely related elymoclavine was less active whereas the lysergic acid amides ergometrine and ergotamine showed almost no influence on the growth of the bacteria. It was demonstrated that no essential metabolization of these clavines took place during incubation so that there was no doubt about the direct antibiotic activity of the two clavines (Schwarz and Eich, 1983; Eichberg, 1983). The effect of agroclavine could be increased by hydrogenation at the 8, 9 position to give festuclavine, another natural compound (Eich and Eichberg, 1982; Eichberg, 1983). Certain 1-alkyl and 6-alkyl-6-*nor* derivatives of festuclavine showed a further enhancement of this effect. This turned out to be also true for human pathogenic bacteria species. For example, 6-allyl-6-*nor*-festuclavine exhibited an inhibitory activity against e.g. *Staphylococcus aureus*

(MIC: 30 μg/ml) and 1-propyl-6-*nor*-festuclavine was active against e.g. *Escherichia coli* (MIC: 60 μg/ml). On the other hand, the natural compounds agroclavine and festuclavine showed only moderate inhibitory effects (MIC: 200 μg/ml) (Eich *et al.*, 1985a). More or less strong bacteriotoxic activity was also found for a series of clavine derivatives in mutagenicity studies with different strains of *Salmonella typhimurium* and one strain of *Escherichia coli* (Glatt *et al.*, 1987, 1992).

In contrast to the remarkable effects of certain clavine derivatives on the growth of bacteria, no comparable influence on fungi could be observed. Yeasts like *Saccharomyces uvarum* or the pathogenic *Candida albicans* as well as the mold *Blakeslea trispora* were not inhibited by agroclavine or its derivatives. Only a modest effect was shown for certain derivatives of festuclavine in experiments with *C. albicans* (MIC: 250–500 μg/ml) (Schwarz and Eich, 1984; Eich *et al.*, 1985a).

15.3.
CYTOSTATIC EFFECTS OF AGROCLAVINE AND FESTUCLAVINE

Using a L5178y mouse lymphoma cell system it was shown that agroclavine and festuclavine are potent cytostatic agents *in vitro* (Eich *et al.*, 1984a, b, c, 1986b). Their EC_{50} values of 6.3μM and 7.1 μM, respectively, were comparable in potency with the EC_{50} value of the therapeutically used cytostatic alkaloid camptothecin (7.2 μM). Surprisingly, other natural 6, 8-dimethylergolenes structurally related to agroclavine (lysergine, isolysergine, setoclavine, isosetoclavine) and other natural 6, 8-dimethylergolines structurally related to festuclavine (pyroclavine, costaclavine) were inactive (Faatz *et al.*, 1989, 1990; Faatz, 1991). Though elymoclavine had shown some bacteriotoxicity neither this alkaloid nor other natural 8-hydroxymethyl-6-methylergolenes (lysergol, isolysergol, penniclavine) including the 8-hydroxymethyl-6-methylergoline dihydrolysergol-I showed cytostatic activity (Eich *et al.*, 1986a). This was also true for the two unusual clavines lysergene and chanoclavine-I as well as for lysergic acid derivatives like methylergometrine, lysergic acid amide, lysergic acid diethylamide (LSD), and isolysergic acid amide (Eich *et al.*, 1984a, 1997).

Thus, agroclavine and festuclavine possess unique biological activities as members of the ergot alkaloid family. This might be not only important for potential drugs of the future but also of relevance concerning aspects of chemical ecology (Eich, 1992). Since agroclavine, a precursor in the biosynthesis of lysergic acid derivatives in *Claviceps purpurea*, is the main alkaloid in the early stage of the sclerotia development, it may be useful as a protecting antimicrobial and cytostatic agent in this colourless, soft and wet mycelium phase.

15.4.
AGROCLAVINE AND FESTUCLAVINE AS LEADS FOR
THE DEVELOPMENT OF ANTITUMOR DRUGS

Agroclavine and festuclavine showed like other ergot alkaloids interactions with α-adrenoceptors, 5-HT receptors, and DA receptors as agonists, antagonists, or partial agonists, respectively (Fuxe *et al.*, 1978; Pertz, 1996). Moreover, festuclavine exhibited direct mutagenicity in the Ames test as well as agroclavine in the presence of a liver xenobiotic metabolizing system (Glatt *et al.*, 1992).

For the development of clavines as potential antitumor agents four main aims were of particular importance: (1) Variations of the ergoline nucleus and structure-activity relationship studies in order to find compounds with increased cytostatic potency *in vitro*. (2) Development of active compounds with enhanced metabolic stability since ergolines usually are metabolized very quickly *in vivo* (pronounced first pass effect). (3) Dissociation between antineoplastic and mutagenic activities presumed the latter is not the mechanism of action for the first. (4) Decrease of their affinity for neurotransmitter receptors mentioned above on the assumption that the cytostatic effect is not

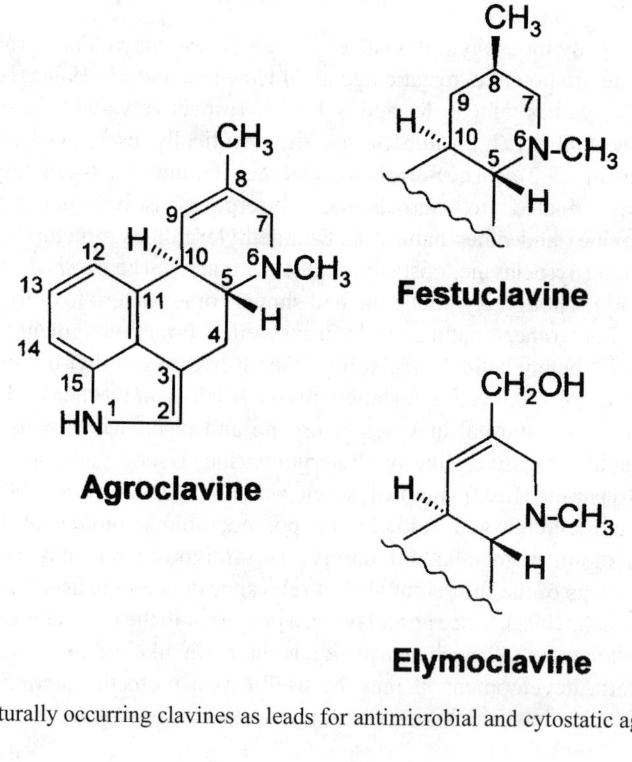

Figure 1 Naturally occurring clavines as leads for antimicrobial and cytostatic agents

based on the interaction with any of these receptors. Otherwise cytostatic
clavines could not be suitable for the therapy of cancer diseases because of
severe adverse effects.

15.5.
INFLUENCE OF DIFFERENT SUBSTITUTIONS ON THE
CYTOSTATIC ACTIVITY OF CLAVINES *IN VITRO*

Substitution at N-1: 1-Alkylation (Eich *et al.,* 1985b) significantly enhanced the
cystostatic effect of agroclavine and festuclavine, respectively. Those
homologues which bear a three to six carbon atoms membered straight-chain
substituent at N-l were especially effective with EC_{50} values between 0.87μM (1-
pentylagroclavine) and 1.6 μM (1-hexylagroclavine) (Eich *et al.,* 1986b). The
cytostatic potential of these clavine derivatives is rather high compared with
clinically used cytostatic antibiotics which reduce cell proliferation under
identical *in vitro* conditions at similar concentrations (EC_{50} of 1.0 μM for
bleomycin; EC_{50} of 1.2 μM for doxorubicin; EC_{50} of 1.9 μM for daunorubicin).
1-Pentylagroclavine was active in 27 different human and non-human cancer cell
systems of the National Cancer Institute, Bethesda/MD (USA) with IC50values
between 2.0 and 8.0 μM (Faatz *et al.,* 1989, 1990).

Surprisingly, 1-alkylation of elymoclavine caused potent inhibition of the cell
proliferation at an order of magnitude similar to that of the agroclavine series in
spite of the fact that the parent compound elymoclavine was inactive by itself
(Eich *et al.,* 1986a). Moreover, the acute toxicity (mouse, i.p.) was diminished
considerably by 1-alkylation of the clavines (Eich *et al.,* 1984b).

Since the n-alkyl substituent is metabolically not very stable, the cycloalkyl
and cycloalkylmethyl groups which are known to be more stable should be of
advantage. Such derivatives were also very active *in vitro,* e.g. 1-
cyclohexylfestuclavine (EC_{50}: 2.1μM) and its 1-cyclohexylmethyl analogue
(EC_{50}: 1.7 μM) (Kasper *et al.,* 1989; Kasper, 1991). Again corresponding
substitutions of the inactive pyroclavine afforded compounds of high potency
indicating the importance of the lipophilic substituent at N-1 (Faatz *et al.,* 1989;
Faatz, 1991).

Substitution at N-1 by *β*-ribofuranosyl or *β*-2-deoxyribofuranosyl residue
should create compounds analogous to nucleotides where more profound
cytostatic and/or antiviral effect can be expected. Series of *β*-ribosides (Křen *et al.,*
1997a, b) and *β*-deoxyribosides (Křen *et al.,* 1997c) were prepared from clavine
alkaloids (agroclavine, elymoclavine, lysergol, lysergene). The compounds were
tested against a large set of viruses including HIV. All compounds tested
displayed higher cytotoxicity towards the host cells and, therefore, their use as
antivirals is excluded. The compounds are now being tested for antineoplastic
activity and especially agroclavine ribosides show promising results.

Substitution at C-2: In order to avoid metabolic attack at the 2,3 double bond,
suitable substituents at C-2 should be useful since other ergolines like lisuride

show metabolic profiles which include oxidations at these positions (Hümpel *et al.,* 1984). However, different substitutions of festuclavine (e.g. -Cl, -Br, -I, -CH$_2$OH, -CHO, -SCH$_3$, -SC$_2$H$_5$) led to compounds with diminished activity or even to its total loss (Faatz, 1991; Milhahn, 1997). But there were some exceptions, e.g. 2-(2-nitro-phenyl)thiofestuclavine (EC$_{50}$: 4.4 μM) and 2-ethinylagroclavine (EC5$_0$: 3.0 μM) (Eich *et al.,* 1997).

Substitution at N-6: Removal of the methyl group at N-6 resulted in a marked loss of activity for agroclavine and festuclavine (Eich *et al.,* 1986b, c). Re-alkylation at N-6 with C$_2$- to C$_7$-membered straight-chain substituents led to compounds with similar potency as agroclavine itself. Interestingly, 1-alkylation of 6-*nor*agroclavine or 6-*nor*festuclavine revealed compounds with a potency which was almost equivalent to the 6-methylated analogues. The ring D opened 6-methyl-6,7-secoagroclavine was inactive (Eich *et al.,* 1997).

Substitution at C-7: The 7-oxo derivative of the inactive alkaloid lysergene was a potent cytostatic agent (Pertz *et al.,* 1989).

Substitution at C-10: Alkylation at C-10 seems to have a similar consequence for the extent of the cytostatic activity as alkylation at N-l, e.g. the EC$_{50}$ values of l0α-pentylagroclavine versus 1-pentylagroclavine were l,3 μM : 0,9 μM (Faatz *et al.,* 1990; Faatz, 1991).

Substitution at C-13: Bromination at C-13 should avoid the metabolic hydroxylation at this position as well as at the vicinal C-12 position. Thus, it was of advantage that 13-bromo-festuclavine was twice as active *in vitro* as festuclavine itself. Once more 1-alkyl and 1-cycloalkyl derivatives were even more active (Eich *et al.,* 1987; Pertz *et al.,* 1987).

15.6.
MECHANISM OF ACTION

In a first approach to determine the mode of action of the cytostatically active clavines, incorporation studies were performed (Eich *et al.,* 1984a). They revealed that the incorporation of [^{3}H]-thymidine into DNA is reduced in the presence of 1-propylfestuclavine. On the other hand, no influence has been determined for [^{3}H]-uridine incorporation into RNA and [^{3}H]-phenylalanine incorporation into protein. The inhibition of the thymidine incorporation rate was only observed after a preincubation period with the clavine derivative for 24 h and was not detectable after a pretreatment period of 2h. The assumption was favoured that clavines interfere with DNA replication processes in L5178y cells rather than an unspecific effect on cell growth. This conclusion was based on the observation that all active clavines caused "unbalanced growth", a property which they share with other cytostatic agents acting selectively by inhibition of DNA synthesis. Experiments with 1-propylelymoclavine revealed that at low concentrations again only the thymidine incorporation rate was significantly reduced. At higher concentrations of the cytostatic agent an additional decrease in the incorporation rates of uridine and phenylalanine was observed (Eich *et al.,*

1986a). From these results, it was concluded that 1-propylelymoclavine also inhibits primarily DNA synthesis by an unknown mechanism. The influence of 1-propylagroclavine on the synthesis of the macromolecules in L5178y cells was comparable to the corresponding festuclavine and elymoclavine analogues. Interestingly, the reduction of cell growth by 50% occurred at a concentration which was very close to that causing a 50% reduction of DNA synthesis. It has been clarified that this inhibition was not due to a direct effect on DNA polymerase α or β. Therefore, an interference with the state of organization of the nuclear reticulum was postulated (Eich *et al.*, 1986b).

The nucleoside uptake and incorporation into DNA and RNA in human lymphoid leukemia Molt 4B cells was also dose-dependently inhibited by higher concentrations of several growth depressing festuclavine derivatives (Hibasami *et al.*, 1990). The most powerful growth inhibiting derivative, 13-bromo-1-cyclopropylmethylfestuclavine, showed complete suppression of nucleoside uptake and incorporation into the macromolecules at a concentration of 50μM. These results were in favour of the suggestion that the inhibitory effects on nucleoside incorporation into DNA or RNA were merely due to inhibition of cellular uptake of nucleosides.

An obvious mechanism of action could involve the interaction with a neurotransmitter receptor since clavines interact with DA receptors, 5-HT receptors, and α-adronoceptors, respectively. But this is apparently not the case (Eich *et al.*, 1986b). The mechanism of action seems to be a fundamentally new one for ergoline compounds.

Agroclavine as well as its 1-propyl- and 1-pentyl derivative, respectively, are not directly mutagenic in the AMES test with different strains of *Salmonellatyphimurium* and one strain of *Escherichia coli* (Glatt *et al.*, 1987). Addition of a subcellular rat liver preparation forming a xenobiotic metabolizing system (S9) resulted in a substantial decrease of cytotoxicity and in the formation of mutagens. The nature of the mutagenic metabolites is unknown. The agroclavine skeleton offers a number of possible sites for oxidative transformations by liver preparations. It is unlikely that a genotoxic mechanism is involved in the antineoplastic activity of the clavines since L5178y lymphoma cells in culture express extremely low levels of cytochrome P450, which are the major active enzymes contained in the liver S9 preparation. This notion is supported by the lack of a correlation between the mutagenic activities in *S. typhimurium* and the cytostatic activities in L5178y cells of a series of clavine derivatives. In addition, there was no correlation between the cytostatic activities of these compounds in L5178y cells and their bacteriotoxic activities in *S. typhimurium* (Glatt *et al.*, 1992). It is not clear therefore whether cytostatic and bacteriotoxic activities are due to independent mechanisms or are mediated by orthologous receptors differing in their affinity spectrum for ligands. The dissociation between cytostatic, bacteriotoxic and mutagenic activities suggests that it may be possible to develop derivatives that are cytostatic but not mutagenic and antimicrobial.

15.7.
ONCOSTATIC ACTIVITY *IN VIVO*

For the development of clavines as potential antitumor agents four main aims were postulated above. The development from agroclavine and its highly active 1-pentyl derivative, both inactive *in vivo,* led via numerous derivatives to 13-bromo-l-cyclopropylmethylfestuclavine and the corresponding 13-bromol-cyclopentyl derivative which showed pronounced antineoplastic activity *invivo* (Eich *et al.,* 1987; Pertz, *et al.,* 1987; Faatz *et al.,* 1989; Hibasami *et al.,* 1990; Kasper, 1991; Glatt *et al.,* 1992). Treatment with 13-bromo-l-cyclopropylmethylfestuclavine increased the median life span (survival time) of mice (lymphoma L5178y; 50mg/kg) more than 2-fold (T/C: 212%; mean dead rate: 35.2 days versus 16.6 days). The T/C value of the 1-cyclopentyl analogue was 223%. However, these compounds showed mutagenicity as well as (an undesired) mitigated cytotoxicity in the Ames test in the presence of the hepatic xenobiotic metabolizing S9 system (Glatt *et al.,* 1992). It was assumed that this could be due to the metabolic formation of certain oxygen species at C-2/C-3 (Milhahn *et al.,* 1993). The 2-methylthio derivative of 13-bromo-l-cyclopropylmethylfestuclavine which should prevent such a metabolization turned out to possess an equivalent cytostatic activity *in vitro* (Milhahn, 1997). Further studies will show if this is also the case *in vivo.* Moreover, an additional metabolic stabilization at N-6 could be of advantage. Such compounds should be promising candidates for *in vivo* experiments.

REFERENCES

Cassady, J.M. and Floss, H.G. (1977) Ergolines as Potential Prolactin and Mammary Tumor Inhibitors. *Lloydia (J. Nat. Prod.),***40,**90–106.

Eich, E. (1992) Ergolin-Derivate—"schmutzige", spezifische und selektive Arzneistoffe. *Pharmaz. Ztg.,***137,**1601–1614.

Eich, E. and Eichberg, D. (1982) Zur antibakteriellen Wirkung von Clavinalkaloiden und deren partialsynthetischen Derivaten. *Planta Med.,***45,**146.

Eich, E., Eichberg, D. and Müller, W.E.G. (1984a) Clavines—new antibiotics with cytostatic activity. *Biochem. Pharmacol.,***33,**523–526.

Eich, E., Decker, C. and Müller, W.E.G. (1984h) Zur zytostatischen Wirkung partialsynthetischer Clavine. *Pharmaz. Ztg.,***129,**2209–2210.

Eich, E., Becker, C. and Müller, W.E.G. (1984c) Zur zytostatischen Wirkung von Festuclavin und seinen Derivaten. *Farmac. Tijdschr. Belg.,***61,**267.

Eich, E., Eichberg, D., Schwarz, G., Clas, F. and Loos, M. (1985a) Antimicrobial activity of clavines. *Arzneim. Forsch./Drug Res.,***35,**1760–1762.

Eich, E., Sieben, R. and Becker, C. (1985b) Partialsynthese neuer Ergolinderivate aus Clavinalkaloiden, 4: *N*-1, *C*-2 und *N*-6-monosubstituierter Agroclavine. *Arch. Pharm. (Weinheim),***318,**214–218.

Eich, E., Becker, C., Mayer, K., Maidhof, A. and Müller, W.E.G. (1986a) Clavines as antitumor agents, 2: Natural 8-hydroxymethyl-ergoline type clavines and their derivatives. *Planta Med.,***52,**290–294.

Eich, E., Becker, C., Sieben, R., Maidhof, A. and Müller, W E.G. (1986b) Clavines as antitumor agents, 3: Cytostatic activity and structure/activity relationships of 1-alkyl agroclavines and 6-alkyl 6-*nor*agroclavines. *J. Antibiot.,***39,**804–812.

Eich, E., Becker, C., Eichberg, D., Maidhof, A. and Müller, W.E.G. (1986c) Ciavines as antitumor agents, 5: *In vitro* activity of semisynthetic festuclavines. *Pharmazie,***41,** 156–157.

Eich, E., Pertz, H. and Müller, W.E.G. (1987) Neue Festuclavin-und Pyroclavin-Derivate, Verfahren zu ihrer Herstellung und deren Verwendung als Arzneimittel, Offenlegungsschrift Deutsches Patentamt, DE 3730124 (16. 3. 1989).

Eich, E., Pertz, H. and Müller, W.E.G. (1997) unpublished results.

Eichberg, D. (1983) Partialsynthese und Pharmakologie substituierter Agro- und Festuclavine. Dissertation Fachbereich Pharmazie, Universität Mainz/Germany.

Faatz, W. (1991) Struktur-Aktivitäts-Beziehungen natürlicher und partialsynthetischer Clavine vom 8-Methyl- und 8-Methylen-Ergolintyp im Hinblick auf deren zytostatische Wirkung, Dissertation Fachbereich Pharmazie, Freie Universität Berlin/Germany.

Faatz, W., Pertz, H., Maidhof, A., Müller, W.E.G. and Eich, E. (1989) Antitumor compounds derived from the ergot minor alkaloid pyroclavine. *Planta Med.,***55,** 653–654.

Faatz, W., Pertz, H., Maidhof, A., Müller, W.E.G. and Eich, E. (1990) Zytostatische Wirkung neuer partialsynthetischer Agroclavine. *Arch. Pharm. (Weinheim),***323,**762.

Fuxe, K., Fredholm, B.B., Agnati, L.F., Ögren, S.-O., Everitt, B.J., Jonsson, G. and Gustafsson, J.-Ä. (1978) Interaction of ergot drugs with central monoamine systems. *Pharmacology,***16**(Supp. 1): 99–134.

Glatt, H., Eich, E., Pertz, H., Becker, C. and Oesch, F. (1987) Mutagenicity experiments on agroclavines, new natural antineoplastic compounds. *Cancer Res.,***47,**1811–1814.

Glatt, H., Pertz, H., Kasper, R. and Eich, E. (1992) Clavine alkaloids and derivatives as mutagens detected in the AMES test. *Anti-Cancer Drugs,***3,**609–614.

Hibasami, H., Nakashima, K., Pertz, H., Kasper, R. and Eich, E. (1990) Inhibitory effects of novel festuclavine derivatives on nucleoside uptake and incorporation into DNA and RNA in human lymphoid leukemia Molt 4B cells. *Cancer Lett.,***50,**161–164.

Hümpel, M., Krause, W., Hoyer, G.-A., Wendt, H. and Pommerenke, G. (1984) The pharmacokinetics and biotransformation of ^{14}C-lisuride hydrogen maleate in rhesus monkey and in man. *Eur. J. Drug Metab. Pharmacokinet.,***9,**347–357.

Kasper, R. (1991) Zur Entwicklung onkostatisch wirksamer Festuclavin-Derivate, Dissertation Fachbereich Pharmazie, Freie Universität Berlin/Germany.

Kasper, R., Pertz, H., Maidhof, A., Müller, W.E.G. and Eich, E. (1989) Struktur-AktivitätsBeziehung zytostatisch wirkender 1-Alkylfestuclavin-Derivate. *Arch. Pharm. (Weinheim),***322,**748.

Kasper, R., Pertz, H., Glatt, H. and Eich, E. (1990) Metabolische Aspekte bei der Entwicklung onkostatisch wirksamer Festuclavin-Derivate. *Arch. Pharm.(Weinheim),* **323,**667.

Křen, V., Pískala, A., Sedmera, P., Havlíček, V., Přikrylová, V., Witvrouw, M. and De Clercq, E. (1997a) Synthesis and antiviral evaluation of *N-β*-D-ribosides of ergot alkaloids. *Nucleosides Nucleotides,***16,**97–106.

Křen, V. (1997b) Enzymatic and chemical glycosylations of ergot alkaloids and biological aspects of new compounds. *Topics Curr. Chem.,***186,**45–65.

Křen, V., Olšovský, P., Havlíček, V., Sedmera, P., Witvrouw, M. and De Clercq, E. (1997c) N-Deoxyribosides of ergot alkaloids: Synthesis and biological activity. Tetrahedron,53,4503–4510.

Milhahn, H.-Chr. (1997) Zytostatische und mutagene sowie α_1- und 5-HT_{2A}-rezeptorantagonistische Wirkungen neuer partialsynthetischer Clavinderivate, Dissertation Fachbereich Pharmazie, Freie Universität Berlin/Germany.

Milhahn, H.-Chr., Pertz, H., Steffen, R., Müller, W.E.G. and Eich, E. (1993) Contribution to the dissociation between antineoplastic and mutagenic activities of the ergot minor alkaloid festuclavine by substitution at C-21. Planta Med.,59,A683-A684.

Pertz, H. (1996) Naturally occurring clavines: antagonism/partial agonism at 5-HT_{2A} receptors and antagonism at α_1-adrenoceptors in blood vessels. Planta Meet.,62, 387–392.

Pertz, H., Maidhof, A., Müller, W.E.G. and Eich, E. (1987) Bromierte 1-Alkylfestuclavine mit Antitumor-Wirkung. Arch. Pharm. (Weinbeim),320,923.

Pertz, H., Maidhof, A., Müller, W.E.G. and Eich, E. (1989) Partialsynthese und zytostatische Aktivität von 9,10-Dihydro-7-oxolysergen-Derivaten. Arch. Pharm. (Weinbetm),322,747.

Schwarz, G. (1978) Einfluß von Agrociavin auf Wachstum und Sekundärstoffproduktion pharmazeutisch interessanter Mikroorganismen, Dissertation Fachbereich Pharmazie, Universität Mainz/Germany.

Schwarz, G. and Eich, E. (1983) Influence of ergot alkaloids on growth of Streptomycespurpurascens and production of its secondary metabolites. Planta Med., 47,212–214.

Schwarz, G. and Eich, E. (1984) Einfluß von Agrociavin auf Wachstum und Sekundärstoffproduktion von Saccharomyces uvarum und Blakeslea trispora.Pharmazie,39,572.

Thorner, M.O., Flückiger, E. and Calne, D.B. (1980) Bromocriptine, a Clinical and Pharmacological Review, Raven Press, New York.

16.
ROLE OF ERGOT ALKALOIDS IN THE IMMUNE SYSTEM

ANNA FIŠEROVÁ and MILOSLAV POSPÍŠIL

Department of Immunology and Gnotobiology, Institute ofMicrobiology Vídeňská 1083, 142 20 Prague 4, Czech Republic

16.1.
INTRODUCTION

The ergolines action on the immune system has been studied scarcely. Here, we will discuss the present state of knowledge on this field, and results concerning ergolines effect on certain subpopulations of lymphoid cells. The immunomodulatory effects of ergolines can be derived from the relationship between immunocytes and cells of the neuroendocrine compartment, the primary targets of ergoline compounds. Some of these data have been demonstrated in various *in vitro* and *in vivo* experimental systems. The recent research has been devoted predominantly to the immunosuppressive effect of dopaminergic and prolactin inhibitory ergopeptines. Studies from our department dealing with wide panel of ergolines (mostly prepared by Prof. Eich), and their direct action on lymphoid cells were provided by Šterzl *et al.* (1987)—modulation of B cells antibody production, and by Fišerová *et al.* (1991, 1995, 1996, 1997)— concerning modulation of cytotoxic (NK, T) cell effector functions. The ergolines cytostatic effect on tumour cells have been described in previous chapter by Eich and Pertz (Chapter 15). We will discuss here the effect of ergot alkaloids on tumour regression process from the immunological point of view, particularly the modulation of natural killer (NK) cell effector functions.

16.2.
CONTEMPORARY KNOWLEDGE OF THE TOPIC— CLINICAL CORRELATES

Since the first observations regarding the central dopaminergic action of substituted ergolines, a number of compounds from this group have been tested using different psychopharmacological and neurochemical assays. Some of them have been introduced into the clinical therapies. Ergolines can serve as a physiologic agonists and/or antagonists at the α-adrenoceptors, dopamine and serotonin receptors (see Chapter 14 of Pertz and Eich).

Some of the clavine type of alkaloids e.g. elymoclavine, chanoclavine, and their analogues influence the monoamine turnover in the brain (Petkov and Konstantinova, 1986). Chanoclavine and its derivatives, as well as bromokryptine stimulating the dopamine D_2R expression (Watanabe et al., 1987), are important for selective pharmacologic intervention in the treatment of various psychosis, hyperprolactinemia and PRL-producing tumours (Wu et al., 1996). On the other hand, effects of ergot alkaloids on vasoconstriction, metabolic influences or smoth muscle cells contraction are mediated by α-adrenoceptors (Abrass et al., 1985; Arnason et al., 1988).

Bromokryptine was the first clinically used central dopaminergic prolactin inhibitor from ergoline-related compounds (Del Pozo et al., 1972). Among the series of ergolines with amino group at the 8-α position the lisurid and tergurid showed considerable activity. The partial dopaminergic properties (D_2R agonist/antagonist) of tergurid, are effective in chronic treatment of Parkinson's disease (Mizokawa et al., 1993; Schulzer et al., 1992) and decreasing of extrapyramidal side effects of neuroleptic drugs (Filip et al., 1992). Tergurid is efficient also in the treatment of stress or drug induced immunodeficiences (Rašková et al., 1987). Tergurid's PRL inhibitory action (Valchar et al., 1989), is utilized in the treatment of hyperprolactinemia (Mizokawa et al., 1993), acromegaly (Dallabonzana et al., 1986) and PRLproducing tumours like prolactinomas or breast cancer (Graf et al., 1986; Anderson et al., 1993). Dihydroergokryptine exhibit similar activity in treatment of microprolactinomas (Faglia et al., 1987). The tumour growth is accompanied incidentally, by inefficient steroid negative feedback on PRL synthesis, downregulation of glucocorticoid receptors, and thus inhibition of the specific immune response. Bromokryptine treatment beside inhibition of PRL levels, induce also the restoration of glucocorticoid receptors (Piroli et al., 1993). The dopamine agonists, except PRL producing tumours, may have an application in the treatment of B cell derived lymphoproliferative disorders arrested in an early differentiation stage (Braesch-Andersen et al.,

---

List of abbreviations: 5-HT—5-hydroxytryptamine (serotonin), CD—cluster of differentiation, CsA—cyclosporin A, CTL—cytotoxic T lymphocyte, cAMP—cyclic adenosin monophosphate, D_1R—dopamine D_1-type receptor, D_2R—dopamine D_2-type receptor, DTH—delayed type of hypersensitivity, GalNAc—N-acetyl-galactosamine, GlcNAc—N-acetyl-glucosamine, GvHR—graft versus host reaction, G-protein—guanine nucleotide binding protein, IFN—interferon, Ig—immunoglobulin, IL—interleukin, LAK—lymphokine activated killer, LPS—lipopolysaccharide, MAPK—mitogen-activated protein kinase, MHC—major histocompatibility complex, MLC—mixed lymphocyte culture, NK—natural killer, NKR-P1—natural killer rat protein 1, NO—nitric oxide, PBMC—peripheral blood mononuclear cells, PHA—phytohaemagglutinin, PKA—protein kinase A, PRL—prolactin, PRL-R—prolactin receptor, SRBC—sheep red blood cells, TCR—T cell receptor, TNF—tumour necrosis factor.

1992). Our preliminary experiments showed also the effectiveness of liposome-encapsulated agroclavine for *in vivo* treatment of syngeneic mouse mastocytoma. Agroclavine delay the tumour incidence, prolonged the survival time of tumour bearing animals (in 75%), and liposome encapsulation eliminate the extrapyramidal side effects (Fišerová *et al.,* 1996a).

16.3.

THE IMMUNE AND NEUROENDOCRINE SYSTEMS COMMUNICATION

16.3.1.

Basic Principles of the Immune Response

The cells of the immune system as a whole are responsible for two different, but interrelated forms of immunity. The nonadaptive, "innate" immunity, including monocytes, macrophages, granulocytes and NK cells, is fully functional before infectious agents or foreign antigens enter the body, and in some cases can be sufficient to destroy the pathogens by phagocytosis or soluble mediators (lymphokines) involved in inflammation. NK cells form a distinct lineage of lymphocytes, that kill a variety of targets including infected or malignantly transformed cells. The adaptive (aquired) immunity is responsible for the immunological memory, specificity and diversity of responses mediated by T and B lymphocytes.

16.3.2.

The Regulatory Role of Biogenic Amines in Neuroendocrine Immune Axis

A variety of immune system products (cytokines, peptides) that function to coordinate the immune response may provide important signals for the neuroendocrine system. Those, currently known to have the most relevance for the neuroendocrine system are interleukins (IL) IL-1, IL-2, IL-6, tumour necrosis factors (TNF) and the interferons (IFN) their major role lies in the physiology of inflammation (infection, tumour growth). Lymphokines enhance the release of dopamine and norepinephrine from the adrenal medulla, the dopamine turnover in the brain and serve as endogeneous neurokines regulating striatal dopaminergic function (Zalcman *et al.,* 1994). Similarly, the dopamine is involved in the modulation of both TNFα and nitric oxide production by peritoneal macrophages, IL-6 release and inhibition of the TNFα synthesized by adrenal cells (Ritchie *et al.,* 1996).

On the other hand, mediators of the neuroendocrine system appear to facilitate or inhibit the functions of immune cells. It is recently well established that the balance between the pituitary prolactin and adrenal glucocorticoids regulate the immune response. The growth stimulatory PRL is a counterregulatory hormone to the anti-inflammatory glucocorticoids, which prevent the immune system overactivation. Their secretion is controled by the hypothalamic releasing or inhibitory hormones underlying the central dopaminergic control (Sinha, 1995). Thus, the nervous, endocrine and immune systems contribute to the maintenance of homeostasis. The dialog underlying this complex network of communication consist of signal molecules (neurotransmitters, neuropeptides, hormones, lymphokines) and related receptors. These are often shared by the immune and neuroedocrine cells and constitutes a biochemical information circuit between and within these systems. The catecholamine synthesis by non-neural cells support the hypothesis of autocrine and paracrine loops in the regulation of lymphocyte activity (Berquist *et al.,* 1994). An impaired hormonal or neurotransmitter function evoke a number of immunopathological situations (immunodeficiency, autoimmunity, malignant cell transformation).

16.4.
ERGOT ALKALOIDS IN EXPERIMENTAL
IMMUNOLOGICAL STUDIES

The proliferation, cell-mediated cytotoxicity, and lymphokine production with respect to antitumour immune response represented our interest. The 30 natural or semisynthetic ergot alkaloids, and 30 glycosylated derivatives of ergolines have been screened. For the preparation of glycosylated derivatives, the dihydrolysergol and elymoclavine has been chosen (Křen *et al.,* 1996). These are the simplest structural representants of 9, 10-dihydroergolines and A-9, 10-ergolines used in the human therapy (e.g. nicergolin, pergolid, tergurid, dihydroergotamine, dihydroergotoxine).

The most important capability of properly functioning immune system is the clonal selection and subsequent proliferation of respective cells. Cell growth and activation in the immune system is regulated to a substantial degree by cytokine family of growth factors in correlation with neurophysiological, neurochemical, and neuroendocrine activities of the brain cells. Catecholamines produced by neuroendocrine or immune cells present in the inflammatory lesions, inhibit the lymphocyte proliferation and differentiation by triggering of apoptosis, and thus counteracting the chronicity of the disease (Josefson *et al.,* 1996). The effect of ergolines on cell proliferation is dependent on the activation (differentiation) state of individual cell subpopulations. In very low concentrations (10^{-7}M-10^{-10}M) ergot alkaloids increased the deprived cell mitotic frequency. In comparison to resting cells, previously activated lymphoblasts taken from IL-2, mitogens or allogeneic stimuli induced cultures, did not respond to the modulation by most of ergoline derivatives (Fišerová *et al.,* 1991). According to our results, synergistic effect was observed only on phytohaemagglutinin (PHA) induced proliferation of human peripheral blood mononuclear cells (PBMC) by dihydrolysergol and purified NK cells by agroclavine. This activation is due to induction of cytokines (IL-2, IFN-γ) release. In contrary, chanoclavine (D_2R agonist) exert a strong inhibitory effect on both IL-2 and PHA stimulated NK cells. The cell surface markers analysis showed that both clavines compete with IL-2 on IL-2 receptor, and thus block the proliferation of cells (Fišerová et al., 1995a, 1997).

16.4.1.
Modulation of Nonadaptive Immunity by Ergot Alkaloids

NK Cells as a Sensitive Targets for Ergot Alkaloids Action

NK cell-mediated cytotoxicity play an important role in the immunosurveillance against invanding pathogens control of tumour growth and regulation of the hematopoiesis in vivo (Phillips et al., 1992). High susceptibility to pituitary hormones and neurotransmitters, make them a suitable targets for ergolines immunomodulatory action. The ergot alkaloid interference with adrenergic, dopamine or serotonin receptors may influence also the immune responsiveness (Hellstrand and Hermodsson, 1987, 1989, 1990).

One of the major mechanisms by which NK activity is regulated in vivo, involves catecholamines released by the sympathetic nervous system, owing to the expression of both D_1 and D_2-type dopamine receptors on lymphocytes (Ovadia et al., 1987; Ricci et al., 1997). Dopaminergic regions in the brain is the effective site for modulation of immune parameters, partially the NK cell functions (Madden and Felten, 1995). Therefore lesions in the striatum or compression of the basal ganglia by brain tumours affect the proliferation of splenocytes and decrease NK cell activity (Imaya et al., 1988; Deleplanque et al., 1994). Central α-adrenergic involvement suppresses the NK cell activity, through the reduction of effector-target conjugates formation by cytolytic effector cells (Carr et al., 1994). The serotoninergic regulation of cell-cell communications between NK cells and monocytes, as well as, the IFN-γ production is mediated through 5-HT$_{1A}$-type receptors. However, the MHC class I antigen expressiveness is regulated via 5-HT$_2$-type receptors (Hellstrand and Hermodsson, 1993).

Natural killer cell activity is suppressed by several dopamine mixed D_1/D_2R or D_2R antagonists (thiothixine, fluphenazine, pimozide, haloperidol) in PRLindependent manner. In comparison, the bromokryptine decreases it via PRL reduction (Won et al., 1995; Nozaki et al., 1996). On the other hand, activation of dopamine (D_1R) neurotransmission has an immunoenhancing effect. The DiR specificity of ergoline compounds is very rare, however, agroclavine exhibit D_1R agonistic/antagonistic activity, as proved by inhibition of specific D_1R antagonist [^{3}H] SCH 23.390 binding (pIC$_{50}$–6.02) on rat striatal neurones. The detection of D_1R specifity of agroclavine on lymphoid cells in our experimental procedure was unsuccessful, even if the specific binding of

The essential role of PRL in NK cell proliferation, lymphokine activated [^{3}H]-agroclavine to lymphocytes was detected (Fišerová et al., 1996b). killer (LAK) cell maturation, IL-2R expression, induction of genes for growth factors, perforins and membrane associated recognition molecules have been described (Matera et al., 1996; Gilmour and Reich, 1994). Concurrently, reduction of NK cell-mediated cytotoxity during pathological hyperprolactinemia, is restored by bromocryptine therapy (Gerli et al., 1986).

The knowledge of the NK cell receptors having affinity to carbohydrates (e.g. NKR-P1 lectin), enable us the search for its ligands involved in cytotoxic effector function with help of glycosylated ergot alkaloids. The carbohydrate part of the ergoline glycosides corresponds to the most common ligands of NKR-P1 receptor expressed on rat NK cytotoxic effectors, represented by either N-acetyl-galactosamine (GalNAc) and N-acetyl-glucosamine (GlcNAc) monosacharides or naturally occurring oligosaccharides containing such terminal moieties (Bezouška et al., 1994a, b).

Using different concentrations of natural and semisynthetic ergolines in 4 hours ^{51}Cr-release human NK cell cytotoxicity assay, a dose-dependent effect was noted. When ergolines applied at high doses (10^{-6}M), inhibition of NK cell activity against sensitive target cells (K562) was observed, whereas at the low doses stimulatory ones. In contrary, when NK-resistant (RAJI) target cells were used, the most potent killing effect of PBMC was detected only in the presence of ergolines concentration from 10^{-6}M. Interestingly, the human PBMC maintained in 5-day cultures in the presence of either agroclavine, 10-α-methoxy-dihydro lysergol or nicergoline enhanced the cytotoxic potency of NK cells against K562 from 4 to 8-fold. However, when IL-2 activated lymphocytes were used, the stimulating effect of ergoline compounds have been found limited (Fišerová et al. in preparation).

Elymoclavine and dihydrolysergol profoundly changed the level of NK cell activity during in vitro assay. The cytotoxicity of human PBMC against NK-resistant (RAJI) target cells was stimulated predominantly in the presence of elymoclavine glycoconjugates, whereas the cytotoxicity against NK sensitive (K562) cells was stimulated in the presence of dihydro-lysergol glycoconjugates. On the contrary, the cytotoxic effector function of mouse splenocytes (Balb/c and Balb/c-nu/nu), is inhibited by glycoconjugates of elymoclavine and particularly by dihydrolysergol, especially in nude mice. It is noteworthy, that strong stimulatory effect of β-glucopyranosid and sialylated N-acetyl-lactosamine derivatives of elymoclavine in human PBMC is completely contradictory to their effects in mouse cell system (Fišerová et al., 1995b; Křen et al., 1996).

Our understanding, if the sensitivity to the immunomodulation by glycosylated ergolines is due to effector or target cells, another experimental system represented by rat NK cell line (RNK16) was used. The rat target cells of astrocyte origin (C-6 glioma), mouse T lymphoma (YAC1) and mouse mastocytoma (P815) were chosen. The highest NK activity in short term (0–4 hours) assay was shown against YAC1 target cells in the presence of GalNAc and GlcNAc derivatives of elymoclavine. The P815 and C-6 target cells were sensitive to killing capability of NK cells after prolonged incubation period (20 hours). This effect was elicited especially by GalNAc derivative of elymoclavine, that was in correlation with its binding specificity to NKR-P1 protein at 10^{-8}M concentration. Generally, we found a weak stimulatory effect of glycosylated ergolines on effector cells, but substantially promoted killing effect with preincubated YAC1 target cells was demonstrated. (Fišerová et al., 1997).

Ergolines Action on Macrophages

Dopamine receptors, present on peritoneal macrophages play an important role in lipopolysaccharide (LPS)-induced TNF-a and nitric oxide (NO) production. Pretreatment with D_2R agonists, bromocriptine or quinpirole, caused the blunting of both the TNF-α and NO responses to LPS injected intraperitoneally. Sulpiride, the D_2R antagonist, has an inhibitory action on both TNF-α and NO production by peritoneal macrophages. However, antagonists of D_1R inhibited only the LPS-induced NO release and did not alter the TNF-α levels (Hasko et al., 1996). These findings may be of clinical relevance, since most of the therapeutically used ergoline compounds posses effect on D_2R. The oxidative burst of macrophages was significantly stimulated also by agroclavine, elymoclavine and nicergoline (Šůla et al., 1991).

16.4.2.
Modulation of Adaptive Immunity by Ergot Alkaloids

Ergolines Action on T Lymphocyte Effector Functions

The neuroendocrine control of the thymus (as a source of T lymphocytes) appears to be extremly complex, with the apparent existence of intrathymic biological circuitry involving the in situ production of pituitary hormones and neuropeptides, as well as, the expression of their respective receptors by thymic cells. Thus, in addition to the classical endocrine pathway, paracrine and autocrine mechanisms are probably implicated in the generation of T cell repertoire. The germline rearrangements of the TCR (T cell receptor) are not completly random, and can be modulated by certain hormones (prolactin, estradiol), cytokines (IL-4) and also drugs like cyclosporine A or bromocriptine (Dardenne and Savino, 1994). Bromocriptine changes the developmental pattern of thymocytes by enhancement of differentiation (Thy1.2, CD4 antigens expression), and inhibition of proliferative rate (Russel et al., 1988; Aboussaouira et al., 1989).

The suppression of pituitary PRL release diminishes a variety of immune responses, including activation of resting T cells by mitogens, IL-2 or alloantigens, T cell-dependent primary IgM response and T cell-dependent macrophage activation. Bromocryptine suppresses all of the mentioned T lymphocyte functions through inhibition of PRL secretion, followed by blocking of IL-2, IFN-γ production and IL-2 receptor (CD25) expression (Bernton et al., 1988, 1992). Moreover, bromocryptine potenciates the immunosuppressive effect of cyclosporin A (CsA) on T cell proliferation and CD25 antigen expression (Bernton et al., 1992). The additive effect is done by two factors: (i) CsA compete with PRL for common binding site on the cell surface, and (ii) bromocryptine decrease the number of circulating PRL molecules (Hiestand et al., 1986; Vidaller et al., 1986; Morikawa et al., 1994). On the other hand, the percentage of PRL-R$^+$lymphocytes is not influenced by bromokryptine, since the PRL-R expression is regulated by other factors, than pituitary PRL level (Leite de Moraes et al., 1995). According to Blank et al. (1995), the immunosuppressive effect of bromocryptine may be added to induction of nonspecific suppressoric factor produced by CD8$^+$ T cells.

The ergolines with various carbohydrate moities strongly inhibit (at 10^{-11} M) the mitotic activity of resting and PHA-activated lymphocytes. Among them the /?-glucopyranosides of all tested ergolines elicited the most pronounced inhibition. These findings are in correlation with carbohydrate binding specificity of lectins important in regulation of cell growth and differentiation. The target population influenced by ergoline β-glucopyranosides are T cells as proved in comparative experiments with athymic nude mice, and their euthymic littermates (Fišerová, unpublished results).

The group of 30 natural and semisynthetic ergot alkaloids were assayed on T cell-dependent immune responses either in vitro (MLC, specific cytotoxic T cell activity—CTL) or in vivo (graft versus host reaction—GvHR). T lymphocyte response to alloantigens in MLC was inhibited only in the presence of ergopeptine alkaloids, α- and β-ergokryptine, nicergoline and lisuride (at 10^{-6}M). The natural and semisynthetic ergolines, or ergopeptines like ergometrine, ergotamine and terguride do not influence the MLC reactivity. The CTL lytic activity against stimulator (allogeneic) cells induced in the presence of ergoline compounds is enhanced by lysergol, dihydrolysergol and nicergoline, and inhibited by β-ergocryptine, ergometrine, ergocristine and nicergoline. GvHR do not show any significant changes, even if high doses (5–20 mg/kg) of ergolines were applied. More sensitive to immunomodulation by ergolines was the delayed type of hypersensitivity (DTH) reaction. The ergopeptine alkaloids—ergometrine, tergurid, lisurid, nicergoline and some of the natural and semisynthetic ergolines (lysergol, chanoclavine, 10α-methoxy-dihydrolysergol) showed significant immunosuppressive effect. Agroclavine, elymoclavine, and dihydrolysergol do not influence the DTH reaction (Šůla et al., 1991). There are contradictory results from the work of Urbanová et al. (1990), who found tergurid as an inducer of the DTH reaction in stressed calves, caused by changes in catecholamine receptor expression and release of neuroendocrine mediators during stress.

Ergolines Action on B Lymphocytes

The presence of D_2-like dopamine receptors on B lymphocytes (Santambrogio *et al.*, 1993), suggesting a functional role of dopamine in mitogen-induced B-cell activation and immunoglobulin synthesis. Experiments provided by KaňkováVančurová *et al.* (1989) have shown that polyclonal activation of splenocytes in the presence of ergot alkaloids (dihydroergocornine, dihydroergocristine, dihydro- α and β-ergocryptine), increased number of antibody producing cells.

Various neurotransmitters and related compounds (e.g. ergot alkaloids) acting in the central nervous system, can affect the B lymphocyte responsiveness and alter the antibody production against T cell-dependent antigens (SRBC—sheep red blood cells and other alloantigens). This is due to the expression of adrenergic (predominantly of β-type), dopamine (D_2R) and sertonin receptors by B cells (Santambrogio *et al.*, 1993; Smejkal-Jagar and Boranic, 1994; Genaro *et al.*, 1986). The conversion of L-DOPA (L-dihydroxyphenylalanine) to dopamine in the organism could selectively affect B lymphocyte effector functions to T-dependent antigens, whereas the antibody production induced by bacterial antigens remains unchanged (Boukhris *et al.*, 1987). Similarly, bromocryptine suppresses through dopamine agonistic, and dihydroergosine *via* serotonin antagonistic properties the immunoglobulin generation by activated B cells to SRBC (Morikawa *et al.*, 1993; Smejkal-Jagar *et al.*, 1993).

The screening of 33 natural and semisynthetic ergot alkaloid provided in Malbrook system for *in vitro* antibody production (to SRBC) was shown by Šterzl *et al.* (1987). Ergopeptines and alkaloids interacting with α-adrenoceptors on lymphocytes negatively affected the number of antibody producing cells. Compounds derived from lysergic acid were completely inactive in this respect. In contrary, the application of natural and semisynthetic ergolines *in vivo* to Balb/c mice showed only slight decrease of B lymphocyte antibody production in most of the tested ergolines. Marked inhibition were seen only in the IgG production by derivatives of lysergic acid—dihydrolysergol, 10-α-methoxy-dihydrolysergol, and lisurid (Šůla *et al.*, 1991). These findings demonstrate that ergolines can influence the B lymphocytes directly (*in vitro* experiments) or by means of neuroendocrine system (*in vivo* experiments) and both of these activities are inhibitory.

The recent investigations fulfil the major regulatory role of PRL also in autoimmune diseases. PRL serves as a co-mitogen of lymphocytes in pathogenesis of some rheumatic diseases e.g. systemic lupus erythematosus, rheumatoid arthritis (collagen type II arthritis), Sjögren's syndrome, Reiter's syndrome, and psoriatic arthritis. High PRL levels in these patients are effectively treated by bromokryptine, that is accompanied also by restoration of immune disturbancies (Gutierez *et al.*, 1994). Another autoimmunities like antiphospholipid syndrome and experimental allergic encephalomyelitis (adequate to multiple sclerosis in man), under PRL control, are also effectively improved by bromocryptine therapy (Shoenfeld, 1994).

16.4.3.
Lymphokine Production by T and NK Cells

The important effect of ergolines on humoral immune mechanisms concerns the induction of lymphokine production mediated by both T lymphocytes and NK cells. The liberation of IFN-γ release following mitogenic stimulation is significantly elevated after treatment of patients by lisurid (Baier and Poehlau, 1994; Poehlau *et al,* 1994).

Our initial experiments indicated, that ergoline compounds modulate in cooperation with other stimulators, as a second signal messengers (IL-2, lectins, alloantigens), the lymphokine production (IFN-γ, TNF-*β*, IL-2) of human PBMC, as well as NK cells. Elevated production of IFN-γ and TNF-*β* has been found in ergoline stimulated cultures in the presence of PHA. Significant enhancement of IFN-γ release was induced preferentialy by agroclavine, and TNF-*β* release by chanoclavine and ergometrine. Agroclavine, elymoclavine and ergometrine, markedly increased the IL-2 production in alloantigen (MLC) stimulated cultures (Fišerová *et al.,* 1995a).

16.5.
INTRACELLULAR EVENTS ASSOCIATED WITH
ERGOLINES ACTION

16.5.1.
Signal Transduction

The lymphoid cells possess membrane receptors for a variety of hormones and neurotransmitters that affect the DNA synthesis and proliferation through adenylate cyclase system (Remaury *et al.,* 1993). Neurotransmitter and peptide hormone receptors, the target structures for ergolines interaction with cell membrane, are coupled to intracellular effector enzymes by GTP-binding proteins (G-proteins). Intracellular induction of adenylate cyclase system by ergolines evoke the transduction of signal to the nucleus with changes in appropriate genes is accompanied by cell membrane receptors expression.

Involvement of PRL and dopamine in the signaling events of lymphoid, as well as, tumour cells is important, in relation to the dopaminergic and PRL inhibitory action of engaged ergolines. PRL can induce a variety of signaling pathways for transmission of the growth signal to cell nucleus, followed by changes of the cell shape and mitotic activity. The common feature, on the top of this cascade, is the activation of the *ras* oncogene. The p21ras-GTP activation by PRL coupled to mitogen-activated protein kinase (MAPK) (Das and Vonderhaar,

1996), accompany the growth and progression of tumour cells. The mechanisms involved in dopamine-mediated inhibition of PRL release and subsequent inhibition of tumour growth, are still not fully understood.

Dopamine receptors are coupled to G-protein α subunits associated with cAMP inhibition (D_2R) or activation (D_1R), and phospholipase C activity (D_1R). Stimulation of mitotic activity is associated with decrease in cAMP content, and the inhibition with cAMP increase. In addition, adenylate cyclase system is involved in protein kinase A (PKA) activation and subsequent PKA phosphorylation of Raf-1, that causes blockage in the Ras pathway (Marx, 1993). It means, that substances elevating cAMP levels in cells (forskolin, D_1R agonists e.g. agroclavine), suppresse *ras* oncogene, therefore can modulate also the carcinogenesis (Suh *et al.,* 1992). Moreover, D_1R is linked to cellular intermediate early gene systems *(c-fos)* producing a phosphoprotein (Fos), involved in the control of class I. MHC antigen expression during the early stages of differentiation. This fact is of great importance in relation to their role in the process of tumour recognition and lysis by cytotoxic (T, NK) cells, and slow the tumour promotion and metastatic spread (Kushtai *et al.,* 1990).

Involvement of Ergolines in Signaling Pathways of NK, T Cells

Biological support for the notion that ergolines participate in the signal transduction of NK cells, was provided by agroclavine. The engagement of appropriate NK cell membrane receptors by agroclavine cause an indirect enhancement of adenylate cyclase through inhibition of G-protein $\alpha i/1,2$subunit (Fišerová *et al.,* 1997). Agroclavine induced in purified population of NK cells the intracellular accumulation of Ca^{2+} and inositolphosphates production, suggesting the phospholipase C-Ca^{2+} pathway activation. Similar activities were not inducible in T cell population (Fišerová *et al.,* 1996b). The involvement of the adenylate cyclase system in the ergolines action was demonstrated by the reduction of both prolactin release and cAMP accumulation produced by dihydroergokryptine and dihydroergocristine (Fiore *et al.,* 1987).

16.5.2.
Nuclear Affinity of Clavines

Many informations concerning the affinity of some ergolines to the nuclear structures, have been obtained from the studies on molecular level. The interaction of lysergic acid diethylamide (LSD) with DNA, followed by chromosomal breaks, is well documented. Juranic *et al.* (1987) described the interaction of other lysergic acid derivatives—ergosine, ergosinine and dihydroergosine with calf thymus DNA. The intrinsic binding constants were higher for the interaction of these alkaloids with single stranded DNA, and is dependent on D-ring steric conformation of the ergoline molecule. Deoxyribose or deoxyribonucleotides are involved in this interaction. The preferential binding to the nuclear structures have been detected also in the case of agroclavine on the series of tumour cell lines and lymphocytes. However, the direct interaction with salmon sperm or calf thymus DNA was not detected (Fišerová et al., 1996b). Festuclavine derivatives, particularly the 13-bromo-1-cyclopropyl-methyl-festuclavine, inhibiting the DNA and RNA syntheses, are due to depressed transport of nucleosides into the human lymphoid leukemia Molt 4B cells (Hibasami et al., 1990).

16.6.
CONCLUSIONS

It is well known that ergot alkaloids and their derivatives exert diverse pharmacological effects in relation to small changes in their chemical structure. Ergoline compounds affect the neuroendocrine system by interaction with membrane receptors for number of biogenic amines. Such a direct interaction of ergolines with neurotransmitter receptors on lymphoid cells till has not been proved. However, we can hypothesise distinct mechanisms of ergot alkaloid interaction with lymphoid and tumour cells, based on indirect experimental findings: (i) expression of dopamine, serotonin or α-adrenoceptors on lymphocytes, that can be involved in action of ergopeptines; (ii) inhibition of signaling pathways through intracellular enzymes, (serin/threonin kinases-Ras-MAPK) leading to mitogenesis by activation of adenylate cyclase system (D_2R agonistic, PRL inhibitory ergolines); (iii) interaction with the nuclear structures or direct binding with DNA, include derivatives of agroclavine, festuclavine, and several ergopeptines—LSD, ergosine, ergosinine, dihydroergosine.

The presented findings showed different effect of ergolines on proliferation of resting and activated lymphoid cells which may be added to their prolactin release regulatory action. Contradictory effects of ergolines on cell-mediated (mostly stimulatory) and humoral (inhibitory) immune response probably involve different type of neuroendocrine regulation in these cell populations.

Generally, we can deduce that the functional activity of glycosylated derivatives of elymoclavine and dihydrolysergol is dependent on both the alkaloid and terminal saccharide part of the molecule. Nevertheless, the status of effector cells (resting activated or stressed), as well as, the sensitivity and origin of the target cells are involved. It could be supposed that ergot alkaloids elymoclavine and dihydrolysergol, play a quantitatively different role in the lytic processes. Increased killing capability of effector cells pretreated by elymoclavine prevails to its protective effect on targets, and glycosylation pronouced these effects. In contrary, dihydrolysergol enhance the susceptibility of pretreated target cells, and cut down the lytic capability of effectors. Again, glycosylation emphasize immunomodulative properties of ergoline molecule.

Thus, the crucial question arising from all these experiments, concerns the nature of receptors, signal transduction and mechanism(s) involved in effector-target interaction. Considering the cytotoxicity and related experiments (binding studies—not shown), appears that ergot alkaloids may influence not only effectors but also tumour targets. Under such a conditions, the mechanism is other, than that involved in direct triggering of effector cells. This is in agreement with the fact, that cytotoxic effect on C-6 targets occurs after 20 hours, i.e. later than on the lymphoid targets.

Using different assays, our studies have confirmed that as many drugs, involving ergot alkaloids in common clinical use, could have potential immunomodulatory properties due to central or peripheral dopaminergic action. Principal evidence leads to the regulation of prolactin release and modification of glucocorticosteroid effects on immune and other target tissues. Apparent involvement of ergot alkaloids in recognition and signaling events of lymphocytes is more complex than appreciated, likely due to the till limited informations.

ACKNOWLEDGEMENT

The immunomodulatory experiments were supported partially by Galena Pharmaceuticals Ltd. Opava (1990, 1991), and by grants No. 310/94/1542, No. 310/98/0347, No. 312/96/1260 of the Grant Agency of Czech Republic (1994–1997).

REFERENCES

Aboussaouira, T., Marie, C. and Idelman, S. (1989) Effect of bromocriptine and bromocriptine combined with prolactin on the proliferation of thymocytes in young rats. *C. R. Acad. Sci. III.,***309,**281–287.

Abrass, C.K., O'Connor, S.W., Scarpace, P.J. and Abrass, I.B. (1985) Characterization of the beta-adrenergic receptor of the rat peritoneal macrophage. *J. Immunol.,***135,** 1338–1341.

Anderson, E., Ferguson, J.E., Morten, H., Shalet, S.M., Robinson, E.L. and Howell, A. (1993) Serum immunoreactive and bioactive lactogenic hormones in advanced breast cancer patients treated with bromocriptine and octreotide. *Eur. J. Cancer.,***29A,** 209–217.

Arnason, B.G., Brown, M., Maselli, R., Karaszewski, J. and Reder, A. (1988) Blood lymphocyte beta-adrenergic receptors in multiple sclerosis. *Ann. N. Y. Acad. Set.,***540,** 585–588.

Baier, J.E. and Poehlau, D. (1994) Is alpha-methyldopa-type autoimmune haemolytic anemia mediated by interferon-gamma?*Ann. Hematol.,***69,**249–251.

Bergquist, J., Tarkowski, A., Ekman, R. and Ewing, A. (1994) Discovery of endogenous catecholamines in lymphocytes and evidence for catecholamine regulation of lymphocyte function via an autocrine loop. *Proc. Natl. Acad. Set. U.S.A.,***91,** 12912–12916.

Bernton, E., Bryant, H., Holaday, J. and Dave, J. (1992) Prolactin and prolactin secretagogues reverse immunosuppression in mice treated with cysteamine, glucocorticoids, or cyclosporin-A. *Brain. Behav. Immun.,***6,**394–408.

Bernton, E.W., Meltzer, M.S. and Holaday, J.W. (1988) Suppression of macrophage activation and T-lymphocyte function in hypoprolactinemic mice. *Science,*239, 401–404.

Bezouška, K., Vlahas, G., Horváth, O., Jinochová, G., Fišerová, A., Giorda, R., Chambers, W.H., Feizi, T. and Pospíšil, M. (1994a) Rat natural killer cell antigen, NKR-P1, related to C-type animal lectins is a carbohydrate-binding protein. *J. Biol.Chem.,*269, 16945–16952.

Bezouška, K., Yuen, C.T., O'Brien, J., Childs, R.A., Chai, W., Lawson, A.M., Drbal, K., Fišerová, A., Pospíšil, M. and Feizi, T. (1994b) Oligosaccharide ligands for NKR-P1 protein activate NK cells and cytotoxicity. *Nature,*372,150–157.

Blank, M., Krause, I., Buskila, D., Teitelbaum, D., Kopolovic, J., Afek, A., Goldberg, I. and Shoenfeld, Y. (1995) Bromocryptine immunomodulation of experimental SLE and primary antiphospholipid syndrome *via* induction of nonspecific T suppressor cells. *Cell. Immunol.,*162,114–122.

Boukhris, W., Kouassi, E., Descotes, J., Cordier, G. and Revillard, J.P. (1987) Impaired T-dependent immune response in L-dopa treated BALB/c mice. *J. Clin. Lab.Immunol.,* 23,185–189.

Braesch-Andersen, S.,Paulie, S. and Stamenkovic, I. (1992) Dopamine-induced lymphoma cell death by inhibition of hormone release. *Scand. J. Immunol.,*36, 547–553.

Carr, D.J., Mayo, S., Gebhardt, B.M. and Porter, J. (1994) Central alpha-adrenergic involvement in morphine-mediated suppression of splenic natural killer activity. *J. Neuroimmunol.,*53,53–63.

Dallabonzana, D., Liuzzi, A., Oppizzi, G., Cozzi, R., Verde, G., Chiodini, P., Rainer, E., Dorow, R. and Horowski, R. (1986) Chronic treatment of pathological hyperprolactinemia and acromegaly with the new ergot derivative terguride. *J. Clin. Endocrinol.Metab.,*63,1002–1007.

Dardenne, M. and Savino, W. (1994) Control of thymic physiology by peptidic hormones and neuropeptides. *Immunology Today,*15,518–523.

Das, R. and Vonderhaar, B.K. (1996) Involvement of SHC, GRB2, SOS and RAS in prolactin signal transduction in mammary epithelial cells. *Oncogene,*13,1139–1145.

Deleplanque, B., Vitiello, S., Le Moal, M. and Neveu, P.J. (1994) Modulation of immune reactivity by unilateral striatal and mesolimbic dopaminergic lesions. *Neurosci. Lett.,* 166,216–220.

del Pozo, E., Brun del Re, R., Varga L. and Friesen, H. (1972) The inhibition of PRL secretion in man by CB 154 (2-Br-α-ergocryptine). *J. Clin. Endocrinol. Metab.,*35, 768–770.

Faglia, G., Conti, A., Muratori, M., Togni, E., Travaglini, P., Zanotti, A. and Mailland, F. (1987) Dihydroergocriptine in management of microprolactinomas. *J. Clin.Endocrinol. Metab.,*65,779–784.

Filip, V., Maršálek, M., Hálková, E. and Karen, P. (1992) Treatment of extrapyramidal side effects with terguride. *Psychiatry Res.,*41,9–16.

Fiore, L., Scapagnini, U. and Canonico, P.L. (1987) Effect of dihydroergocriptine and dihydroergocristine on cyclic AMP accumulation and prolactin release in vitro: evidence for a dopaminomimetic action. *Horm. Res.,*25,171–177.

Fišerová, A., Hajduová, Z., Křen, V., Flieger, M. and Pospíšil, M. (1996a) Induction of antitumor NK cell mediated immunity by liposome encapsulated drugs. *Int. J.Oncology,*9,841.

Fišerová, A., Kovářů, H., Hajduová, Z., Mareš, V., Křen, V., Flieger, M. and Pospíšil, M. (1997) Neuroimmunomodulation of natural killer (NK) cells by ergot alkaloids. *Physiol. Research,* **46,** 119–125.

Fišerová, A., Kovářů, H., Starec, M., Křen, V., Flieger, M., Hajduová, Z. and Pospíšil, M. (1996b) The neuroimmunomodulatory and antineoplastic potential of dopaminergic agents (ergot alkaloids). *Neuroimmunomodulation,* **3,** 185.

Fišerová, A., Křen, V., Augé, C., Šíma, P. and Pospíšil, M. (1995b) Ergot alkaloid derivatives with immunomodulatory activities. *Proceed. Int. Conf. On Exp. Ther.Manip. Host Defence System,* Hradec Králové, 67–73.

Fišerová, A., Pospíšil, M., Šůla, K. and Flieger, M. (1991) Modulation of cell-mediated immunity by ergot alkaloids. *Proceedings 1st Symposium on Consuprene (Cyclosporin A),* Hradec nad Moravici, April 1991, 56–58.

Fišerová, A., Trinchieri, G., Chan, S., Bezouška, K., Flieger, M. and Pospíšil, M. (1995a) Ergot Alkaloids Induced Cell Proliferation, Cytotoxicity and Lymphokine Production. *Adv. Exp. Med. Biol.,* **371A,** 163–166.

Genaro, A.M. and Borda, E. (1989) Alloimmunization-induced changes in beta-adrenoceptor expression and cAMP on B lymphocytes. *Immunopharmacology,* **18,** 63–70.

Gerli, R., Rambotti, P., Nicoletti, I., Orlandi, S., Migliorati, G. and Riccardi, C. (1986) Reduced number of natural killer cells in patients with pathological hyperprolactinemia. *Clin. Exp. Immunol.,* **64,** 399–406.

Gilmour, K.C. and Reich, N.C. (1994) Receptor to nucleus signaling by prolactin and interleukin-2 via activation of latent DNA-binding factors. *Proc. Natl. Acad. Sci. USA,* **91,** 6850–6854.

Graf, K.J., Kohler, D., Horowski, R. and Dorow, R. (1986) Rapid regression of macro-prolactinomas by the new dopamine partial agonist terguride. *Acta Endocrinol.Copenhagen,* **111,** 460–466.

Gutierez, M.A., Anaya, J.M., Cabrera, G.E., Vindrola, O. and Espinoza, L.R. (1994) Prolactin, a link between the neuroendocrine and immune systems. Role in the pathogenesis of rheumatic diseases. *Rev. Rheum. Ed. Fr.,* **61,** 278–285.

Hasko, G., Szabo, C., Merkel, K., Bencsics, A., Zingarelli, B., Kvetan, V. and Vizi, E.S. (1996) Modulation of lipopolysaccharide-induced tumor necrosis factor-alpha and nitric oxide production by dopamine receptor agonists and antagonists in mice. *Immunol. Lett.,* **49,** 143–147.

Hellstrand, K. and Hermodsson, S.J. (1987) The role of serotonin in the regulation of natural killer cell cytotoxicity. *J. Immunol.,* **139,** 869–875.

Hellstrand, K. and Hermodsson, S. (1989) An immunopharmacological analysis of adrenaline-induced suppression of human natural killer cell cytotoxicity. *Int. Arch.Allergy. Appl. Immunol.,* **89,** 334–341.

Hellstrand, K. and Hermodsson, S.J. (1990) Enhancement of human natural killer cell cytotoxicity by serotonin: role of non-T/CD16+ NK cells, accessory monocytes, and 5-HT1A receptors. *Cell. Immunol.,* **127,** 199–214.

Hellstrand, K. and Hermodsson, S.J. (1993) Serotonergic 5-HT1A receptors regulate a cell contact-mediated interaction between natural killer cells and monocytes. *Scand. J. Immunol.,* **37,** 7–18.

Hibasami, H., Nakashima, K., Pertz, H., Kasper, R. and Eich, E. (1990) Inhibitory effects of novel festuclavine derivatives on nucleoside uptake and incorporation into DNA and RNA in human lymphoid leukemia Molt 4B cells. *Cancer. Lett.,* **50,** 61–64.

Hiestand, P.C., Mekler, P., Nordmann, R., Grieder, A. and Permmongkol, C. (1986) Prolactin as a modulator of lymphocyte responsiveness provides a possible mechanism of action for cyclosporine. *Proc. Natl. Acad. Sci. USA,***83,**2599–2603.

Imaya, H., Matsuura, H., Kudo, M. and Nakazava, S. (1988) Suppression of splenic natural killer cell activity in rats with brain tumors. *Neurosurgery,***23,**23–26.

Josefson, E., Bergquist, J., Ekman, R. and Tarkowski, A. (1996) Catecholamines are synthesized by mouse lymphocytes and regulates the function of these cells by induction of apoptosis . *Immunology,***88,**140–146.

Juranic, Z., Kidric, M., Juranic, I. and Petrovic, J. (1987) Interaction of several ergopeptines with calf thymus DNA in vitro. *Studia Biophysica,***120,**163–170.

Kaňková-Vančurová, M., Procházková, J. and Grossman, V. (1989) The effect of hydrogenated ergot alkaloids on immune functions. II. The primary antibody response and proliferative activity in mouse spleen cells in vitro. *Česká Farmakologie,***38,** 351–354.

Křen, V., Fišerová, A., Augé, C., Sedmera, P., Havlíček, V. and Šíma, P. (1996) Ergot alkaloid glycosides with immunomodulatory activities. *Bioorganic & MedicinalChemistry,***4,**869–878.

Kushtai, G., Feldman, M. and Eisenbach, L. (1990) *c-fos* transfection of 3LL tumor cells turns on MHC gene expression and consequently reduces their metastatic competence. *Int. J. Cancer,***45,**1131–1136.

Leite, de-Moraes, M.C., Touraine, P., Kelly, P.A., Kuttenn, F. and Dardenne, M. (1995) Prolactin receptor expression in lymphocytes from patients with hyperprolactinemia or acromegaly. *J. Endocrinol.,***147,**353–359.

Madden, K.S. and Felten, D.L. (1995) Experimental basis for neural-immune interactions. *Physiological Reviews,***75/1,**77–106.

Marx, J. (1993) Two major signal pathways linked. *Science,***262,**988–989.

Matera, L., Bellone, G., Lebrun, J.J., Kelly, P.A., Hooghe Peters, L.E., Francia Di Celle, P., Foa, R., Contarini, M., Avanzi, G. and Asnaghi, V. (1996) Role of prolactin in the in vitro development of interleukin-2-driven anti-tumoural lymphokine-activated killer cells. *Immunology,***89,**619–626.

Mizokawa, T., Akai, T., Nakata, Y., Yamaguchi, M., Nakagawa, H., Hasan, S., Rattig, K.J. and Wachtel, H. (1993) Terguride as a new anti-hyperprolactinemic agent: characterization in rats and dogs in comparison with bromocriptine. *Jpn. J. Pharmacol.,***63,**269–278.

Morikawa, K., Oseko, F. and Morikawa, S. (1993) Immunosuppressive property of bromocriptine on human B lymphocyte function in vitro. *Clin. Exp. Immunol.,***93,** 200–205.

Morikawa, K., Oseko, F. and Morikawa, S. (1994) Immunosuppressive activity of bromocryptine on human T lymphocyte function in vitro. *Clin. Exp. Immunol.,***95,** 514–518.

Nozaki, H., Hozumi, K, Nishimura, T. and Habu, S. (1996) Regulation of NK activity by the administration of bromocriptine in haloperidol-treated mice. *Brain. Behav.Immun.,* **10,**17–26.

Ovadia, H., Lubetzki Korn, I. and Abramsky, O. (1987) Dopamine receptors on isolated membranes of rat thymocytes. *Ann. N.Y. Acad. Set.,***496,**211–216.

Petkov, V.D. and Konstantinova, E. (1986) Effects of the ergot alkaloid elymoclavine on the level and turnover of biogenic monoamines in the rat brain. *Arch. Int.Pharmacodyn. Therapy,***281,**22–34.

Phillips, J.H., Hori, T., Nagler, A., Bhat, N., Spits, H. and Lanier, L.L. (1992) Ontogeny of human natural killer (NK) cells: Fetal NK cells mediate cytolytic function and express cytoplasmic CD3 epsilon, delta proteins . *J. Exp. Med.,***175**,1055–1066.

Piroli, G., Grillo, C., Ferrini, M., Diaz, Torga, G., Libertun, C. and De-Nicola, A.F. (1993) Restoration by bromocriptine of glucocorticoid receptors and glucocorticoid negative feedback on prolactin secretion in estrogen-induced pituitary tumors. *Neuroendocrinology,***58**,273–279.

Poehlau, D., Baier, J.E., Kovacs, S., Galatti, H., Suchy, I., Will, C., Schmutz, T., Neumann, H.A. and Przuntek, H. (1994) Is dopaminergic therapy immunologically rejuvinating? Increased interferon-gamma production with the dopaminergic agent lisuride. *Fortschr. Med.,***112**,174–176.

Rašková, H., Čelada, L., Lavický, J., Vaneček, J., Urbanová, Z., Křeček, J., Priborska, Z., Elis, J. and Krejčí, I. (1987) Pharmacological interventions to antagonize stressinduced immune consequencies. *Ann. N.Y. Acad. Sci.,***496**,436–446.

Remaury, A., Larrouy, D., Daviaud, D., Rouot, B. and Paris, H. (1993) Coupling of α-2-adrenergic receptor to the inhibitory G-protein G(i) and adenylate cyclase in HT29 cells. *Biochem. J.,***292**,283–288.

Ricci, A., Marietta, S., Greco, S. and Bisetti, A. (1997) Expression of dopamine receptors in immune organs and circulating immune cells. *Clin. Exp. Hypertens.,***19**,59–71.

Ritchie, P.K., Ashby, M., Knight, H.H. and Judd, A.M. (1996) Dopamine increases interleukin 6 release and inhibits tumor necrosis factor release from rat adrenal zona glomerulosa cells in vitro. *Eur. J. Endocrinol.,***134**,610–616.

Russell, D.H., Mills, K.T., Talamantes, F.J. and Bern, H.A. (1988) Neonatal administration of prolactin antiserum alters the developmental pattern of T- and B-lymphocytes in the thymus and spleen of BALB/c female mice. *Proc. Nat I. Acad.Sci. USA,***85**,7404–7407.

Santambrogio, L., Lipartiti, M., Bruni, A. and Dal-Toso, R. (1993) Dopamine receptors on human T- and B-lymphocytes. *J. Neuroimmunol.,***45**,113–119.

Schulzer, M., Mak, E. and Calne, D.B. (1992) The antiparkinson efficacy of deprenyl derives from transient improvement that is likely to be symptomatic. *Ann. Neurol.,***32**, 795–798.

Shoenfeld, Y. (1994) Immunosuppression of experimental systemic lupus erythematosus and antiphospholipid syndrome. *Transplant. Proc.,***26**,3211–3213.

Sinha, Y.N. (1995) Structural variants of prolactin: occurence and physiological significants. *Endocrine Reviews,***16**,354–369.

Smejkal-Jagar, L., Pivac, N., Boranic, M. and Pericic, D. (1993) Effect of ergot-alkaloid dihydroergosine on the immune reaction and plasma corticosterone in rats. *Biomed. Phannacother.,***47**,33–36.

Smejkal-Jagar, L. and Boranic, M. (1994) Serotonin, serotoninergic agents and their antagonists suppress humoral immune reaction in vitro. *Res. Exp. Med. Berl.,***194**, 297–304.

Spangelo, B.L. and Gorospe, W.C. (1995) Role of cytokines in neuroendocrine immune axis. *Frontiers in Neuroendocrinology,* **16**,1–22.

Šterzl, J., Řeháček, Z. and Cudlín, J. (1987) Regulation of the immune response by ergot alkaloids. *Czech. Med.,***10**,90–98.

Suh, B.S., Eisenbach, L. and Amsterdam, A. (1992) Adenosine 3',5'-monophosphate suppresses metastatic spread in nude mice of steroidogenic rat granulosa cells transformed by Simian virus-40 and Ha-ras oncogene. *Endocrinology* **131**,526–532.

Šůla, K., Fišerová, A. and Flieger, M. (1991) The preliminary results of ergot alkaloid role on the antibody production, delayed type of hypersensitivity, and oxidative burst of macrophages. *Proceedings 1st Symposium on Consuprene,*April 1991, Hradec nad Moravicí, 50–54.

Urbanová, Z., Rašková, H., Sladkovský, J., Treu, M. and Křeček, J. (1990) Metipranolol and tergurid induced enhancement of delayed hypersensitivity and sedative effect in stressed calves. *Act. Nerv. Super. Praha.,***32,**298–299.

Valchář, M., Dobrovský, K., Krejčí, I. and Dlabač, A. (1989) Comparison of (3H) terguride and (3H)-spiperone binding to dopamine receptors in the rat striatum. *Act. Nerv. Super. Praha.,***31,**77–79.

Vidaller, A., Llorente, L., Larrea, F., Mendez, J.P., Alcocer-Varela, J. and AlarconSegovia, D. (1986) T-cell dysregulation in patients with hyperprolactinemia: effect of bromocriptine treatment. *Clin. Immunol. Immunopathol.,***38,**337–343.

Watanabe, H., Somei, M., Sekihara, S., Nakagawa, K. and Yamada, F. (1987) Dopamine receptor stimulating effects of chanoclavine analogues, tricyclic ergot alkaloids, in the brain.*Jpn. J. Pharmacol.,***45,**501–506.

Won, S.J., Chuang, Y.C., Huang, W.T., Liu, H.S. and Lin, M.T. (1995) Suppression of natural killer cell activity in mouse spleen lymphocytes by several dopamine receptor antagonists. *Experientia,***51,**343–348.

Wu, H., Devi, R. and Malarkey, W.B. (1996) Expression and localization of prolactin messenger ribonucleic acid in the human immune system. *Endocrinology,***137,**349–353.

Zalcman, S., Green-Johnson, J.M., Murray, L., Nance, D.M., Dyck, D., Anisman, H. and Greenberg, A.H. (1994) Cytokine-specic central monoamine alterations induced by interleukin-1, -2 and -6. *Brain. Res.,***643,**40–49.

17.
TOXICOLOGY OF ERGOT ALKALOIDS IN AGRICULTURE

RICHARD A.SHELBY

Department of Plant Pathology, 209 Life Sciences,Auburn University, AL 36849, USA

Ergot is a two edged sword. The sclerotia of *Claviceps* may contain useful pharmaceutical compounds, but as young plant pathologists we were all taught in our introductory course about the impact of ergot on human health. We learned about the misery of "St. Anthony's Fire" and how these mycotoxins from a phytopathogenic fungus can directly impact our society. Like the Irish potato famine, and countless other epiphytotic disasters, this one has shaped the course of human destiny. Early historic records are descriptive in their accounts of the effects of eating ergot bodies in grain, but recent accounts have grown increasingly rare. This is due largely to our understanding of *Claviceps* as a plant pathogen. We can control the fungus with fungicides (McLaren, 1994; *Schultzet et al.,* 1993; Mielke, 1993) or resistant varieties (Pageau *et al.,* 1994) and remove sclerotia from grain with modern equipment. Ergot alkaloids, however, still find their way into our food supply, making the need for testing just as important as ever. Small numbers of sclerotia can still evade the grain cleaning process. In a 5-year survey of food products in Canada, Scott *et al.* (1992) found ergot alkaloids in 118/128 rye flour samples, with the highest amount being almost 4ppm total alkaloids. They found that other commodities were also contaminated: wheat flour was positive in 68/93 samples; bran 29/35; triticale flour 24/26; and baked rye products 52/114. The amounts found were generally insufficient to cause symptoms of ergotism in humans, but the potential is present if the diet is heavy in contaminated grain products. One documented case of human ergotism is described in Fuller (1968). Point-Espirit, France, was exposed to ergot in their bread, due to a local supply of flour which had not been thoroughly inspected. This factual account of the tragedy should be required reading for anyone who doubts the importance of thoroughly testing food products for traces of ergot alkaloids. Several people died, and many more suffered irreversible injuries, some from psychotic episodes similar to those produced by LSD-25.

The ergot alkaloids produced by *C. purpurea* depend to some extent on the host infected. Rye *(Secale cereale)* produces some of the highest amounts of ergot alkaloids, mainly ergocristine, ergotamine, and lesser amounts of ergosine, ergocornine, ergokryptine, and ergonovine (Scott *et al.,* 1992). Wheat *(Triticum aestivum)* ergot contains a similar distribution of alkaloids as does the hybrid of the two (Triticale). Barley, on the other hand, seems to produce more ergotamine than ergocristine (Porter *et al.,* 1987), while other grasses like fescue (infected with *Claviceps* only) produce more ergosine than is found on other host grasses. Of course the story becomes infinitely more complex if the full range of *Claviceps* species and hosts are considered.

17.1.
ERGOT ALKALOIDS IN ANIMAL FEED

Perhaps even more common are ergot alkaloids in animal feeds. We tend to take less care in screening animal feeds for ergot alkaloids, perhaps believing that animals are more resistant to their toxic effects, but this is not the case. A recent survey of ergotism in American livestock (Burfening, 1994) found 655 published reports of ergot in the period from 1970–1992. These included case studies of intoxication of mainly cattle, sheep, and swine. Reports of ergotism in livestock varies from year-to-year in the United States, probably due to varying rainfall and temperature, but in the grain producing areas of the Northern plains, it can be found in most years (Lamey *et al.,* 1982). Several syndromes are associated with the ingestion of ergot alkaloids by livestock.

Nervous ergotism In this type of ergotism, the animal "staggers" as if intoxicated. Later the animal may experience paralysis of the posterior limbs, followed by drowsiness, and/or convulsions. Extreme cases of this type of ergotism sometimes result in mortality. Horses, sheep, and even carnivores can experience this type of ergotism, but rarely cattle.

Gangrenous Ergotism Because some of the ergot alkaloids cause contraction of smooth muscle (vasoconstriction), capillary beds can have reduced blood flow. In cases of chronic exposure, this leads to diminished blood supply, degeneration of the tissues, and eventual gangrene and loss of the afflicted body part. Ears, tails, and often hooves are the target organs. These symptoms are almost identical to those in humans having the worst cases of exposure and in the throes of "St. Anthony's Fire". In animals, these symptoms are exaggerated by cold weather, when capillary circulation in the extremities is further restricted. This type of ergotism is most common in cattle. In several clinical cases, a total alkaloid load amounting to about 0.2% of the diet, caused these symptoms (Burfening, 1994).

Agalactia In dairy animals lactation is severely curtailed by exposure to ergot alkaloids. One documented endocrine effect is reduction in prolactine, which in turn reduces milk production. This malady can not only reduce milk production, but reduce the weaning weight of calves whose mothers were intoxicated by ergot, and in swine, can reduce the birth weight and survivability of piglets in the litter. Schneider *et al.* (1986) reported that 1% ergot in the diet of sows caused a piglet birth weight reduction of about half that of control sows, and 100% mortality of piglets by three days after farrowing, compared to 3% mortality in the control group (Table 1).

This effect can also be demonstrated on sheep reproduction (Table 2). In this experiment, 152 sheep were fed ergot during gestation.

Table 1 Effect of rye ergot on swine reproduction (from Schneider *et al.,* 1986)

Percent ergot in diet	0	0.3	1.0
Live births (%)	89	87	74
Mean birth weight (kg)	1.45	1.14	0.82
Survival day 1 (%)	100	96	79
Survival day 2 (%)	100	75	15
Survival day 3 (%)	97	64	0
Survival day 7 (%)	91	61	0
Survival day 21 (%)	86	61	0
Agalactia in sows (%)	0	33	100

Table 2 Effect of ergot on sheep reproduction (from Burfenicng, 1994)

Percent of ergot in the diet	0	0.1	0.5
Percent lambing	67	47	48
Lambs/ewe	1.02	0.78	0.87

Table 3 Effect of barley ergot on ADG of heifers (from Burfening, 1994)

Percent of ergot in the diet	0	0.1	0.2	0.4	0.8	1.6
ADG (kg/day)	0.83	0.85	0.74	0.73	0.70	0.55
Ergot intake (g)	0	1.14	2.04	4.99	8.17	12.7
Feed intake (kg)	6.36	6.50	6.25	6.13	5.91	10.1

Table 4 Effect of endc ophytic fungi on ADG i in steers (from Sch midt *et al.,* 1982)

	ADG (kg/d)	Feed intake (kg/d)	Feed/gain (kg/kg)
Seed, endophyte −	0.96	6.41	6.7
Seed, endophyte +	0.20	4.14	20.7
Hay, endophyte −	0.66	4.79	7.3
Hay, endophyte +	0.28	4.40	15.4

Reduced Gain Some livestock species, such as beef cattle, have reduced average daily gains (ADG) when ergot alkaloids are present in the diet. This has been demonstrated with both *Claviceps* (Table 3) and with endophytic fungi (Table 4).

This relatively new type of ergot toxicosis was first described as "fescue toxicosis", but other pasture grasses can be affected, too. It is caused by endophytic fungi in the genus *Neotyphodium* (formerly *Acremonium*) which are systemic symbionts in several important genera of pasture grasses, including *Festuca* and *Lolium*. White *et al.* (1993) listed 47 species of grasses in 12 genera in which they had observed endophytic fungi in herbarium specimens. This unusual group of fungi may share a common phylogeny with *Claviceps,*

(Glenn *et al.,* 1996) and indeed produce some of the ergot alkaloids, but in different combinations. Ergovaline is most often the ergot alkaloid observed in plants infected by endophytic fungi (TePaske *et al.,* 1993; Porter, 1995). Ergine is also produced, and lesser amounts of ergosine. The endophytic fungi produce much lower concentrations of these alkaloids than are found in sclerotia of *Claviceps.* It is also common for seed lots to have a simultaneous infection of both *Neotyphodium* and *Claviceps,* resulting in an even more toxic mixture.

Unlike *Claviceps,* which is a phytopathogen, these fungi confer enhanced survivability upon their hosts by mechanisms which are poorly understood at the present time. Certainly involved in this process is reduced feeding by herbivores of all classes, including insects which normally feed upon the grass host. The endophytic fungi generally reproduce vegetatively by spread of fungal mycelium in the seed of host. Like *Claviceps,* these fungi inhabit the floral portion of the host and invade the seed, but do not form sclerotia, or conidia *in vitro.* As the seed germinates, the endophyte resumes growth, colonizes the young host plant, and eventually moves up the bolting culm into the panicle where the life cycle is complete (Figure 1). They are spread exclusively in the viable seed of their host.

17.2.
ROUTES OF CONTAMINATION

When one considers the number of fungal species capable of producing ergot alkaloids and also the broad host range of *Claviceps* and the endophytic *Neotyphodium* species (Flieger *et al.,* 1997), it is hardly surprising that food and feed can become contaminated. The USA National Pest Information Service (NAPIS) database lists 457 records of *C. purpurea* in grain from the period 1984–1995. Almost all of the major grain crops are susceptible to a species of *Claviceps* (Table 5). Some of these diseases are rare, or limited in geographical distribution. *C. gigantea,* for example, is limited in distribution to a small area of Mexico (Fuentes *et al.,* 1964). This may be due to strict habitat requirements of the pathogen, but there have been no attempts to spread the pathogen (nor should there be). *C. purpurea,* on the other hand,

Table 5 Species of *Claviceps* found on important grain crops

Claviceps *species*	Host crops
C. africana	Sorghum
C. fusiformis	Pearl millet
C. gigantea	Maize
C. oryzae-sativae	Rice
C. sorghi	Sorghum
C. purpurea	Barley, Wheat, Rye, Oats, (50 + genera)

is cosmopolitan in distribution, and we may not have discovered its complete host and geographic range. It is by far the most common and troublesome

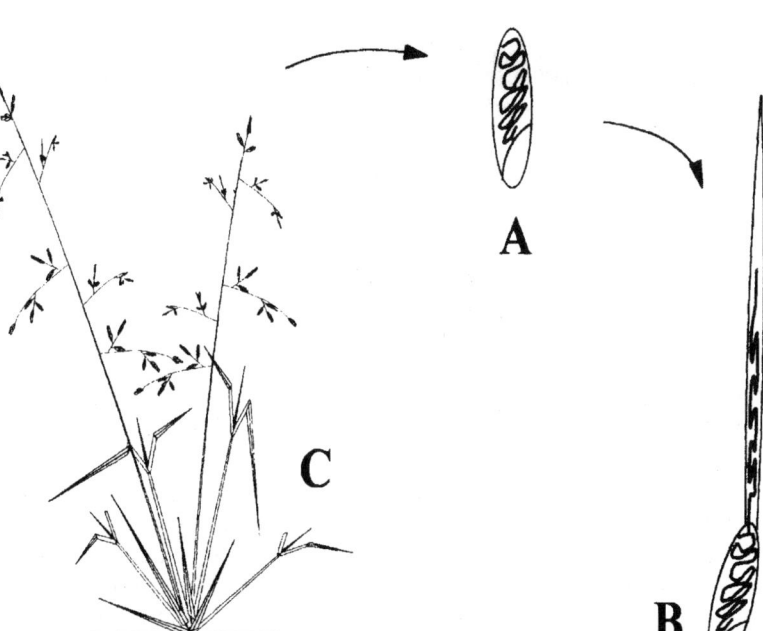

Figure 1 Life cycle of *Neotyphodium coenophialum,* the endophyte of tall fescue. **A** Viable seed of the grass host contains viable mycelium of the fungal endophyte. At this stage it may be quite toxic to herbivores. **B** The seed germinates and the fungus begins to grow intercellularly into the stem of the plant. **C** Stems of the mature plant are completely colonized. When the plant forms an inflorescence, mycelium of the endophyte moves up the culm, into the panicle, and eventually into maturing ovaries, where the cycle is complete

member of the genus. *C. africana* is a species that appears to be expanding into Japan, Thailand, South America, Australia, and the United States (Reis, *et al.,* 1996). Worldwide, at least 40 species of *Claviceps* have been described, but most are parasites of noncultivated grasses. (See also chapter 3 on *Claviceps* taxonomy in this book.)

Contamination of feed and food occurs by two routes: the grain crop itself is infected, or infected grasses growing as weeds in the crop are harvested with the grain and sclerotia become mixed with the grain. Barley, for example, can be contaminated by sclerotia from annual rye grass *(Lolium multiflorum)* which was a weed in the barley field. Similarly, grazing animals can ingest sclerotia from *Claviceps* infected grasses which are harvested with hay, or are ingested from the field in direct grazing common in the USA. Contamination of feeds can also result when infected grains are processed into pelleted feeds, under the mistaken impression that heating will denature the ergot alkaloids. As proven by Scott *et al.* (1992), heat denaturation by baking is not 100%, and some of the active compound may be present in processed feed. Fajardo *et al.* (1995) also found that ergot alkaloids survived not only the process of formation into pasta, but the cooking process as well.

17.3.
TESTING OF FOOD AND FEED

The obvious solution to preventing human and animal intoxication by ergot alkaloids is a thorough regimen of testing agricultural products before they are consumed. Fortunately, there are quite a few published methods designed to detect ergot alkaloids in agricultural commodities (see also chapter 10 Ergot alkaloid analytics by Jegorov in this book).

Thin-Layer Chromatography (TLC) This method has been used for many years to detect ergot alkaloids, and is still used due to its speed, simplicity, and relatively low cost. To use TLC for agricultural commodities, one need only to extract the ergot alkaloids with a suitable solvent, concentrate by evaporation, or solvent partition, spot the sample on a thin-layer plate, and develop in a suitable solvent. Normal phase ergot alkaloid TLC is spotted on silica plates and developed in chloroform : methanol solutions (Svendsen and Verpoorte, 1984). Plates can be viewed in UV light to visualize the fluorescent ergot alkaloids, or they can be sprayed with Ehrlich's reagent which is a positive test for the indole ring, common to most ergot alkaloids. While this method is simple, it does suffer from the problem of low sensitivity. Typically, this method is accurate in the microgram range of detection, and quantitation is by comparison with intensity of spots of control samples run on each plate. TLC works well for *Claviceps* broth cultures, or pure sclerotia, but in cases of low-level contamination of agricultural commodities, the interference of background materials may prevent a good chromatographic analysis.

High-Performance Liquid Chromatography (HPLC) This method takes advantage of the same basic chemistry as TLC: ergot alkaloids have a characteristic affinity for a solid, stationary phase relative to a mobile, liquid phase. The big advantages of HPLC are the rigidly controlled flow rates generated by mechanical pumps as opposed to capillary flow in TLC, and the improved accuracy and sensitivity of UV and fluorescence detectors. A recent review of HPLC methods can be found in Flieger *et al.* (1997). Most HPLC methods for ergot alkaloid analysis in food or feed samples utilize a preliminary cleanup by solvent partitioning (Scott and Lawrence, 1980) or silica columns (Rottinghaus *et al.,* 1993). With fluorescence detection, this method is sensitive in the nanogram range. Of the methods used in routine diagnostic work, HPLC provides the most accurate quantitative and qualitative analysis for ergot alkaloids in agricultural commodities.

Enzyme-Linked Immunoassay (ELISA) Immunological methods to detect ergot alkaloids are based on the ability of an antibody or immunoglobulin to bind with a specific target alkaloid. These antibodies are produced by injecting mice, rabbits, or other animals with the alkaloid which has been conjugated to a strongly antigenic carrier molecule such as bovine serum albumin (Schran *et al.*, 1979; Arens and Zenk, 1980; Shelby and Kelley, 1992). The immunoassay can be in several formats, but the competitive ELISA has largely replaced the radioimmunoassay (RIA). The use of an enzymemediated color change in a chromogenic substrate eliminates the need for radioisotopes. The specificity of the antibody can be manipulated somewhat by utilizing different conjugation chemistries, and thereby different binding sites on the alkaloid molecule. That portion of the alkaloid molecule distal to the point of conjugation to the carrier is the site recognized by the antibody. If that region is C-8 of the ergoline ring, then many more ergot alkaloids can be detected, because this region is common to most ergot alkaloids. If, on the other hand, N-1 is the point of conjugation, the antibody will be more specific, since the distal C-8 is the more variable region of ergot alkaloids. The ELISA methods have the advantage of being quicker and less expensive than other methods, with a level of sensitivity which is at least equal to the HPLC methods. However, they suffer from the problem of quantitative inaccuracy due to cross-reactivity of the antibody with non-target compounds in the matrix. These method are more suited to rapid, large scale screening of agricultural samples for the presence of ergot alkaloids. Detailed accurate quantitation could then be conducted using a more accurate method, such as HPLC.

17.4.
THE FUTURE OF ERGOT

Claviceps will always be around to make life interesting for plant pathologists. This pathogen has shown an ability to adapt to a variety of hosts, many of which form the staple crops essential for human survival. It is unlikely that in the foreseeable future this pathogen will be completely controlled by fungicides, resistant varieties, or phytosanitation. For this reason, it is essential for individuals in the food and feed industries to be kept aware of the potential problem posed by ergot in agricultural commodities. This is made more difficult by the rarity of outbreaks of ergot intoxication, and the lack of public awareness of the problem which could eventually lead to a dangerous situation. From a purely academic standpoint, the phylogenetic relationships between *Claviceps* species and the endophytic fungi are beginning to be understood by modern molecular techniques. Even after more than 100 years of research in ergot alkaloid chemistry we are still finding new compounds from ergot (Cvak *et al.*, 1995). Certainly the future of ergot and the ergot alkaloids will be brighter than the past.

REFERENCES

Arens, H. and Zenk, M. (1980) Radioimmunoassays for the determination of lysergic acid and simple lysergic acid derivatives. *Planta Medica,*39,336–347.

Burfening, P.J. (1994) Ergotism in domestic livestock. Montana Nutrition Conference Special Report SR-50. pp. 8.1–8.11.

Cvak, L., Minář, J., Pakhomova, S., Ondráček, J., Kratochvíl, B., Sedmera, P., Havlíček, V. and Jegorov, A. (1995) Ergoladinine, a new ergot alkaloid. *Photochemistry,*42, 231–233.

Fajardo, J.E., Dexter, J.E., Roscoe, M.M. and Norwiki, T.W. (1995) Retention of ergot alkaloids in wheat during processing. *Cereal Chem.,*72,291–298.

Flieger, M., Wurst, M. and Shelby, R.A. (1997) Ergot alkaloids—sources, structures, and analytical methods. *Folia Microbiol. (Prague),*42,3–30.

Fuentes, S.F., DeLa Isla, M., Ullstrup, A.J. and Rodriquez, A.E. (1964) *Clavicepsgigantea,* a new pathogen of maize in Mexico. *Phytopathology,*54,379–381.

Fuller, J.C. (1968) *The day of St. Anthony's fire.*MacMillan, New York.

Glenn, A.E., Bacon, C.W., Price, R. and Hanlin, R.T. (1996) Molecular phylogeny of Acremonium and its taxonomic implications. *Mycologia,*88,369–383.

Lamey, H.A., Schipper, I.A., Dinusson, W.E. and Johnson, L.J. (1982) *Ergot.*North Dakota State University Circular PP–551.

McLaren, N.W. (1994) Efficacy of systemic fungicides and timing of preventive sprays in the control of sugary disease of grain sorghum. *S. African J. of Plant and Soil,*11, 30–33.

Mielke, H. (1993) Investigations on the control of ergot. *Nachrichtenblatt desDeutschen Pflantzenschutzdienstes,*45,97–102.

Pageau, D., Collin, J. and Wauthy, J.M. (1994) Evaluation of barley cultivars for resistance to ergot fungus, C. purpurea.*Can. J. Plant Sci.,*74,663–665.

Porter, J.K. (1995) Analysis of endophyte toxins: fescue and other grasses toxic to livestock. *J. Anim. Sci.,*73,871–880.

Porter, J.K., Bacon, C.W., Plattner, R.D. and Arrendale, R.F. (1987) Ergot peptide alkaloid spectra of *Claviceps-infected* tall fescue, wheat and barley. *J. Agric. FoodChem.,*35,359–361.

Reis, E.M., Mantle, P.O. and Hassan, H.A.G. (1996) First report in the Americas of sorghum ergot disease, caused by a pathogen diagnosed as *Claviceps africana.Plant Dis.,*80,463.

Rottinghaus, G.E., Schultz, L.M., Ross, P.P. and Hill, N.S. (1993) An HPLC method for the detection of ergot in ground and pelleted foods. *J. Vet. Diag Invest.,*5,242–247.

Schmidt, S.P., Hoveland, C.S., Clark, E.M., Davis, N.D., Smith, A.L., Grimes, H.W. and Holliman, J.L. (1982) Association of an endophytic fungus with fescue toxicity in steers fed Kentucky 31 tall fescue seed or hay. *J. Anim. Sci.,*55,1259–1263.

Schneider, N., Wiernusz, M., Hogg, A. and Fritschen, R. (1986) Ergot and swine reproduction. *Nebraska Research Reports,*pp. 10–12.

Schran, H.F., Schwartz, H.J., Talbot, K.C. and Loeffler, L.J. (1979) Specific radioimmunoassay of ergot peptide alkaloids in plasma. *Clin. Chem.,*25,1928–1933.

Schultz, T.R., Johnston, W.J., Golob, C.T. and Maguire, J.D. (1993) Control of ergot in Kentucky bluegrass seed using fungicides . *Plant Dis.,*77,685–687.

Scott, P.M., Lombaert, G.A., Pellaers, P., Bacler, S. and Lappi, J. (1992) Ergot alkaloids in grain foods sold in Canada. *J. of AOAC International,*15,773–779.

Shelby, R.A. and Kelley, V.C. (1992) Detection of ergot alkaloids from *Claviceps* species in agricultural products by competitive ELISA using a monoclonal antibody. *J. Agric.Food Chem.,* **40,**1090–1092.

Svendsen, A. and Verpoorte, R. (1983) *Chromatography of Alkaloids.*Elsevier, Amsterdam.

TePaske, M.R., Powell, R.G. and Clement, S.L. (1993) Analyses of selected endophyte-infected grasses for the presence of loline-type and ergot-type alkaloids. *J. Agric. Food Chem.,* **41,**2299–2303.

White, J.F., Jr., Morgan-Jones, G. and Morrow, A.C. (1993) Taxonomy, life cycle, reproduction, and detection of *Acremonium* endophytes. *Agric. Ecosystems andEnviron.,* **44,**13–37.

18.
PRODUCERS OF ERGOT ALKALOIDS OUT OF *CLAVICEPS*GENUS

ANATOLY G.KOZLOVSKY

Russian Academy of Sciences, Institute of Biochemistry andPhysiology of Microorganisms, Laboratory of Biosynthesis ofBiologically Active Compounds, 142292, Pushchino,Moscow Region, Russia

18.1.
INTRODUCTION

It is known, that the ability to produce secondary metabolites is connected to the taxonomic position of the producers (Frisvad and Filtenborg, 1990). Traditional source of ergot alkaloids are fungi from the genus *Claviceps*. Up to now many organisms from filamentous fungi to higher plants were identified to be the producers of these biologically active compounds. It was established that strains belonging to fungi imperfecti, *Ascomycetes, Basidiomycetes* and *Phycomycetes* are able to synthesize the ergot alkaloids. In 1960, Hofmann and Tscherter described the occurrence of ergot alkaloids in higher plants. They succeeded in isolating of lysergic acid amide, isolysergic acid amide and chanoclavine from the seeds of *Ipomoea violacea* and *Rivea corymbosa* (Convolvulaceae).

Summary of all ergot alkaloids isolated from filamentous fungi and higher plants is given in Table 1. For the structures of other fungal metabolites not given in this book see e.g., Turner and Aldridge (1983).

Most of ergot alkaloid producers outside of *Claviceps* genus produce clavine alkaloids. They can be classified in three groups:

I. Producers "classical" type of clavines (elymoclavine, agroclavine, festuclavine etc.—for the structures see Chapter 7).
II. Producers of the clavines having different configuration than those from *Claviceps* (Figure 1) (Kozlovsky *et al.*, 1983).

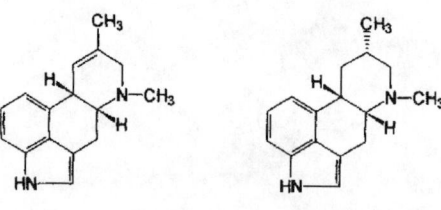

Agroclavine-I   Epicostaclavine

Figure 1

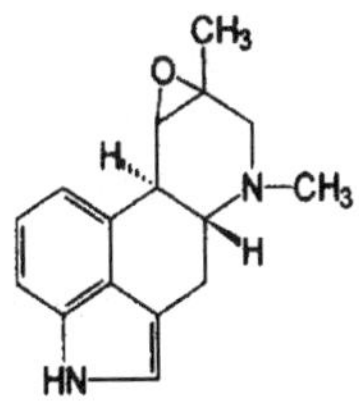

Epoxyagroclavine-I

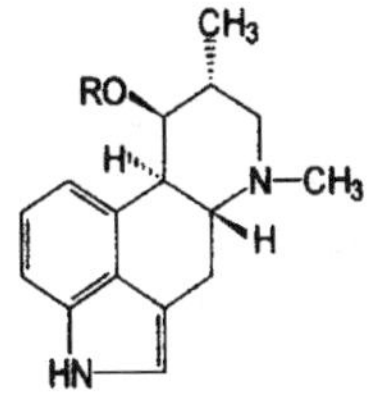

Fumigaclavine A R = CH₃CO
Fumigaclavine B R = H

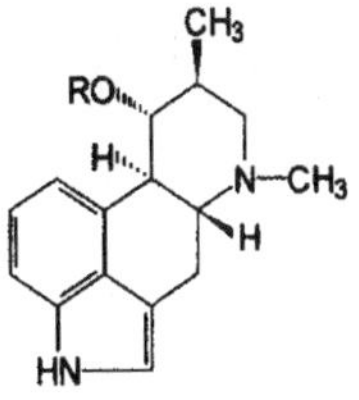

Isofumigaclavine A R = CH₃CO
Isofumigaclavine B R = H

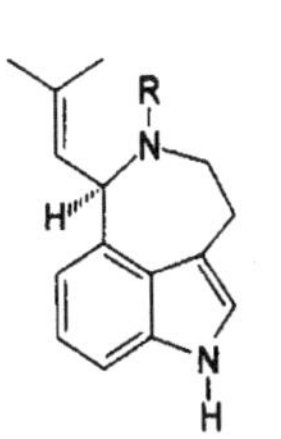

Aurantioclavine R = H
6-N-Ethylaurantioclavine R = C₂H₅

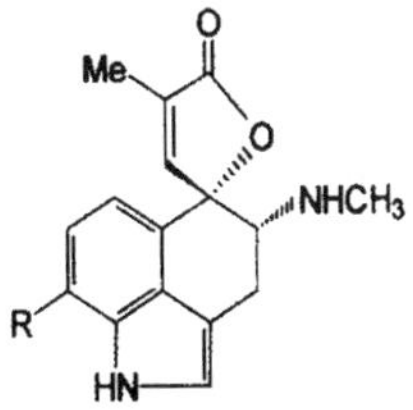

Rugulovasine A R = H
8-Chlororugulovasine A R = Cl

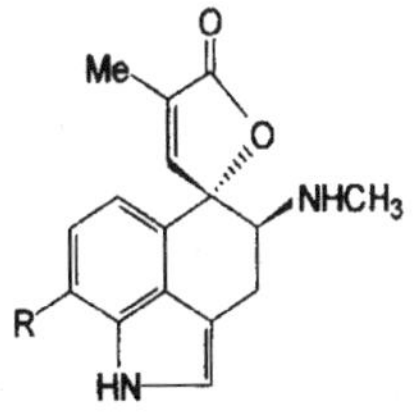

Rugulovasine B R = H
8-Chlororugulovasine B R = Cl

Figure 2

III. Producers of clavines which were never found in Claviceps (Figure 2) and clavine alkaloid dimers (Figure 3).

18.2.
SCREENING OF THE ERGOT ALKALOID PRODUCERS AMONG THE FILAMENTOUS FUNGI

More than 1000 strains of fungi belonging to the *Ascomycetes, Phycomycetes,Basidiomycetes* and *Fungi imperfecti* were screened by Abe *et al.* (1967). Method of the screening included cultivation in submerged and surface culture, the use of Ehrlich reagent for detection of indolic compounds, and paper chromatography with the standard samples of ergot alkaloids. Several strains of fungi belonging to the genera *Penicillium* and *Aspergillus* were identified as producers of clavine alkaloids. The best alkaloid producers were strains of *Aspergillus fumigatus* synthesizing clavine alkaloids (Table 1).

Bekmakhanova *et al.* (1975) screened potentional alkaloid producers among 19 strains from the genus *Penicillium* using TLC. She found that strains *P. gorlenkoanum, P. sizovae, P. roqueforti, P. restrictum* and *P. paxilli* produced metabolites of the ergot alkaloid type. Later it was shown *P. gorlenkoanum, P. sizovae,* and *P. roqueforti* can synthesize ergot alkaloids (see Table 1) (Kozlovsky *et al.,* 1979; Kozlovsky *et al.,* 1981a; Kozlovsky *et al.,* 1986).Vining *et al.* (1982) examined several hundreds isolates of the fungi out of *Claviceps* **Table 1***(Continued)* **Table 1***(Continued)* **Table 1***(Continued)* **Table 1** *(Continued)* genus for the production of ergot alkaloids. Only one, *Pen icillium citreoviride* gave positive reaction for indolic compounds. The main component named as cividiclavine was isolated and its structure was elucidated as a new type of clavine alkaloid dimer. It contains a pyroclavine moiety linked through its indolic nitrogen to a hydroxypyroclavine. The linkage in the latter is tentatively placed at C-13' and the hydroxy group is assigned to C-14'. Free pyroclavine was also isolated from the culture broth (Figure 3). A screening method for alkaloid producing fungi was developed (Kozlovsky and Solov'eva, 1985) including cultivation under the optimal conditions for the secondary metabolite production, sampling during growth, analysis of the culture liquid and mycelium for alkaloids, isolation of metabolites of the alkaloid origin (alkaline, neutral and acid), and TLC of extracts with standard samples. Seven strains of *Aspergillus,* 3 belonging to *Chaetomium,* 10 of *Fusarium,* 6 of *Helminthosporium,* 2 of *Rhizopus* and 36 of *Penicillium* genera were examined. *P. aurantiovirens, P. kapuscinskii,* and *P. palitans* were identified as ergot alkaloid producers (see Table 1) and their structures were elucidated (Kozlovsky *et al.,* 1981; Kozlovsky *et at.,* 1982b; Kozlovsky *et al.,* 1990; Vinokurova *et al.,* 1991).

Ohmomo *et al.* (1989) have screened for indole alkaloid producing fungi and he isolated some new producers from the genus *Aspergillus.* One of them, a thermophilic strain No. 2–18, was identified as *Aspergillus fumigatus* producing mainly fumigaclavine B (Figure 2).

From the culture liquid of *P. sizovae,* known as a producer of agroclavine-I (Figure 1) and epoxyagroclavine-I, a group of new dimeric ergot alkaloids was recently isolated (Zelenkova *et al.,* 1992; Kozlovsky *et al.,* 1995a). Specific feature of these dimers is a linkage between indolic nitrogens of the ergoline moieties. Several derivatives of these dimer were obtained by opening of the oxiran ring of epoxyagroclavine-I (Zelenkova *et al.,* 1992). Later, Kozlovsky *et al.* (1995a) isolated dimer of agroclavine-I and a mixed dimer of agroclavine-I and epoxyagroclavine-I (Figure 3). Based on the preliminary data (retention time in HPLC in comparison with the same for other dimers) they supposed also that chanoclavine dimer can be present in the culture broth.

Clavicipitic acid and its plausible decarboxylation product aurantioclavine were detected as metabolites *P. aurantiovirens (Kozlovsky et al.,* 1981). Later, *N*-6-ethylaurantioclavine has been isolated and it's structure was elucidated *(Kozlovsky et al.,* 1997a). It is the first case of natural ergot alkaloid containing *N*-6-ethyl group (Figure 2).

Rugulovasine A and B (Figure 2) originally found by Abe *et al.* (1967) in *P. concavorugulosum* differs from clavine alkaloid by irregular cyclisation of D

Table 1 Ergot alkaloid producers out of genus *Claviceps*

Organism	Ergot alkaloid	Reference
Fungi imperfecti		
Aspergillus fumigatus	festuclavine, fumiga-clavine A, fumigaclavine B, elymoclavine, chano-clavine-I, ergotamine, ergocristine	Abe *et al.* (1967); Cole *et al.* (1977); Furuta *et al.* (1982); Narayan and Rao (1982); Spilsbury and Wilkinson (1961); Yamano *et al.* (1964)
A. flavus	elymoclavine, cyclopiazonic acid, ergokryptine	El-Refai *et al.* (1970); Hermansen *et al.* (1984); Naim (1980); Yokota *et al.* (1981)
A. japonicus	festuclavine, cycloclavine	Furuta *et al.* (1982); Stauffacher *et al.* (1969)
A. nidulans	clavines	Abe *et al.* (1967)
A. oryzae	cyclopiazonic acid	Hermansen *et al.* (1984); Yokota *et al.* (1981)
A. tamarii	fumigaclavine A, other unidentified clavine alkaloids	Janardhanan *et al.* (1984)
A. versicolor	cyclopiazonic acid, bis-secodehydrocyclopiazonic acid, imine cyclopiazonic acid	Ohmomo *et al.* (1973)
Botrytis fabae	elymoclavine, agroclavine, ergokryptine	Naim (1980)
Curvularia lunata	elymoclavine, agroclavine, ergokryptine	Naim (1980)
Geotrichum candidum	elymoclavine, agroclavine, ergosine, lysergic acid	El-Refai *et al.* (1970)
Penicillium atramentosum	rugulovasine A	Frisvad and Filtenborg (1989)
P. aurantiovirens	clavicipitic acid, chano-clavine-I, agroclavine, elymoclavine, penniclavine, isopenniclavine	Kozlovsky *et al.* (1981)
P. biforme	rugulovasine A, rugulova-sine B, chlororugulova-sine B	Dorner *et al.* (1980)
P. camembertii	cyclopiazonic acid	Hermansen *et al.* (1984); Le Bars (1979); Schoch *et al.* (1983)
P. chermesinum	costaclavine	Agurell (1964)
P. chrysogenum	rugulovasine A, rugulova-sine B, fumigaclavine A, fumigaclavine B, pyro-clavine	Kozlovsky *et al.* (1997); Reshetilova *et al.* (1995)

Organism	Ergot alkaloid	Reference
P. clavigerum	isofumigaclavine A	Frisvad and Filtenborg (1989)
P. commune chemotype-I	rugulovasine A	Frisvad and Filtenborg (1989)
P. commune chemotype-II	cyclopiazonic acid, isofumigaclavine A	Frisvad and Filtenborg (1989); Hermansen et al. (1984)
P. concavo-rugulosum	chanoclavine-I, chanocla-vine-II, isochanoclavine-I, rugulovasine A, chlororugulovasine A, rugulovasine B, chloro-rugulovasine B, dihydro-rugulovasine A	Abe et al. (1969); Cole et al. (1976); Ohmomo et al. (1977)
P. crustosum	cyclopiazonic acid, fumi-gaclavine A, pyroclavine, festuclavine	Filtenborg et al. (1983); Hermansen et al. (1984); Kawai et al. (1992)
P. citreo-viride	cividiclavine, pyroclavine	Vining et al. (1982)
P. cyclopium	cyclopiazonic acid, bissecodehydrocyclo-piazonic acid, imine cyclopiazonic acid	Hermansen et al. (1984); Holzapfel (1968); Holzapfel (1970)
P. expansum	aurantioclavine	Kozlovsky (1990)
P. gorlenkoanum	rugulovasine A, rugulova-sine B, costaclavine, epicostaclavine, chano-clavine-I, isochanoclavine-I	Kozlovsky et al. (1981a); Reshetilova et al. (1995)
P. griseofulvum	cyclopiazonic acid, epicostaclavine	Hermansen et al. (1984); Kozlovsky (1990)
P. implicatum	epoxyagroclavine-I	Kozlovsky et al. (1997)
P. islandicum	rugulovasine A	Cole et al. (1976)
P. kapuscinskii	agroclavine-I, epoxy-agroclavine-I	Kozlovsky et al. (1982a)
P. lanosoviride	cyclopiazonic acid	Hermansen et al. (1984)
P. mononematosum	cyclopiazonic acid	Frisvad and Filtenborg (1989)
P. oxalicum	cyclopiazonic acid, fumiga-clavine A, fumigacla-vine B, pyroclavine, festuclavine, chanoclavine-I	Vinokurova et al. (1991)
P. palitans	cyclopiazonic acid, fumigaclavine A, fumiga-clavine B, pyroclavine, festuclavine, chanocla-vine-I, agroclavine	Reshetilova et al. (1995); Vinokurova et al. (1991)
P. patulum	cyclopiazonic acid	Filtenborg et al. (1983); Hermansen et al. (1984)

Organism	Ergot alkaloid	Reference
P. puberulum	cyclopiazonic acid	Filtenborg et al. (1983); Hermansen et al. (1984)
P. roqueforti	festuclavine, isofumiga-clavine A, isofumigacla-vine B	Frisvad and Filtenborg (1989); Kozlovsky et al. (1979); Ohmomo et al. (1975); Ohmomo et al. (1977); Polonsky et al. (1977); Scott et al. (1976); Scott et al. (1977)
P. rubrum	rugulovasine A, chloro-rugulovasine, chloro-rugulovasine B	Dorner et al. (1980)
P. rugulosum	rugulovasine A, rugulo-vasine B	Abe et al. (1969)
P. sizovae	agroclavine-I, epoxy-agroclavine-I, agroclavine, chanoclavine-III, chanoclavine-I, 6,7-secoagroclavine, elymoclavine, dimer of epoxyagroclavine-I, dimer of agroclavine-I, mixed dimer of agroclavine-I and epoxyagroclavine-I, dimer of chanoclavine	Kozlovsky (1990); Kozlovsky and Reshetilova (1984); Kozlovsky et al. (1995a); Zelenkova et al. (1992); Kozlovsky et al. (1996)
P. solitum	rugulovasine A, rugulo-vasine B, chlororugulovasine A	Reshetilova et al. (1995)
P. viridicatum	cyclopiazonic acid, rugulovasine A, rugulo-vasine B	Filtenborg et al. (1983); Hermansen et al. (1984); Reshetilova et al. (1995)
P. vulpinum	cyclopiazonic acid	Kozlovsky et al. (1997)
Ascomycetes		
Balansia cyperi	ergobalansine, ergobalansinine	Powell et al. (1990)
B. claviceps	chanoclavine-I, ergonovine, ergonovinine	Bacon et al. (1979); Porter et al. (1979)
B. epichloe	chanoclavine-I, isochano-clavine-I, agroclavine, elymoclavine, penniclavine, ergonovine, ergonovinine, 6,7-secoagroclavine	Bacon et al. (1979); Porter et al. (1979); Porter et al. (1981)
B. henningsiana	chanoclavine-I, dihydro-elymoclavine	Bacon et al. (1979); Porter et al. (1981)
B. obtecta	ergobalansine, ergobalansinine	Powell et al. (1990)
B. strangulans	6,7-secoagroclavine, chanoclavine-I	Bacon et al. (1979); Porter et al. (1981)

Organism	Ergot alkaloid	Reference
Epichloe typhina	6,7-secoagroclavine, agro-clavine, elymoclavine, penniclavine, festuclavine, ergovaline, ergovalinine, ergotamine, ergosine, β-ergosine, ergostine, ergoptine, β-ergoptine, ergonine, ergocristine, α-ergokryptine, β-ergokryptine, ergocornine	Porter *et al.* (1979a); Porter *et al.* (1981); Yates *et al.* (1985)
Hypomyces aurantius	ergokryptine, ergokrypti-nine, agroclavine, elymo-clavine, chanoclavine	Yamatodani and Yamamoto (1983)
Sepedonium sp.	elymoclavine, agroclavine, ergocornine	Naim *et al.* (1980)
Basidiomycetes		
Corticium caeruleum	clavines	Abe *et al.* (1967)
Lenzites trabea	clavines	Abe *et al.* (1967)
Pellicularia filamentosa	clavines	Abe *et al.* (1967)
Phycomycetes		
Cunninghamella blakesleana	elymoclavine, agroclavine, ergosine, lysergic acid	Naim (1980)
Mucor hiemalis	ergosine	El-Refai *et al.* (1970)
Rhizopus arrhizus	agroclavine	Sallam *et al.* (1969)
Rh. nigricans	agroclavine, ergosine, ergosinine	El-Refai *et al.* (1970); Sallam *et al.* (1969)
Higher plants Convolvulaceae		
Argyreia (20 species)	agroclavine, chanoclavine-I, chanoclavine-II, rac.chanoclavine-II, elymo-clavine, festuclavine, lysergene lysergine, lysergol, isolysergol molliclavine, penniclavine, setoclavine, isosetoclavine, ergine, isoergine, ergometrine, ergometrinine, lysergic acid 2-hydroxyethylamide, iso-lysergic acid 2-hydroxy-ethyl-lamide, ergosine, ergosinine, cycloclavine	Chao and DerMarderosian (1973)

Organism	Ergot alkaloid	Reference
Cuscuta monogyna	agroclavine	Chao and DerMarderosian (1973)
Ipomoea (11 species)	agroclavine, chanoclavine-I, chanoclavine-II, rac.chanoclavine-II, elymoclavine, festuclavine, lysergol, isolysergol, penniclavine, ergine, isoergine, ergometrine, ergometrinine, lysergic acid 2-hydroxyethylamide, isolysergic acid 2-hydroxyethylamide, ergosine, ergosinine, cycloclavine	Chao and DerMarderosian (1973)
I. rubro-caerulea	chanoclavine-I, lysergic acid amide, lysergic acid 2-hydroxyethylamide, isolysergic acid 2-hydroxyethylamide	Gröger et al. (1963)
I. piurensis	chanoclavine-I, ergine, lysergic avid 2-hydroxyethylamide, ergobalasine, ergobalansinine	Jenettsiems et al. (1994)
Rivea corymbosa	agroclavine, chanoclavine-I, chanoclavine-II, elymoclavine, lysergene, lysergol, molliclavine, setoclavine, isosetoclavine, ergine, isoergine, ergometrine, ergometrinine, lysergic acid 2-hydroxyethylamide, isolysergic acid 2-hydroxyethylamide, ergosine	Chao and DerMarderosian (1973)
Stictocardia tiliifolia	chanoclavine-I, chanoclavine-II, festuclavine, lysergol, ergine, ergometrine, ergometrinine, lysergic acid 2-hyrdoxyethylamide	Hofmann (1961)

ring. Later, also chlororugulovasine A and B were found that are probably single natural ergot alkaloids containing halogen atom (see Table 1).

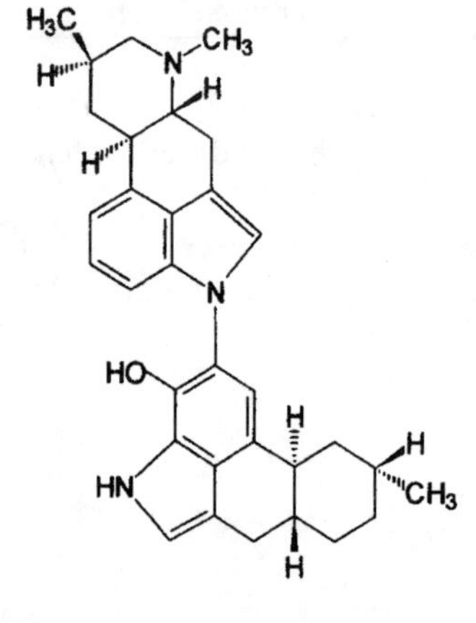

Cividiclavine

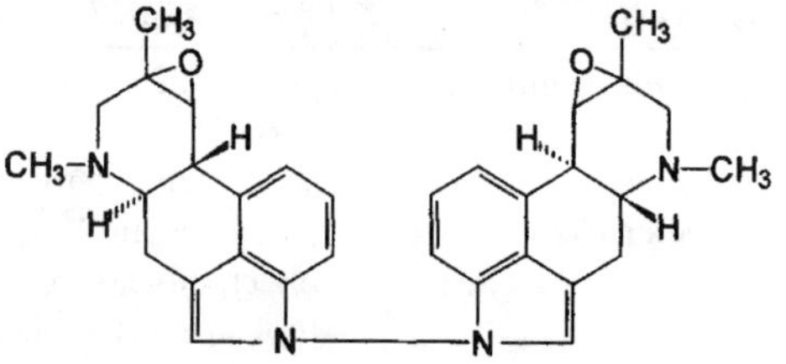

Epoxyagroclavine-I dimer

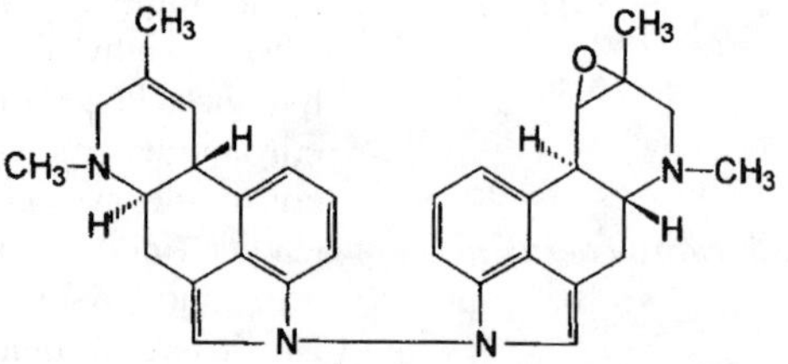

Agroclavine-I-epoxyagroclavine-I mixed dimer

Figure 3

18.3.
PHYSIOLOGY OF THE PRODUCERS AND SOME ASPECTS OF THE REGULATION OF ERGOT ALKALOID BIOSYNTHESIS

The data on the physiology of ergot alkaloid biosynthesis by the fungi, besides *Claviceps* strains are quite scarce. Extensive studies have been done with the strains of *P. sizovae, P. gorlenkoanum, P. kapuscinskii* and *P. aurantiovirens.*

Production kinetics of agroclavine-I and epOxyagroclavine-I, the main components of alkaloid mixture of *P. sizovae,* was studied by Kozlovsky *et al.* (1986). Accumulation and degradation of alkaloids took place in two stages which coincide with two growth phases. During growth in a medium containing succinic acid and mannitol sequential substrate utilization by the culture and biphasic growth was observed. For the alkaloid biosynthesis with *P. sizovae* high residual phosphate concentrations was necessary, compared to *Claviceps* strains, where high phosphate concentrations inhibited the alkaloid production.

An influence of carbon sources on the growth of *P. sizovae* and biosynthesis of agroclavine-I and epoxyagroclavine-I, as well as the activity of key enzymes of the Krebs cycle, the pentose phosphate pathway and glyoxalate cycle were studied (Kozlovsky and Vepritskaya, 1987). The best alkaloid productivity was observed with mannitol and fumaric acid as the carbon sources. A combination of sorbitol with fumaric acid stimulated epoxyagroclavine-I synthesis. A high alkaloid production was accompanied by high activity of the pentose phosphate cycle and low activity of the Krebs cycle.

Tryptophan is a precursor of ergot alkaloids in *Claviceps* and in some cases plays also the role of an inducer and derepressor (Bu'Lock and Barr, 1968; Vining, 1970; Robbers *et al.*, 1972). Isotopically labelled tryptophan shown a high level of incorporation into ergot alkaloids (41%) in the strain of *P. sizovae* indicating that it is also here a direct precursor Kozlovsky *et al.* (1985). In *Claviceps* not only L- but also D-tryptophan are utilised for the ergot alkaloid biosynthesis (Robbers *et al.*, 1972; Floss, 1976). The effect of both L- and D-tryptophan and also of their analogue, D,L-6-methyltryptophan on the alkaloid production kinetics in *P. sizovae* was studied (Kozlovsky *et al.*, 1985a). In most of our experiments L- and D-tryptophan were added at concentrations of 0.1, 0. 25, 0.4 and 2 mM together with the inoculum because their induction effect in *Claviceps* occurs only when fed during the first 24 hours of production cultivation (Bu'Lock and Barr, 1968; Robbers *et al.*, 1978). Feeding of both D- and L-tryptophan to the *P. sizovae* culture at the beginning of the production cultivation did not increase the alkaloid production. The production was even lowered to one half of the control. However, feeding both L- and D-tryptophan on the 6th day of the production stage increased the alkaloid yield 2.4 times with D-tryptophan, and 1.9 times with L-tryptophan. Additions of 6-methyltryptophan did not exert any stimulating effect on the production of ergot alkaloids in *P. sizovae,* moreover, an inhibition of their biosynthesis was observed. Thus it can be concluded, that the induction of ergot alkaloid biosynthesis by tryptophan is absent in *P. sizovae*. The absence of the induction effect in some strains of the *Claviceps* has also been established earlier by some authors (Gröger and Tyler, 1963; Řičicová *et al.*, 1982).

A specific relationship between exogenous tryptophan and alkaloid level was found in *P. roqueforti* (Kozlovsky *et al.*, 1982; Reshetilova and Kozlovsky, 1985). This strain produces two types of alkaloids, clavines (festuclavine, isofumigaclavine A and isofumigaclavine B) (Figure 2) and diketopiperazines (roquefortine and 3,12-dihydroroquefortine) (Figure 4) (Kozlovsky *et al.*, 1979) that have both a common precursors—tryptophan and mevalonic acid. Exogenous tryptophan has different effect on these alkaloid types. Biosynthesis of diketopiperazines was enhanced by the precursor addition, whereas the production of clavines did not depend on the precursor and sometimes it was even inhibited by its addition (Reshetilova and Kozlovsky, 1985).

Effect of various concentrations of inorganic phosphate, microelements, temperature, pH, and sources of carbon, nitrogen on the yield of epoxyagroclavine-I dimer was studied in *P. sizovae* (Kozlovsky *et al.*, 1995). Active biosynthesis of the dimer occurred upon the cultivation on the media containing mannitol, α-ketoglutarate, ammonium, KH_2PO_4 in concentration of 0. 1 g L^{-1}, and microelements (Fe^{2+}, Mn^{2+}) at 28°C and initial pH 7.0.

Active alkaloid production (agroclavine-I and epoxyagroclavine-I) by *P. kapuscinskii* was observed on the media containing mannitol and succinic or

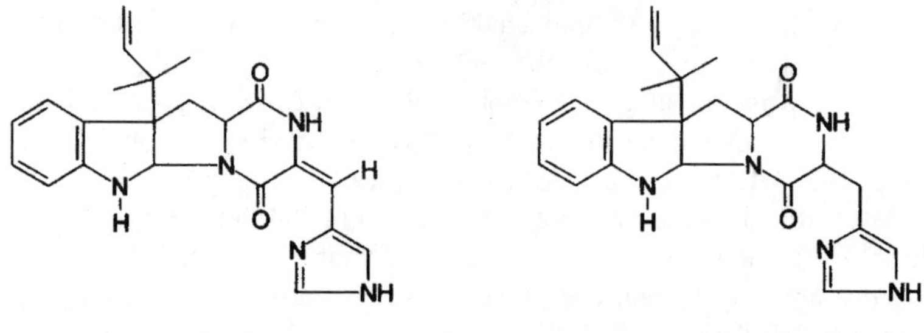

Roquefortine   3,12-Dihydroroquefortine

Figure 4

malic acids as carbon sources, and ammonium sulfate, asparagine, and tryptophan as nitrogen sources (Kozlovsky and Solov'eva, 1986a). The optimum of the phosphate concentration for alkaloid biosynthesis was 1 g L^{-1} KH_2PO_4. Ascorbic acid (1 M) stimulated the yield of epoxyagroclavine-I and agroclavine-I.

The optimum medium for epicostaclavine (Figure 1) synthesis by *P.gorlenkoanum* contained mannitol, succinic acid and 1% KH_2PO_4. Change in the carbohydrate or organic acid concentration, or variation in the phosphate concentration, altered the costaclavine and epicostaclavine ratio (Kozlovsky *et al.*, 1981b). Glucose together with fructose as carbon source inhibited alkaloid synthesis, addition of microelements as well as lowered aeration stimulated the alkaloid biosynthesis (Stefanova-Avramova and Kozlovsky, 1984).

The effect of the composition of the culture medium on the biosynthetic spectrum of ergot alkaloids in *P. aurantiovirens* was studied by Solov'eva *et al.* (1995). Addition of methionine and replacement succinic acid in Abe's medium by asparagic acid caused that besides aurantioclavine also chanoclavine-I was produced. When glucose was used as a sole source of carbon and energy only chanoclavine-I was produced, indicating stimulation of N-6-methyl transferase. Decrease of the dissolved oxygen from 75% to 2% at the end of the exponentional phase, aurantioclavine and intermediates of the ordinary pathway of the ergot alkaloid biosynthesis, e.g., chanoclavine-I, agroclavine, elymoclavine, penniclavine and isopenniclavine were produced. Thus, *P. aurantiovirens* is able to produce clavine alkaloids by two pathways—through clavicipitic acid to aurantioclavine and from chanoclavine-I through agroclavine, elymoclavine to penniclavine and isopenniclavine (Figure 5).

Effect of the culture age, the medium composition, various concentrations of phosphate and possible precursors, tryptophan, mevalonic acid and methionine, on the clavine alkaloids production by *A. fumigatus* was examined more authors (Rao and Patel, 1974; Rao *et al.*, 1977; Narayan and Rao, 1982). Tryptophan, mevalonic acid and methionine stimulated the clavine alkaloid production. Correlation between the alkaloid production and the culture growth was not established. Quality of the results obtained by these authors (Rao and Patel, 1974;

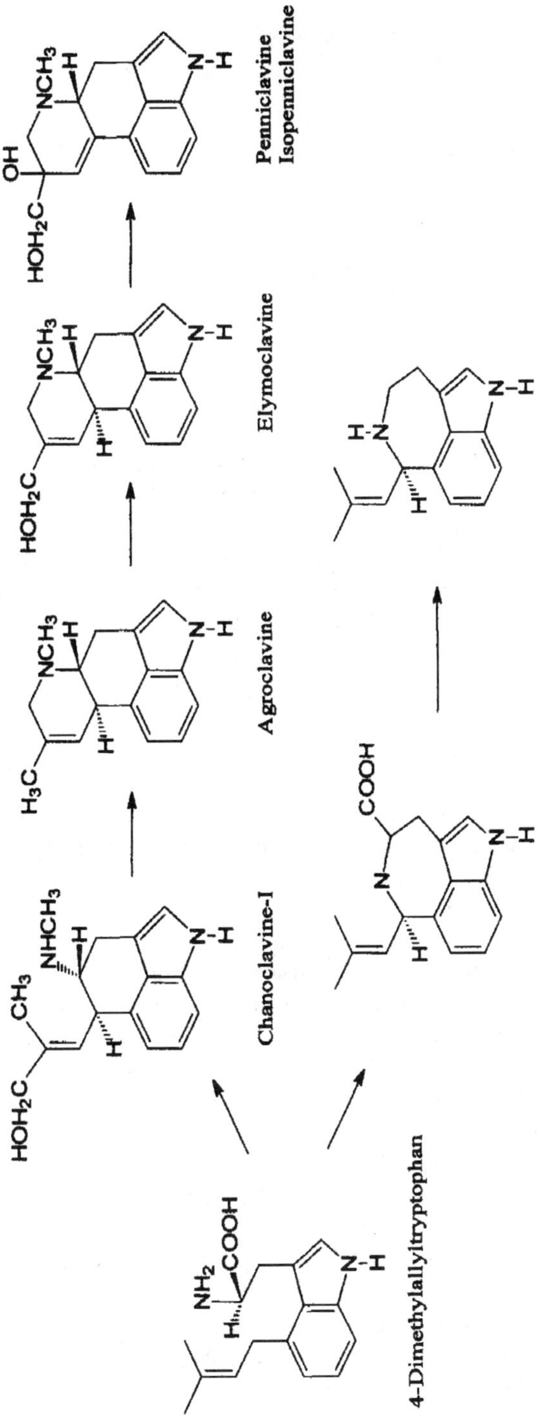

Ergot alkaloid biosynthesis by *Penicillium aurantiovirens*

Figure 5

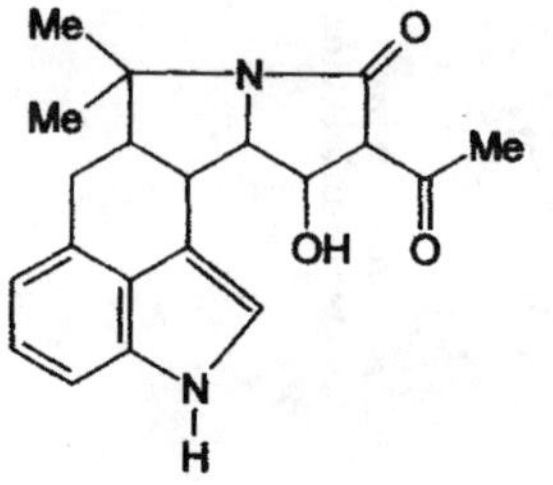

Cyclopiazonic acid

Figure 6

Narayan and Rao, 1982; Ohmomo *et al.*, 1989) can be, however, hampered by questionable methodology.

Ohmomo *et al.* (1989) investigated the effect of carbon and nitrogen sources on the alkaloid synthesis with thermophilic strain *A. fumigatus* producing fumigaclavine B. Optimum medium was the combination of mannitol (5%), glucose (5%) and ammonium succinate (2%). Alkaloids were produced in good yields at 37°C, while the highest growth rate was attained at 41°C. The maximum alkaloid yield—20 mg/L was reached. After 10th day of cultivation, alkaloid degradation started.

18.4.

CYCLOPIAZONIC ACID BIOSYNTHESIS

Cyclopiazonic acid, a toxic metabolite of *P. cyclopium,* which can be belong in some extent to ergot alkaloids, was isolated and identified by Holzapfel (1968) (Figure 6). Stereochemical aspects of D-ring formation of cyclopiazonic acid were investigated by Chalmers *et al.* (1982).

Sixty two isolates of *Penicillium* and *Aspergillus* were screened for cyclopiazonic acid production in surface and submerged culture on different media (Hermansen *et al.,* 1984). The production of this mycotoxin is restricted to *P. camembertii, P. griseofulvum* and *A. flavus* (and its domesticated form *A. oryzae*).

Best yield of cyclopiazonic acid was obtained with *P. griseofulvum* but several strains of *P. camembertii* were also found to be good producers. Submerged cultures gave best yields of cyclopiazonic acid, but in some cases the production occured only in a surface culture. Hermansen *et al.* (1984) described also a simplified procedure for isolation of cyclopiazonic acid.

Effect of carbon and nitrogen sources on the production of cyclopiazonic acid by *P. griseofulvum* was studied by Reddy and Reddy (1988). Glycerol supported cyclopiazonic acid production, while citric acid and lactose were poor substrates. L-Asparagine, potassium nitrate and D,L-alanine supported good production of cyclopiazonic acid, while L-histidine did not.

18.5.
HIGHER PLANTS AS THE PRODUCERS OF THE ERGOT ALKALOIDS

As mentioned above, for the first time ergot alkaloids have been found, in the higher plants by Hofmann and Tscherter (1960). They established that mexican crude drug *"Ololiuqui"* consists of the seeds of *Rivea corymbosa*

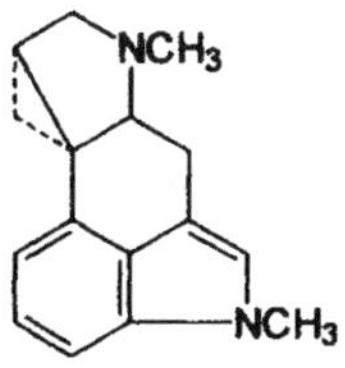

Cycloclavine 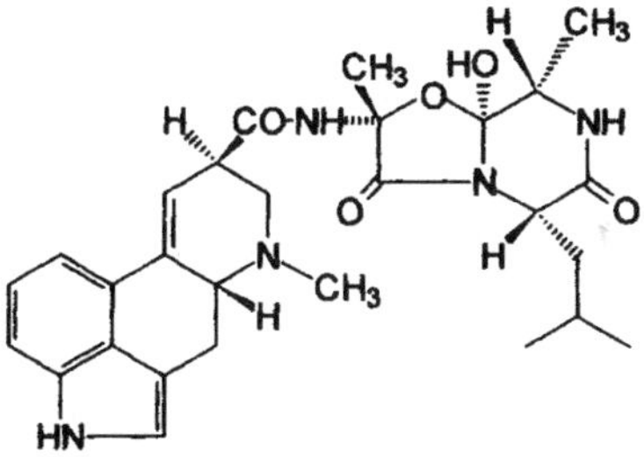   Ergobalansine

Figure 7

and *Ipomoea violacea* belonging to the family Convolvulaceae. They identified three main alkaloid components of the drug to be lysergic acid amide, isolysergic acid amide and chanoclavine-I. Later, Hofmann (1961) found elymoclavine in it. Stauffacher *et al.* (1969) isolated festuclavine and cycloclavine from *Ipomoea hildebrantii* (Figure 7). First ergot alkaloids of the peptide type have been found in higher plants by Stauffacher *et al.* (1965). They have found in the seeds of *Ipomoea argyrophylla* ergosine, ergosinine and also agroclavine (for the structures see Chapter 7). Jenettsiems *et al.* (1994) isolated from *Ipomoea piurensis* and elucidated the structures proline-free peptide ergot alkaloids, ergobalansine and ergobalansinine, and three simple ergoline alkaloids, chanoclavine-I, ergine and lysergic acid 2-hydroxyethylamide.

Gröger *et al.* (1963) have shown that in *Ipomoea rubro-caerulea* producing ergot alkaloids chanoclavine-I, lysergic acid amide, 2-hydroxylysergic acid amide, and 2-hydroxyisolysergic acid amide, tryptophan and mevalonic acid were their precursors analogously as in *Claviceps*.

Topics of the evolutionary relationship of ergoline biosynthesis in fungi and in higher plants has been discussed by Boyes-Korkis and Floss (1992). They raised the questions whether the genetic information coding for ergoline biosynthesis developed two times independently in nature, or it evolved only once and then was passed from the fungus to the plant or *vice versa*. If the latter is the case, are the pathway genes in the plant and/or fungus clustered or scattered throughout the genom? Is the genetic information coding for the ergoline pathway perhaps ubiquitous, but genetically silent in some other organisms? These questions, in their opinion, not yet answered, will be possibly solved in the future by the help of molecular biology.

18.6.
ENVIRONMENTAL AND HAZARD PROBLEMS WITH UNTRADITIONAL PRODUCERS OF ERGOT ALKALOIDS

Safety methods and understanding of the danger from traditional source of ergot alkaloids, fungi of the genus *Claviceps* were developed very well. But now new challenge connected with a change of the relationship between bacterial and fungal microflora in environmental. In many cases it can be explained by a pollution of an antropogenic nature such, as wide using of herbicides, pesticides, another xenobiotics, heavy metals, etc. As a rule, fungi are more resistant than bacteria against the influence of these factors. Fungi of the genera *Penicillium* and *Aspergillus* are widely distributed in our environment. They occur in a soil, on the food and feed, on the plants etc. As a result of these processes an rearrangement and narrowing of the diversity of the fungal community take place. New species of fungi became dominant strains. It is important to know the toxigenic potential of these strains, especially dominant, in the respect to their ability to produce such toxic secondary metabolites as ergot alkaloids, to evaluate the scale of the danger.

Nineteen strains of the twelve species of the genus *Penicillium,* isolated from the polluted soils and extreme places of ihabitation were examined for their ability to produce mycotoxins, included ergot alkaloids (Kozlovsky *et al.,* 1997). It was established that strains *P. chrysogenum* isolated from the city soil, can produce fumigaclavine A, fumigaclavine B and pyroclavine. *P. implicatum* isolated from Turkmenistan soil can synthesize epoxyagroclavine-I. One of the dominant strains, *P. vulpinum,* common in the city soil, can produce in considerable quantities cyclopiazonic acid and its imine. Cyclopiazonic acid belong to mycotoxins that should be strictly controled.

Ability to synthesize ergot alkaloids was tested in 31 fungal strains belonging to the genera *Aspergillus* and *Penicillium* that were isolated from Uzbekistan soils treated with pesticides for a long time (Kozlovsky *et al.,* 1990). It was shown, that one of the examined strains—strain *P. verrucosum* var. *cyclopium* can produce cyclopiazonic acid.

About 15 years ago a limited case of feed contamination by ergot alkaloids in Czechoslovakia was identified (V.Křen—personal communication). Clavine alkaloids (agroclavine, elymoclavine) were found in eggs and hen's meat and later it was found that this was caused by the grain, contaminated by some *Aspergilli* during the ship transport from South America (hot/humid conditions). Hens also suffered from ovaria degradation. They, in the first days of toxications, laid 2–3 eggs per day, however, later within ca 14 days stopped egg production and their ovaria were found severely damaged.

18.7.
OUTLOOK ON USING OF ERGOT ALKALOID PRODUCERS OUT OF THE GENUS *CLAVICEPS* IN PRACTICE

Filamentous fungi, especially those belonging to genus *Penicillium* can be a source of the new ergot alkaloids with "unusual" structures (Skryabin and Kozlovsky, 1984). It is known that variations in the substitution and configuration of the ergoiine moiety may result into the considerable changes in the biological activity (Fluckiger, 1980). From this point of view epoxyagroclavine-I, *N, N'*-dimer epoxyagroclavine-I, mixed dimer epoxyagroclavine-I and agroclavine-I, metabolites of *P. sizovae* and *P. kapuscinskii,* are the most perspective as the base for the obtaining of the new biologically active compounds. Due to the high reactivity of the epoxy-group, these compounds are readily converted into some new ergoline derivatives under mild conditions using classical epoxide chemistry (Kozlovsky *et at.,* 1982a; Kozlovsky *et al.,* 1983; Zelenkova *et al.,* 1992; Kozlovsky *et al.,* 1995a). Analogous reactions were performed with *N, N'*-dimer of epoxyagroclavine-I (Zelenkova *et al.,* 1992) and new derivatives were obtained.

18.8.
CONCLUSIONS

Producers of ergot alkaloids are widely distributed among the various genera belong to various taxons. These organisms, especially filamentous fungi, can be dangerous for people and animals and their level must be controled. They can produce ergot alkaloids with a great variety of structural types including those which have not been found in *Claviceps*. Fungi of genus *Penicillium* are perspective as a source of the new ergot alkaloids and their derivatives, for production of the new biologically active compounds.

In these organisms, precursors of the ergot alkaloids tryptophan and mevalonic acid are the same as in *Claviceps*. Composition of the media and the cultivation conditions are rather specific for the production of the alkaloids by these strains.

ACKNOWLEDGEMENT

I am greatly indebted to Mrs. Anna V. Khrenova for her industriousness and patience in the technical preparation of the manuscript.

REFERENCES

Abe M., Yamatodani, S., Yamano, T., Kozu, Y. and Yamada, S. (1967) Production of alkaloids and related substances by fungi. Examination of filamentous fungi for their ability of producing ergot alkaloids. *J. Agr. Chem. Soc. Japan,* **41,**68–71.

Abe, M., Ohmomo, S., Ohashi, T. and Tabushi, T. (1969) Isolation of chanoclavine-I and two new interconvertible alkaloids, rugulovasine A and B from cultures of *Penicillium concavo-ruguloum.Agr. Biol. Chem.,***33,**469–471.

Agurell, S.L. (1964) Costaclavine from *Penicillium chermesinum.Experientia,***20,** 1, 25–26.

Arushanyan, A.V., Vepritskaya, I.G., Kozlovsky, A.G., Akimenko, V.K. and Kulaev, I.S. (1985) Study of the polyphosphate metabolism and cyanide resistance of the fungus *Penicillium sizovae* in the process of alkaloid formation. *Biokhimiya (in Russian),***50,** 1836–1842.

Bacon, C.W., Porter, J.K. and Robbins, J.D. (1979) Laboratory production of ergot alkaloids by species of *Balansia.J. Gen. Microbiol.,***113,**119–126.

Bacon, C.W., Lyons, P.C., Porter, J.K. and Robbins, J.D. (1986) Ergot toxicity from endophyte-infected grasses: a review. *Agronomy J.,***78,**106–116.

Bekmakhanova, N.E., Kozlovsky, A.G. and Bezborodov, A.M. (1975) Selection of fungi —alkaloid producers by thin-layer chromatography. *Prikl. Biochim. Mikrobiol.* (in *Russian*), **11,**131–135.

Boyes-Korkis, J.M. and Floss, H.G. (1992) Biosynthesis of ergot alkaloids: some new results on an old problem. *Prikl. Biokhim. Mikrobiol.* (in *Russian*), **28,**843–857.

Bu'Lock, I.D. and Barr, J.G. (1968) A Regulation mechanism linking tryptophan uptake and synthesis with ergot synthesis in *Claviceps. Lloydia,***31,**342–354.

Chalmers, A.A., Gorst-Allman, C.P. and Steyn, P.S. (1982) Biosynthesis of cyclopiazonic acid: stereochemical aspects of D-ring formation. *J. Chem. Soc., Chem. Commun.,* 1367–1368.

Chao, J.-M. and DerMarderosian, A.H. (1973) Identification of ergoline alkaloids in the genus *Argyreia* and related genera and their chemotaxonomic implications in the convolvulaceae. *Phytochemistry,***12,**2435–2440.

Cole, R.J., Kirksey, J.W., Clardy, J., Eichman, N., Weinreb, S.M., Singh, P. and Kim, S. (1976) Structures of rugulovasine-A and -B and 8-chlororugulovasine-A and -B. *Tetrahedron Lett.,***43,**3849–3852.

Cole, R.J., Kirskey, J.W., Dorner, J.W., Wilson, D.M., Johnson, J., Bedell, D., Springer, J.P., Chexal, K.K., Clardy, J. and Cox, R.H. (1977) Mycotoxins produced by *Aspergillus fumigatus* isolated from silage. *Ann. Nutr. Alim.,***31,**685–692.

Dorner, J.W., Cole, R.J., Hill, R.A., Wicklow, D. and Cox, R.H. (1980) *Penicilliumrubrum* and *Penicillium biforme,* new sources of rugulovasins A and B. *Appl.Environ. Microbiol.,***40,**685–687.

Dorner, J.W., Cole, R.J., Lomex, L.C., Gosser, H.S. and Diener, U.L. (1983) Cyclopiazonic acid production by *Aspergillus flavus* and its effect on broiler chickens. *Appl. Environ. Microbiol.,***46,**698–703.

El-Refai, A.M.H., Sallam, L.A.R. and Naim, N. (1970) The alkaloids of fungi. I. The formation of ergoline alkaloids by representative mold fungi. *Jap. J. Microbiol.,***14,** 91–97.

Filtenborg, O., Frisvad, J.C. and Svendsen, J.A. (1983) Simple screening method for mold producing intracellular mycotoxins in pure cultures. *Appl. Environ. Microbiol.,***45,** 581–585.

Floss, H.G. (1976) Biosynthesis of ergot alkaloids and related compounds. *Tetrahedron,* **32,**873–912.

Fluckiger, E. (1980) Recent advances in ergot pharmacology. In Phillipson, J.D. and Zenk, M.H. (eds.). *Indole and biogenetically related alkaloids,* Academic Press, London, New York, Toronto, Sydney, San Francisco, pp. 285–291.

Frisvad, J.C. and Filtenborg, O. (1989) Terverticillate Penicillia-. chemotaxonomy and mycotoxin production. *Mycologia,* **81,** 837–861.

Frisvad, J.C. and Filtenborg, O. (1990) Secondary metabolites as consistent criteria in *Penicillium* taxonomy and a synoptic key to *Penicillium* subgenus *Penicillium.* In R.A. Samson and J.I. Pitt, (eds.), *Modern concepts in Penicillium and Aspergillusclassification,* Plenum Press, New York, pp. 313–384.

Furuta, T., Koike, M. and Abe, M. (1982) Isolation of cycloclavine from the culture broth of *Aspergillus japonicus* Saito. *Agr. Biol. Chem.,* **46,** 1921–1922.

Groger, D., Mothes, K., Floss, H.G. and Weygand, F. (1963) Zur Biogenese von ergolinderivaten in *Ipomoea rubro-caerulea* Hook. *Z. Naturforsch.,* **18b,** 157.

Gröger, D. and Tyler, V.E. (1963) Alkaloid production by *Claviceps paspali* in submerged culture. *Lloydia,* **28,** 174–191.

Hermansen, K., Frisvad, J.C., Emborg, C. and Hansen, J. (1984) Cyclopiazonic acid production by sumberged cultures of *Penicillium* and *Aspergillus* strains. *FEMSMicrobiol. Lett.,* **21,** 253–261.

Hofmann, A. und Tscherter, H. (1960) Isolierung von lysergsaurealkaloiden aus der mexikanischen zauberdroge "Ololiuqui" (Rivea corymbosa (L.) Hall.f.). *Experientia,* **16,** 414.

Hofmann, A. (1961) Die Wirkstoffe der mexikanischen Zauberdroge "Ololiuqui". *Planta Med.,* **9,** 354–367.

Holzapfel, C.W. (1968) The Isolation and structure of cyclopiazonic acid, a toxic metabolite of *Penicillium cyclopium* Westling. *Tetrahedron,* **24,** 2101–2119.

Holzapfel, C.W., Hutchison, R.D. and Wilkins, D.C. (1970) The isolation and structure of two new indole derivatives from *Penicillium cyclopium* Westling. *Tetrahedron,* **26,** 5239–5246.

Janapdhanan, K.K., Sattar, A. and Husain, A. (1984) Production of fumigaclavine A by *Aspergillus tamarin* Kita. *Can. J. Microbiol,* **30,** 247–250.

Jenettsiems, K., Kaloga, M. and Eich, E. (1994) Ergobalansine/ergobalansinine, a proline-free peptide-type alkaloid of the fungal genus *Balansia* is a constituent of *Ipomoea piurensis. J. Nat. Prod.,* **57,** 1304–1306.

Kawai, K., Nozawa, K., Yamaguchi, T., Nakajima, S. and Udagawa, S. (1992) Two chemotypes of *Penicillium crustosum* based on the analysis of indolic components. *Proc. Jpn. Assoc. Mycotoxicol.,* **36,** 19–24.

Kozlovsky, A.G., Reshetilova, T.A., Medvedeva, T.N., Arinbasarov, M.U., Sakharovsky, V.G. and Adanin, V.M. (1979) Intracellular and extracellular alkaloids of the fungus *Penicillium roqueforti.Biokhimija* (in *Russian*), **44,** 1691–1700.

Kozlovsky, A.G., Solovyeva, T.F., Sakharovsky, V.G. and Adanin, V.M. (1981) Biosynthesis of the "Unusual" ergot alkaloids by the fungus *Penicillium aurantiovirens. Proceeding of the USSR Academy of Sciences* (in *Russian*), **260,** 230–232.

Kozlovsky, A.G., Stefanova-Avramova, L.N. and Reshetilova, T.A. (1981a) Clavine ergoalkaloids—metabolites of *Penicillium gorlenkoanum.Prikl. Biokhim. Microbiol.* (in *Russian*), **17,** 806–812.

Kozlovsky, A.G., Stefanova-Avramova, L.N. and Reshetilova, T.A. (1981b) Effects of culture age and medium composition on alkaloid biosynthesis by *Penicilliumgorlenkoanum*. *Mikrobiologiya* (in *Russian*), **50**,1046–1052.

Kozlovsky, A.G., Reshetilova, T.A. and Medvedeva, T.H. (1982) The role of tryptophan and histidine in biosynthesis of alkaloids by *Penicillium roqueforti*. *Mikrobiologiya* (in *Russian*), **51**,48–53.

Kozlovsky, A.G., Solovyeva, T.F., Adanin, V.M. and Skryabin, G.K. (1982a) The way of the epoxyagroclavine-I preparation. Avt. svid. 955693 SU Pat., C 12 P. 17/00.

Kozlovsky, A.G., Solovyeva, T.F., Sakharovsky, V.G. and Adanin, V.M. (1982b) Ergoalkaloids agroclavine-I and epoxyagroclavine-I—metabolites of *Penicillium corylophillum*. *Prikl. Biokhim. Mikrobiol.* (in *Russian*), **18**,535–541.

Kozlovsky, A.G., Sakharovsky, V.G., Korpachev, A.V. and Aripovsky, A.V. (1983) Synthesis and structure of epoxyagroclavine-I derivatives. 2nd Int. Conf. *Chemistry andBiotechnology of Biologically Active Nature Products.*, ed T.Kezdy and A.Thomas, Budapest, 83–86.

Kozlovsky, A.G. and Reshetilova, T.A. (1984) Regulation of the biosynthesis of ergot alkaloids by *Penicillium sizovae*. *Folia microbiol.*,**29**,301–305.

Kozlovsky, A.G. and Solov'eva, T.F. (1985) Filamentous fungi—alkaloid producers. *Prikl. Biokhim. Mikrobiol.* (in *Russian*), **21**,579–586.

Kozlovsky, A.G., Vepritskaya, I.G., Gayazova, N.B., H'chenko, V.Ya. (1985a) Effect of exogenous tryptophan on biosynthesis of ergot alkaloids in *Penicillium sizovae.Mikrobiologiya* (in *Russian*), **54**,883–888.

Kozlovsky, A.G., Vepritskaya, I.G. and Gayazova, N.B. (1986) Alkaloid formation by *Penicillium sizovae.Prikl. Biokhim. Mikrobiol.* (in *Russian*), **22**,205–210.

Kozlovsky, A.G. and Solov'eva, T.F. (1986a) Influence of the conditions of fermentation on alkaloid biosynthesis by *Penicillium kapuscinskii*. *Mikrobiologiya* (in *Russian*), **55**, 34–40.

Kozlovsky, A.G. and Vepritskaya, I.G. (1987) Influence of carbon sources on the biosynthesis of ergoalkaloids and the activity of the enzymes of carbon metabolism in *Penicillium sizovae*. *Mikrobiologiya* (in *Russian*), **56**,587–592.

Kozlovsky, A.G. (1990) Nitrogen-containing mycotoxins of the fungi *Penicillium.Proc.Jpn. Mycotoxicol.*,**32**,31–34.

Kozlovsky, A.G., Solov'eva, T.F., Bukhtiyarov, Yu.E., Shurukhin, Yu.V., Sakharovsky, V.G., Adanin, V.M., Nefedova, M.Yu., Pertsova, R.N., Tokarev, V.G. and Golovleva, L.A. (1990) Secondary metabolites produced by new soil strains of microscopic fungi belonging to the genera *Aspergillus* and *Penicillium*. *Microbiologiya* (in *Russian*), **59**, 601–608.

Kozlovsky, A.G., Kuvichkina, T.N., Vinokurova, N.G., Zelenkova, N.F., Solov'eva, T.F. and Arinbasarov, M.U. (1995) The effect of various factors on the synthesis of epoxyagroclavin-I dimer by *Penicillium sizovae* VKM F-1073. *Prikl. Biokhim. Mikrobiol.* (in *Russian*), **31**,207–212.

Kozlovsky, A.G., Zelenkova, N.F., Adanin, V.M., Sakharovsky, V.G. and Arinbasarov, M.U. (1995a) Novel metabolites of *Penicillium sizovae*—dimer of agroclavine-I and mixed dimer of agroclavine-I and epoxyagroclavine-I. *Prikl. Biokhim. Mikrobiol.* (in *Russian*), **31**,540–544.

Kozlovsky, A.G., Zelenkova, N.F., Adanin, V.M., Sakharovsky, V.G. and Arinbasarov, M.U. (1996) Novel metabolites of *Penicillium sizovae*—dimer of agroclavine-I and mixed dimer of agroclavine-I and epoxyagroclavine-I. *Prikl. Biochim. Microbiol.,(in Russian),*31,540–544.

Kozlovsky, A.G., Marfenina, O.E., Vinokurova, N.G., Zhelifonova, V.P. and Adanin, V.M. (1997) Mycotoxins of filamentous fungi of the *genus Penicillium,* isolated from the soils of the natural and antropogenic breached ecosystems. *Microbiologiya* (in *Russian*), 66,112–116.

Kozlovsky, A.G., Vinokurova, N.G., Zhelifonova, V.P. and Adanin, V.M. (1997a) The secondary metabolites of *Penicillium janczewskii* fungi. *Prikl. Biokhim. Mikrobiol.* (in *Russian*), 33,87–91.

Le Bars, J. (1979) Cyclopiazonic acid production by *Penicillium camemberti* thom and natural occurence of this mycotoxin in cheese. *Appl. Environ. Microbiol.,*38, 1052–1055.

Naim, N. (1980) Alkaloid production by some local fungi. *Zbl. Bakteriol.*II. Abt., 135, 715–720.

Narayan, V. and Rao, K.K. (1982) Factors affecting the fermentation of ergot alkaloids by *Aspergillus fumigatus* (Fresenius). *Biotechnol. Lett.,*4,193–196.

Ohmomo, S., Sugita, M. and Abe, M. (1973) Isolation of cyclopiazonic acid, cyclopiazonic acid imine and bissecodehydrocyclopiazonic acid from the cultures of *Aspergillus versicolor* (Vuill) Tarabschi. *J. Agr. Chem. Soc. Jap.,*47,83–89.

Ohmomo, S., Sato, T., Utagawa, T. and Abe, M. (1975) Isolation of festuclavine and three new indole alkaloids, roquefortine A, B and C from the cultures of *Penicillium roqueforti. Agr. Biol. Chem.,*39,1333–1334.

Ohmomo, S., Miyazaki, K., Ohashi, T. and Abe, M. (1977) On the mechanism for formation of indole alkaloids in *Penicillium concavorugulosum. Agr. Biol. Chem.,*41, 1707–1710.

Ohmomo, S., Kaneko, M. and Atthasampunna, P. (1989) Production of fumigaclavine B by a thermophilic strain of *Aspergillus fumigatus. Mircen J.,*5,5–13.

Polonsky, J., Merrien, M.-A. and Scott, P.M. (1977) Roquefortine and isofumigaclavine A, alkaloids from *Penicillium roqueforti. Ann. Nutr. Alim.,*31,693–698.

Porter, J.K., Bacon, C.W. and Robbins, J.D. (1979) Lysergic acid amide derivatives from *Balansia epichloe* and *Balansia claviceps* (Clavicipitaceae). *J. Nat. Prod.,*42, 3, 309–314.

Porter, J.K., Bacon, C.W. and Robbins, J.D. (1979a) Ergosine, ergosinine and chanoclavine-I from *Epichole typhina. J. Agric. Food Chem.,*27,595–598.

Porter, J.K., Bacon, C.W., Robbins, J.D. and Betowski, D. (1981) Ergot alkaloid identification in clavicipitaceae systemic fungi of pasture grasses. *J. Agric. FoodChem.,*29,653–657.

Porter, J.K. (1993) Ergot and other alkaloids associated with toxic syndromes in livestock on endophyte-infected grasses. *Prikl. Biokhim. Mikrobiol.* (in *Russian*), 29,51–55.

Powell, R.G., Planner, R.D., Yates, S.G., Clay, K. and Leuchtmann, A. (1990) Ergobalansine, a new ergot-type peptide alkaloid isolated from *cenchrus echinatus* (Sandbur grass) infected with *Balansia obtecta* and produced in liquid cultures of *B.obtecta* and *B.cyperi. J. Nat. Prod.,*53,1272–1279.

Rao, K.K. and Patel, V.P. (1974) Effect of tryptophan and related compounds on alkaloid formation in *Aspergillus fumigatus. Lloydia,*37,608–610.

Rao, K.K., Gupta, A.R. and Singh, V.K. (1977) Effect of phosphate on ergot alkaloid synthesis in *Aspergillus fumigatus. Folia Microbiol.,*22,415–416.

Reddy, V.K. and Reddy, S.M. (1988) Effect of some carbon and nitrogen sources on the production of cyclopiazonic acid by *Penicillium griseofulvum*. *Nat. Acad. Sci.Letters.,* **11,**133–134.

Reshetilova, T.A. and Kozlovsky, A.G. (1985) The regulation of alkaloid biosynthesis by tryptophan and its analogs in *Penicillium roqueforti*. *Mikrobiologiya* (in *Russian*), **54,** 699–703.

Reshetilova, T.A., Solovyeva, T.F., Baskunov, B.P. and Kozlovsky, A.G. (1995) Study of the alkaloid composition of the foodinfecting penicilles. *Food additives andcontaminants,***12,**461–466.

Řičicová, A., Flieger, M. and Řeháček, Z. (1982) Quantitative changes of the alkaloid complex in a submerged culture of *Claviceps paspali*. *Folia Microbiol.,***27,**433–445.

Robbers, J.E., Robertson, L.W., Hornemann, K.M., Jindra, K. and Floss, H.G. (1972) Physiological studies on ergot. Further studies on the induction of alkaloid synthesis by tryptophan and its inhibition by phosphate. *J. Bacteriol.,***112,**791–796.

Robbers, J.E., Eggert, W.W. and Floss, H.G. (1978) Physiological studies on ergot: time factor influence on the inhibitory effect of phasphate and the induction effect of tryptophan on alkaloid production. *Lloydia,***41,**120–129.

Sakharovsky, V.G. and Kozlovsky, A.G (1981) Stereoisomerie of ergot alkaloids. *Shortcommunications of first international conference on chemistry and biotechnologyof biologically active natural products,*Varna, Bulgaria, **9,**358–361.

Sakharovsky, V.G and Kozlovsky, A.G. (1983) The study of costaclavine and epicostaclavine structure by ^{1}H-NMR. *J. Structumoy Khimii* (in *Russian*), **24,**100–105.

Sallam, L., El-Refai, A.H. and Naim, N. (1969) Detection of indole alkaloids in representative fungi. *Jap. J. Microbiol.,***13,**218–219.

Schoch, U., Luthy, J., und Schlatter, Ch. (1983) Mykotoxine in schimmelgereiften kasen. *Mitt. Gebiete Lebensm. Hyg.,***74,**50–59.

Scott, P.M., Merrien, M.A. and Polonsky, J. (1976) Roquefortine and isofumigaclavine A, metabolites from *Penicillium roqueforti*. *Experientia,***32,**140–142.

Scott, P.M., Kennedy, B.P.C., Harwing, J. and Blanchfield, B.J. (1977) Study of conditions for production of roquefortine and other metabolites of *Penicilliumroqueforti.Appl. Environ. Microbiol.,***33,**249.

Skryabin, G.K. and Kozlovsky, A.G. (1984) Microbial production of alkaloids. In A.A.Baev (ed.), *Biotechnology* (in *Russian*), Nauka, Moscow, pp. 66–70.

Solov'eva, T.F., Kuvichkina, T.N., Baskunov, B.P. and Kozlovsky, A.G. (1995) Alkaloids from the fungus *Penicillium aurantiovirens* Biourge and some aspects of their formation. *Mikrobiologiya* (in *Russian*), **64,**645–650.

Spilsbury, J.F. and Wilkinson, S. (1961) The isolation of festuclavine and two new clavine alkaloids from *Aspergillus fumigatus* Fres. *J. Chem. Soc.,***5,**2085–2091.

Stauffacher, D., Tscherter, H. and Hofmann, A. (1965) Isolierung von ergosin und ergosinin neben agroclavin aus den samen von *Ipomoea argyrophylla* Vatke (Convolvulaceae). *Helv. Chim. Acta,***48,**1379–1380.

Stauffacher, D., Niklaus, P., Tscherter, H., Weber, H.P. and Hofmann, A. (1969) Cycloclavin, ein neues alkaloid aus *Ipomoea hildebrandtii* Vatke—71 Mutterkornalkaloide. *Tetrahedron,***25,**5879–5887.

Stefanova-Avramova, L.N. and Kozlovsky, A.G. (1984) Effect of cultivation conditions on alkaloid biosynthesis by *Penicillium gorlenkoanum*. *Mikrobiologiya* (in *Russian*), **53,**437–441.

Turner, W.B. and Aldridge, D.C. (1983) *Fungal Metabolites II.*Acad. PressLondon, New York.

Vining, L.C. (1970) Effect of tryptophan on alkaloid biosynthesis in cultures of a *Claviceps* species. *Canad. J. Microbiol.,***16,**473.

Vining, L.C., Mclnnes, A.G., Smith, D.G., Wright, J.L.C. and Taber, W.A. (1982) Dimeric clavine alkaloids produced by *Penicillium citreo-viride.FEMS Symp.,***13,** 243–251.

Vinokurova, N.G., Reshetilova, T.A., Adanin, V.M. and Kozlovsky, A.G. (1991) Study of the alkaloid composition of *Penicillium palitans* and *Penicillium oxalicum.Prikl.Biokhim. Microbiol.* (in *Russian*), **27,**850–855.

Yamano, T., Kishino, K., Yamatodani, S. and Abe, M. (1964) Ergot alkaloids. Pat. 10250 Jap., MKU B 4, 39–10250.

Yokota, T., Sakurai, A., Iriuchijima, S. and Takahashi, N. (1981) Isolation and C NMR study of cyclopiazonic acid, a toxic alkaloid produced by muscardine fungi *Aspergillus flavus* and *A.oryzae. Agr. Biol. Chem.,***45,**53–56.

Yamatodani, S. and Yamamoto, I. (1983) Peptide-type ergot alkaloids produced by *Hypomyces aurantius.Nippon Nogeikagaku Kaishi,***57,**453–456.

Yates, S.G., Planner, R.D. and Garner, G.B. (1985) Detection of ergopeptine alkaloids in endophyte infected, toxic Ky-31 Tall fescue by mass spectrometry/mass spectrometry. *J. Agric. Food Chem.,***33,**719–722.

Zelenkova, N.F., Vepritskaya, I.G., Adanin, V.M., Sakharovsky, V.G., Nefedova, M.Yu. and Kozlovsky, A.G. (1992) A new ergoalkaloid of *Penicillium sizovae,* the dimer of epoxyagroclavine-I. *Prikl. Biokhim. Mikrobiol.* (in *Russian*), **28,**738–741.

Other volumes in preparation in Medicinal and Aromatic Plants—Industrial Profiles

Stevia, edited by A.D.Kinghorn

Tea, edited by Y.S.Zhen

Tea tree, edited by I.Southwell and R.Lowe

Tilia, edited by K.P.Svoboda and J.Collins

Thymus, edited by W.Letchamo, E.Stahl-Biskup and F.Saez

Trigonella, edited by G.A.Petropoulos

Urtica, by G.Kavalali

This book is part of a series. The publisher will accept continuation orders which may be cancelled at any time and which provide for automatic billing and shipping of each title in the series upon publication. Please write for details.

INDEX

507